The physiology of excitable cells

FOURTH EDITION

The fourth edition of this highly successful text has been extensively revised and restructured to take account of the many recent advances in the subject and so bring it right up to date. The classic observations of earlier years can now be interpreted with powerful new techniques in cell biophysics and molecular biology. Consequently there is much new material throughout the book, including many new illustrations and extensive reference to recent work. Its essential philosophy, however, remains the same: fundamental concepts are clearly explained, and key experiments are examined in some detail.

This book will be used by students of physiology, neuroscience, cell biology and biophysics and by academics and researchers in these subjects. They will find the text thorough, interesting and clearly written.

David Aidley read Zoology at Cambridge University, and did a Ph.D. there on excitation–contraction coupling in insect muscle. He then spent four years as a Departmental Demonstrator in the Zoology Department at Oxford University. Since 1967 he has been in the School of Biological Sciences at the University of East Anglia, Norwich. He has also worked in Kenya, at the International Centre for Insect Physiology and Ecology, and in Nigeria at Bayero University, Kano.

He has published research papers and reviews on various aspects of physiology and experimental zoology. His books include the three previous editions of the present book (1971, 1979 and 1990), *Animal Migration* (editor, 1981), *Nerve and Muscle* (with R. D. Keynes, 1981 and 1991), and *Ion Channels: Molecules in Action* (with P. R. Stanfield, 1996).

The physiology of excitable cells

FOURTH EDITION

David J. Aidley

School of Biological Sciences
University of East Anglia, Norwich

PUBLISHED BY THE PRESS SYNDICATE OF THE UNIVERSITY OF CAMBRIDGE
The Pitt Building, Trumpington Street, Cambridge, United Kingdom

CAMBRIDGE UNIVERSITY PRESS
The Edinburgh Building, Cambridge CB2 2RU, UK
40 West 20th Street, New York, NY 10011–4211, USA
10 Stamford Road, Oakleigh, VIC 3166, Australia
Ruiz de Alarcón 13, 28014 Madrid, Spain
Dock House, The Waterfront, Cape Town 8001, South Africa

http://www.cambridge.org

First edition 1971
Reprinted 1973, 1974, 1975, 1976
Second edition 1979
Reprinted 1979, 1981, 1982, 1983, 1985, 1986, 1988
Third edition 1990
Reprinted 1991, 1996
Fourth edition 1998
Reprinted 2001

Typeset in Times NR 9/12pt, in QuarkXpress™ [SE]

A catalogue record for this book is available from the British Library

Library of Congress Cataloguing in Publication data
Aidley, David J.
The physiology of excitable cells / David J. Aidley, – 4th ed.
    p.   cm.
Includes bibliographical references and index.
ISBN 0 521 57415 3 (hardcover). – ISBN 0 521 57421 8 (pbk.)
1. Neurophysiology. 2. Neurons. 3. Muscle cells. 4. Sensory
receptors. 5. Excitable membranes. 6. Electrophysiology.
7. Neural transmission. I. Title.
QP356.A46   1998
573.8 – dc21   97-46773   CIP

ISBN 0 521 57415 3 hardback
ISBN 0 521 57421 8 paperback

Transferred to digital printing 2003

To Jessica

Contents

Preface

There has been a great flowering of our knowledge about the physiology of nerve, muscle and sensory cells in recent decades. This book aims to help the reader to learn about the subject by giving an account of some of the experimental evidence on which this knowledge is based. It is intended primarily for use by students taking courses in physiology, neuroscience, cell biology or biophysics, but it should also prove useful to those beginning research and to scientists of related disciplines.

This fourth edition reflects the continuing emphasis on molecular mechanisms that has been such a feature of the biological sciences in recent years. Exciting new developments have continued to flow from the use of the patch clamp technique for examining the currents flowing through single membrane channels, and from the application of recombinant DNA methods for determining the structures of proteins. Hence the book has been extensively revised to take account of these and other advances, and there is much new material throughout. I have also extended the range of the book to cover a wider range of sensory cells and to consider the cellular basis of learning, and I have restructured some of the chapters to maintain a sensible arrangement of the material.

Learning about science is a complicated business. Students are expected to know the phenomena that occur in the natural world and understand the concepts which we use to explain them. But it is not sufficient to stop there, without asking how we arrived at our present consensus. Someone who knows why we believe what we believe is much better educated than someone who merely knows what we believe. So it is essential, it seems to me, that science students should be introduced to the experimental evidence on which our present understanding is based. They need to understand how some of the experiments were done and why they were done. That is why much of this book is so concerned with experimental results and the evidence for conclusions.

I am very grateful to those authors and publishers who have allowed me to reproduce here records and diagrams which originally appeared in their own publications. The sources of each of these are indicated in the captions and detailed in the reference list at the end of the book.

It is a pleasure to thank those of my colleagues who have helped and encouraged me in various ways during the preparation of this and earlier editions. Hence I am particularly grateful to Michael Brown, Alan Coddington, Alan Dawson, George Duncan, Sir Alan Hodgkin, Richard Keynes, Edward Lea, Peter Miller, Eduardo Rojas, Graham Shelton, Peter Stanfield, Paul Taylor, Richard Tregear and David White. I would also like to acknowledge the cheerful assistance that I have received from the Library staff at the University of East Anglia and the editorial staff at Cambridge University Press. Needless to say, the responsibility for the shortcomings that remain is entirely my own.

Part A
Foundations

1
Introduction

Suppose a man has a tomato thrown at his head, and that he is able to take suitable evasive action. His reactions would involve changes in the activity of a very large number of cells in his body. First of all, the presence of a red object would be registered by the visual sensory cells in the eye, and these in turn would excite nerve cells leading into the brain via the optic nerve. A great deal of activity would then ensue in different varieties of nerve cell in the brain and, after a very short space of time, nerve impulses would pass from the brain to some of the muscles of the face and, indirectly, to muscles of the neck, legs and arms. The muscle cells there would themselves be excited by the nerve impulses reaching them, and would contract so as to move the body and so prevent the tomato having its intended effect. These movements would themselves produce excitation of numerous sensory endings in the muscles and joints of the body and in the organs of balance in the inner ear. The resulting impulses in sensory nerves would then cause further activity in the brain and spinal cord, possibly leading to further muscular activity.

A chain of events of this type involves the activity of a group of cell types which we can describe as 'excitable cells': a rather loose category which includes nerve cells, muscle cells, sensory cells and some others. An excitable cell, then, is a cell which readily and rapidly responds to suitable stimuli, and in which the response includes a fairly rapid electrical change at the cell membrane.

The study of excitable cells is fascinating for a number of reasons. These are the cells which are principally involved in the behavioural activities of animals, including ourselves: these are the cells with which we move and think. Yet just because their functioning must be examined at the cellular and subcellular levels of organization, the complexities that emerge from investigating them are not too great for adequate comprehension. So it is frequently possible to pose specific questions as to their properties and to elicit some of the answers to these questions by suitable experiments. It is perhaps for this reason that the subject has attracted some of our foremost scientists. As a consequence, the experimental evidence on which our knowledge

of the physiology of excitable cells is based is often elegant, clear-cut and intellectually exciting, and frequently provides an object lesson in the way a scientific investigation should be carried out. Nevertheless, there are very many investigations still to be done in this field, many questions which have yet to be answered, and undoubtedly very many which have not yet been asked.

Most readers of this book will possess a considerable amount of information on basic ideas in the biological and physical sciences. But it may be as well in this introductory chapter to remind them, in a rather dogmatic fashion, of some of the background which is necessary for a more detailed study: to formulate, in fact, a few axioms.

The biological material
Cells
All large organisms are divided into a number of units called cells, and every cell is the progeny of another cell. This statement constitutes the cell theory. Every cell is bounded by a cell membrane and contains a nucleus in which the genetic material is found. The main part of the living matter of the cell is a highly organized system called the cytoplasm, which is concerned with the day-to-day activity of the cell. The cell membrane separates this highly organized system inside from the relative chaos that exists outside the cell.

In order to maintain and increase its high degree of organization and in order to respond to and alter its environment, the cell requires a continual supply of energy. This energy must be derived ultimately from the environment, usually in the form of chemical energy such as can be extracted by the cell from glucose molecules. We can describe the cell in thermodynamic terms as an open system maintained in a rather improbable steady state by the continual expenditure of energy. Its life is a continual battle against the second law of thermodynamics (which we may state without gross inaccuracy as 'things tend to get mixed up').

The cells of nervous systems are called *neurons*. Their primary function is the handling of information. Within the cell this mainly takes the form of changes in the electric potential across the cell membrane, whereas information is passed between cells largely as chemical messages.

The idea that the nervous system is composed of discrete cells is known as the neuron theory. This view, which is simply a particular application of the cell theory, was developed during the nineteenth century and is now generally accepted (see Shepherd, 1991). The alternative proposal, that nervous systems are not divided into separate membrane-bounded entities (the reticular theory), was difficult to reconcile with the observations of light microscopists, and seems to be conclusively refuted by the evidence of electron microscopy.

Neurons have functional regions specialized for different purposes. Sites where one neuron contacts another cell, usually another neuron, and transmits or receives some information, are known as *synapses*. Synaptic transmission is usually a one-way process, from the presynaptic cell to the postsynaptic cell. Areas of the neuron that receive synaptic contacts from presynaptic neurons form the *input region* (fig. 1.1). The input region commonly consists of branched processes called dendrites, and may include the surface of

the cell body (the soma, containing the cell nucleus) of the neuron. The postsynaptic responses in the input region may be sufficient to produce excitation in the *conductile region* of the neuron, whose activity consists of unitary events called nerve impulses or action potentials. The conductile region is a long process called the *axon*. The axon usually terminates in fine branches that make synaptic contact with other cells, such as other neurons or muscle cells. These terminals are presynaptic and form the *output region* of the neuron. They usually secrete a chemical substance, the neurotransmitter, when an action potential arrives along the axon, and this carries information across the synapses to the postsynaptic cell.

Proteins

So pervasive are the functions of proteins in cells that one way of defining living material is to say that it contains active proteins. Proteins are composed of chains made up from different combinations of twenty different amino acids, and their properties depend critically upon the sequence in which these amino acids are arranged.

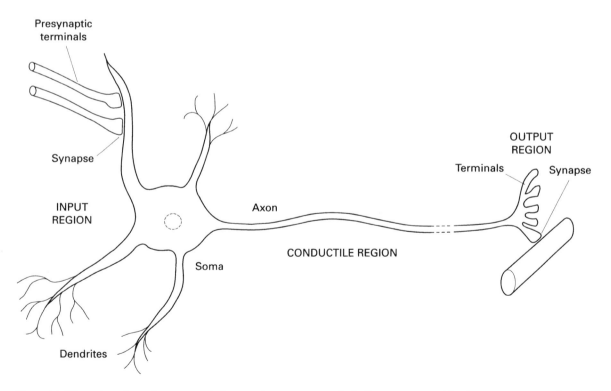

Figure 1.1. The main regions of a neuron. The input region, commonly the dendrites and soma of the cell, receives synaptic inputs from a number of different nerve fibres. The conductile region, the axon, carries all-or-nothing action potentials from the input region to the output region. The output region, the nerve terminals, forms synapses with other neurons or muscle cells. The diagram is based very loosely on a vertebrate spinal motoneuron, where the input region (soma and dendrites) is in the spinal cord, and the axon passes down the ventral root and along peripheral nerves to the output region at the neuromuscular junction.

The protein's amino acid sequence is specified by the nucleotide base sequence in the DNA molecules which form the genetic material of the cell. This means that proteins are the product of evolution, so that present-day proteins largely represent stable and successful sequences. Animals whose cells produced too many unstable or nonfunctional sequences would have died before producing viable offspring, so the genes specifying those sequences have mostly been eliminated. Different animal species may have proteins with only minor variants in amino acid composition. The same animal may produce a number of very similar sequences, perhaps adapted to slightly different roles in different tissues. These variants of essentially the same proteins are called *isoforms*.

The shape of many protein molecules changes when they react with smaller molecules or other proteins. Changes of this type underlie much protein activity, such as enzymic hydrolysis, opening of membrane channels, and muscular contraction. The day-to-day activity of the cell can thus be described largely in terms of the actions of proteins; the reader of modern accounts of cell biology (such as those by Albert *et al.*, 1994, and Lodish *et al.*, 1995) will find ample illustration of this statement.

Animals

Every animal has a history: every animal owes its existence to the success of its ancestors in combating the rigours of life; that is to say, in surviving the rigours of natural selection. Hence every animal is adapted to its way of life, and its organs, its tissues, its cells and its protein molecules are adapted to performing their functions efficiently.

An animal is a remarkably stable entity. It is able to survive the impact of a variety of different environments and situations, and its cells and tissues are able to survive a variety of different demands upon their capacities. An animal is a complex of self-regulating (homeostatic) systems. These systems are themselves coordinated and regulated so that the physiology and behaviour of the animal form an integrated whole.

Nervous systems

A nervous system is that part of an animal which is concerned with the rapid transfer of information through the body in the form of electrical signals. The activity of a nervous system is initiated partially by the input elements – the sense organs – and partially by endogenous activity arising in certain cells of the system. The output of the system is ultimately expressed via effector organs – muscles, glands, chromatophores, etc.

Primitive nervous systems consist of scattered but usually interconnected nerve cells, forming a nerve net, as in coelenterates. Increase in the complexity of responses is associated with the aggregation of nerve cell bodies to form ganglia and, when the ganglia themselves are collected and connected together, we speak of a *central nervous system*. The *peripheral nervous system* is then mainly composed of nerve fibres originating from the central nervous system. Peripheral nerves contain afferent (sensory) neurons taking information inwards into the central nervous system and efferent (motor) neurons taking information outwards. Neurons confined to the central nervous system are known as interneurons. Ganglia which remain or arise outside the central nervous system, and the nerve fibres which lead to and arise from them to innervate the animal's viscera, are frequently described as forming the *autonomic nervous system*.

One of the simplest, but possibly not one of the most primitive, modes of activity of a nervous system is the *reflex*, in which a relatively fixed output pattern is produced in response to a simple input. The stretch reflexes of mammalian limb muscles provide a well-known example (fig. 7.20). Stretching the muscle excites the endings of sensory nerve fibres attached to certain modified fibres (muscle spindles) of the muscle. Nerve impulses pass up the sensory fibres into the spinal cord where they meet motor nerve cells (the junctional regions are called *synapses*) and excite them. The nerve impulses so induced in the motor nerve fibres then pass out of the cord along peripheral nerves to the muscle, where their arrival causes the muscle to contract. Much more complicated interactions occur in the analysis of complex sensory inputs, the coordination of locomotion, the expression of the emotions and instinctive reactions, in learning and other 'higher functions'. These more complicated interactions are outside the scope of this book.

Electricity

Matter is composed of atoms, which consist of positively charged nuclei and negatively charged electrons. Static electricity is the accumulation of electric charge in some region, produced by the separation of electrons from their atoms. Current electricity is the flow of electric charge through a conductor. Current flows between two points connected by a conductor if there is a potential difference between them, just as heat will flow from a hot body to a cooler one placed in contact with it. The unit of potential difference is the *volt*. The current, i.e. the rate of flow of charge, is measured in *amperes*, and the quantity of charge transferred is measured in *coulombs*. Thus one coulomb is transferred by a current of one ampere flowing for one second.

In many cases it is found that the current (I) through a

Table 1.1 *Some electrical quantities and their units*

Quantity	Symbol for quantity	Unit	Symbol for unit	Equivalent
Charge	Q	coulomb	C	A s
Current	I	ampere	A	$C s^{-1}$
Potential difference	V, E	volt	V	$J C^{-1}$
Energy (work)		joule	J	C V
Power		watt	W	$J s^{-1}$, A V
Resistance	R	ohm	Ω	$V A^{-1}$
Conductance	G	siemens	S	Ω^{-1}, A V^{-1}
Capacitance	C	farad	F	$C V^{-1}$

Note:

It is conventional to write the symbols for quantities in italics and the symbols for units in roman type.

conductor is proportional to the potential difference (V) between its ends. This is *Ohm's Law*. Thus if the constant of proportionality, the *resistance* (measured in *ohms*) is R, then

$$V = IR$$

The specific resistance of a substance is the resistance of a 1 cm cube of the substance. The resistance of a wire of constant specific resistance is proportional to its length and inversely proportional to its cross-sectional area. The reciprocal of resistance is called *conductance* (G).

Let us apply Ohm's law to a simple calculation. In chapter 6 we shall see that under certain conditions small channels open to let sodium ions flow through. If we can measure this current flow and we know what the driving voltage is, we can calculate the conductance of the channel. Thus in one experiment the single channel current was 1.6 pA with a driving voltage of 90 mV. (Table 1.1 shows selected electrical units and table 1.2 gives prefix names for multiples and submultiples.) Applying Ohm's law, the conductance of the channel is given by

$$G = I/V$$

$$\text{i.e. conductance (siemens)} = \frac{\text{current (amps)}}{\text{voltage (volts)}}$$

$$= \frac{1.6 \times 10^{-12}}{90 \times 10^{-3}}$$

$$= 17.8 \text{ pS}$$

The total resistance of a number of resistive elements arranged in series is the sum of their individual resistances,

Table 1.2 *Some prefixes for multiples of scientific units*

Multiple	Prefix	Symbol
10^{-2}	centi	c
10^{-3}	milli	m
10^{-6}	micro	μ
10^{-9}	nano	n
10^{-12}	pico	p
10^{-15}	femto	f
10^{3}	kilo	k
10^{6}	mega	M
10^{9}	giga	G

whereas the total conductance of a number of elements in parallel is the sum of their conductances. A patch of membrane containing five channels each with a conductance of 17.8 pS, for example, will have a conductance of 89 pS if all the channels are open.

Two plates of conducting material separated by an insulator form a capacitor. If potential difference V is applied across the capacitor, a quantity of charge Q, proportional to the potential difference, builds up on the plates of the capacitor. Thus

$$Q = VC$$

where C, the constant of proportionality, is the *capacitance* of the capacitor. When the voltage is changing, charge flows away from one plate and into the other, so that we can speak of current, I, through a capacitor, given by

$$I = C\frac{dV}{dt}$$

where dV/dt is the rate of change of voltage with time. The capacitance of a capacitor is proportional to the area of the plates and the dielectric constant (a measure of the ease with which the molecules of a substance can be polarized) of the insulator between them, and inversely proportional to the distance between the plates. The total capacitance of capacitors in parallel is the sum of the individual capacitances, whereas the reciprocal of the total capacitance of capacitors in series is the sum of the reciprocals of their individual capacitances.

Scientific investigation

Science is concerned with the investigation and explanation of the phenomena of the natural world. Any particular investigation usually starts with an idea – a hypothesis – about the relations between some of the factors in the system to be studied. The hypothesis must then be tested by suitable observations or experiments. This business of testing the hypothesis is what distinguishes the scientific method from other attempts at the acquisition of knowledge, and hence it follows that a scientific hypothesis must be capable of being tested. We must therefore understand what is meant by 'testing' a hypothesis.

In mathematics and deductive logic it is frequently possible to prove, given a certain set of axioms, that a certain idea about a particular situation is true or not true. For instance, it is possible to prove absolutely conclusively that, in the system of Euclidean geometry, the angles of an equilateral triangle are all equal to one another. But this absolute proof of the truth of an idea is not possible in science. For example, consider the hypothesis 'No dinosaurs are alive today'. This statement would be generally accepted by biologists as being almost certainly true. But, of course, it is just possible that there are some dinosaurs alive which have never been seen. Some years ago the statement 'No coelacanths are alive today' would also have been accepted as almost certainly correct.

However, in many cases, it *is* possible to prove that a hypothesis is false. The hypothesis 'No coelacanths are alive today' has been proved, conclusively, to be false. If we were to find just one living dinosaur, the hypothesis 'No dinosaurs are alive today' would also have been shown to be false. It follows from this argument that in order to test a hypothesis it is necessary to attempt to disprove it. When a hypothesis has successfully survived a number of attempts at disproof, it seems more likely that it provides a correct description of the situation to which it applies (Popper, 1963).

If we can test a hypothesis only by attempting to disprove it, it follows that a scientific hypothesis must be formulated in such a way that it is open to disproof – so that we can think of an experiment or observation in which one of the possible results would disprove the hypothesis. Any idea which we cannot see how to disprove is not a scientific hypothesis.

But where do the ideas come from? Science is a progressive activity. Advances are usually made step by step. Ideas arise in a controlled imagination: the scientist usually starts from a generally accepted understanding of the situation and makes a small conjecture into the unknown. A high rate of progress follows two particular types of advance: ideas which provide a major reinterpretation of what we know, and new techniques. In 1954, for example, as we shall see in chapter 19, the study of muscular contraction entered a new and highly productive phase as the result of the formulation of the sliding filament theory, which itself arose in the context of advances in X-ray diffraction methods and electron microscopy. More recently, the advent of the patch clamp technique (fig. 2.4) has led to a great flowering of work on the ionic channels of cell membranes.

What implications does the nature of science have for learning about science? Students of any subject must get to grips with its intellectual credentials, if they are to be worth their salt. For the science student, this implies that simply comprehending a proposition that we believe to be true is not enough. It is also necessary to understand why we believe it to be true, what the evidence for the proposition is, and hence what sort of evidence might lead us to revise our opinion about it.

It is for this reason that this book is much concerned with experiments and observations, and not simply with the understanding that has arisen from them. The conclusions from some of these experiments will stand the tests of further investigations in the future, those from others will have to be revised. Science students cannot hope to know everything about their subject, but if they understand just why they believe some of what they know, then they can look future in the face.

2
Electrophysiological methods

Excitable cells can be studied by the great variety of techniques that are available for the study of living cells in general. These include light and electron microscopy, X-ray diffraction measurements, experiments involving radioactive tracers, cell fractionation techniques, cell imaging techniques, biochemical methods, and so on. The techniques which are particular to the study of excitable cells are those involving the measurement of rapid electrical events. So in order to understand the subject, we need to have some idea of how these measurements are made. Here we look briefly at some of the more general methods used. Duncan (1990) gives more detail on a variety of electrical measurement techniques.

Recording electrodes

If we wish to record the potential difference between two points, it is necessary to position electrodes at those points and connect them to a suitable instrument for measuring voltage. It is desirable that these electrodes should not be affected by the passage of small currents through them, i.e. that they should be non-polarizable. For many purposes fine silver wires are quite adequate. Slightly better electrodes are made from platinum wire or from silver wire that has been coated electrolytically with silver chloride. For very accurate measurements of steady potentials, calomel half-cells (mercury/mercuric chloride electrodes) may have to be used.

If the site we wish to record from is very small in size (such as occurs in extracellular recording from cells in the central nervous system), the electrode must have a very fine tip, and be insulated except at the end. Successful electrodes of this type have been made from tungsten wire which is sharpened by dipping it into a solution of sodium nitrate while current is being passed from the electrode into the solution; insulation is produced by coating all except the tip of the electrode with a suitable varnish.

The manufacture of a suitable electrode is rather more difficult if we wish to record the potential inside a cell. Apart from a few large cells such as squid giant axons, this necessitates the use of an electrode which is fine enough to

penetrate the cell membrane without causing it any appreciable damage. The problem was solved with the development of glass capillary microelectrodes by Ling & Gerard in 1949. These are made on a suitable device which will heat a small section of a hard glass tube and then very rapidly extend it as it cools. The heated section gets thinner and cooler as the pulling proceeds, so that finally the pulling force exceeds the cohesive forces in the glass, and the tube breaks to give two micropipettes. If the machine designed to do this has been correctly adjusted, the outside diameter of the micropipette at its tip will be about 0.5 μm.

The micropipette now has to be filled with a strong electrolyte solution such as 3 M potassium chloride solution. This is most readily done if the glass tube from which the electrode is made has a fine filament of glass fused to its inner wall; the angle between the filament and the wall leads to high capillary forces, so that the pipette can be filled to its tip by injecting the solution into the barrel. The connection to the recording apparatus is made via a non-polarizing electrode such as a silver wire coated with silver chloride.

An electrode of this type has a very high resistance, from five to several hundred megohms (MΩ); in fact the suitability of an electrode is usually tested by measuring its resistance, since the tip is too small to be examined satisfactorily by light microscopy. Further details of microelectrode methods are given by Ogden (1994).

Electronic amplification

The potential differences which are measured in investigations on the activity of excitable cells vary in size from just over 0.1 V down to as little as 20 μV or so. Before being measured by a recording instrument, these voltages usually have to be amplified. This is done by means of suitable electronic circuits involving thermionic valves, transistors or integrated circuits, the details of which need not concern us. However, there are three aspects of any amplifier used for electrophysiological recording purposes of which we should be aware: the frequency response, the noise level and the input resistance.

The gain of an amplifier is the ratio of its output voltage to its input voltage. All amplifiers show a higher gain at some frequencies than at others. A typical amplifier for use with extracellular electrodes might have a constant gain over the range 10 Hz to 50 kHz, the gain falling at frequencies outside this range. An amplifier of this type is known as an a.c. amplifier, since it measures voltages produced by alternating current. A d.c. amplifier is one which can measure steady potential differences, i.e. its frequency response extends down to zero hertz. If we wish to measure steady potentials, or slow potential changes without distortion, then obviously we must use a d.c. amplifier. Amplifiers used with intracellular electrodes are usually d.c. amplifiers, so that the steady potential difference across the cell membrane can be measured, as well as the rapid (high frequency) changes involved in its activity.

Any amplifier will produce small fluctuations in the output voltage even when there is no input signal. These fluctuations are known as *noise*, and they are caused by random electrical activity in the amplifier. The existence of noise sets a lower limit to the signal voltage that can be measured, since it is difficult to distinguish very small signals from the noise. Consequently it is necessary to use an amplifier with a low noise level if we wish to measure signals of very small size.

If we connect a potential difference across the input terminals of an amplifier, a very small current flows between them, which is proportional to the potential difference. The proportionality factor is called the *input resistance* of the amplifier. It is determined by application of Ohm's law and is measured in ohms; for instance, the input resistance of a cathode ray oscilloscope amplifier is usually about 1 MΩ. Now suppose we connect an intracellular microelectrode whose resistance is, say, 10 MΩ to an amplifier with an input resistance of 1 MΩ. The equivalent circuit is shown in fig. 2.1*a*. The two resistances form a potential divider, so that the voltage input to the amplifier is only $10^6/(10^6 + 10^7)$ i.e. one-eleventh of the signal voltage. Obviously this is of

little use for measuring the signal voltage. Hence, when using high resistance electrodes, it is necessary to use an input stage with a very high input resistance, such as is given by the use of a junction field effect transistor. Suppose the input resistance of such a device is 1000 GΩ (fig. 2.1*b*); then the voltage recorded is $10^{12}/(10^{12} + 10^7)$, which is effectively equal to the signal voltage.

The cathode ray oscilloscope

The voltage output from an amplifier has to be recorded and measured in some way. If the voltage is steady, or only changing very slowly, we could use an ordinary voltmeter in which the current through a coil of wire placed between the poles of a magnet causes a pointer on which the coil is mounted to move. However, a device of this nature is no use for measuring rapid electrical changes since the inertia of the pointer is too great. What we need, in effect, is a voltmeter with a weightless pointer. This is provided by the beam of electrons in the cathode ray tube of a cathode ray oscilloscope.

The cathode ray tube is an evacuated glass tube containing a number of electrodes and a screen, which is coated with phosphorus compounds so that it luminesces when and where it is bombarded by electrons, at one end (fig. 2.2). The cathode is heated and made negative (by about 2000 V) to the anode; consequently electrons are emitted from the cathode and accelerate towards the anode. Since the anode has a hole in its centre, some of the electrons continue moving at a constant high velocity beyond it; these form the electron beam. The intensity and focusing of the beam can be controlled by other electrodes, not shown in fig. 2.2, placed in the vicinity of the anode and cathode. When the electron beam hits the screen, it produces a spot of light at the point of impact.

Between the anode and the screen, the beam passes between two pairs of plate electrodes placed at right angles to each other, known as the X and Y plates. If a potential difference is connected across one of these pairs, the

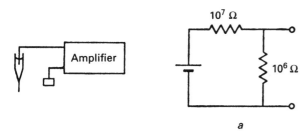

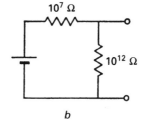

a   *b*

Figure 2.1. Microelectrodes need high resistance input stages. The diagrams show equivalent circuits of a glass capillary microelectrode whose resistance is 10 MΩ, connected to (*a*) an 'ordinary' amplifier with an input resistance of 1 MΩ, and (*b*) an input stage with a resistance of 1000 GΩ. The potential recorded in *a* is only one-eleventh of the source voltage.

electrons in the beam will move towards the positive plate. Thus when the electrons pass out of the electric field of the pair of plates, their direction will have been changed, and so the light spot on the screen will move, by an amount proportional to the potential difference between the plates. The Y plates are connected to the output of an amplifier whose input is the signal voltage to be measured; hence the signal voltage appears as a vertical deflection of the spot on the screen. The X plates are usually connected to a waveform generating circuit (the time-base generator) which produces a sawtooth waveform. This sawtooth waveform thus moves the spot horizontally across the screen at a constant velocity, flying back and starting again at the end of each 'tooth'. As a consequence of these arrangements, the spot on the screen traces out a graph with signal voltage on the *y*-axis and time on the *x*-axis. By making the rise-time of the time-base sawtooth sufficiently fast, it is possible to measure the form of very rapid voltage changes.

Many oscilloscopes have tubes with two beams, each with separate Y plates and amplifiers, so that one can measure two signals at once. The time-base unit can frequently be arranged so that a single sawtooth wave, leading to a single sweep of the beam, can be initiated (or 'triggered') by some suitable electrical signal; this facility is essential for much electrophysiological work. In some cases it is possible to connect the X plates to another input amplifier, instead of to the time-base generator, so that a signal related to some quantity other than time is measured on the *x*-axis. On the screen of a standard oscilloscope the waveform traced out by the electron beam persists for only a fraction of a second. A permanent record of the trace can be obtained by photographing it; many of the diagrams in this book are photographs of oscilloscope traces.

Storage oscilloscopes are devised so that the trace can be held on the screen for some time. Analogue storage devices depend upon rather expensive modifications to the cathode ray tube. Digital storage oscilloscopes store the trace in memory from which it is continually read out and displayed on the screen. An advantage of this type of instrument is that the memory can be read out into other devices, such as a chart recorder for hard-copy production.

Oscilloscopes are always easier to use if one can have a second or third look at what one wishes to display, so it is often useful and sometimes essential to be able to store the recorded signals in some permanent form. Such immediate data storage can be supplied by magnetic tape (using a frequency-modulated tape recorder) or a computer memory.

Electrical stimulation

An electrical stimulus must be applied via a pair of electrodes. Stimulating electrodes may be of any of the types previously described for recording purposes. The simplest way of providing a stimulating pulse is to connect the electrodes in series with a battery and a switch, but this is not satisfactory if brief pulses are required. In the past, stimulating pulses were produced by such means as discharge of condensers or by using an induction coil, but nowadays most investigators use electronic stimulators which produce square pulses of constant voltage, beginning and ending abruptly. A good stimulator unit will be able to produce pulses which can be varied in strength, duration and frequency.

The voltage clamp technique

During most bioelectrical events, the membrane potential and the current flowing through the membrane are both changing at the same time. This makes it difficult to work

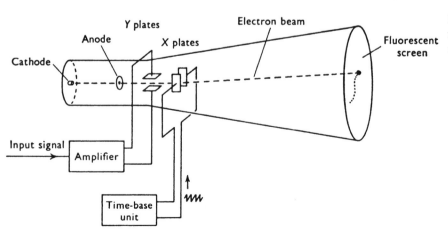

Figure 2.2. Diagram to show the main components of a cathode ray oscilloscope.

out just what is happening in the membrane. It would be much more useful if one could hold the membrane potential constant at some predetermined value and measure the current flow on its own. This capability is provided by a technique known as the *voltage clamp*, first used extensively around 1950 for the analysis of the electrical behaviour of squid axons. The technique was used by Alan Hodgkin & Andrew Huxley in their ground-breaking analysis of the action potential in 1952 (chapter 5) and has been much used by others since then.

The principle of the voltage clamp is to use electronic feedback to maintain the voltage across the membrane at some constant level, and then measure the current flowing through the membrane as a result. Thus it is usual to have two intracellular electrodes, one to measure the membrane potential and the other to pass the current. Figure 2.3 gives an impression of the method. The feedback amplifier is a high gain differential amplifier; it measures the difference between the membrane potential and a reference or command voltage that is set by the experimenter. Its output is proportional to this difference, and provides the current that is fed to the current-passing electrodes; this arrangement ensures that any change in the membrane potential will be opposed by an increase in the membrane current. Consequently, the membrane potential is held constant at a value determined by the experimenter, who thus measures the current flowing through the membrane under these conditions.

The patch clamp

Patch clamping is a technique whereby a very small area of membrane can be voltage clamped, so allowing the current flow through individual channels to be measured directly. The first records of this type were published by Neher & Sakmann in 1976, from the acetylcholine channels of denervated muscle cells (see p. 143). Since then the technique has been used in a large number of investigations so that it now provides a major source of new information about membrane channels.

Figure 2.4 illustrates the method. A glass microelectrode is polished to produce a smooth tip 1 to 2 μm in diameter. It is then coated with a resin to reduce the conductance and capacitance of the glass, and filled with an isotonic electrolyte solution. The electrode is pushed against the cell membrane so that the resistance between the inside of the electrode and the external solution rises to about 100 MΩ. Application of suction pulls the cell membrane into the tip of the electrode where it makes close contact with the glass wall and so forms a seal with a very high resistance, of the order of 50 GΩ (a 'gigaseal'). Such a seal greatly reduces the background noise from the system and so enables the very small currents flowing through the ionic channels to be measured. The gigaseal may not form if there is much extracellular material attached to the cell membrane, hence it is common to use embryonic cells or to 'clean' the cell with an enzyme such as collagenase.

Once the gigaseal has been formed there is a choice of

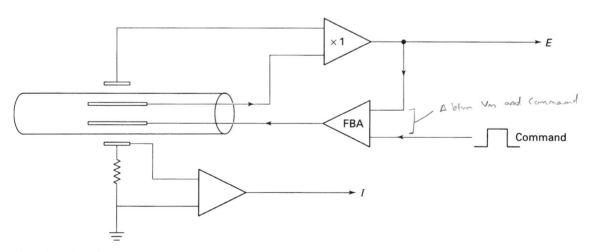

Figure 2.3. The main components of a voltage clamp circuit, such as could be used for a squid axon. Two electrodes are inserted into the cut end of the axon, one to measure the voltage across the membrane, the other to pass current through it. A high impedance follower amplifier (× 1) measures the membrane potential E; its output is fed to the recording system and also to the input of the feedback amplifier, FBA. The other input to the

FBA is the command voltage. Any very small difference between these two is greatly amplified by the FBA and fed to the current electrode. This ensures that current passes across the membrane so as to maintain the membrane potential E at a clamped level equal to the command voltage. The membrane current I is monitored as the current flows to ground.

alternative recording methods, as indicated in fig. 2.5. The pipette electrode may be left as it is, in the 'cell-attached' position, or it may be pulled away, bringing the patch with it in the 'inside-out' arrangement. Application of more suction to the cell-attached arrangement breaks the patch

so that the electrode now records from the whole cell, and pulling the electrode away from this may tear off a patch of membrane in the 'outside-out' configuration. Further details are provided by Hamill *et al.* (1981), Ogden & Stanfield (1994) and Sakmann & Neher (1995).

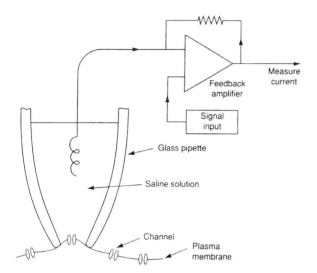

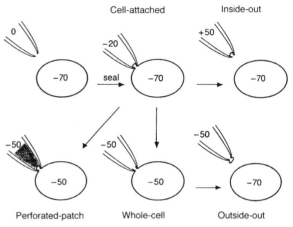

Figure 2.4. The patch clamp method for measuring currents through single ion channels. The patch electrode is a glass micropipette with a smoothed end, filled with a suitable saline solution. Slight suction makes it form a high resistance seal (the 'gigaseal') with the cell membrane, hence any current flowing through the electrode is also flowing through the membrane patch. The feedback amplifier keeps the membrane potential constant and allows the current through the patch to be measured. (From Aidley & Stanfield, 1996.)

Figure 2.5. Different configurations of the patch clamp technique. All begin with the formation of a gigaseal in the cell-attached configuration. Withdrawal of the pipette electrode produces an inside-out patch. Breakage of the patch while the pipette is attached to the cell gives whole-cell recording. Withdrawal from this produces an outside-out patch. A perforated patch (produced with nystatin or amphotericin in the pipette) also produces whole-cell records. The numbers show the electrode voltages required to produce a patch or whole-cell membrane potential of -50 mV if the cell resting potential is initially -70 mV. (From Cahalan & Neher, 1992.)

3
The resting cell membrane

If an intracellular microelectrode is inserted into a nerve or muscle cell, it is found that the inside of the cell is electrically negative to the outside by some tens of millivolts. This potential difference is known as the *resting potential*. If we slowly advance a microelectrode so that it penetrates the cell, the change in potential occurs suddenly and completely when the electrode tip is in the region of the cell membrane; thus the cell membrane is the site of the resting potential. In this chapter we shall consider some of the properties of the cell membrane that are associated with the production of the resting potential.

Membrane structure

Plasma membranes are usually composed of roughly equal amounts of protein and lipid, plus a small proportion of carbohydrate. Human red cell membranes, for example, contain about 49% protein, 44% lipid and 7% carbohydrate. Intracellular membranes tend to have a higher proportion of protein, whereas the protein content of myelin (p. 49) is only about 23%.

Figure 3.1 shows the chemical structure of some membrane lipids. Phospholipid molecules are esters of glycerol with two long-chain fatty acids, the glycerol moiety being attached via a phosphate group to various small molecules. The fatty acid chains thus form non-polar tails attached to polar heads. The fatty acid chains are usually fourteen to twenty carbon atoms long, and some of them are unsaturated, with one or more double bonds in the chain. Sphingolipids contain an amide link between the two fatty acid chains. Glycolipids have one or more sugar residues attached to hydrocarbon chains. Cholesterol is a sterol.

When lipids are spread on the surface of water, they form a monolayer in which the polar ends of the molecules are in contact with the water surface and the non-polar hydrocarbon chains are oriented more or less at right angles to it. This monolayer can be laterally compressed until the lipid molecules are in contact with each other; at this point the lateral pressure exerted by the monolayer has reached a maximum, as can be seen by the use of a suitable surface balance. Using such a balance, Gorter & Grendel (1925)

measured the minimum area of the monolayer produced by the lipids extracted from red blood cells, and compared it with the surface area of the cells. They found that the monolayer area was almost double the surface area of the cells, and concluded that the lipids in the cell membrane are arranged in a layer two molecules thick. (Later work showed that this result was somewhat fortuitous in that an underestimate of the surface area seems to have been compensated for by incomplete extraction of the lipids: see Bar *et al.* 1966.)

The idea that the essential barrier to movement of substances across the cell membrane is a layer of lipid molecules is supported by the observation that lipid-soluble substances appear to penetrate the cell membrane more readily than do many non-lipid-soluble substances. The electric capacitance of cell membranes is usually about $1 \, \mu\text{F cm}^{-2}$; this is what one would expect if the membrane were a bimolecular layer of lipid 50 Å (1 Å = 10 nm) thick with a dielectric constant of 5. ← *why dielectric consta of 5?*

Davson & Danielli (1943) suggested that the lipid bilayer was stabilized by a thin layer of protein molecules on each side of it. High resolution electron microscopy of sections of cells (usually fixed with potassium permanganate) shows that the cell membrane appears as two dense lines separated by a clear space, the whole unit being about 75 Å across (see Robertson, 1960, 1989). This accords well with Davson & Danielli's model, since one would expect electron-dense stains to be taken up by the polar groups of the lipid molecules and the proteins associated with them, but not by the non-polar lipid chains in the middle of the membrane. X-ray diffraction studies (Wilkins *et al.*, 1971) fit well with the bilayer hypothesis, giving a distance of 45 Å between the polar head groups in erythrocyte membranes. This is about what we would expect from the dimensions of phospholipid molecules (fig. 3.2).

About this time it became clear that the Davson–Danielli model, with its thin layer of protein on each side of the lipid bilayer, did not really fit the facts. In red cell membranes there was not enough lipid to account for the observed thickness of the membrane, suggesting that part of its area

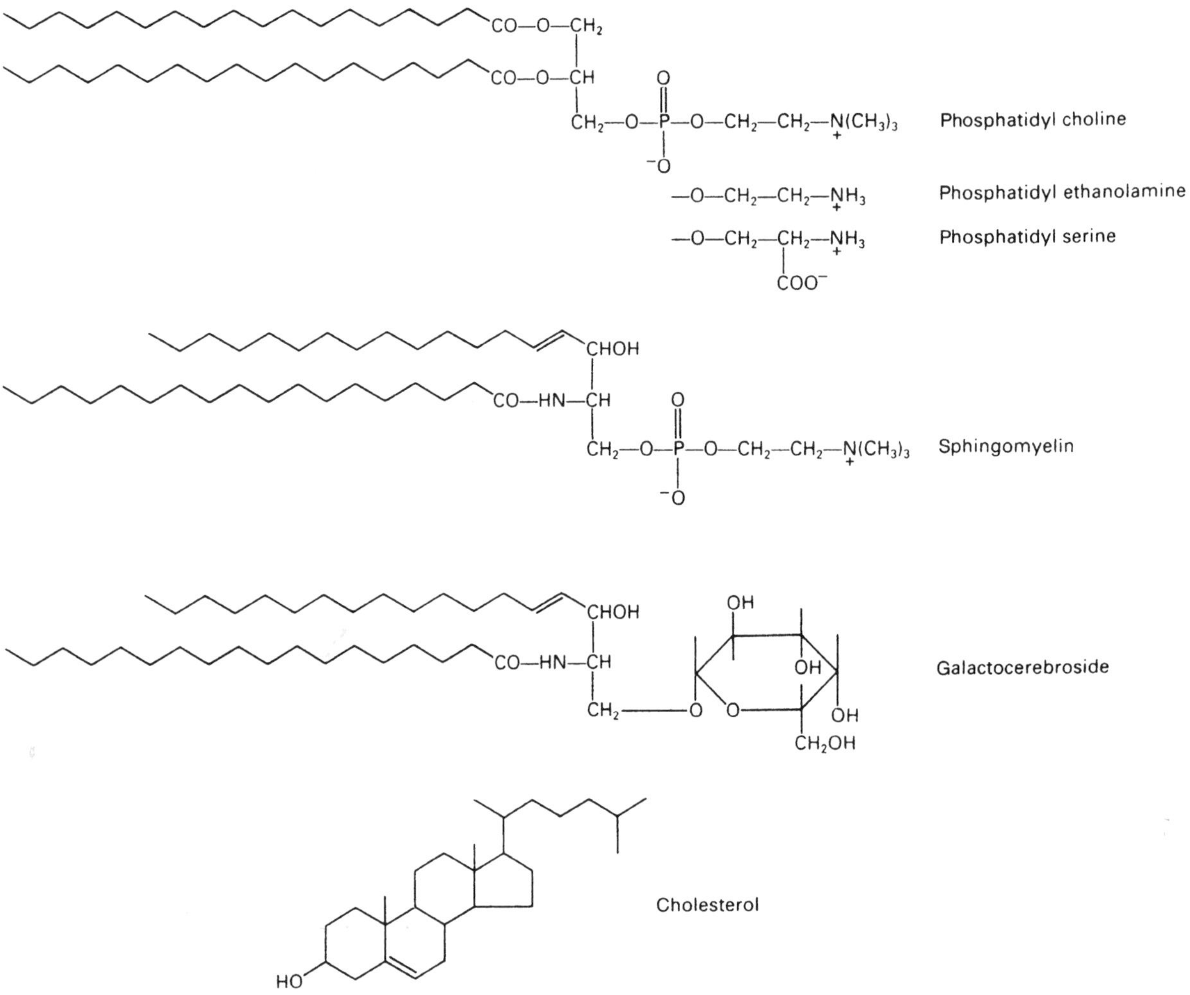

Phosphatidyl choline

Phosphatidyl ethanolamine

Phosphatidyl serine

Sphingomyelin

Galactocerebroside

Cholesterol

Figure 3.1. Chemical structures of some plasma membrane lipids. The fatty acid chains on the left may be of variable length and may have one or more double bonds.

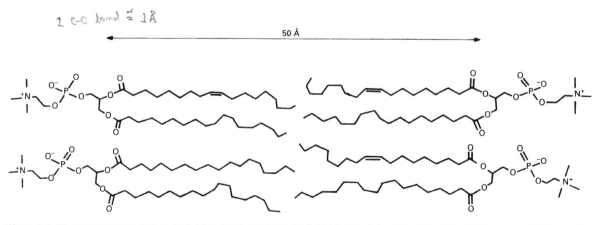

1 C–C bond ≈ 1 Å

50 Å

Figure 3.2. The arrangement of phospholipid molecules to form the lipid bilayer of the plasma membrane. (Based on Griffith *et al.*, 1974, and Marsh, 1975.)

was occupied by protein (Wilkins *et al.*, 1971). Enzymes which would hydrolyse phospholipids were able to attack cell membranes, suggesting that the phospholipids were not protected by an overlying layer of protein. But perhaps the most graphic evidence comes from a special technique of electron microscopy known as freeze fracture. A portion of tissue is frozen and then broken with a sharp knife. The cell membranes then cleave along the middle so as to separate the inner and outer lipid leaflets. A replica of the fractured face is made, 'shadowed' with some electron dense material and then examined in the electron microscope. Small particles are seen projecting from both faces, but especially from the inner one; an example is shown in fig. 10.7. If these are protein molecules (and it is difficult to see what else they could be) then they are clearly embedded within the lipid bilayer, rather than simply applied to its surface as in the Davson–Danielli model.

These results led to the conclusion that much of the protein of the plasma membrane penetrates the lipid bilayer, as is shown in fig. 3.3 (Singer & Nicolson, 1972; Bretscher, 1973). Singer & Nicolson suggested that the hydrophobic parts of these intrinsic membrane proteins are embedded in the non-polar environment formed by the hydrocarbon chains of the lipid molecules, whereas their hydrophilic sections project from the membrane into the polar environment provided by the aqueous media on each side of the membrane.

Singer & Nicolson viewed the membrane as a mosaic in which a variety of protein molecules serve different functions. There is evidence that some of the protein molecules are able to move in the plane of the membrane (rather like icebergs in the surface waters of some arctic sea) and hence the structure illustrated in fig. 3.3 has become known as the fluid mosaic model. The model has been refined somewhat since its original formulation (Singer, 1990, 1992). It now includes peripheral proteins, which are attached to the membrane but not embedded in it, as well as the intrinsic proteins which enter the bilayer. Many of the intrinsic proteins may have their movements restricted by attachment to each other, to peripheral proteins or to proteins of the cytoskeleton. Intrinsic proteins nearly always cross the whole of the bilayer, so that their hydrophilic regions emerge on each side of the membrane.

Plasma membranes are highly asymmetrical structures (Bretscher, 1973; Rothman & Lenard, 1977). Phospholipids are differentially distributed in the two leaflets of the bilayer, so that there is more sphingomyelin and phosphatidylcholine in the outer leaflet and more phosphatidylethanolamine and phosphatidylserine*in the inner one (Verkieij *et al.*, 1973). Proteins show an absolute asymmetry: they are nearly all positioned in their own particular way with respect to the inner or outer surface of the membrane. Thus, for example, transporting enzymes always have their ATP-binding sites on the inner (cytoplasmic)

Exposed phosphatidyl-serine signals death and recruits mø

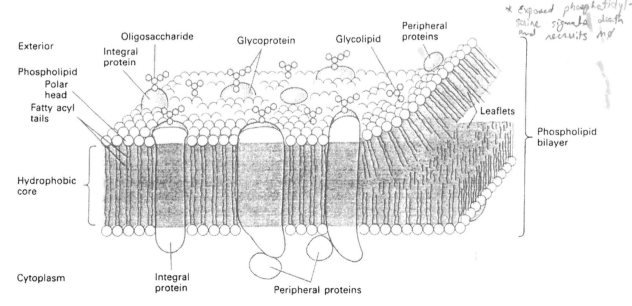

Figure 3.3. Fluid mosaic model of the plasma membrane. The phospholipid bilayer is shown split into two leaflets at the right, as it might be in the freeze fracture technique. Integral proteins traverse the bilayer, peripheral proteins sit on its surface. Oligosaccharides occur on the outer surface of the membrane, attached mainly to proteins but also to lipids, forming glycoproteins and glycolipids respectively. (From Darnell *et al.*, 1986.)

surface, glycoproteins have their sugar residues on the outer surface, and the peripheral proteins on the outer surface have functions quite different from those on the cytoplasmic surface.

Substances that can pass through the lipid bilayer include gases such as oxygen and nitric oxide, some small non-polar molecules such as ethanol, and various lipid-soluble substances such as certain anaesthetics. However the lipid bilayer is largely impermeable to most of the substances in the aqueous media on each side of it. Hence the plasma membrane acts as a barrier separating the contents of the cell from the outside world. But some of the intrinsic proteins embedded in the membrane may act as doors in the barrier through which particular information or specific substances can be transferred from one side to the other.

Membrane proteins
There are three major types of protein concerned with the movement of substances from one side of the cell to the other (Stein, 1990; Yeagle, 1993). *Pumps* act to drive ions against a concentration gradient, and the energy for this is derived from the breakdown of adenosine triphosphate, ATP. *Channels* contain transmembrane pores that can open or close to permit the flow of particular ions or water down their concentration gradients. *Transporters* combine with specific molecules and undergo a conformational change so as to release them on the other side of the membrane, without involving any breakdown of ATP. Some transporters move just one type of molecule down its concentration gradient; these are known as uniporters, or *carriers* for facilitated diffusion. Others, called *coupled transporters*, utilize the energy released by the movement of one substance down its concentration gradient to move another substance up its concentration gradient; the two movements may be either in the same direction (symporters) or in opposite directions (antiporters).

It may be important for the cell to transfer information across the plasma membrane, rather than substances, so that extracellular events can elicit intracellular responses. One way of doing this is via intrinsic proteins known as *receptors*. These combine with a specific molecule on the outer surface of the cell and produce some change in the molecule so as to affect the inside of the cell. Some ion channels are also receptors: the nicotinic acetylcholine receptor, for example, contains an ion channel that opens briefly when the receptor binds two molecules of acetylcholine at its outer surface. Others receptors undergo conformational changes which activate other proteins inside the cell: the β-adrenergic receptor, for example, activates a G protein on the inner surface of the plasma

membrane when it binds noradrenaline at its outer surface.

Information about the structures of proteins is crucial to an understanding of how they work. Proteins are linear polymers each with a unique sequence of amino acid residues, and this sequence is itself determined by the genetic material of the cell. The DNA in the cell nucleus acts as a template for the messenger RNA which in turn determines the amino acid sequence of the protein. A small number of proteins have had their primary structures (their amino acid sequences) determined directly by chemical analysis. But in recent years the development of recombinant DNA technology has allowed the amino acid sequence of a protein to be determined indirectly from the sequence of bases in the nucleic acid that codes for it. This is a much faster and more efficient process than the direct chemical analysis method, and many thousands of proteins have been sequenced in this way.

The twenty amino acid residues in protein chains have different properties. Three of them are basic, with positive charges, two are acidic, with negative charges, seven are polar but uncharged, and eight are non-polar (table 3.1). Different residues have different affinities for aqueous and lipid environments; thus valine and leucine, for example, are non-polar and so hydrophobic (they will tend to be found preferentially in a lipid environment); glutamic and aspartic acids, lysine and arginine are charged and therefore hydrophilic, so they are more likely to be found in an aqueous environment.

The amino acid sequence is an important determinant of the way in which the protein chain is folded. For membrane proteins it is particularly useful to know which parts of the molecule are embedded in the membrane and which parts are in contact with the cytoplasm or the extracellular space. Non-polar residues tend to occur in the middle of the molecule or, in intrinsic membrane proteins, in association with the lipid bilayer. Polar and charged residues are more likely to be found on the outside of the molecule in contact with the aqueous environment. Hydrogen bonds, electrostatic interactions and disulphide links (between pairs of cysteine residues) all serve to hold the chain in its folded conformation.

Inspection of the amino acid sequence of the protein can give some clues towards its secondary or tertiary structure, by which we mean just how the amino acid chain is folded and arranged in space. A common arrangement is known as the α-helix, in which a section of the chain forms a helix that is stabilized by hydrogen bonds between its constituent residues. The helix is right-handed, with a rise per residue of 1.5 Å and a pitch height of 5.4 Å, corresponding to 3.6 amino acid residues per turn (see Creighton, 1993). An

Table 3.1 *The amino acids of proteins. They have the general formula*
$R\text{—}CH(NH_2)COOH$, *where R is the side-chain or residue. Proline is actually an imino acid*

Type	Amino acid	Side chain	Abbreviations		Hydropathy index
Non-polar	Isoleucine	$\text{—CH(CH}_3)\text{CH}_2\text{CH}_3$	Ile	I	4.5
	Valine	$\text{—CH(CH}_3)_2$	Val	V	4.2
	Leucine	$\text{—CH}_2\text{CH(CH}_3)_2$	Leu	L	3.8
	Phenylalanine	$\text{—CH}_2\text{C}_6\text{H}_5$	Phe	F	2.8
	Methionine	$\text{—CH}_2\text{CH}_2\text{SCH}_3$	Met	M	1.9
	Alanine	—CH_3	Ala	A	1.8
	Tryptophan	$\text{—CH}_2\text{C(CHNH)C}_6\text{H}_4$	Trp	W	−0.9
	Proline	$\text{—CH}_2\text{CH}_2\text{CH}_2\text{—}$	Pro	P	−1.6
Uncharged polar	Cysteine	$\text{—CH}_2\text{SH}$	Cys	C	2.5
	Glycine	—H	Gly	G	−0.4
	Threonine	—CH(OH)CH_3	Thr	T	−0.7
	Serine	$\text{—CH}_2\text{OH}$	Ser	S	−0.8
	Tyrosine	$\text{—CH}_2\text{C}_6\text{H}_4\text{OH}$	Tyr	Y	−1.3
	Histidine	$\text{—CH}_2\text{C(NHCHNCH)}$	His	H	−3.2
	Glutamine	$\text{—CH}_2\text{CH}_2\text{CONH}_2$	Gln	Q	−3.5
	Asparagine	$\text{—CH}_2\text{CONH}_2$	Asn	N	−3.5
Acidic	Aspartic acid	$\text{—CH}_2\text{COO}^-$	Asp	D	−3.5
	Glutamic acid	$\text{—CH}_2\text{CH}_2\text{COO}^-$	Glu	E	−3.5
Basic	Lysine	$\text{—(CH}_2)_4\text{NH}_3^+$	Lys	K	−3.9
	Arginine	$\text{—(CH}_2)_3\text{NHC(NH}_2)\text{NH}_3^+$	Arg	R	−4.5

Note:
Hydropathy index from Kyte & Doolittle (1982).

α-helix with twenty or so amino acid residues would be long enough to cross the lipid bilayer from one side to the other. A section of about twenty non-polar residues, therefore, is likely to form a transmembrane α-helix.

Kyte & Doolittle (1982) arranged the twenty amino acids commonly found in proteins on a scale of hydropathicity (also known as hydropathy or hydrophobicity) which measures their affinity for non-polar environments. The scale runs from a hydropathicity index of $+4.5$ for the hydrophobic isoleucine to -4.5 for the hydrophilic arginine; the values for the different amino acids are shown in table 3.1. The scale can be used to tell us something about the structure of a protein whose amino acid sequence is known. For an amino acid at position i, the average hydropathicity index of all the amino acids from position $i-n$ to position $i+n$ (where n is 7 or 9, for example) is plotted. The result is a hydropathicity profile whose peaks show hydrophobic sections of the protein chain and whose valleys show hydrophilic sections. If the hydrophobic sections are long enough, we may suspect that they are embedded in the lipid environment of the membrane and cross from one side of it to the other.

Figure 6.8 shows an example of this type of analysis, for the voltage-gated sodium channel from the electric eel. The hydropathy profile has four similar groups of six peaks, representing four homologous domains each with six putative membrane-crossing segments, labelled S1 to S6. (The word 'putative' means 'commonly supposed to be', and its use indicates the indirect nature of the evidence.)

Singer (1990) classified integral membrane proteins into four main groups according to their transmembrane topography (fig. 3.4). The simpler forms, types I and II in fig. 3.4, have just one transmembrane segment. Type III proteins have a single polypeptide chain which traverses the lipid bilayer a number of times, as in rhodopsin, the β-adrenergic receptor, adenylyl cyclase and many others. Type IV proteins are typical of ion channels; they have a number of domains or subunits that are arranged together around an aqueous pore through which the ion movement occurs.

Ion channels

Ion channels are intrinsic membrane proteins containing aqueous pores that can open or shut. When they are open they permit specific ions to flow down their electrochemical gradients from one side of the membrane to the other. They provide the molecular basis for excitation processes and for many other cellular activities involving changes in the ionic permeability of membranes (Hille, 1992; Aidley & Stanfield, 1996). We shall look in some detail at the structure of a number of different channels later in this book; here is given a brief outline of some of the different types.

Channels vary considerably in their *gating*, by which we mean the factors that make them open or close. Some channels are opened by combination with particular chemicals outside or inside the cell, such as neurotransmitters or cytoplasmic messenger molecules. Others are opened by changes in the voltage across the membrane, and yet others by sensory stimuli of various kinds. They show *selectivity* in the ions to which they are permeable. Some of them will permit only particular ions to pass through, such as sodium, potassium, calcium or chloride ions; others are selective for broader groups of ions, such as monovalent cations, or cations in general.

These two aspects of channel functioning, gating and selectivity, are commonly used to describe the different types of channel, as is illustrated in fig. 3.5. The nerve axon,

for example, possesses *voltage-gated channels*, i.e. channels that are opened by a change in the membrane potential, and these are of two main types, sodium channels and potassium channels. Voltage-gated calcium channels also occur, particularly at the nerve terminals. We can refer to all these voltage-gated channels as a group distinct from others.

Channels gated by neurotransmitters include the nicotinic acetylcholine receptor channel, the γ-amino butyric acid (GABA) receptor channel, and the glycine receptor channel. A third group includes all those gated by combination with internal ligands, such as calcium ions, ATP, cyclic nucleotides and so on. Sometimes these two groups are together described as *ligand-gated* or chemically gated channels, to distinguish them from the voltage-gated channels. Gap junction channels connect the cytoplasmic compartments of adjacent cells by crossing two plasma membranes.

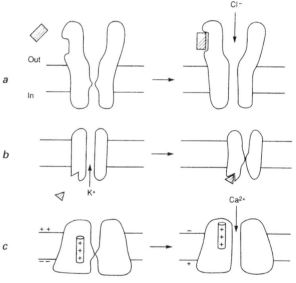

Figure 3.5. Different types of ion channel, illustrating some of the variety in their gating and selectivity. *a* shows a neurotransmitter-gated channel which is selective to anions. *b* shows a channel selective for potassium ions which is closed by the binding of an internal ligand such as ATP. *c* shows a calcium-selective voltage-gated channel; part of the internal structure of the channel is charged and moves when the membrane potential becomes more positive inside, and this acts as the trigger for opening the channel. *a* is based loosely on the glycine receptor channel which mediates synaptic inhibition in the spinal cord, *b* on the ATP-sensitive potassium channel of pancreatic β cells, and *c* on the voltage-gated calcium channels of neurons and secretory cells. The diagrams are much simplified: channels of the *a* and *b* types often require more than one ligand molecule for gating, and channels like *c* usually have four internal gating sections. (From Aidley & Stanfield, 1996.)

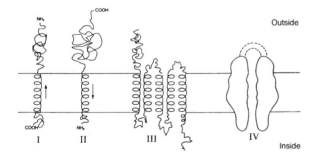

Figure 3.4. Topography of four main types of integral membrane proteins. Type I has a single membrane-crossing segment, with the N-terminus on the external face of the membrane. Type II is similar, but with the N-terminus on the cytoplasmic face. Type III proteins have more than one membrane-crossing segment. Type IV proteins are composed of a number of domains surrounding an aqueous pore; the domains may be either separate subunits or may be connected in a single polypeptide chain as is suggested by the dotted lines. (From Singer, 1990. Reproduced with permission from the *Annual Review of Cell and Developmental Biology* Volume 6, © 1990 by Annual Reviews Inc.)

Concentration cells

Most of the electric potentials across cell membranes arise from the movement of ions from one side of the membrane to the other through ion channels. To understand how this happens we need a little physical chemistry.

Consider the imaginary system shown in fig. 3.6. The two compartments contain different concentrations of an electrolyte X^+Y^- in aqueous solution. They are separated by a membrane which contains ion channels selective for the cation X^+; this means that it will be permeable to the cation X^+ but impermeable to the anion Y^-. The concentration of XY in compartment 1 is higher than that in compartment 2. We start with all the channels closed. Since the concentration of X is equal to that of Y in compartment 1, and the same holds for compartment 2, there will be no excess of electric charge in either compartment and so no potential difference between them.

Now we open the channels. X^+ ions will move down the concentration gradient from compartment 1 to compartment 2, carrying positive charge with them. This movement of charges causes a potential difference to be set up between the two compartments, with compartment 2 positive to compartment 1. The higher the potential gets, the harder it is for the X^+ ions to move against the electrical gradient. Hence an equilibrium position is reached at which the electrical gradient (tending to move X^+ from 2 to 1) just balances the chemical or concentration gradient (tending to move X^+ from 1 to 2). Since the potential

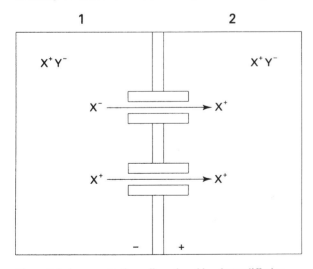

Figure 3.6. A concentration cell produced by electrodiffusion through channels in a membrane. The solute is a salt XY, at higher concentration in compartment 1 than in 2. The channels are permeable to X^+ but not to Y^-. Hence X^+ ions move from 1 to 2 until the potential produced by their excess positive charges in 2 is sufficient to stop further net movement.

difference at equilibrium (known as the *equilibrium potential* for X^+) arises from the difference in the concentration of X^+ in the two compartments, the system is known as a *concentration cell*.

What is the value of this potential difference? Suppose a small quantity, δn moles, of X is to be moved across the membrane, up the concentration gradient, from compartment 2 to compartment 1. Then, applying elementary thermodynamics, the work required to do this, δW_c, is given by

$$\delta W_c = \delta n\, RT \ln \frac{[X]_1}{[X]_2}$$

where R is the gas constant (8.314 J deg^{-1} mol^{-1}), T is the absolute temperature, and $[X]_1$ and $[X]_2$ are the molar concentrations of X in compartments 1 and 2 respectively. (More strictly, we should use the activities of X in the two compartments, rather than their concentrations. The simplification given here is valid as long as the activity coefficient of X is the same in each compartment. For animal cells this is usually the case.)

Now consider the electrical work, δW_e required to move δn moles of X against the electrical gradient, i.e. from compartment 1 to compartment 2. This is given by

$$\delta W_e = \delta n\, zFE$$

where z is the charge on the ion, F is Faraday's constant ($96\ 500$ coulombs mol^{-1}) and E is the potential difference in volts between the two compartments (measured as the potential of compartment 2 with respect to compartment 1). Now, at equilibrium, there is no net movement of X, and therefore

$$\delta W_c = \delta W_e$$

or

$$\delta n/zFE = \delta n/RT \ln \frac{[X]_1}{[X]_2}$$

i.e.

$$E = \frac{RT}{zF} \ln \frac{[X]_1}{[X]_2} \tag{3.1}$$

This equation is known as the *Nernst equation* after its formulator. It is the most important equation in this book. The reason for this is that it helps us to understand the origins of electric potentials in excitable cells, which are in part dependent upon differences in the ionic composition of the cytoplasm and the external fluid.

Equation 3.1 can be simplified by enumerating the constants to become, at 18 °C,

$$E = \frac{25}{z} \ln \frac{[X]_1}{[X]_2} \qquad (3.2)$$

or

$$E = \frac{58}{z} \log_{10} \frac{[X]_1}{[X]_2} \qquad (3.3)$$

where E is now given in millivolts. For instance, in fig. 3.6, suppose $[X]_1$ were ten times greater than $[X]_2$, then E would be $+58$ mV if X were K^+ and $+29$ mV if X were Ca^{2+}; if the channels were permeable to Y and not to X, then E would be -58 mV if Y were Cl^-, and -29 mV if Y were SO_4^{2-}.

How many X ions have to cross the membrane to set up the potential? The answer depends upon the valency of the ion, the value of the potential set up and the capacitance of the membrane. Consider, for example, a squid giant axon with a membrane capacitance C of 1 μF cm^{-2}, and a membrane potential V of 70 mV. Then the charge on 1 cm^2 is given by

$$Q = CV$$

where Q is measured in coulombs, C in farads and V in volts. The number of moles of X moved will be CV/zF, where z is the charge on the ion and F is the Faraday constant. In this case, assuming (to anticipate a little) that the membrane potential is set up primarily by a potassium ion concentration cell,

$$\frac{CV}{zF} = \frac{10^{-6} \times 7 \times 10^{-2}}{96500}$$

$$= 7.3 \times 10^{-13} \text{ mol cm}^{-2}$$

which is a *very* small quantity. If the diameter of the squid axon is 1 mm, then 1 cm^2 of membrane will enclose a volume containing about 3×10^{-5} mol of potassium ions, and thus a loss of 7.3×10^{-13} mol would be undetectable by normal chemical analysis. In this case the difference between the numbers of positive and negative charges inside the axon would be about 0.000002%.

It is important to realize that the movements of ions from one compartment to another which we have been discussing are *net* movements, i.e. the overall movements produced by the sum of all the movements of individual ions crossing the membrane in both directions. At the equilibrium position, although there is now no net ionic movement, X ions will still move across the membrane, but the rate of movement (the *flux*) is equal in each direction. Before the equilibrium position is reached, the flux in one direction will be greater than that in the other. Fluxes can be measured only by using isotopic tracer methods. In

Table 3.2 *Ionic concentrations in frog muscle fibres and in blood plasma*

	Concentration in muscle fibres (mM)	Concentration in plasma (mM)
K^+	124	2.25
Na^+	10.4	109
Cl^-	1.5	77.5
Ca^{2+}	4.9	2.1
Mg^{2+}	14.0	1.25
HCO_3^-	12.4	26
Organic anions	*ca* 74	*ca* 13

Note:
Most of the calcium inside the muscle fibres is not in free solution. Simplified after Conway, 1957.

describing fluxes across cell membranes, movement of ions into the cell is called the influx, and movement out the efflux.

We have seen that an ion will tend to move down its concentration gradient and also down the electrical gradient. The two gradients together form the *electrochemical gradient*. At equilibrium, the electrochemical gradient is zero, so there is no net movement of the ions through their channels. When the potential across the membrane is not equal to the Nernst potential, then the electrochemical gradient is not zero and so there will be a net movement of ions through their open channels and down their electrochemical gradient.

Ionic concentrations in the cytoplasm

Various microchemical methods are available for determining the quantities of ions present in a small mass of tissue. These include various microtitration techniques, flame photometry (in which the quantity of an element is determined from the intensity of light emission at a particular wavelength from a flame into which it is injected) and activation analysis (in which the element is converted into a radioactive isotope by prolonged irradiation in an atomic pile). All estimations made on a mass of tissue must include corrections made for the fluid present in the extracellular space. The size of the extracellular space is usually estimated by measuring the concentration in the tissue of a substance which is thought not to penetrate the cell membrane, such as inulin. Table 3.2 gives a simplified balance sheet of the ionic concentrations in frog muscle determined in this way. In the giant axons of squids, which may be up

Table 3.3 *Ionic concentrations in squid axoplasm and blood*

	Concentration in axoplasm (mM)	Concentration in blood (mM)
K	400	20
Na	50	440
Cl	40	560
Ca	0.4	10
Mg	10	54
Isethionate	250	—
Organic anions	*ca* 74	*ca* 13

Note:
Simplified after Hodgkin, 1958.

Table 3.4 *Application of the Nernst equation to nerve and muscle cells*

	Ion	Ion concentrations		Nernst potential (mV)
		External (mM)	Internal (mM)	
Frog muscle	K	2.25	124	−101
	Na	109	10.4	+59
	Cl	77.5	1.5	−99
Squid axon	K	20	400	−75
	Na	440	50	+55
	Cl	560	40	−66

to 1 mm in diameter, the situation is much more favourable, since the axoplasm can be squeezed out of an axon like toothpaste out of a tube; table 3.3 shows the concentrations of the principal constituents.

The main features of the ionic distribution between the cytoplasm and the external medium are similar in both the squid axon and the frog muscle fibre. In each case the intracellular potassium concentration is much greater than that of the blood, whereas the reverse is true for sodium and chloride. Moreover, the cytoplasm contains an appreciable concentration of organic anions. These features seem to be common to all excitable cells. As we shall see, these inequalities in ionic distribution are essential to the electrical activity of nerve and muscle cells. They are dependent upon active transport processes which are driven by metabolic energy, especially the active extrusion of sodium ions from the cell by the action of the sodium pump.

Active transport of ions

Active transport is the name given to a process that produces a net movement of an ion species or other small molecule across the membrane against its electrochemical gradient. In animal cells the energy required for this movement is derived from metabolic energy by the breakdown of ATP.

Sodium ions

The Nernst potential for any particular ion species shows what the cell membrane potential would be if the distribution of that ion across the membrane had reached equilibrium. If the actual membrane potential is not equal to the Nernst potential, then the ion is not in equilibrium and will

tend to flow down its electrochemical gradient if there are open channels in the membrane permeable to it.

The resting potential of frog muscle cells is usually around −90 to −100 mV, and that for squid axon is typically about −60 mV. Table 3.4 shows the Nernst potentials for the major monovalent ions in these cells. Clearly the distribution of potassium and chloride ions is fairly close to equilibrium in both cases, since their Nernst potentials are not far from the resting potential. Sodium ions, however, have a Nernst potential which is more than 100 mV more positive than the resting potential, and hence there is a large electrochemical gradient for sodium ions, tending to drive them into the cell.

One explanation for this situation might assume that the cell membrane is impermeable to sodium ions. However, this is not so; resting nerve and muscle cells do allow some entry of sodium ions. Experiments with radioactive sodium ions, for example, show that the resting influx of sodium into giant axons of the cuttlefish *Sepia* is about 35 pmol cm^{-2} s^{-1}. During activity each action potential is accompanied by a net inward movement of 3 to 4 pmol cm^{-2} s^{-1}, as we shall see later. If nothing were done about these inward movement of sodium ions, the system would in time run down, producing equal concentrations of all ions on both sides of the membrane. In fact, the cell prevents this by means of a continuous extrusion of sodium ions through the activity of the sodium pump. Since such extrusion must occur against an electrochemical gradient, we would expect this process to be an active one, involving the consumption of metabolic energy.

Let us first consider some experiments by Hodgkin & Keynes (1955a) on the extrusion of sodium from *Sepia* giant axons. They placed an axon in sea water containing radioactive sodium ions and stimulated it repetitively for some time, so that (as we shall see in chapter 5) the interior of the axon became loaded with radioactive sodium. They then put it in a capillary tube through which non-radioactive sea water could be drawn (fig. 3.7), and determined the efflux of sodium ions by measuring the radioactivity of samples of this sea water at intervals. This particular arrangement has two advantages: the efflux from the cut ends of the axon is not measured, and the measured efflux occurs from a known length of axon into a known volume of sea water. They found that the relation between the logarithm of the efflux and time is a straight line of negative slope (as in the first section of the graph in fig. 3.8), indicating that the efflux of radioactive sodium falls exponentially with time. This is just what we would expect to see if both the internal sodium ion concentration and the rate of extrusion of sodium ions are constant, since under these conditions a constant proportion of the radioactive sodium inside the axon will be removed in successive equal time intervals.

When 2,4-dinitrophenol (DNP) was added to the sea water, the sodium efflux fell markedly, and then recovered when the DNP was washed away (fig. 3.8). DNP is an inhibitor of metabolic activity, and probably acts by uncoupling the formation of the energy-rich compound adenosine triphosphate (ATP) from the electron chain in aerobic respiration; hence this experiment implies that the extrusion of sodium is probably dependent on metabolic energy supplied directly or indirectly in the form of ATP. Hodgkin & Keynes also showed that the sodium efflux from *Sepia* axons is dependent upon the external potassium ion concentration; potassium-free sea water reduced the efflux to about a third of its normal value. This implies that the sodium extrusion process is coupled to the uptake of potassium.

More conclusive evidence of the need for ATP is provided by a series of experiments by Caldwell *et al.* (1960), one of which is shown in fig. 3.9. A solution containing radioactive sodium ions was injected into a giant axon of the squid *Loligo* by means of a very fine glass tube inserted longitudinally into the axon, and the efflux was then measured. Soon after the start of the experiment the axon was poisoned with cyanide. Cyanide ions interfere with the energy-producing processes in aerobic respiration (by inhibiting the action of cytochrome oxidase), and so the efflux of sodium fell to a low level. Injections of ATP, arginine phosphate or phosphoenol pyruvate into the axon produced transient increases in the rate of sodium extrusion, so confirming the view that the sodium pump is driven by the energy derived from the breakdown of energy-rich phosphate compounds.

In terms of the fluid mosaic model of the cell membrane (fig. 3.3), we may suppose that sodium ions are pumped at particular sites each consisting of a particular protein molecule or group of molecules. How many of these sites are there in any particular area of membrane? The problem has been approached by measuring the binding of the drug ouabain to nerve axons. Ouabain is a glycoside found in certain plants and used as an arrow poison in parts of Africa. It acts by inhibiting the sodium pump of cell membranes. By measuring the uptake of radioactive ouabain it is possible to estimate the number of sodium-pumping sites in a system, assuming that each site binds one molecule of ouabain. In squid giant axons, Baker & Willis (1972) estimated that there are 1000 to 5000 sites per square micrometre of cell membrane, and Ritchie & Straub (1975) estimated that for garfish olfactory nerve there are 350 sites per square micrometre.

*Review

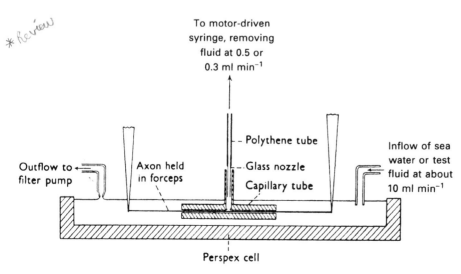

Figure 3.7. The apparatus used to measure the efflux of radioactive sodium ions from giant axons of the cuttlefish *Sepia*. (From Hodgkin & Keynes, 1955a.)

More information about the sodium pump came from experiments on human red blood cells. Thus Post & Jolly (1957) measured the ratio of sodium ions to potassium ions moved by the pump. They stored red cells for two days at 2 °C; this would stop the activity of the pump and allow sodium entry and potassium loss by passive fluxes. On raising the temperature to 37 °C the pump became active again and so after a few hours it was possible to measure the change in the amounts of sodium and potassium ions in the cells. The results showed that two potassium ions were taken up for every three sodium ions extruded. In similar experiments with resealed red cell ghosts (red cells which have had their haemoglobin removed), Garrahan &

Glynn (1967) found that one ATP molecule was split for every three sodium ions extruded.

Our understanding of the sodium pump was much enhanced when Skou (1957, 1989) isolated an ATPase from crab nerves that was stimulated by sodium and potassium ions. An ATPase is an enzyme which splits ATP, usually with the object of utilizing the energy of the ATP molecule for some energy-demanding function of the cell. In this case the energy derived from the splitting of ATP is used to drive sodium ions up their electrochemical gradient and out of the cell. Later work showed that the enzyme is a dimeric protein consisting of an α subunit (molecular weight, M_r, 112 000) which contains the ATPase activity and the ouabain-binding site, and a β subunit (M_r 55 000).

Work by Albers and his colleagues and by Post and his colleagues, using ATP labelled with radioactive phosphorus, showed that sodium ions catalyse the phosphorylation of the pump by ATP and potassium ions catalyse its dephosphorylation (Albers *et al.*, 1963; Charnock & Post, 1963; Post *et al.*, 1965). It seemed likely that during the activity of the pump the Na,K-ATPase molecule cycles between two distinct conformations, E_1 and E_2. The E_1 form would split ATP to become phosphorylated, the reaction being catalysed by sodium ions, and the E_2 form would lose its phosphate in the presence of potassium ions. It was also evident that the phosphorylated E_1 form would change spontaneously to the phosphorylated E_2 form, whereas the dephosphorylated E_2 form would revert spontaneously to the E_1 form.

Putting these reactions together leads to a cycle of events known as the Albers–Post model of the sodium pump,

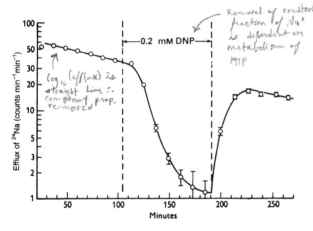

Figure 3.8. The effect of the metabolic inhibitor 2,4-dinitrophenol (DNP) on the efflux of radioactive sodium ions from a *Sepia* giant axon. The sodium pump stops once its fuel source is removed. (From Hodgkin & Keynes, 1955*a*.)

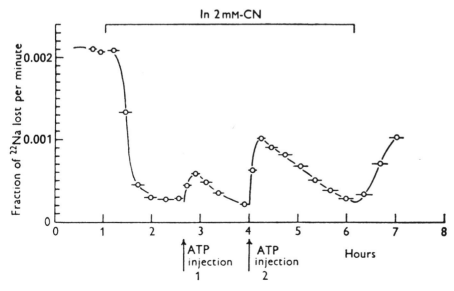

Figure 3.9. Experiment showing that ATP drives the sodium pump in a squid giant axon. The rate of sodium extrusion fell markedly after poisoning with cyanide, an inhibitor of ATP synthesis. Injection of ATP into the axon produced temporary reactivation of the sodium pump. More ATP was introduced in the second injection than in the first. (From Caldwell *et al.*, 1960.)

shown in fig. 3.10. The model also suggests that the the E_1 form has an ion-binding site that is accessible to the interior of the cell and prefers sodium to potassium ions, whereas in the E_2 form the binding site is accessible from the external solution and prefers potassium to sodium ions. Thus the pump molecule binds three sodium ions at the inner face of the membrane, and then changes shape to release them at the outer face, splitting an ATP molecule in the process. It now binds two potassium ions at the outer surface of the membrane, and changes shape again to release them at the inner face, losing an inorganic phosphate group as it does so.

Jørgensen (1975, 1992) produced some neat evidence for the Albers–Post model. He found that the digestive enzyme trypsin cleaves the α chain at different sites according to whether sodium or potassium ions are present. This implies that the molecule changes shape when combined with these two different ions; it seems probable that this shape change is involved in the pumping activity. Further details are discussed by Glynn (1993) and Läuger (1991).

The structure of the α subunit has been determined using complementary DNA cloning techniques. The base sequence of the complementary DNA was determined, and from this the amino acid sequence of the protein could be deduced. Details of such methods are given by Watson *et al.* (1992), for example, and we shall look at their application to the acetylcholine receptor (one of the first large membrane proteins to be examined with these techniques) in chapter 8.

The sequences of the Na,K-ATPase from sheep kidney (Shull *et al.*, 1985) and from the electric organ of the electric ray *Torpedo* (Kawakami *et al.*, 1985) are very similar, with over 85% of the sequence identical in the two enzymes. The α chain in the sheep kidney enzyme consists of a sequence of 1016 amino acid residues. Application of the Kyte–Doolittle analysis shows that there are probably eight or perhaps ten major hydrophobic sequences, suggesting that the protein chain traverses the membrane eight or ten times. ATP hydrolysis seems to involve binding to a lysine residue at position 501 and phosphorylation of an aspartic acid residue at 369; the chain is probably coiled so as to bring these two residues together. Binding of ouabain probably occurs at the tryptophan residue at position 310 in the short external portion between the third and fourth hydrophobic sequences. Figure 3.11 gives a general impression of the structure of the molecule.

Calcium ions

If a small quantity of radioactive calcium ions is injected into a squid axon, the resulting patch of radioactivity does not spread out by diffusion or move in a longitudinal electric field (in contrast to potassium, for example – see fig. 3.13). This suggests that nearly all the calcium is bound in some way, and that only a small proportion is ionized in free solution (Hodgkin & Keynes, 1957). The concentrations of free calcium ions can be determined by using a calcium-sensitive microelectrode or by injecting the protein aequorin, which emits light in the presence of calcium ions. Nowadays it is common to use calcium-sensitive

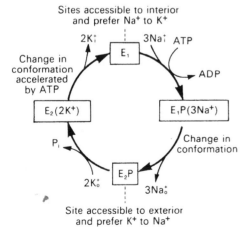

Figure 3.10. The basic cycle of the sodium pump. The pump molecule undergoes conformational change between two forms: the E_1 form, which will combine with sodium ions and ATP on the inside of the membrane; and the E_2 form, which will combine with potassium ions on the outside. (From Glynn, 1993.)

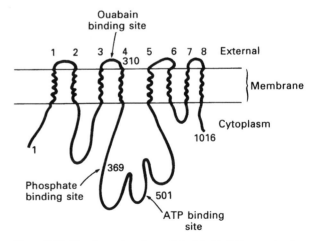

Figure 3.11. Membrane topology of the Na,K-ATPase molecule as suggested by Shull *et al.* (1985). An alternative model (Ovchinnikov *et al.*, 1988) has the eighth hydrophobic region and the C-terminus remaining on the external face of the membrane.

fluorescent dyes such as fura 2, often with calcium imaging techniques that will show the distribution of different calcium ion concentrations within the cell.

Investigations by these means show that most of the calcium inside cells is bound to proteins or sequestered in intracellular compartments such as the endoplasmic reticulum, so that the concentration of free ionic calcium is much lower than the total calcium concentration. In squid axons, for example, the total concentration of calcium in axoplasm is usually in the range 50 to 200 μM, whereas the concentration of ionized calcium is as low as 0.1 μM (Baker & Dipolo, 1984). Such values are typical of animal cells.

This situation means that when relatively small quantities of calcium ions are released inside the cell they produce proportionately large changes in the internal concentration of calcium ions. Thus release of 1% of the bound calcium in squid axoplasm would increase the free ionic concentration forty-fold, from 0.1 to 4 μM. The relative ease with which such changes can be produced by inflow of calcium ions through membrane channels or by release from intracellular compartments allows calcium ions to play key roles in the control of a large number of cellular processes (Dawson, 1990).

With an internal concentration of around 0.1 μM and an external concentration of around 1 mM in mammals, there is clearly a very great concentration gradient of calcium ions across the membrane, and it is therefore not surprising that calcium ions enter the cell whenever appropriate channels are open. This influx is normally balanced by a corresponding outward movement in which the calcium ions are moved up their electrochemical gradient, a process which must consume energy.

There are two components to this efflux. One is an ATP-dependent 'calcium pump', a membrane-bound Ca-ATPase that is active at low internal calcium ion concentrations. The other is a coupled transporter in which internal calcium ions are exchanged for external sodium ions; we shall look at it in the next section.

The Ca-ATPase has an M_r of about 140 000 and is activated by calmodulin. It has some similarities with the Na,K-ATPase in molecular structure. One ATP molecule is split for each calcium ion transported, and there is a cycle involving E_1 and E_2 forms of the enzyme similar to that for the sodium pump. In mammalian cells at 30 °C, each Ca-ATPase molecule transports 25 to 100 calcium ions out of the cell per second (Carafoli & Zurini, 1982; Carafoli, 1991).

Coupled transporter systems

Coupled transporter systems provide an alternative way of moving ions or other small molecules 'uphill' against their electrochemical gradient: the downhill movement of one ion is linked to the uphill movement of another ion or small molecule, and there is no direct involvement of ATP breakdown. The ion gradient that drives the exchanger is itself set up by an active transport process such as the sodium pump. Sodium-coupled transporter systems are used for the uptake of some neurotransmitters in neurons (chapter 10) and for the absorption of glucose in the gut. Here we look at their use in transporting ions.

Sodium–calcium exchange

During the normal operation of the sodium–calcium exchanger, a movement of sodium into the cell down its electrochemical gradient is linked to an outward movement of calcium against its electrochemical gradient. This was first demonstrated in squid axons by Blaustein & Hodgkin (1969) and in heart muscle by Reuter & Seitz (1968); in each case the efflux of ^{45}Ca was dependent upon the presence of external sodium ions. Under suitable conditions a similar exchange working in the opposite direction can be detected; this requires ATP for its activity (Baker *et al.*, 1969; Baker, 1986).

In squid axons the exchange system is not very active at normal calcium ion concentrations, so that most of the calcium extruded from the cell exits via the Ca-ATPase calcium pump; it becomes more important as the internal calcium ion concentration rises (DiPolo, 1989). In heart muscle, however, it is the major route for calcium extrusion from the cell (Philipson *et al.*, 1993).

The sodium–calcium exchanger protein of heart muscle cells has been sequenced by molecular cloning by Nicoll and her colleagues (1990). It is 938 amino acid residues long and probably contains eleven membrane-crossing segments. The corresponding exchanger from retinal photoreceptors (which transports potassium ions as well as sodium and calcium) is different in sequence, but has some similarities in three of the membrane-crossing segments (Reilander *et al.*, 1992).

Chloride movements

In squid giant axons, Keynes (1963) showed that the internal chloride ion concentration is higher than would be expected from a purely passive distribution of the ions. If chloride were passively distributed across the axonal membrane, then the internal chloride concentration would be given by

$$E = \frac{RT}{-F} \ln \frac{[\text{Cl}]_o}{[\text{Cl}]_i} \qquad (3.4)$$

(where E is the resting potential), or, in isolated axons,

$$-60 = -58 \log_{10} \frac{560}{[\text{Cl}]_i}$$

from which $[\text{Cl}]_i = 55$ mM

In fact, however, the average internal chloride concentration in Keynes's experiments was 108 mM, or about twice what one would expect. The chloride influx was halved by treating the axon with DNP, thus suggesting that there is some uptake of chloride ions that is dependent either directly or indirectly on ATP production.

More recent work showed that chloride is taken into squid axons by a sodium–potassium-chloride cotransport system, with an Na:K:Cl stoichiometry of 2:1:3 (Russell, 1983; Russell & Boron, 1990). The sodium ions enter the cell, flowing down their concentration gradient, accompanied by potassium and chloride ions. The energy for the whole process come from the sodium ion concentration gradient, itself set up by the sodium pump. Internal ATP is essential to the process, probably by enabling phosphorylation of the transporter molecule rather than acting as a primary energy supply.

Chloride and hydrogen ions are transported out of the squid giant axon by means of an exchange with sodium and bicarbonate ions. The consequence of this is that the inside of the cell is somewhat more alkaline than it would otherwise be. (If hydrogen ions were passively distributed, then, with a resting potential of -70 mV and an outside pH of 8.0, the internal pH predicted by the Nernst equation would be 6.8; in fact it is about 7.4.) The energy for this cotransport again comes from the sodium ion concentration gradient. The main function of this sodium-dependent chloride/bicarbonate exchanger seems to be in the regulation of intracellular pH (Russell & Boron, 1976, 1990). Thomas (1977) found evidence in snail neurons for a similar system, in which efflux of chloride and hydrogen ions was dependent upon influx of sodium and bicarbonate ions.

In many vertebrate central neurons, E_{Cl} is maintained more negative than the resting potential by means of a potassium–chloride cotransport system; chloride extrusion is linked to outward movement of potassium ions down their electrochemical gradient (Thompson *et al.*, 1988; Thompson & Gähwiler, 1989). This enables inhibitory postsynaptic potentials that are produced by the opening of transmitter-gated chloride channels to be hyperpolarizing (p. 126). In dorsal root ganglion cells, there is an inward accumulation of chloride as a result of a sodium–potassium–chloride cotransport system that operates as the result of the sodium ion concentration gradient (Alvarez-Leefmans, 1990).

The resting potential
The potassium electrode hypothesis

At the beginning of this century, Bernstein produced his 'membrane theory' of the resting potential. At that time there was some doubt as to whether there actually was any resting potential in the intact, uninjured cell. Bernstein held that there was such a potential, and that it arose as a result of the selective permeability of the cell membrane to potassium ions. It seems he was the first to apply the Nernst equation to the living cell. We can restate Bernstein's hypothesis in modern form in terms of ion channels: the resting potential is produced because some of the potassium channels in the plasma membrane are open.

If it is correct that the resting membrane is selectively permeable to potassium ions, then we would expect the membrane potential to be given by the Nernst equation for potassium ions,

$$E_K = \frac{RT}{F} \ln \frac{[\text{K}]_o}{[\text{K}]_i} \qquad (3.5)$$

At 18 °C this may be written as

$$E_K = 58 \log_{10} \frac{[\text{K}]_o}{[\text{K}]_i} \qquad (3.6)$$

The best method of readily testing this hypothesis is to vary the external potassium concentration and observe the change in membrane potential: equation 3.6 predicts that the relation should be a straight line with a slope of 58 mV per unit increase in $\log_{10} [\text{K}]_o$.

Figure 3.12 shows the membrane potential of isolated frog muscle fibres in solutions containing different potassium ion concentrations (Hodgkin & Horowicz, 1959). It is evident that equation 3.6 fits the results very well at potassium concentrations above about 10 mM, but below this value the membrane potential is rather less than one would expect. Similar results have been obtained in squid axons (Curtis & Cole, 1942; Hodgkin & Keynes, 1955*b*) and many other cells (Hodgkin, 1951).

What is the reason for the departure from the simple potassium electrode hypothesis at low external potassium ion concentrations? We have seen that sodium ions are distributed across the cell membrane so as to produce a potential concentration cell with the inside positive, and since the membrane is not completely impermeable to sodium ions, we might expect this concentration cell to play some part in determining the membrane potential. Just how this effect will make itself felt will depend upon the properties of the membrane.

A useful approach to determining what the membrane potential should be when the membrane is permeable to more than one ion was developed by Goldman in 1943. It is known as the *constant field theory*. The theory assumes (1) that ions move in the membrane under the influence of electric fields and concentration gradients just as they do in free solution, (2) that the concentrations of ions in the membrane at its edges are proportional to those in the aqueous solutions in contact with it, and (3) that the electric potential gradient across the membrane is constant. From these assumptions it is possible to show (see Hodgkin & Katz, 1949) that, when there is no current flowing through the membrane, the membrane potential is given by

$$E = \frac{RT}{F} \ln \frac{P_K[K]_o + P_{Na}[Na]_o + P_{Cl}[Cl]_i}{P_K[K]_i + P_{Na}[Na]_i + P_{Cl}[Cl]_o} \quad (3.7)$$

where P_K, P_{Na} and P_{Cl} are permeability coefficients.

Equation 3.7 is often known as the Goldman–Hodgkin–Katz voltage equation. The permeability coefficients are measured in centimetres per second and defined as $u\beta RT/aF$, where u is the mobility of the ion in the membrane, β is the partition coefficient between the membrane and aqueous solution, a is the thickness of the membrane, and R, T and F have their usual significance. Actual values for the permeability coefficients must be determined by

measuring the fluxes of the different ions through the membrane, but their ratios (such as P_K/P_{Na}) can be estimated from membrane potential measurements.

If chloride ions are omitted from the system, or if they are assumed to be in equilibrium so that the equilibrium potential for chloride is equal to the membrane potential, equation 3.7 becomes

Donnan equilibrium

$$E = \frac{RT}{F} \ln \frac{[K]_o + \alpha[Na]_o}{[K]_i + \alpha[Na]_i} \quad (3.8)$$

where α is equal to P_{Na}/P_K. Equation 3.8 is plotted in fig. 3.12, assuming α to be 0.01; it is clear that it provides a good fit for the experimental results. Results similar to those shown in fig. 3.12 have also been obtained from observations on the membrane potentials of nerve axons (Hodgkin & Katz, 1949; Huxley & Stämpfli, 1951). We can therefore conclude that sodium ions play some small part in determining the membrane potential of nerve and muscle cells, their effect being greater when the external potassium ion concentration is low. The potassium channels are not perfectly selective for potassium ions, it seems, and will allow some sodium ions to pass through them.

One further piece of information is required before we can accept the hypothesis that the resting potential is determined mainly by the Nernst equation for potassium ions. It is necessary to show that the internal concentration of free potassium ions (which is involved in equations 3.5 to 3.8) is the same as the total internal potassium concentration (which is what can be measured by chemical methods of analysis), or nearly so. For instance, if half the internal potassium were bound in non-ionic form, the predicted values for membrane potentials would have to be reduced

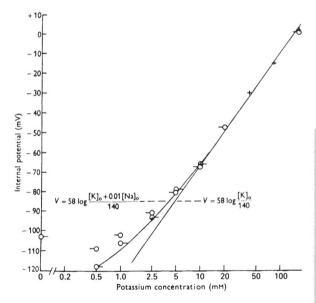

Figure 3.12. How the external potassium ion concentration affects the membrane potential of isolated frog muscle fibres. The external solutions were chloride free, the principle anion being sulphate. The straight line shows the Nernst equation (equation 3.6), but the points at low potassium ion concentrations are better fitted by equation 3.8, a modification of the Goldman–Hodgkin–Katz equation. (From Hodgkin & Horowicz, 1959.)

← Why the deviation from the expected V_m at low $[K^+]$?

Na^+ plays a small role in determining V_m resting. It can go through the K^+ channel at a very low frequency and when $[K^+]$ is low its effects are greater.

in a 0.5 M solution of potassium chloride. Thus the radio-active potassium is present inside the axon in a free, ionic form. Since Keynes & Lewis (1951) had previously shown that radioactive potassium exchanges with at least 97% of the potassium present in crab axons, it seemed very reasonable to conclude that almost all the potassium in the axoplasm is effectively in free solution and so can contribute to the production of the resting potential.

Experiments on internally perfused giant axons

It has been possible to test further the potassium electrode hypothesis by means of the technique of intracellular perfusion (Baker *et al*. 1962*a*; Tasaki & Shimamura, 1962). In the method used by Baker and his colleagues, the axoplasm is squeezed out of a squid axon, leaving a flattened tube consisting mainly of the axon membrane and the Schwann cells and endoneurial sheath. This tube is then reinflated by filling it with a perfusion fluid isotonic with sea water.

When the perfusion fluid is an isotonic potassium chloride solution, the resting potential is about −55 mV. Isotonic solutions of other potassium salts, such as sulphate or isethionate, produce resting potentials a few millivolts greater, indicating that the membrane is slightly permeable to chloride ions (Baker *et al*., 1962*b*). If the solutions inside and outside the membrane were identical in composition, the membrane potential was within 1 mV of zero. Finally, when the internal solution was isotonic sodium chloride and the external isotonic potassium chloride, the inside became positive to the outside by 40 to 60 mV.

By gradually increasing the internal potassium ion concentration, it is possible to see how far the membrane behaves as a potassium electrode. It was found that the membrane potential tended to reach a saturating value of −50 to −60 mV at an internal potassium ion concentration of about 300 mV. This relation is compatible with equation 3.5 if we assume that P_K decreases with increasing (more negative) membrane potentials.

Inward rectifier potassium channels

Katz (1949) discovered that in potassium-depolarized muscle the membrane conductance increased with hyperpolarization and decreased with depolarization. This means that potassium ions could flow into the cell much more easily than they could flow out. Katz named the phenomenon 'anomalous rectification', but the alternative term 'inward rectification' is generally used now. It seems likely that the inward rectifier channels that mediate this rectifying potassium conductance are the main determinants of the resting potential in many cells.

Patch clamp experiments on the inward rectifier channels of rat myotubes (cultured embryonic muscle fibres) indicate a single-channel conductance of about 10 pS with isotonic external potassium ion concentrations (Ohmori *et al*., 1981). The conductance is reduced when the external potassium concentration is lower. The channel density is low, perhaps about one per square micrometre of surface membrane.

Cloning studies on inward rectifier channels show that each subunit contains just two membrane-crossing segments, together with a pore-lining segment which is similar to the H5 or P segment of voltage-gated potassium channels (p. 99) (Ho *et al*., 1993; Kubo *et al*., 1993). It seems likely, by analogy with voltage-gated channels, that each channel is formed from four subunits, as suggested in fig. 3.14. A number of different types are found, forming the Kir family of channels (Doupnik *et al*., 1995).

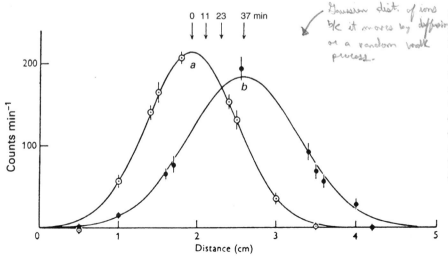

Figure 3.13. Movement of radioactive potassium in a longitudinal electric field in a *Sepia* giant axon. The two curves show the distribution of radioactivity immediately before application of the longitudinal current (*a*), and after 37 min of current flow (*b*). Arrows at 11 and 23 min show the positions of peak radioactivity at those times. (From Hodgkin & Keynes, 1953.)

How is the rectification brought about in these channels? Patch clamp experiments show that it disappears if the patch is removed from the cell and its cytoplasmic surface exposed to a solution with potassium as the only cation. Rectification in inside-out patches can be restored by adding either magnesium or especially spermine (a cytoplasmic polyamine with four ionizable amino groups per molecule) to the cytoplasmic surface (Matsuda *et al.*, 1987; Stanfield *et al.*, 1994; Fakler *et al.*, 1995). This suggests that the channel can normally be blocked by a positively charged cytoplasmic particle (a magnesium ion or spermine) which is drawn into the inner mouth of the channel pore when the membrane is depolarized, but which is displaced from it when the current is inward.

ATP-sensitive potassium channels

Potassium channels which open when the intracellular concentration of ATP falls to a low level are found in muscle and other cells (Standen, 1992). In muscle they open only in conditions of extreme metabolic exhaustion. Their opening leads to some loss of potassium from the muscle cells, but will also tend to stabilize the membrane potential so as to make the muscle less excitable and thus perhaps speed recovery.

Chloride movements in muscle

Vertebrate skeletal muscle cells have a relatively high permeability to chloride ions, with the result that chloride ions are passively distributed in accordance with equation 3.4. If potassium ions are also at equilibrium (equation 3.5), then $E_{Cl} = E_K$ and so

$$\frac{RT}{-F}\ln\frac{[Cl]_o}{[Cl]_i} = \frac{RT}{F}\ln\frac{[K]_o}{[K]_i}$$

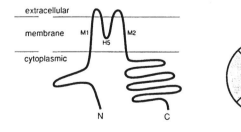

Figure 3.14. Possible structure of inward rectifier potassium channels. Each subunit has two membrane-spanning segments, M1 and M2, and between them a 'membrane-dipping' loop H5. The model suggests that four subunits aggregate to form a channel, with their H5 loops lining the channel pore. (From Kubo *et al.*, 1993. Reprinted with permission from *Nature* **362**, p. 132 Copyright 1993 Macmillan Magazines Limited.)

Hence

$$\frac{[K]_o}{[K]_i} = \frac{[Cl]_i}{[Cl]_o}$$

or

$$[K]_o \times [Cl]_o = [K]_i \times [Cl]_i \qquad (3.9)$$

Equation 3.9 describes a Donnan equilibrium system: the product of the concentrations of the permeant ions on one side of the membrane is equal to their product on the other side. Boyle & Conway (1941) tested the application of this idea to frog muscle by soaking muscles in different potassium chloride concentrations for 24 h and then determining their internal potassium and chloride concentrations. The results were in accordance with equation 3.9, provided the external potassium concentration was above about 10 mM.

Do chloride ions contribute to the resting potential in muscle? This question was extensively investigated by Hodgkin & Horowicz (1959), using intracellular microelectrodes with isolated frog muscle fibres. The great advantage of using single fibres is that, with a suitable apparatus, it is possible to change the ionic concentrations at the cell surface within a fraction of a second, so eliminating the inevitable diffusion delays involved in work with whole muscles. When the external solution was changed for one with different potassium and chloride concentrations but with the same [K][Cl] product, the new membrane potential (given by equations 3.4 and 3.5) was reached within 2 or 3 s, and thereafter remained constant. However, if the chloride concentration was changed without altering the potassium concentration, the membrane potential jumped rapidly to a new value, but this transient effect gradually decayed over the next few minutes and the potential returned to very nearly its original value (fig. 3.15).

The explanation of this effect is based on the Donnan equilibrium hypothesis. The chloride equilibrium potential is given by equation 3.4, or, at 18 °C,

$$E_{Cl} = -58 \log_{10}\frac{[Cl]_o}{[Cl]_i}$$

At the start of the experiment shown in fig. 3.15 (when $[Cl]_o$ is 120 mM) we assume that chloride is in equilibrium, i.e. that E_{Cl} is equal to the membrane potential of -98.5 mV; this gives a value of 2.4 mM for $[Cl]_i$. When $[Cl]_o$ is reduced to 30 mM, E_{Cl} will change by 35 mV to -63.5 mV. The membrane potential changes to -77 mV, which is intermediate between E_K and E_{Cl}, indicating that the membrane is permeable to both potassium and chloride ions. Then, in order to restore the Donnan equilibrium, potassium chloride moves out of the cell until E_{Cl} is equal to E_K. This point is reached when [Cl]$_i$ has fallen to about 0.6 mM and, since there is also some movement of water out of the cell in order to maintain osmotic equilibrium, [K]$_i$ is practically unchanged. Hence the new steady potential (reached after 10 to 15 min in fig. 3.15) is the same as it originally was. When [Cl]$_o$ is returned to 120 mM, there is a similar transient membrane potential change in the opposite direction, and equilibrium is then restored by movement of potassium chloride and water into the fibre. The fact that these experiments can be so readily interpreted in this way provides further evidence in favour of Boyle & Conway's Donnan equilibrium hypothesis.

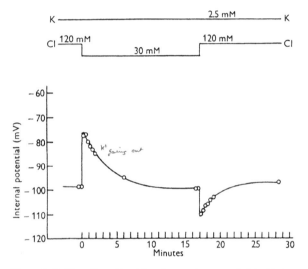

Figure 3.15. The effect of sudden changes in external chloride ion concentration on the membrane potential of an isolated frog muscle fibre. (From Hodgkin & Horowicz, 1959.)

Figure 3.16c shows the results of a similar experiment in which [K]$_o$ is changed from 2.5 to 10 mM without changing [Cl]$_o$. This will change E_K by 35 mV, and so the membrane potential jumps to a new value intermediate between E_K and E_{Cl}. Equilibrium is then reached by entry of potassium chloride (and water) into the fibre, so that the membrane potential gradually moves from -73 to -65 mV. On returning [K]$_o$ to 2.5 mM, there is a small instantaneous repolarization, and then equilibrium (and the normal resting potential) is slowly restored by loss of potassium chloride from the fibre. The larger repolarizations in fig. 3.16a and b are caused by the fact that the internal chloride concentration has not had time to change much from its normal level, so that E_{Cl} is still near the normal resting potential.

Notice that the four-fold increase in [K]$_o$ causes an instantaneous depolarization of 21 mV in fig. 3.16c, whereas the later four-fold decrease causes an instantaneous repolarization of only 3 mV. This must indicate that there is some rectification process in the potassium pathway, so that potassium ions can move inwards very easily but outwards with much more difficulty. Hodgkin & Horowicz calculated that P_K is about 8×10^{-6} cm s^{-1} for inward current but may be as little as 0.05×10^{-6} cm s^{-1} for outward current, whereas P_{Cl} remains at about 4×10^{-6} cm s^{-1} irrespective of the direction of the current flow. We may assume that the potassium movements occur through inward rectifier channels, whereas the chloride channels are not rectifying.

Chloride channels

The cells in the electric organs of the electric ray *Torpedo* are asymmetrical. They are innervated on the upper face but not on the lower face. When they are excited by nervous stimula-

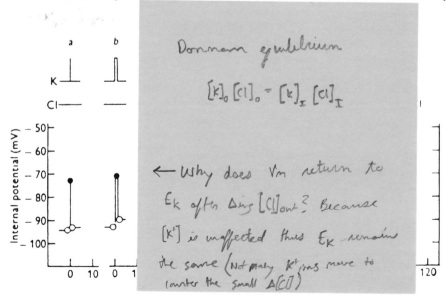

Figure 3.16. The effect of changes in the external potassium concentration on the membrane potential of an isolated frog muscle fibre. (From Hodgkin & Horowicz, 1959.)

Donnan equilibrium

$$[K]_o [Cl]_o = [K]_I [Cl]_I$$

← Why does V_m return to E_k after Δing [Cl]$_{out}$? Because [K$^+$] is unaffected thus E_K remains the same (Not many K$^+$ ions move to counter the small Δ[Cl])

tion, sodium ions pour into the cell through the acetyl-choline-gated channels on the upper face. The lower face has to have a high conductance so as to allow the current to flow readily, and this is brought about by a high density of chloride channels. The cells contain messenger RNA coding for these chloride channels, and cDNA made from this has been cloned and sequenced, so providing the amino acid sequence of the channel protein (Jentsch *et al.*, 1990).

The chloride channel protein (called CLC-0) is 805 residues long and contains perhaps twelve or thirteen transmembrane segments. Perhaps the complete channel is a homotetramer made from four such molecules. Similar proteins (CLC-1 and CLC-2, respectively) have been isolated from skeletal muscle and other mammalian cells. Mutations in the CLC-1 channel produce myotonia (muscle stiffness) and decreased muscle chloride conductance in mice and humans (see Pusch & Jentsch, 1994).

The electrogenic nature of the sodium pump

Consider a membrane ion pump which is concerned solely with the active transport of one species of ion in one direction. There would then be a current flow across the membrane and a change in membrane potential caused by the depletion of charge on one side of the membrane. There would be similar consequences from a pump which coupled inflow of one ion to outflow of another but in which the numbers of ions transported in the two directions were not equal. A pump in which such a net transfer of charge does take place is described as being *electrogenic*. In the sodium pump of mammalian red blood corpuscles, for example, there is very clear evidence that two potassium ions are taken up for every three sodium ions extruded (Post & Jolly, 1957), so there must be a net flow of charge equal to one-third of the sodium ion flow.

The sodium pump of nerve axons was once thought to be electrically neutral, because of a supposed one-for-one exchange of sodium for potassium ions. Later experiments on a variety of cells indicated that this is not so and that the pump is electrogenic (Ritchie, 1971; Glynn, 1984). Good evidence that this conclusion applies to nerves cell comes from experiments by Thomas (1969, 1972) on snail neurons; let us examine them.

The essence of Thomas's method was to inject sodium ions into the cell body of a neuron and observe the subsequent changes in membrane potential. In order to perform the injection two microelectrodes filled with a solution of sodium acetate were inserted into the cell and current passed between them; sodium ions would thus be carried out of the cathodal electrode into the cytoplasm, so raising the internal sodium concentration. A third microelectrode was used to record the membrane potential.

Thomas found that injection of sodium ions into the cell is followed by a hyperpolarization of up to 15 mV, after which the membrane potential returned to normal over a period of about 10 min, as is shown in the first parts of the traces in fig. 3.17. Figure 3.17*a* shows that ouabain greatly

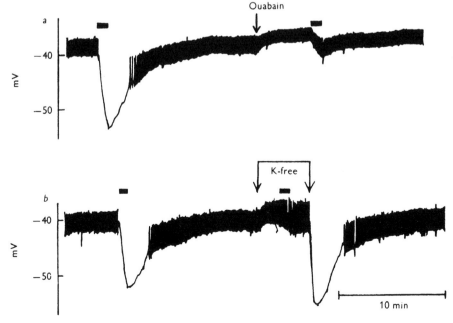

Figure 3.17. Responses of snail neurons to injections of sodium ions, demonstrating the electrogenic nature of the sodium pump. The traces show pen recordings of the membrane potential. Black bars show the timing of the injecting currents. The first response in each trace shows the hyperpolarization that normally follows sodium injection. After treatment with ouabain this is greatly reduced (trace *a*). In a potassium-free external solution (trace *b*) the hyperpolarization does not occur until the external potassium is replaced, indicating that sodium extrusion is coupled to potassium uptake. (From Thomas, 1969.)

reduces this hyperpolarization, suggesting that it is connected with an active sodium extrusion. In fig. 3.17b, the second injection of sodium ions occurs while the cell is in a potassium-free environment, when there is no hyperpolarization, suggesting that sodium extrusion is coupled to potassium uptake. Injections of potassium or lithium ions produced no membrane hyperpolarization.

These observations suggest that sodium extrusion is coupled to potassium uptake, but also that the number of sodium ions moved out of the cell is greater than the number of potassium ions moved in. There is thus a net flow of positive charge outward during the action of the pump, seen as a hyperpolarization of the membrane. Is it possible to estimate the ratio of sodium ion to potassium ion movements? Thomas attacked this problem by using the voltage clamp technique. He found that sodium injection was followed by an outwardly directed current which took about 10 min to fall to zero.

By integrating this current with respect to time it was possible to estimate the amount of charge transferred. Thomas found that this was always much less than the quantity of charge injected as sodium ions. And yet he also found that all of the injected sodium was extruded during the period of membrane current flow; he did this by using an intracellular sodium-sensitive microelectrode (one which produces an electrical signal proportional to the sodium ion concentration in its environment). This means that the sodium outflow must be partly balanced by an inflow of some other cation, for which the obvious candidate is potassium. Thomas calculated that the net charge transfer in the pump was about 27% of the sodium ion flow. This figure is fairly near to the 33% that would be expected from a system like that in red blood cells, where three sodium ions move for every two potassium ions.

Part B
Nervous conduction

4
Electrical properties of the nerve axon

The most striking anatomical feature of nerve cells is that part of the cell is produced into an enormously elongated cylindrical process, the axon. It is this part of the cell with which we shall be concerned in this chapter and the next. The essential function of the axon is the propagation of nerve impulses.

Action potentials in single axons

Let us consider a simple experiment on the giant fibres in the nerve cord of the earthworm. These fibres are anatomically not axons, because they are multicellular units divided by transverse septa in each segment, but physiologically each fibre acts as a single axon. There are three giant fibres, one median and two lateral, which run the length of the worm; the laterals are interconnected at intervals. The experimental arrangement for eliciting and recording impulses in the giant fibres is shown in fig. 4.1. The stimulator produces a square voltage pulse which is applied to the nerve cord at the stimulating electrodes. The recording electrodes pick up the electrical changes in the nerve cord and feed them into the amplifier. Here they are amplified about 1000 times, and then passed to the oscilloscope where they are displayed on

the screen of the cathode ray tube. The output of the stimulator is also fed into the oscilloscope so that it is displayed on the second trace on the screen. The timing of the oscilloscope sweep is arranged so that both traces start at the moment that the pulse from the stimulator arrives.

We begin with a stimulating pulse of low intensity (fig. 4.2*a*). The lower trace shows the stimulating pulse, but nothing appears on the upper trace except a deflection which is coincident with the stimulus and (as we can show by varying the stimulus intensity) proportional to its size.

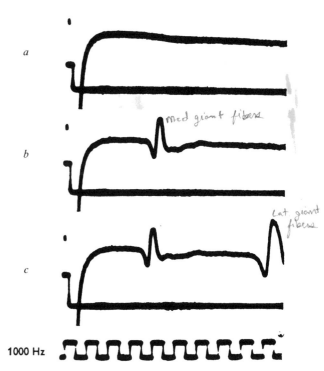

Figure 4.2. Oscilloscope records from the experiment shown in fig. 4.1. In each case the upper trace is a record of the potential changes at the recording electrodes and the lower trace monitors the stimulus pulse. Records *a*, *b* and *c* show responses to increasing stimulus intensity. In *b* the stimulus is sufficient to excite the median giant fibre; in *c* it is sufficient to excite the lateral giant fibres as well.

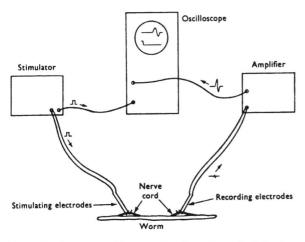

Figure 4.1. Arrangement for recording the action potentials of the giant fibres in the nerve cord of the earthworm, using external silver wire electrodes.

This phenomenon is known as the *stimulus artefact;* it is not a property of the worm, and is due merely to the recording electrodes picking up the electric field set by the stimulus pulse. As we increase the intensity of the stimulus, a new deflection appears on the upper trace (fig. 4.2*b*). This is the nerve impulse, or *action potential*, in the median fibre. Further increase in the stimulus intensity does not change the size of this deflection, and if we turn the stimulus intensity down again it suddenly disappears, without any preparatory decrease in size. In other words, the size of the action potential is independent of the size of the stimulus; it is either there or not there. This phenomenon is known as the *'all-or-nothing' law.* The stimulus intensity which is just sufficient to produce an action potential is called the *threshold stimulus intensity.*

The all-or-nothing principle was first established by K. Lucas and E. D. Adrian in the early years of the twentieth century, using indirect observations on muscle fibres. Following the introduction of electronic amplification in the 1920s, Adrian and his colleagues were able to make direct observations of the constancy in the size of action potentials in individual sensory and motor nerve fibres (Adrian & Zottermann, 1926; Adrian & Bronk, 1928).

The all-or-nothing nature of the action potential has important implications for signalling in nervous systems. The individual action potential cannot carry any information about the amplitude of a signal. To do this we need a series of successive action potentials, so that information can be passed along an axon as the time intervals between them. Thus we find that sensory information, for example, is often in the form of a frequency code: low frequencies represent low intensities of the sensory stimulus, whereas high frequencies represent stronger stimuli. The great advantage of the unitary action potential as a signalling mechanism is that it avoids the need for a complex and sophisticated mechanism such as would be required to conduct signals of different amplitudes without distortion.

Now let us perform another experiment with our earthworm nerve cord. Keeping the stimulus intensity constant, above the threshold for the median fibre, we move the recording electrodes nearer to the stimulating electrodes. We find that the action potential occurs earlier on the trace. If we measure the time between the stimulus and the action potential at a number of different distances between the two pairs of electrodes, we can plot a graph showing the position of the action potential at various times after the stimulus (fig. 4.3). The points lie very nearly on a straight line which passes near the origin. This means that the action potential arises at the stimulating electrodes and is then conducted along the fibre at a constant velocity given by the slope of this straight line.

If we further raise the stimulus intensity in our earthworm experiment, a second deflection may appear at a later time on the trace (fig. 4.2*c*); this is caused by the action potentials of the two lateral fibres. Notice that the lateral fibre action potential occurs later than that of the median fibre; this is because its conduction velocity is slower. We know that the first action potential is from the median fibre and the second from the laterals (and not the other way round) from a delicate experiment by Rushton (1945). He damaged just the median fibre and found that the first action potential disappeared, and in another nerve cord he damaged just the lateral fibres and found that the second action potential disappeared.

Intracellular recording

Experiments with external electrodes, as in fig. 4.2, measure the potential difference between two points on the outside of the axon. Obviously it would be informative to measure

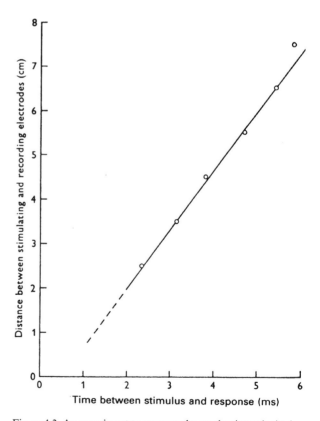

Figure 4.3. An experiment to measure the conduction velocity in the median giant fibre of an earthworm. The latency of the response (the time between the stimulus and the action potential) increases as the stimulating and recording electrodes are moved further apart. The slope of the line gives the conduction velocity, about 12 m s^{-1} in this case.

potential changes across the membrane, using an electrode inside the cell. This was first done in 1939 by Hodgkin & Huxley at Plymouth and by Cole & Curtis at Woods Hole, Massachusetts, using the giant axons innervating the mantle muscle of squids. As was mentioned in the previous chapter, these axons are of unusually large diameter (0.1 to 1 mm) and have been much used by nerve physiologists since their description by J. Z. Young in 1936. The intracellular electrode used in these experiments was a glass capillary tube, about 100 μm in diameter, filled with sea water (Hodgkin & Huxley, 1945) or a potassium chloride solution isotonic with sea water (Curtis & Cole, 1942). It was inserted through the cut end of the axon and pushed along so that its tip was level with an undamaged part of the axon. The potential difference between the electrode and the external sea water was then measured. The axon could be stimulated electrically via a pair of external electrodes.

Figure 4.4 shows a typical record of an action potential obtained by this intracellular recording method. The trace begins with the inside of the axon negative to the outside (this is the resting potential), but during the action potential the membrane potential changes so that the inside becomes positive. The discovery of this 'overshoot' was unexpected at the time; its significance will be considered in the next chapter.

Two useful terms can be introduced here. We can describe the existence of the resting potential by saying that the membrane is 'polarized'. If we reduce the membrane potential, then the membrane becomes *depolarized*. If we increase the membrane potential (i.e. we make it more negative), then the membrane becomes *hyperpolarized*.

In many cases a careful examination of the time course of an action potential shows that the membrane potential does not immediately settle down at the resting level after its completion. As is shown in fig. 4.5, the action potential may be followed by (1) a *positive phase*, evident in fig. 4.4, in which the membrane potential 'underswings' towards a more negative value than the normal resting potential, (2) a *negative after-potential*, in which the membrane potential is less negative than normal, and finally (3) a *positive after-potential*, in which the membrane potential is again more negative than usual. The reason for the apparently paradoxical nomenclature of these after-potentials is that they were first observed using extracellular electrodes, which of course measure the potential on the outside of the axon membrane instead of the inside, so that the directions of all potential changes are reversed; this point is illustrated in fig. 4.6.

Electrical stimulation parameters

What are the requirements for a stimulus to produce an action potential in a nerve fibre? Clearly the membrane must be depolarized so that the membrane potential crosses the threshold. A number of phenomena of the excitation process can be interpreted with this principle in mind.

The strength–duration relation

If an axon is stimulated with square constant current pulses, it is found that the threshold stimulus intensity rises as the pulse length is lessened. This effect is known as the *strength–duration relation* (fig. 4.7). The curve in fig. 4.7 is quite well fitted by the empirical equation

$$\frac{I}{I_0} = \frac{1}{1 - e^{-t/k}} \tag{4.1}$$

where I is the intensity of the pulse, t is the length of the pulse, I_0 is the threshold stimulus intensity when t is large (the *rheobase*) and k is a constant (Lapique, 1907; Hill, 1936). The pulse length when the threshold stimulus intensity is twice the rheobase is called the *chronaxie*.

Equation 4.1 is similar in form to that describing the change of voltage across a circuit consisting of a resistance and capacitance in parallel following the application of a constant current, i.e.

$$V = IR \left(1 - e^{-t/RC}\right) \tag{4.2}$$

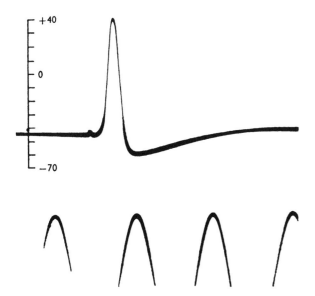

Figure 4.4. An action potential recorded from a squid giant axon with an intracellular electrode. The vertical scale shows the potential (mV) of the internal electrode with respect to the external sea water. The sine wave at the bottom is a time scale, frequency 500 Hz. (From Hodgkin & Huxley, 1945.)

where V is the voltage, I the current, t the time after application of the current, R the resistance and C the capacitance. If V is constant (for instance, if we regard V as the threshold membrane depolarization), this becomes

$$I \times \text{constant} = \frac{1}{1 - e^{-t/RC}} \qquad (4.3)$$

which is equivalent to equation 4.1. The actual situation has some extra complexities (see Noble, 1966), but, despite

this, equation 4.3 provides quite a good fit to the strength–duration curve.

Figure 4.8 may help in understanding how the strength–duration relation comes about. We regard the axon membrane as a resistance and capacitance in parallel. When we pass a constant current through it, the voltage across it will rise exponentially as in equation 4.2, finally reaching a plateau proportional to the strength of the current. For weak currents, the plateau will still be below

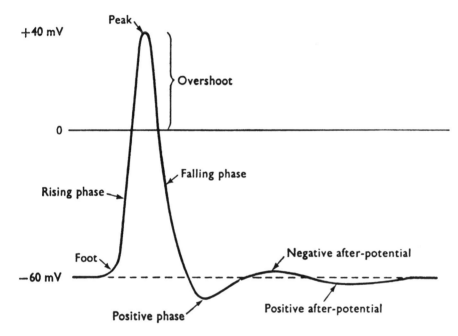

Figure 4.5. Diagram to show the nomenclature applied to an action potential and the after-potentials that follow it.

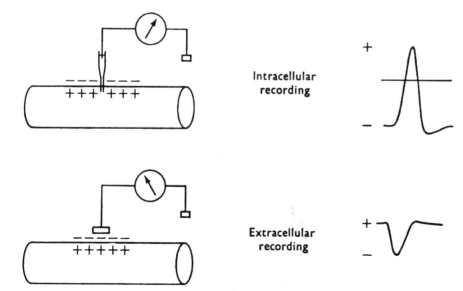

Figure 4.6. Diagram to show the difference in sign of action potentials recorded by intracellular and by extracellular electrodes. Extracellular records are in fact frequently shown with negative potentials upwards.

the threshold membrane potential, as for traces *a* and *b* in fig. 4.8. For somewhat stronger currents, the membrane potential will eventually exceed the threshold value, as in trace *c*. The stronger the current, the more rapid the initial change in voltage and so the sooner the threshold membrane potential is exceeded, as in traces *d* to *f* in fig. 4.8. If the pulse is short, the current strength must be high so that the threshold can be reached before the pulse comes to an end. For longer pulses, the current strength need not be so high, but it must always be above some minimum value.

Latency

The time between the onset of a stimulus and the ensuing action potential is called the *latency* of the response. It is clear from fig. 4.8 that the latency decreases with increasing current strengths.

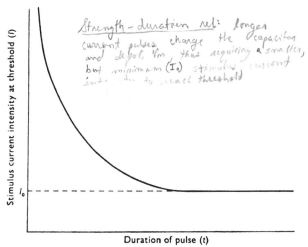

Figure 4.7. The strength–duration curve for a nerve axon. For current pulses of short duration, the threshold stimulus intensity is high. As the duration lengthens, the current required to produce an action potential falls to a minimum level known as the rheobase, I_0.

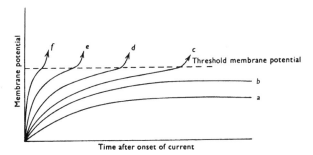

Figure 4.8. How depolarizing currents change the membrane potential of a nerve axon. The current strengths increase from *a* to *f*. Responses *c* to *f* lead to excitation.

Latent addition

For a short time after a brief subthreshold cathodal stimulus, the membrane potential is nearer to the threshold membrane potential than usual. During this time, less current is required to depolarize the membrane by enough to cause excitation, so that stimuli which are normally of subthreshold intensity may be sufficient to elicit an action potential (fig. 4.9). Conversely, if an anodal shock is applied, a cathodal stimulus soon afterwards has to be of a higher intensity than usual in order to cause excitation. These phenomena constitute *latent addition*. They can be used to investigate the time course of subthreshold responses. Katz (1937), for instance, demonstrated indirectly the presence of local responses in this way; his results were very similar to those obtained by Hodgkin (1938) from direct observations on crab nerves.

Accommodation

If a constant subthreshold depolarizing current is passed through an axon membrane, it is found that the threshold slowly rises during the passage of the current, and then falls

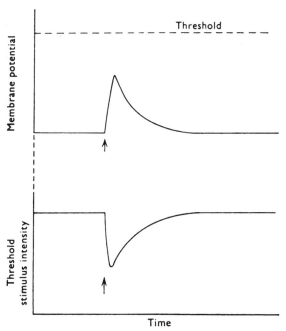

Figure 4.9. To illustrate latent addition. A subthreshold depolarizing current pulse is applied to an axon at the time shown by the arrow. The upper trace shows the consequent change in membrane potential. The lower trace, proportional to the difference between the upper trace and the threshold membrane potential, shows the changes in threshold stimulus intensity that this produces.

again after its cessation. Conversely, when hyperpolarizing current is used, the threshold falls. This delayed dependence of the threshold on membrane potential is known as *accommodation* (fig. 4.10). With hyperpolarizing currents of sufficient strength, the threshold may fall beyond the resting potential, so that when the current is switched off, the membrane potential is temporarily above threshold. This leads to the initiation of an action potential (fig. 4.10*c*), and the phenomenon is therefore known as *anode break excitation*.

If an axon is stimulated by a linearly rising cathodal current, the membrane potential change follows an approximately linear course, and, because of accommodation, the threshold membrane potential also changes slowly. Consequently, the threshold current intensity is lower for rapidly rising currents than it is for slowly rising ones. When

the rate of rise of current is sufficiently low, the change in membrane potential is too slow to be able to gain on the change in threshold level, and therefore no excitation occurs.

The refractory period

For a short time after the passage of an action potential, it is not possible to elicit a second action potential, however high the stimulus intensity is. This period is known as the *absolute refractory period* (fig. 4.11). Following the end of the absolute refractory period, there is a period during which it is possible to elicit a second action potential, but the threshold stimulus intensity is higher than usual. During this *relative refractory period* the second action potential, if it is recorded near the stimulating electrodes, may be reduced in size. The existence of refractoriness places an upper limit on the frequencies at which axons can conduct nerve impulses.

Subthreshold potentials

We will now consider some experiments performed by Hodgkin (1938) on isolated axons from the legs of the shore crab *Carcinus*, with the object of examining the responses to subthreshold electrical stimuli. He stimulated the axon with brief (60 μs) electric shocks, and recorded the response in the vicinity of the stimulating electrodes. Figure 4.12 shows the results of one of his experiments.

The responses to inward current are shown as the deflec-

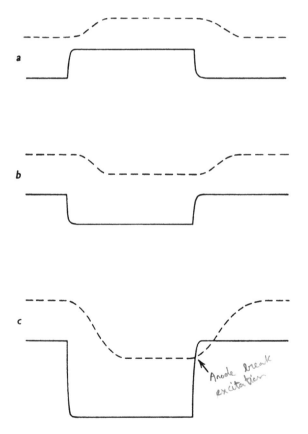

Figure 4.10. Accommodation in response to constant currents. In each case the continuous line indicates the membrane potential and the dashed line indicates the threshold membrane potential. The effect of subthreshold depolarizing current is shown in *a*; effects of hyperpolarizing currents are shown in *b* and *c*. In *c* the current is strong enough to cause 'anode break excitation' after it is switched off, at the point marked by the arrow.

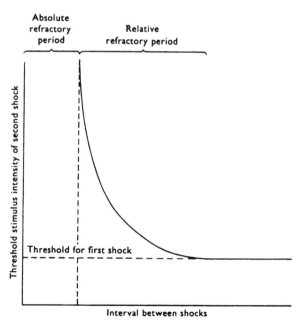

Figure 4.11. Refractory periods.

tions below the baseline in fig. 4.12. These changes would be seen as hyperpolarizations from the resting potential if the experiment had been done with intracellular electrodes. Notice that the voltage does not immediately return to the baseline after the end of the stimulating pulse, but decays gradually. The reason for this is that the nerve membrane behaves electrically as a resistance and capacitance in parallel.

We can model the membrane resistance and capacitance with a simple electric circuit (fig. 4.13). A battery E is connected to a resistance R and capacitance C via a switch S, so that a voltage V appears across the resistance–capacitance network when the switch is closed. As a consequence there is a positive charge on one plate of the capacitance and a negative charge on the other. When the switch is opened, this charge flows through the resistance R as shown by the arrow, so that there is a potential difference across R which falls exponentially as the capacitance discharges. Thus, returning to fig. 4.12, the voltage decay following passage of inward current through the membrane (anodal stimulation) is caused by the discharge of the membrane

capacitance through the membrane resistance. This type of electrical change (i.e. one that can be attributed to current flow through the resistance and capacitance of the resting membrane) is known as an *electrotonic potential*.

The responses above the baseline in fig. 4.12 were obtained with the cathode of the stimulating electrodes next to the recording electrode (i.e. with cathodal stimulation, producing a flow of current outwards through the membrane). They would be seen as depolarizations from the resting potential if the experiment had been done with intracellular electrodes. When the stimulus intensity was very small, the responses were mirror images of the responses to anodal stimulation, and can similarly be described as electrotonic potentials. However, when the stimulus intensity was raised above half the threshold level, the resulting potential did not return to the baseline as rapidly as in the response to anodal stimulation of the same intensity. Hodgkin suggested that these responses were composed of a *local response* added to the electrotonic potential. The time course of the local response could be obtained by subtracting the expected electrotonic response from the whole response. Finally, if the stimulus was large enough to produce a response which crossed a particular level of membrane potential, a propagated action potential was produced. This critical level of membrane potential is called the *threshold membrane potential*.

The results of this experiment are brought together in graphical form in fig. 4.14. For anodal stimuli, and for cathodal stimuli of intensities below about half threshold, the relation between stimulus and response is a straight line, so that the behaviour of the system (assuming the membrane capacitance to be constant) is in accordance with Ohm's law. Above this value this is not so, and the experiment implies that the resistance or capacitance of the membrane changes during local responses and action potentials. (We shall see later that local responses are produced when the membrane depolarization is sufficient to open some voltage-gated sodium channels, and that action potentials

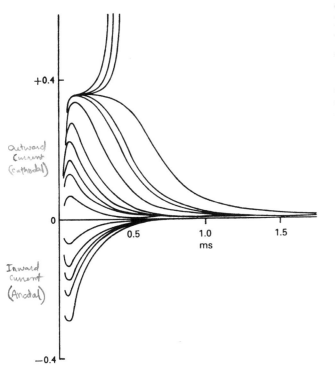

Figure 4.12. Subthreshold responses recorded from a crab axon with extracellular electrodes placed near to the stimulating cathode. The stimulus was very brief. The ordinate is a voltage scale on which the height of the action potential is taken as 1. (From Hodgkin, 1938, by permission of the Royal Society.)

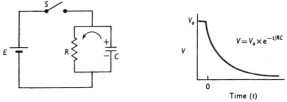

Figure 4.13. Resistance and capacitance in parallel connected to a battery (E). When the switch S is opened the capacitance C discharges through the resistance R. The voltage V across R falls exponentially as C discharges.

occur when many of them are opened; but when Hodgkin did these experiments in 1938 there was no knowledge of these things.)

Impedance changes during activity

If we pass current I through a resistor, the voltage V across the resistor is proportional to the current (Ohm's law again), and the constant of proportionality is called the resistance R, so that

$$V = I R$$

Similarly, if we pass alternating current through a network containing a resistance and capacitance in parallel, the voltage appearing across the network is proportional to the current, and the constant of proportionality is called the *impedance* (Z), so that

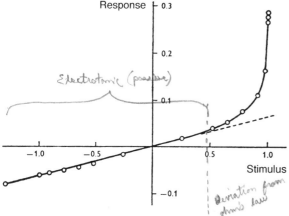

Figure 4.14. The relation between stimulus and response in a crab axon, derived from fig. 4.12. The horizontal axis shows the stimulus intensity, measured as a fraction of the threshold stimulus. The vertical axis shows the recorded potential at 0.29 ms after the stimulus, measured as a fraction of the action potential height. (From Hodgkin, 1938, by permission of the Royal Society.)

$$V = I Z$$

The impedance is a complex quantity containing both resistive and capacitative elements, and dependent upon the frequency of the alternating current.

If we could measure the impedance changes during an action potential, then it should be possible to see what changes there are in the resistance and capacitance of the membrane. Let us see how Cole & Curtis tackled this problem in 1939.

One method of determining the value of an unknown resistance R_x is by means of a simple Wheatstone bridge circuit (fig. 4.15a). The two fixed resistances R_1 and R_2 are known, and R_v is a known variable resistance. R_v is altered until there is no voltage between C and D. In this condition (when the bridge is said to be 'balanced') it is evident that

$$R_1/R_2 = R_x/R_v$$

from which R_x can be calculated. A similar method can be used to measure impedance. In fig. 4.15b the unknown circuit contains a resistance R_x in parallel with a capacitance C_x. If the two fixed resistances are equal, then the bridge is balanced when the variable resistance R_v is equal to R_x and the variable capacitance C_v is equal to C_x.

Cole & Curtis (1939) used this principle to measure the impedance changes in the axon membrane during the passage of an action potential. They placed a squid giant axon in a trough so that it passed between two plate electrodes which were connected to a bridge circuit similar to that shown in fig. 4.15b. A high frequency alternating current (2 to 1000 kHz) was applied to the bridge, and the output was converted to a 175 kHz signal of the same amplitude and then displayed on an oscilloscope. Thus the oscilloscope trace was a thin line when the bridge was in balance, and became a broadened band (the 175 kHz a.c. signal) when it was out of balance. Knowing the values of the variable resistance and capacitance necessary to balance the bridge, and also the geometry of the axon–

Figure 4.15. Wheatstone bridges. *a* shows a simple d.c. bridge for the measurement of resistance. *b* shows an a.c. bridge used to measure resistance and capacitance in parallel.

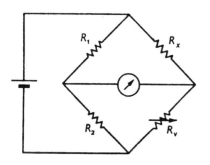

a

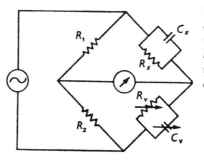

b

electrode system, it was possible to calculate the resistance and capacitance of the axon membrane.

Figure 4.16a shows the effect of stimulation of the axon, with the bridge initially balanced in the resting condition. A short time after the start of the action potential, the oscilloscope trace broadens, indicating that the bridge is out of balance and therefore that the impedance of the system has changed. By starting with the bridge out of balance (as in fig. 4.16b and c), it is possible to find points at which the bridge is in balance at different times during the passage of the action potential and so to determine the resistance and capacitance of the membrane at these times. The results of these experiments showed that the membrane capacitance was 1.1 μF cm^{-2} in the resting condition, and fell by about 2% during activity. The resistance of the membrane, on the other hand, fell very markedly during activity, from its resting value of about 1000 Ω cm^2 to 25 Ω cm^2.

We can now understand this marked fall in resistance in terms of ion channels in the membrane. Voltage-gated sodium channels open during the action potential, allowing sodium ions to rush into the axon; then voltage-gated potassium channels open and potassium ions flow out. The capacitance remains the same during excitation because the lipid bilayer in the membrane remains essentially unchanged. The evidence for this explanation of excitation in terms of ion movements through membrane channels must wait until the next chapter. For the rest of this chapter we consider the cable properties of axons and the way in which action potentials are propagated.

Core-conductor theory

From the results of the experiments so far described, it is evident that the axon membrane has a transverse resistance r_m and a transverse capacitance c_m. If we wish to draw a circuit diagram to show the electrical properties of a length of axon, we must also include two other components, the longitudinal resistance of the external medium r_o and the longitudinal resistance of the axoplasm r_i; the complete network is shown in the upper half of fig. 4.17. The suggestion that the axon can be represented in this way is known as the *core-conductor theory* or *cable theory*, since it implies that the axon behaves as a poorly insulated cable.

If we pass current through a part of the membrane so that a voltage V_0 is set up across the membrane at that point, then the voltage V_x across the membrane at some distant point x must be dependent on the distance x, and (because of the capacitances in the system) it must also vary with time, t. Hence it is necessary for us to know how V varies with x and t. We assume that the system is regular, i.e. that r_o, r_i, r_m and c_m do not change along the length of the axon, and that it is infinitely long. These quantities refer to unit lengths (e.g. 1 cm) of axon. Now the transverse current flowing through the membrane i_m will be the sum of the currents flowing through r_m and c_m, i.e.

$$i_m = \frac{V}{r_m} + c_m \frac{dV}{dt} \tag{4.4}$$

The current (i) flowing through the longitudinal resistances r_o and r_i must get progressively less as we move from its source, since a constant fraction in each unit length is diverted through the membrane; thus it follows that

$$i = -\frac{dV}{dx}\left(\frac{1}{r_m + r_i}\right) \tag{4.5}$$

and

$$i_m = -\frac{di}{dx} \tag{4.6}$$

Hence

$$i_m = \left(\frac{1}{r_o + r_i}\right)\frac{d^2V}{dx^2} \tag{4.7}$$

Then, equating equations 4.7 and 4.4,

$$\frac{V}{r_m} + c_m \frac{\partial V}{\partial t} = \left(\frac{1}{r_o + r_i}\right)\frac{\partial^2 V}{\partial x^2}$$

or

$$V = \left(\frac{r_m}{r_o + r_i}\right)\frac{\partial^2 V}{\partial x^2} - r_m c_m \frac{\partial V}{\partial t} \tag{4.8}$$

We now define two constants, a *space constant* λ and *a time constant* τ, by

$$\lambda^2 = \frac{r_m}{r_o + r_i}$$

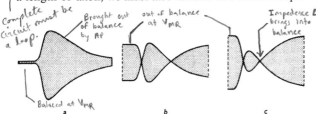

Figure 4.16. Transverse impedance changes in a squid giant axon during the passage of an impulse. In a the bridge was in balance during the resting state; in b and c it was out of balance during the resting state but was brought into balance by the impedance changes associated with the passage of an action potential. (Redrawn after Cole & Curtis, 1939.)

and

$$\tau = r_m c_m$$

Substituting these definitions in equation 4.8, we get

$$V = \lambda^2 \frac{\partial^2 V}{\partial x^2} - \tau \frac{\partial V}{\partial t} \tag{4.9}$$

This is the relation between V, x and t that we have been looking for. The solution of equation 4.9 requires the use of various transform methods (see Hodgkin & Rushton, 1946; Taylor, 1963; Jack *et al.*, 1975); we shall merely consider some relatively simple results that follow from it.

First, consider the situation when a constant current has been applied for a long (effectively infinite) time, to set up the voltage V_0 at that point and the voltage V_x at a distance x from it. Then equation 4.9 simplifies to become

$$V = \lambda^2 \frac{d^2 V}{dx^2}$$

The relevant solution of this equation is

$$V_x = V_0 e^{-x/\lambda} \tag{4.10}$$

This means that the voltage across the membrane falls off exponentially with the distance from the point at which the current is applied. This feature is shown in the lower half of

fig. 4.17. If we make x equal to λ in equation 4.10, V_x becomes $V_0 e^{-1}$, and so λ can be defined as the length over which the voltage across the membrane falls to $1/e$ of its original value.

Now consider the total charge Q on the membrane. This is given by

$$Q = c_m \int_0^\infty V dx \tag{4.11}$$

so that Q is obtained by integrating equation 4.9 with respect to x. If the applied current is constant, beginning at $t = 0$, then the solution of equation 4.11 is

$$Q_t = Q_\infty (1 - e^{-t/\tau}) \tag{4.12}$$

where Q_t is the charge at time t, and Q_∞ is the charge at infinite time (or, effectively, when t is much greater than τ). If a current which has produced a charge Q_0 is suddenly switched off at $t = 0$, then the charge on the membrane decays according to the equation

$$Q_t = Q_0 e^{-t/\tau} \tag{4.13}$$

If we put t equal to τ in equations 4.12 or 4.13 it becomes evident that τ can be defined as the time taken for the change in total charge on the membrane to reach $1/e$ of completion.

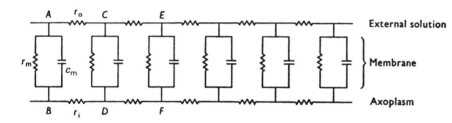

Figure 4.17. Electrical model of the passive (electrotonic) properties of a length of axon. The graph shows the steady-state distribution of transmembrane potential along the model when points A and B are connected to a constant current source. The vertical arrows on the graph are used to suggest the gradual rise of membrane potential to its final value at any point when the current is applied.

The equations describing the time course of the voltage change at any particular point are rather complex, but their general effect is that the potential rises and falls less rapidly the farther that point is from the point at which the current is applied. The responses to the onset and cessation of constant currents are, of course, limited by equations 4.10, 4.12 and 4.13. Figure 4.18 shows the results of Hodgkin & Rushton's calculations of the voltage changes.

The core-conductor equations we have examined so far have been based on quantities referring to a unit length of axon. For many purposes, however, it is useful to possess measurements of membrane properties in terms of unit area; the conversion from one to the other can easily be made

if we know the radius of the axon. The relation between the membrane resistance of a unit length of axon r_m and the resistance of a unit area of membrane R_m is given by

$$r_m = R_m/2\pi a \qquad (4.14)$$

where a is the axon radius. Similarly, the membrane capacitance per unit length c_m is related to the membrane capacitance per unit area C_m by

$$c_m = 2\pi a C_m \qquad (4.15)$$

And the longitudinal resistance of the axoplasm in a unit length of axon r_i is related to the resistivity of the axoplasm R_i by

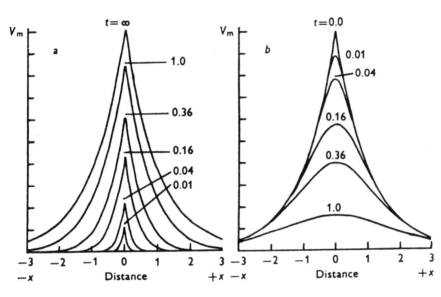

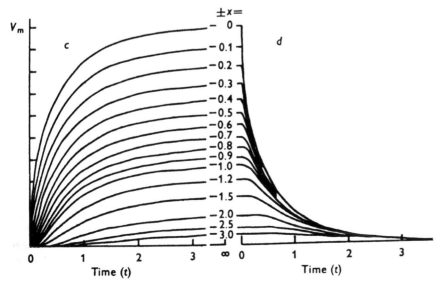

Figure 4.18. Theoretical distribution of potential difference across a passive nerve membrane (i.e. one with no voltage-gated channels active) in response to onset (*a* and *c*) and cessation (*b* and *d*) of a constant current applied at $x = 0$. Graphs *a* and *b* show the spatial distribution of potential difference at different times, and *c* and *d* show the time courses of the potential change at different distances along the axon. Time t is expressed in units equal to the time constant τ, and distance x is expressed in units equal to the space constant λ. (From Hodgkin & Rushton, 1946, by permission of the Royal Society.)

$$r_i = R_i / \pi a^2 \qquad (4.16)$$

To take a numerical example, consider an axon of radius 25 μm, membrane resistance (R_m) 2000 Ω cm^2, axoplasm resistivity (R_i) 60 Ω cm and membrane capacitance (C_m) 1 μF cm^{-2}. Then r_m (membrane resistance of a unit length) is 127 000 Ω cm, r_i (internal longitudinal resistance) is 3 060 000 Ω cm^{-1}, and c_m (membrane capacitance per unit length) is 0.0157 μF cm^{-1}.

These relations can usefully be inserted in some of the core-conductor equations. For simplicity we shall assume that the axon is in a large volume of external medium, so that r_o is very much less than r_i and can therefore be omitted from the equations. First, consider the space constant λ. If r_o is very small compared with r_i, then

$$\lambda = \sqrt{\frac{r_m}{r_i}}$$

$$= \sqrt{\frac{R_m/2\pi a}{R_i/\pi a^2}} = \sqrt{\frac{aR_m}{aR_i}}$$

Hence λ increases with increasing axon radius; if R_m and R_i remain constant, λ is proportional to the square root of the radius. *Space constant*

We have seen that the time constant is defined by

$$\tau = r_m c_m$$

Substituting equations 4.14 and 4.15 in this we get

$$\tau = (R_m/2\pi a)(2\pi a C_m)$$

$$= R_m C_m$$

Thus τ is essentially independent of the axon radius.

A further useful relation follows from equation 4.7; substituting equation 4.16 in this we get

$$i_m = \frac{\pi a^2}{R_i} \times \frac{d^2 V}{dx^2}$$

Now the membrane current density per unit area I_m will be related to the membrane current per unit length i_m by

$$i_m = I_m \, 2\pi a$$

Hence

$$I_m = \frac{a}{2R_i} \times \frac{d^2 V}{dx^2} \qquad (4.17)$$

(This equation, as we shall see later, is of considerable importance in the further analysis of nervous conduction.)

The core-conductor theory was experimentally tested by Hodgkin & Rushton (1946) on large axons from the walking legs of the lobster *Homarus*. Since they used extra-

cellular electrodes, the voltage change measured was that on the outside of the membrane; this will be proportional to the change in potential across the membrane (the proportionality constant will be $r_o/(r_o + r_i)$) so that the core-conductor equations can still be applied. It was found that the observed potential changes were in accordance with equations 4.10, 4.12 and 4.13 and with equations derived from 4.9 describing the change of potential with time at different distances from the point of application of the current. This correspondence between fact and theory is obviously good evidence for the core-conductor theory.

The next step was to evaluate the quantities r_o, r_i, r_m and c_m. Then, knowing the diameter of the axon, the resistance and capacitance of a square centimetre of membrane could be calculated. Average values from experiments on ten lobster axons were 2300 Ω cm^2 for R_m, the membrane resistance (range 600 to 7000 Ω cm^2 – this quantity was rather variable), and 1.3 μF cm^{-2} for C_m, the membrane capacitance. These values are in quite good agreement with those obtained for squid axons by Cole & Curtis using the a.c. bridge method, a fact which further increases our confidence in the applicability of the core-conductor model.

The local circuit theory

We are now in a position to consider how the action potential is propagated. Suppose a small length of axon is stimulated by a depolarizing pulse above threshold intensity, so that an action potential arises in the stimulated area. Regard this action potential as corresponding to the voltage V_0 in fig. 4.17 and assume that V_0 is about four times the threshold change in membrane potential. The potential across CD will now rise towards the point V_1, but at some point before reaching V_1 it will cross the threshold, so that an action potential appears across CD. This means that the potential across EF, originally moving towards V_2, now starts moving towards the potential level of V_1, and so it in turn crosses the threshold and an action potential now arises across EF. Thus an action potential has been propagated from AB through CD to EF, and will of course continue along the chain. Furthermore, unless there is any change along the length of the membrane in the threshold level or in the values of the component resistances and capacitances of the system, the conduction velocity will obviously be constant.

This hypothesis of the mechanism of conduction, proposed by Hermann at the turn of the century, is known as the *local circuit theory*, since it postulates that conduction is dependent on the electrotonic currents across the membrane set up in front of the action potential. The local circuits are shown in fig. 4.19a. Notice that, in order to set up passive (electrotonic) currents at the beginning of the action potential, there must be some inward current flow at the peak of the action potential. This inward current flow is from negative to positive and is therefore analogous to the flow *inside* a battery; in other words, the electrical energy needed to cause the electrotonic currents (which flow from positive to negative) is derived from this inward movement of positive charge at the peak of the action potential. This concept is illustrated in fig. 4.19b.

We shall see in the next chapter that this 'battery' is a sodium ion concentration cell and that the inward current it produces is an inward flow of sodium ions through voltage-gated sodium channels. The 'battery' is brought into action (i.e. the channels open) at any particular point when the membrane potential crosses the threshold at that point. The quantitative distribution of membrane current can be deduced from the form of the action potential by applying equation 4.7.

If the local circuit theory is correct, then it should be possible to observe the local currents associated with the action potential. This was done by Hodgkin (1937) by cooling a short length of nerve so that the action potential could not pass this region. On the other side of the block, small *extrinsic potentials* appeared (fig. 4.20), which diminished in size along the length of the nerve in the same way that electrotonic potentials do.

A less direct method of testing the theory is by consideration of the factors which affect conduction velocity. Reverting to fig. 4.17, it is evident that the conduction velocity must be dependent upon the time taken for the potential at a point a given distance in front of the action potential to cross the threshold. This time, in turn, will be governed by the values of the resistances in the system.

Hence changes in r_o or r_i should produce changes in conduction velocity. Hodgkin (1939) showed that an increase in r_o, produced by immersing an axon in paraffin oil instead of Ringer's solution, decreased the conduction velocity. Conversely, a decrease in r_o, produced by laying the axon across a series of brass bars which could be connected by means of a mercury trough, resulted in an increase in conduction velocity. Later del Castillo & Moore (1959) obtained large increases in conduction velocity by inserting a silver wire down the middle of an axon, so greatly reducing r_i.

Since r_i varies with the axon radius, we would expect conduction velocity to vary with the axon radius also. Theoretical arguments given by Hodgkin (1954) suggest that, other things being equal (a rather important proviso), conduction velocity is proportional to the square root of the axon radius. A simple argument why this should be so is as follows. For an action potential travelling at a constant velocity θ,

$$\frac{dV}{dx} = \frac{1}{\theta} \times \frac{dV}{dt}$$

and

$$\frac{d^2V}{dx^2} = \frac{1}{\theta^2} \times \frac{d^2V}{dt^2}$$

Substituting this relation in equation 4.17, we get

$$I_m = \frac{a}{2R_i\theta^2} \times \frac{d^2V}{dt^2} \qquad (4.18)$$

If the membrane properties remain the same in axons of different radii, then I_m and d^2V/dt^2 will be constant. In this case equation 4.18 becomes

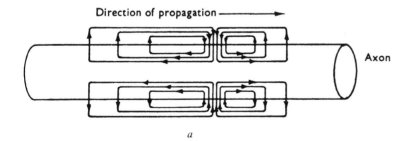

Direction of propagation ⟶

Axon

a

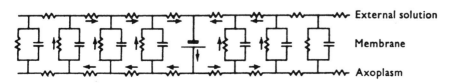

External solution

Membrane

Axoplasm

b

Figure 4.19. Local circuit currents. *a* shows the local circuit currents in a propagating action potential. *b* shows the currents produced in the core conductor model by insertion of a battery; the battery is the analogue of sodium ion flow through open sodium channels at the peak of an action potential.

$$\frac{a}{\theta^2} = \text{constant}$$

or

$$\theta = \text{constant} \times \sqrt{a}$$

i.e. the conduction velocity is proportional to the square root of the fibre radius.

An experimental investigation of this relation was carried out by Pumphrey & Young (1938), using the giant axons of *Sepia* and *Loligo*. The best fit for their results was given by conduction velocity being proportional to (radius)[0.6], which is very near to the expected value. Other investigations have not always provided such good agreement with the square root relation, but it is generally observed that, in a group of axons of the same type, increase in radius is associated with increase in conduction velocity.

The local circuit theory enables us to explain the shapes of action potentials as recorded by external electrodes. In fig. 4.21a two electrodes X and Y are placed a short distance apart on an axon. We shall assume that the input resistance of the recording system used to measure the potential difference between X and Y is very high. As the active region moves under electrode X, X becomes negative to Y, so producing an upward peak in the record, However, when the active region reached electrode Y, X becomes positive to Y and so the potential record shows a downward peak. This type of recording is known as *diphasic*. In fig. 4.21b the axon is crushed between the electrodes so that the action potential is unable to reach Y, and therefore Y never goes negative. This gives a *monophasic* record which is similar in shape to the potential across the membrane at X.

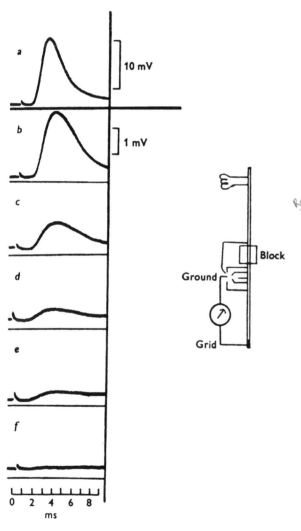

Figure 4.20. Direct observation of local circuits: the extrinsic potentials of frog sciatic nerve. A 3 mm length of nerve was blocked by cooling, and the electrical response of the nerve was recorded with external electrodes 2 mm before the block (a) and from 1.4 to 8.3 mm beyond it (b to f). Notice that the gain for records b to f is five times that for a. (From Hodgkin, 1937.)

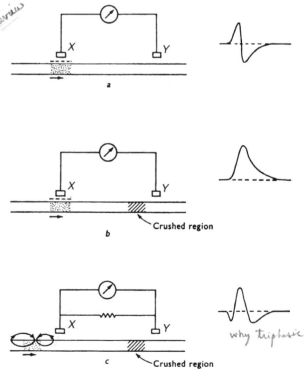

Figure 4.21. The form of action potentials recorded from nerve axons with extracellular electrodes. The active region is stippled in each case. The records on the right show an upward deflection when electrode X is negative to electrode Y.

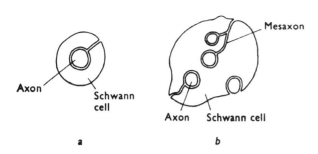

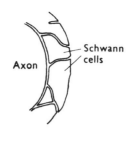

Figure 4.22. The form of Schwann cells surrounding unmyelinated axons. *a* shows the basic arrangement, *b* the situation in many small axons, and *c* the situation with large axons. (Based on Hodgkin, 1964.)

In fig. 4.21*c* there is a shunt resistance between the electrodes (such as might be produced by recording from a nerve in Ringer's solution or *in situ*) so that currents can flow between them. In this case the electrode X records the local currents associated with the action potential and the record is therefore *triphasic*.

Saltatory conduction in myelinated nerves

All peripheral nerve axons are surrounded by accessory cells called *Schwann cells* (fig. 4.22). A connective tissue sheath, the *endoneurium*, which contains small endoneurial cells, surrounds the complex of Schwann cells and axon. Numbers of these Schwann cells and endoneurial cells are found distributed along the length of an axon. Transverse sections of large axons (such as squid giant axons, as shown in fig. 4.22*c*) show that many Schwann cells are needed to surround them.

When the axons are small, the reverse situation may be seen (fig. 4.22*b*), in which a single Schwann cell surrounds a number of axons. In all these cases the axon is separated from the Schwann cell by a space about 150 Å wide, which is in communication with the extracellular fluid via channels (mesaxons) between the membranes of the Schwann cells. Examination of axons of this type by light microscopy does not reveal the presence of a fatty sheath surrounding the axon, and the nerve fibres (we shall regard a 'fibre' as comprising an axon plus its accessory cells) are therefore described as *non-myelinated*.

Many vertebrate and a few invertebrate axons (such as the large diameter axons of prawns; Holmes, 1942) are surrounded by a fatty sheath known as *myelin* and are described as being *myelinated*. The sheath is interrupted at intervals of the order of a millimetre or so, to form the *nodes of Ranvier* (fig. 4.23).

Electron microscope studies by Geren (1954) and Robertson (1960) showed that myelin is formed from many closely packed layers of Schwann cell membrane, produced by the mesaxon being wrapped round and round the axon (fig. 4.24). As a consequence of this arrangement, there are

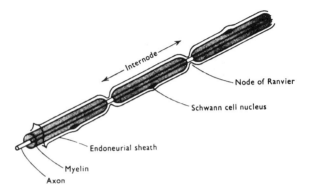

Figure 4.23. Diagram to show the structure of a myelinated nerve fibre.

no extracellular channels between the axon membrane and the external medium in the internodal regions. Such contact is available at the nodes, as is shown in fig. 4.25.

It is obvious that the presence of a myelin sheath will considerably affect the electrical properties of the nerve fibre. The overlapping Schwann cell membranes act as chains of resistances and capacitances in series. This means that the myelin sheath will have a much higher transverse resistance and a much lower transverse capacitance than a normal cell membrane; the actual figures for frog fibres are about 160 000 Ω cm^2 and 0.0025 μF cm^{-2}. The values for the resting membrane at the node are quite different, being of the order of 20 Ω cm^2 and 3 μF cm^{-2}. In view of these figures, we might expect that when current is passed across the membrane the current flow through the nodes would be greater than that at the internodes. If, in addition, the nodes are the regions at which excitation of the membrane occurs, then, applying the local circuit theory, we would expect conduction in a myelinated axon to be a discontinuous (or *saltatory*) process.

Verification of the saltatory conduction theory depends first of all on the ability of the experimenter to dissect out single fibres from a whole nerve trunk. This technique was first developed by Japanese workers in the 1930s (see Kato,

1936; Tasaki, 1953). It was found that blocking agents such as cocaine or urethane were effective only when applied at the nodes, and that the threshold stimulus intensity was lowest at the nodes (fig. 4.26). These results accord well with the predictions of the saltatory theory, but it was possible to argue that the sensitivity of the nodal region was caused merely by the exposure of the axon membrane at this site.

More conclusive evidence that conduction is saltatory would be provided if it could be shown that inward flow of current is restricted to the nodes. Experiments to test this idea were performed in the 1940s by Tasaki & Takeuchi and by Huxley & Stämpfli, and it is worth looking at some of the details of their experiments. The technique used by Tasaki & Takeuchi (1942) is shown in fig. 4.27. The nerve fibre is placed in three pools of Ringer's solution insulated

from each other by air gaps. The two outer pools are earthed, and the middle pool is connected to earth via a resistance. With this arrangement, all the radial current flowing across the axon membrane and myelin sheath of that part of the fibre which is in the middle pool will flow through the resistance, and can therefore be measured by the potential of the middle pool with respect to earth. The results (fig. 4.27b) showed that inward currents occurred only when there was a node in the middle pool, and therefore inward current flow must be restricted to the nodal regions.

The method used by Huxley & Stämpfli (1949) was rather different. The fibre was threaded through a fine hole in a glass slide, and the potential measured on each side of the hole during the passage of an impulse (fig. 4.28). If the

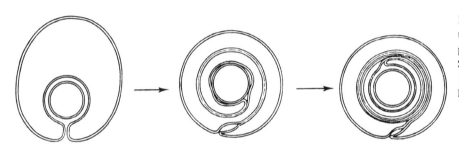

Figure 4.24. The development of the myelin sheath of vertebrate peripheral nerve fibres by Schwann cells. (From Robertson, 1960, by permission of Pergamon Press Ltd.)

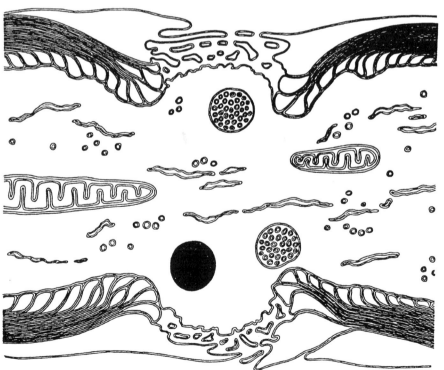

Figure 4.25. Longitudinal section of a node of Ranvier, as seen by electron microscopy. (From Robertson, 1960, by permission of Pergamon Press Ltd.)

resistance of the Ringer's solution surrounding the fibre in the hole is known, then the longitudinal current flow can readily be obtained from the measured potential by application of Ohm's law. By pulling the fibre through the hole, a series of such measurements can be made at intervals along its length, as shown in fig. 4.29. The transverse current across the membrane is then given by equation 4.6,

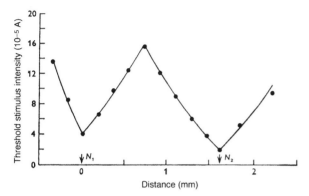

Figure 4.26. The variation in threshold stimulus intensity along the length of a single myelinated nerve fibre. N_1 and N_2 mark the positions of two nodes of Ranvier. (From Tasaki, 1953.)

i.e. by the difference between the longitudinal currents at two adjacent recording sites. The calculated membrane currents are shown in fig. 4.30, from which it is clear that here again inward current flow is restricted to the nodes.

This restriction of the inward current to the nodes implies that, while the transverse current in the internodal regions can be explained as a passive current through a resistance and capacitance in parallel, there must be some 'active' component at the nodes of Ranvier. We shall see later that this current is an inward flow of sodium ions through voltage-gated sodium channels, and that these channels are restricted to the nodal regions.

The conduction velocity of myelinated fibres is normally greater than that of unmyelinated fibres of the same diameter. The reason for this is that the high resistance and low capacitance of the myelin sheath increases the longitudinal spread of the local currents involved in propagation. Theoretical arguments given by Rushton (1951) suggest that conduction velocity should be proportional to the fibre diameter, and not, as in unmyelinated axons, to the square root of fibre diameter. This appears to be so (Hursh, 1939; Tasaki *et al.*, 1943). The conclusion implies that the conduction velocity of very small myelinated fibres should

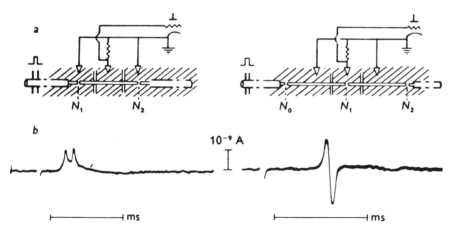

Figure 4.27. The radial currents in a short length of a myelinated fibre during the passage of an action potential. The fibre passes through three pools of physiological saline separated from each other by air gaps. *a* shows the recording arrangements. *b* shows the potential difference across the resistance *R* when the middle pool of saline does (right trace) or does not (left trace) contain a node. (From Tasaki & Takeuchi, 1942.)

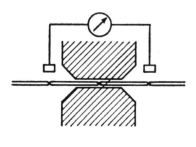

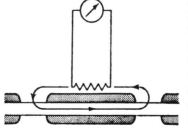

Figure 4.28. The method used by Huxley & Stämpfli to measure longitudinal currents outside a short length of a myelinated nerve fibre. *a* gives a schematic view of the recording arrangement and *b* the equivalent circuit of the system. (After Huxley & Stämpfli, 1949.)

be less than that of unmyelinated fibres of the same diameter (fig. 4.31). Rushton calculated that the critical diameter below which myelination confers no advantage should be about 1 μm, which is near that of the largest unmyelinated fibres in mammalian peripheral nerves. However, Waxman & Bennett (1972) pointed out that myelinated fibres less than 1 μm in diameter are found in the central nervous system, and suggested that the critical diameter there is as low as 0.2 μm.

Why should there be such a difference between central and peripheral fibres? Ritchie (1982) concluded that the answer lies in the different mechanisms of myelination in the two cases. In peripheral nerve fibres each internode is produced by a separate Schwann cell and there is a minimum internodal distance, in the range 200 to 400 μm. This means that at small diameters (less than 4 μm) the internodal distance is longer than the optimum for conduction, and at very small diameters the electrotonic spread along the internode will be insufficient to cause excitation at the succeeding node. Ritchie calculated that the limiting diameter at which conduction will fail is about 0.8 μm. In central fibres, however, the myelin sheaths are produced by

oligodendrocytes, each of which may myelinate up to 50 internodes (Hirano, 1981), and there seems to be no anatomical lower limit to the length of the internodes.

Compound action potentials

The isolated sciatic nerve of the frog has been much used in the study of nervous phenomena. If such a nerve is arranged so that it can be stimulated at one end and recorded from at the other, action potentials can be recorded in the usual manner. These action potentials differ from those of isolated single fibres in that their size is, over

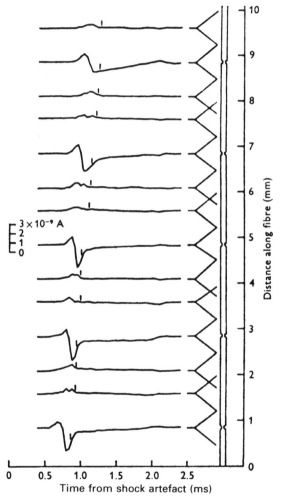

Figure 4.30. Radial (transmembrane) currents at different points along the length of a myelinated nerve fibre during the passage of an action potential, obtained from the difference between the longitudinal currents (fig. 4.29) at two points 0.75 mm apart. Notice that inward current flow (seen as a downward deflection from the baseline) is restricted to sections containing a node. (From Huxley & Stämpfli, 1949.)

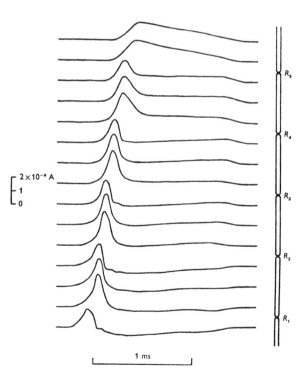

Figure 4.29. Longitudinal currents at different points along the length of a myelinated nerve fibre during the passage of an action potential, measured by the method shown in fig. 4.28. (From Huxley & Stämpfli, 1949.)

Table 4.1 *Classifications of mammalian nerve fibres according to their conduction velocities and other features. The group I to IV classification applies to sensory fibres only*

Type	Group	Diameter (μm)	Conduction velocity (ms⁻¹)	Function
Aα		15–20	50–120	Motor fibres to skeletal muscle
Aα	Ia	15–20	70–120	Primary endings on muscle spindles
Aα	Ib	12–20	70–120	Golgi tendon organ afferents
Aβ	II	5–10	30–70	Secondary endings on muscle spindles, touch, pressure
Aγ		3–6	15–30	Motor innervation of muscle spindles
Aδ	III	2–5	5–25	Pressure/pain receptors
B		3	3–15	Autonomic preganglionic
C		0.5–1	0.5–2	Autonomic postganglionic (non-myelinated)
C	IV	0.5–1	0.5–2	Pain (non-myelinated)

Note:
Based partly on Ottoson, 1983.

a restricted range, proportional to the size of the stimulus. The reason for this is that the recorded action potential is the result of simultaneous activity in a large number of axons (each of which itself obeys the all-or-nothing law) which have different thresholds.

When the distance between the stimulating and recording electrodes is large, it is found that the monophasic compound action potential (recorded as in fig. 4.21*b*) consists of a number of potential waves; this is because the nerve trunk contains fibres of different diameter and hence different conduction velocities. Separation of the different components of the whole response occurs because action potentials with higher conduction velocities reach the recording electrodes before those with lower ones (fig. 4.32).

Erlanger & Gasser (1937) classified vertebrate nerve fibres according to their conduction velocities into three groups, A, B and C, with A being further subdivided into α, β, γ and δ. An alternative classification has been much used for the sensory fibres from mammalian muscles, with groups I to IV designated according to their fibre diameter (Lloyd, 1943; Hunt, 1954). The two systems are summarized in table 4.1.

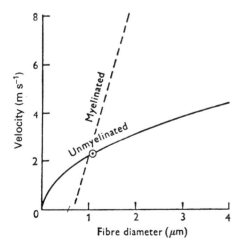

Figure 4.31. Theoretical relations between fibre diameter and conduction velocity in myelinated and unmyelinated nerve fibres. (From Rushton, 1951.)

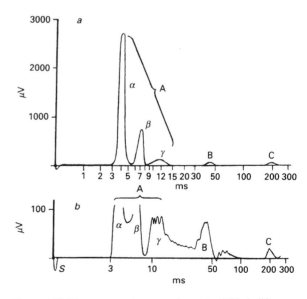

Figure 4.32. The compound nerve action potential in bullfrog sciatic nerve. The time scale is logarithmic, and trace *b* is at a higher amplification than trace *a*. (From Keynes & Aidley, 1991, after Erlanger & Gasser, 1937.)

5
The ionic basis of nervous conduction

The story so far, given in chapter 4, describes the understanding of the nerve action potential as it was in 1945, just after the Second World War. Much was known about the relations between the stimulus and the response in nervous conduction, but the nature of the action potential itself was far from clear. The discovery that the membrane resistance falls during the action potential fitted well with Bernstein's idea that there is an increase in the ionic permeability of the membrane, but the discovery of the overshoot had shown that his proposal of a general increase in permeability to all ions did not fit the facts. What was going on? Over the next few years, work by Hodgkin and his colleagues in Cambridge and Plymouth produced some very convincing answers to this question (Hodgkin, 1958, 1964, 1992). Let us look at the conceptual framework within which they emerged.

The conceptual model used by Hodgkin and his colleagues is shown in the form of an electric circuit in fig. 5.1. In this model, we assume that the concentration gradient of any particular ion in the system acts as a battery, whose electromotive force is given by the Nernst equation for that ion. For example, there is a 'potassium battery', E_K, whose electromotive force is given by

$$E_K = \frac{RT}{F} \ln \frac{[K]_o}{[K]_i} \tag{5.1}$$

In squid axons, E_K is usually about -75 mV (inside negative). Similarly, the electromotive force of the 'sodium battery', E_{Na}, is given by

$$E_{Na} = \frac{RT}{F} \ln \frac{[Na]_o}{[Na]_i} \tag{5.2}$$

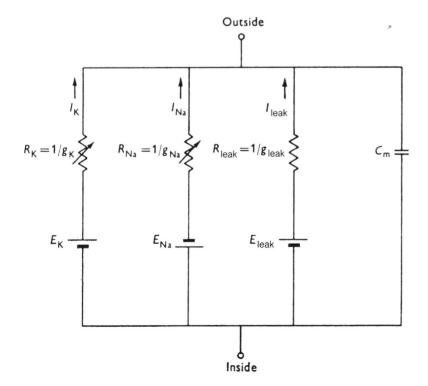

Figure 5.1. Conceptual model of an area of excitable membrane in a nerve axon. The variable conductances g_K and g_{Na} are proportional to the numbers of open voltage-gated potassium and sodium channels. (After Hodgkin & Huxley, 1952*d*.)

Since $[Na]_o$ is usually greater than $[Na]_i$, E_{Na} is positive, typical values being in the region of $+55$ mV. The other ions in the system, of which chloride is usually the most important, contribute to a third battery, E_{leak}, which may produce a 'leakage current'; the value of E_{leak} is near the resting potential, at about -55 mV.

In series with each battery is a resistance, shown as R_K, R_{Na} and R_{leak} in fig. 5.1. However it is more convenient to talk about the corresponding conductances, g_K, g_{Na} and g_{leak}. (Remember that conductance is the reciprocal of resistance.) These conductances represent the ease with which ions can pass through the membrane; they are related to the permeability coefficients mentioned in chapter 3, but of course the units of measurement are different. We assume that the sodium and potassium conductances are variable. These three ionic pathways are arranged in parallel, the current flowing through each pathway being represented by I_K, I_{Na} and I_{leak}. Finally, the membrane capacitance is represented by a fourth element C_m.

A crucial feature of the model shown in fig. 5.1 is the capability of the sodium and potassium conductances to adopt different values. For many years the physical basis of this feature was not at all clear. However, there is now overwhelming evidence that ionic flow through the membrane occurs through the macromolecular structures called ion channels. The channels are usually permeable to only one or a few types of ion, so that we find sodium channels which open to allow the flow of sodium ions, potassium channels which open to allow potassium ions through, and so on. Each channel is normally closed, but can open for short periods of time. Hence the conductance of a single channel is either zero or a particular value such as 18 pS. So changes in the conductance of a whole pathway are brought about by changes in the number of membrane channels which are open.

Let us take an example. Suppose the conductance of a single sodium channel is 18 pS and suppose the sodium conductance of an area of membrane (represented by g_{Na} in fig. 5.1) rises to 36 mS. This increase in total membrane conductance is brought about by a corresponding increase in the number of sodium channels which are in the open state, and in order to get a total conductance of 36 mS we must have 2×10^9 open channels.

A moment's thought will show that the net potential across the membrane, E, will, in the absence of externally applied current, be determined by the relative values of the ionic conductances. If g_K is much higher than g_{Na}, or g_{leak}, for instance, E will be near to E_K. If g_{Na} is then increased, E will move towards E_{Na}, and so on.

Work on the ionic theory of nervous conduction proceeded apace in the years after the Second World War, reaching a peak with the remarkable analysis by Hodgkin & Huxley in 1952. Later work served to fill in the details and expand the boundaries of this analysis. From the 1970s onwards there has been an increasing emphasis on the nature and properties of the individual channels involved, and we shall examine some of that work in the following chapter. For the remainder of this chapter we see how the conceptual model shown in fig. 5.1 leads to an adequate description of the properties of the axon membrane.

The sodium theory

We have seen that the discovery of the 'overshoot', in which the inside of the axon becomes positive to the outside during the peak of the action potential, was the big surprise produced by the intracellular recordings of 1939. It was not immediately evident what the reason for this overshoot was, and various mechanisms that are no longer considered were proposed. Then Hodgkin & Katz (1949) suggested that during the action potential there is a rapid and specific increase in the permeability of the membrane to sodium ions. In terms of the model shown in fig. 5.1, g_{Na} becomes temporarily very much larger than g_K, so that E moves towards E_{Na}.

This suggestion became known as the 'sodium theory' of the action potential. In order to test its validity, Hodgkin & Katz measured the height of action potentials from squid giant axons placed in solutions containing different concentrations of sodium ions. No action potentials could be produced in the absence of sodium ions in the external medium. If the axon was placed in a solution with reduced sodium ion concentration (prepared by mixing sea water with an isotonic glucose solution), the action potential was reduced in height (fig. 5.2). Furthermore the slope of the curve relating the height of the action potential to $\log_{10}[Na]_o$ was close to 58 mV per unit, except at very low sodium ion concentrations, where failure of conduction was imminent. This is just what one would expect from equation 5.2. They also found that the height of the action potential was increased after addition of sodium chloride to the external solution.

Similar results have been obtained from a variety of other nerve cells, including frog myelinated axons (Huxley & Stämpfli, 1951) and insect axons (Narahashi, 1963). Sodium ion currents also form the basis of action potentials in frog 'fast' skeletal muscle fibres (Nastuk & Hodgkin, 1950) and the electrocytes of the electric eel (see later), and are an important component in the action potentials of many mammalian heart muscle fibres (Draper & Weidmann, 1951). In other cases, as in vertebrate smooth muscle, heart pacemaker fibres, and some arthropod muscle fibres, calcium ions carry most of the inward current flow.

Ionic movements during activity

Direct measurements of ionic movements, using radioactive isotopes as tracers and various accurate methods of chemical analysis, are obviously relevant to a study of the nature of the action potential. Let us consider some experiments by Keynes (1951) on the movements of radioactive sodium and potassium ions across the membrane of the giant axons of the cuttlefish *Sepia*. The axon, held in two pairs of forceps, could be loaded with the radioactive isotope by immersion in a pot of sea water containing the isotope. In order to observe its radioactivity, it was transferred to a chamber containing flowing non-radioactive sea water and placed over a Geiger counter. Stimulating pulses could be applied via one pair of forceps, and action poten-

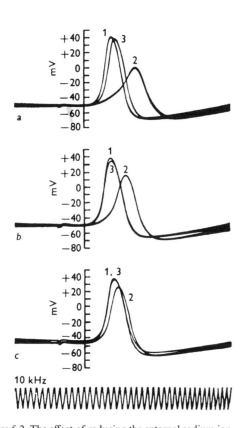

Figure 5.2. The effect of reducing the external sodium ion concentration on the size of the action potential in a squid giant axon. In each set of records, 1 is obtained in sea water, 2 in the experimental solution, and 3 in sea water again. Experimental solutions were made by mixing sea water and isotonic glucose solutions. Proportions of sea water were 33% in *a*, 50% in *b* and 70% in *c*. (From Hodgkin & Katz, 1949.)

tials could be recorded from the other pair so as to provide a check on the excitability of the axon.

Figure 5.3 shows the results of one of Keynes's experiments on the movement of radioactive potassium ions. The axon was initially immersed in ^{42}K-labelled sea water. After 15 min it was removed to the counting chamber and its radioactivity measured at intervals during the next 40 min; when plotted on a logarithmic scale, these values fell on a straight line, and so the radioactivity at 15 min could be estimated by producing this line back. From this value, knowing the potassium ion concentration of the sea water, and its specific activity, the rate of entry of potassium could be calculated. Then the axon was transferred to the radioactive sea water again and stimulated for 10 min; on returning to the measuring chamber its increase in radioactivity could be calculated as before, and from this the extra influx on stimulation could be determined. The rate of loss of radioactivity in the resting condition could be obtained from the slope of the lines through the experimental points in fig. 5.3, and the rate of loss on stimulation could be found by stimulation in the non-radioactive sea water in the counting chamber. By making plausible assumptions as regards the internal potassium concentration, these figures could be converted into absolute values of potassium efflux.

Similar experiments were performed with sodium ions. The main results of the investigation are shown in diagrammatic form in fig. 5.4. The net effect of each impulse was to produce a net entry of sodium ions of 3.7 pmol cm^{-2} and a net exit of potassium ions of 4.3 pmol cm^{-2}.

Entry of sodium ions will cause a build-up of positive charge on the inside of the axon membrane (making the membrane potential more positive), and exit of potassium ions will remove it. Hence it is reasonable to suggest, from Keynes's results, that the rising phase of the action potential is brought about by inward movement of sodium ions, and the falling phase by outward movement of potassium ions. If this suggestion is correct, we should be able to show that the quantity of sodium ions entering the axon is sufficient to cause the observed changes in membrane potential. The charge (Q, measured in coulombs) on a capacitance (C, measured in farads) is given by

$$Q = CV$$

where V is the voltage across the capacitance. If this charge is produced by a univalent ion, the number of moles, n, of the ion moved from one side of the capacitance to the other is given by

$$n = CV/F$$

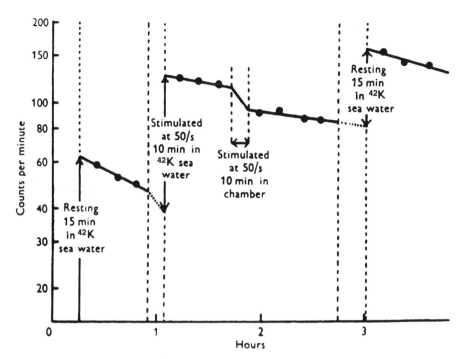

Figure 5.3. Results of an experiment to determine potassium ion movements in a *Sepia* giant axon. The points show the radioactivity of an axon after treatment in various ways. (From Keynes, 1951.)

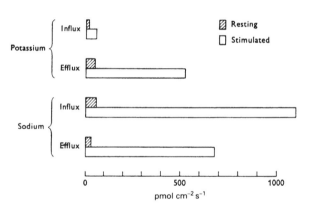

Figure 5.4. Fluxes of sodium and potassium ions in *Sepia* giant axons. Shaded columns show resting fluxes, white columns show fluxes during stimulation at $100 \ s^{-1}$. (Based on data of Keynes, 1951.)

where F is the Faraday constant. In the case of an axon, C is $1 \ \mu F \ cm^{-2}$ and V (the height of the action potential) is about 110 mV. Hence

$$n = \frac{10^{-6} \times 0.11}{10^5}$$

faradays constant

$$= 1.1 \ pmol \ cm^{-2}$$

So the 3.7 pmol of sodium ions that enter the axon are more than sufficient to account for the change in membrane

potential during the action potential. (Why should there be a 'more than sufficient' quantity? Probably because there is some overlap between the timing of the sodium ion entry and the potassium ion exit; clear evidence for this is given in the following section.)

Isotopic experiments provide a very good measure of the gross ionic exchanges occurring on stimulation, but their time resolution is naturally extremely poor in relation to the duration of the action potential. A more detailed picture of ionic movements during activity can be obtained only by using electrical methods, which we must now consider.

Voltage clamp experiments

If the 'sodium theory' is correct, then the initial depolarization which produces an action potential must result in an increase in sodium conductance, and this increase in sodium conductance will itself produce a further depolarization as sodium ions flow in through the open sodium channels. Thus we are dealing with a positive feedback system, as shown in fig. 5.5a, and such systems are very difficult to analyse. These difficulties have been overcome by means of a special technique known as the *voltage clamp* method, which was initiated by Cole (1949) and Marmont (1949) in Chicago and Woods Hole and further developed by Hodgkin *et al.* (1952) in Cambridge and Plymouth. The essence of the method is that the positive feedback effect of sodium conductance on membrane potential is eliminated by passing current through the nerve membrane so as to hold the membrane potential constant at any desired value. The use of this method by Hodgkin & Huxley (1952*a–d*) led to a remarkably complete analysis of the events occurring during the action potential.

Experiments on squid axons

A schematic diagram of the experimental arrangement is shown in fig. 5.6. Two thin silver wire electrodes *a* and *b* are passed down the middle of a squid giant axon. The axon is placed in a trough passing through pools of sea water separated by insulated partitions; on the outside of these pools is an earthed electrode *e*. Current is passed through the nerve membrane by means of a generator connected to the internal electrode *a* and the earth electrode *e*. In the middle pool, this current must pass through the resistance provided by the sea water between electrodes *c* and *d*, and thus (applying Ohm's law) the voltage between *c* and *d* is proportional to the current flowing through the axon membrane in contact with the middle pool. The voltage across this part of the membrane is recorded by means of electrode *c* and the internal electrode *b*. This voltage, after amplification, is fed into a comparator which is also supplied by a signal voltage. The output of the comparator is passed to the current generator so as to increase or decrease the current flowing through the membrane, which makes the voltage across the membrane (after amplification) equal

to the signal voltage. (In practice, a single high gain differential d.c. amplifier acts both as the comparator and the current generator.) Hence this arrangement constitutes a negative feedback control system in which the voltage across the membrane is determined by the externally applied signal voltage. The presence of the outer pools, with part of the current between *a* and *e* passing through them, ensures that the membrane potential is constant over the length of axon in the middle pool.

The current flowing through the axon membrane is assumed to consist of two components, a capacity current (i.e. changes in charge density at the inner and outer surfaces of the membrane) and an ionic current (i.e. passage of ions through the membrane). Hence the total current I is given by

$$I = C_m \frac{dE}{dt} + I_i \qquad (5.3)$$

where C_m is the membrane capacitance, E is the membrane potential and I_i is the ionic current. Thus when the voltage is held constant ('clamped') $dE/dt = 0$, and so the record of current flow gives a direct measure of the total ionic flow.

Figure 5.7*a* shows the type of record obtained by Hodgkin & Huxley. Notice that the voltage measured, *V*, corresponds to the difference between the resting potential, E_r, and the clamped membrane potential, E; the reason for this is that it is difficult to obtain accurate absolute values for the membrane potential with silver wire electrodes. The current record shows three components. First, there is a brief 'blip' of outward current; this is caused by discharge of the membrane capacitance. After this the current is inward for about 1 ms, and finally the current becomes outward and climbs to a steady level which is maintained while the clamp lasts.

Now, applying the 'sodium theory' of the action potential, we might expect that the initial inward current during the clamp is caused by sodium ions flowing inwards. If this is so, then it should disappear if the axon is placed in a

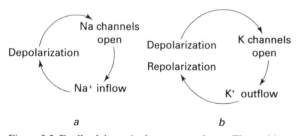

a   b

Figure 5.5. Feedback loops in the axon membrane. The positive feedback sodium loop is shown in *a*. The negative feedback potassium loop is shown in *b*.

Figure 5.6. Schematic diagram of the method used to determine membrane currents in a squid axon under voltage clamp. (Based on Hodgkin *et al.*, 1952.)

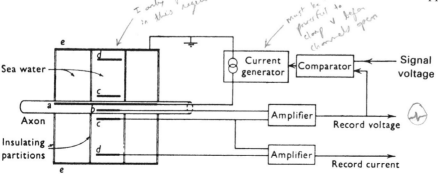

solution containing no sodium ions. In fact, it was found (by using choline chloride as a substitute for sodium chloride) that the inward current is replaced by an outward current under these conditions (fig. 5.7b). This is understandable if we assume that depolarization causes a brief increase in sodium conductance, since the internal sodium concentration is now very much greater than that outside, and we would therefore expect sodium ions to flow outwards.

The direction of sodium ion flow must be dependent upon the membrane potential as well as on the concentration gradient. When the membrane potential is equal to the sodium equilibrium potential E_{Na} (given by equation 5.2),

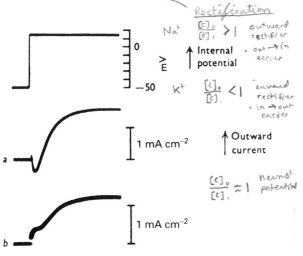

there will be no net flow of sodium ions, so that if the membrane potential is clamped at E_{Na} (produced by a depolarization of V_{Na}, where $V_{Na} = E_{Na} - E_r$) there should be no sodium current. If the depolarization does not reach V_{Na}, the sodium current will be inwards, and if it is greater than V_{Na} it will be outwards. This effect is shown in fig. 5.8, in which it is evident that on this interpretation V_{Na} is approximately 117 mV. An alternative way of eliminating the sodium current is to alter the external sodium ion concentration, so changing E_{Na}; trace *b* in fig. 5.9 was obtained in this way.

These results imply very strongly that the initial current flow under voltage clamp conditions is caused by movement of sodium ions. In order to clinch the matter, it is necessary to show that the potential at which the initial current reverses really is the sodium equilibrium potential, E_{Na}. This could not be done directly in the original experiments since the internal sodium ion concentration was not known exactly, and also because of the uncertainties involved in translating the measured potentials V into membrane potentials E. However, it was possible to test the hypothesis by measuring changes in E_{Na} (measured as changes in V_{Na} plus a small correction for any resting potential change) which occurred when the external sodium ion concentration was changed. The observed changes never differed from those expected by more than 3%.

The slowly rising, maintained outward current is only very slightly affected by changes in the external sodium ion concentration, and is therefore caused by movement of some other ion, probably (it was assumed) potassium. Direct evidence that this was so was provided by an experiment in which the efflux of ^{42}K from the region of an axon membrane under a cathode was measured (Hodgkin & Huxley, 1953). The increase in potassium ion efflux was

Figure 5.7. Typical records of the membrane current during a voltage clamp experiment on a squid *Loligo* giant axon. Records *a* and *c* were obtained with the axon in sea water, record *b* with a sodium-free isotonic choline chloride solution. (From Hodgkin, 1958, after Hodgkin & Huxley, 1952a.)

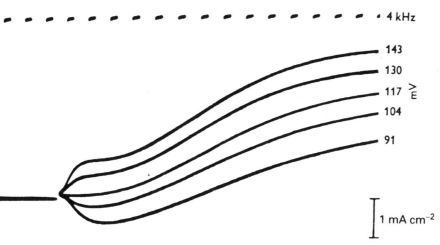

Figure 5.8. Membrane currents from a voltage-clamped squid axon at large depolarizations. Values of V (the amount of depolarization from the resting potential) are shown at the right of each record. (From Hodgkin, 1958, after Hodgkin *et al.*, 1952.)

linearly proportional to the current density, with a slope equal to the Faraday constant, and hence the current was carried by potassium ions.

How the membrane conductances are affected by voltage and time

It was now possible to analyse the ionic current following depolarization into two components, due to the flows of sodium and potassium ions. In fig. 5.9, trace *b* represents the current produced by an axon in 10% sea water (with 90% isotonic choline chloride solution) when *V* was equal to V_{Na}. Trace *a* is the current produced by the same depolarization (56 mV in this case) with the same axon in sea water. Now since there is no sodium current I_{Na} when *V* is equal to V_{Na}, trace *b* must represent the potassium current I_K plus a small constant component, the leakage current I_{leak}. Hence the difference between trace *a* and trace *b* must be equal to the sodium current; this is drawn as trace *c* in fig. 5.9.

The problem of separating the sodium and potassium currents is a little more complicated when the clamped depolarization in the low sodium solution was not equal to V_{Na}, but, by assuming that the sodium current in the low sodium case was always a constant fraction of that in the high sodium case, it is possible to calculate the sodium and potassium currents in each case.

The next step in the analysis was to determine the conductance of the membrane to sodium and potassium ions during a clamped depolarization. Applying Ohm's law again, the ionic current flow through the membrane is equal to the product of the conductance and the electromotive force. In the case of sodium ions, for instance, the electromotive force is given by the difference between the membrane potential and the sodium equilibrium potential, and so the sodium current is given by

$$I_{Na} = g_{Na} (E - E_{Na})$$
$$= g_{Na} (V - V_{Na}) \qquad (5.4)$$

Thus, from trace *c* in fig. 5.9, and knowing V_{Na}, it is a simple matter to calculate g_{Na} throughout the course of the clamp.

By a similar process, the potassium conductance can be calculated from trace *b* and the relation

$$I_K = g_K (V - V_K) \qquad (5.5)$$

The results of these calculations are shown in fig. 5.10. Similar calculations at depolarizations of different magnitudes produced the curves shown in fig. 5.11.

The results of this investigation at this stage can be summarized as follows. Depolarization produces three effects: (1) a rapid increase in sodium conductance, followed by (2) a slow decrease in sodium conductance

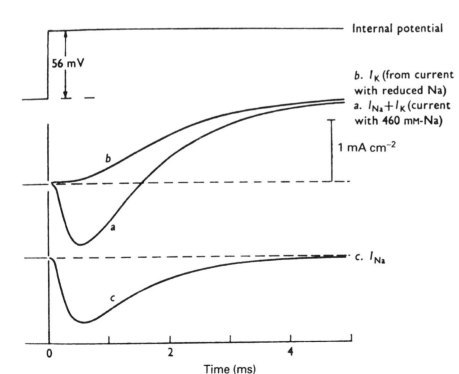

Internal potential

b. I_K (from current with reduced Na)
a. $I_{Na} + I_K$ (current with 460 mM-Na)

1 mA cm^{-2}

c. I_{Na}

56 mV

Figure 5.9. Analysis of the ionic current in a squid axon during a voltage clamp. Trace *a* shows the response to a depolarization of 56 mV with the axon in sea water. Trace *b* is the response in a solution comprising 10% sea water and 90% isotonic choline chloride solution; the membrane potential would be equal to E_{Na} during this trace. Trace *c* is the difference between traces *a* and *b*. (From Hodgkin, 1958, after Hodgkin & Huxley, 1952a.)

Time (ms)

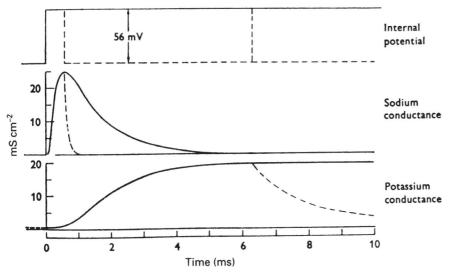

Figure 5.10. Ionic conductance changes during a clamped depolarization of a squid axon, derived from the current curves shown in fig. 5.9. The dashed lines show the effects of repolarization. (From Hodgkin, 1958, by permission of the Royal Society.)

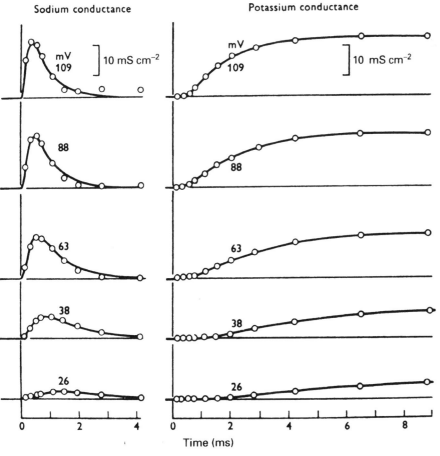

Figure 5.11. Conductance changes in squid axon brought about by depolarizations of different extents. The circles represent values derived from the experimental measurements of ionic currents (as in figs. 5.9 and 5.10), and the curves are drawn according to the equations used to describe the conductance changes. (From Hodgkin, 1958, after Hodgkin & Huxley, 1952*d*.)

(known as the sodium inactivation process) and (3) a slow increase in potassium conductance. The extent of these conductance changes increases with increasing depolarization, reaching a maximum (saturating) level of about 30 mS cm^{-2} at depolarizations of 100 mV and above. If the membrane potential is suddenly returned to its resting level, these changes are reversed, as is shown by the dashed curves in fig. 5.10.

We can restate these conclusions in terms of voltage-gated ion channels, since the conductance of an area of membrane is proportional to the number of its channels that are open. Depolarization produces (1) a rapid opening of the sodium channels, followed by (2) their inactivation, so that they can no longer allow sodium ion flow. Depolarization also (3) opens potassium channels; this is a slower process and is not followed by inactivation on the time scale of these experiments. Larger depolarizations open more channels, up to a maximum at depolarizations of 100 mV or more. When the membrane potential is returned to its resting level again, the channels close.

Calculation of the form of the action potential

If the conceptual model of the membrane shown in fig. 5.1 is correct, then it should be possible to calculate the form of the action potential if we know the values of the fixed components in the system (E_K, E_{Na}, E_{leak}, R_{leak} and C_m) and if equations describing the behaviour of the variables g_K and g_{Na} can be obtained. From their voltage clamp experiments, Hodgkin & Huxley (1952d) were able to provide a series of equations describing the changes of g_K and g_{Na} with depolarization and time (from which the continuous curves in fig. 5.11 were calculated), and so were able to undertake this rather laborious calculation. The results were dramat-ically accurate, with only slight differences between the predicted action potentials and those actually observed (fig. 5.12).

Figure 5.13 shows the theoretical solution for a propagated action potential, together with the associated changes in g_K and g_{Na}. It is instructive to follow the potential and conductance changes during its course. Initially, g_K is small but g_{Na} is much smaller, so that the resting potential is near E_K. Then the presence of an action potential approaching along the axon draws charge out of the membrane capacitance by local circuit action, so causing some initial depolarization. When the membrane has been depolarized by about 10 mV, the sodium conductance begins to increase (some sodium channels open), so that a small number of sodium ions cross the membrane, flowing down their

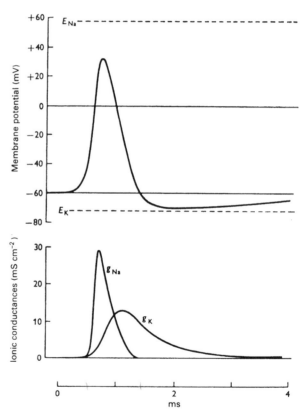

Figure 5.13. Calculated changes in membrane potential (upper curve) and sodium and potassium conductances (lower curves) during a propagated action potential in a squid giant axon. The scale of the vertical axis is correct, but its position may be slightly inaccurate; it has been drawn here assuming a resting potential of -60 mV. The positions of E_{Na} and E_K are correct with respect to the resting potential. In the original calculations, voltages were measured from the resting potential, as in fig. 5.12. (Redrawn after Hodgkin & Huxley, 1952d.)

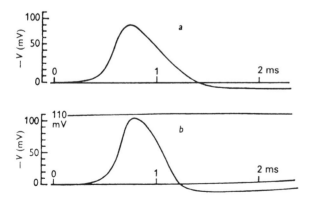

Figure 5.12. Comparison of (*a*) computed and (*b*) observed propagated action potentials in squid axon at 18.5 °C. The calculated conduction velocity was 18.8 m s^{-1}, the observed velocity was 21.2 m s^{-1}. (From Hodgkin & Huxley, 1952d.)

electrochemical gradient into the axon. This transfer of charge results in a further depolarization, which causes a further increase in g_{Na} (more sodium channels open) so more sodium ions cross the membrane, and so on.

The result of this regenerative action between depolarization and sodium conductance is that the sodium battery is relatively much more important than the potassium battery in determining the membrane potential, and so the membrane potential goes racing up towards the sodium equilibrium potential E_{Na}. But now the two slower consequences of depolarization, sodium inactivation (sodium channels stop allowing sodium ions through) and the increase in potassium conductance (potassium channels open), begin to take effect. This means that the potassium battery becomes more important and the sodium battery less important in determining the membrane potential (because sodium ion inflow declines and potassium ion outflow increases), and so the membrane potential begins to fall.

This repolarization closes sodium channels and so further reduces g_{Na} (it also closes potassium channels and so reduces g_K, but more slowly) so that the membrane potential is brought rapidly back to its resting level. At this point, although g_{Na} is extremely low as very few sodium channels are open, g_K is still considerably higher than usual as there are still quite a lot of potassium channels open. So the membrane potential passes the resting level and moves even nearer to the potassium equilibrium potential. Finally, as g_K declines to its normal low value (nearly all the potassium channels close), the membrane potential returns to its resting level.

The equations

Now let us have a brief look at the equations used by Hodgkin & Huxley (1952*d*) in the calculation of the form of the action potential. The first step was to find empirical equations to describe the conductance changes seen, for example, in fig. 5.11. The potassium conductance is given by

$$g_K = \bar{g}_K n^4 \qquad (5.6)$$

where $\bar{g}_K$ is a constant equal to the maximum value of g_K. The idea behind this equation was that potassium ions might be let through the membrane when four charged particles moved to a certain region of the membrane under the influence of the electric field, the quantity n being the probability that one of these particles is in the right position. This idea may or may not be correct; the question is unimportant from the point of view of providing a mathematical description of the conductance changes. (In fact, as we shall see later, there are some remarkable similarities between the molecular structure of the potassium channels and the ideas expressed in equation 5.6.)

The variation of the quantity n with time is given by the equation

$$\frac{dn}{dt} = \alpha_n(1 - n) - \beta_n n \qquad (5.7)$$

in which α_n and β_n are rate constants which, at 6 °C, vary with voltage according to the empirical equations

$$\alpha_n = \frac{0.01(V + 10)}{\exp\left[(V + 10)/10\right] - 1}$$

Also lower driving force

and

$$\beta_n = 0.125 \exp(V / 80)$$

An alternative way of writing equation 5.7 is

$$\frac{dn}{dt} = \frac{(n_\infty - n)}{\tau_n} \qquad (5.8)$$

where n_∞ (the steady state value of n at any particular voltage) is given by

$$n_\infty = \alpha_n/(\alpha_n + \beta_n) \qquad (5.9)$$

and τ_n is a time constant given by

$$\tau_n = 1/(\alpha_n + \beta_n) \qquad (5.10)$$

The sodium conductance is given by

$$g_{Na} = \bar{g}_{Na} m^3 h \qquad (5.11)$$

where $\bar{g}_{Na}$ is a constant equal to the maximum value of g_{Na}. This equation is based on the idea that the sodium channel can be opened by movement of three particles, each with a probability m of being in the right place, and inactivated by an event of probability $(1 - h)$. m is given by the equation

$$\frac{dm}{dt} = \alpha_m(1 - m) - \beta_m m \qquad (5.12)$$

where, at 6 °C,

$$\alpha_m = \frac{0.1(V + 2.5)}{\exp\left[(V + 25)/10\right] - 1}$$

and

$$\beta_m = 4 \exp(V/18)$$

h is given by the equation

$$\frac{dh}{dt} = \alpha_h(1 - h) - \beta_h h \qquad (5.13)$$

where, at 6 °C,

$$\alpha_h = 0.07 \exp(V/20)$$

and

$$\beta_h = \frac{1}{\exp\left[(V+30)/10\right]+1}$$

Equations 5.12 and 5.13 can be written alternatively as

$$\frac{dm}{dt} = \frac{(m_\infty - m)}{\tau_m}$$

and

$$\frac{dh}{dt} = \frac{(h_\infty - h)}{\tau_h}$$

with m_∞, τ_m, h_∞ and τ_h being given by equations analogous to 5.9 and 5.10.

The voltage dependence of the parameters n_∞, m_∞ and h_∞, and their corresponding time constants is shown in fig. 5.14.

The total membrane current, I, is given by equation 5.3, which is expanded to give

$$I = C_m \frac{dV}{dt} + I_K + I_{Na} + I_{leak}$$

$$= C_m \frac{dV}{dt} + g_K(V - V_K) + g_{Na}(V - V_{Na}) + g_{leak}(V - V_{leak})$$

$$= C_m \frac{dV}{dt} + \bar{g}_K n^4(V - V_K) + \bar{g}_{Na} m^3 h(V - V_{Na})$$

$$+ g_{leak}(V - V_{leak})$$

If I is known, it is possible to work out V from this equation by numerical integration. The simplest case occurs when an appreciable stretch of the membrane is excited simultaneously by means of an internal silver wire elec-

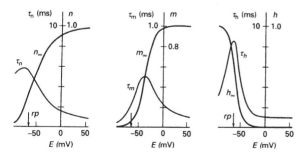

Figure 5.14. How some of the Hodgkin–Huxley parameters vary with membrane potential. See equations 5.7 to 5.9 for the potassium activation parameter n, and the corresponding equations for the sodium activation and inactivation parameters m and h. (From Cole, 1968.)

trode; there is then no propagation and the net current flow through the membrane is zero, so that the ionic current is equal and opposite to the capacitance current. This is known as a 'membrane' action potential. In the case of the propagated action potential, the situation is a little more complicated. We have seen in equation 4.18 that

$$I_m = \frac{a}{2R_i\theta^2} \frac{d^2V}{dt^2}$$

where a is the radius of the axon, R_i is the resistivity of the axoplasm, and θ is the conduction velocity. So the full equation for the propagated action potential is

$$\frac{a}{2R_i\theta^2} \frac{d^2V}{dt^2} = C_m \frac{dV}{dt} + \bar{g}_K n^4(V - V_K) + \bar{g}_{Na} m^3 h(V - V_{Na})$$

$$+ g_{leak}(V - V_{leak}) \qquad (5.14)$$

The right value of θ has to be found by trial and error; wrong values lead to infinite voltage changes. Using these equations, the time course of the action potential can be calculated.

Other predictions

Besides predicting the form of the action potential, the equations derived from analysis of the voltage clamp experiments can account for a number of other features of the physiology of the axon.

By adding g_K and g_{Na}, the impedance change during the course of the action potential can be obtained; the result is very similar to that observed by Cole & Curtis (fig. 4.16). Knowing the time course of the ionic currents during the action potential it is possible, by integrating with respect to time, to calculate the flow of sodium and potassium ions across the membrane during an impulse. For a propagated action potential the calculated net sodium entry was 4.33 pmol cm^{-2} and the calculated net potassium loss was 4.26 pmol cm^{-2}. These values are very close to those obtained by Keynes from radioisotopes measurements on *Sepia* axons.

By calculating the responses to instantaneous depolarizations of varying extent, curves very similar to those of fig. 4.12 were obtained. Thus the local responses produced by membrane potential displacements just less than threshold are caused by small increases in g_{Na} which are soon swamped by increase in g_K. The threshold for production of an action potential is the point at which the increase in g_{Na} is just large and rapid enough to avoid this effect.

At the end of an action potential, g_K is higher than usual, and the sodium inactivation process is well developed, so that g_{Na} cannot be much increased by depolarization. The membrane is thus inexcitable for a time; this corresponds to the duration of the absolute refractory period. A little later, g_K and the extent of the sodium inactivation process have fallen somewhat, so that a submaximal regenerative response can be initiated by sufficient depolarization. This period corresponds to the relative refractory period, where the threshold is higher than normal and the ensuing action potential may be reduced in size. Finally g_K and the sodium inactivation process return to their normal resting levels, and the membrane shows its normal excitability.

Accommodation is explained by the increases in the sodium inactivation process and g_K that are produced by a slowly rising membrane potential; these effects combine to raise the threshold. Conversely, anode break excitation is caused by the lowering of the threshold by means of a decrease in sodium inactivation and potassium conductance.

It has long been known that the conduction velocity of an axon increases with increase in temperature. This is because the time course of the electrical changes constituting the action potential is much faster at higher temperatures, since the rates at which the channels open and close increase with increasing temperature. In terms of the Hodgkin–Huxley equations, the rate constants for the conductance changes (the αs and βs in the equations) increase three-fold for every 10 deg.C rise in temperature.

Myelinated nerves

The myelinated nerve fibres of vertebrates are usually less than 15 μm in diameter, so their membrane potentials cannot be measured with the ease with which those of squid giant axons are. It is not normally possible to insert electrodes into the axon, and so indirect methods have to be used. These depend essentially upon using the axoplasm itself as the intracellular electrode by insulating the cut end of an internode from the external fluid surrounding an adjacent node (Dodge & Frankenhaeuser, 1958).

The results show that the properties of the frog nodal membrane are remarkably similar to those of squid axon membrane (Frankenhaeuser, 1965; Hille, 1971a). The currents produced during a voltage-clamped depolarization consist of an initial sodium current and a delayed potassium current (fig. 5.15), and can be described by relatively minor modifications of the Hodgkin–Huxley equations.

The situation in mammalian myelinated fibres is rather different. Voltage clamp experiments show that the potassium current at the node is negligible (Chiu et al., 1979). The

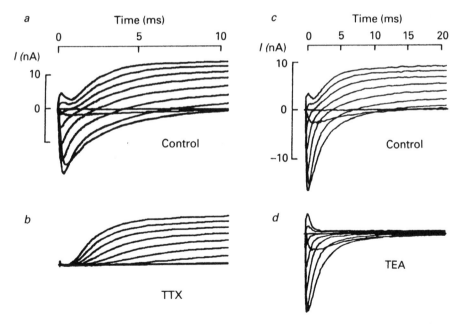

Figure 5.15. Voltage clamp records for frog nodes, showing separation of the sodium and potassium currents by the use of selective blocking agents. The membrane potential was clamped at −120 mV for 40 ms before the start of the records, and then depolarized to various levels ranging from −60 to +60 mV in 15 mV steps. Leakage and capacity currents were subtracted by computer. Records in *a* and *c* show the normal response. In *b* the node shown in *a* was treated with 300 nM tetrodotoxin (TTX), which blocks voltage-gated sodium channels: only the potassium current remains. In *d* the node shown in *c* was treated with 6 mM tetraethylammonium (TEA), which blocks voltage-gated potassium channels: only the sodium current remains. (From Hille, 1984, after Hille, 1966 and 1967.)

rate of inactivation of the sodium conductance is more rapid than in frog nodes. Computer reconstructions of the action potential show that the initial depolarization is brought about by the transient inward sodium current (just as in frog nodes and squid axons), but that the repolarization phase can be accounted for by the leakage current alone.

It is interesting to find that there is some potassium conductance in the internodal region in both frogs and rabbits, so that potassium currents do occur in demyelinated axons (Chiu & Ritchie, 1981, 1982). Chiu & Ritchie (1984) suggested that the function of these channels is to maintain a resting potential under the myelin sheath so as to avoid depolarization of the nodes.

Separation of the ionic currents by drugs

Tetrodotoxin is a virulent nerve poison found in the tissues of the Japanese puffer fish *Spheroides rubripes*. Narahashi and his colleagues (1964), using voltage-clamped lobster axons, found that tetrodotoxin in the external solution blocks the increase in sodium conductance that occurs on depolarization but has no effect on the increase in potassium conductance (see fig. 5.15*b*). This provides further evidence in favour of the idea that sodium and potassium flow through separate channels. Saxitoxin, a poison produced by the dinoflagellate protozoan *Gonyaulax*, also blocks the sodium channels. Both these substances have no effect on nervous conduction if they are injected into the axoplasm. Hence their site of action must be at or near the external surface of the axon membrane.

Tetraethylammonium (TEA) ions produce prolongation of the action potential when injected into squid axons, but external application is ineffective. Voltage clamp experiments have shown that this prolongation is caused by a blockage of the potassium channels, as in fig. 5.15*d*.

Various other drugs prevent or modify the opening of the ion channels; we shall consider some of these actions in the following chapter. Less direct effects are produced by drugs which interfere with the maintenance processes of the cell, so that the ionic gradients across the membrane are altered. Cardiac glycosides, such as ouabain, prevent sodium extrusion directly; metabolic inhibitors do so by removing the energy supply for the process. Poisoning the sodium pump does not immediately affect the electrical properties of the axon. In an experiment by Hodgkin & Keynes (1955*a*), over 200 000 impulses were elicited from a Sepia axon which had been poisoned with dinitrophenol.

More about conductance

One way of representing the results of a typical voltage clamp experiment is shown in fig. 5.16, for the giant axon of the tube-dwelling marine worm *Myxicola*. From records of current flow at different depolarizations, we plot the peak early current (curve *a*, largely sodium current) and the maintained outward current (curve *b*, potassium current) against membrane potential. Treatment with tetrodotoxin abolishes the sodium current, leaving a small outward current (curve *c*) which is composed of leakage current plus a small component of the potassium current. The normal sodium current is then the difference between curves *a* and *c* in fig. 5.16. Notice that curves *a* and *c* meet at the sodium equilibrium potential (E_{Na}), about $+70$ mV in this case.

The curves in fig. 5.16 are relations between current and voltage, and so they can tell us something about the membrane conductance. For any particular point on one of the curves, there are in fact two different measures of conduc-

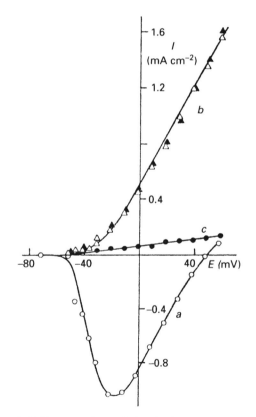

Figure 5.16. Current–voltage relations in the axon membrane. Giant axons of the marine worm *Myxicola* were subjected to clamped depolarizations at different membrane potentials. Curve *a* shows the peak early currents, largely I_{Na}. Curve *b* shows the steady-state currents after 25 ms, largely I_K. Curve *c* shows currents corresponding to those in *a*, but obtained in the presence of tetrodotoxin. (From Binstock & Goldman, 1969. Reproduced from the *Journal of General Physiology* by copyright permission of the Rockefeller University Press.)

tance (fig. 5.17). Firstly, we can draw a line connecting our point to the point where the driving force is zero (this will be at E_{Na} for the early current curve): the gradient of this gives the *chord conductance*. Secondly, we can measure the slope of the curve at our point by drawing the tangent at that point and measuring its gradient: this gives us the *slope conductance*. It is the chord conductance, given by g_{Na} and g_K in equations 5.4 and 5.5, which is utilized in the Hodgkin–Huxley analysis.

The Hodgkin–Huxley equations assume that chord conductances are linear, i.e. that Ohm's law applies strictly to the instantaneous relation between current and voltage. Hodgkin & Huxley (1952*b*) showed that this was so by measuring the ionic currents immediately after a second change in the clamped potential. In fig. 5.18, for example, the membrane was depolarized by 29 mV for 1.53 ms and then set at a new value and the ionic current (the 'tail' current) measured immediately after the change. It is clear from curve *A* in fig. 5.18 that the relation between this instantaneous current and voltage really is a straight line passing through E_{Na}. Curve *A* is in fact the chord conductance at the peak inward current at a depolarization of 29 mV. Instantaneous

current–voltage curves for the potassium current in squid axons were also linear.

Now let us interpret the two curves in fig. 5.18 in terms of ion channels. If the current–voltage relation of a single channel is linear then that of a number of open channels in a patch of membrane will also be linear. Hence we can conclude that the instantaneous current–voltage relation (curve *A* in fig. 5.18) represents the properties of a particular number of open channels. The peak current curve α, however, in which the currents are measured some time after the change of voltage, is far from linear and so must indicate that there are different numbers of channels open at different points on the curve. The slope conductance is a reflection of how these numbers change with changes in membrane potential.

Rectification

When the resistance (or conductance) of a conductor is not independent of the voltage across it, the conductor is said to show rectification. The potassium current curve in fig. 5.16 shows relatively large currents for depolarization and negligible ones for hyperpolarization, hence the potassium pathway is sometimes described as showing 'delayed rectification'. Notice that this form of rectification arises from the increased number of channels which are opened when the membrane is depolarized: the individual channels are not themselves rectifiers.

The constant field theory (chapter 3) predicts a certain amount of rectification, known as 'constant field rectification'. In a system in which this occurs, the instantaneous current–voltage relation will not be a straight line. Rectification of this type must be a property of the individual open channels. There are also a number of ion channels, particularly the inward rectifier potassium channels (chapter 3), which show appreciable intrinsic rectification.

Experiments on perfused giant axons

It is possible to replace the axoplasm of squid giant axons by an internal perfusion solution whose ionic concentration can be altered at will, as is described in chapter 3. This technique has proved to be very useful in confirming and extending our knowledge of the membrane conductance changes involved in nervous conduction.

In the experiment of Hodgkin & Katz shown in fig. 5.2, the action potential was reduced in height by reducing the external sodium ion concentration, so reducing the sodium equilibrium potential as given by equation 5.2. The intracellular perfusion method makes it possible to reduce the sodium equilibrium potential by the alternative method of increasing the internal sodium ion concentration. Baker

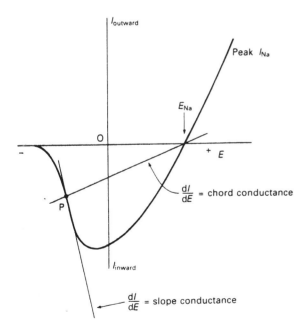

Figure 5.17. Diagram to show the difference between chord and slope conductances. The curve shows a typical current–voltage relation for the peak inward current in an axon membrane. The slope conductance at point P is the gradient of the tangent to the curve at that point. The chord conductance is the slope of the line connecting P to the point on the curve at which the voltage is equal to the sodium equilibrium potential.

and his colleagues (1962b) found that this reduced the over-shoot of the action potential as expected (fig. 5.19).

The voltage clamp technique coupled with internal perfusion was used by Atwater and her colleagues (1969) to measure sodium fluxes during clamped depolarizations. They placed an axon in sea water containing ^{22}Na and perfused it with a potassium fluoride solution. By measuring the radioactivity of this perfusion solution after it had passed through the axon it was possible to calculate the sodium influx. First the sodium influx at rest was measured. Then the axon was subjected to a few thousand clamped depolarizations at a rate of 10 s^{-1}, each pulse lasting about 3 ms. The radioactivity of the emerging perfusion solution rose to reach a new steady level within a few minutes, from which the influx associated with each clamped depolarization could be calculated. The corresponding inward current was calculated from the difference between clamped currents measured in the presence and in the absence of tetrodotoxin, used to block the early inward current.

Figure 5.20 summarizes the results obtained in these experiments. At a number of different membrane potentials the sodium influx accounts almost entirely for the inward current flow, in most convincing agreement with Hodgkin & Huxley's analysis.

Fixed charges and the involvement of calcium ions

Several investigations have suggested that there are fixed negative charges on the outer and inner sides of the axon membrane (see for example Hille *et al.*, 1975, and Gilbert & Ehrenstein, 1984). We would expect such charges to occur at the polar ends of the membrane lipids and also on the protein molecules. Their existence would modify the electric potential field in the membrane and in its immediate vicinity, perhaps in the manner shown in fig. 5.21a. We would also expect that the excitability mechanisms in the membrane (which must presumably be dependent upon the electric field in the membrane) would be affected by any alterations in the density of these fixed charges. Such altera-

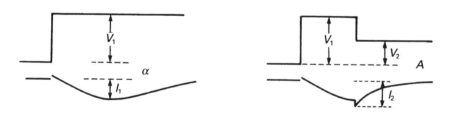

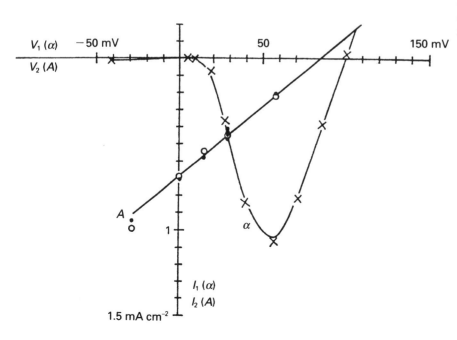

Figure 5.18. Sodium current–voltage relations in squid axon. Curve α shows the peak early currents in a voltage clamp, corresponding to curve *a* in fig. 5.16. Curve *A* shows the instantaneous current–voltage relation obtained by measuring the currents produced on changing the membrane potential again at 1.53 ms after the onset of a clamped deplarization of 29 mV. (After Hodgkin & Huxley, 1952b.)

tions would be produced by changes in the divalent ion concentration, pH or ionic strength of the solutions on either side of the membrane.

As an example of these effects, let us look at the action of calcium ions. Reduction of the calcium ion concentration in the external medium produces an increase in the excitability of axons, seen as a reduction in threshold and a tendency towards spontaneous and repetitive activity.

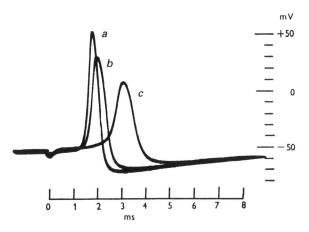

Figure 5.19. Action potentials from an internally perfused squid giant axon, showing the effect of increasing the internal sodium ion concentration. Record *a* shows the response with an isotonic potassium sulphate solution as the perfusion fluid, records *b* and *c* show responses with respectively a quarter and a half of the potassium replaced by sodium. (From Baker *et al.*, 1961.)

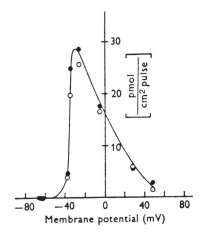

Figure 5.20. The nature of the early current during voltage clamp in a squid giant axon. White circles show the sodium ion influx as measured with a radioactive tracer method. Black circles show the charge movement (coulombs divided by Faraday's constant) attributable to the early inward current, measured from the difference between current flows in the presence and absence of tetrodotoxin. (From Atwater *et al.*, 1969.)

Frankenhaeuser & Hodgkin (1957) used the voltage clamp technique to investigate these effects. They concluded that the increases in excitability were caused by an increase in the sodium and potassium conductances of the membrane, such that a five-fold decrease in external calcium ion concentration produces effects similar to a depolarization of 10 to 15 mV.

Figure 5.21 shows one way in which these results can be interpreted in terms of fixed charges. With low external calcium ion concentrations, there are many more fixed negative charges on the outer surface of the membrane than on its inner surface. These charges will not affect the total potential across the membrane as measured at short distances away from its surfaces, but they will have marked effects on the detailed form of the potential gradient within the membrane and in its immediate vicinity. In particular,

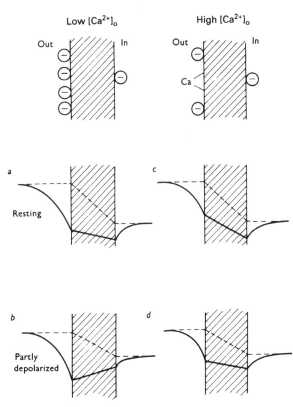

Figure 5.21. Schematic diagrams to show how fixed surface charges may affect the potential distribution in the membrane and its immediate vicinity (heavy lines). Dashed lines show the potential distribution assumed by a constant field model and disregarding fixed charges. The diagrams show low and high calcium concentrations, with membrane potentials at the resting level and partly depolarized. Notice how calcium ions alter the potential gradient within the membrane.

the greater number of negative charges on the outer surface ensures that the gradient of the potential field within the membrane will be less than it would otherwise be (fig. 5.21*a*). We may assume that the voltage-dependent ionic conductances are determined by the gradient of the field within the membrane. An increase in external calcium concentration will reduce the negative charge density on the outer surface of the membrane, and so will increase the gradient of the potential field within the membrane and hence raise the membrane potential at which the channels become opened.

In an internally perfused squid axon, dilution of the internal medium with a solution of sucrose or some other non-electrolyte results in a reduced resting potential, as we would expect. But a similar reduction in excitability does not occur: large and often prolonged action potentials may be seen under these conditions. Using the voltage clamp technique it was possible to show that the relation between sodium conductance and membrane potential was moved to a less negative position; the sodium inactivation relation was moved similarly. In other words, the sodium channels behaved as though the membrane potential was more negative than it actually was. Chandler and his colleagues (1965) explained this in terms of fixed charges on the inside of the membrane; when the ionic strength of the internal solution is low, such charges will have a much larger effect on the potential field within the membrane. Figure 5.22 gives an indication of how this system might work.

Gating currents

We have seen that the Hodgkin–Huxley equations were based on the idea that the ion channels are opened by the movement of charged particles within the membrane. Indeed it seems almost inevitable that a voltage-dependent change should be initiated by some sort of movement of charge. There should therefore be some current flow in the membrane just prior to the increase in ionic permeability. The search for such 'gating currents' (so called because they would open the 'gates' for ionic flow) was eventually successful (Armstrong & Bezanilla, 1973, 1974; Keynes & Rojas, 1974; see also Armstrong, 1981, and Keynes, 1983).

The main difficulty in detecting these gating currents is that they are very small in comparison with the ionic and capacity currents. The ionic currents can be blocked pharmacologically (by using external tetrodotoxin to block the sodium current and internal caesium fluoride to block the potassium current, for example). The capacity current is symmetrical with respect to the direction of voltage change, so it can be set aside by subtracting the response to

a hyperpolarizing voltage clamp pulse from the response to a depolarizing pulse of the same magnitude; the difference is then the gating current.

The reason for the asymmetry of the gating current is as follows. Suppose the charges that are moved during the gating current are positively charged particles that are held near to the inside of the membrane by the resting potential (fig. 5.23). During a hyperpolarization they will not be able to move any further towards the inside of the membrane simply because they are at the limit of their possible movement. There will therefore be no gating current on hyperpolarization. On depolarization, however, the particles will move towards the outer side of the membrane and this movement will constitute the gating current. Notice that the arguments are similar in principle if the gating particles are negatively charged and move inwards.

(It is possible, of course that the gating particles will

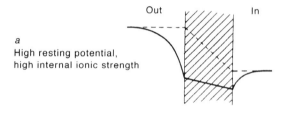

a
High resting potential,
high internal ionic strength

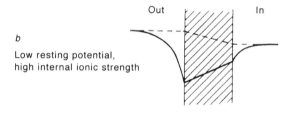

b
Low resting potential,
high internal ionic strength

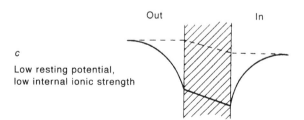

c
Low resting potential,
low internal ionic strength

Figure 5.22. Schematic diagrams to show how the gradient of potential across the membrane at low membrane potentials may be altered by internal perfusion with a solution of low ionic strength. Such a solution allows the fixed charges on the inner surface of the membrane to exert a much greater effect.

reach the inner limits of their movement at membrane potentials more negative than the resting potential, but it would seem unlikely that the difference would be very great. It is for this reason that many measurements of gating currents have been made from 'holding potentials' of $-100\,\mathrm{mV}$ or so, rather more negative than the resting potential.)

Figure 5.24 shows gating currents of a squid axon for clamped depolarizations, and also the ionic currents obtained from the same axon. Notice that the 'on' gating current is outward and has a much faster time course than the ionic current. The total quantity of charge moved (obtained by integrating the current with respect to time) is about 30 nC cm^{-2} in a squid axon, which corresponds to about 2000 electronic charges per square micrometre of membrane (Keynes & Rojas, 1974).

On repolarization at the end of a voltage clamp pulse there is an inward gating current, as is evident in fig. 5.24. For short pulses (less than 1 ms), the total charge movement during the 'off' gating current is equal and opposite to that at the beginning of the pulse. For longer pulses there is some reduction in this 'off' charge movement in parallel with the sodium inactivation process; this phenomenon is sometimes called charge immobilization (Armstrong & Bezanilla, 1977).

What is the implication of the very small size of the gating currents? Clearly there are just a few charges associated with each ion channel, but large numbers of ions flow through the channel when it is open. We shall discuss the nature of gating in the next chapter.

Electrical excitability in some electric organs

A number of different types of fish are capable of producing appreciable electric currents in the water surrounding them. These currents are produced by special organs known as *electric organs*.

Electric organs (with one exception, in *Sternarchus*, considered below) are composed of columns of cells called *electrocytes* (sometimes called electroplaques, electroplates or electroplaxes), each of which is innervated by an excitor nerve. These electrocytes appear to be modified muscle cells which have lost their contractile function. In life the electrocytes are excited initially by synaptic action, which then (in freshwater species) initiates an all-or-nothing response involving voltage-gated sodium channels. Here we consider only the electrically excited responses in two different electric fish, leaving the synaptic components of the responses to chapter 7.

The electrocytes of the electric eel

The electric eel *Electrophorus electricus* produces an electric discharge consisting of about half a dozen pulses each lasting 2 to 3 ms, which it uses to stun its prey and to repel

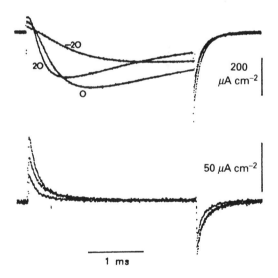

Figure 5.24. Sodium ionic and gating currents in squid axon, produced during depolarizations under voltage clamp. The upper traces show currents recorded in a solution containing only one-fifth of the normal sea water sodium ion concentration, for depolarizations to -20, 0 and $+20$ mV. The initial brief outward current is the gating current, followed by the much larger inward sodium ionic current. The lower traces show the gating currents alone, after blockage of the sodium channels with tetrodotoxin. Potassium ionic currents were eliminated by using potassium-free solutions for both internal and external media. (From Bezanilla, 1986.)

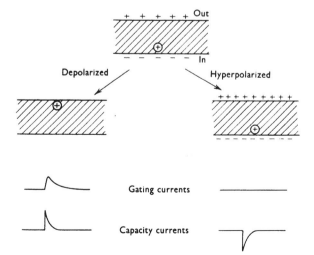

Figure 5.23. Diagram to show one way in which gating currents could arise by movement of charges within the membrane. The capacity currents are much larger than the gating currents.

predators. In large specimens the voltage may be more than 600 V; the record seems to be 866 V. The gross structure of the electric organ is shown in fig. 5.25. Each electrocyte is about 100 μm thick (longitudinally), 1 mm wide (vertically) and 10 to 30 mm long (radially). The nerve endings are restricted to the posterior face. The anterior (non-innervated) faces are much folded, giving rise to numerous papillae. The high voltage discharges are produced by the main organ; Sachs' organ gives much smaller discharges. In the main organ, up to 6000 rows of electrocytes are arranged in series with each other. This series arrangement of the electrocytes leads to addition of the voltages produced by each one of them; for example, if a discharge of 600 V were produced by an organ containing 4000 rows of electrocytes, each electrocyte would have to produce a potential of 150 mV across it.

The electrical properties of the electrocyte were investigated by Keynes & Martins-Ferreira (1953), using intracellular electrodes. They used the electrocytes from the organ of Sachs for their experiments, since these electrocytes can be more easily isolated than can those of the main organ. In the resting condition, there is a resting potential of about −80 to −90 mV. Electrical stimulation of the electrocyte results in an all-or-nothing action potential

appearing across the innervated (posterior) face, but there is no potential change across the non-innervated (anterior) face. The experimental evidence for this statement is as follows. In fig. 5.26, trace *a* is recorded from two microelectrodes placed just outside the innervated face; there is, of course, no potential difference between them either at rest or during activity. For trace *b*, one of the microelectrodes was inserted into the cell; this then records the resting potential, and, on stimulation, an 'overshooting' action potential of about 150 mV is observed. This means that the innervated face is electrically excitable, like the cell membranes of nerve axons and vertebrate twitch muscle fibres. In trace *c*, one of the electrodes is pushed right through the electrocyte so that the potential across the whole electrocyte is recorded. There is no steady potential at rest, indicating that the resting potential across the non-innervated face is equal to that across the innervated face. On stimulation, an action potential of the same size and shape as that across the innervated face appears across the whole electrocyte. This suggests that the non-innervated face is electrically inexcitable, and that the whole of the discharge is accounted for by the activity of the innervated face.

Confirmation of this view is provided by the experiment

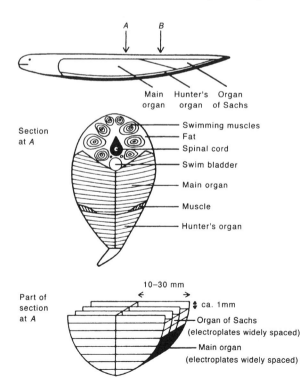

Figure 5.25. The electric organs of the electric eel *Electrophorus*. (From Keynes & Martins-Ferreira, 1953.)

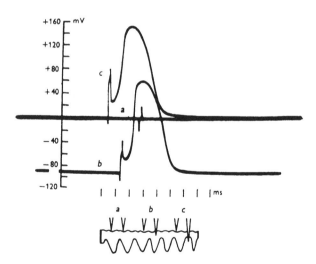

Figure 5.26. Responses of an electrocyte from the Sachs' organ of the electric eel to electrical stimulation. The potential is measured between two microelectrodes, positioned as shown in the lower diagram with the innervated face of the electrocyte uppermost. For trace *a* both electrodes are external and on the same side of the electrocyte, in trace *b* one microelectrode is intracellular so that the potential across the innervated face is recorded, and in trace *c* the potential across the whole electrocyte is recorded. (From Keynes & Martins-Ferreira, 1953.)

shown in fig. 5.27. Here, the potential across the non-innervated face (trace *b*) is not affected by stimulation (except for some small electrotonic changes), whereas the potential across the whole electrocyte (trace *c*) shows the familiar action potential. In these experiments Keynes & Martins-Ferreira were investigating only the all-or-nothing action potential of the electrocytes; in life this is preceded by synaptic activation of the electrocyte, and we shall look at this in chapter 7.

Thus the voltage produced by a stimulated electrocyte is caused by the asymmetry of the responses of its two faces, as is shown diagrammatically in fig. 5.28. In the complete electric organ, the electrocytes are arranged in series so that, as we have seen, their voltages are additive. Notice that the innervated face becomes negative to the non-innervated face during the discharge; so, since it is the posterior faces of the electrocytes that are innervated, the head end becomes positive to the tail during the electric discharge.

The ionic basis of the electrocyte action potential is much the same as that of the action potentials of nerve axons and twitch muscle fibres. Keynes & Martins-Ferreira showed that its size is dependent on the external sodium ion concentration, and Schoffeniels (1959), from radioactive trace measurements on isolated electrocytes, found that the sodium inflow across the innervated membrane increases greatly during activity.

Tetrodotoxin combines with the sodium channel protein and renders the electrocytes inexcitable. Antibodies to the tetrodotoxin-binding protein can be prepared. When these are allowed access to thin tissue sections, they bind entirely to the innervated faces of the electrocytes and not at all to the non-innervated faces (Ellisman & Levinson, 1982). So the voltage-gated sodium channels are confined to the innervated faces of the electrocytes, and it is the massive inflow of

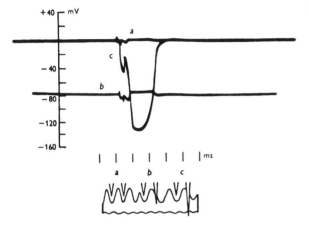

Figure 5.27. Results of an experiment similar to that of fig. 5.26, but this time the non-innervated face of the electrocyte is uppermost. Trace *b* shows that there is no action potential across the non-innervated face. (From Keynes & Martins-Ferreira, 1953.)

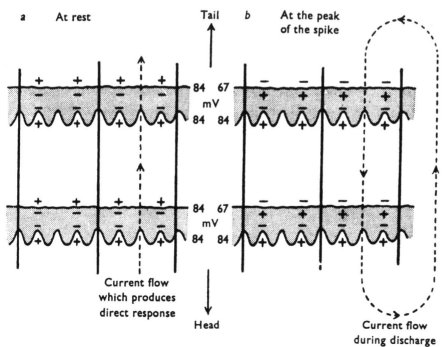

Figure 5.28. Diagram to show potentials across the two faces of the electrocytes in the Sachs' organ of the electric eel at rest (*a*) and at the peak of the action potential (*b*). The electrocytes are stippled, their innervated faces are uppermost (tailward). (From Keynes & Martins-Ferreira, 1953.)

sodium ions through them that produces the discharge current when they are excited.

The *Electrophorus* electric organ is one of the richest sources of sodium channel protein; this can be isolated from it (Agnew *et al.*, 1978, 1983) and sequenced by isolating the messenger RNA and then complementary DNA cloning (Noda *et al.*, 1984). Thus the electric eel has been of major importance in the molecular study of voltage-gated sodium channels, with results that we shall look at in the next chapter.

Electric organs of Sternarchus

The electric organs of *Sternarchus* and its immediate relatives are unique in that they are composed of tissue derived from nerve axons rather than from muscle cells (Couceiro & de Almeida, 1961). The electromotor axons leaving the spinal cord are swollen distally to form 'electrotubes' which are analogous in function to the electrocytes of other electric fishes. After entering the electric organ each axon

passes forwards for several millimetres, and then turns back on itself and runs backwards for an approximately equal distance. The diameter of the axon is about 100 μm in the swollen portions but only about 20 μm elsewhere. Each arm of the loop has three very large nodes (about 50 μm long) in the region just distal to the swollen portion (Waxman et al., 1972).

Waxman and his colleagues investigated the functioning of these remarkable axons (fig. 5.29). The greatly enlarged nodes on the distal side of each swollen portion are electrically inexcitable, whereas those on the proximal side behave in the normal manner. Thus inward current flow through the voltage-gated sodium channels at the proximal nodes will pass outwards at the enlarged distal ones. The net external current produced by the whole axon reverses in direction as the second arm of the loop is excited. The swollen portions will allow the internal current to pass along to the enlarged distal nodes without much attenuation, and the enlarged area of the distal nodes will facilitate its outward flow.

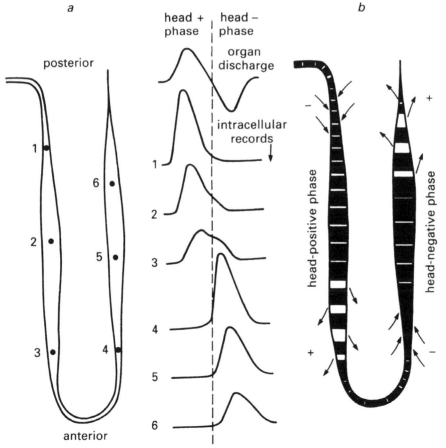

Figure 5.29. The mode of action of *Sternarchus* electrocytes, which are modified myelinated nerve fibres. *a* shows intracellular records from different sites during the discharge. Notice that the potentials in the regions with large nodes are smaller than elsewhere, suggesting that these nodes are not electrically excitable. *b* shows the current flow deduced from these observations. (From Waxman *et al.*, 1972. Reproduced from the *Journal of Cell Biology*, 1972, **53**, 222, by copyright permission of The Rockefeller University Press.)

The voltages produced by *Sternarchus* are far too small to be used for stunning prey. Like those of other weakly electric fish (most of which possess electrocytes which operate by mechanisms similar to those of the electric eel, although there are far fewer of them), they are used as part of an ingenious sensory system that can locate objects in the environment of the fish, as we shall see in chapter 17.

6
Voltage-gated channels

The Hodgkin–Huxley analysis of nervous conduction, which we examined in the chapter 5, showed that the sodium and potassium conductances of the axon membrane are switched on and off (or 'gated') by changes in membrane potential. Hence the channels through which the sodium and potassium ions flow are themselves gated by changes in membrane potential. In this chapter we examine some of the properties of these voltage-gated channels.

Voltage-gated sodium channels

Voltage clamp studies of the Hodgkin–Huxley type can provide information about the overall actions of large numbers of channels, but they cannot tell us how the individual channels behave. Two methods have been developed for this: fluctuation analysis and the patch clamp technique. Fluctuation analysis was developed first, but the patch clamp method allows more direct observation if the individual channel currents are large enough.

Patch clamping

The patch clamp technique has the very great advantage that the current flow through individual ion channels can be measured. An outline of the method is given in chapter 2. For voltage-gated channels, the stimulus for channel opening is a clamped depolarization applied to the patch of membrane containing the channel.

The first single-channel records of sodium channel currents were obtained by Sigworth & Neher (1980) by patch clamping myoballs, which are spherical cells prepared by tissue culture of embryonic rat muscle in the presence of colchicine. They used pipettes containing tetraethylammonium (TEA) to block any potassium channels present and α-bungarotoxin to block acetylcholine channels (p. 114). The records in fig. 6.1 show successive responses to a clamped depolarization by 40 mV. Most of the traces show square inward current pulses. These are always the same amplitude at the same membrane potential but they differ in duration.

How do we know that these unitary pulses of current

are carried by sodium ions? Sigworth & Neher showed that (1) they were blocked by tetrodotoxin, (2) they decreased in size at less negative membrane potentials and disappeared around $+ 20$ to $+ 40$ mV, which is presumably near E_{Na}, and (3) they were reduced when the sodium ion concentration in the pipette was reduced (fig. 6.1d). Trace b in fig. 6.1 shows the average of 300 patch clamp records; it is equivalent to the current flow which we would expect to see in a single record from an area 300 times as large as the experimental one and so corresponds to the current

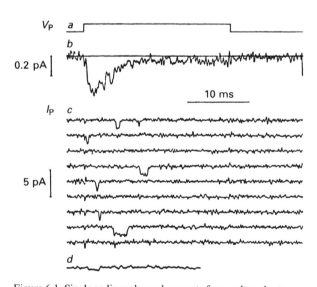

Figure 6.1. Single sodium channel currents from cultured rat muscle cells, recorded with the cell-attached patch clamp technique. Trace a shows the imposed membrane potential, held at $V = -30$ mV (where $V = 0$ is the resting potential) and depolarized by 40 mV to $V = +10$ mV for about 23 ms at 1 s intervals. Trace b shows the average of a set of 300 of current records elicited by these pulses. c shows nine successive individual records from this set. Square pulses of inward current (average size 1.6 pA) can be seen in most of the records; these correspond to the opening of individual channels. Trace d shows a record taken when two-thirds of the sodium ions in the pipette had been replaced with tetramethylammonium ions; the single-channel current is reduced accordingly. (From Sigworth & Neher, 1980.)

flow produced in a conventional large-area voltage clamp system.

The individual pulses in fig. 6.1 are much the same size but they vary considerably in duration and timing. This means that the opening and closing of the channels are stochastic processes: we cannot predict precisely when any particular channel will open or close, but we can in principle determine the probability that it will do either of these things in any one time interval. In voltage-gated channels, the probability of opening is greatly affected by the membrane potential.

We can now give a molecular interpretation of the sodium conductance changes during a voltage clamp of a large area of membrane. At any particular instant a channel is either open or closed: for practical purposes there is no halfway stage at which the conductance of the channel is at some intermediate value. This means that changes in the conductance of an area of axon membrane are produced by changes in the number of channels that are in the open state. When the membrane is depolarized the probability of any particular channel opening is increased so the total number of open channels rises and the conductance of the membrane as a whole rises. During inactivation or after repolarization the number of channels that are open falls and so the overall conductance falls also.

Computer simulations of these effects were performed by Clay & DeFelice (1983), and programs with similar capabilities have since become commercially available (e.g. Heitler, 1992). In the Hodgkin–Huxley equations the rate constant α determines the increase of the parameters n, m and h, and the rate constant β determines their decrease. In stochastic terms α and β are also rate constants determining the opening and closing of channels, hence they can be used in equations describing the probabilities of these events at any particular time. So computer simulations can indicate the changes of state of individual channels, and these can be added together to show the overall effect of large numbers of channels. Figure 6.2 shows some of the results of these simulations. Although the current flow through any single channel is a series of discrete events of constant magnitude and variable duration, the total current through large numbers of channels shows the continuous and determinate relations described by the Hodgkin–Huxley equations.

The mean current during the single-channel pulses in fig. 6.1 is 1.6 pA. We can calculate what this means in terms of sodium ions moved as follows:

$$\text{number of ions per second} = \frac{\text{current in amps} \times \text{Avogadro's number}}{\text{Faraday constant}}$$

$$= \frac{1.6 \times 10^{-12} \times 6 \times 10^{23}}{9.6 \times 10^4}$$

$$= 10^7 \text{ ions s}^{-1}$$

The mean channel lifetime is 0.7 ms so on average 7000 sodium ions pass through each channel before it closes. A transfer rate of 10^7 ions s^{-1} is much higher than the rates of carrier transport systems and enzyme actions and so provides excellent evidence that the open channel really is an aqueous pore through which sodium ions can diffuse.

We can calculate the conductance γ of the open channel from the equation

$$i = \gamma (V - V_{Na}) \tag{6.1}$$

where i is the single-channel current and $(V - V_{Na})$ is the difference between the clamped membrane potential and the sodium equilibrium potential. Notice that this equation is the single-channel equivalent of equation 5.4. In this case

$$\gamma = \frac{1.6 \text{ pA}}{90 \text{ mV}}$$

$$= 18 \text{ pS}$$

Patch clamping provides us with perhaps the most direct evidence for the existence of discrete ion channels in that we can actually observe and measure the currents flowing through them. It is a particular exciting technique in that it allows us to see changes in the behaviour of individual channel molecules as they actually happen.

Fluctuation analysis

Fluctuation analysis is a technique whereby we can gain information about the nature of unit electrical events by examining the fluctuations (or 'noise') in the gross currents produced by the sum of a larger number of such events. It was developed before patch clamping and has been largely superseded by that method. However, it is still useful where the membrane is not accessible to a patch clamp electrode (as in nerve axons surrounded by Schwann cells) or where channel densities are very high (as at the node of Ranvier or the neuromuscular junction). Details of the method are given by Stevens (1972), Neher & Stevens (1977) and DeFelice (1981). Sigworth (1980a) used fluctuation analysis to determine single sodium channel currents at the node of Ranvier; let us have a look at some of his experiments.

Sigworth's method depends essentially on a simple relationship between the variance and the mean in a binomial distribution. Suppose that under some specified situation (such as the time of peak sodium current during a clamped depolarization) the probability of any particular channel being open is p, and that there is a total of N channels of which r are open on any particular occasion when the specified situation occurs. The sodium current will be proportional to the proportion of channels open, r/N. The variance of r/N, determined from a large number of occasions, is given by:

$$\mathrm{var}(r/N) = \mathrm{var}(r)/N^2 = p(1 - p)/N$$

Hence the variance of r/N enables us to estimate N; a high variance means that a small number of channels are involved.

We can illustrate this by considering the finances of two analogical characters, Harry and Tom. Each day they are presented with a bag containing one dollar's worth of coins, always of the same denomination, and our problem is to know how many coins are in the bag, i.e. what denomination they are. Our characters share the coins between them by tossing: Harry takes those that fall head and Tom takes the tails. The amount each receives will fluctuate from day to day, and the variance of these amounts will be greater the higher the denomination of the coins. For example, if Harry tells us at the end of the year that he several times received the whole dollar's worth of coins we can be confident that the coins were 25 or 50 cent pieces, not 10 cents or less. More precisely, we can expect the mean daily income for each man to be near to 50 cents whatever the coinage used, but the standard deviations (the square roots of the variances) will be 25 cents for quarters ($N = 4$), 16 cents for 10-cent pieces ($N = 10$), 11 cents for 5-cent pieces ($N = 20$) and 5 cents for single cents ($N = 100$).

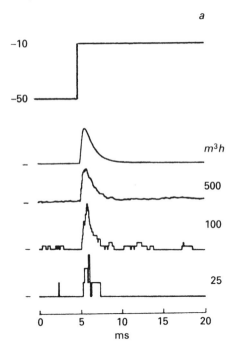

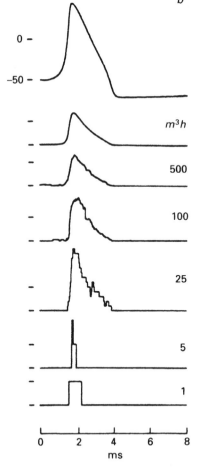

Figure 6.2. Computer simulations of sodium channel gating in patches of axon membrane, based on the Hodgkin–Huxley equations. Responses are shown during (*a*) a voltage-clamped depolarization by 40 mV, and (*b*) an action potential. The upper trace in each case shows the membrane potential. The m^3h curve shows the probability of an individual channel being open and is proportional to the sodium current in a large area of membrane. The lower graphs show the fraction of open channels in patches of 1 to 500 channels. Compare the 100 channels trace in *a* with trace *b* in fig. 6.1. (From Clay & DeFelice, 1983.)

Sigworth's results are illustrated in fig. 6.3. Trace *a* shows six successive current records produced by depolarizations to −5 mV. By taking the mean of such a group of records and subtracting it from each individual record, we get the deviations of the individual records from the mean; this is shown in trace *b*. From a large number of such records we can calculate the variance of the current, as in *c*, which was computed from 65 groups of six records. (The use of local means to get traces as in *b* largely eliminates problems of long-term drift. Potassium currents were eliminated with TEA.) The variance shown in trace *c* contains a component caused by thermal noise in the system; this has to be subtracted to give var*I*, the variance in the current due to fluctuations in conductance.

We assume that each of the sodium channels at the node can be either open or closed and that this is independent of what state other channels are in. *N* is the number of channels, *p* is the probability of any particular channel being open, *i* is the current flowing through an open channel and *I* is the total current. Then

$$\text{mean of } I = Npi$$

and

$$\text{var}I = Np(1 - p)i^2 \qquad (6.2)$$

Hence, substituting for *p*,

$$\text{var}I = iI - I^2/N \qquad (6.3)$$

Equation 6.3 is a parabola which hits the *I* axis at *iN* and has a slope of *i* where it hits the var*I* axis. Hence by plotting var*I* against *I* from records such as that in fig. 6.3, Sigworth was able to estimate *i* and *N*.

Figure 6.4 shows one of Sigworth's variance–mean plots; let us see how the various parameters can be determined from it. The curve is a parabola fitted by eye; ideally it should go through the origin of the graph, but a small residual variance has been added. The slope of the curve is given by differentiating equation 6.2:

$$\frac{\text{d}(\text{var}I)}{\text{d}I} = i - \frac{2I}{N} \qquad (6.4)$$

so when *i* = 0 the slope of the curve is the single channel current *i*. At the maximum value of var*I*, d(var*I*)/d*t* = 0 and so *i* = *Ni*/2, which enables *N* to be calculated. In fig. 6.4 *i* = −0.34 pA and *N* = 42 000. If all the channels were open at once the current would therefore be 42 000 × 0.34 pA, i.e. 14.2 nA. The maximum current produced during the clamp was 8.2 nA and so at this time 58% of the sodium channels were open (the corresponding value at +125 mV was 93%). The clamp potential was −5 mV and E_{Na} was +54 mV, so the driving force was 59 mV and hence the single-channel conductance γ was 0.34/0.059 i.e. 5.8 pS.

The method of fluctuation analysis described in this section is named non-stationary or ensemble analysis since it utilizes a large number of records, in each of which the probability of channels being open changes with time. A simpler method, known as stationary fluctuation analysis,

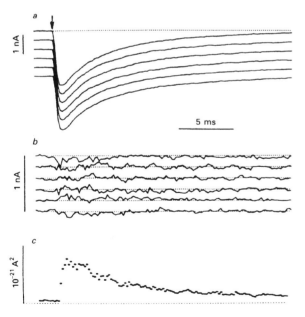

Figure 6.3. Sodium current fluctuations at the frog node of Ranvier. Trace *a* shows six successive current records produced by clamped depolarizations to −5 mV. *b* shows the deviations of the individual currents in *a* from their mean. *c* shows the variance of 65 such groups of records. (From Sigworth, 1980*a*.)

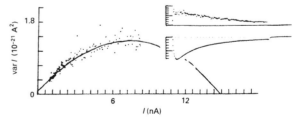

Figure 6.4. Variance–mean plot from non-stationary fluctuation analysis of the sodium currents at a frog node of Ranvier. The inserts show the mean sodium current *I* (lower trace, 1 nA per small division) and its variance var*I* (upper trace, 2 × 10⁻²² A² per small division) measured as in fig. 6.3, in response to 20 ms clamped depolarizations to −5 mV. The points on the main graph show these two quantities plotted against each other and the curve is equation 6.3 with *i* = 0.34 pA and *N* = 42 000. (From Sigworth, 1980*a*.)

is used for situations where this is not so; we shall examine this form of the technique in chapter 8.

Selectivity

Chandler & Meves (1965) carried out voltage clamp experiments on squid giant axons perfused internally with different solutions. When an axon was perfused with a 300 mM potassium chloride solution, they found that the initial inward current (which is attributed entirely to sodium ions in the Hodgkin–Huxley analysis) became an outward current at a reversal potential in the region of +70 mV. But under these conditions, with no internal sodium ions, the sodium equilibrium potential is infinite and so a purely sodium ion current should be inward. Chandler & Meves concluded that potassium ions can flow through the sodium channels. By applying the constant field equation it is possible to estimate the relative permeabilities of the channels to sodium and potassium ions. Thus, with no internal sodium and no external potassium, and assuming that the channels are impermeable to anions, equation 3.5 becomes

$$E = \frac{RT}{F} \ln \left(\frac{P_{Na}[Na]_o}{P_K[K]_i} \right) \qquad (6.5)$$

At 0 °C, with 320 mM potassium ions inside and 472.5 mM sodium ions outside, the mean reversal potential for the early current was +67.8 mV. Substituting these values in equation 6.5 we get

$$67.8 = 54.2 \log_{10} \left(\frac{472.5 \, P_{Na}}{320 \, P_K} \right)$$

whence

$$P_{Na} / P_K = 12.1$$

By repeating this type of experiment with other monovalent cations it was possible to estimate their permeabilities through the sodium channel. They fell in the order Li > Na > K > Rb > Cs, with relative values of 1.1, 1, 1/12, 1/40 and 1/61.

Similar experiments were carried out by Hille (1972), on the selectivity of the nodal sodium channels in myelinated fibres. Here it is not possible to change the internal solution, but one can compare the reversal potential for the current through the sodium channels under normal conditions with that when a test ion X is substituted for sodium in the external solution. With sodium as the only cation in the external solution, equation 3.5 becomes

$$E_{rev,Na} = \frac{RT}{F} \ln \frac{P_{Na}[Na]_o}{P_{Na}[Na]_i + P_K[K]_i}$$

When the sodium is replaced by the test ion X in the external solution, we have

$$E_{rev,X} = \frac{RT}{F} \ln \frac{P_X[X]_o}{P_{Na}[Na]_i + P_K[K]_i}$$

whence

$$P_X/P_{Na} = \frac{[Na]_o}{[X]_o} \exp(\Delta E_{rev} F/RT) \qquad (6.6)$$

where ΔE_{rev} is the change in reversal potential, $(E_{rev,X} - E_{rev,Na})$.

With lithium replacing sodium in the external solution, Hille found that the reversal potential changed by only −1.6 mV. This gives a value of 0.93 for the ratio P_{Li}/P_{Na}, so the permeability to lithium ions is only slightly less than that to sodium ions. With potassium substituted for sodium, however, the change in reversal potential was much larger, as is shown in fig. 6.5. The difference was −59 mV, giving a P_K/P_{Na} ratio of 0.086, so the permeability of the channel to potassium ions is only one-twelfth of that to sodium ions, just as in squid axons. Permeabilities to rubidium, caesium, magnesium and calcium ions were too small to measure.

The sodium channel is thus highly selective of the ions that will pass through it. Hille (1971b) suggested that at some point the channel pore must be narrow enough to bring the permeant ion directly into contact with its walls, since otherwise it is difficult to see how the selectivity can arise if the ion is 'hidden' inside a shell of water molecules. He called this region of the channel the *selectivity filter*.

This idea fits well with what we know about the permeability of sodium channels to organic cations (Hille, 1971b). Those which will pass through the channels will all fit snugly into a space 3.2 Å × 5.2 Å, as will a sodium ion plus one water molecule. Ions larger than this, such as trimethylammonium (fig. 6.5), will not pass through the channel. Hence it is reasonable to suggest that this space gives the dimensions of the selectivity filter.

It is notable that guanidinium ions will pass through the sodium channel and that both tetrodotoxin and saxitoxin have guanidinium moieties in their molecules; it may be that these toxins cause block by getting stuck in the narrow part of the channel. Potassium ions are somewhat larger than sodium ions, and the size of a potassium ion attached to one water molecule is too large to pass through a 3.2 Å × 5.2 Å space.

Selectivity may well not be simply a question of size alone. It seems very likely that the filter includes what is in effect an ion-binding site, so that there is a sequence of very fast reactions such as

$$Na_o + S \rightarrow Na-S \rightarrow Na_i + S$$

where S is the binding site in the selectivity filter and the subscripts o and i indicate ions on the outside and inside of the filter. There may be two or more such binding sites (Hille, 1975; Begenisich & Cahalan, 1980).

A physiologist's model for the sodium channel

Figure 6.6 gives an impression of some of the features that we might expect to find in the sodium channel. This model is based on physiological evidence. It reflects views current in the 1970s and 1980s, before information about the molecular structure of the channel became available. We shall see later that knowledge of the molecular structure has refined the detail of what we know about the channel; nevertheless the main features of the model remain as conceptual components which will need to be identified within the molecular structure.

The channel protein sits in the lipid membrane so that its *aqueous pore* connects the external medium to the internal cytoplasm. At the narrowest part of the pore is the *selectivity filter*, which determines which ions can pass through the open channel. The pore is closed at rest by an *activation gate*. The gate is opened following movement of electric

charge within a *sensor* as a result of a change in membrane potential; this charge movement can be detected as the gating current.

Internal perfusion of a squid axon with pronase, a proteolytic enzyme, removes the sodium inactivation process (Armstrong *et al.*, 1973). So inactivation must be brought about by a part of the molecule which is readily accessible from the cytoplasmic side of the membrane. The model as drawn here suggests that an *inactivating particle* can swing into the pore to block it when the activation gate is open.

Neurotoxins and other agents acting on sodium channels

Voltage-gated sodium channels are crucial to the movements of animals, since the action potentials of nerves and muscles cannot occur without them. Hence they are obvious targets for toxins produced by animals and plants for attack or defence. The actions of these and other substances are interesting both in themselves and in relation to their possible uses in medicine or agriculture. They are also of much use as tools in the study of sodium channel function (see Strichartz *et al.*, 1987; Catterall, 1992; Adams & Olivera, 1994).

Neurotoxins which act on voltage-gated ion channels are

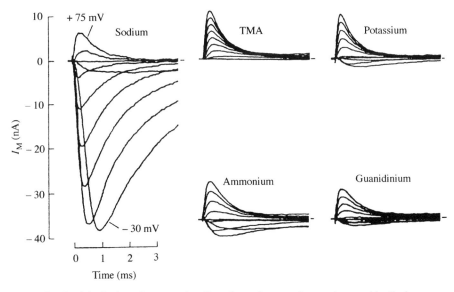

Figure 6.5. Ionic selectivity in the voltage-gated sodium channel of frog nerve. Records show currents through the sodium channels in a node in response to different clamped depolarizations; potassium channels were blocked with tetraethylammonium (TEA). The sodium curves were obtained in normal physiological saline. Other families of records show responses when the sodium in the external solution was substituted by the less permeant ions ammonium, guanidinium and potassium, and by the impermeant tetramethylammonium (TMA). Inward currents would be carried by the external cation, outward currents would be carried mainly by internal sodium ions. Reversal potentials were determined from these curves, and permeability ratios P_X/P_{Na} were calculated from equation 6.6. (From Hille, 1971*b*, 1972 and 1992. Reproduced from the *Journal of General Physiology*, by copyright permission of The Rockefeller University Press.)

of two types: those which block the channels and those which modify the gating process so as to keep them open for longer than usual.

The water-soluble substances *tetrodotoxin* and *saxitoxin* block the passage of sodium ions, probably by binding near the mouth of the channel. The dose–response relation of the effect of saxitoxin on sodium currents suggests that each saxitoxin molecule combines with one sodium channel so as to block it completely (Hille, 1968). A similar conclusion arises from fluctuation analysis experiments: in the presence of a low concentration of tetrodotoxin the number of open channels at a frog node fell but their single-channel conductance did not (Sigworth, 1980*a*).

Conotoxins are active agents in the venom of cone shells (*Conus*). These predatory molluscs inject the venom into their prey (other molluscs, worms or fish) by means of a disposable tooth that acts like a harpoon. Conotoxins are small peptides with unusually short chain lengths and two or three disulphide cross-links. There are a variety of different types, acting on various channels and receptors (Olivera *et al.*, 1990; Myers *et al.*, 1993). The μ-conotoxins selectively block voltage-gated channels, acting from the outside of the cell.

Other peptide neurotoxins act on other features of sodium channel activity. Scorpion α toxins and sea anemone toxins prevent inactivation and so produce pro-longed action potentials in muscle and nerve. Scorpion β toxins make nerve cells more excitable by shifting the activation–voltage curve to more negative membrane potentials.

A number of alkaloids also produce persistent activation and so make nerve cells more excitable. They include *batrachotoxin* from a Columbian arrow poison frog, *veratridine* from a lily, and *aconitine* from the ornamental flower monk's hood. Since they are lipid soluble they can penetrate the lipid membrane and so access the channel at sites embedded in it.

Pyrethrins are natural insecticides produced from a *Chrysanthemum* species farmed in East Africa. They are useful because they have a rapid 'knock-down' effect on insects and because they are largely non-toxic to mammals. They prolong sodium channel activation and so produce repetitive firing of nerve axons. Their synthetic analogues (pyrethroids) and the organochlorine insecticide DDT have similar effects.

A number of *local anaesthetics* such as procaine and lidocaine act by blocking sodium channels. These are all lipid-soluble compounds and it seems likely that they reach their binding site in the channel by first dissolving in the lipid phase of the membrane. A number of quaternary ammonium compounds of otherwise similar structure (such as QX-314, pancuronium and *N*-methylstrychnine) probably

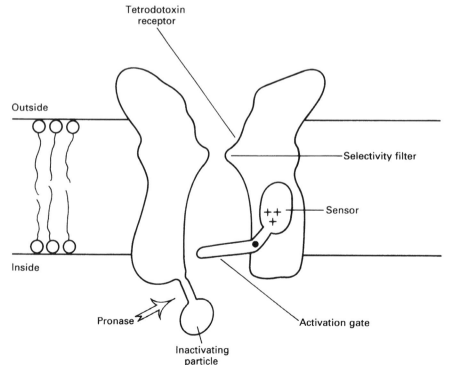

Figure 6.6. A sketch to show the main features of the sodium channel, as deduced from physiological experiments. (Based partly on diagrams by Armstrong, 1981, and Hille, 1984.)

Table 6.1 *Estimates of voltage-gated sodium channel densities in various membranes*

Preparation	Channel densities (μm^{-2}) estimated from		
	Toxin binding	Gating currents (charge/6)	Single-channel conductance
Garfish olfactory nerve	35 [a]		
Lobster walking leg nerve	90 [a]		
Squid giant axon	110 [b]	317 [c, d]	330 [e]
			92 [f]
Squid fin nerve (16 μm)	94 [b]		
Frog node		3000 [g]	400–900 [h]
Mammal node		2000–3000 [i]	700 [j]
Rabbit central fibre node	400–700 [k]		
Frog muscle	371–557 [l]	650 [m]	140 [n]

Note:

References: *a*, Ritchie *et al.*, 1976; *b*, Keynes & Ritchie, 1984; *c*, Armstrong & Bezanilla, 1974; *d*, Keynes & Rojas, 1974; *e*, Conti *et al.*, 1975; *f*, Bekkers *et al.*, 1983; *g*, Nonner *et al.*, 1975a,b; *h*, Sigworth, 1980a; *i*, Chiu, 1980; *j*, Neumke & Stämpfli, 1982; *k*, Pellegrino & Ritchie, 1984; *l*, Hansen Bay & Strichartz, 1980; *m*, Collins *et al.*, 1982b; *n*, calculated from Collins *et al.*, 1982a, and Sigworth & Neher, 1980.

act at the same site. These quaternary compounds are charged and so are not lipid soluble. They are only effective when applied from the inside of the cell, which suggests that their site of action is on the inner side of the selectivity filter (Strichartz, 1973; Schwarz *et al.*, 1977).

Peak sodium currents in voltage clamp experiments are reduced in acid environments. Using fluctuation analysis, Sigworth (1980b) showed that the single-channel conductance in frog node sodium channels is reduced at pH 5 to about 40% of the normal value. His explanation for this was that *hydrogen ions* compete with sodium ions for the ion binding site in the selectivity filter (as suggested by Woodhull, 1973) but that the duration of the consequent block is so short that individual events cannot be resolved by the recording apparatus.

Counting the channels

Clearly it is of considerable interest to estimate how many sodium channels there are in an area of axon membrane. There are various ways of doing this. One method is to measure the binding of radioactive tetrodotoxin or saxitoxin to the nerve. The estimates vary from 35 sodium channels per square micrometre for the thin unmyelinated fibres of the garfish olfactory nerve to about 300 μm^{-2} for squid giant axons, and rather larger values for the nodes of myelinated axons (see table 6.1).

An alternative method is provided by the analysis of gating currents. Thus Keynes & Rojas (1974) found that the average maximum gating charge displacement was 1900 electronic charges per square micrometre. To utilize this figure we need to know how many gating charges there are per sodium channel. A minimum value for this can be calculated from the Boltzmann relation, expressed here as

$$P_o = \frac{1}{1 + \exp\left[-ze_0(V - V_{0.5})/kT\right]}$$

where P_o is the fraction of channels open, e_0 is the elementary unit charge, z is the number of charges per channel, V is the membrane potential and $V_{0.5}$ is the value of V when half the channels are open, k is the Boltzmann constant and T is the absolute temperature. At 5 °C when V is measured in millivolts this becomes

$$P_o = \frac{1}{1 + \exp\left[-z(V - V_{0.5})/24\right]}$$

which means that when P_O is small it increases e-fold for every $24/z$ mV change in membrane potential. Hodgkin & Huxley (1952a) found that the initial slope of the curve of peak sodium conductance against voltage was 3.9 mV per log unit, implying that $z = 6$. With 1900 charges moved per square micrometre, the channel density is then 316 μm^{-2}. Later work, however, suggests that the situation may be more complicated, and that uncertainties in the appropriate value of z lead to corresponding uncertainties in estimating channel density (Bezanilla & Stefani, 1994; Keynes, 1994).

The third method of estimating channel density is to estimate the maximum sodium conductance per unit area and divide it by the single-channel conductance as determined by patch clamping or fluctuation analysis. In frog myelinated fibres, for example, Conti *et al.* (1976*a,b*) found a typical maximum sodium conductance of 683 nS per node and a mean single channel conductance of 7.9 pS, giving about 86 500 channels per node. If the nodal membrane area was 50 μm^2 (and there is some uncertainty here) then the channel density would be 1730 μm^{-2}. Sigworth's (1980*a*) results give rather lower values, with 20 000 to 46 000 channels per node, corresponding to 400 to 920 channels per square micrometre.

Table 6.1 shows some estimates of sodium channel densities in various cells. There is evidently some variation in the various estimates, and the figures should not be regarded as being very precise. Nevertheless it is clear that we can make two useful generalizations: (1) densities are appreciably greater at nodes than in non-myelinated axons, and (2), in non-myelinated axons, densities increase with increasing fibre diameter.

Is there an optimum density of the sodium channels? Clearly, conduction velocity will increase with increasing sodium conductance. But it will decrease with increasing membrane capacitance, and each channel, with its little store of moveable gating charges, must act as a small capacitance. At very high sodium channel densities, therefore, the effective membrane capacitance will be large enough to reduce the conduction velocity. Hence there is an optimum sodium channel density at which conduction velocity is maximal. This argument was developed by Hodgkin (1975), who calculated that the optimum density is in the range 500 to 1000 channels μm^{-2}, in reasonable agreement with the figures derived from experiments on squid axons.

Why should the sodium channel density be lower in axons of smaller diameter? Perhaps it is simply to reduce the quantity of sodium entering per impulse, all of which must later be pumped out using metabolic energy. This is much more of a problem for small diameter fibres because of their higher surface-to-volume ratio.

An important corollary of the differences in channel density in different axons is that calculations which assume that membrane characteristics are the same for fibres of different diameters (as on p. 47) are somewhat oversimplified.

It is interesting to notice that the density of sodium channels in non-myelinated axons is about one-tenth of that of sodium pumping sites (see p. 22). This ratio is presumably related to the much greater rate of flow of sodium ions through the channels than through the pumping sites.

Molecular biology of the sodium channel

The biochemical isolation of voltage-gated sodium channels is much dependent upon the great specificity of their binding of tetrodotoxin and saxitoxin. Using radioactive toxins, all the biochemist has to do is to isolate the fraction of the cell protein which is radioactive. The proteins can be extracted by using non-ionic detergents with some phospholipid present (Agnew *et al.*, 1978, 1983).

The electric organ of the electric eel *Electrophorus electricus* contains a large number of electrically excitable cells and so it is a good source of sodium channels. Agnew and his colleagues found that the sodium channel protein is a single large peptide with a number of carbohydrate residues attached, and an M_r of about 260 000. The carbohydrate accounts for 29% of the mass, half of it being sialic acid residues, which are negatively charged (Miller *et al.*, 1983). Partial sequencing of the amino acid chain was achieved, but the protein was too large for a full determination of the primary structure.

Reconstitution in model membranes

Artificial phospholipid bilayers, in the form of either planar bilayers or liposomes, have been much used as models for cell membranes. Planar bilayer membranes are produced by coating a small hole in an insulating partition separating two chambers. Liposomes are multilayered bodies formed when phospholipids are suspended in aqueous solutions; sonication produces small vesicles bounded by a single membrane.

Ion channel proteins can sometimes be inserted into these artificial membranes and their properties studied (Miller, 1984, 1986). Thus Kreuger *et al.* (1983) fused membrane vesicles from rat brain with planar bilayers and observed current flow through single channels, and Hartshorne *et al.* (1985) did similar experiments with the purified protein. Tanaka *et al.* (1983) inserted purified rat muscle sodium channels into unilamellar vesicles and observed ion fluxes by fast reaction techniques. All these studies used batrachotoxin to prevent inactivation of the sodium channels.

Some very pleasing experiments by Rosenberg *et al.* (1984) involved patch clamping of liposome membranes. Purified sodium channel protein from *Electrophorus* electric organ was incorporated into liposomes. Then a patch clamp electrode was applied to the outside of a large vesicle. After the seal had formed, the electrode was briefly withdrawn into the air, with the result that the body of the liposome fell away to leave a small patch of the bilayer membrane on the end of the electrode.

Rosenberg and his colleagues were thus able to measure

single-channel currents in response to brief depolarizations, just as in the experiments of Sigworth & Neher (p. 76), and with very similar results. The single-channel conductance was 11 pS, and the mean open time was about 1.9 ms. Inactivation was evident as a reduction in the number of channel openings at late times during a depolarizing pulse. In the presence of batrachotoxin there was no inactivation, the mean open time was 28 ms and the single-channel conductance increased to 25 pS. These experiments show conclusively that the single 260 000 M_r molecule, isolated from *Electrophorus* electric organ on the basis of its tetrodotoxin binding, is all that is required to make a functional sodium channel.

Primary structure of the sodium channel

Numa and his colleagues at Kyoto University used recombinant DNA ('gene cloning') techniques to determine the full amino acid sequence of the electric eel voltage-gated sodium channel (Noda *et al.*, 1984, 1986a). They extracted messenger RNA from the electric organ, made complementary DNA (cDNA) from it by using the enzyme reverse transcriptase, made a library of all the different cDNA fragments produced in this way, and then probed this library with DNA that would code for the partial amino acid sequences already determined from the channel protein. The methods were generally similar to those used for the earlier determination of the nicotinic acetylcholine receptor amino acid sequence, which we shall look at in chapter 8.

The protein chain is 1820 amino acid residues long and contains four homologous regions which have very similar amino acid sequences (fig. 6.7); they are named domains I to IV. A hydropathy analysis (p. 17) of the protein chain is shown in fig. 6.8; it shows six hydrophobic peaks within each domain. These six peaks correspond to six segments (S1 to S6), each consisting of 18 to 28 amino acid residues that are largely (but not entirely) hydrophobic and are probably arranged in an α-helix. These properties suggest that the six segments each traverse the membrane from one side to the other. The N- and C-termini of the chain are both on the cytoplasmic side of the membrane. The N-terminal end has 110 residues before domain I and the C-terminal end has 315 after domain IV.

In rat brain the channel consists of three subunits: a large α chain similar to the *Electrophorus* chain at 260 000 M_r, and two smaller ones called the β1 and β2 chains, with molecular weights of about 39 000 and 37 000 respectively (Hartshorne & Catterall, 1984). Noda *et al.* (1986a) determined the amino acid sequence of the α chain by recombinant DNA methods. They found two different molecular

forms, with 2009 and 2005 amino acid residues per chain. The degree of sequence similarity (i.e. the proportion of identical amino acids at corresponding positions in two protein chains) is 87% between the two forms and 62% between each of them and the *Electrophorus* sodium channel. The four homologous domains are similar in all three sodium channels, and all have hydropathy profiles with peaks corresponding to the S1 to S6 segments.

Segments S1 to S3 have some negatively charged residues in their helices, whereas segments S5 and S6 are entirely non-polar. Segment S4 is particularly highly conserved in the different domains and the different sodium channel proteins. Its structure is unusual in that it contains four to eight arginine or lysine residues located at every third position; arginine and lysine are the most positively charged of all amino acid residues, whereas the other residues in the S4 sequences are mostly non-polar. Perhaps the S4 sequence achieves stability in the membrane by forming ion-pairs with the negative charges on the S1 to S3 segments.

Figure 6.9 shows the probable transmembrane topography of the voltage-gated sodium channel. The four domains are probably bundled together so as to bound the transmembrane pore, as is suggested in fig. 6.10. The section of the chain between S5 and S6 in each domain is called the SS1–SS2 section, the H5 loop or the P loop. It is probably looped into the membrane so as to form the lining of the pore, at least in its outer half. This idea was first proposed in modelling studies (Greenblatt *et al.*, 1985; Guy & Seetharamulu, 1986), and has since had much confirmation from mutagenesis work, as we shall see later.

Expression in oocytes

During protein synthesis in animal cells, the nucleotide base sequence in the DNA of the appropriate gene is transcribed to form a complementary sequence in messenger RNA molecules, and this then acts as a template for translation to form the amino acid sequence of the protein. The oocytes of the African clawed toad *Xenopus laevis* provide an efficient translation system, and they can manufacture the corresponding proteins when messenger RNA is injected into them (Gurdon *et al.*, 1971). Since ion channels can be detected by their electrophysiological effects, *Xenopus* oocytes can be used to investigate their properties in a readily accessible membrane system.

We saw earlier how the Kyoto University group sequenced cloned cDNA to determine the structure of the α chain of rat brain sodium channels. They extended this work to make specific messenger RNA from the cDNA and injected it into *Xenopus* oocytes (Noda *et al.*, 1986b). Some

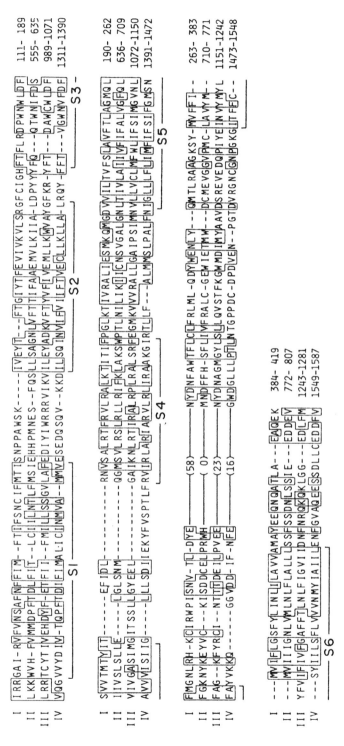

Figure 6.7. Parts of the amino acid sequence of the *Electrophorus* sodium channel, to show the four homologous domains. The four domains are aligned, with gaps introduced to maximize homology. The one-letter amino acid code (table 3.1) is used. Boxes enclose identical or similar amino acids. The segments S1 to S6 are thought to be regions where the chain forms α-helices which traverse the membrane. (From Noda *et al.*, 1984.)

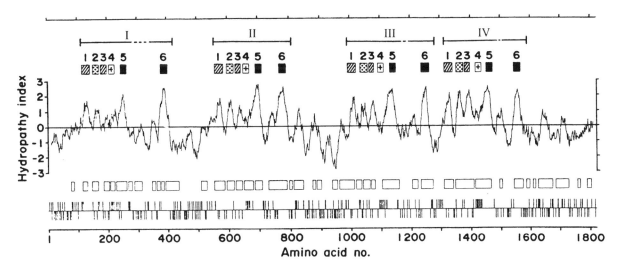

Figure 6.8. Hydropathy profile of the amino acid sequence of the *Electrophorus* sodium channel. The averaged hydropathic index of amino acid residues $i - 9$ to $i + 9$ is plotted against i, where i is the position number of the amino acid residue in the sequence. The bars show the positions of the homologous domains I to IV, and the boxes below them show the positions of the sections S1 to S6 that are thought to cross the membrane. The line of open boxes shows all sections of predicted α-helix or β-strand structure. The bottom line shows the positions of the positively charged lysine and arginine residues as upward lines and the negatively charged aspartate and glutamate residues as downward lines. (From Noda *et al.*, 1984.)

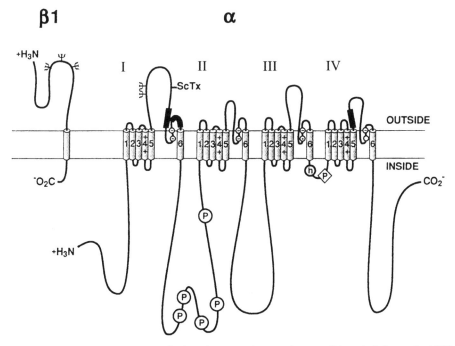

Figure 6.9. Membrane topology of the rat brain voltage-gated sodium channel. The four domains in the α chain are labelled I to IV, and the S1 to S6 transmembrane segments are shown in each of them. The SS1–SS2 (H5 or P) sections are shown dipping into the lipid bilayer between S5 and S6; small circles in them represent residues involved in tetrodotoxin binding. N-linked glycosylation sites are shown as ψ. Sites of phosphorylation are shown as P in a circle for cyclic-AMP-dependent phosphorylation and P in a diamond for protein kinase C phosphorylation. The inactivation particle is shown by h in a circle. Black rectangles and ScTx show sites involved in binding α-scorpion toxins. The β2 chain (not shown in this diagram) is similar to the β1 chain in its topology. (From Catterall, 1992.)

days later the oocytes were tested for the presence of voltage-gated sodium channels, which are not normally present in the oocyte membrane. Voltage clamped depolarizations gave transient inward sodium currents sensitive to tetrodotoxin (fig. 6.11).

These results imply that the α chain alone is sufficient to form the voltage-gated channel. However their expression is enhanced (i.e. more channels are produced) and their properties are altered (rates of activation and inactivation are higher) if the β1 subunit is also present (Isom *et al.*, 1992). The β subunits thus serve an accessory modulating function.

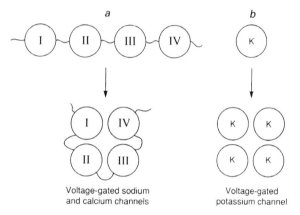

Figure 6.10. Subunits and domains in voltage-gated channels. *a* shows how the four homologous domains of the voltage-gated sodium channel molecule are arranged to form a unitary structure with the transmembrane pore passing through the middle. Voltage-gated calcium channels have a similar structure. In voltage-gated potassium channels (*b*) four separate protein chains are brought together as subunits to form the whole channel. (From Aidley & Stanfield, 1996).

The molecular basis of gating

A crucial question about the sodium channel must be the nature of its voltage sensor, whose movement provides the gating current. Segment S4, with its unusual array of arginine and lysine residues and their positive charges, looks a very attractive candidate. In the electric eel sodium channel there are five such residues in S4$_I$ (the S4 segment of domain I) and S4$_{II}$, six in S4$_{III}$ and eight in S4$_{IV}$. Similar patterns occur in other voltage-gated sodium channels.

Site-directed mutagenesis a powerful technique in molecular biology. It enables experimenters to introduce specific changes into cloned DNA molecules, so as to produce proteins with particular amino acid residues changed or removed (Smith, 1994). Stühmer *et al.* (1989) used the method to replace some of the arginine and lysine residues in the S4 segments of rat brain sodium channels. They injected messenger RNA made from the altered cDNA into *Xenopus* oocytes and investigated the gating of the channels after they had been expressed in the oocyte membrane. They found that the steepness of the relation between channel opening and depolarization was progressively reduced as the positively charged residues of the S4$_I$ segment were progressively replaced by neutral or negatively charged residues.

How does the S4 segment move? Catterall (1986, 1992) has developed a prescient idea by Armstrong (1981) in proposing a 'sliding helix' model of gating. Each positive charge in S4 is separated from its neighbours by about 5 Å, measured along the axis of the helix, and we may suppose that each one is loosely bound to a corresponding negative charge, probably on the S1, S2 or S3 segments. Catterall suggested that the S4 segment responds to a depolarization of the membrane by rotating by about 60° and moving

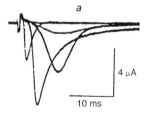

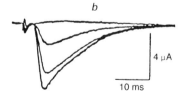

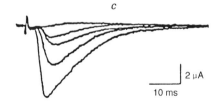

Figure 6.11. Expression of voltage-gated sodium channels in *Xenopus* oocytes. Oocytes were injected with mRNA transcribed from cDNA coding for rat brain sodium channel II, and then tested 3 to 6 days later by depolarizing under voltage clamp and measuring the membrane currents. Trace *a* shows a family of inward currents produced by depolarizations from −100 mV to −51 mV (the flat trace), −25, −21, −6 mV (the largest peak current) and +59 mV. Trace *b* shows responses to depolarizations from −100 to −10 mV in the presence of different external sodium ion concentrations, from 118 mM (the bottom trace) down to 2.75 mM (the top one). Trace *c* shows the response to tetrodotoxin: depolarizations were all from −100 to −10 mV; the largest trace was obtained before adding tetrodotoxin, the remainder, in decreasing order of size, were obtained 15, 20, 35 and 70 s after washing with 0.3 μM tetrodotoxin. (From Noda *et al.*, 1986*b*. Reprinted with permission from *Nature* **322**, p. 827. Copyright 1986 Macmillan Magazines Limited.)

outward by 4.5 Å, so as to bring each positive charge into alignment with the next negative charge (fig. 6.12). The net effect of this would be to expose a negative charge at the inner side of the membrane and a positive charge at the outer side, which is equivalent to moving one charge across the whole membrane.

Since there are four S4 segments in the molecule, Catterall's model implies a movement of four charges per channel. However, the S4 segments of domains III and IV contain six or more positive charges, whereas those of I and II contain four or five. The S4 segments of III and IV might move two steps up the ladder, as it were, so producing a total gating current of six charges per channel, the figure suggested by Hodgkin & Huxley (1952d). If all the S4 segments were to move two or three steps, we would get a gating charge movement of eight to twelve per channel, in line with more recent investigations (see Keynes, 1994). Sigworth (1994) suggested that such a large movement is

unlikely, and instead proposed that the S4 segment undergoes some conformational change in association with its charge movement, so as to keep it more within the bounds of the lipid bilayer.

Some evidence for movement of the S4 segment has been provided by an ingenious mutagenesis experiment by Yang & Horn (1995). They substituted a cysteine residue for the outermost arginine residue in the S4 segment of domain IV. The characteristics of the mutant channel could be changed by exposure to sulphydryl agents (which would attach to the cysteine residue), but only when the membrane was depolarized. This suggests that depolarization moves the cysteine into the extracellular space where the sulphydryl agents can reach it. Further work with the other charged residues in S4 showed that depolarization exposed the three outer charged residues in S4, and that two of these were accessible to sulphydryl agents from the inside in the hyperpolarized condition (Yang *et al.*, 1996).

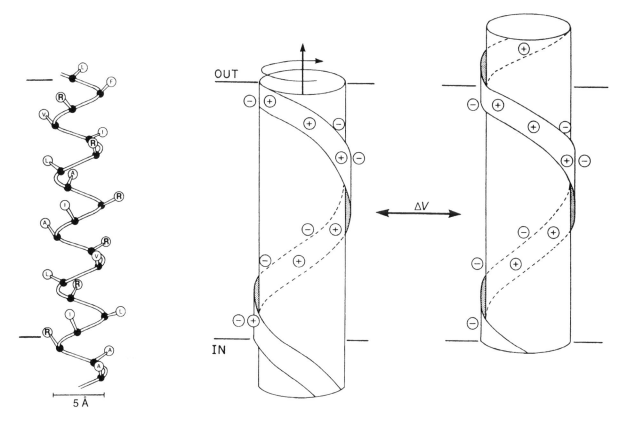

Figure 6.12. Catterall's sliding helix model of gating charge movement in the sodium channel. Segment S4 is presumed to form an α-helix crossing the membrane as shown on the left. Here the black circles represent the α-carbon atoms of the different amino acid residues and the white circles represent their side-chains. Residues are indicated by the single-letter code, with R showing the positively charged arginine; the rest are non-polar. The arginine residues thus form a helix of positive charges, shown on the right. The model proposes that these form ion-pairs with an array of negative charges on other segments, and that depolarization allows an outward movement of the S4 segment by one step along this array. (From Catterall, 1986.)

Keynes (1990, 1994) has produced a model of sodium channel gating that is compatible with a number of the complexities of channel gating, and arising out of his extensive experimental work on gating currents. He suggested that changes can occur simultaneously in the four separate S4 segments, as is shown in fig. 6.13.

We can conclude from all this that the gating charge movement, which can be measured, is brought about by some change in the position of the S4 segments, which thus act as the sensor in the model shown in fig. 6.6. The sliding helix model suggests a plausible way in which this might happen. Much less certain is how the movement of the S4 segments leads to opening of the channel.

Inactivation

During a maintained depolarization the sodium current rises to a peak and then falls back to a low level, as is shown for squid axon in fig. 5.9, for frog nerve in fig. 5.15, and for sodium channel cDNA expressed in *Xenopus* oocytes in fig. 6.11. Inactivation is the fall in current after the peak; it refers to the channels becoming non-conductive while the membrane is still depolarized, and should be distinguished from the closure that occurs when the membrane is repolarized.

Site-directed mutagenesis of some of the amino acid residues in the cytoplasmic loop connecting domains III and IV reduces the rate of inactivation. This region contains a cluster of positively charged residues, and also three adjacent hydrophobic residues (Ile-Phe-Met) that appear to be crucial to the inactivation process. Three more hydrophobic residues that seem to be important occur at the inner end of S6. Perhaps the III–IV loop acts as a 'hinged lid', with the two hydrophobic trios forming a catch to hold it in place (Stühmer *et al.*, 1989; West *et al.*, 1992).

Phosphorylation

Phosphorylation is a major method of modifying the activity of proteins in cells. A phosphate group is transferred from ATP to the protein by means of an enzyme called a kinase, of which there are a number of different types. It can be removed by a different enzyme called a phosphorylase. The phosphate group is usually attached at a serine or threonine residue. The voltage-gated sodium channel can be phosphorylated by cyclic-AMP-dependent protein kinase (sometimes called protein kinase A) at a number of sites on the cytoplasmic loop between domains I and II, as indicated in fig. 6.9. Overall sodium currents are smaller after phosphorylation; patch clamp experiments show that this is caused by a reduced probability of the individual channels opening, not by a reduction in their conductance when they are open (Li *et al.*, 1992).

Protein kinase C phosphorylates the sodium channel at a separate site, in the III–IV cytoplasmic loop (West *et al.*, 1991). This again produces a fall in the probability of the channel opening on depolarization, but there is also a fall in the inactivation rate, so that channel open lifetimes are longer. Perhaps the presence of the negatively charged phosphate group in the III–IV loop interferes with its operation as a 'hinged lid' in the inactivation of the channel.

The two separate ways of phosphorylating the sodium channel, with somewhat different effects on its function, implies that there are complex possibilities for control of its functions in the cell. Phosphorylation by cyclic-AMP-dependent protein kinase will be dependent on the presence of cyclic AMP. This is produced from ATP by the action of

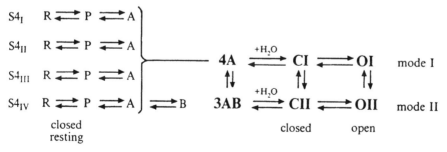

Figure 6.13. The Keynes kinetic model of the voltage-gated sodium channel. S4 segments in the four homologous transmembrane domains undergo independent transitions from the resting state R, via the primed state P to the activated state A. When all four S4 segments have reached A the channel takes up twenty to thirty water molecules (Rayner *et al.*, 1992) to move to the CI state, and from there to the open OI state. Inactivation is associated with a third transition of the $S4_{IV}$ segment to the B state, and this converts the channel to the mode II condition, in which opening is much less probable. Charge movements are associated with the R→P, P→A, A→B and C→O transitions (and their reversals), but not with the hydration step. (From Keynes, 1994.)

the enzyme adenylyl cyclase, which is itself activated by a number of membrane receptors for various hormones and neurotransmitters, and inhibited by others (fig. 9.5). Phosphorylation by protein kinase C is dependent upon the enzyme being activated by diacylglycerol, itself produced from the membrane phospholipid phosphatidylinositol by the action of a different set of receptors for hormones and neurotransmitters (fig. 9.6). In this way it is clear that agents external to the cell could produce changes in the activity of the sodium channels. Such changes might not be of much importance in the all-or-nothing propagation of nerve action potentials along the axon, but they could have marked effects on thresholds and frequencies of firing at impulse initiation sites or on transmitter release at nerve terminals (Catterall, 1992).

Dysfunctional sodium channels

Three somewhat similar inherited diseases are produced by mutations in the skeletal muscle sodium channel gene (Barchi, 1995; Hoffman *et al.*, 1995; Cannon, 1996). They are hyperkalaemic periodic paralysis (HYPP), paramyotonia congenita (PC) and potassium-aggravated myotonia (PAM). The paralysis in HYPP is brought on by heavy physical work and is associated with raised plasma potassium concentrations. Patients with the myotonias (PC and PAM) show intermittent muscle stiffness or involuntary contraction, often triggered by cold (PC) or by potassium-rich food such as bananas (PAM). Electrophysiological analysis of muscle biopsies from affected patients show that their voltage-gated sodium currents do not inactivate fully. The consequent slight depolarization might well explain both the paralysis (a rise in threshold) and the muscle stiffness (a slight activation of the contractile apparatus).

Genetic analysis of affected families, and sequence determination of the mutant genes showed that the faulty sodium channel proteins have just single amino acid changes, as is summarized in table 6.2. The mutations in the III–IV loop and at the cytoplasmic end of the $S6_{IV}$ segment are clearly connected with the operation of the 'hinged lid' in inactivation. The glycine residue at position 1306 is one of a pair that may well be concerned with providing the flexibility need for the 'hinge'; it is not surprising that its replacement by larger residues is associated with a reduced degree of inactivation.

Voltage-gated potassium channels

So far we have designated the potassium channels opened by depolarization in squid axons as 'the' potassium channel. However, it turns out that there are potassium channels with a variety of somewhat different properties in other cells, and

Table 6.2 *Some naturally occurring mutants of the human skeletal muscle sodium channel α subunit. The hereditary diseases they produce are hyperkalaemic periodic paralysis (HYPP), paramyotonia congenita (PC) and potassium-aggravated myotonia (PAM)*

Mutation	Position	Domain	Disease
Thr → Met	704	II S5	HYPP
Ala → Thr	1146	III S4-5	HYPP-PAM
Gly → Val ⎫ Gly → Glu ⎬ 1306 Gly → Ala ⎭	1306	III-IV	PAM
Thr → Met	1313	III-IV	PC
Met → Val	1360	IV S1	HYPP
Phe → Leu	1419	IV S3	HYPP (horse)
Arg → His ⎫ Arg → Cys ⎬ 1448 Arg → Pro ⎭	1448	IV S4	PC
Val → Met	1589	IV S6	PAM
Met → Val	1592	IV S6	HYPP

Note:
Data mainly from Brown, 1993, and Hoffmann *et al.*, 1995.

hence we need to distinguish between these different types. The maintained potassium conductance increase which follows depolarization in nerve axons has been called 'delayed rectification' and so the channels involved are sometimes known as delayed rectifier potassium channels.

Patch clamping

Figure 6.14 shows some patch clamp records from delayed rectifier potassium channels in frog muscle (Standen *et al.*, 1985). Depolarization produces opening of the channel, seen as current pulses averaging 1.5 pA at a membrane potential of 0 mV and with 2.5 mM potassium in the patch pipette. Applying the equivalent equation to 6.1,

$$\gamma = \frac{i}{V - V_K}$$

we get a value of about 15 pS for the single channel conductance γ.

Single-channel records of delayed rectifier potassium currents in squid giant axons were first obtained by Conti & Neher (1980). It was difficult to apply a patch clamp electrode to the outside of the axon because it is closely surrounded by Schwann cells, so they made a special L-shaped pipette electrode to apply to the inner surface of the axon

membrane. Other tricks included using a short internal perfusion with pronase to remove inactivation, external
tetrodotoxin to eliminate sodium currents, and a perfusion
solution of low ionic strength so as to give a good seal for
the patch electrode.

Conti & Neher calculated that the single-channel conductance of the potassium channels in squid axons is about
10 pS. This is close to the value of 12 pS deduced from
earlier investigations using noise analysis (Conti *et al.*,
1975). Using Hodgkin & Huxley's value of 36 mS cm^{-2} (i.e.
360 pS μm^{-2}) for the maximum potassium conductance, we
get a channel density of 36 μm^{-2}. This is much lower than
the corresponding figure for sodium channel density.

Molecular structure
Proteins forming the voltage-gated sodium channel and the
nicotinic acetylcholine receptor channel (chapter 8) were
first isolated from tissues where they occur in large quantities, such as the electric organs of some electric fish. Highly
specific neurotoxins (tetrodotoxin and α-bungarotoxin,
respectively) were essential to the isolation procedure.
Partial amino acid sequences then provided a key to the

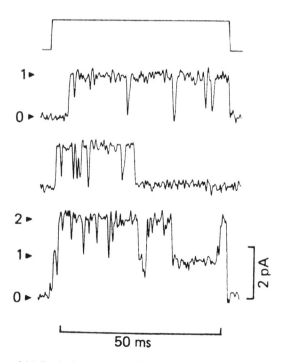

Figure 6.14. Patch clamp records of the opening of voltage-gated
potassium channels in frog muscle plasma membrane. The
stimulus was a depolarization from -100 mV to 0 mV. There
were at least two channels in the patch; in the first two records
only one of them opened, but in the lower record both channels
were open for much of the time. (From Standen *et al.*, 1985.)

production of complementary DNA for the whole protein.
A different approach had to be adopted for the determination of potassium channel structure, since there were no
tissues similarly rich in potassium channels and there were
until recently no similarly specific neurotoxins available.

The genetic structure of the fruit fly *Drosophila* has been
under detailed scrutiny for much of the twentieth century.
Shaker is a behavioural mutant which can be readily
detected: the flies shake their legs when anaesthetized with
ether. Voltage clamp studies on flight muscle cells showed
that the potassium currents are affected by mutations at the
Shaker locus (Salkoff & Wyman, 1981). The Jans and their
colleagues at the University of California, San Francisco,
therefore cloned genomic DNA from the *Shaker* locus so as
to get at the structure of the potassium channel proteins.

The *Shaker* locus is in the 16F region of the X chromosome. Papazian *et al.* (1987) cloned genomic DNA from
this region and used it to probe cDNA libraries made from
fly head messenger RNA. They were thus able to prepare
clones of *Shaker* genes and of cDNA coding for *Shaker*
locus gene products. From this cDNA, Tempel *et al.* (1987)
deduced the amino acid sequence of a protein which was
probably a component of the potassium channel.

The putative potassium channel protein was 616 amino
acid residues long with an M_r of 70 200. It had a hydrophobicity profile just like that of one of the four domains
of the voltage-gated sodium channel, suggesting a pattern
of transmembrane folding as in fig. 6.15. The S4 segment
was remarkably similar to the S4 section of the sodium
channel: it had a positively charged arginine or lysine
residue at every third position (fig. 6.16). A similar S4
segment occurs in voltage-gated calcium channels, but does
not occur in ligand-gated channels such as the acetylcholine or glycine receptors. This fits very well with the idea
that the S4 segment acts as the voltage sensor in voltage-
gated channels, and suggests further that these channels are
derived from a common evolutionary ancestor.

By analogy with the sodium channel we would expect the
complete voltage-gated potassium channel to be a tetramer
made up of four individual subunits. When messenger
RNAs coding for two subunits with somewhat different
properties are injected into *Xenopus* oocytes, channels with
hybrid properties are formed. This indicates that they must
consist of more than one subunit, and a detailed analysis
showed that they are indeed tetramers (Ruppersberg *et al.*,
1990; MacKinnon, 1991). Confirmation came from the
manufacture of an artificial tetramer (a protein chain containing four subunits linked together) with properties just
like those of a normal potassium channel (Liman *et al.*,
1992).

Sequence diversity

A striking feature of voltage-gated potassium channels is the degree of diversity that they show in their molecular make-up. This arises in three different ways. Firstly, different proteins can be made from the same gene. Most genes of multicellular organisms contain appreciable sections of DNA that do not code for the amino acids of the protein chain. These sections are called introns, and they separate the coding sections, called exons. The whole DNA sequence is represented in the primary RNA transcript, and then the non-coding sections are cut out so that the coding sections can be spliced together to form the messenger RNA. Sometimes different exons can be put together to form partly different proteins; this process is called alternative splicing.

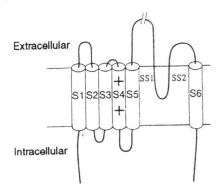

Figure 6.15. Characteristic structures in the voltage-gated cation channel superfamily. Each subunit or domain contains six transmembrane α-helices, shown here as cylinders, with the rest of the nearby peptide chain drawn as a line connecting them. The membrane-associated segment SS1–SS2 (also called H5 or P) occurs between S5 and S6 and probably forms part of the lining of the pore. The S4 segment contains the positively charged amino acid residues arginine or lysine at every third position. The complete channel (see fig. 6.10) contains four subunits of this type in potassium channels, four domains in a single molecule in sodium and calcium channels. (From Catterall, 1993. Reproduced from: Structure and function of voltage-gated ion channels, *Trends in Neurosciences* **16**, 500, with permission from Elsevier Trends Journals.)

The *Shaker* gene in *Drosophila* contains at least twenty-three exons, and alternative splicing of these leads to ten different channel variants. There are five different N-terminal regions and two different C-termini, but the central section containing most of the membrane-crossing region is invariant, as shown in fig. 6.17. Particular splice variants can be indicated by the nomenclature *Shaker* A1, G2, etc. (Stocker *et al.*, 1990).

A second source of diversity in *Drosophila* voltage-gated potassium channels is the presence of sister genes similar to but not identical with *Shaker*. They are known as *Shab*, *Shaw* and *Shal*, and occur at particular loci on chromosomes 2 and 3; the proteins of any two them show about 40% identity in their amino acid sequences. Homologous genes occur in mice and other mammals. The mammalian genes also fall into four subfamilies which are homologous with the four *Drosophila* genes, as indicated in table 6.3 (Wei *et al.*, 1990; Salkoff *et al.*, 1992). The mouse versions of *Shaker* (called mKv1) show on average 70% identity with *Shaker* but only about 40% with the mouse versions of the other genes. The mouse genes usually do not possess introns in their coding regions and so cannot display alternative splicing. Perhaps as a result of this there are numerous slightly different copies of the genes present in mice; thus there are at least twelve varieties of the mouse version of *Shaker*, called Kv1.1 to Kv1.12. Almost identical sequences have been determined for rat (rKv1.1 etc.) and human (hKv1.1 etc.) potassium channels. Exons and alternative splicing do occur in the rat Kv3 family.

Autoradiography using nucleotide probes for specific mRNAs shows that these different potassium channel varieties have different distributions in the mammalian brain. In the rat cerebellar cortex, for instance, Kv3.1 is densely present in the granule cell layer, Kv3.2 is present predominantly in the Purkinje cell layer and to a lesser extent in the granule cell layer, Kv3.4 is thinly present throughout, and Kv3.3 is absent (Vega-Saenz de Meira *et al.*, 1994; Weiser *et al.*, 1994).

Further diversity arises because individual potassium channels can be heteromultimers, assembled from subunits

ShA –Arg–Val–Ile–Arg–Leu–Val–Arg–Val–Phe–Arg–Ile–Phe–Lys–Leu–Ser–Arg–His–Ser–Lys–Gly–Leu–Gln–

Na⁺ –Arg–Val–Val–Arg–Val–Phe–Arg–Ile–Gly–Arg–Ile–Leu–Arg–Leu–Ile–Lys–Ala–Ala–Lys–Gly–Ile–Arg–

Ca²⁺ –Lys–Ile–Leu–Arg–Val–Leu–Arg–Val–Leu–Arg–Pro–Leu–Arg–Ala–Ile–Asn–Arg–Ala–Lys–Gly–Leu–Lys–

Figure 6.16. Amino acid sequences of the S4 segments of three voltage-gated channels: clone ShA of the potassium channel from the *Shaker* locus in *Drosophila* (Tempel *et al.*, 1987), the *Drosophila* sodium channel (Salkoff *et al.*, 1987), and the rabbit muscle calcium channel (Tanabe *et al.*, 1987). Positively charged amino acids are indicated by shading. (From Miller, 1988.)

Table 6.3 *Homologies and nomenclature of some* Drosophila *and mammalian voltage-gated potassium channel genes*

| Drosophila | | Mammal | | |
	Mouse	Rat	Human
Shaker	Kv1.1	MBK1	RCK1, RBK1	HBK1, HK1
		MK1	RMK1, RK1	
	Kv1.2	MK2	RCK5, BK2	HBK4
			RK2, NGK1	
	Kv1.3	MK3	RCK3, RGK5	HGK5, HPCN
			KV3	
	Kv1.4		RCK4, RHK1	HBK4, HK2
			RK3	hPCN2
	Kv1.5		RCK7, KV1	hPCN1
			RK4	
	Kv1.6		RCK2, KV2	HBK2
	Kv1.7	MK4		
	Kv1.8		RCK9	
Shab	Kv2.1	M.Shab	DRK1	DHK1
	Kv2.2		cdrk	
Shaw	Kv3.1	NGK2	KV4, Raw2	
			Raw2a	
	Kv3.2		RkShIIIA, Raw1	
	Kv3.3	MK5		KCNC3
	Kv3.4	MK6	Raw3	KCNC4
Shal	Kv4.1	M.Shal	R.Shal	
	Kv4.2		RK5	

Note:

From Aidley & Stanfield (1996), partly after Pongs (1992), with information from Strong *et al.* (1993).

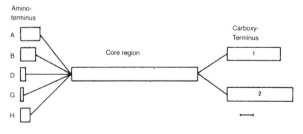

Figure 6.17. Alternative splicing of *Shaker* potassium channels. The core region, containing segments S1 to S5 and H5, is common to all transcripts. This is combined with any one of five different N-terminal regions and with either of two different S6 and C-terminal regions. The scale bar is twenty amino acid residues long. (From Pongs, 1992.)

that are not identical to one another. Oocytes injected with messenger RNAs for just one subunit produce channels which must be homomultimers. Those expressing more than one subunit can produce channels with hybrid properties, as long as the subunits all come from the same subfamily. However, subunits from different subfamilies will not combine to form channels (Covarrubias *et al.*, 1991). Heteromultimers have been demonstrated in brain nerve cells from mice (mKv1.1 and mKv1.2) and rats (rKv1.1 and rKv1.4); in each case the two subunits could be immunoprecipitated together by antibodies that would only recognize one of them (H. Wang *et al.*, 1993).

Gating

There seems little doubt that the S4 segments constitute the sensor for membrane potential changes in voltage-gated potassium channels, just as they do in the sodium channel. They have the same arrangement of positively charged arginine and lysine residues, as is evident in fig. 6.16. Here also, site-directed mutagenesis experiments show that replacement of the charged residues alters the relationship between channel opening and membrane potential (Papazian *et al.*, 1991). It seems very likely that the gating currents reflect movement of the S4 segments in response to changes in membrane potential.

The cysteine-substitution method has been used by Larsson *et al.* (1996) to investigate the accessibility of various residues in the S4 segment to a sulphydryl agent. The agent would not pass through the membrane and so could be applied either from the outside or the inside; it would combine with the cysteine residues (producing channel block) only when they were exposed at the appropriate surface of the membrane. The results suggested that the inner and outer ends of the S4 segment were only separated by five residues when the channel was closed, and that the S4 segment moves bodily outwards when the channel opens (fig. 6.18).

Potassium channel gating currents proved much more difficult to detect than did the gating currents for sodium channels. They were investigated in squid axons by Gilly & Armstrong (1980) and by White & Bezanilla (1985). The maximum amount of charge moved was around 490 e_0 μm^{-2}. This is about a quarter of the corresponding figure for the sodium gating current, and probably reflects a lower density of channels in the membrane.

It is an interesting feature of the potassium gating current that the time constants for on and off currents are similar to those for the potassium ionic currents at the same voltage. This is not what one would expect from the Hodgkin–Huxley scheme, which predicts (from equation

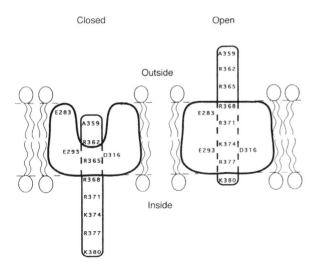

Figure 6.18. Movement of the S4 segment across the membrane in *Shaker* voltage-gated potassium channels, as deduced from cysteine-substitution experiments. The S4 segment is shown as a rod with its eight charged residues (A359 to K380) indicated. In the closed state at the resting membrane potential, five of these residues are, after mutation to cysteines, accessible to the sulphydryl reagent MTSET applied to the inside of the membrane, and two are accessible from the outside. In the depolarized membrane, when the channel is open, only one residue is accessible to MTSET applied at the inside and three are accessible from the outside. Also shown are negatively charged residues in the S2 segment (E283 and E293) and S3 segment (D316), thought to act in stabilizing the S4 segment by forming ionic bonds with its positively charged residues. (From Larsson *et al.*, 1996. Reproduced with permission from *Neuron*. Copyright Cell Press.)

5.6) that the time constant for potassium conductance should be related to the fourth power of that for gating.

A very thorough model of the gating of *Shaker* channels has been produced by Aldrich and his colleagues (Hoshi *et al.*, 1994; Zagotta *et al.*, 1994a). Each subunit undergoes a change from a resting state via an intermediate state to a permissive state. When all four units are in the permissive state, the channel opens. Each movement from one state to the next is associated with some gating charge movement. Total charge movement is about 13 electronic charges per channel, with 3.3 charges attributable to each subunit. This is a parallel-activation scheme similar to that produced by Keynes for the sodium channel.

Inactivation

In the experiments by Hodgkin & Huxley on squid axons there was no evidence for any reduction of the potassium currents with time. The use of longer depolarizations showed that this did happen, so that the potassium conductance was reduced to about a third over a period of some seconds (Ehrenstein & Gilbert, 1966). Much faster inactivation occurs in neurons of the sea slug *Archidoris*, and in some crab axons that show repetitive activity in response to depolarizing currents; the potassium channels involved are sometimes called A channels (Connor & Stevens, 1971; Quinta-Ferreira *et al.*, 1982). The different *Drosophila* potassium channels have different degrees of inactivation. This is illustrated in fig. 6.19, where they are expressed in *Xenopus* oocytes: *Shaker* is rapidly inactivated, *Shal* rather more slowly, *Shab* very slowly and *Shaw* not at all (Wei *et al.*, 1990).

A molecular model for inactivation was suggested by Armstrong and his colleagues (see Armstrong, 1992),

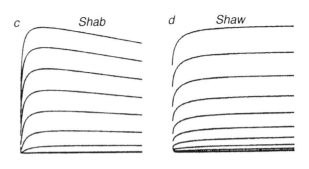

Figure 6.19. Different degrees of inactivation in different *Drosophila* voltage-gated potassium channels. *a* to *d* show the responses of *Xenopus* oocytes injected with cRNAs from the sister genes *Shaker*, *Shal*, *Shab* and *Shaw*. Each family of curves shows the currents produced by depolarizations reaching −80 to +20 mV in 10 mV steps, from a holding potential of −90 mV. Notice that the currents inactivate fairly rapidly in *a*, more slowly in *b* and *c*, and not at all in *d*. (From Wei *et al.*, 1990. Reprinted with permission from *Science* **248**, 599–603. Copyright 1990 American Association for the Advancement of Science.)

initially for the sodium channel of squid axons, but later for potassium channels. This is known as the 'ball-and-chain' model: inactivation is produced by a mobile part of the channel protein that swings into the inner mouth of the open channel pore so as to block it, as is shown in fig. 6.20*a*.

Excellent evidence for the ball-and-chain model has been provided by Hoshi and his colleagues (1990), using site-directed mutagenesis on *Shaker* potassium channels. They prepared deletion mutants in which various sections of the N-terminal cytoplasmic region of the molecule had been removed, and expressed them in oocytes. They found that deletions in the first twenty-two residues slowed or removed inactivation. Deletions of sufficient length in the sequence between residues 23 and 83 tended to speed up inactivation. Examples of these effects are shown in fig. 6.21.

These results fit the ball-and-chain model very well: deletions from the ball alone (residues 1 to 19), or from the ball and part of the chain, disrupt inactivation. Deletions from the chain alone tend to speed it up, as if a longer chain gives more freedom of movement to the ball so that it takes longer to find the open channel. The ball itself contains a concentration of positively charged residues, which we might suppose to be important in holding it in contact with the negatively charged pore of a cation-selective channel. There must be four balls with chains in each *Shaker* potassium channel, since there are four subunits, but presumably only one of them blocks the channel pore at any one time.

Further experiments showed that the ball on its own is sufficient to block the channel (Zagotta *et al.*, 1990). A synthetic peptide with the same amino acid sequence as the first twenty residues of the normal channel was prepared. Mutant *Shaker* channels with part of the ball sequence removed were expressed in oocytes and their activity was measured using the inside-out patch clamp method; the

channels opened on depolarization and there was little or no inactivation. Then a solution of the synthetic peptide was brought into contact with the cytoplasmic side of the patch; this restored the inactivation process.

The S4–S5 cytoplasmic loop contains a number of residues that are conserved in a wide variety of *Drosophila* and mammalian voltage-gated potassium channels. Mutations to some of these in the *Shaker* channel reduced the degree of inactivation (Isacoff *et al.*, 1991). This suggests that the S4–S5 loop is near to the internal channel mouth and forms part of the receptor for the inactivation ball when it blocks the channel pore.

The inactivating ball peptide from *Shaker* channels will also block other voltage-gated potassium channels, and mammalian calcium-activated channels (Toro *et al.*, 1992), all of which are in the same molecular superfamily. It will not, however, block the ATP-dependent channels of mammalian muscle, which are sufficiently different in molecular structure to lack the receptor for the peptide (Beirão *et al.*, 1994).

A variant of the ball-and-chain inactivation system has been discovered in some voltage-gated potassium channels from mammalian brain. They can be isolated from brain tissue as α-dendrotoxin (DTX) receptors, and they contain two types of subunit, α and β. The α subunits are of the familiar *Shaker*-related Kv1 (RCK) family, but the β subunits are quite different in structure, with no obvious hydrophobic membrane-crossing segments. It looks as though they are attached to the cytoplasmic end of the channel, probably with $\alpha_4\beta_4$ stoichiometry. The β subunits would not form channels when expressed in *Xenopus* oocytes, but they greatly enhanced the inactivation of channels formed from the α subunits (Rettig *et al.*, 1994). Deletion of part of the N-terminal region from the β

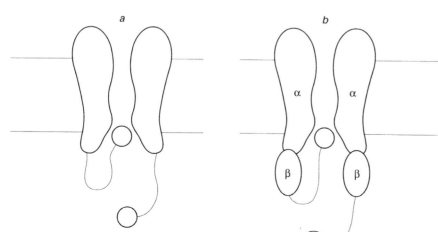

Figure 6.20. Ball-and-chain models of inactivation in voltage-gated potassium channels. *a* shows the situation in the *Shaker* channels of *Drosophila*, where the ball and chain are intrinsic to the subunit polypeptide chain. *b* shows the situation in some mammalian channels, such as Kv1, where separate β subunits carry the inactivating particle.

subunit removed its inactivation capability, and a peptide with the same sequence as its first twenty-four amino acid residues would produce inactivation in the absence of the rest of the β subunit. All this suggests strongly that the β subunit carries a ball-and-chain section which can block the channel in the same way as the N-terminal ball and chain of the pore-forming subunits does in the *Shaker* B channels of *Drosophila*. This idea is illustrated in fig. 6.20*b*.

The ball-and-chain inactivation system in potassium channels has become known as N-type inactivation

because it is identified with part of the N-terminal region of the molecule. When N-type inactivation is absent in *Shaker* deletion mutants, a slower inactivation system remains. This second type of inactivation is associated the C-terminal region, and so is called C-type (Hoshi *et al.*, 1991).

The long-pore effect

Ussing (1949) showed that, if an ion X crosses the cell membrane solely under the influence of the chemical and

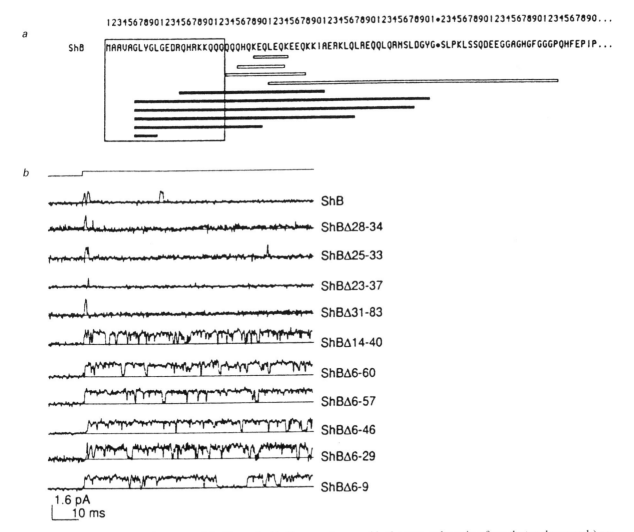

Figure 6.21. Effects of removing parts of the N-terminal ball-and-chain on inactivation of *Shaker* B potassium channels expressed in *Xenopus* oocytes. *a* shows the first ninety amino acid residues of a subunit. The bars show the sections removed in different deletion mutants; black bar deletions removed inactivation, white bar ones did not. Sample single-channel records from the *Shaker* B channel and the deletion mutants (arranged in the same order as in *a* from the top downwards) are shown in *b*. Notice the presence of inactivation in the first five records and its absence in the last six. The stimulus in *b* was a voltage step from −100 to +50 mV. (From Hoshi *et al.*, 1990. Reprinted with permission from *Science* **250**, 533–8. Copyright 1990 American Association for the Advancement of Science.)

electrical gradients across the membrane, and if the movement of any individual ion is independent of the movement of its neighbours, the ratio between the influx (M_i) and the efflux (M_e) is given by

$$\frac{M_i}{M_e} = \frac{[X]_o}{[X]_i} \exp\left(\frac{-EzF}{RT}\right) \qquad (6.7)$$

where the symbols in the exponent are as in equation 3.1.

Hodgkin & Keynes (1955*b*) measured the potassium fluxes of *Sepia* giant axons which had been poisoned with dinitrophenol so as to eliminate active transport effects. The axon was arranged so that its membrane potential could be altered by passing current through it at the same time as the potassium fluxes were measured. Thus it was possible to find the potential at which the fluxes were equal in each direction. If the flux ratio is given by equation 6.7, i.e. if

$$\frac{M_i}{M_e} = \frac{[K]_o}{[K]_i} \exp\left(\frac{-EF}{RT}\right)$$

or

$$\frac{M_i}{M_e} = \exp\left(\frac{-(E - E_K)F}{RT}\right) \qquad (6.8)$$

then the influx should be equal to the efflux when E is equal to E_K. This expectation was fulfilled. This means that, in poisoned axons, potassium ions move solely under the influence of the concentration and electrical gradients across the membrane.

A most interesting discovery was made when the flux ratios at membrane potentials not equal to the potassium equilibrium potential were measured. Equation 6.8 can be written as

$$\log_{10}\frac{M_i}{M_e} = \frac{-(E - E_K)}{58} \text{ mV}$$

This implies that the flux ratio should change ten-fold for a 58 mV change in membrane potential. But the relation turned out to be much steeper than this, with a ten-fold change in flux ratio for about a 23 mV change in membrane potential, as is shown in fig. 6.22. The results could be described by a modification of equation 6.8 such that

$$\frac{M_i}{M_e} = \exp\left(\frac{-n(E - E_K)F}{RT}\right) \qquad (6.9)$$

where n is 2.5. What does this mean? Hodgkin & Keynes suggested that potassium ions move through the membrane via pores whose length is appreciably greater than their diameter, so that a number of ions (n) can occupy a pore at the same time. Any one ion will then only pass through if it

is hit n times in succession from the same side. This is much more likely to happen if the ion is moving in the same direction as the majority of the other ions. The phenomenon became known as 'the long-pore effect'.

Selectivity

If only potassium ions are moving through the potassium channels, then the reversal potential for the ionic current is equal to the equilibrium potential for potassium ions, as in equation 3.4. If external potassium is now replaced by another ion X^+ then the amount of change in the reversal potential gives an estimate of the relative permeabilities of X^+ and K^+; there will be no change if $P_X = P_K$ and a large change if P_X is much different from P_K. Using a modification of equation 3.5 similar to that which gave us equation 6.6, it can be shown that

$$E_{rev,X} - E_{rev,K} = \frac{RT}{F} \ln \frac{P_X[X]_o}{P_K[K]_o}$$

where $E_{rev,K}$ is the reversal potential with external potassium ions and $E_{rev,X}$ is the reversal potential with external X^+.

Hille (1973) used this technique to measure the relative permeabilities of different ions in the delayed rectifier potassium channels in frog nodes. His results were rather

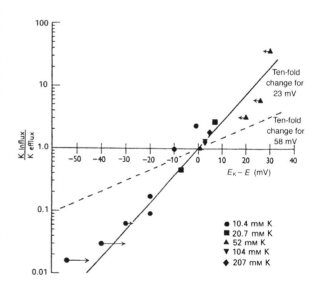

Figure 6.22. The effect of membrane potential (plotted here as the difference between the membrane potential and the potassium equilibrium potential) on the potassium flux ratio in *Sepia* giant axons. The results are better described by equation 6.9 than by 6.8, implying that the potassium ions pass through their channels in single file. The phenomenon is known as the long-pore effect. (From Hodgkin & Keynes, 1955*b*.)

striking: only four ions, potassium, thallium, rubidium and ammonium, with diameters in the range 2.66 to 3.0 Å, could readily pass through the channel. Organic cations and caesium, all with diameters of 3.3 Å or more, could not. This suggests that the selectivity filter in the channel is a pore 3 to 3.3 Å in diameter. We can imagine a potassium ion sitting snugly in this pore, perhaps with the oxygen atoms lining the pore taking the place of the oxygen atoms of the water molecules that surrounded the ion in free solution. Lithium and sodium are smaller than potassium ions, so why do they not pass through the channel? Perhaps the walls of the pore cannot make close contact with them so as to provide an adequate substitute for the water molecules of their hydration shells.

The selectivity of an ion channel must be greatly dependent upon the particular amino acids that line the transmembrane pore. The pore of voltage-gated potassium channels is probably lined at least in part by the H5 segment (also called the P region and equivalent to the SS1–SS2 section of sodium channels). This stretch of twenty-one amino acid residues probably forms a hairpin loop so that the proline residues at each end are on the outer side of the membrane and the methionine and threonine residues in the middle are towards the inner side, as is shown in fig. 6.23. The inner part of the pore appears to be lined with the cytoplasmic S4–S5 loop and the cytoplasmic end of the S6 segment (Slesinger *et al.*, 1993).

Convincing evidence that the H5 segment forms the lining of the most selective part of the channel pore comes from a neat experiment in molecular engineering. Kv3.1 and Kv2.1 are two voltage-gated potassium channels from different mammalian subfamilies. The single-channel conductance of Kv3.1 is almost three times that of Kv2.1. Hartmann and her colleagues (1991) replaced a section of Kv2.1 that contained most of the H5 segment with the corresponding section of Kv3.1. They found that the resulting chimaera had the conductance properties of Kv3.1, even though most of its amino acid sequence was still that of Kv2.1.

If the selectivity filter is relatively localized within the channel, then we might expect it to be associated with a restricted subset of amino acid residues within the P region. Figure 6.24 shows the amino acid sequences of the P region in a number of different potassium channels. Heginbotham and her colleagues (1994) identified a stretch of eight of these (TXXTXGYG, in the single-letter code) as being very similar in all of them. This 'signature sequence' appears as TMTTVGYG (residues 9 to 16) in the *Shaker* and Kv2.1 channels shown in fig. 6.23. *Shaker* channels lose their specificity for potassium ions if subject to mutations at positions 6 and 8 of the signature sequence (the glycine

residues in the GYG triplet), and some of the mutations at position 5 also result in non-selective channels (Heginbotham *et al.*, 1994). The YG doublet at 7 and 8 is absent from the P region sequences in the non-selective cyclic-nucleotide-gated channels of visual and olfactory sensory cells and in voltage-gated calcium channels; and mutant *Shaker* channels from which it has been removed are permeable to sodium as well as to potassium ions (Heginbotham *et al.*, 1992).

An ingenious mutagenesis technique has been used to indicate which P region residues line the pore of the channel (Lü & Miller, 1995; Kürz *et al.*, 1995; Pascual *et al.*, 1995). Cysteine residues are substituted one at a time for each of the residues in turn, and the mutant channels are then exposed either to silver ions or to sulphydryl agents (both of which will form covalent links with any −SH groups accessible to them) and tested to see whether the channel has been blocked. Block by silver ions occurred when residues 0, 1, 4, 5, 8, 13, 15 or 17 to 20 were changed to cysteine, suggesting that these residues line the channel pore.

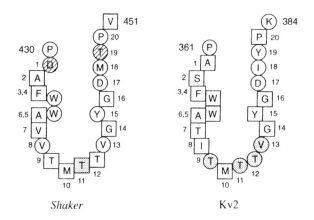

Shaker   Kv2

Figure 6.23. The pore-lining P region (also called the H5 or SS1–SS2 loop) in two voltage-gated potassium channels, *Shaker* from *Drosophila* and Kv2 from the mouse. Amino acid residues are shown by the single-letter code, and numbered from 0 to 20, beginning at the homologous residues proline-430 in *Shaker* and proline-361 in Kv2. For the *Shaker* channel, residues in circles are those that line the pore as determined by the method of individual cysteine substitution and subsequent block by silver. Mutations at D1 and T19 (cross-hatched) affect block by external tetraethylammonium (TEA) ions, whereas mutations at T11 (shaded) affect block by internal TEA. For the Kv2 channel, residues in circles are those that combine with sulphydryl agents after individual mutagenesis to cysteine: white circles for extracellular agents and shaded circles for agents applied to the cytoplasmic side of the membrane. (From Aidley & Stanfield, 1996, courtesy of Dr M. J. Sutcliffe, data from Lü & Miller, 1995, Kürz *et al.*, 1995, and Pascual *et al.*, 1995.)

With the sulphydryl agents, which are too large to enter the pore, block occurs when residues 0, 17 to 19 or 21 are changed and the agent is applied on the external side of the membrane, whereas block occurs with internally applied agents when residues 9 or 11 to 13 are changed. These results (shown by the circled residues in fig. 6.23) provide convincing evidence for the membrane-dipping nature of the loop, but show that it is not a regular structure such as an α-helix or a β-sheet.

Blocking agents

Delayed rectifier channels in squid axons are readily blocked by caesium and barium ions, and by TEA and other quaternary ammonium ions when applied from the inside of the membrane. The lipid-soluble compounds 4-aminopyridine, strychnine and quinidine cause block whether applied from the inside or from the outside, pre-sumably because they can diffuse through the lipid bilayer to reach their site of action if that is necessary.

TEA has a diameter of about 8 Å, which is about the same as that of a hydrated potassium ion. Experiments by Swenson (1982), using an extensive range of quaternary ammonium compounds, imply that the critical size above which block cannot occur is at a diameter of about 12 Å. This suggests that perhaps the channel has an inner 'mouth' of this size, which is large enough to receive hydrated potassium ions, and that TEA and the other quaternary ammonium ions fit into this. We assume that potassium ions enter the narrowest part of the channel from the mouth by losing their hydration shells.

TEA blocks many potassium channels from the outside as well as from the inside; this applies to *Shaker* channels expressed in *Xenopus* oocytes and to the potassium channels of vertebrate myelinated fibres. Mutations at positions 1 and 19 of the *Shaker* channel P region produce changes in the concentrations required to block by external TEA, and mutations at position 11 do so for internal TEA (MacKinnon & Yellen, 1990; Yellen *et al.*, 1991). This provides further strong evidence for the view that the P region really does form a loop in which its two ends are on the outer side of the channel and its middle is accessible from the inner side, as is shown in fig. 6.23.

Charybdotoxin is a small peptide found in the venom of the scorpion *Leiurus quinquestriatus*. It blocks voltage-gated potassium channels when applied to the outside of the membrane, and is interesting in that it binds only to the fully assembled tetrameric channel. Mutation studies show that the residues at positions 1 and 19 in the P region loop are important for binding of the toxin to the channel (MacKinnon *et al.*, 1990). The structure of charybdotoxin is quite well known, so we can get some idea of the shape of the potassium channel region that it binds to. The toxin molecule is an ellipsoidal structure about 15 Å wide by 25 Å long; it binds only to the fully assembled tetrameric potassium channel. This suggests that it fits into a rather flat wide vestibule on the outer side of the channel mouth (Miller, 1995).

Other blocking agents effective on many voltage-gated potassium channels include 4-aminopyridine, noxiustoxin from the scorpion *Centruroides noxius*, and dendrotoxin from the green mamba snake.

```
        ← - - - - - - P-REGION- - - - - - →

                  12345678
         _____          ___
Sh       DAFWWAVVTMTTVGYGDMT
Kv1.1    DAFWWAVVSMTTVGYGDMY
Shab     EAFWWAGITMTTVGYGDIC
Kv2.1    ASFWWATITMTTVGYGDIY
Shaw     LGLWWALVTMTTVGYGDMA
Kv3.1    IGFWWAVVTMTTLGYGDMY
Shal     AAFWYTIVTMTTLGYGDMV
Kv4.1    AAFWYTIVTMTTLGYGDMV
mSlo     ECVYLLMVTNSTVGYGDVY
fSlo     TCVYFLIVTMSTVGYGDVY
eag      TALYFTMTCMTSVGFGNVA
AKT1     TSMYWSITTLTTVGYGDLH
KAT1     TALYWSITTLTTTGYGDFH
ROMK1    SAFLFSLETQVTIGYGFRF
IRK1     AAFLFSIETQTTIGYGFRC
GIRK1    SAFLFFIETEATIGYGYRY
```

Figure 6.24. Sequence alignment of the P region in various cloned potassium channels, using the single-letter code for amino acids. The first eight sequences are from voltage-gated channels of four *Drosophila* genes and their mammalian homologues; mSlo and fSlo are from calcium-activated potassium channels of mouse and *Drosophila*; eag is a cyclic-nucleotide-regulated potassium channel; AKT1 and KAT1 are inward rectifying potassium channels from the plant *Arabidopsis*; ROM1 and IRK1 are two mammalian inward rectifying potassium channels; and GIRK1 is a G-protein-modulated potassium channel. Shading shows residues identical with those in the *Shaker* (SH) channel. The section labelled 1 to 8 is the 'signature sequence' thought to be common to most potassium-selective channels. (From Heginbotham *et al.*, 1994.)

Calcium-activated potassium channels

Some potassium channels are opened by an increase in the intracellular calcium ion concentration as well as, or instead of, a change in membrane potential. Meech (1972) discovered that injection of calcium ions into molluscan neurons causes an increase in potassium conductance. Experiments by Gorman & Thomas (1978, 1980) showed how this calcium-activated potassium current is involved in the bursting behaviour of these cells (fig. 6.25). They injected the calcium-sensitive dye arsenazo III into the cell to monitor its intracellular calcium concentration. During a burst of action potentials calcium ions enter the cell (via the calcium channels to be described in the next section) and so open the calcium-activated potassium channels. The resulting outward current terminates the burst and produces an inter-burst hyperpolarization. This dies away as the calcium ion concentration falls, presumably as a result of uptake by the endoplasmic reticulum and other intracellular organelles.

Calcium-activated potassium channels are found in a wide variety of cells (Latorre *et al.*, 1989; Latorre, 1994). They are of two types. High conductance K(Ca) channels (also called maxi-K or BK channels) have single-channel conductances in the range 100 to 250 pS. Barrett *et al.* (1982) showed that they are activated both by a rise in internal calcium ion concentration and by depolarization. They are blocked by charybdotoxin. Small conductance K(Ca) channels (also called SK channels) have conductances as low as 6 pS; they are show little or no voltage dependence.

Mutations at the *slowpoke* (*slo*) locus in the fruit fly *Drosophila* abolish a calcium-activated potassium current, with the result that the muscle action potentials take longer than usual to repolarize. Cloning studies show that the *slo* gene product is a protein that has much similarity with the *Shaker* potassium channels in its N-terminal sequence, with the usual S1 to S6 and P segments present (Adelman *et al.*,

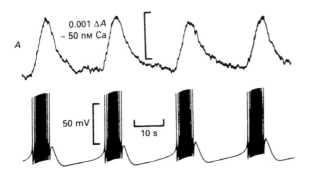

Figure 6.25. Calcium ion concentration changes in an *Aplysia* pacemaker neuron. Electrical activity of the cell (bursts of action potentials at intervals of about 24 s) is shown on the lower trace. Changes in light absorbance (*A*), which reflect changes in intracellular calcium ion concentration since the cell was injected with the calcium-sensitive dye arsenazo III, are shown in the upper trace. The calcium ion concentration rises during the burst of action potentials and falls during the quiescent period. (From Gorman & Thomas, 1978.)

1992). A similar gene *mSlo* encodes for BK calcium-activated potassium channels in mice (Butler *et al.*, 1993). The C-terminal third of the protein seems to be concerned with calcium binding (Wei *et al.*, 1994).

Voltage-gated calcium channels

Appreciable calcium ion currents occur in many different types of cell, notably the muscle fibres of crustaceans and many other invertebrates, vertebrate heart muscle cells, the cell bodies of many invertebrate neurons (fig. 6.25 provides an example), nerve terminals, secretory cells, some sensory receptor cells and many egg cells (Hagiwara & Byerly, 1981). There may be some inflow of calcium ions through sodium channels (Baker *et al.*, 1971) and by the operation of the sodium–calcium exchange system (Requena *et al.*, 1986), but any appreciable inflow usually takes place through specific calcium channels (see Stanfield, 1986; Tsien & Tsien, 1990).

Crustacean muscle fibres provided some of the first evidence for action potentials based on calcium ion flow. Hagiwara & Naka (1964), for example, used the large fibres of the barnacle *Balanus nubilis*, perfused internally with an isotonic potassium sulphate solution. They showed that the size of the all-or-nothing action potential was unaffected by removal of external sodium ions but was increased by a raised external calcium ion concentration.

Molecular structure

Mammalian skeletal muscles have calcium channels located in the T tubules. These are invaginations of the cell surface membrane which are concerned with carrying excitation from the cell surface into the interior of the muscle fibre (see chapter 20). The calcium channels bind dihydropyridines (DHP), hence it is possible to use their DHP-binding capability as an assay during their isolation and purification. For this reason it has sometimes been considered more precise to refer to the purified calcium channel protein as the DHP receptor.

The channel consists of a large principal subunit, called $\alpha1$, which contains the channel pore, and a number of accessory subunits, $\alpha2$, β, γ and δ. Complementary DNA coding for the main $\alpha1$ subunit was prepared and sequenced by Numa and his colleagues at Kyoto University, using their usual molecular cloning techniques (Tanabe *et al.*, 1987). The deduced protein sequence has 1813 amino acid residues and a calculated molecular weight of 212 018. The most striking feature of the sequence is its similarity to that of the sodium channel; 29% of the positions are occupied by identical residues and a further 26% by ones showing only conservative changes.

The hydropathy plot of the deduced calcium channel amino acid sequence is shown in fig. 6.26. Just as in the sodium channel, there are four domains of high internal homology, each with six sections which appear to be membrane-crossing helices. The S4 section of each domain showed an unusual sequence, with arginine or lysine every third residue, just like those of the S4 sections in the sodium and potassium voltage-gated channels (fig. 6.16). It looks very much as though the voltage-gated sodium and calcium channels are closely related in evolution.

Selectivity

Under normal circumstances, voltage-gated calcium channels are highly selective for calcium ions over sodium and potassium ions; P_X/P_{Ca} values of less than 0.001 have been reported (Tsien *et al.*, 1987). Permeabilities for strontium and barium ions are not much less than for calcium. At very low external calcium ion concentrations, however, the channels become much less selective, and allow monovalent ions to flow through relatively freely, as is shown in fig. 6.27 (Almers *et al.*, 1984).

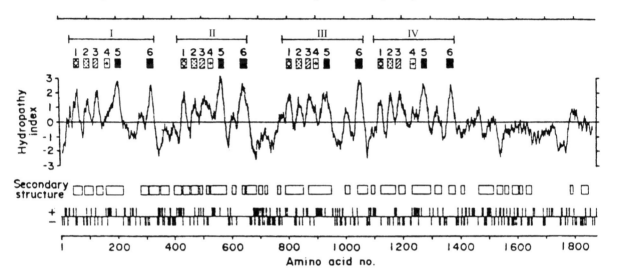

Figure 6.26. Hydropathicity plot of the amino acid sequence of the DHP receptor from rabbit muscle, thought to be a voltage-gated calcium channel protein. Compare with the voltage-gated sodium channel plot shown in fig. 6.8. (From Tanabe *et al.*, 1987.)

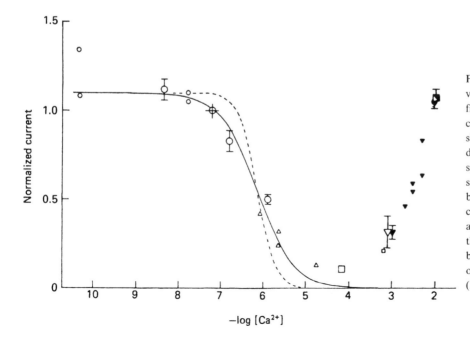

Figure 6.27. Currents through voltage-gated calcium channels in frog muscle at different external calcium concentrations. Points show peak inward currents on depolarization to −20 mV (open symbols) or −5 to +7 mV (black symbols). Sodium channels were blocked by tetrodotoxin. The curves were calculated by assuming that sodium ion flow through a channel could be blocked by one (continuous line) or two (dashed line) calcium ions. (From Almers *et al.*, 1984.)

Almers & McCleskey (1984) produced a persuasive model of the calcium channel which accounts for these properties. They suggested that within its pore each channel has two binding sites with a high affinity for calcium ions. If neither site is occupied by a calcium ion, then sodium and other monovalent ions can pass readily through the channel. If one of them is occupied by a calcium ion, then sodium and other ions are repelled and so cannot pass through the channel; the flow of calcium ions will also be low. But if both sites are occupied by calcium ions, the electrostatic repulsion between them will greatly increase the probability of one of the ions leaving the pore. Hence the rate of flow of calcium ions through the membrane is roughly proportional to the number of doubly occupied channels.

The pores of voltage-gated calcium channels are probably lined by their P segments (H5 or SS1–SS2 segments), just as in other voltage-gated channels. The amino acid sequences of calcium channel P segments are generally similar to those of sodium channels, except that they always have a glutamic acid residue at the same position, about half-way up the second arm of the hairpin loop. This must produce a ring of four negative charges in the pore, and it seems likely that these form the two binding sites postulated by Almers & McCleskey. Further evidence for this idea comes from mutagenesis experiments, with the mutant channels expressed in *Xenopus* oocytes. Sodium channels with glutamate residues introduced at the appropriate points into their P segment loops were much more permeable to calcium ions (Heinemann *et al.*, 1992), and substitution of uncharged residues for the glutamate in calcium channel P regions allowed sodium ion currents at much higher calcium concentrations (Yang *et al.*, 1993). Figure 6.28 shows how the glutamate ring might bind calcium ions.

Diversity

Patch clamp records from chick sensory neurons in tissue culture suggest that there are three types of channel in these cells (Nowycky *et al.*, 1985). L channels produce long-lasting currents at large depolarizations, T channels provide transient currents at small depolarizations, and N channels require very negative potentials for removal of inactivation and large depolarizations for activation. P channels are found in cerebellar Purkinje cells, and Q and R channel types have also been described (Zhang *et al.*, 1994). The different channel types have different blocking agents: L channels are blocked by DHP, N channels by ω-conotoxins (from cone shell venom), P channels by ω-AgaIVA (from funnel-web spider venom) and T channels by nickel ions. These differences are related to differences in the amino

acid sequences; at least six different α1 subunit genes and four different β subunit genes have been described (Hofmann *et al.*, 1994; McCleskey, 1994).

What is the value of this diversity? In some tissues, such as crustacean muscles, vertebrate heart muscle and some secretory cells, it is clear that calcium ion flow through these channels is a major vehicle of cell depolarization. In others this is not so. In the squid giant axon terminal, for example, the peak calcium current is less than one-twentieth of the peak sodium current and occurs largely while the membrane potential is returning to its resting level; its purpose is rather to increase the calcium ion concentration inside the terminal, which is essential for the release of the transmitter substance, as we shall see in chapter 10. The intracellular calcium ion concentration is generally of great importance in the regulation of cellular activity, controlling such activities as secretion, neurotransmitter release, muscular contraction, enzyme activity, calcium-activated potassium channel opening, and many others. Perhaps the complex and subtle control of intracellular calcium ion concentration requires different types of calcium channel for different circumstances.

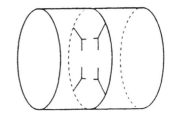

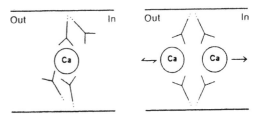

Figure 6.28. How the ring of four glutamate residues in the voltage-gated calcium channel pore might bind one or two calcium ions. Forks represent negatively charged carboxyl groups. With a single ion in the ring, all four glutamates might be involved in high affinity binding. Entry of a second calcium ion into the ring would reduce the binding affinity and so promote flow of ions through the pore. (From Yang *et al.*, 1993. Reprinted with permission from *Nature* **366**, 161. Copyright 1993 Macmillan Magazines Limited.)

Part C
Synaptic transmission

7
Fast synaptic transmission

Synapses are junctional regions between neurons, or between a neuron and another cell, where information is passed rapidly from one cell to the other. Electrical activity in the first (presynaptic) cell produces a change in the electrical activity of the second (postsynaptic) cell. There are two general types of synaptic transmission, chemical and electrical. In *chemical transmission*, by far the commoner mechanism, the presynaptic cell releases a chemical transmitter substance which diffuses across the intercellular space between the two cells and then binds to receptor molecules on the surface of the postsynaptic cell. If the postsynaptic receptors are also ion channels, this produces rapid changes in the flow of ions across the cell membrane; such events

imply *fast* synaptic transmission (fig. 7.1*a*), the subject of this chapter and the next. Sometimes the postsynaptic receptors are not themselves ion channels, but their activation produces further chemical changes before there are any changes in the electrical activity of the cell; events of this type occur in *slow* synaptic transmission (fig. 7.1*b*), examined in chapter 9. In *electrical transmission* (fig. 7.1*c* and chapter 12), current flows directly from the presynaptic cell into the postsynaptic cell so as to alter its membrane potential significantly.

The nature of synaptic transmission
With the realization, towards the end of the nineteenth century, that the nervous system was composed of individual

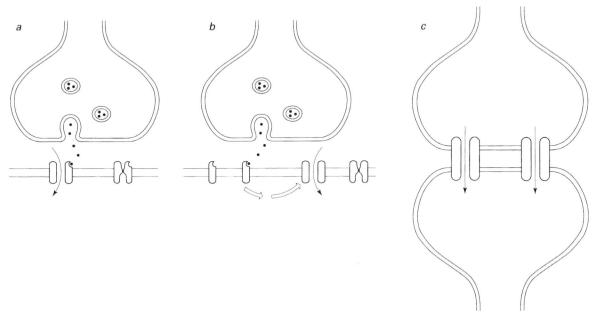

Figure 7.1. Different types of synaptic transmission. In fast chemical transmission (*a*), the neurotransmitter combines with ion channels in the postsynaptic membrane, so as to open them and allow current flow across the postsynaptic membrane. In slow chemical transmission (*b*), the neurotransmitter binds to receptors which produce changes in other proteins, and often second-messenger molecules, leading to the opening or closing of ion channels in the postsynaptic membrane. Chemical transmission allows the postsynaptic currents to be much greater than the presynaptic ones. In electrical transmission (*c*), currents flow directly from the presynaptic cell to the postsynaptic one, usually via gap junction channels that bridge the pre- and postsynaptic membranes. The diagrams are not drawn to scale.

cells, the neurons, the problem arose as to how excitation was transmitted across the gap between two of them. Du Bois Reymond suggested in 1877 that the presynaptic cell could influence the postsynaptic cell either by electric currents or by chemical mediators. We now know that transmission at the majority of synapses is brought about by release of a chemical *transmitter substance* or *neurotransmitter* from the presynaptic cell, although there are some cases of electrical transmission. The history of the establishment of this generalization provides a fascinating example of the development of a scientific theory; let us look at some of the main steps in the story.

The first tentative evidence for the chemical transmission theory came from the work of Elliott in 1904. He showed that those mammalian muscles which are innervated by the sympathetic nervous system are responsive to adrenaline, whereas muscles which are not so innervated are not. In order to explain these observations he suggested that sympathetic nerves act by release of adrenaline from the nerve endings when a nerve impulse arrives at the periphery.

In 1914, Dale investigated the effects of the substance acetylcholine, originally isolated from ergot (a fungal infection of cereals), on various body functions. He found that it lowered blood pressure in the cat, inhibited the heart beat in the frog, and caused contractions of frog intestinal muscles. He suggested that acetylcholine occurred naturally in the body, possibly acting as an antagonist to the effects of adrenaline, and, as an inspired guess, suggested that it was normally rapidly broken down by a hydrolytic enzyme. This would account for the fact that it had so far been impossible to isolate acetylcholine from the body.

The next stage in this story came from a classic experiment by Loewi (1921) on the frog heart. The heart normally beats spontaneously, but it can be inhibited by stimulation of the vagus nerve. Loewi found that the perfusion fluid from a heart which was inhibited by stimulation of the vagus would itself reduce the amplitude of the normal beat in the absence of vagal stimulation. Perfusion fluid from a heart beating normally did not have this effect. This means that stimulation of the vagus results in the release of a chemical substance (initially called the *vagusstoff*), presumably from the nerve endings. The inhibitory action of vagal stimulation and of the vagusstoff was prevented in the presence of the alkaloid atropine. In this and other respects the pharmacological properties of the vagusstoff seemed to be similar to those of acetylcholine. It was later shown that the actions of the vagusstoff and acetylcholine could be potentiated by the alkaloid eserine, and that this substance acted by preventing the hydrolysis of acetylcholine by the enzyme acetylcholinesterase.

These experiments were leading to the conclusion that acetylcholine and the vagusstoff were one and the same substance; all that was needed for the general acceptance of this conclusion was a demonstration that acetylcholine does in fact occur in the body. This was provided by Dale & Dudley (1929), who isolated it from the spleens of cows and horses. Further advances came with the demonstrations that acetylcholine is released on stimulation from the sympathetic ganglia (Feldburg & Gaddum, 1934) and from the motor nerve endings in skeletal muscles (Dale *et al.*, 1936).

However, as the evidence for the chemical transmission theory accumulated, some difficulties arose in its application to all synapses. A number of neurophysiologists remained convinced that many cases of synaptic transmission must involve electric current flow across the synapse. They found it difficult to see how the transmitter substance could be released, produce its action and be destroyed all within the space of a few milliseconds, and there was conflicting evidence about the action of acetylcholine in the central nervous system. These doubts were not finally resolved until the early 1950s, when it became possible to insert intracellular electrodes into the postsynaptic region (Fatt & Katz, 1951; Brock *et al.*, 1952), as we shall see in the following pages.

The molecular basis of chemical transmission has become more firmly established in recent years. The activity of individual acetylcholine-gated ion channels was first observed by Neher & Sakmann when they invented the patch clamp technique in 1976. The chemical isolation of acetylcholine receptors began in the 1970s and the determination of their amino acid sequences using recombinant DNA technology by Numa and his colleagues occurred in 1982. Since then a great deal of further information has been forthcoming on the structure and behaviour of a variety of different receptors involved in synaptic transmission.

Eccles (1976) gives the reasons why he originally believed in the electrical transmission theory, and provides a stimulating account of how he came to change his mind, and Bennett (1994) surveys the interaction between concepts and experiments in our understanding of the synapse from Sherrington (who introduced the term in 1897) to Katz. An ironic footnote on the controversy was provided by the discovery of some synapses in which the transmission process *is* by means of electric currents, as we shall see in chapter 12.

Neuromuscular transmission

We begin our account of synaptic transmission by considering the structure and physiology of the neuromuscular junction in the 'twitch' fibres of vertebrate skeletal muscle. The situation here differs in some details from that found at other neuromuscular junctions and at synapses between neurons. Nevertheless, the detailed study of this particular synapse provides an excellent introduction to the physiology of chemically transmitting synapses in general.

The structure of the neuromuscular junction

A vertebrate 'twitch' skeletal muscle fibre, such as is found in the sartorius muscle of the frog, is a long cylindrical multinucleate cell, which is innervated at one or sometimes two points along its length by branches of a motor nerve axon. The structure of the neuromuscular junction in frogs and toads is indicated in fig. 7.2. In the region of contact, the muscle fibre is modified to form the motor end-plate, which contains numbers of mitochondria and nuclei. The motor nerve axon loses its myelin sheath near the ending, and the terminal branches of the axon are partly sunken into 'synaptic gutters' as they spread over the surface of the end-plate. Schwann cell nuclei are associated with the terminals, but the precise distribution of their cytoplasm cannot be determined with the light microscope. A series of 'subneural lamellae' project into the end-plate cytoplasm under the terminal branches of the axon.

Electron microscopy (fig. 7.3) shows that there is a space about 500 Å wide, the *synaptic cleft*, between the nerve terminal and the muscle cells (Birks *et al.*, 1960; Heuser & Reese, 1977), The 'subneural lamellae' are seen to be infoldings of the postsynaptic membrane. There is some extra-cellular material in the cleft, especially in these folds, which forms the basal lamina. The Schwann cell covers the nerve terminal in places where it is not in contact with the muscle cell. The terminal branches of the axon contain large numbers of *synaptic vesicles* about 500 Å in diameter, which may be concentrated in regions opposite the folds in the postsynaptic membrane.

We shall see later that the synaptic vesicles contain the transmitter substance acetylcholine, and that it is released from them when the nerve action potential arrives at the terminal, to diffuse across the synaptic cleft and bind to the acetylcholine receptors on the postsynaptic membrane. Observations on other synapses show that the synaptic cleft and synaptic vesicles are found in all chemically transmitting synapses.

Amphibian neuromuscular junctions are somewhat unusual in having the fine terminals run longitudinally along the surface of the muscle cell. In reptiles and mammals the terminals are more compact in form, with the fine nerve branches and the postsynaptic gutters curled to form a junctional area that is oval in shape (see Salpeter, 1987).

The release of acetylcholine by the motor nerve endings

In a very elegant series of experiments, Dale and his colleagues (1936) showed that acetylcholine is released when the motor nerves of various vertebrate skeletal muscles are stimulated. Their experiments on the cat's tongue serve to illustrate some of their methods and conclusions. They

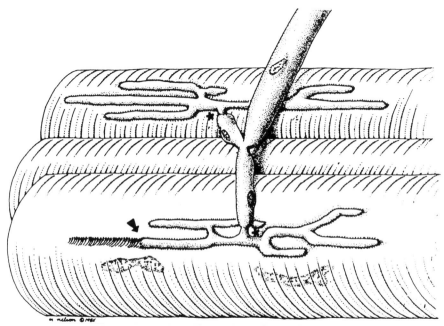

Figure 7.2. A schematic drawing of the frog neuromuscular junction, based mainly on light microscopy and scanning electron microscopy. The nerve terminal branches to contact a number of muscle fibres. The asterisks show where the myelin sheath ends. Part of a terminal branch has been pulled away at the arrow to show the postsynaptic gutter traversed by postsynaptic folds. (From Salpeter, 1987 in *The Vertebrate Neuromuscular Junction*, ed. M. M. Saltpeter, © 1987 Alan R. Liss. Reprinted by permission of John Wiley & Sons, Inc.)

perfused the arteries supplying the tongue with a suitable saline solution and collected it for analysis from the jugular vein after its passage through the tongue capillaries.

The quantities of acetylcholine released in an experiment of this kind were too small to be identified and measured by ordinary chemical methods, and so some type of bio-assay had to be used. Acetylcholine in very dilute concentrations causes contraction of the dorsal longitudinal muscles of leeches (after sensitization by eserine) and produces a fall in the blood pressure of anaesthetized cats. Thus if the perfusate also produced these effects, and particularly if the same concentration of acetylcholine was needed to mimic the effects of the perfusate in different

tests, then there is a strong indication that the perfusate contained acetylcholine.

Now consider fig. 7.4, which shows the results of one of the experiments on the cat's tongue. Each record in the diagram is a smoked drum trace of the contraction of a strip of leech muscle in response to perfusate from the tongue (diluted by 50%) or a solution containing a known concentration of acetylcholine. Trace *a* shows the absence of response to perfusate obtained before stimulation of the hypoglossal nerve. Trace *b* shows the contraction produced by acetylcholine at a concentration of 10^{-8} w/v. Trace *c* shows the response to perfusate obtained during a period of stimulation at 5 shocks s^{-1}, traces *d* and *e* the responses

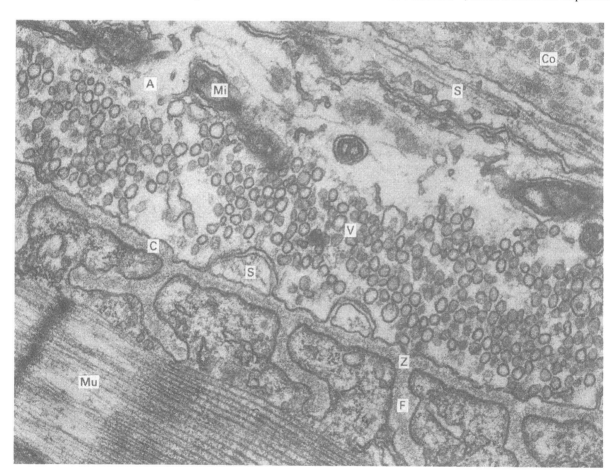

Figure 7.3. Electron micrograph of a frog neuromuscular junction. The axon terminal (A) is seen in longitudinal section. It contains mitochondria (Mi) and numerous synaptic vesicles (V), and is covered by a Schwann cell (S). Collagen fibres (Co) can be seen over the Schwann cell. The muscle fibre (Mu) is separated from the axon terminal by the synaptic cleft (C), which contains some darkly staining material. The muscle fibre postsynaptic membrane is indented to form junctional folds (F), and is underlaid by dense material at the top of them, where the acetylcholine receptors are concentrated. Presynaptic active zones (Z) occur opposite some of the junctional folds; notice the slight protrusion of the presynaptic membrane, the dense cytoplasmic material and the concentration of synaptic vesicles there. Magnification 43 000×. i.e. 1 mm is equivalent to 230 Å. (Photograph kindly supplied by Professor J. E. Heuser.)

to perfusate obtained 2 and 20 min later. Trace *f* shows the response to perfusate obtained during a second period of stimulation; trace *g*, 20 min later. Finally trace *h* shows a smaller response to perfusate obtained from a third period of stimulation, which can be compared with trace *j*, the contraction produced by acetylcholine at a concentration of 2.5×10^{-9} w/v.

This experiment shows that a substance which causes leech muscle to contract is released on stimulation of the nerves supplying the cat's tongue. It was also found that the perfusate from a period of stimulation reduced the blood pressure in the cat, and that this effect was increased in the presence of eserine and abolished by atropine; in all these respects the released substance was similar to acetylcholine. Furthermore, the concentrations of acetylcholine required to mimic the effects of the perfusate on cat blood pressure and on the leech muscle were the same. It follows that acetylcholine must have been present in the perfusion fluid obtained during a period of stimulation.

Where does this acetylcholine come from? At this stage of the investigation there were three possibilities: the acetylcholine could be released from the motor nerves, from the sensory nerves, or from the muscle fibres. In order to resolve this question, Dale and his colleagues used a similar perfusion technique on leg muscles of the dog and the frog. Stimulation of the sensory nerves via the dorsal roots (which were cut proximally to the electrodes so as to avoid reflex excitation of the motor nerves) did not induce release of acetylcholine, whereas stimulation of the motor nerves via the ventral roots did. Direct stimulation of the muscles also produced acetylcholine release, but of course it is impossible to stimulate a normal muscle without also stimulating the motor nerve endings in the muscle. However, stimulation of a muscle which had been denervated ten days previously, or stimulation of a muscle after transmission had failed following prolonged stimulation of the motor nerves, did not result in release of acetylcholine. Hence acetylcholine is released from the motor nerve endings on stimulation.

Was this release of acetylcholine essential to the excita-tion of the muscle, or was it merely some incidental by-product of nervous activity? Earlier work had shown that some contraction could be obtained from muscles after injection of acetylcholine, but only when the amount injected was very high. Brown and his colleagues (1936) reinvestigated the problem, paying particular attention to the details of the perfusion technique, so as to ensure that the administered doses of acetylcholine came into contact with all the fibres in the muscle as rapidly as possible. In this way they were able to show that injections of as little as 2 µg of acetylcholine could produce contractions reaching as high as that in a twitch produced by stimulation of the motor nerves. This response was abolished in the presence of curare, a South American poison that blocks neuro-muscular transmission. Injections of eserine caused the contractions produced by single stimuli to the motor nerves to be enhanced and prolonged, as in repetitive stimulation. Hence it was concluded that the action of acetylcholine on the muscle fibre is an essential step in the transmission process.

The end-plate potential

What is this action of acetylcholine on the muscle fibre? Electrical activity in the region of the end-plate, produced by nervous stimulation, was first observed by Göpfert & Schaefer in 1938 and was subsequently investigated by a number of other workers (e.g. Kuffler, 1942). The general conclusion to be reached from these initial experiments was that the first postsynaptic electrical event is a depolariza-tion (recorded as a negative potential with external elec-trodes) of the muscle fibre membrane, known as the *end-plate potential* (EPP) or postsynaptic potential (PSP). This initial depolarization is restricted to the end-plate region of the muscle fibre, but is normally of sufficient size to elicit an all-or-nothing action potential which propa-gates along the length of the fibre.

The end-plate potential was first investigated with intra-cellular electrodes by Fatt & Katz (1951), using the frog sartorius muscle. Figure 7.5 gives an idea of their experi-mental arrangement. The microelectrode is filled with a

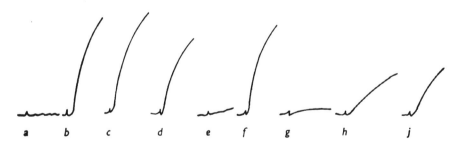

a   b   c   d   e   f   g   h   j

Figure 7.4. Contractions of eserinized leech muscle in response to acetylcholine (traces *b* and *j*) and to perfusates from cat tongue muscle under various conditions, including periods of motor nerve stimulation (traces *c*, *f* and *h*). For further explanation, see the text. (From Dale *et al.*, 1936.)

concentrated solution of potassium chloride and is inserted into the muscle fibre in the end-plate region. A suitable amplifier then measures the voltage between the tip of the electrode and another electrode in the external solution, so giving the membrane potential. It is much easier to study the nature of the EPP if it is reduced in size by partial block with curare, so that the complicating effects of the propagated action potential are absent. Under these conditions, the EPP is a brief depolarization with a rapid rising phase and a much slower falling phase, as is shown in the uppermost record of fig. 7.6.

When the electrode is inserted at increasing distances from the end-plate, the electrical change becomes reduced in size and its time course is lengthened (fig. 7.6). This is what one would expect if the EPP is a result of (1) a brief 'active' phase of ionic current flow across the membrane at the junctional region, produced by the synchronous opening of a large number of ion channels located in the subsynaptic membrane, followed by (2) a passive electronic spread and decay of the charge across the membrane. The form of this second phase of the EPP should be in accordance with the equations used by Hodgkin & Rushton (p. 43) in the application of the core-conductor theory to nerve axons. A relatively simple way of testing this idea is to see whether the total charge on the membrane (which is proportional to the integral of voltage with respect to distance) decays exponentially, as is predicted by equation 4.13. Fatt & Katz found that it does.

The ionic current at the end-plate was later measured directly by Takeuchi & Takeuchi (1959), using a voltage clamp technique (fig. 7.7). The results, shown in fig. 7.8, were fully in agreement with the scheme suggested by Fatt & Katz.

Initiation of the action potential

If a frog sartorius fibre is stimulated electrically so as to depolarize the membrane sufficiently, an action potential arises in the depolarized region and is propagated along the length of the fibre. This action potential is similar to that of non-myelinated axons (although the time course is longer) in that it shows the familiar phenomena of threshold, all-or-nothing response and refractoriness. It is produced by a specific regenerative increase in the permeability of the cell membrane to sodium ions, mediated by voltage-gated sodium channels (Nastuk & Hodgkin, 1950).

In the normal muscle, the depolarization produced by the EPP is sufficient to cross the threshold for production of an action potential. However, Fatt & Katz showed that the form of an action potential produced in response to nervous stimulation and recorded at the end-plate (the N response) is different from that recorded at distances away from the end-plate (fig. 7.9). It is also different from responses to direct stimulation of the muscle fibre (M responses) recorded at the end-plate (fig. 7.10). Figure 7.9 shows the transitional stages between the two types that occur at points near the end-plate.

What is the reason for these differences between the N and M responses? It is obvious that the inflection in the rising phase of the N response at about -50 mV is caused by the action potential 'taking off' from the rising phase of an EPP as the membrane potential crosses the threshold. The first part of the N response is mediated by acetylcholine-gated channels restricted to the subsynaptic membrane, and the depolarization so produced is eventually sufficient to activate the voltage-gated sodium channels found all over the muscle fibre surface. The later differences between N and M responses were at first more difficult to

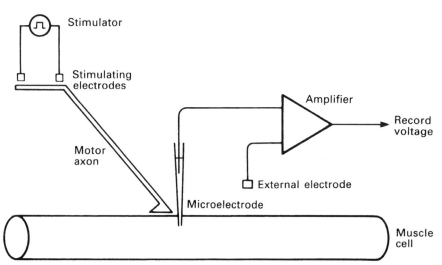

Figure 7.5. Schematic diagram showing the arrangement for intracellular recording of postsynaptic responses at the vertebrate neuromuscular junction.

✷ interpret; the peak of the N response is lower than that of the M response and there is a hump on the falling phase of the N response. Fatt & Katz suggested that these phenomena were caused by the effect of the conductance change at the end-plate on the form of the normal action potential. We shall see later that, when the acetylcholine-gated channels are open as well as the sodium channels, the membrane potential does not approach the sodium equilibrium potential as closely as it would if only the sodium channel were

↑

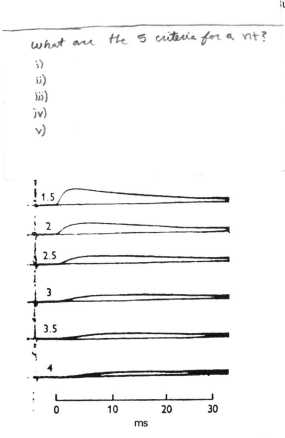

what are the 5 criteria for a nt?

i)

ii)

iii)

iv)

v)

Figure 7.6. The end-plate potential in curarized frog sartorius muscle, recorded by means of an intracellular microelectrode inserted at different distances from the neuromuscular junction. The figures to the left of each record indicate the distance (mm) between the electrode and the motor nerve ending; the time scale at the bottom is in milliseconds. *S* shows the time of the stimulus to the nerve axon. (From Fatt & Katz, 1951.)

how is the acetylcholine to be applied to the end-plate? The normal action of the transmitter substance is a brief impulsive event lasting no more than a few milliseconds, but it is not possible to limit the action of externally applied acetylcholine to less than a few seconds when using the usual perfusion techniques. Such relatively long applications of acetylcholine tend to produce neuromuscular desensitization (p. 116) rather than excitation.

The answer to this problem was provided by an ingenious technique developed by Nastuk (1953) and further used by del Castillo & Katz (1955) and others. Acetylcholine ionizes in solution so that the acetylcholine ion is positively charged. Consequently it will move electrophoretically if placed in an electric field. Thus, after filling a glass micropipette (of the same order of size as a glass microelectrode) with acetylcholine, it is possible to eject it from the tip of the pipette for very brief times by passing a pulse of electrical current outward through the pipette. This technique is known as *ionophoresis* or iontophoresis. It is usually necessary to pass an inward 'braking' current through the pipette in order to prevent outward diffusion of acetylcholine in the absence of an ejecting pulse.

The experimental arrangement used by del Castillo & Katz is shown in fig. 7.11. When the acetylcholine pipette was brought near to the end-plate, an ejecting pulse caused a depolarization of the muscle fibre membrane, which could initiate propagated action potentials if it was sufficiently large (fig. 7.12). On the other hand, if the pipette was inserted into the muscle fibre, the acetylcholine-ejecting pulse merely caused an electrotonic potential change, which is what would occur even if there were no acetylcholine in the pipette.

These experiments show that acetylcholine, when applied to the outside of the end-plate membrane, causes a depolarization similar to that produced during neuromuscular transmission. The longer time course of the rising phase of the acetylcholine potential was ascribed to the much longer distance from the sensitive sites: a few micrometres instead of 500 Å. Later workers have obtained responses that are much more rapid.

Acetylcholine acts via particular molecular sites known as *receptors*, which occur on the outside of the postsynaptic membrane in the end-plate region. The receptors have been isolated in recent years and we now know a great deal about their structure and properties, as we shall see in chapter 8. Probably two acetylcholine molecules are bound to each receptor in order to activate it. When this happens a channel is momentarily opened, so permitting ionic current flow across the membrane. End-plate potentials, and the responses to acetylcholine applied by ionophoresis, are

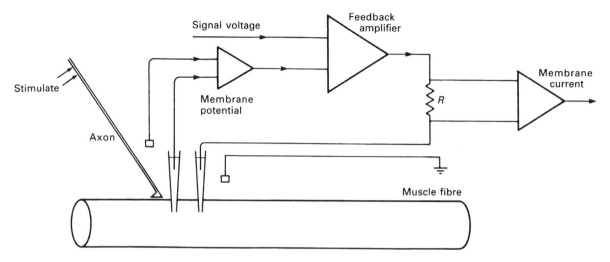

Figure 7.7. Arrangement for voltage clamp measurements on neuromuscular transmission. The output of the feedback amplifier passes through a microelectrode so as to keep the membrane potential (measured via the other microelectrode) constant at a level set by the signal voltage. The potential across the resistor R is proportional to the current flowing through the membrane. The muscle fibre is in Ringer solution containing curare.

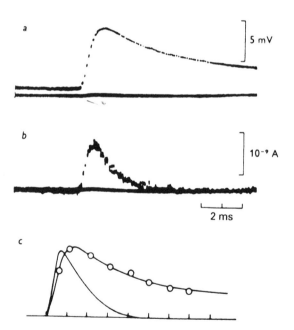

Figure 7.8. Voltage clamp analysis of the end-plate potential in a curarized frog muscle fibre. Trace *a* shows the EPP, recorded in the usual manner. Trace *b* shows the end-plate current (EPC) recorded from the same end-plate with the membrane potential clamped at its resting level. The continuous lines in *c* show the actual EPP and EPC superimposed; the circles show the expected EPP as calculated from the EPC, assuming that the muscle fibre has a time constant of 25 ms and an effective resistance of 320 kΩ. (From Takeuchi & Takeuchi, 1959.)

produced by currents flowing simultaneously through large numbers of channels.

Localization of the acetylcholine receptors
Responses to acetylcholine in ionophoresis experiments can be obtained only when it is applied to the end-plate region of the muscle fibre. This suggests that the acetylcholine receptors are localized in or near the subsynaptic region. Just how localized they are has been shown in some elegant experiments by Kuffler & Yoshikami (1975a). They used the proteolytic enzyme collagenase to free the nerve terminals from the muscle fibres, so enabling acetylcholine to be applied directly to the subsynaptic membrane. They found, as expected, that the subsynaptic membrane had a high sensitivity to acetylcholine; but the sensitivity fell to as little as 2% of this within a distance of 2 μm once the ionophoresis pipette was moved over the extrasynaptic membrane.

The localization of acetylcholine receptors has also been determined histochemically. A most useful substance for this is α-bungarotoxin, a polypeptide constituent of the venom of the Formosan snake *Bungarus multicinctus*; it causes neuromuscular block by binding tightly to the acetylcholine receptors (see Lee, 1970). Using toxin made radioactive with either tritium or iodine-125 it is possible to show by autoradiography that the toxin rapidly binds to the postsynaptic membrane in the end-plate region (Barnard *et al.*, 1971; Fertuck & Salpeter, 1974, 1976). Fertuck &

Salpeter found that binding occurs at the crests of the post-synaptic membrane between the junctional folds and for about 2000 Å into the folds. This distribution of binding sites corresponds with the occurrence of electron-dense thickenings of the postsynaptic membrane.

By counting the grains of silver in the autoradiographs, Fertuck & Salpeter calculated that there are about 30 000 sites per square micrometre (±27%) in these regions in mouse muscle. If there are two binding sites per ion channel (see chapter 8) then the channel density will be half this at about 15 000 per square micrometre. This density is considerably higher than that of sodium channels in nerve axon

membranes. We look further at some of the quantitative aspects of neuromuscular transmission in chapter 10 (p. 173).

Acetylcholine receptors have also been seen by electron microscopy of neuromuscular junctions and electric organs. A variety of techniques has been used, including negative staining of membrane fragments, freeze-fracture and freeze-etching of quick-frozen material (Cartaud *et al.*, 1973, 1978; Rosenbluth, 1974; Heuser & Salpeter, 1979; Hirokawa & Heuser, 1982). The results (fig. 7.13) show arrays of particles on the subneural regions (but not in the depths of the junctional folds of the neuromuscular junction). The particles

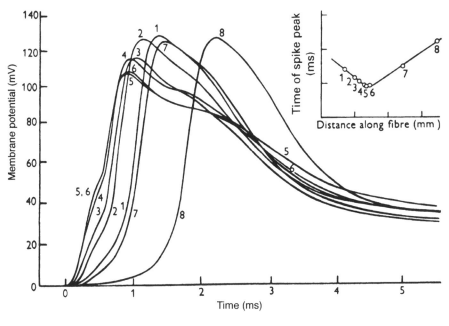

Figure 7.9. The initiation of the muscle action potential at the motor end-plate. Traces were obtained by inserting a microelectrode at different distances (shown in the inset) along the length of a frog muscle fibre, and recording the response to stimulation of the motor nerve. Membrane potentials are shown as changes from the resting level. (From Fatt & Katz, 1951.)

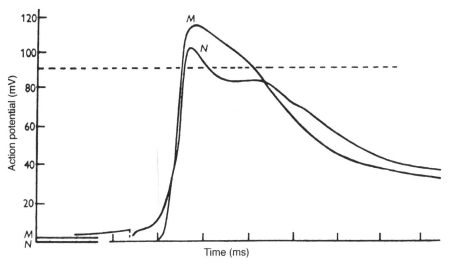

Figure 7.10. Muscle action potentials recorded intracellularly from the end-plate region in response to direct electrical stimulation of the muscle fibre (trace M) and in response to stimulation of the motor nerve (trace N). The voltage shows the change with respect to the resting potential; the dashed line shows where the potential across the membrane is zero. (From Fatt & Katz, 1951.)

are closely packed together at densities in the region of 10 000 per square micrometre. Hirokawa & Heuser showed that the subsynaptic density at the neuromuscular junction is in fact a submembranous meshwork which appears to anchor the acetylcholine receptors in the membrane to underlying filaments of the cytoskeleton.

Desensitization

If a muscle is soaked in a solution containing acetylcholine, the initial period of excitation is followed by neuromuscular block which can be removed only by washing away the acetylcholine. Thesleff (1955) showed that such application causes an initial depolarization which then dies away, and introduced the term *desensitization* to describe this effect. The situation was further investigated by Katz & Thesleff (1957a) using the ionophoretic application technique. Both barrels of a double-barrelled pipette were filled with an acetylcholine solution; steady ('conditioning') currents were passed through one barrel and brief pulses of current through the other. When acetylcholine was continuously applied by switching on the 'conditioning' current, the resulting depolarization gradually faded, and the response to 'pulses' of acetylcholine also declined, as is shown in fig. 7.14. Cessation of the conditioning current was followed by a recovery in the response to acetylcholine pulses.

The ionic basis of the end-plate potential

Fatt & Katz suggested that the EPP is produced by a general increase in the membrane conductance. But it was not clear whether this conductance increase occurred for all ions or whether only certain specific ions were involved. What is the ionic selectivity of the acetylcholine-gated channel?

The problem was solved by Takeuchi & Takeuchi (1960), using a voltage clamp method on frog sartorius muscle fibres. With two intracellular microelectrodes, the membrane potential can be clamped at the end-plate region, as long as no large currents flow through the adjacent unclamped membrane. This technique allows the end-plate current (EPC) to be measured, as in fig. 7.8. The principle of the Takeuchis' experiments was to determine the relations between the size of the EPC and the membrane potential in solutions containing different ionic concentrations.

First, consider fig. 7.15a. In a solution containing 3×10^{-6} g curare ml^{-1}, the EPC varies linearly with membrane potential over the measured range of -50 to -120 mV (it was not possible in these experiments to clamp the membrane at depolarizations greater than about -50 mV). When the line drawn through these points is extrapolated, it cuts the membrane potential axis at -15 mV. This means that the EPC would be zero at this value, and negative (i.e. in the opposite direction) beyond it; hence this point (-15 mV) is called the *reversal potential*. If the curare concentration is increased, the values of EPC are reduced, but the reversal potential is not altered.

Figure 7.15b shows the results of a similar experiment, in which the relation between the EPC and the membrane potential in a curarized muscle fibre was compared in two solutions containing different sodium ion concentrations. It is evident that reduction of the sodium ion concentration makes the reversal potential more negative, and thus it follows that the sodium conductance of the end-plate membrane must increase during the EPC. Similar experiments were carried out on the effect of variations in potassium ion concentration (fig. 7.15c). Again, it was found that the reversal potential is dependent upon the potassium ion concentration, and therefore the potassium conductance of

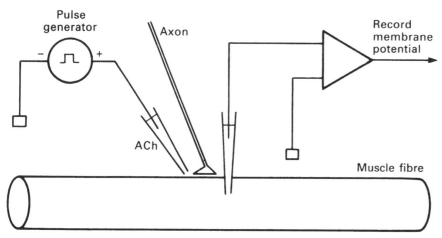

Figure 7.11. Arrangement for ionophoretic application of acetylcholine at the neuromuscular junction. The intracellular microelectrode is inserted in the end-plate region and connected to the circuit on the right to record the membrane potential. The ionophoresis circuit is shown on the left: a pulse generator is connected to an extracellular micropipette which contains a solution of acetylcholine (ACh).

the end-plate membrane must also increase during the EPC. However, with drastic changes in chloride concentration, using glutamate as a substitute, there was no change in reversal potential, and therefore there is no change in the chloride conductance of the end-plate membrane during the EPC (fig. 7.15*d*). (Although the reversal potential is unchanged in fig. 7.15*d*, it is obvious that the EPCs recorded were much higher in the absence of chloride. The reasons for this were trivial: the glutamate solution contained a higher calcium concentration, which would increase the amount of transmitter released, and also a lower curare concentration.)

We can conclude that the transmitter substance, acetylcholine, acts by increasing the permeability of the subsynaptic membrane to cations, so that sodium ions flow inward and potassium ions flow outward. Notice that, in contrast to the situation in a propagated action potential, the ratio of the sodium and potassium conductances does not vary with time, and moreover the conductances are not initiated by changes in membrane potential, so that the potential change constituting the EPP is not regenerative.

These conclusions tell us some important things about the ion channels which are opened during neuromuscular transmission. They differ fundamentally from the channels involved in the nerve action potential in that they are opened not by changes in membrane potential but by binding of the transmitter substance: they are ligand gated, not voltage gated. Thus the linear relations shown in fig. 7.15 show that, for any one of the straight lines in that diagram, the number of channels opened is constant at different membrane potentials. Curare blocks the channels, so the difference between the two lines in fig. 7.15*a* is that there are fewer channels open at the higher curare concentration. Each individual channel allows both sodium and potassium ions to pass through, but not chloride ions. Many channels are open at the peak of the end-plate current, but the number falls steadily to zero over the next few milliseconds.

Single channel currents

The patch clamp technique, which we have met in chapter 2, was invented by Neher & Sakmann (1976) and first used by them to examine single-channel currents through acetylcholine receptors. It had been known for some time that denervation results in a remarkable change in vertebrate muscle fibres: the whole surface becomes sensitive to acetylcholine, as if the occurrence of receptors outside the

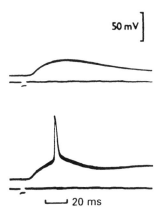

50 mV

20 ms

Figure 7.12. Responses of a muscle fibre to ionophoretic application of acetylcholine to the end-plate. The upper traces show the membrane potential of the fibre, the lower traces monitor the current in the ionophoresis circuit. (From del Castillo & Katz, 1955.)

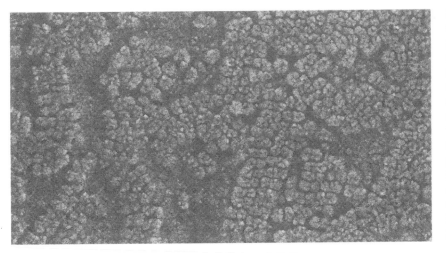

Figure 7.13. Array of acetylcholine receptors on the electrocyte postsynaptic membrane in the electric ray *Torpedo.* Notice the tendency for the receptors to form rows of four abreast, and that each receptor consists of a number of subunits around a central hollow. The picture is of a platinum replica of the surface of a fragment of postsynaptic membrane, quick-frozen and freeze-etched. Magnification 296 000×, i.e. 1 mm is equivalent to 34 Å. (Photograph kindly supplied by Professor J. E. Heuser, from Heuser & Salpeter, 1979.)

end-plate region were normally inhibited in some way by the presence of the motor axon (Axelsson & Thesleff, 1959). Neher & Sakmann pushed the polished tip of a micropipette electrode containing a low concentration of acetylcholine against the membrane of a denervated muscle fibre; they used a voltage clamp circuit to measure the current flow across the small patch of membrane enclosed by the wall of the electrode. Since the density of the channels is much lower than at the end-plate, only a relatively small number of channels would be included in the patch and so the activity of individual channels could be recorded.

Neher & Sakmann found that each channel produces a square pulse of current lasting for up to a few milliseconds. Figure 7.16 shows similar channel openings recorded with an improved technique. The single-channel currents here are about 3 pA. We can calculate the single-channel conductance γ from the relation

$$\gamma = i/(E - E_r)$$

where i is the single-channel current, E is the membrane potential and E_r is the reversal potential. Here the membrane potential was -70 mV and the reversal potential would be near to 0 mV, so

$$\gamma = 3/0.07$$
$$= 43 \text{ pS}$$

The end-plate current in a frog muscle fibre is approaching 10^{-6} A in magnitude. The individual ion channel currents are about 3×10^{-12} A, which means that about 3×10^5 channels are open during the EPP.

Synaptic transmission in electric organs

We have seen in chapter 5 how the discharge of the electric organ in the electric eel *Electrophorus* is dependent upon the production of action potentials in the electrocytes of the electric organ, and that the asymmetry of these cells is an essential feature of their functioning. In the living animal, discharge of an electrocyte is not, of course, initiated by direct electrical stimulation, but by excitation via the efferent nerves innervating it. Altamarino and his colleagues (1955) found that stimulation of the motor nerves elicits an excitatory postsynaptic potential across the innervated surface of the electrocyte, which then induces the action potential (fig. 7.17).

The postsynaptic potential in fig. 7.17 is equivalent to the end-plate potential of vertebrate twitch muscles: it is produced by release of acetylcholine from the electromotor nerve ending and represents the summed activity of many thousands of acetylcholine receptor channels. An important difference between the two systems is that each electrocyte is innervated by a number of nerve fibres, whereas the twitch muscle is innervated by only one. The postsynaptic response (the end-plate potential) in the muscle is easily large enough to cross the threshold for production of an action potential. In the electric organ, however, the response to a single nerve fibre is much smaller, and not sufficient on its own to raise the membrane potential past the threshold. But if a high proportion of the nerve fibres are stimulated at once, their postsynaptic potentials add up to produce a response that *is* large enough to cross the threshold, as is shown in fig. 7.17*b*. This phenomenon is known as *spatial summation*.

Electric organs occur in a variety of different fish. They have probably arisen independently at least six times in evolution: in (1) the rajoid and (2) torpedoid rays among the cartilaginous fish; in (3) the stargazer *Astroscopus* ((1) to (3) are marine); and, in freshwater forms, (4) the electric catfish *Malapterurus*, and in two large groups separated by the Atlantic ocean, (5) the six gymnotiform families of Central and South America and (6) *Gymnarchus* and the mormyrids of tropical Africa. Much detail is now known about the nature of their various electric discharges (see

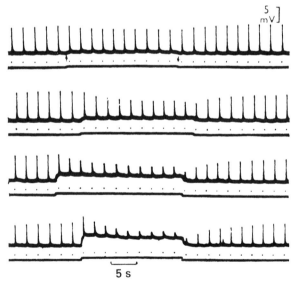

Figure 7.14. A series of records showing desensitization to acetylcholine in a frog muscle fibre. In each case the lower trace monitors the ionophoretic current and the upper trace shows the membrane potential of the muscle fibre. Acetylcholine was applied ionophoretically from a double-barrelled micropipette. Brief 'test' pulses were applied through one barrel (producing the dots on the lower traces and the vertical deflections on the upper traces), and constant 'conditioning' currents were applied through the other barrel at the time indicated. The strength of the 'conditioning' current is lowest in the top record and highest in the bottom one. (From Katz & Thesleff, 1957*a*.)

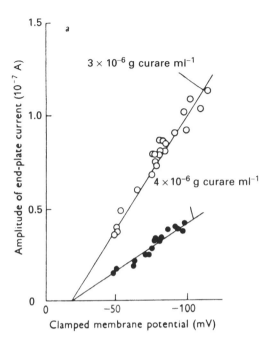

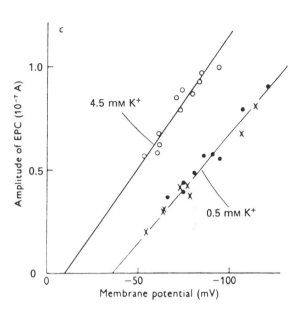

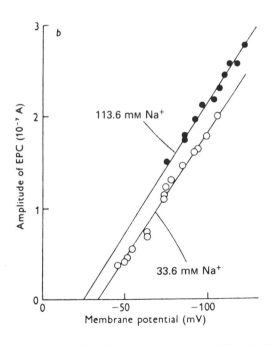

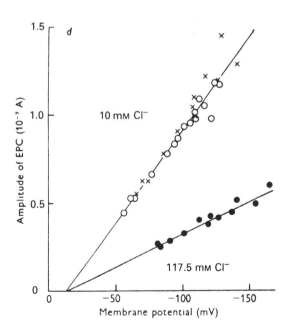

Figure 7.15. Relations between membrane potential and end-plate current in a frog muscle fibre under various conditions, determined by means of the voltage clamp technique as in figs. 7.7 and 7.8. Graphs show the effects of variation in the external concentrations of *a*, curare; *b*, sodium ions; *c*, potassium ions; and *d*, chloride ions. Notice particularly the changes in the reversal potential brought about by changes in sodium and potassium ion concentrations, and the absence of such changes when the curare or chloride concentrations are altered. (From Takeuchi & Takeuchi, 1960.)

Bennett, 1971; Bass, 1986). Most of the gymnotids and mormyrids have relatively weak discharges, insufficient to stun prey or deter predators, that are used instead as components of a sophisticated object-location system, as we shall see in chapter 17.

The marine forms include the electric rays *Torpedo* and *Raia*. Here the innervated faces of the electrocytes are not electrically excitable, and the response to nervous stimula-

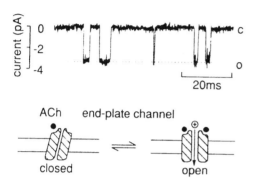

Figure 7.16. Currents through a single acetylcholine-gated channel in rat muscle, recorded with the patch clamp technique. Opening of the channel is seen as a downward deflection of the trace, indicating an inward current of about 3 pA. The membrane potential was −70 mV. The sketch indicates that normally two acetylcholine molecules have to be bound before the channel opens. (From Sakmann, 1992.)

tion consists solely of a depolarizing postsynaptic potential (Brock *et al.*, 1953; Bennett *et al.*, 1961), as is shown for *Raia* in fig. 7.18. Synaptic transmission is cholinergic, with large numbers of nerve endings and high densities of nicotinic acetylcholine receptors on the innervated face of the electrocyte. Ionic current flow across the innervated face occurs via the channels of the acetylcholine receptors and there are no voltage-gated sodium channels. Hence the maximum potential across the active electrocyte cannot be as large as it is in *Electrophorus*; it is usually less than 70 mV. The non-innervated face of the electrocyte has a high density of CLC chloride channels (see p. 30), leading to a high conductance and thus allowing the ready flow of current through the electric organ (White & Miller, 1979).

The electric organ of *Torpedo* has been much used as a rich source of acetylcholine receptors and their messenger RNA (Raftery *et al.*, 1976; Noda *et al.*, 1982*a*), with the results that we shall see in chapter 8. Volta's invention of the voltaic pile in 1800, a crucial step in the science of electricity, was described by him as 'an artificial electric organ' and may well have been inspired by the structure of the *Torpedo* electric organ (Wu, 1984).

Fast synaptic excitation in mammalian spinal motoneurons

Motoneurons are neurons which directly innervate skeletal muscle fibres. In mammals, the cell bodies of the motoneu-

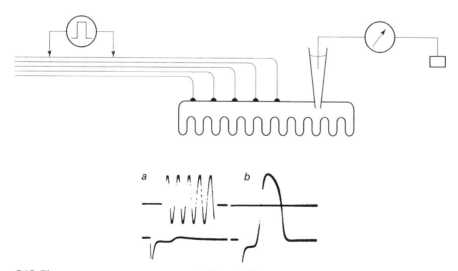

Figure 7.17. Electrocyte responses to nervous stimulation in the electric eel. Record *a* shows the response to nervous stimulation at a low intensity, exciting just a few of the presynaptic nerve fibres. A small EPSP is seen, insufficient to cross the threshold for action potential production. Record *b* shows the response to a higher intensity stimulus, exciting more nerve fibres. Spatial summation of the postsynaptic responses produces a large EPSP which immediately produces an action potential. Upper traces show the 0 mV level and, in *a*, a 100 mV 1000 Hz calibration signal. The diagram above indicates the sites for stimulation and recording. (From Altamirano *et al.*, 1955.)

rons innervating the limb muscles lie in the ventral horn of the spinal cord, and their axons pass out to the peripheral nerves via the ventral roots. The cell body, or soma, is about 70 μm across, and extends into a number of fine branching processes, the dendrites (fig. 7.19). The surface of the soma

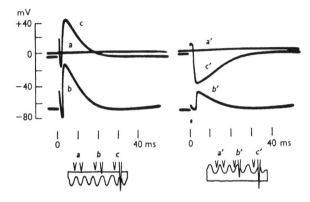

Figure 7.18. Responses of *Raia* electrocytes, as recorded by electrodes in various positions. (From Brock *et al.*, 1953.)

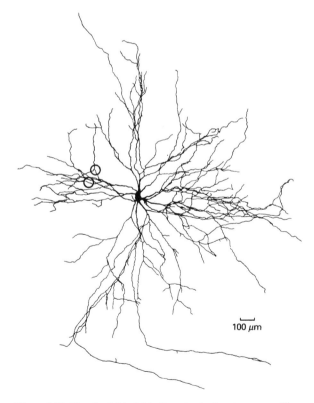

Figure 7.19. The dendritic field of a cat spinal motoneuron. The motoneuron was injected with horseradish peroxidase and its structure then reconstructed from stained serial sections. The circles show sites of synaptic contact with one particular presynaptic fibre. (From Redman & Walmsley, 1983*a*.)

and dendrites is covered with small presynaptic nerve endings (terminal boutons), showing the typical features of a chemically transmitting synapse. Some of these terminal boutons are the endings of group Ia fibres from stretch receptors (muscle spindles) in the muscle which the motoneuron innervates. Stretching the muscle excites these group Ia fibres, which may then excite the motoneurons supplying the muscle so that it contracts. This system is known as a *monosynaptic reflex* (fig. 7.20).

Much of our knowledge of the synaptic responses of motoneurons comes from investigations in the lumbar region of the spinal cord of the cat, using intracellular electrodes. This technique was first developed by Eccles and his colleagues (Brock *et al.*, 1952; Eccles, 1964). They used anaesthetized cats with the spinal cord transected in the thorax. The spinal cord was exposed, a patch of the tough sheath round the cord removed, and the microelectrode inserted through this region until it reached a motoneuron. Maps of the positions of various motoneuron groups were available, prepared by observing the chromatolytic changes following section of the nerve supplying a particular muscle (e.g. Romanes, 1951), so that the microelectrode could be placed in approximately the desired position. The identity of a motoneuron was accurately established by stimulating the nerves supplying particular muscles and observing the response in the motoneuron. In many experiments the ventral roots were then cut so that stimulation of a peripheral nerve did not result in antidromic stimulation of the motoneurons.

Motoneurons have a resting potential of about −70 mV. Depolarization of the membrane by about 10 mV results in the production of an action potential which propagates along the axon to the nerve terminals. The results of a number of experiments on the effects of injection of various ions into motoneurons via microelectrodes indicate that the ionic basis of the resting and action potentials is much the same as in squid axons. That is to say, the resting potential is slightly less than the potassium equilibrium potential, the action potential is caused primarily by a regenerative increase in sodium permeability, and the ionic concentration gradients necessary for these potentials are maintained by an active extrusion of sodium ions.

Excitatory postsynaptic potentials

The responses shown in fig. 7.21 were recorded by means of a microelectrode inserted into a motoneuron supplying fibres in the biceps or semitendinosus muscles. They were produced by stimulating the nerve supplying these muscles with single shocks. The size of the mass response of the sensory nerves was recorded from the dorsal roots, as shown in the inset records of fig. 7.21; the form of these dorsal root responses shows that all the sensory fibres stimulated were of about the same conduction velocity – all group Ia fibres, in fact. The ventral roots were cut so as to avoid antidromic stimulation of the motoneurons. Each record was obtained by superimposition of faint traces from about forty separate responses to stimuli of the same intensity. The responses are known as *excitatory postsynaptic potentials*, or EPSPs for short.

The form of the EPSP is much the same as that of the end-plate potential in a curarized frog sartorius muscle fibre: a fairly rapid rising phase followed by a slower decay which follows an approximately exponential time course. The neurotransmitter is glutamate, not acetylcholine, and glutamate molecules combine with glutamate receptors on the postsynaptic membrane, opening ion channels that allow the cations sodium, potassium, and to some extent calcium to flow down their electrochemical gradients. There are two types of glutamate receptor with intrinsic ion channels at many excitatory synapses in the mammalian nervous system; one has somewhat slower kinetics than the other, so that the EPSC (the excitatory postsynaptic current) has two components, one slower than the other. We shall look at this situation more closely in the next chapter (p. 152).

The size of the responses in fig. 7.21 is proportional to the stimulus intensity, and therefore to the number of presynaptic fibres which are active. This is another example of *spatial summation*. If the EPSP is large enough, a propagated action potential is set up, as in fig. 7.22. If a second EPSP is produced a short time later, the total response is greater; thus two successive EPSPs may be able to produce an action potential whereas either alone could not do so. This phenomenon is called *temporal summation*.

If the cell membrane is progressively depolarized, the EPSP decreases in size and eventually becomes reversed in sign; the reversal potential in cat motoneurons is about 0 mV (fig. 7.23). This is just what we would expect if the EPSP is produced by the simultaneous openings of a large number of chemically gated channels in the subsynaptic membrane. The channels involved are activated by glutamate, the neurotransmitter released at group Ia fibre terminals. They allow sodium and potassium ions to flow down their electrochemical gradients when they are open.

Initiation of the action potential

As is shown in fig. 7.22, an EPSP of sufficient size elicits an action potential in the motoneuron. The form of this action potential is rather more complicated than that in the axon at some distance from the site at which it is initiated. As recorded in the soma, the action potential consists of three components. These are: (1) the M spike, representing activity in the myelinated region of the axon and recorded by a microelectrode placed in the soma as a small depolarization of 1–3 mV; (2) the IS spike, representing activity in the initial segment of the axon, which is not surrounded by a myelin sheath, and recorded in the soma as a depolarization of 30–40 mV; and (3) the SD spike, representing activity in the soma and dendrites, and therefore recorded as a full-sized action potential of 80–100 mV.

This analysis (due to Coombs *et al.*, 1955a, 1957a,b) was derived mainly by examination of the effects of the soma membrane potential on the form of an antidromic action potential produced by stimulation of the ventral roots. The results of such an experiment are shown in fig. 7.24a. At the resting potential (−80 mV in this case) only the IS spike is seen. It is reasonable to assume that the size of the IS spike *in the initial segment* is about 100 mV, but the electrotonic

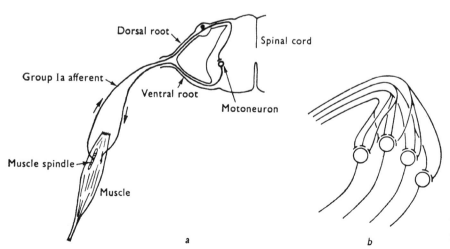

Figure 7.20. Anatomical organization of the monosynaptic stretch reflex system. Group Ia afferents from muscle spindles in a muscle synapse with the motoneurons that make that muscle contract. Diagram *a* is greatly simplified: there are very many stretch receptors and afferent and efferent neurons associated with each muscle. *b* is somewhat less simplified, showing how the afferent fibres branch to synapse with different members of the motoneuronal pool.

currents set up by this activity produce a depolarization of only about 40 mV in the soma, where it is recorded; since this is still below the threshold for the regenerative response of the soma and dendrites, no SD spike arises. However, if the soma is depolarized below about -78 mV, the electrotonic currents produced by activity in the initial segment are now sufficient to excite the soma, so an SD spike arises from the IS spike, with decreasing latency at lower membrane potentials (compare the traces obtained at -77 and -63 mV in fig. 7.24a). If the soma is hyperpolarized beyond about -82 mV (a procedure which must also hyperpolarize

the initial segment to some extent), the IS spike disappears, and we are left with a very small deflection, the M spike, shown in the traces obtained at -87 mV in fig. 7.24a. This seems to be caused by the electrotonic currents produced in the soma by the action potential in the myelinated region of the axon.

How does this analysis relate to the action potentials initiated orthodromically by excitatory synaptic action? An initial IS component can be seen at the beginning of the SD spike produced by an EPSP of sufficient size. This implies that the initial segment is the site of impulse initiation in

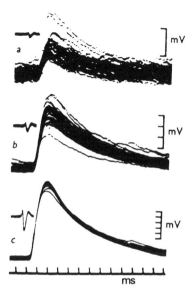

Figure 7.21. Excitatory postsynaptic potentials recorded intracellularly from a cat biceps-semitendinosus motoneuron in response to stimuli applied to the group Ia afferent fibres from the muscle. Each trace shows a number of superimposed responses. Stimulus intensity increases from a to c; notice the change in the voltage scale and the increase in size of the responses. The inset records, taken at constant amplification, show the size of the dorsal root responses, i.e. they monitor the number of afferent fibres that are active. (From Coombs *et al.*,

Figure 7.22. Initiation of an action potential by the EPSP in a cat gastrocnemius motoneuron. The stimulus intensity to the afferent nerve was increased in the order a to d, with the result that the EPSP is of sufficient size to produce an action potential in b to d, and does so progressively earlier in these cases. (From Coombs *et al.*, 1957b.)

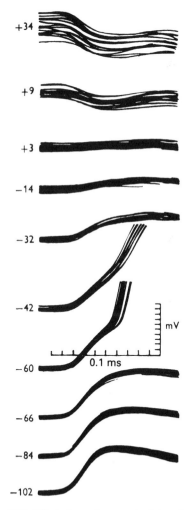

Figure 7.23. Effect of membrane potential on the size of the EPSP in a cat motoneuron. The membrane potential was set at the values shown to the left of each set of records by passing current through one barrel of a double-barrelled intracellular electrode. The other barrel was used to record the membrane potential. The traces at -42 and -60 mV show the initiation of action potentials. (From Coombs *et al.*, 1955c.)

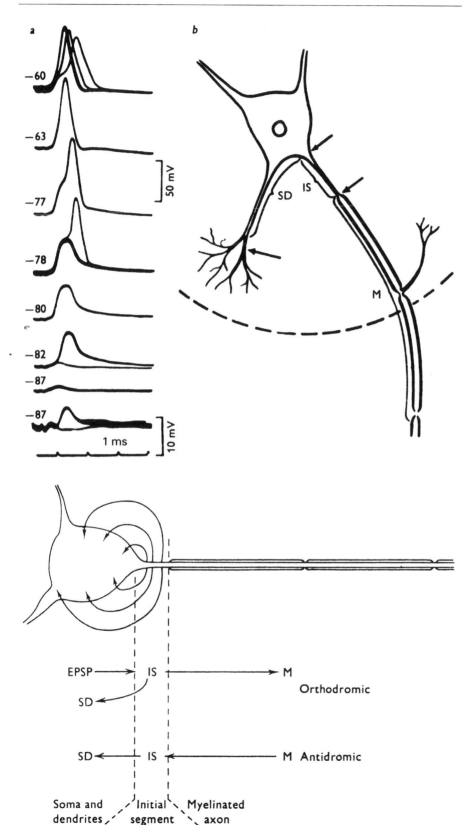

a

−60

−63

−77

−78

−80

−82

−87

−87

50 mV

10 mV

1 ms

b

SD IS

M

Figure 7.24. Analysis of action potentials recorded in the motoneuronal soma following antidromic stimulation. The traces in *a* show the effect of soma membrane potential on the response. The diagram of a motoneuron in *b* shows the probable sites of the SD, IS and M spikes. Below −78 mV the antidromic action potential propagates into the soma and dendrites, so the SD spike is recorded. At −80 mV the hyperpolarization of the soma stops the action potential at the initial segment, and at more negative values it may be stopped in the myelinated region. (From Eccles, 1957.)

Figure 7.25. Spike initiation in a vertebrate spinal motoneuron. The arrows in the diagram show how the initial segment is depolarized by the local circuit currents set up by excitatory synaptic action in the soma and dendrites. Also shown are the sequences of events in orthodromic and antidromic activation.

EPSP ⟶ IS ⟶ M   Orthodromic

SD ⟵

SD ⟵ IS ⟵ M   Antidromic

Soma and dendrites / Initial segment / Myelinated axon

orthodromic excitation (fig. 7.25). The reason for this is that the threshold is very much lower in the initial segment, so that it can be preferentially excited by the depolarization produced by the EPSP. The IS spike then itself excites both the soma and the myelinated axon.

Postsynaptic inhibition in mammalian spinal motoneurons

If the contraction of a particular limb muscle is to be effective in producing movement, it is essential that those muscles which oppose this action, the *antagonists,* should be relaxed. In the monosynaptic stretch reflex systems of mammals, this is brought about by *inhibition* of the motoneurons of the antagonistic muscles, through a system known as the direct inhibitory pathway. Figure 7.26 shows the arrangement of neurons in this pathway. Group Ia afferent fibres from the stretch receptors in a particular muscle (an extensor in fig. 7.26) synapse in the ventral horn with motoneurons innervating that muscle and, to a lesser extent, with its synergists (i.e. other muscles acting in a similar way). These afferents also synapse, in a region called the intermediate nucleus, with small interneurons which themselves innervate the motoneurons of antago-

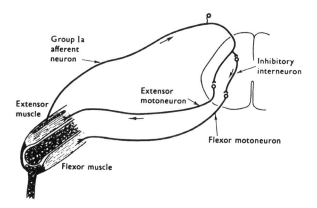

Figure 7.26. The direct inhibitory pathway. This diagram is much simplified, in that many afferent, inhibitory and efferent neurons are involved at each stage. Each inhibitory interneuron innervates several motoneurons, and is itself innervated by several afferents.

nistic muscles. It is these interneurons which exert the inhibitory action on the motoneurons of antagonistic muscles.

This inhibitory action can thus be examined by inserting a microelectrode into a motoneuron and stimulating the group Ia afferents from an antagonistic muscle. Figure 7.27 shows the results of such an experiment: the records were taken from a biceps-semitendinosus (flexor) motoneuron in response to group Ia afferent volleys from the quadriceps muscle (an extensor). The responses consist of small hyperpolarizing potentials known as *inhibitory postsynaptic potentials,* or IPSPs.

Apart from the fact that it is normally hyperpolarizing rather than depolarizing, the shape of the IPSP is similar to the shapes of the EPSP and the EPP. Hence we might expect it to be produced by a similar mechanism, i.e. by the opening of neurotransmitter-gated channels so as to produce a brief increase in the ionic conductance of the membrane, which causes the initial hyperpolarization, followed by a passive decay of the charge on the membrane capacitance. Observation of the effects of the membrane potential on the size of the IPSP provides a useful test of this hypothesis. It is found (Coombs *et al.,* 1955*b*) that the IPSP increases on depolarization, but decreases and then reverses in sign on hyperpolarization (fig. 7.28).

The question now arises to which ions does the postsynaptic membrane alter its permeability during the action of the inhibitory transmitter substance? As we have seen, this type of problem is most readily solved by observing the effects of altering the ionic concentration gradients across the membrane. This was done by Coombs and his colleagues (1955*b*); since it is very difficult to change the external environment of spinal motoneurons, they altered the ionic concentrations inside the cell by injecting various ions through one barrel of a double microelectrode, the other barrel being used to measure membrane potentials. After chloride injection, the reversal potential always moved to a more depolarized level, so that the IPSP at the resting potential became a depolarizing response of increased size (fig. 7.29). This result implies that the

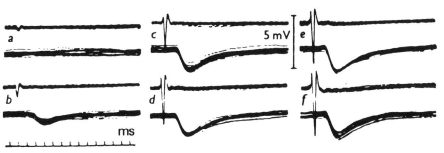

Figure 7.27. Inhibitory responses in a cat motoneuron (posterior biceps-semitendinosus) to afferent volleys from the antagonistic muscle (quadriceps). Stimulus intensity increases from *a* to *f.* The upper trace shows the afferent volley recorded from the dorsal root. (From Coombs *et al.,* 1955*d*.)

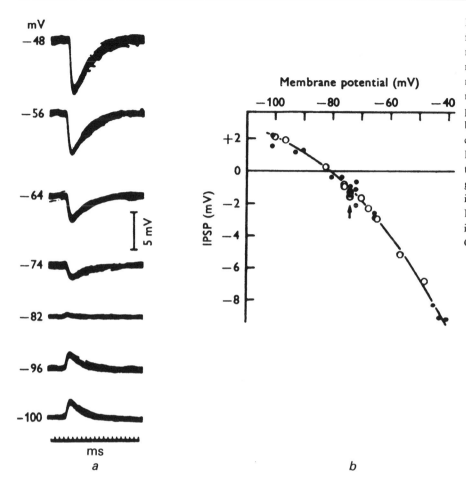

Figure 7.28. The effect of membrane potential on the magnitude of the IPSP in a cat motoneuron. The series of records in *a* show the IPSP recorded at different membrane potentials, using a double-barrelled microelectrode to pass current through the membrane. In *b* these results and others from the same cell are expressed in graphical form; the arrow indicates the resting potential. Notice that the reversal potential is at about −80 mV. (From Coombs *et al.*, 1955*b*.)

Figure 7.29. The effect of increasing the internal chloride concentration on the IPSP of a cat motoneuron. The records were obtained by inserting a micropipette electrode filled with 3 M KCl into the cell. Record *a* was obtained immediately after insertion, records *b* and *c* at successively later times. Notice the change in the IPSP following diffusion of chloride out of the electrode. Record *d* was obtained immediately after *c*, but with the membrane potential set at a lower level (−27 mV, instead of −59 mV). (From Coombs *et al.*, 1955*b*.)

potential at the peak of the IPSP is at least partially determined by the Nernst equation for chloride ions, i.e. the inhibitory transmitter substance (glycine in this case) opens channels in the postsynaptic membrane that are permeable to chloride ions.

Interactions between inhibitory and excitatory postsynaptic potentials

The peak depolarization during an EPSP is reduced if there is an overlap in time with an IPSP. If the EPSP was just large enough to elicit an action potential in the absence of an IPSP, then the IPSP may reduce the membrane potential change during the EPSP so that it no longer crosses the threshold for production of an action potential (fig. 7.30). When the motoneuron is prevented from producing an action potential in this way it cannot induce muscular contraction and so it is effectively inhibited.

There are two components in postsynaptic inhibition. One is caused by the hyperpolarization of the membrane, and lasts for the whole of the IPSP. The other is caused by a 'clamping' of the membrane near to the chloride equilibrium potential; this lasts only for the time that the glycine-gated channels are open during the rising phase of the IPSP. This analysis accords very well with the effect of an inhibitory volley on the monosynaptic response recorded from a ventral root. It is found (fig. 7.31) that the compound action potential in the ventral root is very much reduced in size when the excitatory volley occurs up to 2 ms after the inhibitory volley, but thereafter the degree of inhibition is such as one might expect from the hyperpolarization constituting the IPSP.

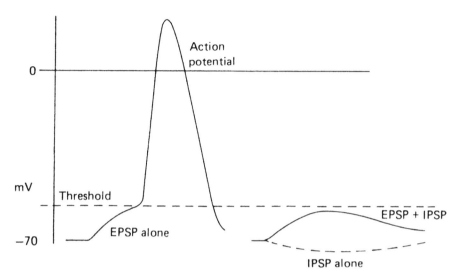

Figure 7.30. Interaction between excitatory and inhibitory PSPs in the motoneuron. The diagram shows an EPSP that is just large enough to cross the threshold for excitation of an action potential. When an IPSP occurs at the same time, the combined result is insufficient to cause excitation, and so no action potential propagates out along the axon to the muscle. (From Keynes & Aidley, 1991.)

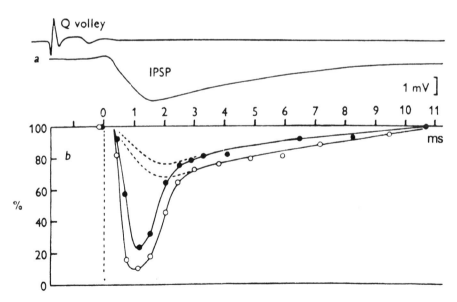

Figure 7.31. The effect of an inhibitory volley on the size of the monosynaptically excited compound action potential recorded in the ventral root. *a* shows the inhibitory (quadriceps, Q) volley recorded in the dorsal root (upper trace), and also the IPSP that this produces in the biceps-semitendinosus motoneuron (lower trace). The time scale of these records is the same as the abscissa of the graph *b*. The ordinate of *b* shows the size of the compound action potential in the ventral root, with the response in the absence of inhibition taken as 100%. The two curves in *b* were obtained from different sizes of afferent volley. The dashed lines indicate the time courses of that part of the inhibitory process which can be attributed directly to the hyperpolarization of the IPSPs. The extra effect in the first 2 ms is coincident with the time that the chloride channels gated by the inhibitory transmitter are open, and demonstrates the extra stabilizing effect that this has. (From Araki *et al.*, 1960.)

The motoneuron is in a sense a decision-making device. The decision to be made is whether or not to 'fire'; that is to say, whether or not to send an action potential out along the axon towards the muscle. The mechanism for this is spatial summation of the incoming excitatory and inhibitory inputs. If the incoming excitatory synaptic action is sufficiently in excess of the incoming inhibitory action, the resulting depolarization will cross the threshold for production of an action potential and the motoneuron will 'fire'. But a reduction in synaptic excitation or an increase in synaptic inhibition will make the membrane potential more negative so that it drops below the threshold and the motoneuron ceases firing. We should remember that the motoneuron receives excitatory and inhibitory inputs from many sources, so that, for example, a 'decision' based on inhibition from group Ia fibres from an antagonistic muscle may be 'overruled' by excitatory inputs from neurons descending from the brain.

8
Neurotransmitter-gated channels

The fast synaptic responses that we looked at in the previous chapter are mediated by ion channels that open when they bind to the neurotransmitter active at the synapse. Thus transmission at the vertebrate neuromuscular junction or electric organ is mediated by acetylcholine-gated channels, excitatory transmission at spinal motoneurons is mediated by glutamate-gated channels, and inhibitory transmission there is mediated by glycine-gated or GABA-gated channels. These channels are also *receptors* since they are each specifically activated by their own particular neurotransmitter.

Structure of the nicotinic acetylcholine receptor

There are two main types of acetylcholine receptor. The type we have met previously, which mediates fast synaptic transmission at vertebrate neuromuscular junctions and electric organs, is known as the *nicotinic* acetylcholine receptor, or nAChR for short. The second type, known as the *muscarinic* acetylcholine receptor, is involved in some slow synaptic transmission processes and contains no ion channel. Nicotinic receptors are activated by nicotine and blocked by curare, whereas muscarinic receptors are activated by muscarine and blocked by atropine. We will look at muscarinic receptors in chapter 9.

The richest known source of nAChRs is the electric organ of the electric ray *Torpedo*, 1 kg of which contains over 100 mg of the receptor protein. Lower concentrations are present in the electric organ of *Electrophorus*, the electric eel, and in vertebrate muscle fibres. Acetylcholine receptors have been isolated from all these sources, but the *Torpedo* electric organ remains the biochemists' favourite.

In order to isolate receptors from *Torpedo* electric organ it is first necessary to homogenize the tissue. Then fractionation by centrifugation in a sucrose density gradient separates the receptor-rich subsynaptic membranes, which are relatively dense. Further purification leads to small vesicles in which the receptors are embedded outside-out in the vesicular membrane. Treatment with detergents releases the receptors from the membrane components, and they can then be further purified by affinity chromatography on α-bungarotoxin columns.

The molecular mass of the acetylcholine receptor is in the region of 285 to 290 kDa. Electrophoresis on poly-acrylamide-sodium dodecyl sulphate (SDS) gels shows that the receptor contains four different polypeptide chains, with apparent molecular masses of 40, 50, 60 and 65 kDa, known respectively as the α, β, γ and δ subunits. There are two α chains and one of each of the others in each receptor. The binding sites for acetylcholine are located on the α chains, suggesting that there are two binding sites per receptor (see Raftery *et al.*, 1980; Popot & Changeux, 1984; Changeux, 1995).

Primary structure

Knowledge of the amino acid sequence of the protein chains is clearly necessary before we can understand how the receptor works. Raftery and his colleagues (1980) determined the first fifty-four amino acid residues in each of the four subunits. They found that each subunit was distinct from the others, but that there was considerable sequence identity between them: if any two subunits were compared then 35% or more of the residues in corresponding positions were the same.

The next step was to determine the full amino acid sequences of the receptor subunits. Direct analysis – working along the protein chain and identifying each amino acid residue in turn – is a difficult and laborious business. With the advent of DNA cloning techniques it became much easier to obtain the sequence of bases in the nucleic acids coding for the protein and then to deduce the amino acid sequence from this, than to determine the amino acid sequence directly by analysis of the protein. Laboratories in Japan, Britain, France and the United States attacked the problem more or less simultaneously (Noda *et al.*, 1982*a*; Sumikawa *et al.*, 1982; Claudio *et al.*, 1983; Devillers-Thiéry *et al.*, 1983). Work by the Kyoto University group was particularly productive, so that they had soon published the amino acid sequences of all four subunits (Noda *et al.*, 1982*a*, 1983*a*,*b*; Numa *et al.*, 1983; see also Numa, 1986). Let us have a brief look at how this was done; further details of recombinant DNA methods are given by Watson *et al.* (1992), among others.

The normal sequence of information transfer in protein synthesis by cells can be stated as 'DNA makes RNA makes protein'. The genetic information in the appropriate section of the nucleotide base sequence of the DNA in the nucleus is first transcribed to form a complementary sequence of RNA bases on a primary transcript. This is then modified to give messenger RNA (mRNA), which in turn is translated into the amino acid sequence forming the protein. The first step in the analysis of the nAChR, therefore, was to isolate mRNA from the electric organ tissue. The extract was a mixture of different mRNAs, less than 1% being of the desired type.

Next, a complementary DNA (cDNA) strand was formed on each mRNA molecule by the action of reverse

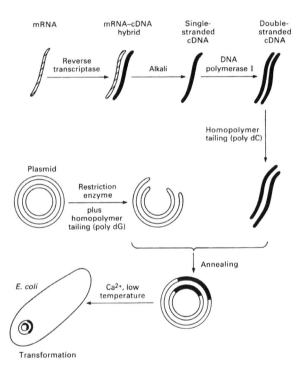

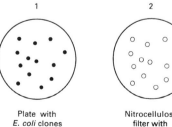

Figure 8.1. An outline of the method used to produce clones of cDNA in *E. coli* from *Torpedo* electric organ mRNA.

transcriptase, an enzyme found in retroviruses. The mRNA was removed with alkali and the cDNA made double-stranded with the enzyme DNA polymerase I from the bacterium *Escherichia coli*. The cDNA molecules were inserted into plasmids (loops of double-stranded DNA which self-replicate in bacteria) that contained genes for antibiotic resistance, and the plasmids were taken up by *E. coli* cells (fig. 8.1). The bacteria were then plated on a medium containing antibiotic, which ensured that only those cells containing plasmids would grow. Thus a large number of small colonies of bacteria were produced, each one being a clone which contained many copies of its own particular piece of the cDNA. In their determination of the primary structure of the α subunit, Noda and his colleagues (1982) used 2.3 µg of mRNA to produce a library of about 200 000 clones of cDNA.

Only a few of these clones would have contained cDNA coding for the desired protein, so there was a need to select these particular ones out of the mass, using a probe which had something in common with the sequence that was sought. Noda and his colleagues synthesized small pieces of DNA corresponding to the amino acid residues 25 to 29 and 13 to 18 of the α subunit in the sequence determined by Raftery *et al.* (1980), and labelled them with radioactive phosphorus. These probes would hybridize only with complementary sequences in the DNA library. A nitrocellulose filter was pressed onto each plate of *E. coli* clones, bringing samples of the clones away with it on removal (fig. 8.2). The samples were lysed and heated to expose single-stranded cDNA and the radioactive probes were applied to them: binding occurred only with complementary sequences in the cDNA, and excess probes could be washed away. The position of the probes on the filter was then detected by autoradiography; this indicated which clones contained the required cDNA and so they could be removed from the original plate and cultured so as to produce sufficient cDNA for sequencing (a few nanograms is enough). Noda and his colleagues found twenty clones that would

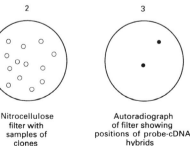

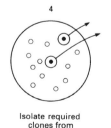

Figure 8.2. Screening of cDNA clones from *Torpedo* electric organ for the nicotinic acetylcholine receptor α subunit. For further details, see the text.

hybridize with both their probes, and they picked the two largest of these for sequencing.

DNA can be cut by restriction endonucleases, bacterial enzymes that act only at particular sites. Using a number of different endonucleases it is possible to cut the cDNA into a series of smaller units whose position on the DNA strand is approximately known. These pieces can themselves be cloned using a bacteriophage vector in *E. coli* cells. The nucleotide sequences of these various pieces are then determined by one of the standard methods (Maxam & Gilbert, 1980). Then by matching up the sequences of the different pieces the nucleotide sequence of the whole cDNA strand can be worked out.

The results of this endeavour are shown in fig. 8.3, which gives the base sequence of the mRNA coding for the α subunit of the *Torpedo* electric organ acetylcholine receptor, and the amino acid sequence of the protein deduced from this. In protein synthesis the mRNA is read from the 5′-terminal end until the sequence AUG is reached: this indicates the beginning of the protein chain and also codes for the amino acid methionine. From then on successive base triplets act as codons for the various amino acids in the sequence until a stop codon (UAA in this case) is reached, when the protein chain is complete.

The protein chain of the α subunit, as determined by amino acid analysis by Raftery *et al.* (1980), begins with the sequence Ser-Glu-His-Glu-Thr, which does not occur until the twenty-fifth residue in fig. 8.3. The first twenty-four residues probably form a 'signal sequence' which serves to start the insertion of the protein chain into the membrane and is removed after it has reached its final position. Hence in numbering the residues the first twenty-four are given negative numbers. The mature protein thus contains 437 residues with serine at the N-terminus (position 1) and glycine at the C-terminus. It has a calculated M_r (neglecting glycosylation and other modifications of the amino acid chain) of 50 116.

The mature β, γ and δ subunits contain 469, 489 and 501 amino acid residues, with calculated M_r values of 53 681, 56 279 and 57 565, respectively. Their sequences show marked similarity to that of the α subunit. This suggests that the four subunits are similar in structure and that the genes encoding for them are descended from a single common ancestor (Noda *et al.*, 1983*a*).

Transmembrane topology

The hydrophobicity profiles for each subunit of the nAChR are very similar to each other. They show four regions of eighteen or more residues which look as though they would form hydrophobic α-helices (fig. 8.4). These segments,

named M1 to M4, are long enough to cross the membrane and it seems highly likely that they do just that.

Figure 8.5*a* shows the probable transmembrane topology of the receptor. The long section from the N-terminus to the beginning of M1 is probably all on the outside of the cell. It contains sites for glycosylation (attachment of sugars) and disulphide crosslinking. In the α subunit it contains the α-bungarotoxin- and acetylcholine-binding sites. Then, proceeding along the chain, we have the M1, M2 and M3 putative membrane-crossing segments, with short connecting sections between them. There is a fairly extensive cytoplasmic section between the M3 and M4 segments, which contains sites for the attachment of phosphate groups. Finally there is a short extracellular section finishing at the C-terminus. Transmembrane topologies different from that shown in fig. 8.5*a* have been proposed from time to time, but the weight of the evidence is in its favour.

Particularly good evidence for the consensus model has been provided by Chavez & Hall (1992). They made fusion proteins in which a fragment of the hormone prolactin replaced various C-terminal lengths of nAChR subunits, and expressed them in pancreas microsomes using an *in vitro* translation system. The microsomes are made by homogenizing pancreatic secretory cells and are vesicles formed from endoplasmic reticulum membrane; the lumen (inside) of the microsome is topologically equivalent to the extracellular space. Thus in some fusion proteins the prolactin section would be on the cytoplasmic face of the membrane, and accessible to a proteolytic enzyme, whereas in others it would be inside the microsome and so inaccessible to proteolysis (see fig. 8.5). Recovery of a polypeptide containing the prolactin fragment (identified by immunoprecipitation with specific antibodies) would thus show that it had not been attacked by the enzyme and therefore that the point of attachment of the prolactin fragment was on the extracellular face of the membrane. The method is sometimes called an epitope protection assay, since the prolactin fragment acts as an epitope (a recognition and binding region) for the antibody.

The results of this experiment were quite clear. If the prolactin domain were attached just after the M2 or M4 segments, it could be recovered, whereas if it were attached just after the M1 or M3 segments, it could not. So the model shown in fig. 8.5*a* really does seem to be the correct one.

There is strong evidence that the M2 α-helix lines the channel pore at its narrowest. It is possible to label the nAChR with radioactive molecules of the non-competitive blocking agent chlorpromazine, whose action appears to be due to binding to the walls of the channel pore. The label

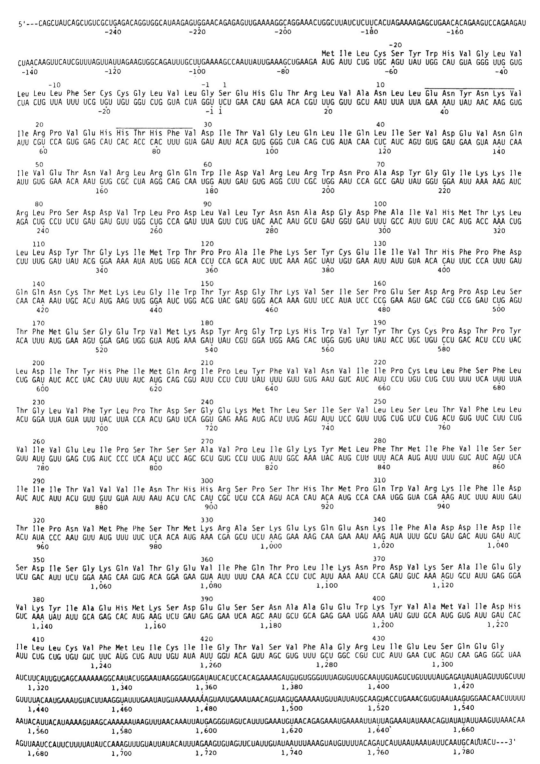

Figure 8.3. Base sequence of the mRNA coding for the α subunit of the nicotinic acetylcholine receptor in the electric organ of *Torpedo californica*, and the amino acid sequence predicted from this. Numbering of the amino acid sequence starts at the first residue (serine) of the mature protein. (From Noda *et al.*, 1982a.)

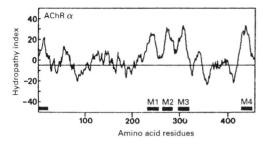

Figure 8.4. Hydropathy profile of the nicotinic acetylcholine receptor α subunit. The horizontal axis shows the position of amino acid residue i in the chain, the vertical axis shows the sum of the hydropathy indices (table 3.1) from $i − 8$ to $i + 8$. Black bars show the positions of the signal sequence at the left and the four putative membrane-crossing segments (M1–M4). (From Schofield *et al.*, 1987. Reprinted with permission from *Nature* **328**, 223. Copyright 1987 Macmillan Magazines Limited.)

becomes attached to certain amino acids in the M2 segments, suggesting that they line the pore (Revah *et al.*, 1990; see Changeux *et al.*, 1992). Point mutations of amino acid residues in the M2 sequence can produce marked effects on the permeability of the ion channel, as we shall see later.

Quaternary structure

Postsynaptic membranes of *Torpedo* electric organ show arrays of receptors which seem to consist of a number of subunits arranged round a central pit (Cartaud *et al.*, 1978; fig. 7.13). The $\alpha_2\beta\gamma\delta$ structure suggests that there are five subunits and that the receptor should therefore show an approach to pentameric symmetry.

The structure of biological macromolecules can be determined by X-ray diffraction if three-dimensional crystals of sufficient size can be prepared. This has been done for haemoglobin, lysozyme and a number of other proteins. Two-dimensional (2-D) arrays may have a sufficient degree of order in them for analogous methods to be used. Electron micrographs of such '2-D crystals' can be subjected either to optical diffraction or to computerized image analysis. In order to obtain some idea of the 3-D structure of the molecules, the crystalline array must be viewed at different angles in the electron microscope, and the image digitized and subjected to Fourier analysis. Details of the process are given by Amos *et al.* (1982).

This technique has been applied to nAChRs from *Torpedo* electric organ using artificial tubular crystalline arrays of receptors embedded in ice (Brisson & Unwin, 1985; Toyoshima & Unwin, 1988; Unwin, 1993, 1995). The results are in the form of 3-D contour maps of image density, as is shown in fig. 8.6. Maps parallel to the plane of the cell membrane show that the receptor is a pentameric

structure with its axis of pseudosymmetry at right angles to that plane. Clearly the five subunits form a rosette embedded in the membrane with the channel pore passing down the middle.

Much of the mass of the receptor is clearly on the outside of the cell, projecting into the synaptic cleft. This outer part is hollow, with a central entrance tube 20 to 25 Å in diameter, leading into a transmembrane pore that is too narrow to be resolved by the image analysis method. There is a dense rod in each subunit very close to the wall of the transmembrane pore. This must be the M2 segment. The maps show that it is kinked at its middle, and the kink apparently forms the narrowest part of the pore..

The other material at the level of the bilayer, as displayed by the imaging technique, does not appear to be in the form of α-helices, which would appear as dense rods in the contour maps. Unwin suggests that the M1, M3 and M4 segments form a β-barrel structure of antiparallel strands holding the whole receptor together. There are, however, other possibilities (Popot, 1993; Hucho *et al.*, 1994).

A large part of the nAChR projects into the synaptic cleft on the outside of the cell. It is made up of the long N-terminal regions of all five subunits, and must contain the two acetylcholine-binding sites in the α subunits. The high resolution image analysis revealed three dense rods in the synaptic part of each subunit, and these are presumably α-helices (Unwin, 1993*a*). It seems likely that in the α subunits they are involved in the acetylcholine-binding site.

We can regard the nAChR as an allosteric protein, one that changes its shape in response to combination of acetylcholine molecules with its α subunits (see Changeux, 1995). Further work by Unwin (1995) has provided some remarkable detail on how this happens. He did this by comparing the structures of the receptors before and after they bound acetylcholine; to eliminate the problem of desensitization he applied acetylcholine to the receptors by means of an atomizer spray just 5 ms before freezing them in liquid ethane at − 178 °C. This timing ensured that all the receptors would be frozen with their channels open.

Unwin found that activation produces movement of the dense rods surrounding the binding sites in the two α subunits. There was also some clockwise twisting of both the α subunits by about 4°, seen from the outer side, and a displacement of the subunit between them, probably the γ subunit. It seems likely that this is the means whereby changes at the binding sites produce effects 30 to 50 Å away at the membrane level.

At rest in the closed state the kinks in the five M2 α-helices form the narrowest part of the pore, and it seems likely that the conserved leucine residues (L251 in the α subunits) at the kink form a hydrophobic ring which acts as a block to any ion movement through the pore. Activation leads to a marked change in the position of these helices: the kink is withdrawn from the axis of the pore and the 'lower' halves (the halves in the inner leaflet of the bilayer) of the helices swing round to become much more tangential to the pore, as is shown in fig. 8.7. This removes the large hydrophobic leucine residues from near the axis of the pore and replaces them with a line of smaller polar residues.

The narrowest part of the pore is now a ring of threonine residues (T244 in the α subunits) at the lower ends of the M2 helices, with a diameter of 9 to 10 Å.

Expression in the oocyte membrane

We have seen in chapter 6 that oocytes from the African clawed toad *Xenopus laevis* are very useful in investigating the properties of various membrane proteins. Messenger RNA coding for the protein is injected into the oocyte, and

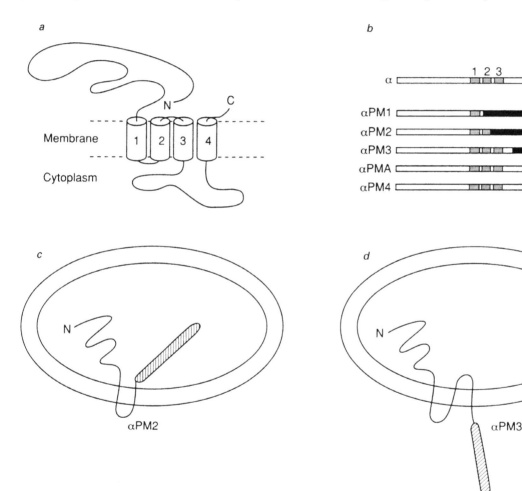

Figure 8.5. The epitope protection method for determining transmembrane topology of the nicotinic acetylcholine receptor. *a* shows the generally accepted transmembrane topology as predicted by the hydrophobicity analysis. *b* shows the fusion proteins, in which a prolactin fragment (black) is attached downstream from part of the α subunit chain, just after the four putative membrane-crossing segments M1 to M4 and also after the amphipathic MA segment. These were inserted into pancreatic microsomes, in which the inside of the microsome corresponds to the extracellular face of the channel; *c* shows the expected topology for αPM2 and *d* shows that for αPM3, with the prolactin fragment cross-hatched. So a proteolytic enzyme would remove the prolactin fragment from αPM3 but not from αPM2, indicating that the C-terminal end of M3 is cytoplasmic whereas that of M2 is extracellular. In the experiments, prolactin fragments were still present after proteinase digestion for the αPM2 and αPM4 fusion proteins, but not for αPM1, αPM3 and αPMA. These results (and similar ones for the δ subunit) show that the model in *a* is correct. (From Aidley & Stanfield, 1996, based on Chavez & Hall, 1992.)

then the membrane can be tested for protein activity a few days later. Figure 8.8 shows the results of one experiment with mRNA coding for nAChRs from *Torpedo* electric organ (Barnard *et al.*, 1982). Two days after injecting the mRNA, the oocyte responds to ionophoretic application of acetylcholine by a smooth depolarization that begins very soon after the start of the electrophoresis current.

The *Xenopus* oocyte system has been used extensively to look at proteins produced by specific products of gene cloning. Thus Mishina *et al.* (1984) injected mRNAs coding for the four different subunits into oocytes in

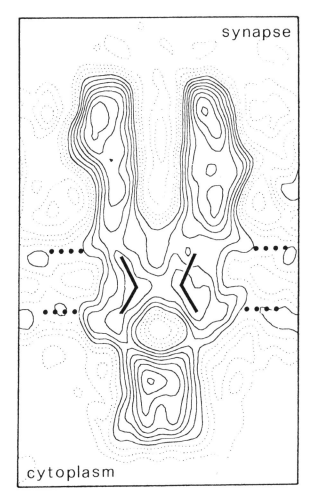

Figure 8.6. Quaternary structure of the nicotinic acetylcholine receptor as determined by Fourier transform analysis from diffraction patterns of multiple electron microscope images. Continuous contours show electron density of the receptor. Large dots show the limits of the membrane bilayer, 30 Å apart. The bent transmembrane rods, thought to be the M2 segment, are shown. The square structure at the bottom is a protein attached to the receptor but not part of it. (From Unwin 1993*a*.)

different combinations. Starting from their clones of *E. coli* containing plasmids with the cDNA coding for one of the subunits, they constructed an artificial recombinant plasmid which was capable of entering a cell line of tissue-cultured monkey cells and which also contained the receptor subunit cDNA. Following this process of *transfection*, the cultured cells produced relatively large quantities of the mRNA coding for the receptor subunit. In this way, mRNAs specific for the four different subunits were obtained, and they could then be injected into oocytes in various combinations.

The results were clear-cut. After injection of mRNAs specific for all four subunits, the oocytes produced appreciable inward currents when acetylcholine was applied by ionophoresis in the presence of atropine. With only three subunits, there was normally no response, or sometimes a small response when either the γ or δ subunit was absent. This indicates that all four subunits are required for normal acetylcholine receptor function. In contrast, only the α subunit was required for α-bungarotoxin binding.

Single channel currents can be measured by applying the patch clamp technique to injected oocytes. Sakmann and his colleagues (1985) used this method to look at the roles of the different acetylcholine receptor subunits in ion transport and gating. Messenger RNAs from the α, β, γ and δ subunits were injected in the proportions 2:1:1:1. Single-channel currents from bovine (cow) receptors were of longer average duration than those from *Torpedo* receptors. Then the group produced hybrid receptors by injecting oocytes with bovine mRNA for one of the subunits and *Torpedo* mRNA for the others. They found that the hybrid with the bovine δ chain had a long average opening time like that of the whole bovine receptor; the corresponding α hybrid had an intermediate average opening time. Thus it seems that the δ chain is particularly important in determining the duration of channel opening.

The next step was to use genetic manipulation techniques to construct chimaeric δ subunit cDNAs which contained partly the *Torpedo* sequence and partly the bovine sequence (Imoto *et al.*, 1986). These were used to make chimaeric δ subunit mRNAs which were then injected into oocytes together with mRNAs for the other *Torpedo* subunits. At low divalent cation concentrations (less than 1 mM) the single-channel conductance for *Torpedo* receptors is higher than that for bovine ones (mean values were 87 and 62 pS, respectively). With a bovine δ chain and the other chains from *Torpedo*, the lower value was observed. With chimaeric chains, single channel conductances were either in the higher (*Torpedo*) or lower (bovine) ranges. The crucial part of the molecule seemed to be the M2 helix and the external sequence connecting the M2 and M3 helices: low conductances were found when this section was bovine and high ones when it was from *Torpedo*.

Receptors at different sites

The amino acid sequences of corresponding subunits of nAChRs from related animals show a high degree of similarity or identity (Numa, 1986). The α subunit chains from humans and calves, for example, show 97% identity (i.e. the same amino acid residue occurs in both chains at 97% of the positions), whereas those from humans and *Torpedo* show only 80% identity. This reflects the fact that the evolutionary divergence between humans and cows is much more recent than that between humans and cartilaginous fish. Corresponding figures for the γ subunit are 92% and 55%, and the figures for the β and δ subunits are similar, suggesting that evolutionary changes in the α subunit have been slower than in the other subunits. The presence of the acetylcholine-binding site must make the α subunit less tolerant of mutational changes than the other subunits.

The different subunits in one species differ from each other more than do the same subunits in different species. The *Torpedo* α and β subunits, for example, show only 42% identity. This suggests that they diverged (probably as a result of gene duplication) at an early stage in the evolutionary history of animals. A rather different subunit, the ε (epsilon), occurs in adult mammals; it takes the place of the γ subunit, which is present in foetal mammals. It produces single-channel currents that are somewhat larger and on average briefer than those of foetal mammals, as is shown in fig. 8.9 (Takai *et al.*, 1985; Mishina *et al.*, 1986).

Nicotinic acetylcholine receptors occur in the nervous system as well as in electric organs and muscle (Sargent, 1993). These neuronal nAChRs seem to have just two general types of subunit: those which bind acetylcholine, called α subunits, and those which do not, called non-α or β subunits. The distinction is usually made on the basis of the presence of adjacent cysteines a few residues before the start of the M1 segment, corresponding to cysteine 192 and cysteine 193 in the *Torpedo* α subunit. These paired cysteines are assumed to be diagnostic of the acetylcholine-binding site.

Neuronal nAChR sequences have been found by probing either cDNA libraries made from brain mRNAs or libraries of genomic DNA. At least seven different α subunits exist (named α2 to α8, with α1 as the corresponding muscle subunit), and at least three different β subunits. Messenger RNAs coding for these can be localized in sections of nervous tissue by using autoradiography after *in*

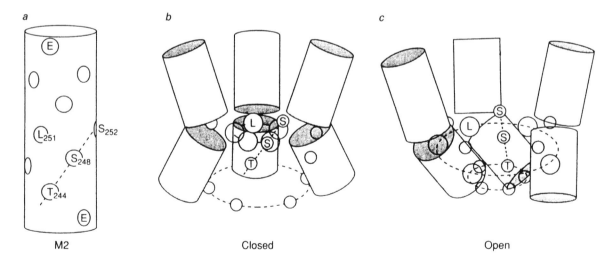

Figure 8.7. Movement of the pore-lining M2 segment of the nAChR during gating, as deduced from image processing of arrays of receptors from *Torpedo* electric organ fast-frozen before and immediately after application of acetylcholine. *a* shows pore-facing amino acid residues on an α subunit M2 α-helix; corresponding residues are found on the other subunits. In the closed condition (*b*) the five M2 segments are kinked so that their leucine residues (L251 and corresponding positions) line the pore at its narrowest part. On opening (*c*), the M2 segments move so as to withdraw the leucine residues and line the 'lower' (inner) half of the pore with a row of smaller polar residues; the lowest of these (T244 and corresponding ones) now form the narrowest part of the open pore. (From Unwin, 1995. Reprinted with permission from *Nature* **373**, 42. Copyright 1995 Macmillan Magazines Limited.)

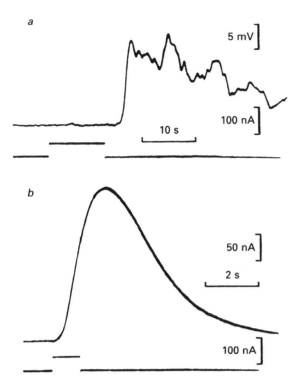

Figure 8.8. Responses of *Xenopus* oocytes to acetylcholine applied by ionophoresis. Trace *a* shows the muscarinic response, a delayed and rather irregular depolarization in an oocyte which had been injected with a little water as a control. Trace *b* is a voltage clamp record of the nicotinic response, a relatively rapid and smooth inward current, in an oocyte which had been injected with *Torpedo* electric organ mRNA two days previously. (From Barnard *et al.*, 1982.)

situ hybridization with appropriate radioactive antisense oligonucleotide probes; different subunits have different distributions in the brain (Wada *et al.*, 1988). It seems likely that neuronal receptors are usually made up of two α and three β subunits (Cooper *et al.*, 1991), but α7 can form functional homo-oligomer channels just on its own (Couturier *et al.*, 1990). A further subunit, α9, forms homomeric receptors which appear to mediate the cholinergic efferent nerve action on the sensory hair cells of the cochlea (Elgoyhen *et al.*, 1994).

Permeability and selectivity of the nAChR channel

The experiments of the Takeuchis, referred to in chapter 7, showed that the nAChR channel was selectively permeable to cations and that both sodium and potassium ions could pass through. Later work confirmed these conclusions and extended them to calcium and magnesium ions (Takeuchi, 1963*a,b*; Linder & Quastel, 1978; Lewis, 1979).

Some organic cations can also pass through the acetylcholine channel, and this opens the possibility of investigating its dimensions. Hille and his colleagues (Dwyer *et al.*, 1980) measured the permeability of frog muscle end-plate channels to a wide variety of organic cations with this end in view. They used frog muscle fibres with their ends cut and open to sodium fluoride solutions so that sodium was the principal internal cation. The end-plate region was in a separate pool isolated with vaseline (petroleum jelly) from the cut ends; it was perfused with an external solution containing either 114 mM NaCl or 114 mM XCl, where X$^+$ is the organic cation. Acetylcholine was applied to the end-plate via a micropipette by ionophoresis. A voltage clamp circuit held the membrane potential at various desired levels.

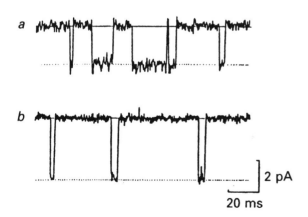

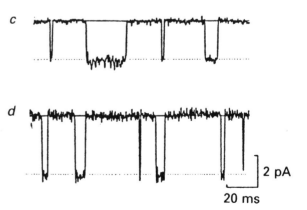

Figure 8.9. Responses of foetal and adult acetylcholine receptors of bovine muscle. The traces on the left show single-channel currents from (*a*) foetal and (*b*) adult muscle. Those on the right show currents from *Xenopus* oocytes which had been previously injected with mRNA coding for (*c*) the α, β, γ and δ subunits or (*d*) the α, β, δ and ε subunits. Clearly the difference between foetal and adult muscle receptors lies in the substitution of the ε for the γ subunit. (From Mishina *et al.*, 1986.)

By applying acetylcholine at different membrane potentials it is easy to measure the reversal potential. In fig. 8.10, for example, the reversal potential for sodium is $+4$ mV and that for Tris is -39 mV, a difference of -43 mV. Let us see how such measurements can tell us something about the permeabilities of the channel to different ions.

Reversal potentials for currents flowing through the nAChR channels will be given by appropriate use of the Goldman–Hodgkin–Katz voltage equation, 3.7, with anions omitted, since we know the channels are cation selective. With X^+ as the cation in the external solution, the reversal potential for acetylcholine-induced currents is given by

$$E_{r,X} = \frac{RT}{F} \ln \frac{P_X[X]_o}{P_{Na}[Na]_i}$$

where P_{Na} and P_X are permeability coefficients as in equation 3.7. When sodium is the external cation, the reversal potential is given by

$$E_{r,Na} = \frac{RT}{F} \ln \frac{P_{Na}[Na]_o}{P_{Na}[Na]_i}$$

Hence the change in reversal potential on substituting X^+ for sodium ions is

$$\Delta E_r = E_{r,X} - E_{r,Na}$$
$$= \frac{RT}{F} \ln \frac{P_X[X]_o}{P_{Na}[Na]_o}$$

Thus, if $[X]_o$ is equal to $[Na]_o$,

$$\Delta E_r = 24.66 \ln (P_X / P_{Na}) \text{ in mV}$$

or

$$P_X/P_{Na} = \exp (\Delta E_r / 24.66)$$

If we apply this equation to our value of ΔE_r for Tris, for example, we get

$$P_{Tris}/P_{Na} = \exp (-43 / 24.66)$$
$$= 0.175$$

Dwyer and his colleagues measured the values of P_X/P_{Na} for a large number of organic cations, with the results shown by the open circles in fig. 8.11. Cations with permeability coefficients greater than P_{Na} were all smaller than 5 Å in diameter, and there was a fairly steady drop in permeability with increasing diameter, reaching zero at about 7 Å. They concluded that the nAChR channel pore must have a cross-sectional area of at least 40 Å^2 and suggested that a square 6.5 Å $\times$ 6.5 Å, as in fig. 8.12, would provide a suitable model. Since we now know that the acetylcholine receptor is composed of five subunits, a pentagonal cross-section might be more appropriate; a regular pentagon 7 Å across would have an area of 41 Å^2.

Hille and his colleagues also measured the relative permeabilities of the end-plate membrane to various metal ions, using essentially the same techniques (Adams *et al.*, 1980). The results are shown as the black symbols in fig. 8.11. Permeability ratios for the alkali metal ions fell in the sequence Cs $>$ Rb $>$ K $>$ Na $>$ Li from $P_{Cs}/P_{Na} = 1.42$ to $P_{Li}/P_{Na} = 0.87$. This order is the same as that of the mobil-

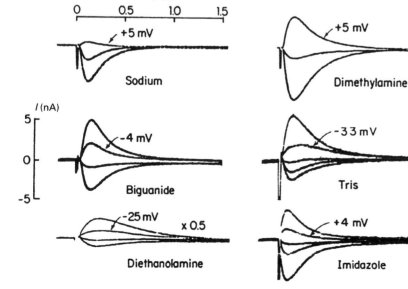

Figure 8.10. Reversal potential measurements at the neuromuscular junction with organic cations substituted for external sodium. The records show currents under voltage clamp in response to brief pulses of acetylcholine at different membrane potentials, separated by 7 mV steps. Reversal potentials (i.e. the potential at which the membrane current is zero) can be estimated by interpolation. (From Dwyer *et al.*, 1980. Reproduced from the *Journal of General Physiology*, by copyright permission of The Rockefeller University Press.)

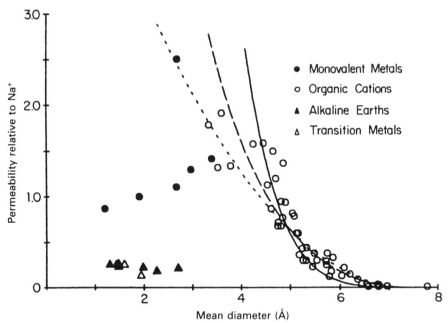

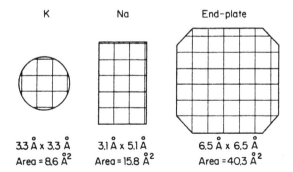

Figure 8.11. Relations between ionic diameter and the relative permeability of nicotinic acetylcholine receptor channels at the frog muscle end-plate. The three curves represent different theoretical models, all of them assuming a cylindrical pore with a diameter of 7.4 Å. (From Adams *et al.*, 1980. Reproduced from the *Journal of General Physiology*, by copyright permission of The Rockefeller University Press.)

Figure 8.12. Hypothetical cross-sections of three types of ion channel in frog nerve and muscle, based on their permeabilities to ions of different sizes. The voltage-gated sodium and potassium (delayed rectifier) channels occur in nerve axons at the nodes of Ranvier. (From Dwyer *et al.*, 1980. Reproduced from the *Journal of General Physiology*, by copyright permission of The Rockefeller University Press.)

and the relatively high values for single-channel conductance suggest that the ion channel is a water-filled pore. The reduction of divalent cation permeability at higher concentrations might be explained if there are negative charges on the external surface near the mouth of the channel (Lewis, 1979; Dwyer *et al.*, 1980). Such charges could also account for the impermeability of the channel to anions.

The molecular basis of selectivity

Can we relate the selectivity of the nAChR channel for cations to its molecular structure? We have seen that the membrane-crossing pore of the channel is very probably lined by the M2 α-helices of its five subunits. Their amino acid sequences are very similar to one another, so the pore is lined by rings of similar or identical residues, occurring at every third or fourth position along the length of the α-helices (figs. 8.13 and 8.14). Three rings of negatively charged residues seem to be particularly important in determining the selectivity of the channel. There is one on the extracellular end of the M2 region at position 20' in fig. 8.14, another one (the 'intermediate ring') just at the inner end (position −1') of the M2 segment, and a third on the cytoplasmic side of the pore at position −4'.

Imoto and his colleagues used point mutations to look at the importance of these rings for channel permeability (Imoto *et al.*, 1988). Reduction of the negative charge in any of these rings causes a corresponding reduction in

ities of the ions in free solution, and so reflects the larger numbers of water molecules surrounding the ions with the smaller atomic masses. The thallous ion has an unexpectedly high P_{Tl}/P_{Na} ratio of 2.51.

The permeabilities for divalent cations are lower, and decrease in the order Mg > Ca > Ba > Sr. They are affected by the concentration of the ion; P_{Ca}/P_{Na} is 0.22 at 20 mM $CaCl_2$ but only 0.16 at 80 mM.

The relatively low selectivity between different metal ions

single-channel conductance. The effect is greater for changes in the intermediate ring, and furthermore changes here produce changes in the relative permeabilities to different cations (Konno *et al.*, 1991). This suggests that the intermediate ring may be part of the selectivity filter. Perhaps the extracellular and cytoplasmic rings serve to attract cations and repel anions.

Point mutations in another ring (the 'central ring' or 'hydroxyl ring', at position 2′) also affect single channel conductance (Imoto *et al.*, 1991; Villaroel & Sakmann, 1992). This central ring contains serine or threonine residues, both relatively small residues whose hydroxyl groups make them relatively polar. It also may be part of the selectivity filter. We have seen earlier that Unwin (1995) suggests that a ring of leucine residues near the middle of the pore (position 9′) is the narrowest part of the pore in the closed channel and so acts as a hydrophobic barrier to ion flow, but that it moves out of the way when the channel opens. This movement would leave the central ring at 2′ as the narrowest part of the pore, so presumably it would form the selectivity filter.

Mutations in the M2 region of the neuronal α7 nAChR can change the channel from cation selective to anion selective (Galzi *et al.*, 1992). The crucial changes are the introduction of a proline or alanine residue between −2′ and −1′, the replacement of the negatively-charged glutamate residue at −1′ by the neutral alanine, and the replacement of the hydrophobic valine residue at 13′ by the polar threonine. Surprisingly, it was not essential to remove the negative charges at −4′ and 20′ in order to make the channel anion-selective.

Kinetics of nAChR channel state changes

Channels can exist in different states. We may distinguish two different *conductive states*, open or closed. A simple channel will be either open or closed at any instant, not half-open (we ignore for the moment the existence of sub-conductance states). Within each conductive state there may be different *conformational states*, in which the shape or behaviour of the molecule is modified. *Kinetics* is the study of these different states and the rates at which the channel changes from one state to another.

Let us start with some simple assumptions. del Castillo & Katz (1957) suggested that receptor activation is a two-stage process in which the acetylcholine (A) first combines with the receptor (R) to form a complex AR and that this then undergoes a conformational change which opens the ionic channel. The channel-open state is represented by AR*. We assume that these reactions are reversible and are governed by different rate constants (k_1, k_{-1}, α and β), to give an overall reaction scheme

$$A + R \underset{k_{-1}}{\overset{k_1}{\rightleftharpoons}} AR \underset{\beta}{\overset{\alpha}{\rightleftharpoons}} AR* \qquad (8.1)$$

closed   closed   open

		-4′	-1′	2′		9′	13′		20′
nAChR	α1	D S G -	E K M T L S I S V L L	S L T V F L L V I V E
	β	D A G -	E K M G L S I F A L L	T L T V F L L L L A D
	γ	K A G G Q	K C T V A T N V L L A Q	T V F L L L L A D
	δ	D C G -	E K T S V A I S V L L A Q	S V F L L L I S K
	α7	D S G -	E K I S L G I T V L L	S L T V F M L L V A E
5HT₃R		D S G -	E R V S F K I T L L L	G Y S V F L I I V S D
GABA_AR	α	E S V P A R T V F G V T T	V L T M T T L S I S A R N
GlyR		D A A P A R V G L G I Y T	V L Y M T T Q S S G S R A

Figure 8.14. Amino acid sequences of the M2 segments and adjacent regions in subunits of the family of neurotransmitter-gated channels related to the nicotinic acetylcholine receptor (nAChR). The upper six sequences are from the mouse muscle nAChR (α1 to δ), chick neuronal α7 nAChR, and an ionotropic 5-hydroxytryptamine receptor, all of which are cation selective; notice the negatively charged residues (D and E) at positions −4′, −1′ and 20′, the positively charged residues (K and R) at position 0, and the polar (S and T) residues at position 2′. The two lower sequences are from the α subunits of the GABA_A and glycine receptors, which are selectively permeable to anions; notice the positively charged R residues at position 19′ and the absence of negative charge at 20′, although the negative charge at −4′ remains. Notice also, in all the channels, the large hydrophobic L residue at 9′, and the positively charged K and R residues at 0′. (Redrawn from Lester, 1992. Reproduced with permission from the *Annual Review of Biophysics and Biomolecular Structure* Volume 21, (1992 by Annual Reviews Inc.)

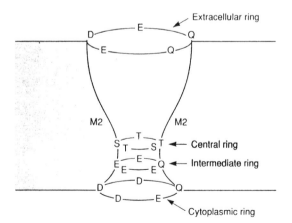

Figure 8.13. Rings of similar amino acid residues in the M2 segments lining the membrane-crossing pore of the nicotinic acetylcholine receptor channel. Note the three anionic rings with negatively charged glutamate (E) and aspartate (D) residues, and the central ring with uncharged but polar serine (S) and threonine (T) residues. Q, glutamine. (From Imoto, 1993.)

Later it became clear that two molecules of acetylcholine combine with each receptor molecule. The reaction scheme then becomes

$$2A + R \underset{k_{-1}}{\overset{2k_1}{\rightleftharpoons}} A + AR \underset{2k_{-2}}{\overset{k_2}{\rightleftharpoons}} A_2R \underset{\beta}{\overset{\alpha}{\rightleftharpoons}} A_2R^* \qquad (8.2)$$

   closed   closed   closed   open

The factor of 2 for the rate constants k_1 and k_{-2}, which both lead to the intermediate AR, arises because there are two possible forms of AR according to which of the two binding sites is occupied.

Schemes 8.1 and 8.2 imply that the ion channel has just two conductive states: it can be either open or closed, and the conductance when it is open is constant. On the other hand, there are three conformational states in scheme 8.1 and four in 8.2. We cannot predict what the duration of any particular opening by a single channel will be since the rate constants are probabilistic rather than determinate. But the schemes do imply that the probability of an open channel closing in any one time interval is constant. This means that if we measure a large number of channel openings, their durations (the time between opening and closing) will be exponentially distributed with a mean value of $1/\alpha$.

Fluctuation analysis

The depolarization produced by acetylcholine applied to a neuromuscular junction is the sum of the effects of a very large number of random collisions between acetylcholine molecules and receptor molecules. We would expect there to be some random fluctuations in the rate of these molecular collisions, which would produce corresponding fluctuations – or 'noise' – in the resulting depolarization. Katz & Miledi (1970, 1972) found that such fluctuations are indeed detectable, as can be seen in fig. 8.15. The phenomenon was also investigated by Anderson & Stevens (1973)

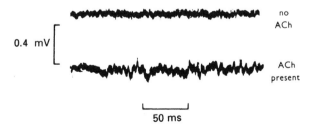

Figure 8.15. End-plate 'noise' produced by acetylcholine. The upper trace shows the membrane potential recorded at high gain from the end-plate region of a frog muscle fibre, with no acetylcholine (ACh) present. The lower trace is a similar record obtained when acetylcholine was applied by diffusion from a nearby micropipette. (From Katz & Miledi, 1972.)

with the voltage clamp technique; this allowed the noise (measured as fluctuations in current) to be measured at different membrane potentials, independently of the acetylcholine concentration. Let us examine some of their results.

We can estimate the single-channel conductance by measuring the variance of the current. This conclusion depends upon a simple relation between the mean and the variance in a binomial distribution, as described in equation 6.2, for example. Suppose there are a total of N channels of which r are open at any one time. Suppose the probability of any particular channel being open is p. Then

mean value of $r = Np$

and

variance of $r = Np(1-p)$.

The membrane conductance G will be the number of open channels (r) times the single-channel conductance γ, i.e.

$$G = \gamma r$$

so

mean value of $G = \gamma Np$

and

$$\text{variance of } G = \gamma^2 \text{ (variance of } r) = \gamma^2 Np(1-p)$$

hence

$$\frac{\text{variance of } G}{\text{mean of } G} = \gamma(1-p)$$

If p is very low (as it would be in low acetylcholine concentrations) this simplifies to give

$$\frac{\text{variance of } G}{\text{mean of } G} = \gamma \qquad (8.3)$$

Figure 8.16 shows some of the results obtained by Anderson & Stevens. There is clearly a linear relation between the mean and the variance of the membrane conductance. The slope of this gives a value for the single channel conductance of 19 pS. In another fibre where a particularly good voltage clamp was achieved, a value of 32 pS was obtained.

Information about the duration of channel openings can be derived from a frequency analysis of the noise. The current fluctuations are digitized and subjected to Fourier analysis by computer. This leads to a spectral density curve (or 'power density spectrum') in which the density at any particular frequency is the sum of the squared amplitudes of the sine and cosine components of the current fluctuations at that frequency. The spectral density curve then shows the contribution of different frequencies to the noise, as is shown in fig. 8.17. It can also – and this is its really useful property – give us some information about the time characteristics of the individual events which cause the fluctuations.

(Should there be any readers who regard the last sentence as nonsensical magic, I invite them to consider the Analogy of the Fleas. Suppose the platform of a small rapid-reading balance is placed in a box in which there are a large number of fleas jumping about at random. Each flea remains still for a short period of time (for an average of 100 ms, let us say) between jumps. Clearly the number (and hence weight) of fleas on the platform will fluctuate, and so the continuous record of weight will be 'noisy'. If we measure the weight of fleas on the platform every second we will find that successive measurements are essentially independent of one another because almost all the fleas on the platform at one particular time will have jumped off when we look again one second later. Hence the variance of the difference between successive weights recorded at 1 s intervals will be maximal. The same will apply for measurements taken at 10 s intervals. Another way of describing this situation is to say that the low frequency components of the noise are large.

However, if we measure at intervals of 1 ms, each successive measurement will be very close to the previous one, because only about one-hundredth of the fleas will have jumped off in the meantime. Hence the variance of the differences between successive weights recorded at 1 ms intervals will be low; that is to say; the high frequency components of the noise are small. At intermediate frequencies we will see corresponding intermediate values for the components of the noise, and so a graph of these components against frequency might well look like that in fig. 8.17. Notice that the contribution of any particular high frequency to the noise depends upon the time for which the fleas are still between jumps. If we warm them up, for example, so that this time is reduced to 10 ms, then measurements at 1 ms intervals will show larger differences between successive measurements because now about one-tenth of the fleas will have jumped off in the meantime instead of only about one-hundredth. Clearly there must be information in the frequency characteristics of the noise which can tell us about the length of time between jumps. For fleas read acetylcholine molecules, for platform read postsynaptic membrane and for weight read current. Now read on.)

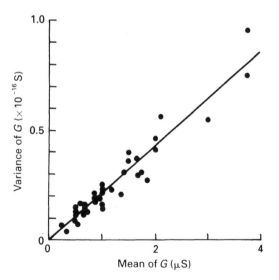

Figure 8.16. Determination of the single-channel conductance from acetylcholine noise measurements. Each point is derived from a voltage clamp record of the current through the membrane, from which the mean and the variance of the membrane conductance (G) can be determined. The slope of the relationship gives the single-channel conductance, from equation 8.3. (From Anderson & Stevens, 1973.)

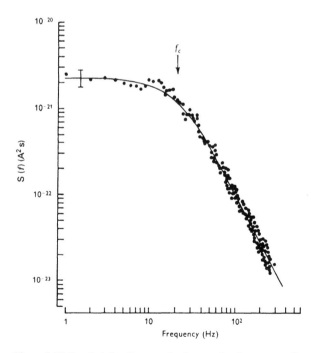

Figure 8.17. Spectral density curve (or 'power density spectrum') of current fluctuations produced at the end-plate by acetylcholine. The membrane potential was clamped at -60 mV. The curve is drawn according to equation 8.4 with $\alpha' = 132$ s^{-1}. f_c is the frequency at which spectral density is half maximal. (From Anderson & Stevens, 1973.)

at the acetylcholine noise is produced by that described in scheme 8.1, in which ings occur at random with an exponential pening times, then it can be shown that the tral density curve should be given by

$$\frac{S_0}{+ (2\pi f/\alpha')^2} \qquad (8.4)$$

:equency, S_0 is a constant (the value of $S(f)$ 1 α' is a rate constant. (Anderson & Stevens constant α from scheme 8.1 here, implying measured from the spectral density curve, but iter that this view is probably not correct, and so w~ ~ iere use α' instead.) This type of relation is known as a Lorentzian curve: $S(f)$ is maximal and the curve is flat at low frequencies, whereas at high frequencies $S(f)$ falls in proportion to $1/f^2$, as in fig. 8.17. Further details of the theory of fluctuation analysis are give by Stevens (1972), Neher & Stevens (1977) and DeFelice (1981).

We can measure the apparent mean duration of channel opening from the spectral density curve. Let f_c be the frequency at which the spectral density falls to half its maximum value. Then

$$S(f_c) = \frac{S_0}{2} = \frac{S_0}{1 + (2\pi f_c/\alpha')^2}$$

Hence

$$(2\pi f_c/\alpha')^2 = 1 \longrightarrow \quad f_c = \sqrt{1} \cdot \alpha'$$

and so

$$f_c = \alpha'2\pi$$

In fig. 8.17, for example, the value of f_c is 21 Hz, giving a rate constant α' of 0.13 ms^{-1}. The apparent mean channel opening time is $1/\alpha'$, i.e. 7.6 ms in this case. Anderson & Stevens found that f_c became larger (and thus the apparent mean opening time became shorter) at more positive membrane potentials and at higher temperatures.

Patch clamp records

The patch clamp technique, which we have met in chapter 2, was invented by Neher & Sakmann (1976) and first used by them to examine single-channel currents through nAChR channels. They found that each channel produces a square pulse of current lasting for up to a few milliseconds. The single-channel currents were 2.2 pA at a membrane potential of -80 mV and 3.4 pA at -120 mV. This change of 1.2 pA per 40 mV implies that the current would be zero at -7 mV, which is thus the reversal potential. We

can calculate the single-channel conductance γ from the relation

$$\gamma = \frac{i}{(E - E_r)} \qquad y = \frac{j}{(\epsilon - \epsilon_{rev})}$$

where i is the single-channel current, E is the membrane potential and E_r is the reversal potential. Here, at -80 mV,

$$\gamma = 2.2/0.073$$
$$= 30 \text{ pS}$$

The first patch clamp records, then, were in pleasing agreement with the results of fluctuation analysis. But, with improvements in techniques (Hamill et al., 1981) and more extensive observations, it was not long before complications arose. Openings occurred in *bursts* (and sometimes the bursts were grouped in *clusters*), there were brief closings and brief openings, and sometimes the single-channel conductance was not always the same size.

One of the first of the new observations was that single 'openings' of the channel are frequently interrupted by brief closed periods or 'gaps' (Colquhoun & Sakmann, 1981), such as is shown in fig. 8.18. This means that the events are actually bursts of openings. The phenomenon had actually been predicted by Colquhoun & Hawkes (1977, 1981) on the basis of scheme 8.1: if β is not much less than k_{-1} then AR is quite likely to revert to AR* instead of dissociating to A + R, as is shown in fig. 8.19. A similar argument can be applied to scheme 8.2. This means that the results of noise analysis must be reinterpreted. If the gaps within bursts are short, then the apparent opening times measured by noise analysis are actually burst length times, and the rate constant α in scheme 8.1

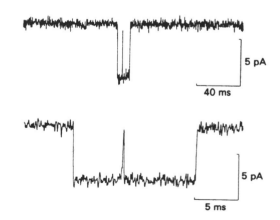

Figure 8.18. Single-channel current at the frog neuromuscular junction, produced by acetylcholine, showing a burst of two openings separated by a brief gap. The two traces show the same event at low and high time resolution. (From Colquhoun & Sakmann, 1985.)

or 8.2 cannot be equated to the parameter α' in scheme 8.3.

Colquhoun & Sakmann (1985) made some careful high resolution measurements of the time characteristics of the end-plate channel openings in frog muscle fibres, and used these to estimate some of the rate constants in scheme 8.2. An open channel A_2R^* can close only by reverting to A_2R, and the rate constant for this is α. Hence

$$\text{mean duration of channel opening} = 1/\alpha \qquad (8.5)$$

The rate constant for the departure from the A_2R state in scheme 8.2 will be the sum of the rate constants for conversion to A_2R^* and to AR, hence the time constant will be the reciprocal of this, i.e.

$$\text{mean gap duration} = 1/(\beta + 2k_{-2}) \qquad (8.6)$$

The number of gaps in a burst will be dependent upon the relative values of β and k_{-2}. If β is greater than k_{-2} the chances of A_2R reverting to A_2R^* will be relatively high and so there will be on average more openings (and so more gaps) in the burst; in fact

$$\text{mean number of gaps per burst} = \beta/k_{-2} \qquad (8.7)$$

Equations 8.5 to 8.7 show that the rate constants in scheme 8.2 can be calculated from measurements of the time characteristics of the single-channel currents. There is a technical difficulty here since we cannot measure all of the channel closings: some of them are too brief for the recording apparatus. This leads to two errors. Firstly, the apparent mean duration of gaps in bursts will be too long. Secondly, two openings separated by a very brief gap will be recorded as a single long opening, and so the mean value for the open time will be too long. Colquhoun & Sakmann

met this problem by fitting an exponential distribution to all the brief gaps that they could record, and assuming that the number of even shorter gaps was as predicted from this distribution. They could thus work out the total number of brief gaps and so, knowing the total time during bursts, calculate the mean duration of channel closing.

After applying these corrections, Colquhoun & Sakmann produced values of 1.4 ms for the mean open time duration, 20 µs for the mean gap duration and 1.9 for the mean number of gaps per burst. From these they calculated that $\alpha = 714 \text{ s}^{-1}$, $\beta = 30\,600 \text{ s}^{-1}$, and $k_{-2} = 8150 \text{ s}^{-1}$.

These values for the rate constants imply that the conformational change from A_2R to A_2R^* is energetically favoured, so that a channel which has two molecules of acetylcholine bound to it will spend most of its time in the open condition. This, of course, is just what is required for effective functioning (Colquhoun & Sakmann, 1983).

There are other aspects of patch clamp records which show that scheme 8.2 is not the whole story. Colquhoun & Sakmann (1985) found that a small proportion of the gaps in a burst are appreciably longer than the rest; they were named intermediate gaps. A proportion of the channel openings are much shorter than the rest. They suggested that these could be caused by channel opening sometimes occurring after only one acetylcholine molecule has been bound. Jackson (1988) has formalized this situation as follows, with a scheme in which there are three open and three closed states, corresponding to binding of 0, 1 or 2 molecules of acetylcholine or other agonist:

$$
\begin{array}{ccccc}
C & \underset{}{\overset{K_1}{\rightleftharpoons}} & A_1C & \underset{}{\overset{K_2}{\rightleftharpoons}} & A_2C \quad \text{closed} \\
\alpha_0 \updownarrow \beta_0 & & \alpha_1 \updownarrow \beta_1 & & \alpha_2 \updownarrow \beta_2 \\
O & \underset{J_1}{\rightleftharpoons} & A_1O & \underset{J_2}{\rightleftharpoons} & A_2O \quad \text{open}
\end{array}
\qquad (8.8)
$$

Here the αs and βs are rate constants, the Ks and Js are equilibrium constants. Estimates for the rate constants for

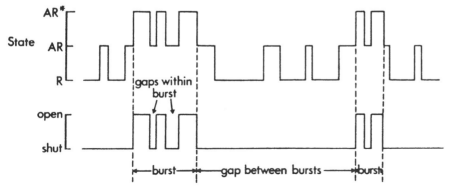

Figure 8.19. Diagram to show how the transitions between three states, as in scheme 8.1, give rise to bursts of channel openings. (From Colquhoun & Hawkes, 1982.)

opening, β_0, β_1 and β_2, were, respectively, 0.0028, 1.1 and 2800 s^{-1}, using carbachol as the agonist, in accordance with the low chances of a receptor opening with no or one molecule of agonist bound.

Jackson's estimates of the equilibrium constants in scheme 8.8 suggested that the two binding sites are not precisely equivalent, so that the first acetylcholine molecule is bound more tightly to the closed channel than the second, and also that both molecules are bound more tightly when the channel is in the open configuration. He argues that these are essential features of the functioning of the channel, allowing rapid opening in the presence of high acetylcholine concentrations (as when acetylcholine is released from the nerve terminal) and rapid termination of the response afterwards. He also suggested that the conformational change involved in opening of the channel requires an amount of energy that is normally only released by binding two molecules of acetylcholine. So the allosteric properties of the nAChR channel make it well adapted to its function as a rapidly activated neurotransmitter-gated channel (Jackson, 1989, 1994).

Another departure from simple models is shown by the existence of *subconductance states*, in which the single-channel conductance is some fraction of the normal value (Hamill & Sakmann, 1981; Auerbach & Sachs, 1983). These states seem to be common in cultured cells, but do also occur in mature frog muscle; Colquhoun & Sakmann (1985) found that 1% to 2% of all bursts contained a partial closure to either 18% or 71% of the normal open channel conductance level.

In conclusion, we can explain many of the features of the patch clamp records of nAChR single-channel currents by means of the relatively simple model of receptor behaviour described by scheme 8.2, but there are other features which imply that the actual situation is rather more complicated.

Pharmacology of the nAChR

Many different drugs and toxins affect the opening of nAChR channels. They are known as *competitive agents* if they attach to the acetylcholine-binding site and compete with the acetylcholine for it; they are known as *non-competitive agents* if they do not. Competitive agents are known as *agonists* if their binding results in opening of the channel, as *antagonists* if it does not. Agonists can sometimes cause block by inducing desensitization. Non-competitive agents include channel-blocking agents.

Nicotine, found in tobacco leaves, is the classical competitive agonist for the nAChR. It is an insecticide, and pre-

sumably this is its value to the tobacco plant. Other agonists include the acetylcholine analogues carbachol and succinylcholine, and the alkaloid anatoxin A from the blue-green alga *Anabaena flos-aquae*.

It has been known since the work of Claude Bernard in 1857 that *curare*, the substance used by South American Indians to poison the tips of their arrows, acts at the neuromuscular junction as an antagonist to acetylcholine. Crude curare is a mixture of substances obtained from various plants including *Chondodendron tomentosum*; there are three varieties, known as pot-curare, tube-curare and calabash-curare. One of the active constituents of tube curare is the alkaloid D-tubocurarine, which is the most widely used of the various curare extracts. *β-Erythroidine* is another alkaloid, derived from the American leguminous plant *Erythrina*. Unlike most neuromuscular blocking agents it is active when taken orally.

One of the most effective nAChR antagonists is α-bungarotoxin, obtained from the venom of the banded krait *Bungarus multicinctus*. Similar α toxin venoms are found in related snakes. They are peptides sixty-one to seventy-four amino acid residues long. α-Bungarotoxin binds firmly to the acetylcholine binding site on each of the two α subunits. It does not bind to most neuronal nAChRs, but a different constituent of *Bungarus* venom, κ-bungarotoxin, does.

Another group of peptides that act as competitive antagonists are the α-conotoxins, produced by *Conus* shells (Myers *et al.*, 1993). These are only thirteen to fifteen amino acid residues long, with just two loops of the peptide chain held in position by two disulphide links. Some of the α-conotoxins are remarkable in showing much greater action on the neuromuscular junction nAChRs of fish than on those of mammals.

Non-competitive antagonists include the alkaloid histrionicotoxin from the arrow poison frog *Dendrobates histrionicus* (which increases desensitization), the diterpene lactone lophotoxin from gorgonian corals and the glycoside neosurugatoxin from the Japanese ivory mollusc *Babylonia japonica*, both of which block the channel pore. In addition to its effects on dopamine receptors, the tranquillizer chlorpromazine blocks the nAChR channel pore, and has been used to help to determine the regions lining the pore (Revah *et al.*, 1990).

The 5-hydroxytryptamine-gated channel

5-Hydroxytryptamine (5-HT, serotonin) is released from blood platelets and from chromaffin cells of the gut lining, as well as acting as a neurotransmitter. It exerts its varied actions by activating a nmber of different receptor types.

Most of these produce relatively slow responses mediated via G proteins and second-messenger systems, as we shall see in chapter 9, but the $5HT_3$ receptor is a ligand-gated ion channel and so produces fast synaptic responses (Derkach *et al.*, 1989). The channel is permeable to both sodium and potassium ions, and its activation leads to a rapid depolarization of the cell.

Molecular cloning of the $5HT_3$ receptor shows that its structure is generally similar to that of the *Torpedo* nAChR α subunit, with the usual set of four transmembrane α-helices and an amino acid sequence showing 27% identity (Maricq *et al.*, 1991). Figure 8.14 shows the sequence similarity of the putative pore-lining M2 segment with those of other members of the nAChR-like superfamily of receptor/channels. Expression in *Xenopus* oocytes shows that a single subunit type is sufficient to produce functioning channels (probably as homopentamers, we may

suppose), but it could be that under normal conditions there are other subunits as well.

The GABA$_A$ and glycine receptor channels

γ-Aminobutyric acid (GABA) and glycine act as fast inhibitory transmitters in the mammalian central nervous system, whereas glutamate and probably aspartate and others are responsible for fast excitatory transmission. The depressant and excitatory actions of these amino acids were first discovered by using the ionophoresis technique on central neurons (Curtis *et al.*, 1960). Nevertheless it was first thought that they were not neurotransmitter substances released from nerve terminals and acting on postsynaptic receptors, but that they acted via some more generalized effect on membrane excitability. This view was revised with the discovery of particular blocking agents for amino acid action, such as strychnine for glycine and bicu-

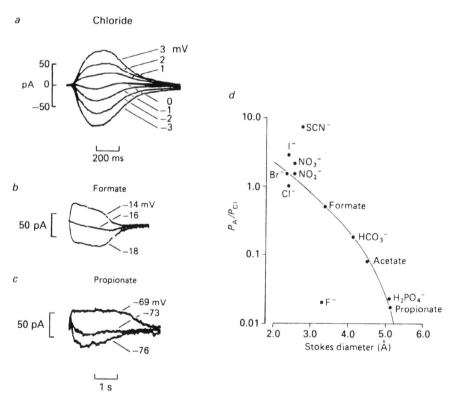

Figure 8.20. Permeability of the GABA-activated chloride channel to various anions. The oscilloscope traces *a* to *c* show whole-cell currents of cultured mouse spinal neurons following application of GABA while held under voltage clamp at different membrane potentials. *a* shows the responses when the extracellular and intracellular solutions had an identical chloride concentration of 145 mM; the reversal potential is very close to 0 mV. *b* and *c* show similar responses when the cell was internally

dialysed with a solution containing 140 mM formate (*b*) or propionate (*c*) and 4 mM chloride; the reversal potential is lower for formate and much lower for propionate, showing that the channels are less permeable to these ions than to chloride. In *d* the permeabilities of the different ions with respect to chloride, as calculated from the reversal potential experiments, are plotted against their diameters, so giving an estimate of the size of the channel. (From Bormann *et al.*, 1987.)

culline for GABA, suggesting that there are specific receptors for these substances. Krnjević (1986) gives a pithy summary of the development of research in this field.

Krnjević & Schwarz (1967) applied GABA by ionophoresis to cells in the cat cerebral cortex. This produced hyperpolarizations similar to IPSPs. Both responses were converted to depolarizations by injection of chloride ions into the cell, and the reversal potential of the GABA response was always the same as that of the IPSP. They concluded that GABA was probably the inhibitory transmitter substance and that it acted by opening ion channels permeable to chloride ions. The receptors mediating this fast response are known as GABA$_A$ receptors; GABA$_B$ receptors produce slower responses and have no intrinsic ion channel.

The two types of receptor differ in their distribution in the nervous system. GABA is the neurotransmitter for most postsynaptic inhibition in the brain and for presynaptic inhibition in the spinal cord. Glycine mediates postsynaptic inhibition in the spinal cord.

Permeability and selectivity

Much information about the nature of these neurotransmitter-gated chloride channels has been obtained by application of patch clamp techniques to mammalian spinal neurons in tissue culture (Bormann *et al.*, 1987),. The whole-cell clamp (fig. 2.5) allows the internal ionic concentrations to be varied because they equilibrate with the solution in the recording electrode. Bormann and his colleagues found that the reversal potential of the response to GABA is given by the chloride equilibrium potential, hence the GABA$_A$ channel is permeable to chloride ions. They also determined the reversal potential with different anions internally, and from this they calculated the permeability ratios for a variety of an ions with respect to chloride. The results (fig. 8.20) show that permeability falls with increasing ionic diameter, reaching zero at about 5.6 Å. This suggests that the internal diameter of the GABA channel is slightly smaller than that of the nicotinic acetylcholine receptor channel (7.4 Å), as determined in the essentially similar experiments by Dwyer *et al.* (1980) described above. Results from the glycine receptor channel were very similar.

Bormann and his colleagues also obtained single GABA channel records using the patch clamp technique. Different conductance states were evident, with conductances of 44, 30, 19 and 12 pS, the 30 pS state being by far the most common. With thiocyanate as the permeant anion, the main conductance level was reduced from 30 to 22.5 pS. With mixtures of chloride and thiocyanate, however,

channel conductance was still lower, reaching a minimum of 12 pS at 16% thiocyanate (fig. 8.21). This rather surprising phenomenon is known as the anomalous mole fraction effect; it has been seen in other channels and is interpreted as evidence that there are at least two interacting ion-binding sites in each channel.

Analysis of how we might expect two-ion channels to behave involves Eyring rate theory and some complex calculations with an uncomfortably large number of constants to which values have to be assigned (see Hille, 1992; Aidley & Stanfield, 1996). But the results are in good agreement with actuality (fig. 8.21) and provide some justification for the model shown in fig. 8.22.

Pharmacology

The pharmacology of the GABA$_A$ receptor channel is quite complicated (Macdonald & Olsen, 1994). There are at least five different binding sites, for GABA, benzodiazepines, barbiturates, certain steroids and picrotoxin. Muscimol, a constituent of the mushroom *Amanita muscaria*, is an agonist of GABA (i.e. it binds to the GABA-binding site and opens the channel). The alkaloid bicuculline and a number of synthetic compounds are competitive antagonists; they probably attach to the GABA-binding site, but without opening the channel.

A number of agents act as *modulators* of GABA action,

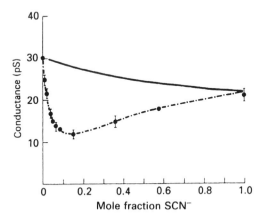

Figure 8.21. The anomalous mole fraction effect in GABA-activated chloride channels of cultured mouse spinal neurons. The points show the single-channel conductance at -70 mV with different proportions of thiocyanate in a chloride–thiocyanate mixture. Solutions were the same on each side of the membrane and the total anion concentration was always 145 mM. The continuous line shows the expected relation for a channel with a single binding site for which the two anions compete. The dashed line shows that for a channel with two such sites. (From Bormann *et al.*, 1987.)

i.e. they make the receptor more (or less) responsive to GABA without themselves being the primary agent for opening the channel. They include the benzodiazepines and the barbiturates, two groups of drugs widely used in medical practice as central nervous system depressants.

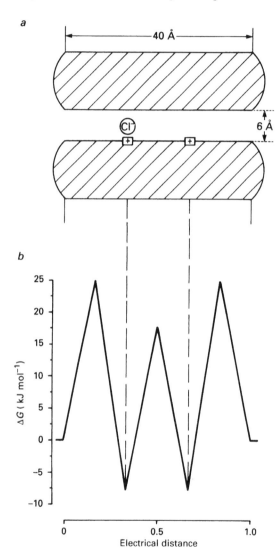

Figure 8.22. Schematic model of the GABA-activated chloride channel in cultured mouse spinal neurons. The diagram in *a* is drawn to scale and shows two fixed positive charges to which the chloride ion can bind. The profile in *b* shows the free-energy changes for a chloride ion passing through the open channel in the absence of a potential across the membrane. Energy barriers have to be crossed on entering and leaving the channel. Wells correspond to the positions of the two binding sites, and there is another barrier between them. The profile is calculated according to Eyring rate theory using values suitable to fit the experimental results. (From Bormann *et al.*, 1987.)

The benzodiazepines are used to produce sedation and sleep; diazepam is marketed as the tranquilliser Valium. On combination with the binding site on the $GABA_A$ receptor, they increase the effectiveness of the transmitter substance GABA. Single-channel records show that this is done by increasing the frequency of channel opening, probably by enhancing the rate at which GABA binds; the single-channel conductance remains the same (Rogers *et al.*, 1994). Compounds which have the opposite effect, the β-carbolines, are known as inverse agonists; they reduce the frequency of channel opening in the presence of GABA. There are also compounds which bind to the benzodiazepine site and have no action other than to prevent benzodiazepine agonists from being effective; they are known as benzodiazepine antagonists. The action of all these compounds on GABA binding is described as allosteric, since an event in one part of the molecule produces a response in another.

Barbiturates are also allosteric modulators of $GABA_A$ receptor channels. They bind at a site separate from that which binds GABA. This increases the proportion of time that the channel is open when it has been activated by GABA (Macdonald *et al.*, 1989). Similar effects are produced by the steroids androsterone and pregnenolone (Twyman & Macdonald, 1992). The inhalational general anesthetics, such as halothane and isoflurane, also potentiate the response of $GABA_A$ receptors to GABA, and this is probably the means whereby they produce anaesthesia (Franks & Lieb, 1994).

Picrotoxin is a convulsant substance from the seed of *Anamirta cocculus*, a climbing shrub from southeast Asia. It is a non-competitive inhibitor of GABA action at $GABA_A$ receptors. It may block the channel directly by binding to a site in the channel pore, or it may be that it interferes with the allosteric link between GABA binding and channel opening so as to reduce the probability of long openings. Penicillin G reduces the average open time of $GABA_A$ receptor channels, probably by blocking the open channel (Macdonald & Olsen, 1994).

Molecular structure

Sequencing of subunits from the $GABA_A$ receptor, using appropriate probes with brain cDNA libraries, showed that they are similar in structure to those of the nAChR, with four putative membrane-crossing segments named M1 to M4 (Schofield *et al.*, 1987). The glycine receptor has a similar structure, so that the three receptors (plus also the 5-HT$_3$ receptor, which is similar in general structure to the nAChR), form a family of related proteins. Since there is good evidence that the glycine receptor has five subunits

(Langosch *et al.*, 1988), it seems very likely that all members of the nAChR-related family have a similar pentameric arrangement with five subunits grouped round a central pore.

GABA$_A$ channels show considerable diversity in their subunit make-up (Barnard, 1992; Macdonald & Olsen, 1994). There are at least four types of subunit in channels in the central nervous system, α, β, γ and δ. A fifth type, ρ (rho), is found in the retina. Furthermore there are at least six different varieties of the α chain, four of the β chain, three of the γ chain and two of the δ chain. The amino acid sequences show 70% to 80% homology between different members of the same subunit type, 30% to 40% between different types. Complete GABA$_A$ receptors in the nervous system probably contain both α and β chains together with either γ or δ. Different combinations of subunits are found in different parts of the nervous system. Motor neurons in the spinal cord, for example, have GABA$_A$ receptors with the α2, β3 and γ2 subunits, whereas many brain neurons show the α1, β2 and γ2 combination.

The M2 pore-lining segments of GABA$_A$ and glycine receptor channels are similar to each other in their amino acid sequences, and show some similarities to those of the nAChR (fig. 8.14). They differ from the nAChR in having a hydrophobic ring of alanine residues instead of the negatively charged glutamate ring at position −1, and there is a ring of proline residues between this and position −2. At the outer end of the pore there is a ring of positively charged arginine residues at position 19 in place of the negatively charged glutamate at 20 in the nAChR. Overall, then, the pore has an excess of five positive charges in these chloride channels, whereas the cation-selective nAChR has an excess of ten negative charges.

Glutamate receptor channels

Curtis and his colleagues (1960) found that glutamic, aspartic and cysteic acids cause excitation of spinal motoneurons when applied to them by ionophoresis. Later studies extended these results to a wide variety of neurons and a considerable number of other acidic amino acids. It is now clear that glutamate is the neurotransmitter at the great majority of synapses for fast excitatory transmission in the vertebrate central nervous system, acting via ionotropic glutamate receptor channels. The channels are permeable to monovalent cations, so that under physiological conditions the reversal potential is about 0 mV and depolarization is largely attributable to an outflow of sodium ions. Some channels are also permeable to calcium ions. Metabotropic glutamate receptors, which produce slow responses and have no intrinsic ion channel, also exist.

Pharmacology

The three major types of ionotropic glutamate receptor are named AMPA, kainate and NMDA receptors according to the agonists which will activate them (Monaghan *et al.*, 1989). AMPA receptors are activated by α-amino-3-hydroxy-5-methyl-4-isoxazolepropionate and also by quisqualate (hence they were previously known as quisqualate receptors); they show rapid responses and are permeable to sodium and potassium ions when their channels are open. Kainate receptors are similar to them in general properties, so the two groups are sometimes put together as AMPA-kainate receptors. NMDA receptors are activated by *N*-methyl-D-aspartate; they are permeable to sodium, potassium and calcium ions when activated and are blocked by external magnesium ions.

The synthetic compounds known as quinoxalinediones are competitive antagonists at AMPA-kainate receptors and have no effect on NMDA receptors. They include 6-cyano-7-nitroquinoxaline-2,3-dione (CNQX). A number of spider and wasp toxins are effective blocking agents for AMPA-kainate receptor channels. Glutamate is the neuromuscular transmitter in insects and the glutamate receptors of insect muscle are similar to the AMPA-kainate receptors of vertebrates; the spiders paralyse their prey by injecting neuromuscular blocking agents into them (Usherwood & Blagbrough, 1991).

NMDA channels can be blocked by conantokin-T and conantokin-G, *Conus* peptides that have the unusual amino acid γ-carboxyglutamate in their sequences (Olivera *et al.*, 1990). A number of synthetic drugs are effective as NMDA channel-blocking agents. They may be of some use in preventing the potentially toxic effects of receptor overactivity (see below) in conditions such as epilepsy and neurodegenerative diseases (Rogawski, 1993). The general anaesthetic ketamine probably acts by inhibiting NMDA receptors (Franks & Lieb, 1994). The synthetic compounds D-2-amino-5-phosphonovalerate (AP5) and 3-(2-carboxypiperazine-4-yl)propylphosphonate (CPP) are competitive antagonists of glutamate action at NMDA receptors. The non-competitive agents phencyclidine, MK-801 and zinc ions all act by blocking the channel pore (Hollmann & Heinemann, 1994).

NMDA responses are considerably potentiated by glycine, which is, of course, an inhibitory transmitter elsewhere in the central nervous system (Johnson & Ascher, 1987). Single-channel currents are unchanged in amplitude, but the probability of a channel opening in response to NMDA is greatly increased. The effect is evident in *Xenopus* oocytes expressing the cloned NMDA receptor, showing that glycine binds directly to the NMDA receptor molecule (Moriyoshi *et al.*, 1991).

Molecular structures

Molecular cloning shows that the protein chains making up glutamate receptor channels are much longer than those of the nAChR and its relatives; they are usually over 800 residues long rather than in the range 400 to 500. Molecular modelling suggests that there are five subunits forming each receptor, as in the nAChR-related family, but a tetrameric structure cannot be ruled out (Sutcliffe *et al.*, 1996).

Hydropathy analysis suggests that glutamate receptor subunits have four hydrophobic membrane-crossing segments, plus a signal sequence that is removed in forming the mature protein. The pattern is not unlike that of the nAChR at first sight, and it was initially thought that the topology of the protein chain would be similar to that of the nAChR, as shown in fig. 8.5a. A closer look, however, showed that this was not so. Bennett & Dingledine (1995) used prolactin as a reporter in an epitope protection assay, as in the experiments shown in fig. 8.5 for the nAChR. It turned out that each end of the M2 segment was on the cytoplasmic side of the membrane, so it cannot cross the membrane from one side to the other, but must loop into it so that both its ends are cytoplasmic (fig. 8.23). Confirmation of this model has been provided by introducing sites for glycosylation into the molecule at different points; only if these were extracellular would glycosylation occur. The results suggested that only M1, M3 and M4 were completely transmembrane, that the M3–M4 link was extracellular, and that the N-terminus was extracellular and the C-terminus cytoplasmic (Hollmann *et al.*, 1994).

Some similarities have been detected between the M2 loop of glutamate channels and the H5 loop thought to line the pore of potassium channels (Wo & Oswald, 1995). It is as if the potassium channel H5 loop had been turned through 180° to enter from the cytoplasmic side of the membrane. It may be relevant to this idea that magnesium ions block the NMDA channel from the outside whereas they block the inward rectifier potassium channel from the inside.

The diversity of glutamate receptor subunits is considerable (Seeburg, 1993; Hollmann & Heinemann, 1994; Nakanishi & Masu, 1994). Mammalian AMPA receptor subunits are named GluR1 to GluR4 (or GluR-A to -D). Kainate receptor subunits show 35% to 40% amino acid sequence identity with them; they are named GluR5 to 7. KA1 and KA2 are high affinity kainate receptors. Related clones have been isolated from the fruit fly *Drosophila*; glutamate is the excitatory neuromuscular transmitter in insects.

NMDA receptors have subunits that show 22% to 24% sequence identity with those of AMPA or kainate receptors. They are composed of NMDAR1 chains (also called NR1), which exist in at least seven splice variants, and sometimes contain NMDAR2A to NMDAR2D chains (also called NR2A etc.), sometimes not. The four NMDAR2 subunits are rather larger in size (they have an extensive C-terminal region) and cannot form functional NMDA receptors without combining with NMDAR1 chains (McBain & Mayer, 1994). Heteromers of NMDAR1 with different NMDAR2 chains show differences in gating and sensitivity to magnesium block when expressed in cultured cells. Messenger RNAs coding for the different chains can be detected and localized in the brain by means of *in situ* hybridization experiments, using radioactive antisense oligonucleotide probes. The results show that NMDAR1 chains are found throughout the brain, but the different NMDAR2 chains are more localized: NMDAR2B is present in the rat cerebral cortex but not in the cerebellum, for example, whereas the reverse is true for NMDAR2C (Monyer *et al.*, 1992).

Diversity is further increased by alternative splicing. In the GluR1 to GluR4 subunits a section thirty-eight residues long situated just before the M4 segment is coded for by either one of two adjacent exons; these alternatives have been called the 'flip' and 'flop' segments. They alter the rel-

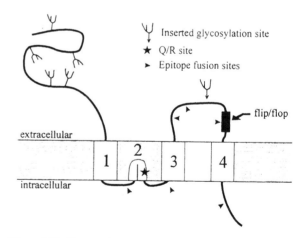

Figure 8.23. The membrane-crossing topology of the glutamate-activated channel GluR3 subunit as determined by the prolactin fusion protein method. The tree-like structures are N-linked glycosylation sites, so they must be extracellular. The tree with an arrow shows where a glycosylation site can be inserted, and so is probably extracellular. The M1, M3 and M4 transmembrane segments are shown. The star shows the Q/R site in the M2 membrane-dipping segment. Arrowheads show prolactin fusion sites, leading to the results that justify this model. (From Bennett & Dingledine, 1995. Reproduced with permission from *Neuron*. Copyright Cell Press.)

ative responses of the receptors to glutamate or AMPA and to kainate, and subunits containing them have different distributions in the brain (Sommer *et al.*, 1990). Alternative splicing of three exons in NMDAR1 subunits generates eight different variants, which are also expressed differentially in different parts of the brain (Zukin & Bennett, 1995).

A remarkable further source of diversity occurs at a particular site in the M2 segment of the GluR2, GluR5 and GluR6 subunits (Sommer *et al.*, 1991). Here the genomic DNA codes for glutamine (Q) but the residue in the protein (as determined from cDNA made from mRNA) is often arginine (R). The position is thus called the Q/R site or (since asparagine occurs at the homologous position in NMDA receptors) the QRN site. The change appears to arise from the process of RNA editing, whereby the mRNA sequence is altered by enzymic action so that the RNA codon for glutamine (CAG) is converted to that for arginine (CGG).

Permeability and block

MacDermott and her colleagues (1986) showed that calcium ions can pass through NMDA channels. They used voltage clamp methods with cultured mouse embryo spinal neurons, and measured the intracellular calcium ion activity with the calcium-sensitive dye arsenazo III. At -60 mV, application of NMDA caused an increase in internal calcium ion activity, and this was blocked by magnesium ions. Does this mean that the calcium entered via the NMDA channels? It could not have entered via voltage-gated calcium channels because they are not activated at this membrane potential. Another possibility is that the calcium might be released from internal stores as a result of inositol trisphosphate production. But application of NMDA at $+60$ mV (where the ionic current through the channel is outward) did not produce a rise in internal calcium, showing that activation of the NMDA receptor on its own is not sufficient to produce the response. Further evidence for calcium flow through the channel came from experiments on the reversal potential for the ionic current produced by NMDA: it is altered by changes in external calcium ion concentration, whereas the response to kainate is not.

The QRN site in the M2 segment seems to be important in the calcium permeability of glutamate channels. AMPA receptor channels are not permeable to calcium ions when arginine is the residue at the site, whereas with glutamine there they are (Hume *et al.*, 1991). Replacement of the asparagine residue at the QRN site in NR1 subunits of NMDA receptors makes them less permeable to calcium ions (Burnashev *et al.*, 1992).

NMDA channels can act as rectifiers. This first became evident when some ionophoresis experiments on spinal motoneurons *in vivo* led to the surprising conclusion that glutamate apparently produced a decrease in membrane conductance (Engberg *et al.*, 1979). Experiments on neurons in tissue culture showed that the current–voltage curve of the glutamate-induced current had a region of negative slope conductance (MacDonald *et al.*, 1982). The explanation for this came when magnesium was removed from the bathing solution: the current–voltage relation was now much more conventional, suggesting that magnesium ions can block the channels at negative membrane potentials but that this block is removed by depolarization (Nowak *et al.*, 1984). Patch clamp experiments are in line with this explanation: magnesium causes 'flickering' (suggesting intermittent blockage of the channel) at negative membrane potentials (fig. 8.24). Non-NMDA channels do not show this sensitivity to magnesium ions.

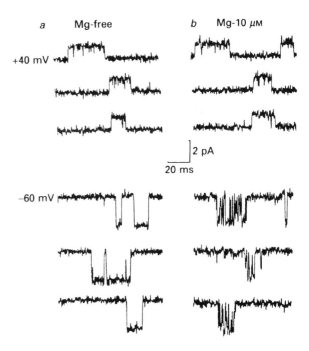

Figure 8.24. Effect of magnesium ions on single-channel currents induced by glutamate in cultured embryonic mouse brain cells. In the absence of magnesium (*a*) the timing of the currents is largely unaffected by membrane potential. With a low magnesium ion concentration (*b*) the currents are similar to the Mg-free situation at $+40$ mV, but at -60 mV they are much interrupted. This suggests that a magnesium ion can block a glutamate-activated channel when it is held in position by a negative membrane potential. (From Nowak *et al.*, 1984.)

Two-component synaptic responses

Excitatory synapses in the mammalian central nervous system often involve both non-NMDA and NMDA receptors. The EPSPs produced by glutamate action have two components, fast and slow, as has been demonstrated in spinal neurons in cell culture (Forsythe & Westbrook, 1988), and in pyramidal neurons of the hippocampus (Collingridge *et al.*, 1988). The dual nature is revealed by the use of selective blocking agents: APV (D-2-amino-phosphonovalerate) blocks NMDA receptors but not non-NMDA receptors, and so reveals the fast component, whereas CNQX (6-cyano-7-nitroquinoxalene-2,3-dione) blocks non-NMDA receptors and so reveals the slow component (fig. 8.25).

Why should the two components have different time courses? The non-NMDA component has a fast rise time of less than 1 ms, and decays rapidly with a time constant of 0.2 to 8 ms. The NMDA response, on the other hand, rises slowly to a peak at around 20 ms ands decays slowly,

with two time constants of the order of tens and hundreds of milliseconds. The decay constants are very sensitive to temperature, with Q_{10} values of 2.7 to 3.5 for the various components (the Q_{10} is the factor by which a rate increases when the temperature rises by 10° C); these values are much too high for diffusion processes, and so the prolonged decay cannot be attributable to slow diffusion of glutamate out of the synaptic cleft (Hestrin *et al.*, 1990*a,b*). Rapid application of APV during the decay phase does not speed it up, again suggesting that NMDA receptors do not release and rebind glutamate (Lester *et al.*, 1990*a,b*). The overall conclusion is that NMDA receptor channels have a relatively long latency to first opening after they have bound their two glutamate molecules, and that they will then open and close a number of times before they release them (Edmonds *et al.*, 1995).

The size of the NMDA-dependent slower component of the EPSC is much affected by the voltage-dependent magnesium block to which NMDA receptor channels are

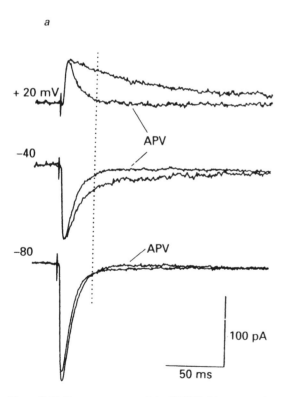

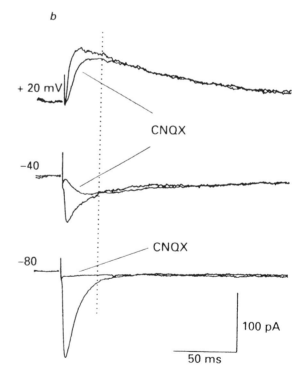

Figure 8.25. Two components of the EPSC in hippocampal neurons revealed by the use of selective blocking agents. *a* shows the effect of APV, which blocks NMDA receptor channels. *b* shows the effect of CNQX, which blocks AMPA receptor channels. Currents were measured with patch clamp electrodes in the whole-cell configuration. The NMDA-dependent part of the response has a slower time course than the AMPA-dependent response. Notice also that it is much larger at +20 mV than at lower membrane potentials; this effect disappears in magnesium-free solutions, and so can be attributed to the blockage of NMDA receptor channels by magnesium at negative membrane potentials. (From Hestrin *et al.*, 1990*a*.)

subject. Because of this, it seems likely that NMDA receptors are active only when the membrane potential is depolarized, such as might occur during a burst of synaptic activation of non-NMDA channels. This feature might well be involved in the cellular basis of learning and memory, as we shall see in chapter 11. Perhaps opening of NMDA channels allows calcium to enter the cell and this calcium then brings about long-term changes in cellular responsiveness.

Not all glutamatergic synapses show this two-component phenomenon: the EPSCs of cerebellar Purkinje cells are completely blocked by CNQX and unaffected by APV, hence NMDA receptors are not involved (Perkel *et al.*, 1990).

Glutamate neurotoxicity
Injection of glutamate into newborn mice causes obesity, brain lesions and other developmental disturbances (Olney, 1969, 1995). Similar effects, involving the death of brain cells, are shown by other substances which activate glutamate receptors. Such agents are described as *neurotoxic* or *excitotoxic*. It may be that the brain damage associated with oxygen lack, stroke, hypoglycaemic coma, epilepsy or various neurodegenerative diseases is brought about in part by excessive activation of NMDA receptors and, to a lesser extent, other glutamate receptors (Rothman & Olney, 1987, 1995; Choi, 1992*a,b*). Neurons in tissue culture die when they are exposed to high concentrations (0.1 to 1 mM) of glutamate, as also do non-neural kidney cells transfected with cloned NMDA receptors; these effects can be prevented by drugs that block the NMDA receptors.

The precise mechanisms of glutamate neurotoxicity are of great interest and some complexity. It seems evident that excessive rises in internal calcium ion concentrations are a major step in the process (Choi, 1988, 1995). The relatively high permeability of NMDA channels to calcium ions is clearly an important factor here. In the transfected kidney cell experiments, NMDA receptor mutants that were less permeable to calcium ions produced less cell death (Anegawa *et al.*, 1995).

One way in which the calcium might act is via the production of nitric oxide through stimulation of the enzyme nitric oxide synthase; nitric oxide is a free radical and therefore highly reactive, and may produce other free radicals. Inhibition of nitric oxide synthase prevents NMDA-mediated neurotoxicity in cultured neurons (Dawson *et al.*, 1991), and use of agents that trap free radicals reduces glutamate neurotoxic lesions in rats (Schulz *et al.*, 1995).

Calcium has wide effects on cellular control systems, however, and it seems likely that neurotoxicity involves more than one second-messenger pathway (Rothman & Olney, 1995).

The P$_{2X}$ ATP receptor channel
We have seen that adenosine triphosphate (ATP) forms part of the contents of cholinergic synaptic vesicles and that it is released into the synaptic cleft along with the acetylcholine, so it is not too surprising to find that in some cases ATP itself may be a neurotransmitter. Burnstock and his colleagues (1970) found that some smooth muscle responses to nerve stimulation persist in the presence of both cholinergic and adrenergic blocking agents. These non-adrenergic non-cholinergic (NANC) responses could be imitated by the application of ATP, leading to the concept of 'purinergic' transmission. It is now clear that a number of other neurotransmitters are also involved in the control of the gut musculature, but the concept of purinergic transmission remains viable (Burnstock, 1995*a*). ATP acts as a cotransmitter with noradrenaline in the sympathetic nervous system, and is responsible for the fast components of the postsynaptic responses (Edwards *et al.*, 1992; Brock & Cunnane, 1993).

A neurotransmitter-gated channel of quite different design from the nAChR-related and glutamate receptor families has been detected in cDNA libraries prepared from smooth muscle mRNA (Brake *et al.*, 1994; Valera *et al.*, 1994; Surprenant *et al.*, 1995). Extracellular ATP acts on smooth muscle cells via a group of receptors known as the P$_2$ purinoceptors, of which the P$_{2X}$ receptors are ion channels.

The P$_{2X1}$ receptor protein is 399 amino acid residues long; other subtypes are a little larger. It has two putative membrane-crossing segments separated by a large extracellular loop. An H5 region next to M2 shows considerable similarity to the H5 regions of various potassium channels. When expressed in *Xenopus* oocytes it forms channels that allow sodium and calcium ions to enter when they are activated by extracellular ATP. It seems reasonable to assume that the channel is made of a number of subunits surrounding a central pore.

The P$_{2X3}$ subtype is apparently expressed only in certain sensory neurons, including those concerned with peripheral pain sensations (C.-C. Chen *et al.*, 1995; Lewis *et al.*, 1995). It forms channels in oocytes when P$_{2X2}$ is also present. It seems likely that it is important in the initiation of pain sensations, so the hunt is on for specific blocking agents that could be used for the alleviation of pain (Burnstock, 1996).

9
Slow synaptic transmission

The preceding two chapters have examined the mechanisms of fast synaptic transmission, which involves the opening of ion channels by the direct action of the neurotransmitter. In this chapter we consider the slower responses produced by indirect mechanisms, where the neurotransmitter receptor does not contain its own intrinsic channel. Activation of the receptor, we shall see, sets in train a series of changes in one or more other proteins, leading eventually to the opening or closing of a particular set of ion channels. Let us look first of all at the electrical phenomena that need to be explained.

Slow synaptic potentials
In the sympathetic nervous system of vertebrates there is a chain of ganglia lying near to the spinal cord. These contain the cell bodies of the postganglionic fibres which terminate on smooth muscle or gland cells. The preganglionic fibres arise in the spinal cord and form synapses with the cell bodies of the postganglionic fibres in the ganglia.

Bullfrog sympathetic ganglia contain B cells and C cells, the B cells being the larger. Each is innervated by preganglionic fibres which form numerous synaptic boutons on the neuronal soma. They show a variety of different types of synaptic activity (Kuffler, 1980; Adams *et al.*, 1986), as is shown in fig. 9.1.

A single stimulus applied to the preganglionic fibres produces a fast EPSP in both B and C cells, and this may be large enough to produce an action potential in the postganglionic fibres (fig. 9.1*a*). The response is blocked by curare and can be mimicked by acetylcholine. The size of the fast EPSP is linearly related to membrane potential, with a reversal potential at about −5 mV (Nishi & Koketsu, 1960). All this suggests that the mechanism of production of the fast EPSP is very similar to that at the neuromuscular junction: it is mediated by nicotinic acetylcholine receptors in which a cation-selective ion channel opens when acetylcholine is bound to it.

In B cells a slow EPSP with a much longer time course occurs after the fast EPSP (fig. 9.1*c*). Similar responses are seen after application of acetylcholine to the ganglion. The slow EPSP occurs in the presence of curare or other nico-

tinic blocking agents but is itself blocked by atropine; hence the receptors which mediate it are muscarinic. Further investigation showed that the mechanism of the slow EPSP is quite different from that of the fast EPSP in that it is produced by a closing of ion channels, not an opening of them (Kobayashi & Libet, 1970; Weight & Votova, 1970). Evidence for this conclusion is shown in fig. 9.2, in which fast and slow EPSPs are elicited at different membrane potentials. The fast EPSP decreases in size with depolarizing (positive) currents, reversing at a potential near to −5 mV. The slow EPSP, however, increases in size with depolarizing currents, and is reduced by hyperpolarizing currents, with a reversal potential at about −88 mV.

It is very striking in fig. 9.2 that the fast EPSP is always a

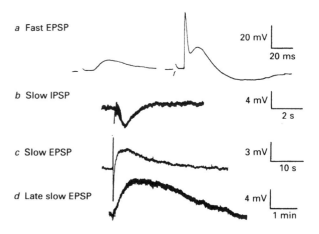

Figure 9.1. Fast and slow synaptic responses in frog tenth sympathetic ganglion neurons. *a* shows a fast EPSP produced by a single preganglionic stimulus, and (right) a stronger stimulus excites more preganglionic fibres giving a larger EPSP which is sufficient to produce an action potential. *b* shows a slow IPSP in a C neuron, produced in response to a burst of stimuli applied to the spinal nerves; fast EPSPs were blocked with the curare-like compound dihydro-β-erythroidine. *c* shows a slow EPSP produced by brief repetitive stimulation of the sympathetic chain. *d* shows a late slow EPSP produced by repetitive stimulation of the spinal nerves. Note the different time scales. (From Kuffler, 1980.)

voltage change towards the reversal potential (just as in figs. 7.23 and 7.28) whereas the slow EPSP is always a movement away from it. This implies that, while there is an increase in membrane conductance during the fast EPSP, there is a decrease in conductance during the slow EPSP. The reversal potential for the slow EPSP is very near to the equilibrium potential for potassium ions and is unaffected by changes in chloride equilibrium potential. Hence we can conclude that the slow EPSP is brought about by a temporary closure of some of the potassium channels in the cell.

Later experiments showed that the particular potassium channels that are closed are responsible for a potassium ion current (the M current) which is activated rather slowly by depolarizations beyond -60 mV (Brown & Adams, 1980; Adams & Brown, 1982; Caulfield *et al.*, 1994). The functional purpose of the slow EPSP may be to increase the excitability of the neuron over relatively long time periods.

In C cells the fast EPSP is followed by a slow, hyperpolarizing IPSP. This IPSP is muscarinic with a latency of 50 ms or more and lasts for a few seconds (fig. 9.1*b*). It has a reversal potential around -102 mV; this is unaffected by chloride ion concentration but strongly dependent upon the external potassium ion concentration with a slope of 55 mV per ten-fold change (Dodd & Horn, 1983). Hence the slow

IPSP is produced by the opening of channels selective for potassium ions.

The final event in this complex sequence of postsynaptic responses is a depolarization which lasts for a few minutes after a prolonged period of repetitive stimulation of the preganglionic fibres. This was discovered by Nishi & Koketsu, (1968) and called by them the late slow EPSP (fig. 9.1*d*). It persists in the presence of both nicotinic and muscarinic blocking agents and is thus not cholinergic. There is good evidence that the transmitter substance is a peptide similar to luteinizing hormone-releasing hormone, LHRH (Kuffler, 1980; Jan & Jan, 1982). The ion mechanism appears to be similar to that of the slow EPSP: closure of the M current potassium channels and, in some cells, an additional inward current which may involve sodium ion flow (Jones, 1985).

The late slow EPSP is found in both C and B cells but only C cells are contacted by presynaptic boutons that show immunoreactivity to LHRH. Hence it seems likely that the LHRH-like substance which is released from terminals on C cells can diffuse to the B cells many micrometres away. In accordance with this idea the time course of the late EPSP is faster in C cells than in B cells (Jan & Jan, 1982).

Slow potentials are widely distributed. Their time course,

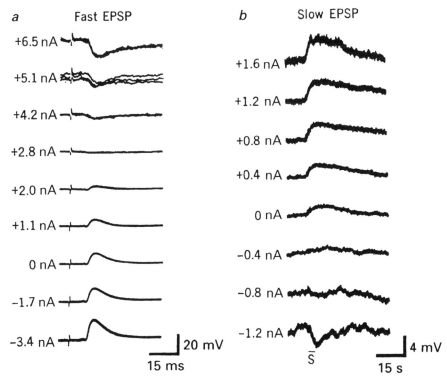

a   Fast EPSP

+6.5 nA

+5.1 nA

+4.2 nA

+2.8 nA

+2.0 nA

+1.1 nA

0 nA

-1.7 nA

-3.4 nA

20 mV

15 ms

b   Slow EPSP

+1.6 nA

+1.2 nA

+0.8 nA

+0.4 nA

0 nA

-0.4 nA

-0.8 nA

-1.2 nA

S̄

4 mV

15 s

Figure 9.2. Effects of applied currents on fast and slow EPSPs in frog sympathetic ganglion cells. *a* shows a series of fast EPSPs each produced by a single preganglionic stimulus, with depolarizing ($+$) or hyperpolarizing ($-$) currents applied to the postsynaptic cell. *b* shows a similar series of slow EPSPs, produced by a burst of repetitive stimuli; the fast EPSPs were blocked with nicotine. The resting potential was -65 mV; reversal potentials were approximately 0 mV for the fast EPSP and -88 mV for the slow EPSP. The directions of voltage change indicate that the fast EPSP is produced by an increase in membrane conductance, whereas the slow EPSP is produced by a decrease; this implies that channels open for the fast EPSP but close for the slow EPSP. (From Weight & Votova, 1970.)

and especially their long latency, could be explained if channel opening or closing is mediated not by a direct combination of the transmitter with the channel but by an indirect process involving intermediate steps between binding at the receptor and the response of the channel. In most cases it is clear that cytoplasmic 'second messengers' are involved. In others there may be a more direct link involving only membrane-delimited components. Figure 9.3 compares in outline the three main ways in which neurotransmitter receptors are linked to the ion channels through which the postsynaptic membrane currents flow.

Second messengers and G proteins

The second-messenger concept was first introduced to describe the role of cyclic adenosine monophosphate (cyclic AMP or cAMP) in hormone action (Sutherland & Rall, 1960; Sutherland, 1971). In skeletal muscle, for example, adrenaline causes an increase in intracellular cyclic AMP levels which leads ultimately to the breakdown of glycogen. Binding of the hormone to its receptor leads to activation of the membrane-bound enzyme adenylyl cyclase, which converts ATP to cyclic AMP. The cyclic AMP combines with a cyclic-AMP-dependent protein kinase which then

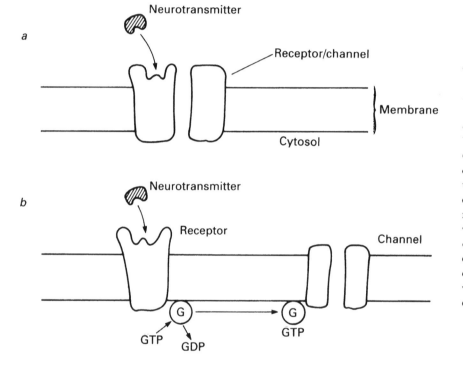

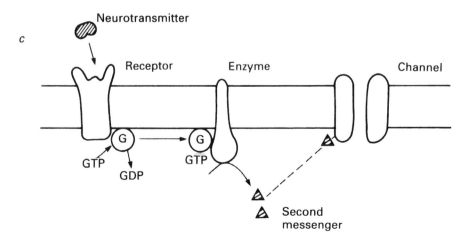

Figure 9.3. Direct and indirect actions of neurotransmitters on ion channels. *a* shows the direct action that occurs when the ion channel is an integral part of the receptor, as in the nicotinic acetylcholine receptor. In *b* and *c* the receptor molecule is not directly linked to an ion channel but acts via a G protein. In *b* the G protein acts directly on the channel to open or close it. In *c* the G protein activates an enzyme which generates a second messenger such as cyclic AMP which itself then alters the state of the channel, either directly by combination with it or, more commonly, by activating a kinase which phosphorylates the channel protein.

serves to phosphorylate target proteins, so altering their activity and producing some physiological or metabolic response. Protein phosphorylation provides a major general mechanism for the control of intracellular events.

At first it was thought that the adenylyl cyclase was itself the receptor for hormone action. It soon became evident, however, that many different hormones and neurotransmitters could activate the enzyme, suggesting that each one had its own receptor. Did each receptor act also as an adenylyl cyclase, or did they all act by stimulating a single separate enzyme? Birnbaumer & Rodbell (1969) found that enzyme activities in response to maximal effective concentrations of different hormones were not additive, as they would be if there were a number of separate cyclases. Then Rodbell and his colleagues (1971) found that adenylyl cyclase could not be activated by the activated hormone receptor unless GTP (guanosine triphosphate) was present. Further work by

Gilman, Rodbell and others led to the detection and isolation of a number of GTP-binding proteins that became known as G proteins, which act as intermediate links in the chain of events from receptor to adenylyl cyclase (Gilman, 1987). The details of these discoveries have been fascinatingly described by Gilman (1995) and Rodbell (1995) in their Nobel Prize lectures.

The G protein cycle

G proteins are activated by a set of receptors with a common structure, with seven transmembrane segments. The receptors themselves are widely diverse: they act individually with particular hormones, neurotransmitters, and other intercellular messengers, and also with various sensory stimuli such as odorant molecules and photons. We consider some of them in more detail later.

What G proteins do is indicated in fig. 9.4. They are

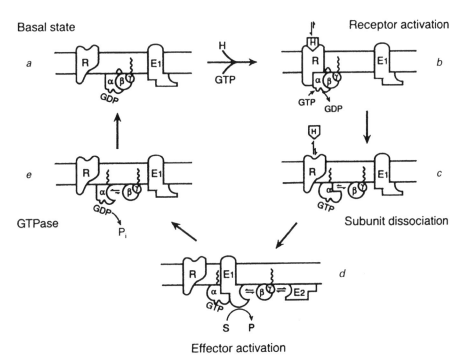

Effector activation

Figure 9.4. The G protein cycle for transmembrane signalling. In the basal state (*a*) G proteins exists as heterotrimers with GDP bound to the α-subunit. The hormone or neurotransmitter receptor R is unoccupied and the effector molecule E is inactive. When the neurotransmitter or hormone H binds to the receptor (*b*), the G protein binds to the inner side of the receptor and releases its bound GDP, taking up GTP in its place. This GTP binding makes the G protein dissociate from the receptor and itself dissociate into its α and βγ subunits (*c*). Now the α-GTP subunit binds to the effector molecule E_1 (usually an enzyme such as adenylyl cyclase) and activates it so that it produces its second-

messenger product P (cyclic AMP, for example) (*d*). The βγ subunit may also modulate the activity of the same or different effector molecules. The GTPase activity of the α subunit returns it to α-GDP (*e*), leading to reassociation with the βγ subunit and return to the basal state. Steps *b* and *c* may repeat many times, leading to activation of many G proteins molecules per activated receptor, and further amplification occurs at step *d*, where many P molecules are produced by each activated effector. (From Hepler & Gilman, 1992. Reproduced from: G proteins. *Trends in Biochemical Sciences* **17**, 383–7, with permission from Elsevier Trends Journals.)

heterotrimeric in structure; that is to say, they are composed of three different subunits, α, β and γ. The α subunit separates from the other two when it binds GTP, but the β and γ subunits are always bound tightly together. Fatty acid chains (palmitic or myristic acids) attached to the α and γ subunits tend to keep them in contact with the cytoplasmic face of the plasma membrane, so that they can shuttle between the hormone or neurotransmitter receptor (R in fig. 9.4) and the effector protein or proteins which they activate (E_1 and E_2 in fig. 9.4).

At rest the G proteins are in the trimeric $\alpha\beta\gamma$ form with GDP (guanosine diphosphate) bound tightly to the α subunit (stage *a* in fig. 9.4). When a neurotransmitter or hormone molecule (H) binds to the receptor, the receptor interacts with the G protein so that it releases the GDP and replaces it with GTP (stage *b*). This binding of GTP to the α subunit leads to dissociation of the G protein from the receptor and separation of the α subunit from the $\beta\gamma$ subunit (stage *c*). The receptor may then interact with other G protein molecules, until it becomes inactive after separation of the neurotransmitter or hormone. The separate α subunit now binds to an effector molecule E_1 and activates it, promoting the conversion of the substrate S into the product P (stage *d*). In some cases the $\beta\gamma$ subunit also binds to an effector molecule, either the same one (E_1) or a different one (E_2). This activity is brought to an end by the intrinsic GTPase activity of the α subunit (stage *e*); the bound GTP is hydrolysed leaving GDP in the binding site. When this happens the α and $\beta\gamma$ subunits reassociate and the system returns to its original state (stage *a*).

A system of this type clearly involves some considerable amplification. One receptor molecule binding a single neurotransmitter molecule (the first messenger) can activate a number of G protein molecules, and each of the activated effector molecules will convert many S molecules into the second messenger P.

There are twenty-three or more different G protein α subunits in mammals, with at least five different β subunits and ten different α subunits. The amino acid sequences of the different α subunits all show at least 50% identity with each other, and some are much more similar than that. Sequence analysis (Simon *et al.*, 1991) shows that they fall into four subfamilies: (1) the G_s group (G_s and G_{olf}) act as activators of adenylyl cyclase and are sensitive to cholera toxin; (2) the G_i group, mostly sensitive to pertussis toxin, including the ubiquitous α_i subgroup, the α_o group from the brain, and G_t (transducin) and G_g (gustducin) from photoreceptors and taste buds respectively; (3) the G_q group, activators of phospholipase Cβ; and (4) the G_{12} group.

Each G protein has its own effector protein or proteins, indicated by E_1 and E_2 in fig. 9.4. Some examples are the enzymes adenylyl cyclase, phospholipase A_2, phospholipase C and phosphodiesterase (table 9.1). When they are activated by the G protein, these enzymes act to alter the concentrations of second-messenger molecules such as cyclic AMP and inositol trisphosphate, and these in turn may affect the opening of ion channels. In other cases, ion channels are themselves directly activated by G proteins. Let us have a look at some of these signalling systems activated by G protein action.

Cyclic AMP

Adenylyl cyclase (also called adenylate cyclase and previously adenyl cyclase) is a membrane-bound enzyme that converts ATP into the second messenger cyclic AMP when it is activated by an appropriate activated G protein. Cloning studies show that there are at least eight different isoforms of the enzyme, with protein chains 1064 to 1248 amino acid residues long. There are probably two groups of six membrane-crossing segments in the molecule, each followed by a large cytoplasmic loop. Parts of the large cytoplasmic loops show considerable sequence identity in the different isoforms, and it seems likely that they include the catalytic sites for production of cyclic AMP and the regulatory sites for binding of G proteins (Taussig & Gilman, 1995).

Regulation of adenylyl cyclase activity was originally thought to be quite simple: it was switched on by G_s-linked receptors and switched off by G_i-linked receptors. Later work has shown that the situation is more complicated than this, with different isoforms subject to different patterns of activation and inhibition. All isoforms are activated by the α subunit of G_s. The α and $\beta\gamma$ subunits of G_i, G_o and G_z inhibit isoform I, and the α subunits of G_i and G_z inhibit V and VI also, but the $\beta\gamma$ subunits of G_i and G_o will activate isoform II. G_q will activate most isoforms via its activation of protein kinase C, and activates isoform I via calcium–calmodulin, but inhibits isoforms V and VI via its action on calcium ion concentrations. Adenylyl cyclases are thus focal points for the convergence of a great deal of regulatory information (Gilman, 1995; Sunahara *et al.*, 1996).

A major target for the cyclic AMP produced by adenylyl cyclase action is protein kinase A, an enzyme whose activity is dependent on cyclic AMP binding. Kinases are enzymes that phosphorylate their own target proteins by transfer of a phosphate group from ATP to the protein:

$$\text{protein} + \text{ATP} \xrightarrow{\text{protein kinase A}} \text{protein–P} + \text{ADP}$$

Table 9.1 *Properties of some mammalian G protein α subunits*

Family/subunit	Toxin	Tissue	Receptors	Effector
G_s				
α_s	CT	Ubiquitous	β-Adrenergic etc.	Adenylyl cyclase (+)
α_{olf}	CT	Olfactory organs	Odorant	Adenylyl cyclase (+)
G_i				
α_{i1}	PT	Nearly ubiquitous		Adenylyl cyclase (−)
α_{i2}	PT	Ubiquitous	M_2 muscarinic etc	Opens K channels
α_{i3}	PT	Nearly ubiquitous	Various	Phospholipase A_2 (+)
α_{oA}	PT	Brain, etc.	Met-enkephalin	Calcium channels (−)
α_{oB}	PT	Brain, etc.	α_2-adrenergic	Adenylyl cyclase (−)
α_T	CT, PT	Photoreceptors	Rhodopsin	Phosphodiesterase
α_g	PT	Taste buds	Sweet taste etc.	Adenylyl cyclase (+)
G_q				
α_q		Nearly ubiquitous	M_1 muscarinic, etc	
α_{11}		Nearly ubiquitous		
α_{14}		Lung, kidney, liver	Various	Phospholipase C
G_{12}				
α_{12}		Ubiquitous		
α_{13}		Ubiquitous		

Note:
CT, cholera toxin; PT, pertussis toxin. From Gilman, 1995, somewhat simplified.

The phosphorylation usually modifies the properties of the protein, which is not surprising in view of the negative charges on the phosphate group.

There are a number of different types of protein kinase, distinguished by the different ways in which their activity is controlled. Protein kinase A is one of the best known (Taylor, 1989). The enzyme molecule is a complex of two regulatory subunits, each of which will bind two molecules of cyclic AMP, and two catalytic subunits which catalyse the phosphorylation of the substrate protein. The complex is inactive in the absence of cyclic AMP, but it dissociates to release the active catalytic subunits when the cyclic AMP is bound. Figure 9.5 summarizes the cascade of events involving cyclic AMP as a second messenger.

Phosphorylated proteins can be dephosphorylated by means of a different enzyme, a phosphorylase:

$$\text{protein--P} + H_2O \xrightarrow{\text{phosphorylase}} \text{protein} + \text{phosphate}$$

Consequently the state of many cellular activities is controlled by the degree of phosphorylation of enzymes and other proteins, and that in turn may be controlled by the activity of a protein kinase that is activated by cyclic AMP or some other second messenger. Synaptic transmission can then be effected or modulated by the phosphorylation of an ion channel.

An example of cyclic AMP-dependent phosphorylation associated with synaptic transmission is given by the adrenergic stimulation of the heart beat. Sympathetic cardioaccelerator nerves release noradrenaline, which binds to the β_1-adrenergic receptor. This activates a G protein (G_s), so that the Gα subunit binds GTP and is released from the receptor and the βγ subunit. The Gα subunit in turn acts as an activator for adenylyl cyclase, so producing cyclic AMP. Cyclic AMP activates protein kinase A and this then phosphorylates L-type calcium channels in the heart muscle plasma membrane. Phosphorylation increases the open probability for the calcium channels, so more calcium ions enter the cell when it is next depolarized and the contraction force is increased (Trautwein & Hescheler, 1990; Hartzell *et al.*, 1991).

Some ion channels are directly activated by cyclic

nucleotide second messengers (see Yau, 1994*a*). In olfactory neurons, cation-selective ion channels that are similar in structure to voltage-gated channels are opened by the action of internal cyclic AMP. Similar channels in photoreceptor cells are opened by cyclic GMP, whose concentration is reduced by the action of phosphodiesterase, an enzyme activated by G_t (transducin), which is itself activated by the visual pigment rhodopsin after absorption of light.

The phosphatidylinositol system
Phosphatidylinositol is a phospholipid found largely in the inner leaflet of the cell membrane lipid bilayer. It consists of two fatty acyl chains linked to a glycerol moiety whose third ester linkage is connected via a phosphate group to inositol, a six-carbon sugar. The sugar moiety is readily phosphorylated, to give phosphatidylinositol-4,5-bisphosphate (PIP_2). This can be hydrolysed by the membrane-bound enzyme phospholipase C to give two second

messengers, inositol 1,4,5-trisphosphate (IP_3) and diacylglycerol (DAG). IP_3 is water soluble and so diffuses through the cytoplasm, whereas DAG remains in the membrane. The phospholipase Cβ1 isoform is activated by the α subunits of certain G proteins, particularly G_q and G_{11}, and the β2 isoform is probably activated by some βγ subunits (see Berridge, 1993; Watson & Arkinstall, 1994).

IP_3 promotes release of calcium ions from the endoplasmic reticulum in a variety of cell types (Streb *et al.*, 1983; Berridge, 1993, 1997). It does this by binding to a receptor which is also a calcium release channel in the endoplasmic reticulum membrane. The IP_3 receptor is a large tetrameric molecule, each subunit of which is typically 2749 amino acid residues long (Furuichi *et al.*, 1994).

DAG activates the membrane-bound enzyme protein kinase C, which will then phosphorylate a target protein and so alter its activity. There at least ten different isoforms of protein kinase C; their targets include various ion channels and other membrane components, and also other pro-

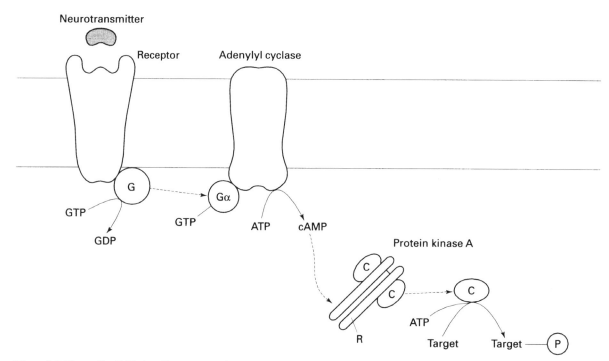

Figure 9.5. The cyclic AMP signalling system. The neurotransmitter or hormone combines with the 7TM receptor and so converts it to the active form which will activate G protein molecules of the G_s group. These then bind GTP and their α subunits can then activate the membrane-bound enzyme adenylyl cyclase. This activates the adenylyl cyclase so that numbers of ATP molecules are converted to cyclic AMP (cAMP). The cyclic AMP may activate a channel directly (as in olfactory neurons), but more commonly serves as an activator for protein kinase A by binding to its regulatory subunit. The catalytic subunits (C) of protein kinase A are thereby released to promote the phosphorylation of a target protein, which may be a channel or another enzyme. Adenylyl cyclase can be inhibited by the action of other G proteins such as G_{i1} and G_{oB}, themselves activated via other 7TM receptors.

teins involved in various cell responses (Nishizuka, 1984, 1992; Tanaka & Nishizuka, 1994). DAG can also be produced by hydrolysis of phosphatidylcholine; this source tends to be more important for longer term changes in protein kinase C activity (Exton, 1990). Figure 9.6 summarizes the cascade of event in the phosphatidylinositol signalling system.

Direct action of G proteins on channels

In heart muscle it is clear that the muscarinic acetylcholine receptors are coupled to inward rectifier potassium channels via a G protein without the intervention of a second-messenger system (Pfaffinger *et al.*, 1985; Yatani *et al.*, 1987). The channel has been cloned as GIRK1. Combination of the $G_{\beta\gamma}$ subunit with the potassium channel then opens it (Logothetis *et al.*, 1987; Reuveny *et al.*, 1994; Wickman *et al.*, 1994). This is the means whereby acetylcholine, released on stimulation of the vagus nerve as Loewi showed in his classic experiments in the 1920s, produces inhibition of the heart beat.

Systems of this type (fig. 9.1*b*) probably occur elsewhere. Clapham (1994) has discussed some of the problems of distinguishing a direct action of the G protein of the channel from a less direct action that involves only membrane-delimited components.

G-protein-linked receptors

Let us now have a look at the receptors that act by activating G proteins. The first neurotransmitter receptor in this group to have its primary structure determined was the β_2-adrenergic receptor (Dixon *et al.*, 1986). The receptor is a single protein chain 413 amino acid residues long. It was immediately evident that the amino acid sequence showed considerable identity with that of the visual pigment rhodopsin. Hydrophobicity analysis suggested that, just as in rhodopsin, there are seven hydrophobic membrane-spanning sections in the molecule. The muscarinic acetylcholine receptor was cloned soon afterwards, and it too showed considerable sequence identity with the β-adrenergic receptor and with rhodopsin and had seven putative membrane-crossing segments (Kubo *et al.*, 1986).

Large numbers of receptors of this superfamily have since been discovered, all possessing the same general structure with seven hydrophobic transmembrane segments, as is shown in fig. 9.7 (Dohlman *et al.*, 1991; Strosberg, 1991; Watson & Arkinstall, 1994). They are thus sometimes named seven-transmembrane-segment receptors, 7TM receptors, R7G receptors or heptahelical receptors. Different members of the superfamily respond to various neurotransmitters, neuropeptides, hormones, olfactory stimulants, and light-induced isomerizations of retinal.

Evidence that the membrane topography of the β_2-adrenergic receptor is like that shown in fig. 9.7 comes from a number of different investigations. Glycosylation sites (which would be external) are found near the N-terminus, and phosphorylation sites (which would be cytoplasmic) are found on the third internal loop and the C-terminal region (Dohlman *et al.*, 1987). Hamster ovary cells in tissue

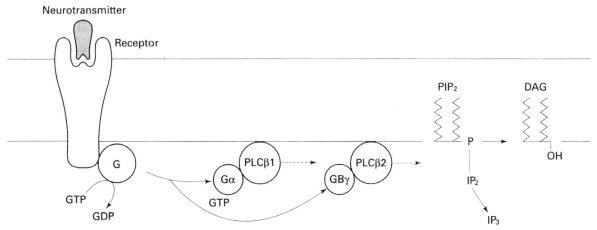

Figure 9.6. The phosphatidylinositol signalling system. Combination of a neurotransmitter or hormone molecule with the appropriate 7TM receptor activates G proteins of the G_q group. Each G protein molecule binds GTP and splits into its α and βγ subunits. These in turn activate phospholipases Cβ1 and Cβ2. These PLCβ molecules then hydrolyse the membrane phospholipid PIP_2 (phosphoinositol 4,5-bisphosphate), giving two second-messenger molecules, IP_3 (inositol 1,4,5-trisphosphate) and DAG (diacylglycerol). IP_3 binds to the calcium release channel and so promotes release of calcium ions from the endoplasmic reticulum. DAG activates protein kinase C, which phosphorylates a variety of target proteins..

culture can be transfected with cDNA coding for the receptor so that it is expressed densely in the plasma membrane. Antibodies to short sections of the protein chain in the N-terminus or to the putative external loops will bind to the cells, whereas antibodies to sections of the chain in the internal cytoplasmic loops will not, unless the cells are first permeabilized so that the antibodies can enter them (Wang *et al.*, 1989).

The high concentration of rhodopsin in photoreceptor membranes has allowed electron cryomicroscopy and image reconstruction techniques to be used (Unger & Schertler, 1995). The results agree well with the seven transmembrane α-helix model; four of the helices are almost perpendicular to the plane of the membrane, the other

three are more tilted (fig. 9.8). It seems reasonable to assume that other members of the superfamily have a similar arrangement of their transmembrane segments. Baldwin (1993) compared 105 different sequences and demonstrated appreciable structural concordances in their transmembrane segments, such that it seemed very likely that all members of the superfamily do indeed have a similar topographical arrangement.

The cysteine residue at position 341 in the C-terminal region of the β_2-adrenergic receptor binds covalently to palmitic acid (O'Dowd *et al.*, 1989). The lipid palmitoyl chain apparently becomes embedded in the lipid bilayer so that a fourth cytoplasmic loop is formed.

The ligand-binding sites for those 7TM receptors that

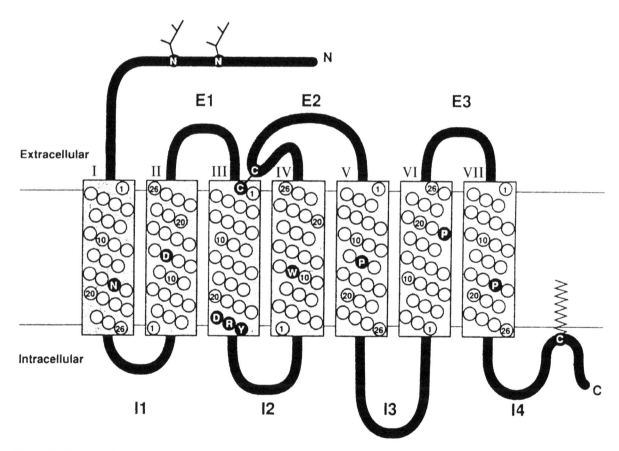

Figure 9.7. Transmembrane topology of a generalized 7TM receptor of the type binding small ligands, as in rhodopsin, the β_2-adrenergic receptor, or muscarinic receptors. The seven membrane-spanning α-helices (I to VII) are connected by three extracellular loops (E1 to E3) and three intracellular ones (I1 to I3). A fourth intracellular loop (I4) occurs when a conserved cysteine residue in the C terminal region is palmitoylated. *N*-glycosylated residues occur in the N-terminal region, and a

disulphide link between two cysteine residues connects E2 to the E1–III junction. Highly conserved residues (i.e. those found in most members of the superfamily) are shown as black circles. Numbering of the residues in the transmembrane helices is as given by Baldwin (1993, 1994). Residues at the base of III and at the ends of I2, I3 and I4 are involved in binding to G proteins. (From Shenker, 1995.)

bind small molecules are situated in the membrane between the membrane-crossing helices. Evidence for this is provided by site-directed mutagenesis experiments. Thus changes in amino acid residues in the hydrophilic regions of the β-adrenergic receptor have little effect on binding, but the presence of the aspartate-113 residue with its ionized carboxyl group on the third transmembrane helix, and of two serine residues (at 204 and 207) with their hydroxyl groups on the fifth helix, are crucial (Dixon *et al.*, 1987; Strader *et al.*, 1988, 1989). Aspartate-113 corresponds to position III-7 in the generalized diagram of fig. 9.7, and serine-204 and -207 correspond to positions V-7 and V-10. These residues are found in all 7TM receptors that bind catecholamines, and the aspartate at III-7 is also found in those that bind small biogenic amines such as acetylcholine or 5-hydroxytryptamine.

The fluorescent antagonist carazolol binds firmly to the β-adrenergic receptor, and its fluorescence emission spectrum in this state suggests it is in an extremely hydrophobic environment (Tota & Strader, 1990). Thus we have a picture of the ligand sitting in the space between some of the transmembrane helices, with its positively charged amino group forming an ionic bond with the carboxyl group of arginine-113, and the hydroxyl groups of its catechol ring forming hydrogen bonds with serine-204 and -207. Figure 9.9 suggests how adrenaline might be bound in this way.

Some of the peptide-binding 7TM receptors have a much longer N-terminal segment than do those for the majority of neurotransmitters and hormones, and it seems likely that in these receptors the peptide-binding sites are located in the N-terminal region. They can be regarded as a separate group within the 7TM receptor superfamily (Ishihara *et al.*, 1992). Metabotropic glutamate receptors form another subfamily; they have an even larger N-terminal region, and here also it contains the ligand-binding site (see Hollmann & Heinemann, 1994).

Site-directed mutagenesis can be used to investigate interactions with G proteins. It is found that the cytoplasmic end of segment III and the intracellular loops I2, I3 (especially its two ends) and I4 are all involved in G protein activation (see Ostrowski *et al.*, 1992; Shenker, 1995). We may assume that binding of the neurotransmitter or other ligand produces a conformational change at the binding site, and that this then produces further conformational change in the cytoplasmic parts of the receptor that interact with the G protein.

Many different G-protein-linked receptors have been described, and undoubtedly there are more to be discovered. Each type is activated by its own particular hormone, neurotransmitter or other messenger molecule. Each forms the first link in a three-protein chain (receptor, G protein, effector) of sequential activation in the cell signalling process. Since each receptor may activate more than one type of G protein, and since each G protein may activate or inhibit more than one effector protein, the whole system is a complex signalling network with divergent and convergent pathways (Gudermann *et al.*, 1996). It is a major challenge for cell biologists to work out how this extensive cross-talk between signalling pathways can occur without loss of precision in the control systems of the cell.

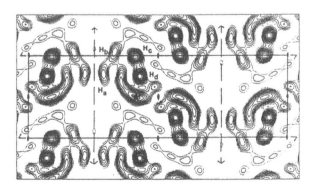

Figure 9.8. Electron density of part of the photoreceptor membrane in a cow's eye, suggesting that the rhodopsin molecule has seven transmembrane α-helices. Four of these (H_a, H_b, H_c and H_d) are nearly perpendicular to the plane of the membrane, whereas the other three are tilted so that they form an arc in the diagram, indicated by the asterisk (*). (From Unger & Schertler, 1995.)

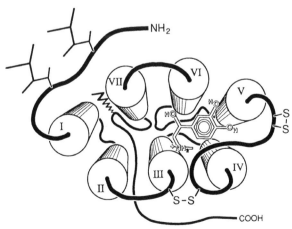

Figure 9.9. The probable arrangement of the transmembrane helices in the β_2-adrenergic receptor, with adrenaline in the binding site. The molecule is drawn as seen from the outer side of the plasma membrane. (From Ostrowski *et al.*, 1992. Reproduced with permission from the *Annual Review of Pharmacology and Toxicology* Volume 32, © 1992 by Annual Reviews Inc.)

Table 9.2 *Some neurotransmitters and their receptors*

Transmitter	Ionotropic	Metabotropic
Acetylcholine	Nicotinic receptors	Muscarinic receptors M_1 to M_5
GABA	$GABA_A$ receptor	$GABA_B$ receptor
Glycine	Glycine receptor	—
5-Hydroxytryptamine	5-HT_3 receptor	$5\text{-HT}_{1,2,4}$ receptors
Glutamate	AMPA-kainate receptors NMDA receptors	$mGluR_1$ to $mGluR_5$ receptors
ATP	P_{2X} receptor	P_{2Y} receptor
Noradrenaline	—	α_1, α_2, β_1 and β_2 receptors
Dopamine	—	D_1-like and D_2-like receptors
Neuropeptides	—	Rhodopsin-like (e.g. substance P, enkephalin) Glucagon-receptor-like (e.g. VIP)

Notes:

GABA, γ-amino butyric acid; 5-HT, 5-hydroxytryptamine; AMPA, α-amino-3-hydroxy-5-methyl-4-isoxazolepropionate; NMDA, *N*-methyl-D-asparate; VIP, vasoactive intestinal polypeptide.
Ionotropic receptors have their own intrinsic ion channel and mediate fast synaptic transmission. Metabotropic receptors activate ion channels indirectly via G proteins and (usually) second-messenger systems. Many receptors have a number of different subtypes or isoforms not indicated here.

The rest of this chapter is concerned with slow synaptic transmission by particular neurotransmitters. Sources for further information include Watson & Arkinstall (1994), Cooper *et al.* (1996) and Hardman *et al.* (1996). Table 9.2 outlines some of the different types of neurotransmitter receptor, both ionotropic (for fast transmission) and metabotropic (for slow).

Muscarinic receptors

Muscarine is a constituent of the poisonous mushroom *Amanita muscaria*. It acts as an agonist of acetylcholine at postganglionic parasympathetic synapses and elsewhere. Dale (1914) distinguished between the nicotinic and muscarinic actions of acetylcholine according to whether the effects were mimicked by nicotine or muscarine. We now know that nicotinic receptors have intrinsic ion channels whereas muscarinic receptors activate G proteins. Hence the postsynaptic potentials produced by the activation of muscarinic receptors are much slower than those involving nicotinic receptors, as is illustrated in figs. 8.8 and 9.1.

Atropine is a well-known antagonist of muscarinic actions. It is an alkaloid, largely responsible for the poisonous properties of the deadly nightshade *Atropa belladonna*. The deadly nightshade's Latin name, given to it by Linnaeus, reflects the properties of atropine: Atropos was

the lady in Greek mythology who cut the thread of life, and *belladonna* ('beautiful woman') refers to the cosmetic effect of the dilation of the pupils produced by atropine. (We tend to feel attracted to those whose pupils dilate when they look at us. Candlelight is cheaper than atropine and has fewer side-effects.)

Not all muscarinic receptors have precisely the same pharmacological properties. Those in sympathetic ganglia have a high affinity for the antagonist pirenzepine, whereas those in heart muscle have a much lower affinity (Hammer & Giachetti, 1982). Another distinction appears in the action of agonists on chick embryonic heart cells: both carbachol and oxotremorine inhibit the production of cyclic AMP, but only carbachol stimulates phosphoinositol hydrolysis (Brown & Brown, 1984). These results suggested that different types of receptor occurred, a conclusion substantiated more recently by cloning methods.

Muscarinic receptors purified from pig brain are single proteins with an M_r of about 70 000. Good evidence for the association of the muscarinic receptor with G proteins was provided by reconstitution experiments by Haga and his colleagues (1985, 1986). They mixed purified receptors from pig brain and purified G proteins with phospholipid vesicles. This would allow the proteins to be reconstituted in a membrane environment free from other membrane

Table 9.3 *Some characteristics of cloned muscarinic receptors*

	M_1	M_2	M_3	M_4	M_5
Amino acid residues	460	466	590	479	532
Selective antagonists	Pirenzipene	Himbacine	Hexahydro-siladifenidol	Tropicamide	—
Location	CNS, autonomic ganglia	Heart, smooth muscle	CNS, ganglia, glands	CNS	CNS
G protein	$G_{q/11}$	$G_{i/o}$	$G_{q/11}$	$G_{i/o}$	$G_{q/11}$
Main second messenger pathway	Phospho-inositide	Inhibition of adenylyl cyclase and L-type Ca channels, excitation of K channels	Phospho-inositide	Inhibition of adenylyl cyclase and L-type Ca channels, excitation of K channels	Phospho-inositide

Notes:

CNS, central nervous system.

The agonists do not show great selectivity between the different receptors. The number of amino acid residues per receptor is given for the human isoforms.

proteins. Reconstituted G_i alone has some GTPase activity, which is not affected by the muscarinic agonist carbachol. Reconstituted receptor alone has no GTPase activity. When both proteins are reconstituted together, the GTPase activity is much increased by carbachol. This shows that G_i is indeed activated when the muscarinic receptor is activated. We would therefore expect that activation of muscarinic receptors linked to G_i would inhibit the production of cyclic AMP.

The molecular structure of muscarinic receptors from pig brain was first deduced from the complementary DNA sequence by the Kyoto University group (Kubo *et al.*, 1986). The protein consisted of 460 amino acid residues, giving an M_r of 51 416. The difference between this figure and the 70 000 derived from SDS-polyacrylamide gel measurements is probably largely accounted for by the carbohydrate content of the receptor protein. Further cloning studies showed that there are at least five different types of muscarinic receptor, named M_1 to M_5 (Hulme *et al.*, 1990; Wess, 1993). Some details of these different types are given in table 9.3.

Hydrophobicity plots of muscarinic receptors show seven hydrophobic peaks just as in other 7TM receptors. Binding studies with cloned mutants have produced some interesting conclusions about how they bind acetylcholine. Just as in the β-adrenergic receptor, the aspartate residue at III-7 is crucial, its charged carboxyl group forming an ionic bond with the quaternary nitrogen atom of the acetylcholine. In addition, the hydroxyl groups on the tyrosine

residue at III-8 and the threonine residues at V-3 and V-6 seem to be involved in binding, probably via the formation of hydrogen bonds (Wess *et al.*, 1991; Wess, 1993; Baldwin, 1994).

Catecholamines

The catecholamines are an important group of physiologically active compounds which contain the catechol group, a benzene ring with two adjacent hydroxyl groups. The hormone *adrenaline* (epinephrine) is released by the adrenal glands of mammals and serves to prepare the animal for action; the rate and amplitude of the heart beat increase, constriction of the blood vessels (except in the muscles) occurs, the spleen contracts, the pupils dilate, intestinal movements are inhibited, etc. Adrenaline is synthesized from the amino acid tyrosine via a series of other catecholamines, dopa, dopamine and noradrenaline, as is shown in fig. 10.24.

As mentioned earlier, Elliott in 1904 suggested that adrenaline was released from postganglionic sympathetic nerves on stimulation. In later years, a number of discrepancies between the action of adrenaline and sympathetic stimulation were discovered, so the sympathetic transmitter substance became known by the non-committal name 'sympathin'. Later, von Euler (see von Euler, 1955) established that in most cases the sympathetic transmitter is not adrenaline but noradrenaline (norepinephrine).

Neurons in which noradrenaline is the transmitter substance are fairly widespread in the mammalian central

Table 9.4 *Agonists and antagonists at adrenergic receptors*

Receptor type	Agonists	Antagonists
α_1 and α_2		Phentolamine
		Phenoxybenzamine
α_1	Phenylephrine	Prazosin
		Corynanthine
α_2	Clonidine	Yohimbine
		Rauwolscine
β_1 and β_2	Isoproterenol	Propranolol
β_1	Dobutamine	Practolol
β_2	Salbutamol	Butoxamine

nervous system. They can be demonstrated by a useful technique in which freeze-dried sections of brain tissue are exposed to hot formalin vapour; the catecholamines are then converted to substances which fluoresce strongly under ultraviolet light (Carlsson *et al.*, 1962; see Carlsson, 1987). Neurons in which dopamine appears to be the transmitter are also present.

Adrenergic receptors

Ahlquist (1948) compared the activity of the adrenergic agonist isoproterenol (isoprenaline) with those of adrenaline and noradrenaline on a variety of tissues. He found that in some cases (some blood vessels, the pregnant uterus, nictitating membrane, ureter) isoproterenol was much less active than the other compounds; he suggested that the adrenergic receptors in these tissues should be called α-receptors. For other tissues (other blood vessels, non-pregnant uterus, heart) isoproterenol was more active; these were described as having β-receptors. There are different competitive blocking agents (antagonists of noradrenaline) at the two types: α-receptors are blocked by phentolamine whereas β-receptors are blocked by propranolol, for example.

Later work resulted in further subdivisions of these receptor types, largely on the basis of their responsiveness to different agonists or antagonists. Thus β_1-receptors are responsive to dobutamine and not responsive to salbutamol, whereas the reverse holds for β_2-receptors. α_1-receptors are blocked by prazosin and corynanthine, whereas α_2-receptors are blocked by yohimbine and rauwolscine (see table 9.4). Cloning studies have shown that there are further subtypes of α_1- and α_2-adrenergic receptors, called α_{1A-D} and α_{2A-C} (see Watson & Arkinstall, 1994).

There is good evidence that α_1-receptors are linked to the phosphatidylinositol second-messenger system via G_q. In

rat liver cells, for example, treatment with adrenaline causes rises in the cytoplasmic concentrations of IP_3 and calcium ions (Charest *et al.*, 1985; Exton, 1985). α_2-Receptors, on the other hand, act in a variety of tissues to inhibit adenylyl cyclase via activation of the inhibitory protein G_i (Bylund & U'Prichard, 1983).

β-Receptors are also linked to adenylyl cyclase, but via the stimulatory protein G_s (see Levitzki, 1986). Thus activation of β-receptors leads to an increase in the cytoplasmic concentration of the second messenger cyclic AMP. A nice test of this conclusion is provided by the reconstitution experiments by May *et al.* (1985). They purified the three necessary proteins for this system: the β_1-adrenergic receptor (from turkey erythrocytes), the G_s protein (from rabbit liver) and the enzyme adenylyl cyclase (from cow brain). These were then restored to their membrane lipid environment by reconstitution in phospholipid vesicles. Adenylyl cyclase activity (the ability to make cyclic AMP from ATP) was stimulated 2.6-fold by β-adrenergic agonists. This stimulation was dependent upon GTP and was blocked by β-adrenergic antagonists.

Table 9.4 gives some examples of adrenergic agonists and antagonists. Propranolol and other 'β-blockers' are clinically useful in the treatment of high blood pressure. Clonidine produces a brief rise in blood pressure, probably by stimulating α_2-receptors in the arterioles, followed by a prolonged fall which is thought to be caused by activation of the presynaptic α_2-receptors on adrenergic neurons in the brain (Katzung, 1995). Isoproterenol and salbutamol are β-receptor agonists which are useful as bronchodilator agents in the treatment of asthma. Isoproterenol may produce side-effects by stimulating the β_1-receptors in the heart, whereas salbutamol does not since it is selective for β_2-receptors.

Dopamine

Dopamine is a precursor of noradrenaline (fig. 10.24), but it also acts as a neurotransmitter. The discovery in 1961 that particular neurons in the brain contain appreciable quantities of dopamine and that these same neurons were the ones which degenerate in Parkinson's disease, set the scene for a dramatic advance in the treatment of the disease (see Hornykiewicz, 1973; Carlsson, 1993). Administration of L-dopa, the immediate precursor of dopamine, produced a rapid relief of the symptoms. Malfunction of dopaminergic neurons has also been implicated in schizophrenia, but the evidence here is much less conclusive (Owen *et al.*, 1985).

Kebabian & Calne (1979) divided dopamine receptors into two types, D_1 and D_2, on the basis of their properties. Stimulation of D_1 receptors activates adenylyl cyclase via

G_s and so increases cyclic AMP formation. Stimulation of D_2 receptors inhibits adenylyl cyclase and increases potassium channel activity, probably via G proteins of the G_i/G_o group. D_1 and D_2 receptors are both found on postsynaptic membranes at dopaminergic synapses, but D_2 receptors also occur presynaptically on the terminals of dopaminergic and other neurons, where they must act as a negative feedback system controlling the amount of dopamine released by the terminal. Antipsychotic drugs such as chlorpromazine and others act as antagonists at D_2 receptors.

The primary structure of the D_2 receptor was determined by cDNA cloning (Bunzow *et al.*, 1988), and the use of probes based on this led to cloning of the D_1 receptor and also to further receptors named D_3 to D_5 (see Grandy & Civelli, 1992; Gingrich & Caron, 1993). D_5 shows appreciable sequence identity to D_1 and has similar pharmacology, indeed some authors prefer the nomenclature D_{1A} and D_{1B} for D_1 and D_5, respectively. D_3 and D_4, on the other hand, are more like D_2. The D_1-like receptors D_1 and D_5 have shorter I3 cytoplasmic loops than do the D_2-like receptors, and much longer C-terminal regions. It is interesting that the genes coding for the D_1-like receptors have no introns in the coding sequence (as is the case for many 7TM receptor genes), whereas those coding for the D_2-like receptors do.

Some other 7TM receptors for small molecules

5-Hydroxytryptamine

5-Hydroxytryptamine (5-HT, serotonin) is an indoleamine (fig 9.10). Indoleamines and catecholamines are sometimes grouped together as monoamines. Relatively large quantities of 5-HT are found in blood serum and platelets, and in the chromaffin cells of the gut lining. It causes potentiation of the heart beat in molluscs, and this effect was used by Twarog & Page (1953) to show that there is an appreciable concentration of 5-HT in the brain.

5-HT has a wide variety of effects on nerve cells. Inhibitory effects depend largely on increases in membrane potassium conductance, and many excitatory effects seem to be caused by a decrease in potassium conductance. Some effects are produced via changes in acetylcholine or noradrenaline release, suggesting that presynaptic 5-HT receptors are involved.

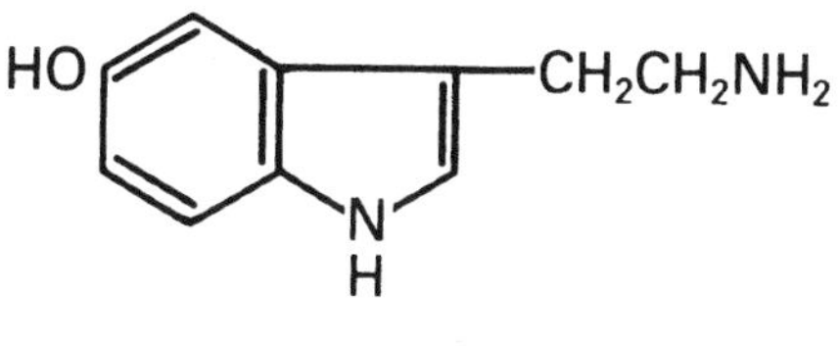

Figure 9.10. Serotonin.

Pharmacological studies have demonstrated the presence of different types of 5-HT receptors. Peroutka *et al.* (1981) measured the ability of various different drugs to displace radioactive 5-HT and radioactive spiperone from binding sites on rat brain membrane fractions. They found that inhibition of 5-HT binding was proportional to the degree of inhibition of 5-HT-sensitive adenylyl cyclase activity. However, inhibition of spiperone binding was not related to adenylyl cyclase activity but was proportional to the inhibition of 5-HT-induced head twitches in rats. They suggested that the two types of response involved two different receptor types, which they called 5-HT_1 and 5-HT_2.

Later work confirmed this view and has added at least two more receptor types (Hen, 1992; Hoyer *et al.*, 1994). 5-HT_1 and 5-HT_2 receptors are 7TM receptors coupled to G proteins. 5-HT_3 receptors are ion channels of similar structure to the nicotinic acetylcholine receptor, as described in chapter 8. 5-HT_4 receptors are further 7TM receptors, but with drug responsiveness different from that of 5-HT_1 and 5-HT_2 receptors (Bockaert *et al.*, 1992; Gerald *et al.*, 1995). Further details are given in table 9.5.

There are further subtypes of these four main groups. Thus the 5-HT_1 group has A, B and D subtypes, and the 5-HT_2 group has subtypes A to C. Sequences isolated by cloning only have been named 5-HT_5, 5-HT_6 and 5-HT_7.

Some of the substances active at 5-hydroxytryptaminergic synapses are useful in medical practice. Methysergide, a $5\text{-HT}_{2A/2C}$ antagonist, and sumitriptan, a 5-HT_{1D} agonist, are used in the treatment of migraine. Cisapride, a 5-HT_4 agonist, can be useful in promoting gastrointestinal motility. Lysergic acid diethylamide (LSD) is a hallucinogen; it is an agonist at $5\text{-HT}_{2A/2C}$ and other receptors.

5-HT is synthesized from the amino acid tryptophan and can be deaminated by the enzyme monoamine oxidase. In the pineal organ (which contains large quantities of 5-HT) it is converted into the hormone melatonin.

Histamine

The main concentrations of histamine in the body are in the mast cells and basophil leucocytes, from which it is released to act as a 'local hormone'. Histamine has a marked effect on some smooth muscles, increases capillary permeability, and stimulates sensory nerve endings, leading to sensations of pain and itching. It also occurs as a neurotransmitter in the brain (see Schwartz *et al.*, 1991). It is synthesized from the amino acid histidine, using the enzyme histidine decarboxylase.

Postsynaptic and effector cell receptors are of two types, H_1 and H_2 (Black *et al.*, 1972). Mepyramine and the 'antihistamines' (such as promethazine) used in the treatment of

Table 9.5 *Properties of 5-hydroxytryptamine receptors*

	5-HT$_1$	5-HT$_2$	5-HT$_3$	5-HT$_4$
Molecular structure	7TM	7TM	Multimeric ion channel	7TM
Effect of activation	Inhibition of AC via G$_i$/G$_o$	Activation of PLC via G$_q$	Cation flow→depolarization	Activation of AC via G$_s$
Agonists	8-OH-DPAT Sumatriptan	α-Methyl-5-HT	2-Methyl-5-HT	Cisapride
Antagonists	Spiperone Isamoltane	Ketanserin Spiperone	Ondansetron	GR 113808

Note:

AC, adenylyl cyclase; PLC, phospholipase C. The antagonists and agonists listed may act specifically only on certain subtypes; e.g. sumatriptan is specific for 5-HT$_{1D}$ receptors.

allergic reactions and in preventing motion sickness are antagonists at H$_1$ receptors, whereas burimamide is an antagonist at H$_2$ receptors. Presynaptic histamine receptors form a third class (H$_3$), both in the brain and at peripheral autonomic nerve endings.

H$_1$ and H$_2$ receptors are both 7TM receptors coupled to G proteins. H$_1$ receptors activate the phosphoinositide signalling pathway, probably via G$_q$/G$_{11}$ G proteins. H$_2$ receptors are coupled to G$_s$ so as to activate adenylyl cyclase. At the time of writing less is known about the molecular biology of H$_3$ receptors.

Purines

The phenomenon of purinergic transmission has been discussed briefly in the chapter 8. G-protein-linked receptors for adenosine and ATP occur in the central nervous system. Adenosine receptors were at first known as P$_1$ purinoceptors, but are now called A receptors. The A$_1$ and A$_3$ receptors inhibit adenylyl cyclase, probably via G$_i$/G$_o$ proteins. The A$_2$ subtypes stimulate adenylyl cyclase via G$_s$. It may be that the stimulants caffeine (from coffee) and theophylline (from tea) act by blocking adenosine receptors.

P$_2$ purinoceptors are activated by ATP. P$_{2X}$ receptors are ion channels, as we have seen in the chapter 8. P$_{2Y}$ receptors are G protein coupled, producing activation of the phosphoinositide pathway.

The cannabinoid receptor

Extracts of the hemp plant *Cannabis sativa* have been used as an intoxicant for thousands of years. Marijuana smokers seek sensations of euphoria and an altered perception of time; they will also experience sedation, delayed reaction time, lack of motor coordination and memory impairment. The active components of the drug have been isolated and are known as tetrahydrocannabinols.

Binding sites for synthetic cannabinoid agonists were found in brain membranes, and the receptor was eventually isolated and cloned (Matsuda *et al.*, 1990). It is a member of the 7TM superfamily, and inhibits adenylyl cyclase activity probably via a G protein of the G$_i$/G$_o$ group. The natural ligand for the receptor appears to be anandamide, a derivative of arachidonic acid, so it seems likely that this compound is used a messenger in the brain in some way (Devane *et al.*, 1992).

Metabotropic glutamate receptors

We have seen in previous chapters that glutamate is the major neurotransmitter for fast synaptic action in the central nervous system, and that this is mediated via AMPA, KA and NMDA receptors, all of which are ionotropic, having intrinsic ion channels. A different type of glutamate receptor was discovered by Sugiyama *et al.* (1987). They found that inward currents could be induced by glutamate and quisqualate in *Xenopus* oocytes injected with rat brain mRNA, but that these currents were slow and irregular in form with the delayed onset typical of the slow responses mediated by second messengers. The responses were much reduced by pertussis toxin, suggesting the involvement of G proteins. Injection of inositol trisphosphate into the oocyte produced similar responses. Injection of EGTA (a calcium-chelating agent) inhibited both the quisqualate–glutamate responses and the IP$_3$ responses, suggesting that the currents are produced by calcium-activated chloride channels. Receptors of this type have since become known as metabotropic glutamate receptors.

Houamed and his colleagues (1991) isolated mGluR1 by expression cloning in *Xenopus* oocytes. They grouped cDNA clones from a rat cerebellum library in pools of 100 000 to make mRNA, injected it into oocytes, and then looked for calcium-activated chloride currents in response to glutamate application to the outside of the oocyte. They then subdivided the pools showing positive responses and repeated the process until a single positive clone was identified, and then they sequenced its cDNA. Knowledge of this sequence enabled others to use low stringency hybridization screening with probes from mGluR1, with the result that other members of the mGluR family were soon discovered.

Cloning shows that at least eight different subtypes of metabotropic glutamate receptors exist, known as mGluR1 to mGluR8 (Hollmann & Heinemann, 1994; Pin & Bockaert, 1995). They all have seven transmembrane segments but show virtually no sequence identity with other 7TM receptors. Their N-terminal sections are all over 500 amino acid residues long, much longer than in most other 7TM receptors. It seems very likely that the glutamate-binding site is in this region, rather than in the transmembrane segment region as in adrenergic and muscarinic receptors. This part of the molecule shows some similarity with bacterial periplasmic binding proteins, and also, to some extent, with the presumed glutamate-binding region of ionotropic glutamate receptors (O'Hara *et al.*, 1993).

The different subtypes have been grouped according to their various properties. Group I (mGluR1 and mGluR5) are linked to the phosphoinositide signalling system, probably via q/11 G proteins. The other subtypes are linked to inhibition of adenylyl cyclase, probably via i/o G proteins. Groups II (mGluR2 and mGluR3) and III (mGluR4, mGluR6-8) are distinguished by their different responses to agonists and antagonists; thus group III receptors respond to L-AP4 (L-amino-4-phosphobutanoate) whereas group II receptors do not.

Metabotropic glutamate receptors appear to function largely as neuromodulators. Located on both pre- and post-synaptic membranes at glutamatergic synapses, their activation may increase or decrease cytoplasmic calcium ion concentrations as well as perhaps having more direct effects on ion channels (Pin & Bockaert, 1995). Some of these effects, as we shall see in chapter 11, may be important in learning and memory.

Neuropeptide transmitters

Neuropeptides are single chains of amino acids produced in nerve cells by the standard methods of protein synthesis. Some years ago it was generally believed that all of the few secreted neuropeptides were released into the blood stream and so acted as hormones. Later it became evident that there are many more neuropeptides than was originally suspected and that the majority of them act as neurotransmitters, often in addition to a role as hormones (table 9.6). Let us see how this change of viewpoint happened.

Table 9.6 *Some neuropeptides in the mammalian nervous system*

Pituitary peptides
  Corticotropin (ACTH)
  Growth hormone (GH)
  Lipotropin
  α-Melanocyte-stimulating-hormone (α-MSH)
  Oxytocin
  Vasopressin

Circulating hormones
  Angiotensin
  Calcitonin
  Insulin

Gut hormones
  Cholecystokinin (CCK)
  Gastrin
  Motilin
  Pancreatic polypeptide (PP)
  Secretin
  Substance P
  Vasoactive intestinal polypeptide (VIP)

Opioid peptides
  Dynorphin
  β-Endorphin
  Met-enkephalin
  Leu-enkephalin
  Kyotorphin

Hypothalamic releasing hormones
  Corticotropin-releasing factor (CRF)
  Luteinizing hormone-releasing hormone (LHRH)
  Somatostatin
  Thyrotropin-releasing hormone (TRH)

Miscellaneous peptides
  Bombesin
  Bradykinin
  Calcitonin gene-related peptide (CGRP)
  Carnosine
  Neuropeptide Y
  Neurotensin
  Proctolin
  Substance K

Certain cells of the hypothalamus produce secretory materials which pass along their axons to be released into the circulation at their terminals in the pituitary gland. This process is called neurosecretion, and the released substances are neurohormones (Scharrer & Scharrer, 1945; Bargmann & Scharrer, 1951). The first of the hypothalamic neurohormones to have their structures determined were vasopressin and oxytocin (see du Vigneaud, 1956); they are each peptides with nine amino acid residues (fig. 9.11). Vasopressin increases the blood pressure and oxytocin produces contraction of the uterus.

von Euler & Gaddum (1931) obtained an alcoholic extract from horse brain and gut which produced marked lowering of the blood pressure when injected into rabbits and made rabbit intestinal muscle contract. von Euler later showed that the activity of the extract was destroyed by trypsin, suggesting that the active component (by now called 'substance P') was a peptide. It was not until many years later that the precise structure of substance P was determined (Chang *et al.*, 1971).

Relatively high levels of substance P are found in the dorsal horn of the spinal cord. The possibility that it might be the transmitter at some afferent endings was discussed in the 1950s and 60s (e.g. McLennan, 1963); further evidence for this idea was later provided from a variety of sources and it is now well established (Go & Yaksh, 1987). In the early 70s, then, substance P appeared to be the only neuropeptide likely to be a neurotransmitter. This view did not survive investigations arising from the awesome human effects of the products of the opium poppy.

Opiates are drugs that have powerful actions on the nervous system; they include morphine, which is of great importance in the relief of pain, and heroin, well known as an addictive narcotic. Pert & Snyder (1973) used a tritium-labelled antagonist which could be displaced by competition with opiates to demonstrate the presence of specific opiate receptors in the brain. But what normally activates these receptors? This question initiated a search for endogenous opioid substances, which was fulfilled by the discovery and sequencing of two of the opioid peptides, Met-enkephalin and Leu-enkephalin, by Hughes *et al.* (1975) and Simantov & Snyder (1976). Then followed a great burst of research on brain neuropeptides, so that the number of them to be considered as possible candidates for neurotransmitter status is now well over 50 (Iversen, 1995).

The detection of neuropeptides in the nervous system was greatly aided by the development of new techniques. Immunohistochemistry uses antibodies to detect particular peptides. Radiolabelled peptides can be used to localize their receptors. Molecular cloning methods have produced amino acid sequences for peptide precursors, and specific antibodies can be made against peptides synthesized from them. Nucleic acid probes can be used for hybridization with mRNAs coding for particular peptides, or for their receptors

The concentrations of neuropeptides are typically in the range of some picomoles per gram of brain, perhaps 1000 times less than the monoamines and about 100 000 times less than the amino acid neurotransmitters (Hökfelt *et al.*, 1980). Their synthesis occurs in the cell body rather than in the nerve terminal, and they are commonly released at the same time as other small-molecule transmitters; these topics are discussed in chapter 10.

Three main types of opiate receptors have been dis-

Met-enkephalin	Tyr-Gly-Gly-Phe-Met
Leu-enkephalin	Tyr-Gly-Gly-Phe-Leu
Angiotensin II	Asp-Arg-Val-Tyr-Ile-His-Pro-Gly-NH$_2$
Arg-vasopressin	Cys-Tyr-Phe-Gln-Asn-Cys-Pro-Arg-Gly-NH$_2$
Lys-vasopressin	Cys-Tyr-Phe-Gln-Asn-Cys-Pro-Lys-Gly-NH$_2$
Oxytocin	Cys-Tyr-Ile-Gln-Asn-Cys-Pro-Leu-Gly-NH$_2$
LHRH	pGlu-His-Trp-Ser-Tyr-Gly-Leu-Arg-Pro-Gly-NH$_2$
Dynorphin	Tyr-Gly-Gly-Phe-Leu-Arg-Arg-Ile-Arg-Pro-Lys-Leu-Lys-Trp-Asp-Asn-Glu
Substance P	Arg-Pro-Lys-Pro-Gln-Gln-Phe-Phe-Gly-Leu-Met-NH$_2$
β-Endorphin	Tyr-Gly-Gly-Phe-Met-Thr-Ser-Glu-Lys-Gln-Thr-Pro-Leu-Val-Thr-Leu-Phe-Lys-Asn-Ala-Ile-Val-Lys-Asn-Ala-His-Lys-Lys-Gly-Gln
CGRP	Ser-Cys-Asn-Thr-Ala-Thr-Cys-Val-Thr-His-Arg-Leu-Ala-Gly-Leu-Leu-Ser-Arg-Ser-Gly-Gly-Val-Val-Lys-Asp-Asp-Phe-Val-Pro-Thr-Asp-Val-Gly-Ser-Glu-Ala-Phe-NH$_2$

Figure 9.11. The amino acid sequences of some neuropeptides. The NH$_2$ groups indicate amidation of the C-terminal residue. pGlu indicates pyroglutamate, in which an N-terminal glutamate residue has been modified to form an internal peptide bond.

tinguished, differing in their affinity for various opioid agent and in their distribution in the nervous system (North, 1986; Mansour *et al.*, 1995). μ (mu) and δ (delta) receptors bind the enkephalins and endorphins most readily, and κ (kappa) receptors are specific for dynorphin. The μ receptors are most specific for morphine. The antagonist naloxone is effective at all three types, but especially so at δ receptors. All three types are coupled to G proteins, probably of the G_i/G_o group, and produce inhibition of adenylyl cyclase. Cloning showed that they are members of the 7TM superfamily with a high degree of mutual sequence similarity (Reisine & Bell, 1993; Zaki *et al.*, 1996).

Other neuropeptide receptors are also members of the G-protein-coupled 7TM superfamily, with structures broadly similar to the opioid receptors. They include receptors for substance P, neurotensin, neuropeptide Y, somatostatin, and many others (Watson & Arkinstall, 1994). As mentioned earlier, some of the peptide-binding 7TM receptors have a much longer N-terminal segment than do those for the majority of neuropeptides and peptide hormones, and it seems likely that in these receptors the peptide-binding sites are located in the N-terminal region. This group includes receptors for secretin, calcitonin, vasoactive intestinal polypeptide (VIP) and others, and forms a separate subfamily of 7TM receptors (Ishihara *et al.*, 1992).

Gaseous transmitters

Nitric oxide

One of the most surprising physiological discoveries of recent years has been that nitric oxide, a diffusible highly reactive gas, can act as a neurotransmitter. The story begins with the demonstration by Furchgott & Zawadzki (1980) that arterial smooth muscle cannot be relaxed by acetylcholine unless the endothelial cell layer lining the artery is present. They suggested that activation of muscarinic acetylcholine receptors releases a diffusible substance from the endothelial cells and that this produces relaxation in the adjacent arterial smooth muscle. The substance was later called endothelium-derived relaxing factor, or EDRF.

A clue to the nature of EDRF came from the knowledge that certain nitro-compounds cause vasodilation; they include amyl nitrite and the explosive nitroglycerin, once used to treat Alfred Nobel, its inventor, for high blood pressure. Could the EDRF be nitric oxide? Palmer and his colleagues (1987) tested this idea by comparing the effects of EDRF on arterial muscle with those of nitric oxide. They found that the two substances had similar lifetimes of a few seconds, and that the relaxations they each produced were inhibited to the same extent by haemoglobin and were enhanced to the same extent by the enzyme superoxide dis-

mutase. (Haemoglobin combines with nitric oxide. Superoxide dismutase converts the superoxide ion O_2^- to hydrogen peroxide and oxygen, so making it unavailable for the removal of nitric oxide.) Similar results were obtained by Ignarro and his colleagues (1987), together with the demonstration that both EDRF and nitric oxide increased the cyclic GMP content of the vascular smooth muscle, which is probably the prime cause of the relaxation.

Knowing that glutamate can increase the cyclic GMP content of brain tissue, Garthwaite and his colleagues (1988) had the idea that this effect might be mediated by nitric oxide. They applied NMDA to a suspension of cerebellar cells and looked for the release of EDRF, using rings of rat aorta as a bioassay. The results were clear-cut (fig. 9.12): activation of the NMDA receptors of the cerebellar cells produced relaxation of the vascular smooth muscle just as EDRF did. In other words, nitric oxide is released from the nerve cells and acts as a messenger for the production of cyclic GMP in the muscle cells.

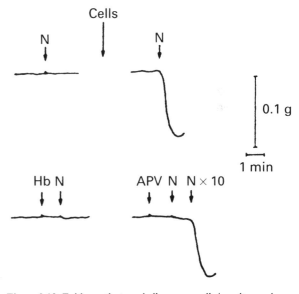

Figure 9.12. Evidence that cerebellar nerve cells in culture release nitric oxide when their NMDA receptors are activated. Nitric oxide was detected as endothelium-derived relaxing factor by bioassay using endothelium-free rat aortic muscle contracted with phenylephrine: nitric oxide would cause relaxation (seen as a downward deflection on the traces here). The upper traces show that NMDA (N) has no relaxing effect on the aortic muscle alone, but does produce a relaxation when cerebellar cells are added. Haemoglobin (Hb), which inactivates nitric oxide, abolished this action. APV, an antagonist of NMDA, blocks the action unless the NMDA concentration is increased ten times. (From Garthwaite *et al.*, 1988. Reprinted with permission from *Nature* **336**, 386. Copyright 1988 Macmillan Magazines Limited.)

Evidence that nitric oxide can be released from presynaptic nerves and acts as a neurotransmitter was provided by Bult and his colleagues (1990), using the inhibitory non-adrenergic non-cholinergic (NANC) innervation of the gut smooth muscle. Stimulation of the NANC nerves to the isolated ileocolonic junction released a factor that caused rings of rabbit aorta from which the endothelial cells had been removed to relax. Neither noradrenaline nor acetylcholine produced such relaxation. The relaxation could be mimicked by nitric oxide and nitroglycerin, inhibited by haemoglobin, and enhanced by L-arginine (the precursor of nitric oxide) and superoxide dismutase. Tetrodotoxin, which would block conduction in the NANC nerves, abolished the response. All this suggests strongly that the NANC nerves act by releasing nitric oxide from their presynaptic terminals, and that this nitric oxide causes relaxation of the postsynaptic smooth muscle cells.

Nitric oxide is produced by oxidation of the amino acid arginine to citrulline by the enzyme nitric oxide synthase:

$$HOOC.CH(NH_2).(CH_2)_3.NH.C(NH_2)NH$$
$$+ 1.5O_2 \rightarrow HOOC.CH(NH_2).(CH_2)_3.NH.CONH_2$$
$$+ NO + H_2O$$

The enzyme requires NADPH, FAD and FMN as redox-active cofactors, contains haem as a prosthetic group, and is activated by calcium and calmodulin. Cloning shows that it has binding sites for these cofactors, and shows considerable homology with the enzyme cytochrome P450 reductase (see Bredt & Snyder, 1994; Griffith & Stuehr, 1995). It can be inhibited by various ω-N substituted analogues of arginine, of which N^{ω}-amino-L-arginine is a much-used example.

Nitric oxide is a highly reactive substance; it can be regarded as a free radical since the shell of valence electrons around the nitrogen atom is incomplete. Its main target in the nervous system seems to be the enzyme soluble guanylyl cyclase. This contains a haem group to which the nitric oxide probably attaches, and is responsible for the production of cyclic GMP. The cyclic GMP may then activate ion channels of the type found in the retina, certain protein kinases and phosphodiesterases, and perhaps other second-messenger systems (Garthwaite & Boulton, 1995).

Southam & Garthwaite (1993) used histochemical methods to compare the distribution of nitric oxide synthase in the brain with that of cyclic GMP after perfusion with the nitric oxide donor nitroprusside. They found that the two were closely associated but not identical; in some areas nitric oxide synthase was found in postsynaptic structures and cGMP in adjacent presynaptic structures, in others the nitric oxide synthase was presynaptic and the cGMP postsynaptic. Cyclic GMP also appeared in glial cells. All this suggests that nitric oxide acts mainly as a mediator of cell-to-cell signalling rather than simply as an intracellular messenger within the cell in which it is produced. Calculations on the diffusional spread of nitric oxide suggest that a point source of nitric oxide lasting for a few seconds would distribute the gas to a sphere about 200 μm across, enclosing perhaps 2 million synapses (Wood & Garthwaite, 1994).

Other possible targets for nitric oxide action exist, in addition to soluble guanylyl cyclase. Details of its presumed neurotransmitter role have still to be worked out. Much interest has been aroused by the involvement of nitric oxide in neurotoxity (see chapter 8) and in long-term potentiation and depression (see chapter 11). There is clearly still much to be discovered in this area (see Garthwaite, 1991, 1995; Jaffrey & Snyder, 1995; Vincent, 1995).

Carbon monoxide

Carbon monoxide is another free-radical-like small gaseous molecule that may act as a messenger between nerve cells (Verma *et al.*, 1993; Dawson & Snyder, 1994). It is formed by the action of the enzyme haem oxygenase and, like nitric oxide, can act as an activator of guanylyl cyclase. There is also some evidence that it may interfere with the activation of guanylyl cyclase by nitric oxide (Ingi *et al.*, 1996).

10
Synthesis, release and fate of neurotransmitters

So far in our consideration of the mechanisms involved in synaptic transmission, we have been concerned very largely with postsynaptic structures and events. It is time to look at what happens in the presynaptic terminal and the synaptic cleft. How is the neurotransmitter released from the terminal when a nerve impulse arrives there? How is it synthesized and packaged for release? What happens to it after it has dissociated from the postsynaptic receptors? We begin by considering some classic experiments by Bernard Katz and his colleagues on the release of acetylcholine at the frog neuromuscular junction.

Transmitter release from synaptic vesicles
During the controversy in the 1930s as to whether synaptic transmission was essentially an electrical or a chemical process, those who held to the electrical hypothesis could not see how a chemical process could account for the speed of synaptic transmission. The advent of the intracellular microelectrode demonstrated, as we have seen in chapter 7, that the transmission process is dependent in most cases upon the release of a chemical substance, the neurotransmitter, from the presynaptic nerve terminal. It also showed that this release is a very rapid process: fast synaptic transmission really is fast, so that the action of the neurotransmitter begins within a millisecond or so of the nerve impulse reaching the presynaptic terminal. How is this rapid release of the neurotransmitter brought about?

Miniature end-plate potentials
In 1952 Fatt & Katz found a completely unexpected phenomenon when they used a high gain amplifier with their microelectrode inserted into a frog sartorius muscle fibre in the end-plate region. In the resting muscle there were small discrete fluctuations in membrane potential, as is shown in fig. 10.1. The potentials were similar in time course to normal end-plate potentials (EPPs), but only about 0.5 mV in height, so Fatt & Katz called them *miniature end-plate potentials*.

These miniature EPPs could be found only at the end-plate region, and were not found in muscle that had been denervated two weeks previously, so they were clearly associated with the presence of the nerve terminal. They were reduced in size by curare, and increased in the presence of prostigmine (an anticholinesterase similar to eserine in its action), suggesting that they were produced by the action of acetylcholine on nicotinic receptors. Statistical analysis of the time intervals between successive miniature EPPs showed that their occurrence was randomly distributed in time. These observations strongly implied that the miniature EPPs were caused by the action of acetylcholine released spontaneously from the motor nerve ending.

How much acetylcholine is involved in the production of a single miniature EPP? The most obvious suggestion is perhaps just one molecule. Fatt & Katz argued that this could not be so, however, for the following reasons. Firstly, if one molecule could produce a depolarization of about 0.5 mV, then not more than 1000 at the most would be needed to produce an EPP of 50 mV, and therefore the

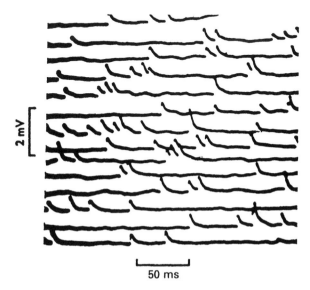

Figure 10.1. A series of membrane potential records from a frog neuromuscular junction showing miniature end-plate potentials. (From Fatt & Katz, 1952.)

amount of acetylcholine released per impulse would be no more than 10^{-21} mol. In fact it is much larger than this (Krnjević & Mitchell (1961), estimated that each nerve terminal in rat diaphragm muscle releases about 10^{-17} mol acetylcholine per impulse), so that each miniature EPP must be produced by several thousand molecules of acetylcholine. Secondly, the response to externally applied acetylcholine is a smooth depolarization, not a series of unitary events about 0.5 mV in height.

Kuffler & Yoshikami (1975b) provided a more direct estimate of the number of molecules of acetylcholine per miniature EPP. They used preparations in which the subsynaptic membrane was exposed, by removing the nerve terminals from frog or snake muscles which had been treated with the enzyme collagenase so as to loosen their connective tissue sheaths. They then measured the size of the responses to ionophoretic applications of acetylcholine. The amount of acetylcholine ejected from the pipette during each ionophoretic pulse was determined by means of an ingenious bioassay: several thousand such pulses were passed into a droplet of Ringer solution which was then applied to an end-plate under a layer of fluo-

rocarbon oil, and its depolarizing effect was compared with that of a droplet of the same volume of an acetylcholine solution of known concentration. They found that about 6000 molecules of acetylcholine were required to produce a depolarization of 1 mV. The average size of a miniature EPP in their experiments was about 1.5 mV, and hence somewhat less than 10 000 molecules were involved in its production.

A most interesting property of these miniature EPPs is that their frequency can be altered by procedures which we would expect to change the membrane potential of the presynaptic nerve terminals. Liley (1956) investigated these effects at the neuromuscular junctions in the rat diaphragm. If depolarizing currents were passed through an external electrode placed near the nerve terminals, the frequency of discharge increased; conversely, it decreased if the current was hyperpolarizing (fig. 10.2a). The frequency also increased if the external potassium ion concentration was raised above 10 mM; in this case the increase can be related to an estimate of the presynaptic membrane potential, which cannot be measured directly since the terminals are too fine (fig. 10.2b).

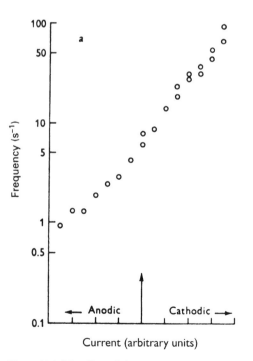
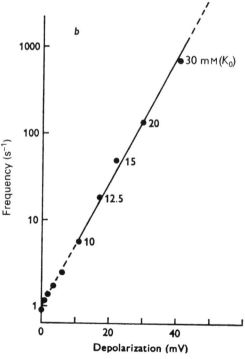

Figure 10.2. The effect of the presynaptic membrane potential on the frequency of miniature end-plate potentials in a rat diaphragm muscle fibre. a shows the effect of applying various electrical currents (measured in arbitrary units) to the nerve ending. b shows the effect of increasing the external potassium ion concentration; the abscissa shows the estimated depolarizations of the presynaptic nerve membrane produced by this treatment. (a from Liley, 1956; b from Katz, 1962.)

We have seen that the size of the miniature EPPs can be altered by curare and anticholinesterases; it also changes with postsynaptic membrane potential in the same way as does the EPP. In the frog sartorius, the size of miniature EPPs varies inversely with the diameter of the muscle fibre; thus, since large fibres have a smaller total membrane resistance than small fibres, it would seem that the currents associated with the miniature EPPs (and therefore the amounts of acetylcholine released) are more or less constant (Katz & Thesleff, 1957*b*). The discharge frequency can be altered by changing the osmotic pressure of the bathing solution or (in mammals) the magnesium or calcium ion concentrations, as well as by polarization of the nerve terminals. These observations lead to an important generalization: as Katz (1962) put it, 'the *frequency* of the miniature potentials is controlled entirely by the conditions of the presynaptic membrane, while their *amplitude* is controlled by the properties of the postsynaptic membrane'.

The quantal nature of the end-plate potential

It has long been known that an excess of magnesium ions blocks neuromuscular transmission. del Castillo & Engbaek (1954) showed that this effect was due to a reduction of the quantity of acetylcholine released per impulse, so that the EPP was very much smaller. Calcium ions antagonized this action, i.e. the degree of reduction of acetylcholine release was dependent upon the ratio of the magnesium and calcium ion concentrations. When the EPPs were greatly reduced by magnesium block, it was found that the size of successive EPPs fluctuates in a stepwise manner, as is evident in fig. 10.3. This led to the suggestion that acetylcholine is released from the motor nerve terminals in discrete 'packets' or *quanta*, that the normal EPP is the response to some hundreds of these quanta, and that miniature EPPs are the result of spontaneous release of single quanta.

An extension of the quantal release hypothesis is that there is a large population of quanta in each nerve terminal, each one of which has a small probability of being released by a nerve impulse. This idea was tested by del Castillo & Katz (1954*a*) by means of a statistical analysis applied to a large number of nervous stimuli. The experiments were performed on a frog toe muscle in a solution containing a high magnesium ion concentration so as to reduce the quantal content of each response. Let x be the number of quanta comprising any one response, and let m be the mean of all values of x. Let P_0, P_1, P_2, etc. be the proportion of events in the classes $x = 0, 1, 2$, etc. (An analogy may help with the terminology here. Suppose a large class of students attends a long series of lectures, and each

student has a constant low probability of falling asleep during a lecture. Then x is the number of students asleep in any particular lecture, m is the mean of x, P_0 the proportion of lectures with no students asleep, P_1 the proportion with one asleep, and so on.)

In this situation the proportions P of the different values of x should be given by the terms of a Poisson distribution, i.e.

$$P_x = e^{-m} \times \frac{m^x}{x!} \tag{10.1}$$

The easiest way of finding m in this equation is to take the case when $x = 0$, when we get

$$P_0 = e^{-m}$$

or

$$m = \ln(1/P_0) \tag{10.2}$$

i.e. m is given by the natural logarithm of the reciprocal of the proportion of events in which transmission fails. There is also another way of finding m. Since we suspect that each miniature EPP corresponds to one quantum, the mean number of quanta per response should be given by

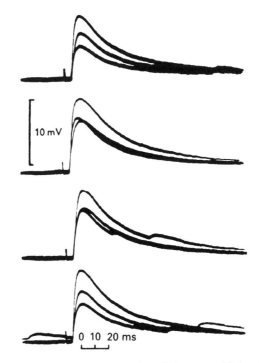

Figure 10.3. Quantal fluctuations in end-plate potential size after magnesium poisoning. Spontaneous miniature potentials can be seen in some of the records. (From del Castillo & Katz, 1954*b*.)

$$m = \frac{\text{mean amplitude of EPP}}{\text{mean amplitude of miniature EPPs}} \quad (10.3)$$

Thus we have two independent ways of determining m, from equations 10.2 and 10.3, and if the hypothesis is correct, they should agree. del Castillo & Katz found that they do.

Now consider the situation when x is 1 or more. The proportions of EPPs of different sizes should be described by a Poisson distribution, as in equation 10.1. A slight complication arises here, however, since the size of the spontaneous miniature EPPs varies somewhat (fig. 10.4a), and this has to be taken into account when calculating the frequency distribution of EPP amplitudes. Figure 10.4b shows that the curve calculated in this way is in good agreement with the actual distribution of EPP amplitudes.

Can we calculate m for normal EPPs? Equation 10.3 has

to be modified to take account of the non-linear addition of quantal changes. For example, it follows from Ohm's law that a change in conductance which produces a depolarization of 0.5 mV when the membrane potential is −95 mV will produce a depolarization of only three-quarters of that amount (i.e. 0.375 mV) at −75 mV, if we assume that the reversal potential is −15 mV. From these considerations, Martin (1955) calculated that a normal EPP of about 50 mV is produced by the action of about 250 quanta.

The Poisson distribution is a limiting case of the binomial distribution when the probability of any individual event is low, as in quantal release in high magnesium solutions. Consequently one might expect that quantal release would be distributed binomially under more normal conditions. For a simple binomial, we assume that there are n identical release sites each with a probability p of releasing

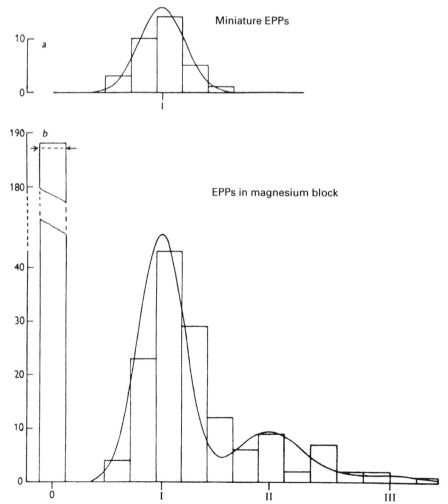

Figure 10.4. Statistical analysis of the quantal components of transmission at a frog neuromuscular junction. The histogram in *a* shows the variation in size of spontaneous miniature end-plate potentials; the continuous curve is a normal distribution. The histogram in *b* shows the frequency distribution of the sizes of responses to nervous stimulation in the same muscle fibre as for *a*, after magnesium block. The ordinate shows the number of occurrences of potentials in a particular size range; the abscissa shows the size of the potentials, the unit of measurement being the mean size of the spontaneous potentials from *a*. The arrows in *b* (to show the expected number of failures of transmission) and the continuous curve were calculated from a Poisson distribution, modified to take account of the variability in size of the quantal units as seen in *a*. (From del Castillo & Katz, 1954b.)

a quantum of transmitter. Then the probability P_x of releasing x quanta from the whole terminal is given by

$$P_x = \left(\frac{n!}{(n-x)!x!}\right)p^x(1-p)^{n-x} \tag{10.4}$$

The mean quantal content m is the product of the number of release sites n and the probability of release from any one of them p. Some investigations have found results in agreement with equation 10.4 (e.g. Wernig, 1975; Bennett *et al.*, 1977), but others have implied more complex behaviour and there are difficulties in relating the binomial parameters to our understanding of what is happening in the nerve terminal (Silinsky, 1985; van der Kloot & Molgó, 1994).

We can now see the importance of the experiments showing that the frequency of spontaneous discharge of miniature EPPs rises when the presynaptic terminal is depolarized. If the line in fig. 10.2*b* is extrapolated to, say, 90 mV (so that, to put it another way, we are assuming that the probability of discharge of a quantum of acetylcholine is logarithmically proportional to the depolarization of the presynaptic terminal membrane), the discharge frequency would be about 10^6 per second. Since the presynaptic action potential lasts only for a fraction of a millisecond, this is about the right order of magnitude to account for the release of the number of quanta needed to produce an EPP.

The vesicle hypothesis
But why should acetylcholine be discharged from the nerve ending in quantal units of 10^3 to 10^4 molecules? Figure 7.3 shows that the nerve terminal contains large numbers of vesicles about 500 Å in diameter. These vesicles were first observed by de Robertis & Bennett (1954) and Palade & Palay (1954), and have since been found in the presynaptic terminals of all chemically transmitting synapses where they have been looked for. del Castillo & Katz (1956) suggested that they contain the transmitter substance, and that the discharge of the contents of one vesicle into the synaptic cleft corresponds to the release of one quantum of the transmitter. This suggestion is sometimes known as the 'vesicle hypothesis'.

The vesicle hypothesis has been contested on occasion. An alternative suggestion is that the quanta represent the release of a group of seven to ten synaptic vesicles from the same active zone (see below) in the nerve terminal (Kriebel & Gross, 1974; Wernig & Stirner, 1977). This idea seems to have been conclusively refuted by some very precise measurements by Hurlbut and his colleagues (1990). They treated frog muscle fibres with α-latrotoxin, a constituent of black widow spider venom that causes massive quantal release of acetylcholine from the nerve terminals. At room

temperature it takes some 20 min or so for the store of vesicles to be exhausted, and the frequency of the miniature EPPs declines during this time. They counted the numbers of vesicles remaining in the terminals fixed for electron microscopy at different times during the treatment and compared this with the total number of quanta released. The results are shown in fig. 10.5: the relation between quanta released and vesicles remaining has a slope very

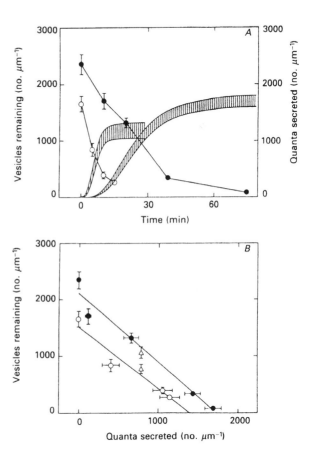

Figure 10.5. The one-to-one relation between quantal release and vesicle loss at the frog neuromuscular junction. Black widow spider venom (or α-latrotoxin purified from it) was added at time zero, causing massive release of acetylcholine from the nerve terminal. Hatched curves in *a* show the cumulative release of transmitter quanta as measured with an intracellular microelectrode in the muscle fibre, and symbols show how the counts of vesicles in the terminals fell, as measured from electron micrographs. Measurements are given per micrometre of nerve terminal. White circles and the left-hand hatched curve show experiments at 22–23 °C, black circles and the right-hand hatched curve show those at 9–10 °C. The two variables are plotted against each other in *b*: the regression lines have slopes of −1.1 and −1.2 at the upper and lower temperatures, respectively. (From Hurlbut *et al.*, 1990.)

near to −1, giving strong evidence that one quantum of transmitter is equivalent to the contents of one vesicle.

Contents of the vesicles

If the vesicle hypothesis is correct it should be possible to isolate vesicles and show by chemical means that they actually do contain acetylcholine. Much work on this problem has been done by Whittaker and his colleagues. Homogenization of a suitable tissue (such as guinea pig brains) followed by centrifugation enables a subcellular fraction containing fragments of nerve endings to be isolated. The fragments are known as *synaptosomes* and they contain synaptic vesicles. The vesicles can be released from the synaptosomes by rupturing them in a hypo-osmotic solution. Such extracts from homogenized brains contain quantities of a variety of transmitter substances, including acetylcholine (Whittaker *et al.*, 1964; Whittaker, 1993).

The electric organ of the electric ray *Torpedo* is very richly innervated with cholinergic nerve terminals and provides an excellent source of nearly pure fractions of acetylcholine-containing vesicles (Whittaker, 1984). The acetylcholine content of each vesicle is in the region of 200 000 molecules. Electric organ vesicles are larger than those at the frog neuromuscular junction: the outside diameter is about 84 nm instead of 50 nm and so the volume they contain (allowing 4 nm for membrane thickness) is nearly six times greater. Since the osmotic concentration of frog tissues is about one-fifth of that in rays, we might expect frog vesicles to contain only one-thirtieth the amount of acetylcholine in *Torpedo* vesicles, i.e. about 7000 molecules. This figure is in good agreement with the estimates of quantal content from physiological measurements or chemical determinations (Kuffler & Yoshikami, 1975*b*; Miledi *et al.*, 1983).

The vesicles also contain adenosine triphosphate (ATP), about one molecule for every seven molecules of acetylcholine (Dowdall *et al.*, 1974). This ATP is released into the synaptic cleft along with the acetylcholine, as is shown by collecting perfusates from rat muscle during stimulation or by using patch-clamped ATP-gated channels as an assay system (Silinsky, 1975; Silinsky & Redman, 1996).

Preparations of vesicles also contain phospholipids and membrane-bound proteins, including proteoglycans, which are proteins to which large quantities of acid polysaccharides are attached. The polysaccharide chains may serve to balance the ionic charges of the vesicle contents.

Vesicle discharge and recycling

del Castillo & Katz (1956) suggested that each synaptic vesicle discharges its contents by fusion of its membrane with the presynaptic plasma membrane as is shown in fig. 10.6. Membrane fusion of this type has been seen in secretory cells and the process is called *exocytosis*. Electron microscopy showed that the discharge sites are not simply distributed at random but are localized to 'active zones' (Couteaux & Pécot-Dechavassine, 1970). These are seen as bands of dense material and double rows of intramembranous particles, with a row of vesicles on each side, opposite the junctional folds of the postsynaptic membrane. This means that the vesicles are positioned where release of their contents will be most effective, since the acetylcholine receptors are concentrated near the openings of the junctional folds.

Direct evidence for the fusion of the vesicular and presynaptic membranes comes from quick-freezing experiments (Heuser *et al.*, 1979; Heuser, 1989). Heuser and his colleagues built a device that would slam the muscle onto a block of copper, cooled to −269 °C by liquid helium, just a few milliseconds after the nerve had been stimulated. They enhanced transmitter release with 4-aminopyridine, which blocks potassium channels and so produces a much pro-

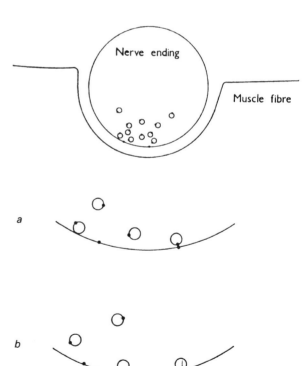

Figure 10.6. del Castillo & Katz's hypothesis about the quantal release of transmitter. Reactive sites on the vesicles and the presynaptic membrane are shown as dots. Release of transmitter occurs when two sites meet, and the probability of this happening increases greatly on depolarization. (From Katz, 1969.)

longed action potential in the nerve terminal. Their freeze-fracture electron micrographs showed vesicles caught in the act of exocytosis (fig. 10.7), in numbers comparable to the enlarged quantal content of the EPP.

What happens to the membrane of the vesicle after the release of its contents? Heuser & Reese (1973) found that the population of synaptic vesicles in the nerve terminal became temporarily depleted after a period of repetitive stimulation. After 1 min of stimulation of the nerve at 10 s^{-1} there was a 30% loss of vesicles together with a corresponding increase in the area of the presynaptic membrane. However, after 15 min of such stimulation numerous membrane-bound vacuoles (cisternae) appeared in the terminal, whose surface area accounted largely for the 60% loss in synaptic vesicles. After a period of rest these cisternae disappeared and the vesicles reappeared. Heuser & Reese concluded that there is a recycling of the vesicular membrane after exocytosis.

More evidence for this view came from the use of horseradish peroxidase (HRP) as a marker in the extracellular space. This enzyme oxidizes a variety of aromatic compounds to produce reaction products which combine with osmium tetroxide to produce electron-dense material, so its position in the cell can be detected by electron microscopy. After stimulation it appeared inside the nerve terminal (Holtzmann *et al.*, 1971). This indicates uptake of plasma membrane and some extracellular material, a process known as 'endocytosis'. Some of these vesicles are coated with fibrous material (probably the protein clathrin), and endocytosis probably begins with the formation of invaginations of the presynaptic membrane called coated pits. There may also be some invagination of non-coated membrane when stimulus rates are high, as is shown in fig. 10.8 (Miller & Heuser, 1984).

It is common to think of the vesicles in the motor nerve terminal as forming a 'pool' from which a small number become attached to the active zones in preparation for release. How structured is this pool? Do the recycled vesicles mix freely with the other vesicles in the pool or do they go to the front or the back of the queue? Betz & Bewick (1992) studied recycling in living nerve terminals by using fluorescent dyes that are taken up into the terminal during

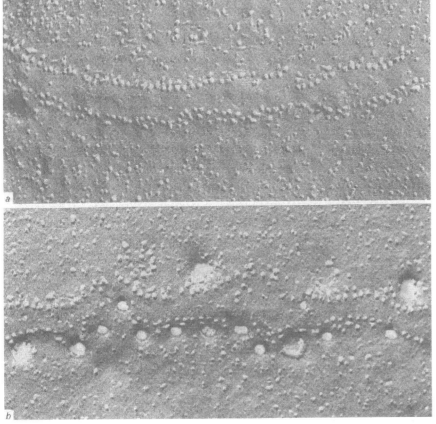

Figure 10.7. Synaptic vesicle exocytosis caught in the act by quick freezing of a frog neuromuscular junction. The muscle was slammed into a copper block cooled to −269 °C by liquid helium, then freeze-fractured at −105 °C, etched (i.e. ice on the surface was allowed to evaporate), and then a replica of the exposed surface was made by evaporation of platinum and carbon onto it. The photographs show transmission electron micrographs of such replicas, showing the cytoplasmic face of the presynaptic membrane. The double row of membrane particles in each case represents an active zone. The nerve was stimulated just before cooling: 3 ms before in *a* and 5 ms before in *b*. The 'holes' in *b* are thought to be openings into synaptic vesicles that were frozen while they were discharging their contents into the synaptic cleft. Magnification 121 000×, i.e. 1 mm is equivalent to 83 Å. (Photographs kindly supplied by Professor J. E. Heuser, from Heuser *et al.*, 1979.)

endocytosis. The release of the vesicles containing the dye could then be followed by measuring the fluorescence changes in the terminal in response to stimulation. The results implied that the recycled vesicles mix thoroughly with the other vesicles in the pool. One complete cycle, from vesicle release, recycling and then release again during a period of continuous stimulation at 20 s^{-1}, took about 1 min.

By treatment of the neuromuscular junction with albumin solution before fixation with osmium tetroxide, Gray (1978) was able to show that the synaptic vesicles in the nerve terminals are closely associated with microtubules. These run either parallel to the active zones or across them, hence it looks as though they may be involved in the transport of vesicles to their release sites.

Quantal transmission at central synapses

Small excitatory postsynaptic potentials (EPSPs) in spinal motoneurons undergo stepwise fluctuations in size, suggesting a unitary basis for transmission (Kuno, 1964). The normal motoneuronal EPSP (as in fig. 7.21, for example) arises from the activity of several afferent fibres. What would the response to individual afferent fibres be? Could we distinguish responses at single terminal boutons, or see quantal responses arising from the release of single

vesicles? How many ion channels are opened by such responses?

Some answers to these questions were produced by Redman and his colleagues (Redman, 1990). Jack *et al.* (1981) looked at the responses to activity in single presynaptic axons: they cut away the dorsal root to leave just a thin filament in which only one group Ia fibre from a particular muscle was active. The response to stimulation of a single presynaptic fibre is called a *unitary EPSP*. These unitary EPSPs were only a few hundred microvolts in size and were superimposed upon a noisy background (fig. 10.9*a*). A large number (800) of these EPSPs were recorded and stored in digital form for computer analysis following stimulation of the same presynaptic fibre. Figure 10.9*b* shows the average form of the EPSPs and fig. 10.9*e* is a histogram showing their frequency distribution. This can be compared with the frequency distribution of the background noise voltage (fig. 10.9*f*).

The histogram in fig. 10.9*e* is wider than that of the noise and somewhat skewed. This means that there must be some variation in the EPSPs in addition to that attributable to the background noise. If we assume that this variation is discrete rather than continuous, then it is possible to determine its frequency distribution by using a rather sophisticated statistical computing technique known as

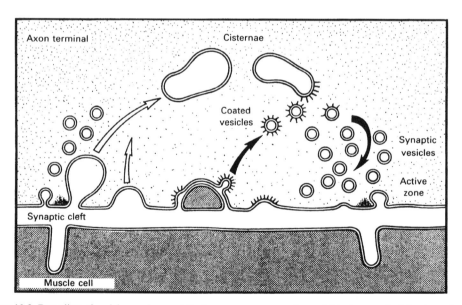

Figure 10.8. Recycling of vesicle membrane at the frog neuromuscular junction. Vesicles discharge their contents at the active zones, where their membranes fuse with the presynaptic membrane. Vesicular membrane, including its intramembrane particles, is retrieved via coated pits to form coated vesicles, which may fuse with cisternae from which synaptic vesicles are reformed (black arrows). After intensive stimulation uncoated pits may form, retrieving excess membrane but not selecting intramembrane particles (white arrows). (After Miller & Heuser, 1984. Reproduced from the *Journal of Cell Biology*, 1984, **98**, 697, by copyright permission of The Rockefeller University Press.)

deconvolution,. The result, shown in fig. 10.9g, implies that the EPSPs in this experiment fluctuate between four discrete peak voltages.

The amplitudes of the four components in this experiment differ by about 100 μV and the smallest one is 302 μV, so it looks as though they consist of three, four, five and six quantal units, respectively. Other unitary EPSPs consisted of fewer or (rarely) more quantal units. We might expect that these quantal units correspond to the release of single vesicles, or the activity of single synaptic boutons, or perhaps (if one vesicle is released per bouton) both. We need some more detailed anatomical information to resolve this question, and this was provided by Redman & Walmsley (1983a,b).

Redman & Walmsley recorded unitary EPSPs from single motoneurons and analysed them in the same way as in the earlier experiments. However, here the recording microelectrode was filled with a solution containing HRP. This could be injected into the motoneuron by ionophoresis and later stained for light or electron microscopy, so as to reveal the dendritic field of the motoneuron. HRP was also used to stain the terminals of the group Ia fibre that produced the unitary EPSP, by applying it to the cut end of the dorsal root filament. The results showed that the quantal content of a unitary EPSP was always equal to or smaller than the number of synaptic contacts made by the presynaptic fibre via its terminal boutons.

In one experiment, for example, deconvolution implied

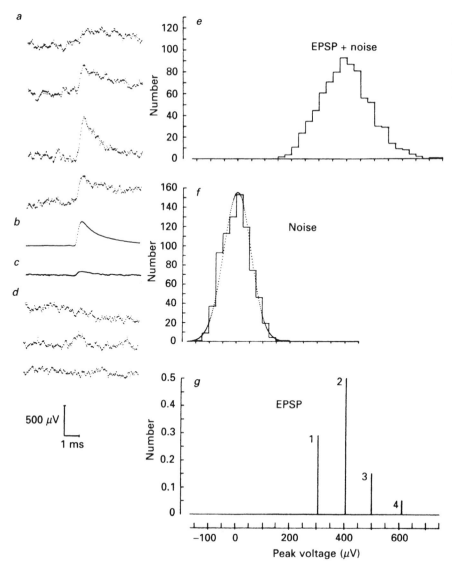

Figure 10.9. Components of unitary EPSPs in a spinal motoneuron. Eight hundred responses to stimulation of a single group Ia fibre were recorded in digitized form; the records in *a* show four of them. *b* shows the averaged time course for all 800 responses and *c* is the standard deviation of these responses. Eight hundred similar traces were obtained in the absence of stimulation, so as to measure the background noise of the system; *d* shows three of these records. *e* is a histogram of the sizes of the 800 EPSPs. *f* is a similar histogram showing the frequency distribution of the noise, with the dotted line as a Gaussian curve fitted to it. *g* shows the deconvolved probability graph by which *e* can be modelled as a combination of four discrete events combined with the Gaussian curve. (From Jack *et al.*, 1981.)

that the unitary EPSPs were made up of three quanta in some cases and four in others. The HRP technique showed that the presynaptic terminal made contact with the motoneuron at four boutons, three on one branch of a dendrite and one on the other. Redman & Walmsley concluded that transmission never failed at three of the boutons, but that it sometimes did at the fourth. The all-or-nothing nature of the response at individual boutons suggested that the contents of just one vesicle were released at each bouton. It was very pleasing that these delicate and time-consuming experiments provided such agreement between the microcircuitry of the synaptic contact and the physiological response in the motoneuron.

The patch clamp technique, used in the whole-cell clamp configuration, gives a more precise method of looking at unitary responses. Sakmann and his colleagues (1989) have used this method to access central neurons in brain slices. Tissue covering a particular neuron in the slice is washed away by a gentle stream of saline solution, allowing access to the cell membrane by the patch pipette. A second pipette can then be brought into contact with a nearby cell to apply a stimulus current sufficient to produce a unitary synaptic response in the first cell.

This method was used by Edwards *et al.* (1990) to investigate inhibitory transmission in rat hippocampal slices. The inhibitory postsynaptic currents (IPSCs) showed considerable variation in size, and their frequency distributions could be fitted by a number of overlapping gaussian curves. In fig. 10.10*a*, for example, the IPSCs appear to be consist of from one to eight quantal components. Perhaps the most likely interpretation of this is that when the presynaptic cell is stimulated, transmission occurs at between one and eight of its boutons terminating on the postsynaptic cell. The responses in a high magnesium environment look as though they were composed of just one of these quantal components (fig. 10.10*b*).

There is an interesting difference between the size distribution results shown in fig. 10.10*a* and the corresponding results from the neuromuscular junction, as in fig. 10.4*b*. In the neuromuscular junction case, we assume that the size distribution for a single quantum can be described by a gaussian curve with mean x and variance σ^2, so the curve for two quanta would have mean $2x$ and variance $2\sigma^2$, that for three quanta would have mean $3x$ and variance $3\sigma^2$, and so on. In other words the gaussian curves fitted to succeeding peaks should get wider and wider as the number of their component quanta increases. This does not happen in fig. 10.10*a*, where the variance is much the same in all the fitted gaussians. A similar effect is evident in other central synapses (Jack *et al.*, 1981, 1994; Stern *et al.*, 1992).

What does this mean? One possibility is that the size of individual quantal responses is limited not by the number of transmitter molecules in each vesicle but by the number of receptors on the postsynaptic membrane opposite each terminal bouton. Transmission at a bouton would then produce saturation of the receptors on the subsynaptic membrane, so that the contents of one vesicle would be more than sufficient to open all the postsynaptic transmitter-activated channels at any particular bouton. Variation in vesicle size would not then be reflected in variation in the postsynaptic response. Even if two vesicles were released from one bouton there would be no increase in the postsynaptic response (Redman, 1990; Edwards *et al.*, 1990).

Simulation experiments suggest that saturation is unlikely to be complete, however, because of the stochastic behaviour of the diffusing transmitter molecules and the receptors with which they combine (Faber *et al.*, 1992). The

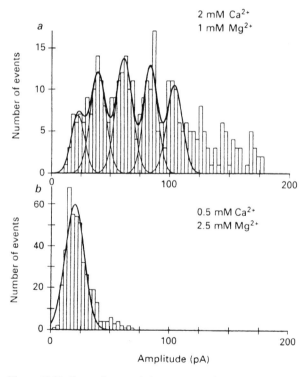

Figure 10.10. Quantal transmission at a central nervous synapse. IPSCs were recorded with the whole-cell patch clamp from cells in rat hippocampus slices, following stimulation of nearby cells. The histogram *a* gives the responses of 362 events, showing a mean peak separation of about 21 pA. Notice that the breadth of the gaussian curves used to fit the various peaks does not increase with increasing quantal content. The lower histogram *b* shows responses in a high magnesium solution; only a single peak is evident. (From Edwards *et al.*, 1990.)

question of the origin of the variation in quantal size at central synapses seems still to be open (Stevens, 1993; Bekkers, 1994; Jack *et al.*, 1994; Korn *et al.*, 1994). Computer simulations by Walmsley (1995) suggest that quite large differences between the amplitudes of quantal currents at different release sites may still produce a regular series of peaks in the frequency distribution of current amplitudes, implying that it is dangerous to draw too many conclusions from diagrams such as fig. 10.10*a*.

Spontaneous postsynaptic potentials are commonly observed in central neurons. These could be caused either by the spontaneous release of quanta just as in the miniature EPPs at the neuromuscular junction, or by the 'spontaneous' activity of individual presynaptic fibres, or by a mixture of the two. Spontaneous potentials persisting in the presence of tetrodotoxin are unlikely to be produced by presynaptic action potentials. Two further difficulties arise in trying to compare these spontaneous 'miniature potentials' with the quantal content of postsynaptic potentials evoked by electrical stimulation: their size and time course may depend on the distance of the release site from the recording site, and they may occur at sites which are different from the sites of evoked postsynaptic potentials.

These difficulties have been circumvented by Isaacson & Walmsley (1995) by looking at a particular synapse where auditory nerve fibres contact neurons in the cochlear nucleus by means of large endbulbs on the cell soma. When stimulating a single presynaptic fibre, they found that they could see and count individual quanta in excitatory postsynaptic currents (EPSCs) that were reduced in size by cadmium ions. The size distribution of these quanta was the same as that of spontaneous miniature EPSCs recorded in the same cell. Furthermore the size distribution of the evoked EPSCs followed Poisson statistics just as in experiments on the neuromuscular junction (fig. 10.11).

Presynaptic events producing transmitter release

We have seen that the arrival of a nerve impulse at the nerve terminal results in the release of a number of quantal packets of neurotransmitter. What happens in the terminal between the arrival of the nerve impulse and the release of the transmitter?

Synaptic delay

When an impulse arrives at the motor nerve ending, there is a short delay before the EPP arises in the muscle fibre. This delay between the presynaptic action potential and the postsynaptic response is a general property of chemically transmitting synapses. Katz & Miledi (1965*a*,*b*) used an ingenious technique to measure synaptic delay at the frog

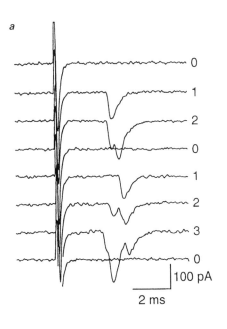

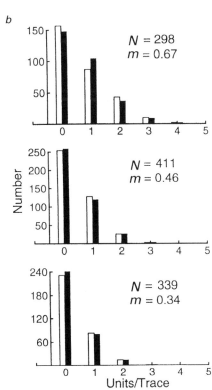

Figure 10.11. Counting quanta at a central synapse. The traces at the left (*a*) show individual responses recorded by whole-cell patch clamp from neurons in slices of the cochlear nucleus to stimulation of single presynaptic auditory fibres, in the presence of 100 μM cadmium ions. The number of quanta is shown at the right of each trace. At the right (*b*) are frequency distributions of the number of quanta in the responses (white columns) compared with that expected from the Poisson distribution (black columns). Here N is the total number of traces and m is the mean number of quanta per response. The upper graph is from the same cell as in *a*, the lower ones are from two other cells. (From Isaacson & Walmsley, 1995. Reproduced with permission from *Neuron*. Copyright Cell Press.)

sartorius neuromuscular junction. They applied an extracellular microelectrode closely to a point on the terminal so as to be able to make 'focal recordings' of the presynaptic action potential and the postsynaptic EPP from localized points at the junction.

One possible explanation of synaptic delay was that the action potential in the motor nerve axon does not invade the terminal endings, so that the depolarization there would be a slower process, caused by electrotonic spread from the myelinated region of the axon. However, Katz & Miledi found that the presynaptic spike recorded from the myelinated terminals occurred at later times after the stimulus as the recording electrode was moved distally. They also found that antidromic impulses (ones travelling in the direction opposite to that which normally occurs in the animal, usually as a result of electrical stimulation) could be set up in the axon by focal stimulation of the terminals. Hence they concluded that the action potential in the axon does invade the terminals.

The focal recording technique was then applied to measuring accurately the extent of the delay at any particular point; the method used for this is extremely ingenious. The preparation in Katz & Miledi's experiments was kept in a calcium-free Ringer solution containing 1 mM magnesium ions; this completely blocked neuromuscular transmission but did not prevent presynaptic impulse conduction. The recording microelectrode contained a 0.5 M solution of calcium chloride, so that diffusion of calcium ions from the electrode (which could be controlled electrophoretically) allowed transmission to occur at the point of recording but not anywhere else. Thus the focal electrode recorded both the presynaptic action potential at a point on the nerve terminal and the EPP produced immediately below it; there could be no interference from transmitter action at distances away from this point, where presynaptic depolarization would occur at different times.

Figure 10.12a shows the form of the record obtained by this means, and fig. 10.12b shows a number of successive actual records. The size of the EPPs produced fluctuates in a stepwise manner and may fail completely, as is to be expected from the quantal release hypothesis. More interesting is the fact that the synaptic delay is shown to be variable. By measuring the delays of a large number of responses it was possible to measure the probability of transmitter quanta being released at various times after the presynaptic impulse. The minimum synaptic delay is about 0.5 ms at 17 °C.

What is happening during this 0.5 ms? Katz & Miledi considered three possibilities: delay between depolarization of the axon terminals and release of acetylcholine, the time

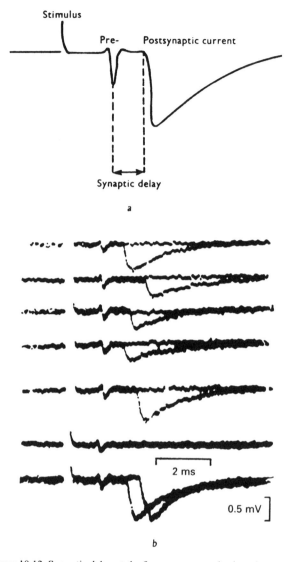

Figure 10.12. Synaptic delay at the frog neuromuscular junction. Records were obtained with the focal recording technique, using an external micropipette electrode filled with a calcium chloride solution. This picks up the currents produced both by the presynaptic action potential in the nerve terminal and the postsynaptic current in the muscle fibre. The saline solution was calcium free, so transmission occurred only in the immediate vicinity of the electrode. The form of a typical record is shown in *a*, indicating the components of the response and the method of measuring the synaptic delay. Part of a series of such responses is shown in *b*; two traces are superimposed on each record. (From Katz & Miledi 1965b, by permission of the Royal Society.)

taken for the acetylcholine to diffuse across the synaptic cleft, and the time between the impact of acetylcholine on the postsynaptic membrane and the subsequent depolarization. They showed that the time needed for acetylcholine molecules to diffuse the distance of 500 Å across the synaptic cleft must be very small, and certainly less than 0.05 ms. When acetylcholine was applied ionophoretically to the end-plate, the consequent postsynaptic depolarization began within 0.15 ms of the start of the ejecting current pulse. So at least 0.3 ms, the major part of the synaptic delay, is needed for the release of acetylcholine from the presynaptic nerve terminals.

The depolarization of the presynaptic terminal

As mentioned in chapter 5, the puffer fish poison tetrodotoxin blocks nervous conduction. A frog sartorius nerve–muscle preparation treated with tetrodotoxin will not therefore show the normal transmission process when the nerve is stimulated in the normal manner, because the axon and terminal membranes have been made inexcitable. However, Katz & Miledi (1967*a*) showed that if the terminals of such a preparation are briefly depolarized by a current pulse applied via a microelectrode, an EPP is elicited in the muscle fibre. Thus the depolarization of the terminal membrane, however it is produced, appears to be the effective agent in producing transmitter release.

The giant synapse in the stellate ganglion of the squid has very distinct advantages for the study of presynaptic events in that the presynaptic axon is large enough for an intracellular microelectrode to be inserted into it. Under normal conditions, an action potential in the presynaptic fibre produces an action potential in the postsynaptic fibre, with a delay of about 0.4 ms at 21 °C. However, if the preparation is fatigued by repetitive stimulation, or synaptic block is induced by magnesium ions or by hyperpolarization of the postsynaptic fibre, the EPSP in the postsynaptic fibre can be seen. Katz & Miledi (1966) determined the input–output relations of the squid giant synapse using tetrodotoxin to eliminate the nerve action potentials. Presynaptic depolarizations of 40 mV or more produced postsynaptic responses, as is shown in fig. 10.13.

The role of calcium ions

Synaptic transmission is dependent upon calcium ions in the external medium and is, as we have seen, much reduced when the magnesium/calcium ratio is high. This suggests that calcium ions may be involved in the release of transmitter. Direct evidence for this idea was provided by Miledi (1973), using the squid giant synapse. Because of the large size of the presynaptic axon he was able to inject calcium ions into it via an intracellular micropipette. This procedure caused a depolarization of the postsynaptic axon, indicating that raising the presynaptic intracellular calcium ion concentration causes release of the transmitter substance.

Axon terminals are rich in voltage-gated calcium channels (Llinas *et al.*, 1981; Smith & Augustine, 1988). One might therefore suppose that transmitter release is triggered by calcium flowing in through these channels, which are themselves opened by the presynaptic action potential. Katz & Miledi (1967*b*) tested this idea by applying calcium ions electrophoretically to the terminals of a tetrodotoxin-treated preparation in calcium-free Ringer solution at various times before and after the application of depolarizing pulses. They found that application of calcium ions as little as 50 μs before the depolarizing pulse resulted in transmission, whereas application of calcium ions after it did not.

Cell imaging methods have been used to show that the calcium ion concentration in the presynaptic terminal rises when the neurotransmitter is released. The calcium-sensitive dye fura-2 is very useful here. When it is injected into the presynaptic terminal of the squid stellate ganglion giant synapse, stimulation produces marked fluorescence changes indicating high calcium ion concentrations (fig. 10.14). These are initially localized next to the immediate presynaptic membrane, suggesting that the calcium channels are next to the active zones (Smith & Augustine, 1988; Smith *et al.*, 1993).

Changes in cellular calcium ion concentrations can also be monitored with the protein aequorin, extracted from a bioluminescent jellyfish; aequorin luminesces in the presence of calcium ions. Llinas and his colleagues (1995) used a synthetic analogue, *n*-aequorin-J, to measure the time course of calcium ion concentration changes. They found that rise in calcium ion concentration triggered by a presynaptic action potential peaks within 200 μs and lasts about 800 μs. This is just what one would expect from the measurements on calcium entry as determined earlier by voltage-clamp experiments (Llinas *et al.*, 1981). Furthermore the light emission occurred at discrete spots on the presynaptic membrane, suggesting that the calcium ion concentration rises markedly in the immediate vicinity of individual calcium channels.

This agrees with some ingenious experiments by Stanley (1991, 1993) on transmission in the chick ciliary ganglion. The presynaptic terminal here forms a large 'calyx' that wraps around much of the postsynaptic cell. Patch clamping of enzymically dissociated terminals showed that calcium channels occur only on the inner side of the terminal from which the acetylcholine is released. Individual channels have been visualized by atomic force microscopy after labelling with ω-conotoxin GVIA linked to colloidal gold particles; they form clusters in which the channels are spaced about 40 nm apart (Haydon *et al.*, 1994).

Stanley (1993) used a patch electrode containing a chemiluminescent acetylcholine assay solution, so that release of a vesicle from the membrane patch produced a brief burst of photons, detected by means of a photomultiplier. He could thus measure the opening of calcium channels and the release of transmitter in the same patch of presynaptic membrane. He found that opening of a single calcium channel in the patch was sufficient to induce vesicle release in the same patch; one of his records is shown in fig. 10.15. Calculations imply that as few as 200 calcium ions enter the terminal through a single channel, and this in turn suggests that the target for calcium action must be very close to the calcium channel. It looks as though each

calcium channel is associated with its own synaptic vesicle just prior to release.

del Castillo & Katz (1956) supposed that quantal release occurs when calcium ions react with a substance X and that the release rate should be proportional to the concentration of the intermediate CaX. However, the size of the EPP is a non-linear function of the external calcium concentration, so Dodge & Rahamimoff (1967) proposed that several calcium ions needed to act cooperatively to produce release, perhaps by the formation of an intermediate Ca_nX, in which case

$$\text{EPP amplitude} = K([Ca^{2+}]_o)^n$$

or

$$\log \text{EPP amplitude} = \log K + n\log [Ca^{2+}]_o$$

where K is a constant. Dodge & Rahamimoff found that the maximum value of n at the neuromuscular junction was

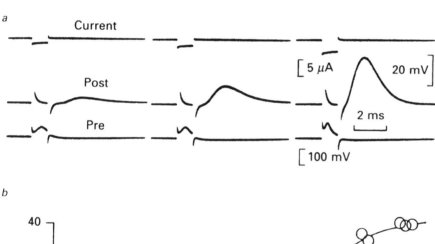

a

Current

Post

Pre

$\left[\, 5\ \mu A \right.$ 20 mV

2 ms

$\left[\, 100\ mV \right.$

b

Post (mV)

40

30

20

10

50   100

Pre (mV)

Figure 10.13. The relation between presynaptic membrane potential and postsynaptic response in the squid giant synapse. Currents of various strengths were passed through the presynaptic terminal, while microelectrodes recorded the pre- and postsynaptic membrane potentials. The preparation was treated with tetrodotoxin to eliminate action potentials. Sample records are shown in *a*, and these form part of the graph shown in *b*. (From Katz & Miledi, 1966.)

3.8, suggesting that four calcium ions cooperate to cause release of a quantum.

Further quantitative information about the role of calcium ions in transmitter release has been provided by some experiments on the bipolar neurons of the goldfish retina by Heidelberger and her colleagues (1994). Bipolar neurons (see chapter 15) have large terminals and can be isolated from the retina by treatment with the enzymes hyaluronidase and papain. A patch clamp electrode applied to the terminal in the whole-cell clamp configuration allowed the membrane capacitance of the cell to be measured. They changed the calcium ion concentration inside the cell by using a 'caged' calcium compound, DM-nitrophen. This is a chelating agent which has a high affinity for calcium in the dark but a much lower affinity when illuminated with ultraviolet light. Consequently a flash of UV light will produce a very rapid increase in the calcium ion concentration (the change is complete within 0.1 ms). The DM-nitrophen was introduced into the terminal via the patch electrode, which also contained a calcium-sensitive dye, furaptra, so that the calcium ion concentration in the terminal could be measured.

Figure 10.16 shows some results of these experiments. Calcium is released inside the terminal by the flash of UV light. When the concentration change is large enough, the membrane capacitance of the cell increases as number of synaptic vesicles fuse with the presynaptic membrane and so increase its area. The higher the calcium ion concentration, the more rapid this change is. The probability of any particular vesicle being released is effectively zero below a threshold calcium ion concentration of about 10 to 20 μM, and becomes very high at concentrations above 100 μM. The steepness of this relationship suggests that at least four calcium ions must bind to the trigger for vesicle fusion, and the relatively high concentrations required confirm the view that the trigger must be close to the calcium channel.

Similar experiments have been carried out on adrenal chromaffin cells (Heinemann *et al.*, 1994). Here the secretory granules are larger than the clear synaptic vesicles involved in fast synaptic transmission and are broadly

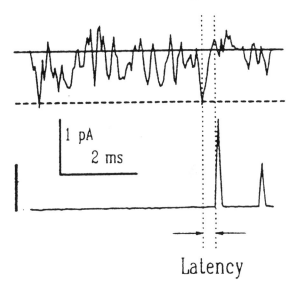

Figure 10.15. Evidence that opening of a single calcium channel can elicit transmitter release from a single nearby synaptic vesicle. The upper trace shows a cell-attached patch clamp record from the synaptic face of a presynaptic terminal from chick ciliary ganglion, during a voltage step to -30 mV. The full horizontal line represents zero current and the dashed line the inward current associated with opening of one calcium channel; although the trace is rather noisy, two individual calcium channel openings are easily seen. The lower trace shows the light output from the patch pipette, which contained a luminescent acetylcholine assay system; the bar represents one photon. Bursts of two or more photons indicate release of acetylcholine. They were associated with calcium channel openings 0.25 to 0.5 ms previously, and one example of this is shown here. The cells had been disassociated with collagenase and other enzymes. (From Stanley, 1993. Reproduced with permission from *Neuron*. Copyright Cell Press.)

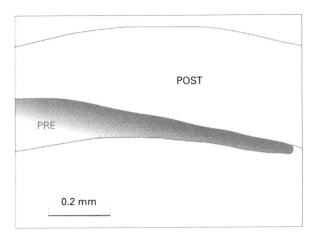

Figure 10.14. Calcium entry into squid giant presynaptic terminals during activity. The diagram is based on experiments using the fluorescent calcium-sensitive dye fura-2; shading indicates higher calcium ion concentrations. Presynaptic action potentials produce marked increases in ionic calcium concentration next to the presynaptic membrane. This suggests that voltage-gated calcium channels are clustered at the active zones, adjacent to the vesicle release sites. (After Smith & Augustine, 1988. Redrawn from: Calcium ions, active zones and synaptic transmitter release. *Trends in Neurosciences* **11**, 458, with permission from Elsevier Trends Journals.)

similar in structure and positioning to the dense-core vesicles of many nerve terminals. There is a longer delay between calcium release and vesicle fusion with the plasma membrane, but the calcium ion concentration required for release is somewhat less than for clear synaptic vesicles; the rate of exocytosis is half-maximal in the range 10 to 20 μM calcium. Since dense-core vesicles do not appear to be associated with active zones, and since they are released only when stimulation rates are relatively high, it seems likely that their release is controlled by the calcium concentration in the bulk cytoplasm of the terminal, as is indicated in fig. 10.17.

Molecular components of the synaptic vesicle cycle

A great deal of information has accumulated in recent years about the molecules involved in the release and recycling of synaptic vesicles; it is possible to make sensible suggestions about the functions of some of these proteins, but there are noticeable gaps in what is still an unfolding story (Jahn & Südhof, 1994; Südhof, 1995; Calakos & Scheller, 1996). The average vesicle in the mammalian central nervous system has a diameter of 40 to 50 nm. This small size must restrict the membrane components to about 8000 to 10 000 phospholipid molecules and to perhaps about a hundred protein molecules with a total mass of not more than 5000 kDa.

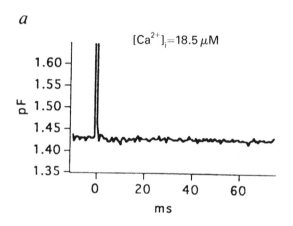

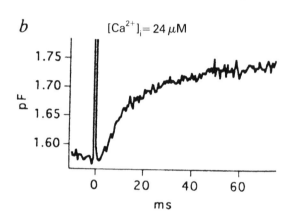

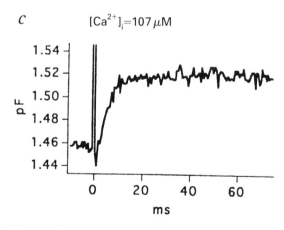

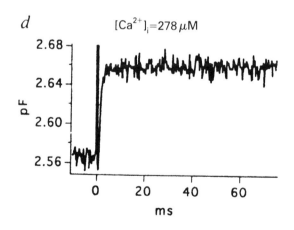

Figure 10.16. Vesicle fusion produced by release of caged calcium inside the presynaptic terminals of goldfish retinal bipolar cells. Cells were patch clamped in the whole cell configuration; each trace shows the membrane capacitance of a different cell, measured with a 1600 Hz sine wave (compare equation 4.4). The terminal contained the caged calcium compound DM-nitrophen, and the fluorescent dye furaptra that was used to measure the internal calcium ion concentration; both of these were introduced via the patch electrode. The artefact at time zero shows the timing of the ultraviolet flash used to release calcium ions from the DM-nitrophen. With a low intensity flash (*a*) the calcium ion concentration was insufficient to produce any vesicle release. At progressively higher intensities (*b* to *d*) the capacitance change, reflecting the increase in plasma membrane area as vesicles fuse with it, took place more and more rapidly. (From Heidelberger *et al.*, 1994. Reprinted with permission from *Nature* **371**, 514. Copyright 1994 Macmillan Magazines Limited.)

The vesicle membrane proteins are of two types: those concerned with the uptake and storage of the transmitter substance, and those concerned with membrane trafficking. Vesicles appear to be released from particular sites on the presynaptic membrane. The process whereby they become positioned there is called *docking*. Before the docked vesicle is released by the action of calcium ions, it is likely that further preparatory changes have to occur: this is called *priming*. Membrane fusion, via the formation of a *fusion pore*, follows very rapidly after calcium entry.

Some of the proteins involved in these processes are indicated in fig. 10.18. The proton pump complex and the neurotransmitter transporter (which is different for different neurotransmitters) and perhaps SV2 are all concerned with transport of the neurotransmitter from the cytoplasm into the vesicle. Synapsin I probably connects vesicles to the cytoskeleton, releasing them when it is phosphorylated by calcium–calmodulin-dependent protein kinase II.

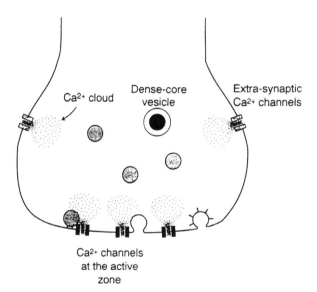

Figure 10.17. Calcium involvement in two types of neurotransmitter release. Small clear vesicles used in fast responses (as with acetylcholine or amino acid neurotransmitters) are docked next to voltage-gated calcium channels in the active zones of the presynaptic terminal. Calcium entry through one of these channels produces a high calcium ion concentration of limited extent, sufficient to trigger release of the adjacent vesicle. Dense-core vesicles (where the transmitter is a catecholamine or neuropeptide) are not associated with active zones, and require lower calcium concentrations in the bulk cytoplasm of the terminal for their release, such as would be produced by repetitive stimulation of the nerve fibre. (From Burgoyne & Morgan, 1995. Reproduced from: Ca^{2+} and secretory vesicle dynamics. *Trends in Neurosciences* **18**, 191, with permission from Elsevier Trends Journals.)

The roles of the various proteins in vesicle release are not fully established at the time of writing, but it may be useful to refer to them in the context of one model of the release process, shown in fig. 10.19. The synaptic vesicle membrane contains at least two isoforms of VAMP ('vesicle-associated membrane proteins', also called synaptobrevins). Docking occurs when these bind to two proteins on the presynaptic plasma membrane, syntaxin, a protein which binds to N-type calcium channels, and SNAP-25 ('synaptosomal associated protein', M_r 25 000). Soluble proteins α-SNAP ('soluble NSF-attachment protein') and NSF (*N*-ethylmaleimide-sensitive factor) also become attached to the docking link, and the complex then splits ATP to become primed ready for release. The soluble protein n-sec1 (also called munc-18) may be attached to syntaxin at rest and may have to leave before docking can take place. The role of the vesicle membrane protein synaptotagmin is not too clear; it binds four calcium ions per molecule so may be involved in the calcium-sensitivity of the release process.

A number of natural neurotoxins target the various proteins involved in vesicle release. Tetanus and botulinum toxins are produced by the soil-dwelling anaerobic bacteria *Clostridium tetani* and *C. botulinum*. There are seven different versions of the botulinum neurotoxins (BoNT/A to BoNT/G), but just one version of the tetanus toxin. They are all proteins of M_r about 150 000, readily split on proteolysis into an L chain (M_r about 50 000) and an H chain (M_r about 100 000), with the two chains then held together by a disulphide link. The L chains act as metalloproteases, targeting specific protein components of the vesicle cycle (see Niemann *et al.*, 1994).

Tetanus toxin attacks synaptobrevin and prevents transmitter release, particularly in the terminals of neurons that inhibit motoneurons, and so causes convulsions. The botulinum toxins prevent transmitter release in motoneurons, so they cause paralysis by neuromuscular block. Botulinum toxins A and E target SNAP-25, toxins B, D and F target synaptobrevin, and toxin C1 targets syntaxin. Botulinum toxins are the most powerful toxins known: the dose required to kill a mouse is 10^{-10} g kg^{-1}, so the amount required to kill a human would be about 5×10^{-9} g, which means that about 30 g would be sufficient to eliminate the population of the world.

A component of black widow spider venom, α-latrotoxin, causes a massive release of neurotransmitter vesicles. It binds to neurexins and also to synaptotagmin.

Ideas on how membrane fusion occurs have been influenced by experiments on exocytosis in secretory cells, where the secretory vesicles are rather larger than synaptic vesicles. When a secretory vesicle fuses with the plasma membrane the area of the plasma membrane must increase, and this can be detected, using a whole-cell patch clamp and applied high frequency alternating current, as an increase in the membrane capacitance of the cell (see Almers, 1990; Neher, 1992). There is a brief outflow of current associated with each vesicle fusion, as is shown in fig. 10.20. This is associated with an abrupt increase in membrane conductance followed by a slower further increase. Perhaps a rosette of fusion pore monomers connecting both membranes forms, then expands a little so as to open the pore, and finally the lipids of the two bilayers run together between the protein

monomers as the pore expands further, as suggested in fig. 10.20. An alternative view is that nearby proteins form a 'scaffold' which produces dimpling of the vesicle membrane so that the two bilayers are brought close together, and that the pore then forms spontaneously by their fusion and consists entirely of lipid molecules (Monck & Fernandez, 1994).

Facilitation and depression

If a curarized rat diaphragm muscle is repetitively stimulated through its motor nerve, it is found that successive

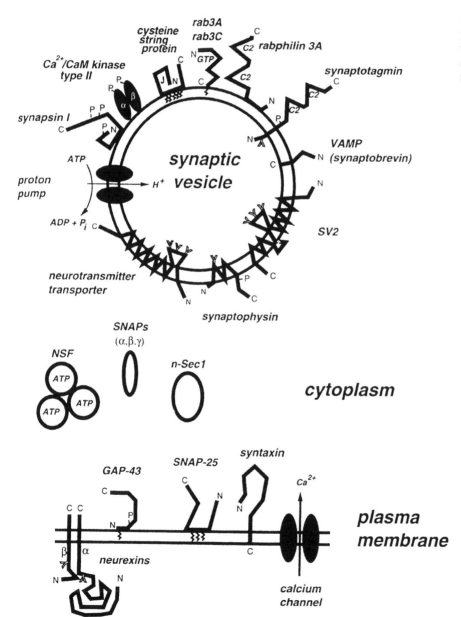

Figure 10.18. Some of the proteins involved in synaptic vesicle release. For abbreviations, see the text. (From Calakos & Scheller, 1996.)

EPPs decline in size, as is shown in fig. 10.21*a*. This phenomenon is known as *neuromuscular depression*. However, if the calcium ion concentration is lowered, so that the number of vesicles release per impulse is much reduced, the reverse effect, known as *neuromuscular facilitation*, occurs (fig. 10.21*b*).

The mechanisms of these effects were investigated by del Castillo & Katz (1954*b*), using a frog toe muscle. In solutions with high magnesium concentrations, they found that the proportion of failures of transmission decreased during a train of stimuli. If the stimuli were paired, the number of failures following the second stimulus was fewer than that following the first stimulus. These experiments indicate that facilitation is a presynaptic phenomenon; it is produced by an increase in the probability of discharge of acetylcholine quanta. When transmission was blocked by curare (which, of course, does not itself reduce the number of quanta released per impulse), depression was observed, and it was found that the later EPPs in a train showed fluctuations in amplitude. Here, again, it is evident that depression is mainly a presynaptic phenomenon, although it is possible that desensitization may be important in some cases after prolonged stimulation.

What is facilitation caused by? Katz & Miledi (1968) suggested that some of the calcium entry during the first nerve impulse remains active and adds to that entering with the second impulse so as to produce a larger response. They tested this idea using the ionophoretic application of calcium ions, and found that facilitation is much enhanced if calcium is present during the first nerve impulse.

The fourth-power relation between calcium entry and release has an interesting implication for the 'residual calcium' hypothesis. Suppose the first impulse raises the calcium level of the active zones to some value q, from which it declines rapidly to $0.2q$ and then much more slowly. The release rate at $0.2q$ will be proportional to $(0.2)^4q$, i.e. about 1/600 of its value at q. A second impulse will raise the calcium level to $1.2q$ and so the release rate will be proportional to $(1.2)^4q$, approximately twice its value for the first impulse (Katz & Miledi, 1968).

More recent analyses suggest that there are at least four different facilitatory processes, separable by their time courses and their differential responses to strontium and barium ions (Magleby & Zengel, 1982). Facilitation proper is seen as a two-stage process, with recovery time constants of tens and hundreds of milliseconds, respectively. Augmentation has a recovery time constant of a few seconds, and potentiation (previously known as post-tetanic potentiation: Lloyd, 1949; Hubbard, 1963) lasts for several seconds to some minutes. Augmentation and

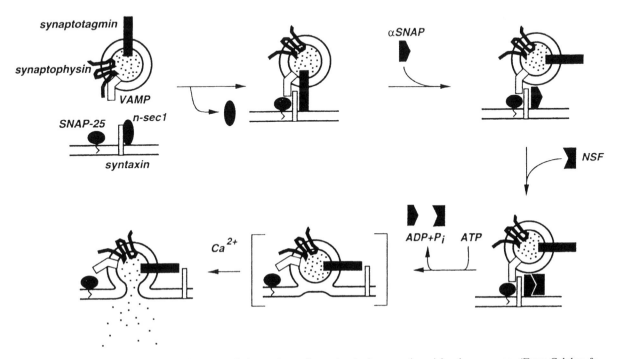

Figure 10.19. A possible model for the docking, priming and pore formation in the synaptic vesicle release process. (From Calakos & Scheller, 1996.)

potentiation are normally seen only after a series of presynaptic impulses.

Long-term potentiation may be defined as a potentiation lasting for an hour or more. We will consider this and other long-term changes at synapses in the next chapter.

Presynaptic receptors and presynaptic inhibition

Noradrenergic neurons have their activity modulated by noradrenaline and a number of other substances via presynaptic receptors (Langer, 1974, 1981). Evidence for this conclusion comes from experiments in which the amount of noradrenaline released at a synapse is measured in the presence of drugs which act at adrenergic receptors. In the dog heart, for example, Yamaguchi *et al.* (1977) found that the amount of noradrenaline released after stimulating sympathetic nerve fibres is increased in the presence of phenoxybenzamine (an α-receptor antagonist) and reduced in the presence of clonidine (an α_2-receptor agonist). Thus there appear to be α_2-adrenergic receptors on the presynaptic

varicosities, whose action is to inhibit noradrenaline release. It looks as though there is some negative feedback control of the amount of transmitter released.

Other presynaptic receptors occur. Inhibition of noradrenaline release may be brought about by acetylcholine (via muscarinic receptors), dopamine, 5-hydroxytryptamine or opioid peptides. Noradrenaline release can be increased via β-adrenergic receptors, angiotensin II receptors (angiotensin II is a peptide hormone involved in the control of blood pressure) and nicotinic acetylcholine receptors. The mechanisms of these effects could perhaps involve the control of the presynaptic intracellular calcium ion concentration.

If there were some mechanism in the nervous system whereby the amount of transmitter released from the presynaptic terminal could be reduced by the activity of a second neuron, then this second neuron would be capable of reducing the responses in the postsynaptic cell elicited by the action of the first neuron. Such a mechanism would thus be inhibitory, and since it would act presynaptically, we could call the phenomenon *presynaptic inhibition*, to contrast it with the postsynaptic inhibition mechanism which we have examined in chapter 7.

There is good evidence that presynaptic inhibition does in fact occur in the mammalian spinal cord (Frank & Fuortes, 1957; Eccles *et al.*, 1961). Consider the experiment whose results are shown in fig. 10.22. The EPSPs shown in records *a–d* were obtained from a gastrocnemius motoneuron in response to a single stimulus exciting the group Ia fibres in the gastrocnemius-soleus nerve. When the stimulus was preceded by stimulation of the group I fibres in the posterior biceps-semitendinosus nerve, as in records *b–d*, the EPSP was depressed in size, although there was no evidence of any inhibitory synaptic action on the motoneuron itself. This depression, or inhibitory action, began about 2.5 ms after the inhibitory volley entered the cord, reached a maximum after about 15 ms, and lasted for over 200 ms (fig. 10.22*e*).

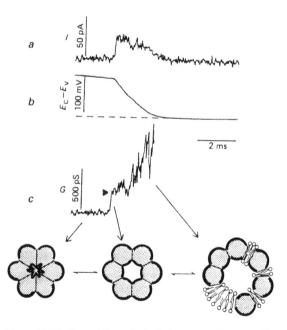

Figure 10.20. Current through the fusion pore of a mast cell during secretory vesicle fusion with the plasma membrane. *a* shows the current transient through the fusion pore, recorded from a mast cell during secretion. *b* shows the potential driving the current, calculated from the time integral of *a*. *c* is the conductance of the fusion pore, obtained by dividing *a* by *b*. The diagram below interprets these changes as a fusion pore formed by a number of monomers. This opens abruptly by conformational change in the pore protein, and then gets larger as pore monomers separate so that bilayer lipid molecules can flow in between them. (From Almers *et al.*, 1988.)

Figure 10.21. Facilitation and depression in curarized rat diaphragm muscle, stimulated at 120 shocks s⁻¹. In *a* the external calcium ion concentration was 2.5 mM, allowing release of many quanta per stimulus, although the size of the responses was reduced with curare; depression is evident. In *b* the external calcium ion concentration was reduced to 0.28 mM, greatly reducing the probability of release of individual quanta; this record shows facilitation. (From Lundberg & Quilisch, 1953.)

This experiment shows that it is possible for the monosynaptic EPSP to be depressed by neuronal action without there being any IPSP in the motoneuron, and therefore suggests that the inhibitory action takes place presynaptically. Presynaptic inhibition occurs elsewhere in the central nervous system, and also peripherally at the neuromuscular junction in crustacea (Dudel & Kuffler, 1961; see fig. 21.22). There is more than one mechanism for the process.

Volleys of action potentials in spinal sensory fibres are followed by a depolarization, the dorsal root potential, which spreads electrotonically along the same or adjacent dorsal roots. Eccles and his colleagues (1962) showed that dorsal root afferent fibres subjected to presynaptic inhibition are in fact depolarized, and that the depolarization follows a time course which is apparently identical with that of the inhibitory action. Thus it seems that presynaptic inhibition is brought about by depolarization of the afferent nerve terminals in the way suggested at the beginning of this section. The dorsal root potential is a reflection of this depolarization. A central delay of 2.5 ms is longer than would be expected in a pathway with a single synapse only, so it is probable that the presynaptic inhibitory pathway involves one or two interneuronal stages (fig. 10.23a). Figure 10.23b shows the suggested anatomical basis of presynaptic inhibition; 'serial synapses' of this type have in fact been seen in electron micrographs of the spinal cord (E. G. Gray, 1962).

It seems likely that the depolarization seen as the dorsal root potential is brought about by chloride channels constituting GABA$_A$ receptors (Nicoll & Alger, 1979). The increased chloride conductance would tend to 'clamp' the membrane potential near to the chloride equilibrium potential and so reduce the depolarization produced by the presynaptic action potential. A system of this type is well established for the peripheral presynaptic inhibition at crustacean neuromuscular junctions (Takeuchi & Takeuchi, 1966).

A second and probably more important mechanism involves GABA$_B$ receptors (Dunlap & Fischbach, 1981; Wu & Saggau, 1995). These are members of the 7TM family (see p. 161), acting via G proteins. In the spinal cord they probably act to reduce the sensitivity of voltage-gated calcium channels, so reducing the amount of calcium entering the terminal so that less transmitter can be released. At other sites, as in the hippocampus for example, there may be an increased activity of potassium channels as well, leading to reduction in the size of the presynaptic action potential (Nicoll *et al.*, 1990; Wu & Saggau, 1997).

Synthesis and packaging of neurotransmitters

Small-molecule neurotransmitters such as acetylcholine, noradrenaline, GABA, and so on, are synthesized in the presynaptic terminals by specific enzymes. Synthesis occurs in the cytosol of the terminal, and from there the transmitter is moved across the vesicular membrane by one of a family of vesicular neurotransmitter transporters. Neuropeptides, on the other hand, since they are derived directly from transcripts of nuclear genes, have to be synthesized in

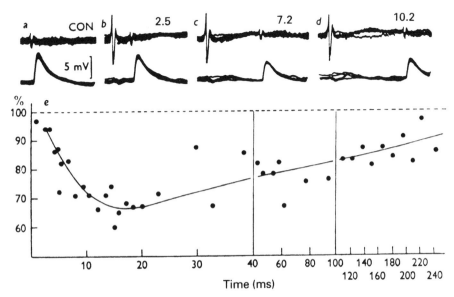

Figure 10.22. Presynaptic inhibition in the spinal cord of the cat. The lower traces in *a* to *d* show EPSPs recorded from a gastrocnemius motoneuron, produced by stimulation of the gastrocnemius-soleus nerve. The upper traces show extracellular records from one of the dorsal roots. In *b* to *d*, the excitatory volley was preceded by a volley in the posterior biceps-semitendinosus nerve; the figures above each trace give the time intervals between the two volleys. The graph *e* shows the depression of the EPSP (expressed as a percentage of its size in the absence of inhibition) at various times after the inhibitory volley. (From Eccles *et al.*, 1961.)

the cell body and transported along the axon to the terminal, usually packaged in large granular storage vesicles. Let us have a look at some examples.

Acetylcholine

Acetylcholine is synthesized by the transfer of an acetyl group from the acetyl carrier acetyl coenzyme A to choline:

$$acetyl\ CoA + choline \rightarrow acetylcholine + CoA$$

The reaction takes place in the cytosol of the terminal and is catalysed by the enzyme choline acetyltransferase (see

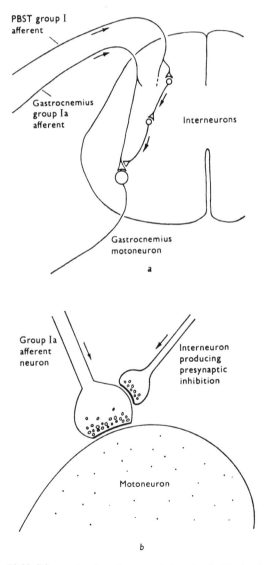

Figure 10.23. Diagram to show the suggested anatomical basis of presynaptic inhibition: *a*, neuronal connections; *b*, synaptic structure.

Cooper *et al.*, 1996). Acetyl CoA is formed in the mitochondria, and is used as an acetyl carrier in a variety of metabolic reactions, including especially that whereby carbon is fed into the citric acid cycle. The choline is supplied partly by reuptake from the synaptic cleft (see later); it may also be derived from breakdown of membrane phospholipids such as phosphatidylcholine or be taken up from the blood.

Acetylcholine in the cytosol is moved into the synaptic vesicles by the action of a specific transporter molecule, the vesicular acetylcholine transporter (see Parsons *et al.*, 1993; Usdin *et al.*, 1995). This action can be blocked by the drug vesamicol. The amino acid sequence of the transporter suggests that there are twelve membrane-crossing segments, with the N- and C-termini on the cytoplasmic side of the vesicular membrane, and with a large loop projecting into the vesicular lumen between transmembrane segments 1 and 2. Remarkably, the gene coding for the transporter is situated within the first intron in the gene coding for choline acetyltransferase, both in rats and humans and in the nematode worm *Caenorhabditis elegans* (Erickson *et al.*, 1994). Perhaps this arrangement facilitates joint regulation of the expression of the two genes.

The vesicular acetylcholine transporter is an antiporter in which acetylcholine movement into the vesicle is driven by an outward movement of hydrogen ions. The hydrogen ion gradient is set up by an active transport of hydrogen ions into the vesicle; the energy for this proton pump is provided by the breakdown of ATP. Studies on *Torpedo* electric organ suggest that the acetylcholine concentration in the cytoplasm is about 4 mM, while that in the vesicle is 400 to 800 mM. The vesicle lumen has a pH of 5.2 to 5.5, whereas the cytoplasmic pH is 7.2 to 7.5. We would expect the activity of the proton pump, pushing hydrogen ions into the vesicle, to set up a potential gradient across the vesicular membrane, with the inside of the vesicle positive to the cytoplasm. If two hydrogen ions move outward for every acetylcholine ion that moves inward, then this potential gradient plus the 100-fold hydrogen ion concentration gradient would easily provide sufficient energy to drive the process (see Parsons *et al.*, 1993; Schuldiner *et al.*, 1995).

The vesicles contain ATP as well as acetylcholine, and this is probably transported across the vesicular membrane by a specific transporter.

Noradrenaline and other monoamines

The noradrenergic neurons of the sympathetic nervous system have non-myelinated axons which terminate in various effector organs (see Fillenz, 1990). The terminals branch, and have a large number of swollen portions called *varicosities* which are packed with mitochondria and vesicles. Most of the vesicles have electron-dense cores and are hence known as 'granular vesicles'. They are of two sizes, large (diameter up to 100 nm) and small (diameter 40 to 50 nm). The small ones are predominant in the varicosities, whereas the large ones are also found in the cell body and axon.

Both types of vesicle contain noradrenaline and ATP. The large vesicles contain the proteins chromogranin A and dopamine β-hydroxylase (the enzyme which converts dopamine to noradrenaline) and a number of others, together with enkephalins and other opioid peptides which presumably are acting as cotransmitters. Klein & Lagercrantz (1981) calculated that small vesicles contain 700 to 1000 molecules of noradrenaline whereas large ones contain 8000 to 16 000. In some electron micrographs there are also some 'agranular vesicles' present; it seems likely that these are derived from small granular vesicles by loss of some of their contents.

Catecholamines are synthesized from the amino acid tyrosine in the sequence of reactions shown in fig 10.24,

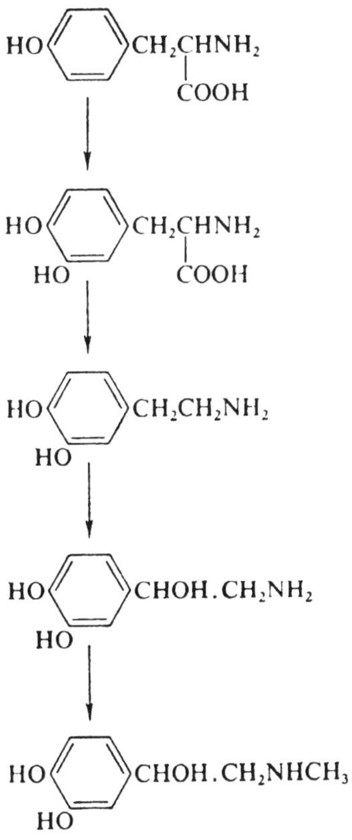

Figure 10.24. The synthesis of adrenaline (epinephrine) from tyrosine.

which gives rise to the neurotransmitters dopamine, noradrenaline (norepinephrine) and adrenaline (epinephrine). The enzymes catalysing the first stages of this sequence (tyrosine hydroxylase and DOPA decarboxylase) are cytoplasmic, but the next one, dopamine β-hydroxylase, is found in the vesicles, so it seems that dopamine has to be taken into the vesicles before it can be converted to noradrenaline.

Dopamine is transported into the vesicles by a vesicular monoamine transporter of molecular structure similar to that of the vesicular acetylcholine transporter (Erickson *et al.*, 1992; Henry *et al.*, 1994). There is a high concentration of hydrogen ions inside the vesicle because of the action of a proton pump. The transporter exchanges one monoamine molecule for two hydrogen ions, and the uptake is driven by the hydrogen ion gradient (Johnson, 1988).

Molecular cloning shows that there are two closely related vesicular transporters, called VMAT1 and VMAT2. VMAT1 is found in chromaffin cells of the adrenal gland, but not in nerve cells. VMAT2 is found in sympathetic ganglion cells and enteric neurons, in central nervous neurons using dopamine, noradrenaline, adrenaline, 5-hydroxytryptamine or histamine (all of them monoamines) as their neurotransmitters, and also at low concentration in chromaffin cells (Peter *et al.*, 1995).

Amphetamine promotes the release of transmitter at noradrenergic and dopaminergic synapses in the brain. Cocaine blocks the reuptake mechanism (p. 198) at the same group of synapses. Both these substances produce sensations of euphoria and so have become drugs of abuse. Drugs used as antidepressants include imipramine (a reuptake inhibitor) and iproniazid (an inhibitor of monoamine oxidase). The drug reserpine, used to treat high blood pressure, blocks the vesicular transporter.

Amino acid neurotransmitters

Glycine and glutamate are dietary amino acids and can also be readily synthesized from various precursors. GABA is formed from glutamic acid by the action of glutamic acid decarboxylase, an enzyme restricted to the central nervous system and the retina.

The concentrations of these amino acids in the nerve terminal cytoplasm are in the millimolar range, which implies that the concentration gradient across the vesicular membrane is not as high as for monoamines. They are probably all transported into the vesicles by an antiport system in exchange for hydrogen ions, but this may be a one-for-one rather than a one-for-two exchange (Schuldiner *et al.*, 1995). The vesicular glutamate transporter is sensitive to chloride ions and may require their inward movement (Hartinger & Jahn, 1993).

Neuropeptides

Non-peptide neurotransmitters such as acetylcholine and noradrenaline are synthesized in the nerve terminal by specific enzymes, and there are usually mechanisms for their reuptake from the extracellular space. Neither of these systems is available for neuropeptides, which are synthesized, usually as precursors, by the normal ribosomal protein synthesis system in the cell body. This means that the peptide or its precursor must be transported along the axon to the terminal, usually in large granular storage vesicles, before it can be released (fig. 10.25). The concentrations of neuropeptides are typically in the range of some picomoles per gram of brain, perhaps 1000 times less than the monoamines and about 100 000 times less than the amino acid neurotransmitters (Hökfelt *et al.*, 1980).

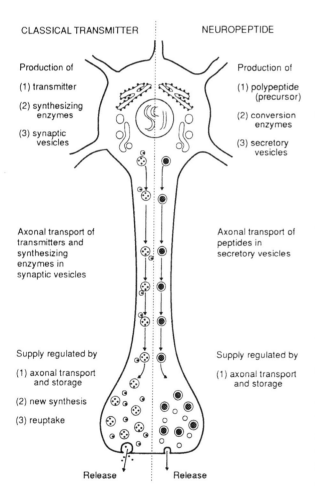

CLASSICAL TRANSMITTER   NEUROPEPTIDE

Production of

(1) transmitter

(2) synthesizing
    enzymes

(3) synaptic
    vesicles

Production of

(1) polypeptide
    (precursor)

(2) conversion
    enzymes

(3) secretory
    vesicles

Axonal transport of
transmitters and
synthesizing
enzymes in
synaptic vesicles

Axonal transport of
peptides in
secretory vesicles

Supply regulated by

(1) axonal transport
    and storage

(2) new synthesis

(3) reuptake

Supply regulated by

(1) axonal transport
    and storage

Release   Release

Figure 10.25. Diagram to show the differences between a peptidergic neuron (right) and one releasing a non-peptide ('classical') neurotransmitter. (From Hökfelt *et al.*, 1980, as redrawn in Brown, 1994.)

It seems to be the common pattern that neuropeptides and peptide hormones are formed from longer-chain protein precursor molecules. Molecular cloning methods have allowed the sequences of many of these precursors to be determined (see Lynch & Snyder, 1986; Sossin *et al.*, 1989). Preproenkephalin A, for example, is the translation product of a gene which is expressed in the adrenal gland and in the brain. It contains 268 amino acid residues of which the first twenty-four probably form a signal sequence which is removed once the proenkephalin is formed in the endoplasmic reticulum (Noda *et al.*, 1982*b*). There are four sequences of Met-enkephalin and one of Leu-enkephalin in the complete chain (fig. 10.26).

Each of the enkephalin sequences in proenkephalin is bounded at each end by a pair of basic residues, lysine or arginine, which form the sites at which endopeptidases of the trypsin type will be able to break the peptide chain. Some rather larger sequences correspond to peptides found in the adrenal gland; it seems likely that processing proceeds further in the brain to give the smaller opioids as end-products. Two brain peptides of intermediate size, metorphamide and amidorphin, are amidated at their C-terminal ends (Weber *et al.*, 1983; Seizinger *et al.*, 1985); this is done by the enzymic breakage of a terminal glycine residue. Similar results have been obtained for other neuropeptide precursors. Thus pro-opiomelanocortin contains the sequences for γ-melanocyte-stimulating hormone (γ-MSH), adrenocorticotropic hormone (ACTH) and β-lipotropin (β-LPH), and further splitting can occur so that ACTH produces α-MSH and β-LPH produces γ-LPH and β-endorphin. The large peptides are produced in the anterior lobe of the pituitary, but processing proceeds farther in the brain, to give the smaller ones (see Douglass *et al.*, 1984).

Another way of generating diversity occurs when the same gene produces different mRNAs as a result of alternative RNA splicing. In the thyroid the calcitonin gene produces an mRNA which codes for the hormone calcitonin, but in the hypothalamus most of the mRNA has a partly different sequence and so codes for another peptide, referred to as the calcitonin gene-related peptide or CGRP (Amara *et al.*, 1982). The existence of CGRP was predicted from the mRNA sequence, and it is highly satisfactory that it was later shown to have many of the properties expected of a neurotransmitter (Poyner, 1992). It was localized by immunocytochemistry to particular parts of the brain such as the trigeminal ganglion, using antisera to a synthetic peptide based on the predicted sequence (Rosenfeld *et al.*, 1983).

Substance P is another peptide whose gene can produce two different messenger RNAs by alternative splicing. The precursor gives rise either to substance P alone or to substance P and another predicted peptide called substance K because of its similarity in sequence to the amphibian peptide kassinin (Nawa *et al.*, 1984).

Neuropeptides are commonly released along with small-molecule non-peptide neurotransmitters; the phenomenon is known as *cotransmission*. It was discovered by Hökfelt and his colleagues (1977) in sympathetic ganglion cells, using the fluorescent antibody technique. A thin section of tissue was washed first with a solution containing sheep antibodies to somatostatin, and then with fluorescent rabbit antibodies to sheep antibodies. An adjacent section, mostly passing through the same cells was similarly stained for dopamine β-hydroxylase, the enzyme which converts dopamine to noradrenaline. Hökfelt and his colleagues found that many of the cells (a majority in the inferior mesenteric ganglion) were fluorescent in both sections and must therefore have contained both somatostatin and noradrenaline. Since then a considerable number of such cases of coexistence have been described (Hökfelt *et al.*, 1982; Campbell, 1987); some of them are listed in table 10.1.

The terminals of peptidergic neurons commonly contain small vesicles that contain only the small-molecule transmitter, together with larger ones that contain both neuropeptide and small-molecule transmitter. The neuropeptide is usually released only when firing rates in the presynaptic neuron are high, perhaps in response to higher calcium ion concentrations in the terminal as a whole, as is suggested in

Table 10.1 *Some examples in the mammalian brain of the coexistence of peptide and non-peptide ('classical') neurotransmitters in the same neuron*

Classical transmitter	Peptide	Location
Dopamine	Neurotensin	Ventral midbrain
	CCK	Ventral midbrain
Noradrenaline	Neuropeptide Y	Medulla oblongata
5-HT	Substance P	Medulla oblongata
	Enkephalin	Medulla oblongata
Acetylcholine	VIP	Cerebral cortex
	Substance P	Pons
GABA	Somatostatin	Thalamus, cerebral cortex
	CCK	Cerebral cortex
	Enkephalin	Retina

Notes:
CCK, cholecystokinin; 5-HT, 5-hydroxytryptamine; VIP, vasoactive intestinal polypeptide; GABA, γ-aminobutyric acid.
Simplified after Hökfelt *et al.*, 1986.

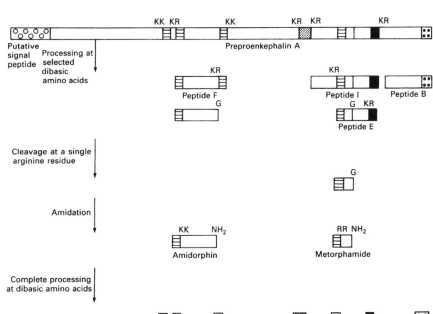

Figure 10.26. Processing of preproenkephalin A, the precursor of the enkephalins and a number of other opioid peptides. Proteases usually split the molecule at sites where there are two basic residues, lysine (K) or arginine (R). Sometimes cleavage can occur at a single arginine residue next to a glycine (G) residue, the latter being then fractured to leave an amide (NH_2) group. (From Lynch & Snyder, 1986.)

fig. 10.17 (see Hökfelt, 1991). In some cases, more than one neuropeptide occurs in the same neuron; this is particularly so in the enteric nervous system, where in some cells acetylcholine coexists with as many as six different neuropeptides (Steele & Costa, 1990).

Non-quantal release of acetylcholine

In addition to the relatively massive release of acetylcholine from motor nerve endings on stimulation, there is also a continuous low level release in the resting state (Mitchell & Silver, 1963; Fletcher & Forrester, 1975). In rat diaphragm muscles, for example, Fletcher & Forrester found values of 0.65 pmol min^{-1} over a 30 min period.

Can the spontaneous release of quanta, producing miniature EPPs, account for this effect? Assume that each quantum contains 10 000 molecules, that the frequency of miniature EPPs is 3 s^{-1}, and that there are 1000 end-plates per muscle. Then the amount of acetylcholine in spontaneous quantal release should be $(3 \times 60 \times 10000 \times 1000)$, i.e. 1.8×10^9 molecules min^{-1} or about 0.03 pmol min^{-1}. This is less than one-twentieth of the actual value. Using a slightly different approach, Vizi & Vyskocil (1979) concluded that quantal release represents not more than 1% of the total release of acetylcholine from the resting muscle.

Does this steady 'leak' of acetylcholine come from the resting presynaptic terminal and would it produce any postsynaptic depolarization? Katz (1969) considered that any such depolarization would be very small and likely to be undetected, but a closer look by Katz & Miledi (1977) did produce results. They used muscles treated with the anticholinesterase DFP (diisopropylphosphorofluoridate) to enhance the effect, and applied a massive dose of curare to the end-plate by ionophoresis. This produced a small hyperpolarization of about 40 μV for a few seconds, which we can attribute to a temporary block of the acetylcholine receptors.

Non-quantal release of acetylcholine is much reduced by vesamicol, which also blocks the action of the vesicular acetylcholine transporter (Edwards *et al.*, 1985). It seems likely that the transporter molecule becomes incorporated into the plasma membrane when vesicles are discharged, so that it transports acetylcholine from the cytoplasm into the synaptic cleft instead of into the vesicle lumen. This suggests that the effect may be an accidental consequence of the vesicle discharge. Nevertheless it may also serve some functional purpose in the development or maintenance of the nerve–muscle interconnection.

Removal of neurotransmitter from the synaptic cleft

Synaptic transmission is necessarily a brief event: the response of the postsynpatic cell must be brought to an end very soon after the presynaptic signal stops. It would be no use, for example, to have a neuromuscular junction in which a single presynaptic nerve impulse produced a massive maintained depolarization of the muscle cell. Consequently we find that there are various methods for removing the neurotransmitter from the synaptic cleft after it it has dissociated from the receptor. The most widespread of these is a reuptake mechanism so that the transmitter is transported back into the presynaptic terminal or, in some cases, into neighbouring glial cells. Cholinergic synapses are exceptional in that the acetylcholine is rapidly broken down by enzymic action. Diffusion of transmitter out of and away from the cleft may be important in some cases.

Neurotransmitter transporters

Many tissues, and especially adrenergic nerve endings, are able to take up catecholamines from the extracellular fluid. Thus Whitby *et al.* (1961) found that tritium-labelled noradrenaline was rapidly concentrated in adrenergic nerve endings. Iversen (1963) found that the adrenergic nerve terminals in a perfused rat heart were capable of clearing noradrenaline from the entire extracellular space of the heart in about ten seconds. He calculated that this uptake system would therefore remove noradrenaline from the immediate vicinity of the terminals in a matter of milliseconds, thus constituting a most effective method of inactivating the transmitter (Iversen, 1971). It is also an economical method, since the noradrenaline is accumulated intracellularly in the synaptic vesicles, from which it can be released again in further synaptic transmission.

Such uptake systems have since been found for dopamine, 5-HT, glutamate, GABA (γ-aminobutyric acid), and glycine, and also for choline at cholinergic synapses. Uptake is tightly coupled to entry of sodium ions into the cell, so the energy for the process is derived from the sodium ion electrochemical gradient, which is itself set up by the sodium pump. The GABA transporter is dependent on chloride as well as sodium ions. For each GABA molecule transported into the terminal, two sodium ions and one chloride ion are also transported; since GABA is probably in the form of a zwitterion with no overall charge, this means that one positive charge is moved for each GABA molecule and so the transport is electrogenic (Keynan & Kanner, 1988).

Experiments with the glutamate transporter expressed in *Xenopus* oocytes suggest that three sodium ions and one hydrogen ion are transported into the cell with each glutamate ion, while one potassium ion is moved out (Zerangue & Kavanaugh, 1996). Here again the transporter is electrogenic, with a net movement of positive charge into the terminal. The coupling of each glutamate ion movement to three sodium ions implies that the transporter can move glutamate up a concentration gradient in which the internal concentration is 10^6 times the external concentration; this highly effective system may be crucially important in preventing the adverse effects of extracellular glutamate in the brain.

The GABA transporter was the first of these neurotransmitter transporters to be cloned (Guastella *et al.*, 1990). The cDNA sequence predicted a protein chain of 599 amino acid residues with twelve transmembrane segments and a large extracellular loop between segments 3 and 4. Cloning of the noradrenaline transporter revealed a protein of similar structure with 48% similarity in the amino acid sequence (Pacholczyk *et* al., 1991). Other members of this molecular family include transporters for glycine, 5-HT, dopamine, and also for choline, the result of acetylcholinesterase activity (Kanner, 1993*a*; Worrall & Williams, 1994). The glutamate transporter, however, shows no sequence similarity to these and has a different structure, with probably eight or nine transmembrane segments (Pines *et al.*, 1992; Kanner, 1993*b*).

The energy for the process of uptake from the synaptic cleft is provided by the sodium ion electrochemical gradient, itself set up by the action of the ATP-driven sodium pump in the plasma membrane. This is in contrast with the energy source for the transport of neurotransmitter from the cytoplasm into the vesicles, which, as we have seen, is provided by the electrochemical gradient of hydrogen ions across the vesicular membrane that is set up by the ATP-driven vesicular membrane proton pump. Figure 10.27 summarizes this situation.

Acetylcholinesterase

The enzyme acetylcholinesterase hydrolyses acetylcholine to form choline and acetic acid. It occurs in very high concentration at the neuromuscular junction. The first evidence for this came from the work of Marnay & Nachmansohn (1938), using a manometric estimation technique on frog sartorius muscle. They calculated that there was sufficient acetylcholinesterase at each neuromuscular junction to hydrolyse about 10^{-14} mol of acetylcholine in 5 ms; this is more than enough to deal very rapidly with the amount of acetylcholine released per impulse. Acetylcholinesterase is a remarkably efficient enzyme: later studies have shown that each active site hydrolyses 1.4×10^4 acetylcholine molecules per second, and it seems likely that nearly all contacts between the enzyme and the substrate molecules result in hydrolysis, so that the reaction rate is limited largely by the rate of

diffusion of the substrate (Rosenberry, 1975; Quinn, 1987).

The breakdown of acetylcholine by acetylcholinesterase can be inhibited by a number of compounds known as anticholinesterases. The most well known of these is eserine (physostigmine), an alkaloid extracted from the calabar bean. The EPP is greatly prolonged in the presence of eserine and may initiate a series of action potentials in the muscle fibre. Many organophosphorus compounds also inactivate cholinesterases, a capability which has formed the basis for their use as war gases and insecticides; examples are DFP and tetraethylpyrophosphate (TEPP), and such insecticides as parathion and malathion.

Salpeter and her colleagues (1978) used tritiated DFP as a marker to localize acetylcholinesterase at the neuromuscular junction, using electron microscope autoradiography. They found that the silver grain density was high over the synaptic cleft and especially high over the junctional folds. In mouse fast extraocular muscle they estimate that there are about 2500 acetylcholinesterase sites per square micrometre of postsynaptic membrane. There is

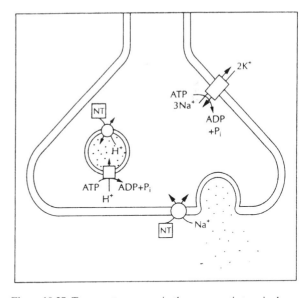

Figure 10.27. Transport processes in the presynaptic terminal. Reuptake of neurotransmitter (NT) from the synaptic cleft is carried out by cotransport with sodium ions flowing down their concentration gradient. The sodium concentration gradient is maintained by the sodium pump, which requires the breakdown of ATP as its energy source. Transport of neurotransmitter from the terminal cytoplasm into the synaptic vesicles is carried out by an antiporter which utilizes the hydrogen ion gradient across the vesicular membrane. To maintain this gradient, hydrogen ions are pumped into the vesicles by the action of a proton pump, which also uses ATP as its energy source. (From Amara & Pacholczyk, 1991.)

good evidence that the acetylcholinesterase molecules are in the extracellular material (the basal lamina) of the synaptic cleft and junctional folds. Hall & Kelly (1971) found that treatment of end-plates with proteolytic enzymes would release acetylcholinesterase into the perfusion fluid. After collagenase treatment neuromuscular transmission still occurred but the postsynaptic responses had a longer time course, comparable with that after eserine treatment.

Acetylcholinesterase is usually a multimeric protein containing a number of catalytic subunits. These have been sequenced and their structure has been investigated by X-ray diffraction techniques. The protein chain is 575 amino acid residues long, and the catalytic active site seems to be located at the bottom of a deep gorge in the molecule, where the histidine-440 and serine-200 residues are brought close together (Schumacher *et al.*, 1986; Sussman *et al.*, 1991; Taylor, 1991; Taylor & Radic, 1994).

The subunits may be aggregated in various ways. The nomenclature for these is based on the number of catalytic subunits and whether they form a globular (G) or asymmetric (A) complex. Thus G_1, G_2 and G_4 forms are globular with one, two and four catalytic subunits, whereas A_4, A_8 and A_{12} forms are asymmetric with four, eight and twelve catalytic subunits. G_4 forms may include an extra P subunit which forms a hydrophobic tail anchoring the complex in the plasma membrane. The A forms all have collagen-like Q subunits, each connected to a tetramer of catalytic subunits; the collagen-like tails serve to anchor them in the extracellular material. The A_{12} form seems to be predominant in the synaptic cleft at the neuromuscular junction, with G_4 attached to the surrounding perijunctional membrane (see Massoulié *et al.*, 1993).

Adrenaline and noradrenaline are enzymically inactivated in two main ways: by oxidative deamination (by the mitochondrial enzyme monoamine oxidase) or by methylation of a hydroxyl group (by the cytoplasmic enzyme catechol-*O*-methyl transferase). Both these enzymes are intracellular and hence can act only on the catecholamines after they have been taken up by cells. This situation is in marked contrast with that at cholinergic synapses, where the extracellular localization of acetylcholinesterase in the synaptic cleft enables it to inactivate the transmitter substance immediately and effectively. Clearly enzymic inactivation cannot be the prime method of terminating the effects of noradrenaline on the postsynaptic membrane; that role, as we have seen, is played by the noradrenaline transporter.

11
Learning-related changes at synapses

One of the remarkable characteristics of animals is that much of their complex behaviour can be modified as a result of experience. In ourselves we see this in such diverse activities as avoiding foods that we do not like, learning to ride a bicycle, being able to identify new faces and new voices, memorizing a new telephone number, remembering what happened last week or many years ago, and so on. Our learned capabilities and our memories are the basis of our individual personalities.

What happens in the nervous system when such changes take place? What is the cellular basis of learning and memory? Since the work of Ramón y Cajal (1911), it has seemed likely that modifications of the effectiveness of transmission at synapses might provide the answer to this question. Hebb (1949) produced a particular model for this: he assumed that repetitive activity at a particular synapse could produce lasting cellular changes. More precisely, as Hebb put it, 'when an axon of a cell *A* is near enough to excite a cell *B* and repeatedly or persistently takes part in firing it, some growth process or metabolic change takes place in one or both cells such that *A*'s efficiency, as one of the cells firing *B*, is increased'.

Learning-related changes in nerve cells have been investigated in a number of model systems. We begin by considering in some detail one system that has proved particularly fruitful, and this is followed by a briefer treatment of some other aspects.

Simple learning in *Aplysia*

Aplysia is a large and graceful opisthobranch mollusc, sometimes called the sea-hare. It grazes peacefully on seaweeds, repelling predators by ejecting clouds of purple ink into the water if disturbed. Like many other molluscs, it has large nerve cells in a number of separate ganglia. These are individually identifiable and readily accessible to microelectrodes, and so they have been much used by electrophysiologists interested in the cellular basis of behaviour (Kandel, 1976).

The gill of *Aplysia* is a delicate structure held in a fold of that part of the body surface called the mantle. The mantle also forms a short tube called the siphon, through which sea water is drawn over the gill. These organs can readily be displayed in the living animal by holding back the mantle folds that cover them. If the siphon or the nearby mantle is touched or stimulated with a jet of water, then both gill and siphon are contracted and withdrawn. This is a simple reflex action: the sensory tactile stimulus is followed directly by an automatic motor response. The size of the response is very roughly proportional to the size of the stimulus.

If the animal is subjected to a whole series of tactile stimuli, the responses rapidly get smaller, as is shown in fig. 11.1. This phenomenon is called *habituation*. Kandel and his colleagues found that the excitatory postsynaptic potentials in the motor neurons are correspondingly reduced in size, and that this is due to a reduction in the amount of transmitter released in response to each presynaptic nerve impulse. The habituated response can be restored to its original size by means of a strong stimulus (an electric shock) applied to the head or tail of the animal, also shown in fig. 11.1. This is known as *sensitization* (see Kandel & Schwartz, 1982; Kandel *et al.*, 1987; Kandel, 1991; Kennedy *et al.*, 1992). The decrease of transmitter release in habituation is called synaptic depression, and the increase on sensitization is called facilitation.

Habituation and sensitization have both short-term and long-term time-scales. A single series of ten tactile stimuli at 1 min intervals, for example, produces a habituation that lasts only for several hours. But if the training series are repeated on four successive days, then the habituation lasts for three weeks or more (Carew *et al.*, 1972). The sensitization produced by a single stimulus to the tail lasts for minutes or hours, but repeated sensitizing stimuli will sensitize the response for days or weeks, as is shown in fig. 11.2 (Pinsker *et al.*, 1973; Frost *et al.*, 1985). The distinction between these two time-scales is strongly reminiscent of the two phases of memory, short-term and long-term, revealed by psychologists in behavioural tests on people and other vertebrates (see Atkinson & Shiffrin, 1971; Baddeley, 1976; Squire & Zola-Morgan, 1988).

Habituation and sensitization are both rather simple forms of learning; they are described as non-associative. They demonstrate modification of behaviour as a result of experience, but there seems to be little intellectual dimension to them. More 'intelligent' behaviour is shown in *associative learning*, where previously separate stimuli are brought together to form a new behaviour pattern. A classic example is provided by Pavlov's dogs (Pavlov, 1906). Dogs normally produce saliva when they see food. Pavlov found that if a bell was rung just before the food was brought in, then the dogs learned to associate the bell with the food and would salivate when it rang. The response is called a conditioned reflex and the whole behaviour pattern is an example of *classical conditioning*; the sound of a bell is called the conditioned stimulus, the sight of food is the unconditioned stimulus and salivation is the unconditioned response.

Classical conditioning occurs in *Aplysia*. A light stimulus to the mantle shelf or the siphon will produce a small contraction of the gill. A strong stimulus to the tail (the unconditioned stimulus) will produce a large contraction of the gill (the unconditioned response). If the shock to the tail is preceded by a light stimulus to the mantle shelf (the conditioned stimulus), then the animal soon shows strong gill contraction in response to the light stimulus to the mantle (Carew *et al.*, 1983).

The molecular basis of sensitization

The neural circuitry for the gill-withdrawal reflex is shown in fig. 11.3. Sensitization involves an increase in the amount of transmitter released by the sensory neurons (Castelluci & Kandel, 1976); this is sometimes called heterosynaptic facilitation, since it follows stimulation of the facilitatory interneurons rather than the sensory neurons themselves. The facilitatory interneurons form synapses with the termi-

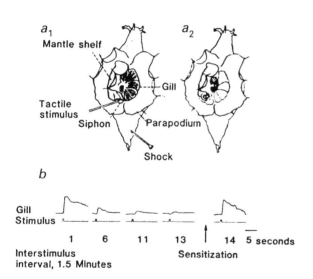

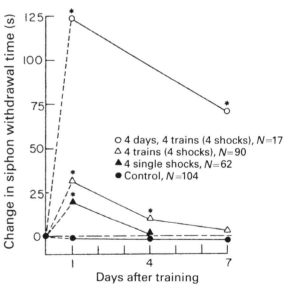

Figure 11.2. Long- and short-term sensitization of the siphon-withdrawal reflex in *Aplysia*. Animals received sensitizing shocks to the tail, either as four single shocks on one day (black triangles), four trains of four shocks on one day (white triangles), or four trains of shocks a day for four days (white circles). Control animals (black circles) received no sensitizing shocks. Responses to a light tactile stimulus to the siphon were measured one, four and seven days after the end of training; the results are shown as the difference between the duration of siphon withdrawal before and after training. Asterisks show statistically significant differences from the controls. Notice that the four days of training produced a large sensitization that was still well maintained after a week, whereas the single shocks produced smaller, short-term changes only. (From Frost *et al.*, 1985. *Proceedings of the National Academy of Sciences, USA* **82**, 8267. Copyright 1985 National Academy of Sciences, USA.)

Figure 11.1. The gill-withdrawal reflex in *Aplysia*, showing habituation and sensitization. a_1 shows the experimental arrangement: the parapodia (folds of the mantle) are drawn back so as to expose the gill. When the tactile stimulus (a jet of water) is applied to the siphon, both it and the gill are withdrawn (a_2). The size of the gill can be monitored with a photocell arrangement. *b* shows photocell recordings of the response of the gill to stimuli. The response is initially large (stimulus 1), but decreases (habituates) with succeeding stimuli. Between stimulus 13 and 14 an electric shock was delivered to the tail; the response is now restored to its initial size (sensitization). (From Kandel & Schwartz, 1982. Reprinted with permission from *Science* **218**, 433–43. Copyright 1982 American Association for the Advancement of Science.)

nals of the sensory neurons, with 5-hydroxytryptamine (5-HT) as the transmitter in some cases. Stimulation of the facilitatory interneurons, or treatment with 5-HT, closes a particular subset of potassium channels in the sensory neuron terminal. This makes the presynaptic action potentials longer, so voltage-gated calcium channels are open for longer, more calcium enters the terminal with each action potential, and so more transmitter is released (Hochner *et al.*, 1986). Remarkably, the synapses between the sensory and the motor cells can be reconstituted after isolating them in cell culture, and the facilitatory changes can still then be induced by applying 5-HT to the sensory terminals (Montarolo *et al.*, 1986).

Klein & Kandel (1980) suggested that the 5-HT produces these effects by a G protein cascade that activates adenylyl cyclase and results in the production of cyclic AMP. This then activates cyclic-AMP-dependent protein kinase (protein kinase A), which phosphorylates the potassium channels and closes them. Injection of the catalytic subunit of the protein kinase A (p. 159) into the sensory cells mimics sensitization in that it broadens the action potential

and increases the amount of transmitter released (Castellucci *et al.*, 1980), and patch clamp records of potassium channels from the cells show that they are closed by external application of 5-HT or injection of cyclic AMP into the cell (Siegelbaum *et al.*, 1982). The particular potassium channel type has been called the S–K channel, to indicate its indirect closure by serotonin (another name for 5-HT). Cell imaging techniques show that 5-HT produces marked increases in cyclic AMP concentration in the finer processes of the sensory cells (Bacskai *et al.*, 1993).

The clearest effect of the protein kinase A activation is to phosphorylate the S–K channel and thus close it. Other mechanisms may also be involved. The G protein that activates adenylyl cyclase in this cascade is probably a member of the G_s group. There is some evidence that a member of the G_o group is also activated by the 5-HT receptor, and that this leads via diacylglycerol to the activation of protein kinase C. Protein kinases A and C may then act together to promote increased transmitter release by phosphorylation of proteins involved in mobilization of vesicles in the terminal.

The second-messenger cascade involved in short-term sensitization is summarized in fig. 11.4. In the similar changes during classical conditioning, there seems to be an increased activation of adenylyl cyclase as a result of calcium–calmodulin activation following calcium entry

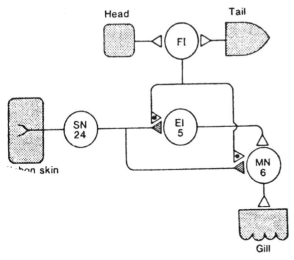

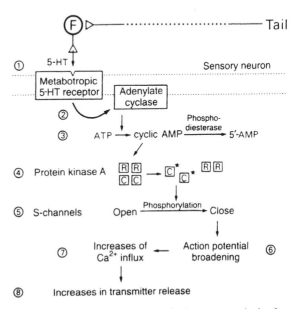

Figure 11.3. Simplified diagram of ... circuit of the gill-withdrawal reflex in *Aplysia*. Touching the siphon skin excites up to twenty-four sensory neurons (SN), and these in turn excite up to six motor neurons (MN) that supply the muscle cells in the gill; they also contact a number of different excitatory (EI) and inhibitory interneurons that also innervate the motor neurons. The amount of transmitter released by the sensory neurons is reduced when habituation occurs. Noxious stimuli to the tail or head excite the facilitating interneurons (FI) which contact the terminals of the sensory neurons. (From Kandel & Schwartz, 1982. Reprinted with permission from *Science* **218**, 433–43. Copyright 1982 American Association for the Advancement of Science.)

Figure 11.4. The molecular cascade in the nerve terminals of *Aplysia* sensory neurons during short-term sensitization. (From Kuno, 1995. Reproduced from M. Kuno, 1995, *The Synapse: Function, Plasticity and Neurotrophism*, by permission of Oxford University Press.)

into the terminal during the nerve impulses produced by the conditioning stimulus (Kandel *et al.*, 1987; Kandel, 1991).

Long-term memory storage

All the changes we have so far mentioned are characteristic of short-term changes in behaviour, by which we mean changes that last for minutes or hours rather than days or weeks. What happens in the long-term changes such as those illustrated in fig. 11.2? Bailey & Chen (1983, 1988*a*,*b*; Bailey *et al.*, 1994) found that long-term sensitization in *Aplysia* is accompanied by appreciable morphological changes in the sensory neurons. The active zones in the synaptic contacts with the motor neurons become larger and have more vesicles in them. The terminal arborizations are enlarged (fig. 11.5) and the numbers of synaptic varicosities in them are increased. Experiments on cells in culture show that these changes occur only when the sensory cell makes synaptic contact with the motoneuron (Glanzman *et al.*, 1990).

We might expect such morphological changes to be accompanied by protein synthesis. Flexner and her colleagues (1963) showed that puromycin, an antibiotic that inhibits protein synthesis, caused memory loss in mice when it was injected into their brains. Further work from a wide variety of experimental situations led to the conclusion that protein synthesis, during or soon after the training period, is an essential step in long-term memory (Davis & Squire, 1984).

Protein and RNA synthesis is also necessary for long-term sensitization changes in *Aplysia*, but not for short-term ones (Montarolo *et al.*, 1986). Sweatt & Kandel (1989) applied 5-HT briefly to isolated sensory neurons and found that seventeen different proteins were transiently phosphorylated. Repeated applications produced long-term phosphorylation of the same seventeen proteins, still evident 24 h later. The long-term effect was prevented in the presence of inhibitors of protein or RNA synthesis, whereas the short-term effect was not. This suggests that the persistent phosphorylation of the seventeen proteins requires some new protein or proteins to be synthesized as a result of switching on their genes.

Protein synthesis associated with long-term facilitation was measured more directly by Barzilai *et al.* (1989). They found that 5-HT applied for 1.5 h leads to synthesis of specific proteins on different time-scales. Synthesis rates for ten proteins were transiently increased in the first hour after 5-HT exposure, whereas those for five other proteins were transiently decreased. Rates for four more proteins increased after 1 h, rose to a peak at 3 h and fell away by 8 h. Two more proteins were still being synthesized at 24 h after training.

What initiates this protein synthesis? Does cyclic AMP act as the trigger for long-term changes as it does for short-term ones, and if so how does it do it? Injection of cyclic AMP into sensory neurons does produce long-term facilitation, and inhibitors of protein kinase A prevent it (Schacher *et al.*, 1988; Bailey *et al.*, 1994). Cell imaging using protein kinase A labelled with fluorescent markers shows that protein kinase A will not enter the nucleus as the complete enzyme, but that its catalytic subunit does so when it is elicited by increases in cytoplasmic cyclic AMP concentration (Bacskai *et al.*, 1993). All this suggests that the relevant genes could be switched on by phosphorylation of some nuclear protein by protein kinase A following its activation by cAMP.

The transcription of eucaryotic genes is controlled partly by enhancer elements, sections of the DNA some way upstream from the protein-coding section. Genes whose

Control   Sensitized

Figure 11.5. Effects of long-term sensitization on sensory neurons in *Aplysia*. The neurons were injected with horseradish peroxidase, a marker that can be histochemically processed to reveal their branching patterns. The diagrams show superimposed serial sections in the middle parts of sample neurons. The two neurons on the right came from long-term sensitized animals, that on the left from a control animal. (From Bailey & Chen, 1988*a*. *Proceedings of the National Academy of Sciences, USA* **85**, 2375. Copyright 1988 National Academy of Sciences, USA.)

transcription is enhanced by cytoplasmic cyclic AMP depend on an element with the nucleotide sequence TGACGTCA, known as the cyclic AMP-responsive element or CRE. A specific protein CREB (CRE-binding protein) increases transcription when it is phosphorylated by protein kinase A (see Comb *et al.*, 1986, 1987; Alberts *et al.*, 1994). In human cultured cells this activator action can be repressed by another protein; the activator protein is then called CREB1 and the repressor CREB2 (Karpinski *et al.*, 1992).

Kandel's group found a CREB protein in *Aplysia*, later called apCREB1. They were able to prevent its action by injecting short pieces of DNA with the CRE sequence (the CREB protein would be swamped by binding to the many injected CRE sequences rather than the genomic one, it would seem). With the CREB protein out of action, they found that long-term facilitation did not occur, although there was no effect on short-term facilitation (Dash *et al.*, 1990). There is also a repressor protein, apCREB2. If this is put out of action by injection of antibodies to it, then a single pulse of 5-HT will produce long-term facilitation in cultured *Aplysia* neurons (Bartsch *et al.*, 1995). It may be that CREB2 normally holds CREB1 in the inactive state, and that it is phosphorylated when stimuli eliciting long-term sensitization occur.

To summarize the argument so far, long-term facilitation depends on an increase in cytoplasmic cyclic AMP, which activates protein kinase A, allowing its catalytic subunit to separate and enter the cell nucleus. Here it phosphorylates the CREB1 protein which then binds to CRE, a short enhancer sequence of the genomic DNA, and this change then initiates transcription of some gene or genes that are involved in long-term facilitation. These genes are known as immediate-early genes (IEGs) or early response genes (ERGs).

What proteins do these IEGs produce? It seems likely that they are of two types: early effectors, which may produce early changes in the cytoplasm, such as a maintained activation of protein kinase A, and regulators for further genes known as late response genes (LRGs). (The nomenclature for these genes is derived from that used for the comparable signal cascades initiated in cells by growth factors – see Cross & Dexter, 1991). The consolidation phase of long-term memory, it is thought, covers the period of time between the initial stimulus and the synthesis of the various proteins coded for by the late response genes (Alberini *et al*, 1994; Frank & Greenberg, 1994).

A number of candidates for early effector genes in *Aplysia* have been suggested (Bailey *et al.*, 1994). One is a protein that enhances the breakdown of the regulatory subunit of protein kinase A, whose concentration is reduced by a third during long-term sensitization, so leading to a maintained higher level phosphorylating action even when cAMP concentrations are not raised (Bergold *et al.*, 1990; Hegde *et al.*, 1993). Another possibility is the light chain of clathrin, a protein involved in endocytosis; clathrin's role in this case is to remove a cell adhesion molecule (apCAM) from the surface of the presynaptic terminal, probably to allow the terminal to grow (Bailey *et al.*, 1992; Mayford *et al.*, 1992; Hu *et al.*, 1993). Figure 11.6 gives an impression of how this might happen.

One early regulator gene involved in long-term facilitation in *Aplysia* has been detected by Alberini and her colleagues (1994). They looked for a gene similar to the

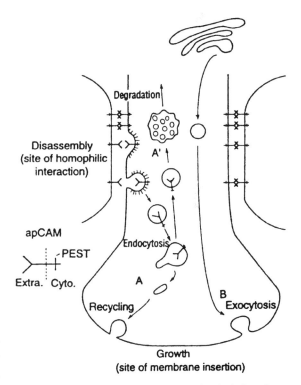

Figure 11.6. Turnover of cell adhesion molecules in learning-related neuron growth in *Aplysia*. Y-shaped symbols represent the cell adhesion molecule apCAM. Long-term sensitization leads to clathrin production, hence to endocytosis via clathrin-coated vesicles. The apCAM molecules may be either recycled to different parts of the membrane (path A') or broken down (path A'). The apCAM amino acid sequence contains regions rich in proline, glutamate, serine and threonine (PEST) residues, that are known to be sites readily accessible to enzymic degradation. New membrane may also be supplied from the Golgi network (path B). (From Bailey *et al.*, 1992. Reprinted with permission from *Science* **256**, 645–9. Copyright 1992 American Association for the Advancement of Science.)

CCAAT enhancer binding protein (C/EBP) transcription factors that are known to be involved in nerve terminal differentiation in various mammalian cells. They found such a clone (called ApC/EBP) in an *Aplysia* cDNA expression library, and were then able to show that it contained a CRE sequence (so it could be activated by CREB) and that its protein product would bind to and activate various known late response genes. Crucially, they found that it was induced in *Aplysia* sensory cells during the consolidation phase following prolonged exposure to 5-HT.

Proteins appearing in the later stages of the long-term response, presumably coded for by LRGs activated by IEG regulator proteins, include BiP and calreticulin (Kennedy *et al.*, 1992). BiP is a chaperone protein that is commonly found in cells undergoing protein synthesis. Calreticulin is a calcium-binding protein found in the lumen of the endoplasmic reticulum; change in the amount present in the cell would probably be related to changes in the levels of intracellular calcium signalling. Since the long-term changes include the morphological ones mentioned earlier, we would also expect proteins related to growth, such as cytoskeletal proteins, adhesion molecules, and proteins involved in neurotransmitter synthesis and release, to be synthesized. Figure 11.7 gives a summary of how long-term sensitization is thought to act on the sensory neuron via control of gene expression.

Long-term potentiation in the hippocampus

Long-term potentiation (LTP) is an activity-dependent increase in synaptic efficacy which is brought about by a brief high frequency train of stimuli to the presynaptic neurons and lasts for several hours. It was discovered in the hippocampus of rabbits by Bliss & Lømo (1973) and since then has been a subject of major interest in the search for the cellular basis of learning and memory (see Bliss & Collingridge, 1993; Bear & Malenka, 1994; Larkman & Jack, 1995; Martinez & Derrick, 1996).

The hippocampus plays an important role in many memory processes in mammals (see L. R. Squire, 1992). It is a paired structure lying deep within the cerebral cortex beneath the corpus callosum, and forms part of the limbic system. The main neurons in it are the granule cells of the dentate gyrus, and the pyramidal cells of areas CA1 to CA4. The principle synapses, shown in fig. 11.8, are (1) those between the perforant path and the granule cells, (2) those between the mossy fibres (axons of the granule cells) and the CA3 pyramidal cells, and (3) those between axons of the CA3 cells and the pyramidal cells of the CA1 region, via the Schaffer collaterals and the commissural fibres from contralateral CA3 cells. The dendrites of the granule and pyramidal cells are covered with small protuberances called

dendritic spines, and it is here that the incoming fibres make synaptic contact. The total number of cells in the rat hippocampus is well over a million, with each one receiving some thousands of synaptic inputs (see Brown & Zador, 1990). The arrangement of the cells is such that many of the

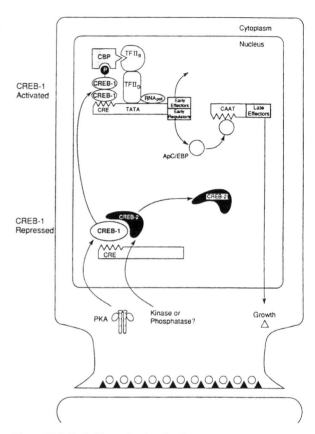

Figure 11.7. Probable mechanism for the control of gene expression in *Aplysia* sensory neurons during long-term sensitization. Production of cyclic AMP leads to activation of protein kinase A (PKA), as shown in fig. 11.4. The catalytic subunit of PKA enters the nucleus and phosphorylates the CRE-binding protein CREB1. This dissociates from the repressor protein CREB2 (it may be that phosphorylation or dephosphorylation of CREB2 is also necessary for this dissociation, perhaps following activation of another cytoplasmic signalling pathway), and is now attached to the CRE (cyclic-AMP-responsive element) region of the IEGs (initial early genes) and bound by the CREB-binding protein (CBP). This leads to activation of the general transcription factors TFII$_B$ and TFII$_D$ attached to the promoter region of the IEGs, so that the IEG gene products are synthesized. IEGs are of two types: early effectors, producing proteins exported to the cytoplasm, and early regulators. apC/EBP is an early regulator gene product: it acts as an enhancer for some of the various late effector genes whose products are involved in synaptic growth. (From Carew, 1996. Reproduced with permission from *Neuron*. Copyright Cell Press.)

properties of the synapses can be studied in slices of hippocampus maintained in saline or culture medium.

LTP was discovered by Bliss & Lømo (1973) in the hippocampus of rabbits. They used extracellular electrodes to record the postsynaptic response of granule cells in the dentate gyrus of an anaesthetized rabbit, and stimulated the presynaptic perforant path fibres. After conditioning trains of a few hundred stimuli at 20 s^{-1} or 100 s^{-1}, responses to single stimuli were markedly increased for periods of up to 10 h. Similar experiments with chronically implanted electrodes in intact animals showed that LTP might last for several weeks (Bliss & Gardner-Medwin, 1973).

An example of LTP at the synapses on CA1 cells in the hippocampal slice is shown in fig. 11.9 (Bliss & Collingridge, 1993). Figure 11.10 gives a diagrammatic summary of what is happening in the experiment. The graphs show the size of the 'field' EPSP produced by single stimuli and recorded by an extracellular electrode (this is a summed response from many pyramidal cells, equivalent to the mean response of the individual cells), and arrows show the times of tetanic bursts of high frequency stimuli. There are two stimulus sites, S1 (for weak stimuli, exciting few presynaptic fibres) and S2 (for strong stimuli, exciting many

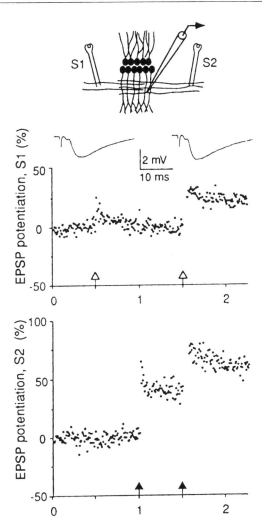

Figure 11.9. LTP at synapses on CA1 cells in a hippocampal slice. Stimuli were delivered to incoming fibres at two separate sites, S1 and S2, and the population excitatory postsynaptic potential (EPSP) was recorded with an extracellular electrode. Dots on the graphs show the size of the response to single stimuli, as a percentage increase on the average response in the first 30 min of the experiment. Stimuli were delivered at 15 s intervals alternately to S1 and S2; stimuli to S1 were weaker than those to S2, so would have excited fewer presynaptic fibres. Conditioning bursts of high frequency stimuli are shown as arrows on the graphs. At 30 min, a high frequency burst to S1 (the 'weak' pathway) produces a brief increase in EPSP size, but this is not maintained and so is not LTP. At 1 h, a similar burst to S2 (the 'strong' pathway) produces a marked and maintained increase in the response to S2 stimulation, but no change in the response to S1 stimulation. At 90 min, high frequency bursts are applied simultaneously to both S1 and S2; this time LTP is produced in the S1 pathway as well as in the S2 pathway. See also fig. 11.10. (From Bliss & Collingridge, 1993. Reprinted with permission from *Nature* **361**, 32. Copyright 1993 Macmillan Magazines Limited.)

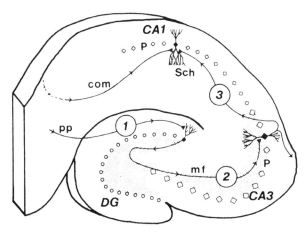

Figure 11.8. Excitatory pathways and synapses in the hippocampus, as seen in a transverse slice. (1) Axons from cells in the nearby entorhinal cortex, which themselves receive various sensory inputs, enter via the perforant pathway pp and synapse with granule cells in the dentate gyrus *DG*. (2) The axons of the granule cells (mossy fibres mf) synapse with pyramidal cells P in the CA3 region. (3) These each send an axon branch (the Schaffer collateral Sch) to pyramidal cells in the CA1 region, which also receive inputs from contralateral CA3 cell along the commissural pathway (com). (From Brown & Zador, 1990. From *The Synaptic Organization of the Brain*, third edition, edited by Gordon M. Shepherd. Copyright © 1990 by Oxford Universtiy Press, Inc. Used by permission of Oxford University Press, Inc.)

presynaptic fibres). A tetanic burst to the weak stimulus site S1 (at 30 min on the graphs, *A* in fig. 11.10) does not produce LTP, although there is some increase in EPSP size (short-term potentiation, STP) for a few minutes subsequently. However, a tetanic burst to the strong stimulus site S2 (at 1 h, *B*) does produce LTP in response to subsequent S2 stimuli, but not to subsequent S1 stimuli.

The experiment so far demonstrates two general characteristics of LTP: cooperativity and input specificity. *Cooperativity* is shown by the response to the strong stimulus alone (*B* in fig. 11.10) when compared with that to the weak stimulus (*A*): a number of presynaptic inputs have to be simultaneously active in order to produce LTP. *Input specificity* is shown in *B*: other inputs not active at the time of the tetanic burst do not show LTP. At 90 min in the experiment show in fig. 11.9 (*C* in fig. 11.10), tetanic bursts are given to both the weak and the strong inputs; this time the response to the weak input also shows LTP. This demonstrates a third general characteristic of LTP, *associativity*, whereby a weak input can be potentiated if it is active at the same time as a separate but convergent strong input. Associativity clearly has much in common with classical conditioning.

These general characteristics lead to a straightforward hypothesis about LTP: that a synapse will be potentiated if, and only if, it is active at a time when the region of the dendrite on which it terminates is sufficiently depolarized. This idea can be tested by delivering depolarizing pulses to the postsynaptic cell through an intracellular electrode; LTP then occurs with low frequency, low intensity stimuli pro-

vided they are paired with the depolarizing pulses (Kelso *et al.*, 1986; Sastry *et al.*, 1986).

Cellular mechanism of LTP at CA1 synapses

We have seen in chapter 8 that there are two main types of ionotropic glutamate receptor on many mammalian nerve cells. The AMPA receptor channel is readily opened to allow sodium and potassium ions to flow through. The NMDA receptor channel is blocked by magnesium ions at negative membrane potentials, but can be opened by glutamate if the postsynaptic cell is depolarized, and when this happens there is an appreciable inflow of calcium ions through the channel. Collingridge *et al.* (1983) found that LTP is prevented when the NMDA receptor is blocked by the selective antagonist AP5.

This discovery led to an influential model of how LTP works, illustrated in fig. 11.11. The two main types of ionotropic glutamate receptor, AMPA and NMDA, are present in the postsynaptic membrane. The NMDA receptors are blocked by magnesium ions at membrane potentials near to the resting potential, so the EPSP in response to a single presynaptic action potential is produced entirely by opening of the AMPA receptor channels. A high frequency burst, however, will depolarize the postsynaptic membrane and so remove the magnesium ions from the NMDA receptor channel and allow it to open. Calcium ions enter through the NMDA channel, where they activate various processes leading to enhanced transmission. Notice that the NMDA receptor channel requires two factors to open it: glutamate, released from the presynaptic terminal,

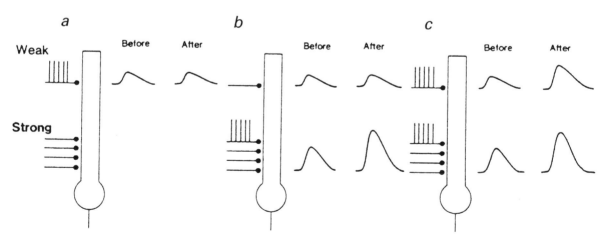

Figure 11.10. Diagram illustrating what is happening in fig. 11.9. A single pyramidal cell is shown receiving a weak synaptic input (few presynaptic fibres) and a strong one (many). EPSPs are shown before and after conditioning high frequency bursts of stimuli. *a* corresponds to the burst at 30 min in fig. 11.9, *b* at 1 h, and *c* at 90 min. Notice that *b* illustrates cooperativity and input specificity, and *c* illustrates associativity. (From Nicoll *et al.*, 1988. Reproduced with permission from *Neuron*. Copyright Cell Press.)

and depolarization, produced by the action of glutamate on other nearby receptors. Metabotropic glutamate receptors are probably also involved in LTP, which is reduced in the presence of the blocking agent MCPG (Bortoletto *et al.*, 1994; Riedel & Reymann, 1996).

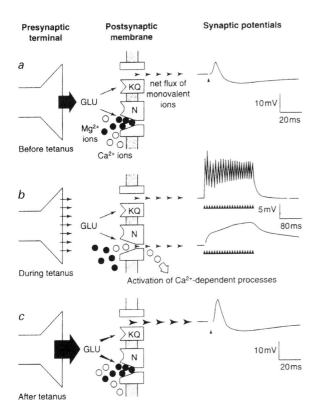

Figure 11.11. Collingridge's model for the initiation of LTP. At rest the NMDA receptor channels are blocked by magnesium ions, which are held there by the negative resting membrane potential. Single presynaptic nerve impulses (*a*) release glutamate, which binds to both AMPA (labelled KQ in this diagram) and NMDA (N) receptors, but ion inflow (largely sodium ions) occurs only through the AMPA receptor channels. A high frequency tetanus (*b*) produces depolarization (especially if adjacent areas are also depolarized by extra presynaptic fibres), which removes magnesium ions from the NMDA receptor channels; these then open so that calcium ions flow into the cell. The raised intracellular calcium ion concentration acts in some way as a trigger for long-lasting changes in the response to single stimuli (*c*), perhaps, as implied here, by enhanced release of glutamate from the presynaptic terminal, or alternatively by means of an increase in the number of active AMPA receptor channels. (From Collingridge & Bliss, 1987, 1995. Reproduced from: Memories of NMDA receptors and LTP. *Trends in Neurosciences* **18**, 54, with permission from Elsevier Trends Journals.)

The importance of raised calcium ion concentrations in the postsynaptic region is shown by a number of experiments. Injection of the chelating agent EGTA into the postsynaptic cell prevents LTP (Lynch *et al.*, 1983). Release of caged calcium inside a CA1 pyramidal cell, by injecting the cell with the photochemically labile chelating agent calcium nitr-5 and then releasing the calcium by an intense flash of light, enhances EPSPs in that cell but not in others (Malenka *et al.*, 1988).

Calcium imaging shows that tetanic stimulation elevates calcium levels in the dendritic spines of the pyramidal cells (Regehr *et al.*, 1989; Denk *et al.*, 1996). In one investigation using the fluorescent indicator dye mag-Fura 5, the calcium ion concentration following tetanic stimulation rose to 20 to 40 μM, easily sufficient to activate a number of different enzymes (Petrozzino *et al.*, 1995). These rises are largely prevented by the NMDA antagonist AP5. They are also reduced in the presence of dantrolene and thapsigargin, substances that reduce the calcium content of the internal calcium stores in the endoplasmic reticulum, so it may be that the calcium signal is boosted by release from these stores. Such calcium release might be triggered via a phosphoinositol cascade initiated by activation of metabotropic glutamate receptors.

Just how the increased postsynaptic calcium levels produce LTP is not entirely clear. One general possibility is that glutamate-gated channels, both AMPA and NMDA, may be made more active in some way by phosphorylation, and that calcium acts as a trigger for the activation of the protein kinases that will do that. NMDA receptor activation increases cyclic AMP levels in area CA1, presumably via activation of protein kinase A (Chetcovich *et al.*, 1991). Inhibitors of protein kinase C and of calcium–calmodulin-dependent protein kinase II (CaMKII) both block LTP when injected into postsynaptic CA1 cells (Malinow *et al.*, 1989). Mice that have been genetically engineered not to express the α-CaMKII gene do not show LTP and have much reduced spatial learning capabilities (Silva *et al.*, 1992*a*,*b*).

A crucial question, still not solved to everyone's satisfaction, is whether LTP at CA1 synapses involves a change in the presynaptic terminals. Attempts to ascertain whether or not the amount of glutamate released per presynaptic impulse increases in LTP have produced conflicting results (e.g. Manabe & Nicoll, 1994; Clark & Collingridge, 1995). One study used minimal stimulation intensities and recorded from single pyramidal cells, so that the number of failures of transmission could be measured; after a tetanic burst the proportion of failures fell (Stevens & Wang, 1994). The most likely explanation for this is that the probability of release of transmitter vesicles rises during LTP, implying that something changes in the presynaptic terminal. Another possibility, however, is that LTP involves the activation of a set of AMPA receptors that were previously 'silent'; if this were so then the number of apparent failures would decrease (Kullmann & Siegelbaum, 1995).

If there is a presynaptic component to LTP at CA1 synapses, how does the increase in calcium ion concentration in the postsynaptic cell influence what goes on in the presynaptic terminal? A possible answer is provided by the idea of a *retrograde messenger*, some substance produced in the postsynaptic terminal which diffuses or is transported to the presynaptic terminal and produces some effect there.

One candidate for the role of the retrograde messenger is arachidonic acid, produced by the action of phospholipase A_2, an enzyme activated by phosphoinositide signalling. It is released from cultured neurons following NMDA receptor activation and LTP, and LTP is blocked by inhibitors of phospholipase A_2. Another possibility is platelet-activating factor, another product of phospholipid hydrolysis.

Perhaps the most exciting candidate for the retrograde messenger is nitric oxide. As mentioned in chapter 9, activation of NMDA receptors in brain cells leads to release of nitric oxide from them. Hence the nitric oxide can diffuse to its target guanylyl cyclase in adjacent cells, acting as a messenger for the production of cyclic GMP there, as well as perhaps having effects within the cell in which it is produced. Experiments to determine the role of nitric oxide in LTP have led to conflicting conclusions (Schuman & Madison, 1994; Hawkins, 1996). Figure 11.12 illustrates one way in which this situation may have arisen; a nitric

oxide synthase inhibitor greatly reduces LTP at low conditioning stimulus intensities, but has relatively little effect at high intensities (O'Dell *et al.*, 1994). This may indicate that nitric oxide is an important retrograde messenger, but not the only one.

Further possible actors in the LTP play are the neurotrophins. These are substances such as nerve growth factor (NGF), brain-derived neurotrophic factor (BDNF), and neurotrophin-3 (NY-3), all proteins that are released from neurons to affect the growth of other neurons. Nerve growth factor was discovered by Levi-Montalcini & Hamburger in 1951, as a substance which promoted the growth of nerve cells in tissue culture, and which has since been shown to have a key role in the survival and development of sympathetic and sensory neurons (Levi-Montalcini, 1987). Other growth factors have been discovered; they are all proteins that have marked effects on cell development, usually acting via tyrosine kinase receptors (Heath, 1993).

Induction of LTP produces changes in neurotrophin mRNA levels in the hippocampus: increases for BDNF and NT-3 in CA1, increases for BDNF and NGF but a decrease for NT-3 in the dentate gyrus (Patterson *et al.*, 1992; Carstén *et al.*, 1993). Application of BDNF or NT-3 (but not NGF) to hippocampal slices produced a rapid and sustained increase in synaptic transmission at CA1 synapses (Kang & Schuman, 1995). LTP at CA1 synapses is significantly reduced in genetically engineered ('knockout') mice lacking the BDNF gene, although hippocampal morphology seems to be normal; LTP can be restored in hippocampal slices from these mice by application of BDNF (Korte *et al.*, 1995; Patterson *et al.*, 1996). All this suggests that

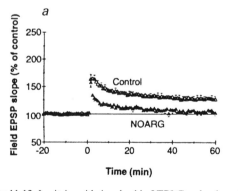
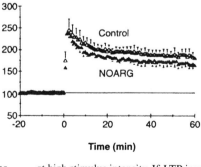

Figure 11.12. Is nitric oxide involved in LTP? Graphs show the effect of the nitric oxide synthase inhibitor *N*-nitro-arginine (NOARG) on field EPSP slope following a high-frequency burst at low (*a*) and high (*b*) stimulus intensities. Notice that there is little or no LTP in the presence of the inhibitor at low stimulus intensities, but that LTP is only slightly reduced by the inhibitor at high stimulus intensity. If LTP involves retrograde messengers, then perhaps nitric oxide is the main one at low stimulus intensities but others become important at higher intensities. (From O'Dell *et al.*, 1994. Reprinted with permission from *Science* **265**, 542–6. Copyright 1994 American Association for the Advancement of Science.)

neurotrophins may have some role to play in LTP, although the details have still to be worked out (Lo, 1995; Berninger & Poo, 1996; Lewin & Barde, 1996). One possibility is that neurotrophins act as retrograde messengers, released by the postsynaptic cells to affect the presynaptic terminals (Thoenen, 1995).

For LTP to be a really long-term affair, lasting several hours or days, it would seem necessary that protein synthesis, and probably morphological change at synapses, should take place. The later phases of LTP, several hours after the conditioning stimulus, are indeed blocked by inhibitors of protein synthesis and RNA transcription if they are present during the conditioning stimuli and for an hour or so afterwards (Frey *et al.*, 1988; Nguyen *et al.*, 1994). Figure 11.13 shows an experiment which illustrates this. It is clear that there is a critical 'window' of time when the establishment of late-phase LTP can be disrupted by the

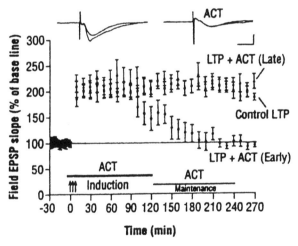

Figure 11.13. Induction of late-phase LTP is blocked by inhibition of mRNA synthesis. Graphs show the slopes of field EPSPs recorded from area CA1 of a rat hippocampus slice. Repeated tetanic stimuli at zero time produced a doubling of the rate of rise of the EPSP produced by single shocks. Three experiments are shown. (1) In the control experiment, LTP is maintained for at least 4.5 h. The sample EPSPs shown at the top left are before the tetanic stimulation and (the larger one) 270 min after. (2) The transcription inhibitor actinomycin D (ACT) is applied during the tetanic stimulation and for 2 h afterwards. LTP is maintained for about 90 min, but then falls steadily, and is no longer evident after about 3 h. The EPSPs at the top right were obtained before and 270 min after the tetanic stimulation. (3) Actinomycin D is applied for the period 2 to 4 h after the tetanic stimulation. The late-phase LTP is unaffected. (From Nguyen *et al.*, 1994. Reprinted with permission from *Science* **265**, 1104–7. Copyright 1994 American Association for the Advancement of Science.)

prevention of transcription, but that after this the LTP is impervious to interference with protein synthesis.

What genes are switched on during this critical period? Just as in *Aplysia*, it seems likely that they are immediate early genes coding for proteins that regulate the expression of other genes. Thus expression of the IEG *zif/268* and the proto-oncogenes *jun* and *fos* is associated with LTP (Cole *et al.*, 1989; Abraham *et al.*, 1991; Richardson *et al.*, 1992). Another IEG, an effector rather than a regulator, is tissue-plasminogen activator. This is an extracellular serine protease, whose release is correlated with morphological differentiation. It seems likely that it is involved in the structural changes that accompany LTP, perhaps by altering the adhesive contacts between neurons (Qian *et al.*, 1993).

We saw that in *Aplysia* some IEGs are activated when a specific protein CREB (CRE-binding protein) binds to the CRE (cAMP-responsive element) enhancer element, a short length of DNA upstream from the gene coding sequence. A similar system occurs in LTP. Thus Bourtchuladze *et al.* (1994) found that mice with targeted mutations of the α and δ isoforms of CREB ('CREB knockout' mice) are deficient in long-term memory, whereas short-term memory (up to an hour) is unaffected. LTP in hippocampal slices from such mice lasted no more than 90 min.

An ingenious method of demonstrating CRE-mediated gene expression in LTP was used by Impey *et al.* (1996). They made transgenic mice containing the reporter gene *CRE-LacZ*, which contained the CRE sequence attached to the *E. coli* gene β-galactosidase. Any binding to CRE will then lead to synthesis of β-galactosidase, and this can be detected by immunocytochemistry. They found that tetanic stimuli that produced LTP in hippocampal slices also resulted in β-galactosidase production in the pyramidal cell layer. Increased β-galactosidase expression was detectable 2 h after the tetanic stimulus, and reached a maximum after 4 to 6 h.

Electron microscopy of the hippocampus has shown that LTP in both the CA1 and dentate gyrus regions is accompanied by an increase in the number of synaptic contacts on the cells involved (Chang & Greenough, 1984; Geinisman *et al.*, 1993). The changes are remarkably rapid, occurring within 15 min to 1 h of the induction of LTP.

Mossy fibre LTP

The synapses between mossy fibres and CA3 pyramidal cells show a form of LTP that is different in a number of ways from that at synapses on CA1 cells (Nicoll & Malenka, 1995). It is not dependent on activation of NMDA receptors and seems to be a purely presynaptic phenomenon. Indeed, mossy fibre LTP can occur even if the synapse is completely blocked by glutamate receptor antagonists during the conditioning tetanus. However, removal of external calcium ions during the tetanus abolishes LTP. It seems likely that the first stage in mossy fibre LTP is the entry of calcium ions into the presynaptic terminal during the conditioning stimulus, probably via voltage-gated calcium channels.

There is good evidence for the involvement of a cyclic AMP signalling system in mossy fibre LTP. The diterpene compound forskolin activates adenylyl cyclase and so increases the amount of cyclic AMP in cells. Application of forskolin to hippocampal slices produced considerable increase in the size of EPSPs produced by mossy fibre action on CA3 cells, and this increase was reduced by antagonists of the cAMP-dependent protein kinase A. It seems likely that LTP is brought about by a cascade beginning with the activation of adenylyl cyclase I by calcium–calmodulin, leading to production of cyclic AMP and hence activation of protein kinase A. The protein kinase A may then phosphorylate some protein or proteins concerned with the vesicular release process (Weisskopf *et al.*, 1994; López-García *et al.*, 1996).

Long-term changes in mossy fibre LTP can be prevented by RNA and protein synthesis inhibitors. This long-term LTP is also blocked by inhibitors of protein kinase A, suggesting that it has features in common with the long-term changes in CA1 area LTP (Huang *et al.*, 1994).

Neurogenetics of fruit fly learning

The fruit fly *Drosophila* has been used for investigating genetic systems ever since T. H. Morgan began his experiments with it in 1908. Consequently we know a great deal about its genes, and there are many mutant strains available. Fruit flies are able to learn, and mutant flies lacking certain genes show deficits in learning and memory (Quinn *et al.*, 1974; Dudai *et al.*, 1976; Dudai, 1988; Greenspan, 1995).

Let us have a look at some elegant experiments by Tully & Quinn (1985), in which they demonstrated strong learning in wild-type *Drosophila* by using classical Pavlovian conditioning. The flies were trained to associate a particular odour with an electrical shock; fig. 11.14 shows how this was done. Typically 95% of wild-type flies could do this when tested immediately afterwards, around 65% still retained the memory 7 h later, and even at 24 h (quite a long time in the life of a fly) there was still some retention. This was not so, however, with the mutants *amnesiac, rutabaga* and *dunce*. These all showed appreciable capability for learning, but their memories decayed rapidly during the first 30 min after training, as is shown in fig. 11.15.

Tully and his colleagues (1994) later showed that fly memories would last much longer with more intensive training, and they used various interventions and mutants to investigate the consolidation processes involved in this. With ten training sessions spaced at 15 min intervals, flies were still

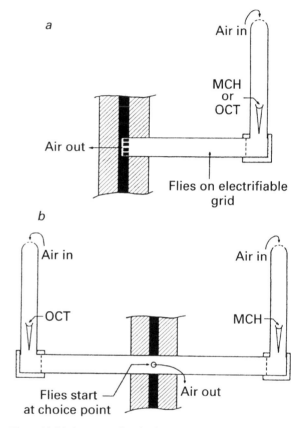

Figure 11.14. Apparatus for classical conditioning in *Drosophila*. Flies are trained to associate a particular odour with electric shocks, so that they avoid it in future. *a* shows the training tube, with its inner surface covered with an electrifiable printed circuit grid. Snapped on to the end of this is an odour tube containing either 3-octanol (OCT) or 4-methylcyclohexanol (MCH), air is sucked in through holes in the top of the odour tube and over the flies. Electric shocks are delivered to a group of about 150 flies while they are exposed to one of the odours but not to the other. The group of flies is later tested for learning in the choice chamber shown in *b*. Here air is drawn through both odour tubes so that flies entering the chamber from a small centre compartment had to move either one way or the other according to their choice of odour. After 2 min a barrier is interposed between the two halves of the choice chamber and the number of flies in each half is counted. (From Tully & Quinn, 1985. Reproduced with permission from Classical conditioning and retention in normal and mutant *Drosophila melanogaster. Journal of Comparative Physiology* **157**, 265, fig. 1, © Springer-Verlag 1985.)

showing the learned response seven days later; they called this long-term memory or LTM. If the ten sessions were held one after the other, however, with no spacing between them ('massed'), retention was longer than with a single session, but fell to zero by four days. Anaesthesia by cooling for 2 min had no effect on this retention if it was done more than an hour after the end of the training sessions, so this was called anaesthesia-resistant memory, ARM. The brief retention shown by the *dunce* and *rutabaga* mutants was described as information acquired during learning (LRN) and the rather longer retention shown by *amnesiac* as short-term memory (STM).

The protein synthesis inhibitor cycloheximide reduced retention at one day after the spaced training, but not after massed training, so LTM is protein-synthesis sensitive whereas ARM is not. However the mutant *radish* eliminates ARM but leaves LTM intact. This suggests that ARM and LTM are reached by independent paths, but since *amnesiac* affects both of them there must be some other common point on the path after STM; it is called medium-term memory or MTM. These ideas are put together in fig. 11.16.

We have some information about what the various genes involved in this consolidation sequence do, but it is far from complete. The *dunce* mutant has much reduced cyclic AMP phosphodiesterase activity, and its cDNA sequence shows a high degree of identity with a mammalian cyclic AMP phosphodiesterase (Byers *et al.*, 1981; Chen *et al.*, 1986). The consequence of this is that *dunce* mutants have cyclic AMP levels much higher than normal. Remarkably, the *dunce* gene has several introns that contain coding sequences for quite separate genes, as well as producing several different phosphodiesterase isoforms by alternative splicing (Davis & Dauwalder, 1991; Qiu *et al.*, 1991). The *rutabaga* gene, on the other hand, codes for a calcium–calmodulin-dependent adenylyl cyclase, again indicating some lack of regulation of cyclic AMP levels (Levin *et al.*, 1992). Larvae of both mutants show impaired facilitation and post-tetanic potentiation at their neuromuscular synapses (Zhong & Wu, 1991).

Sequencing of the *amnesiac* gene indicates that it codes for a neuropeptide showing some homology with the mammalian neuropeptides adenylyl cyclase activating peptide (PACAP) and growth hormone releasing hormone (GHRH). Perhaps this neuropeptide promotes cyclic AMP synthesis (Feany & Quinn, 1995). The learning-deficit gene *linotte* probably encodes a receptor tyrosine kinase (Dura *et al.*, 1995).

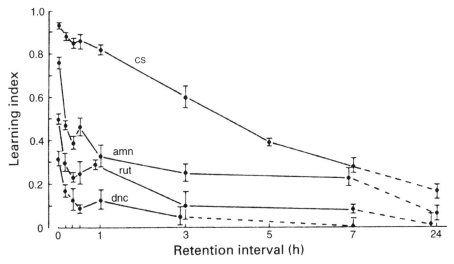

Figure 11.15. Memory retention in normal and mutant *Drosophila*. Flies were trained to associate a particular odour with electric shocks and tested at intervals up to 24 h later, using the method shown in fig. 11.14. The graphs show results for normal flies (cs) and three different mutants, amn (*amnesiac*), rut (*rutabaga*) and dnc (*dunce*). The learning index is calculated as the fraction of flies avoiding the shock-associated odour minus the fraction avoiding the unshocked odour; this is averaged over two groups of flies, one shocked in the presence of OCT and the other shocked in the presence of MCH. If 93% of flies avoided the shocked odour, for example, and 4% avoided the unshocked odour, with 3% left in the central chamber, then the learning index would be 0.89. Each point represents four experiments. (From Tully & Quinn, 1985. Reproduced with permission from Classical conditioning and retention in normal and mutant *Drosophila melanogaster. Journal of Comparative Physiology* **157**, 271, fig. 10, © Springer-Verlag 1985.)

Binding proteins for CRE enhancers are found in *Drosophila*. Just as in *Aplysia*, they seem to be of crucial importance in eliciting expression of immediate-early genes in long-term memory. Expression of a CREB repressor blocks LTM, and induced expression of the activator isoform enhances it (Yin *et al.*, 1994, 1995).

The cellular basis of memory

The three model systems that we have looked at, sensitization in *Aplysia*, LTP in the mammalian hippocampus, and learning genetics in *Drosophila*, are very different in many ways. But it is evident that there are a number of features in common between them. The activation of intracellular signalling systems by neurotransmitters is clearly an essential first step in the establishment of learning and short-term memory. To what extent intercellular retrograde messengers are active is not altogether clear, and it may well be that some systems require them and others do not. Long-term memory seems to require protein synthesis leading to morphological change. Genes have to be switched on to do this. CRE-mediated gene expression following a rise in cytoplasmic cyclic AMP levels seems to be a common feature of our model systems. But clearly there is much yet to learn in this field and many more experiments to be done.

Figure 11.16. Genetic pathway for memory formation after olfactory learning in *Drosophila*. Flies initially learn to associate a specific odour with electric shock at stage LRN. This is then passed into short-term memory (STM) of about an hour and medium-term memory (MTM) of several hours. With multiple training sessions, further memory consolidation takes place, so that memories last for two or three days with massed sessions (ARM) or at least seven days with spaced trials (LTM). ARM stands for anaesthesia-resistant memory, since it is not abolished by a 2 min cold shock at 1 h or later after the end of training. The protein synthesis inhibitor cycloheximide (CXM) interferes with LTM but not with ARM, and the transcription factor dCREB2-b is also involved in LTM. The various mutants act at different points in this pathway, so that *latheo* and *linotte* prevent learning, *dunce* and *rutabaga* show learning but no retention, and *amnesiac* allows retention for the time-span of STM only. *radish* shows no ARM but has no effect on LTM, hence ARM is not a precursor of LTM. (From Tully *et al.*, 1994. Reproduced with permission from *Cell*. Copyright Cell Press.)

12
Electrotonic transmission and coupling

We have seen in chapter 7 that the controversy as to whether synaptic transmission was electrical or chemical in nature seemed to have been resolved by the advent of intracellular recording in the early 1950s. It was a great surprise, therefore, when intracellular recordings at the end of that decade provided clear evidence that some synapses do operate by the direct flow of current from one cell to another (see Bennett, 1985). Such synapses are described as electrotonic or electrically transmitting.

Synapses operating by electrotonic transmission

An excitatory electrotonic synapse is one in which the postsynaptic cell is directly excited by the electrotonic currents accompanying an action potential in the presynaptic axon. Electron microscopy of electrotonic synapses shows regions where the intercellular space between the two cells is much narrower than usual. These regions are known as *gap junctions*. As we shall see later, they contain channels which provide direct connections between the pre- and postsynaptic cells, so that current can flow readily from one cell to the other.

In order for electrotonic transmission to be effective, the electrical and geometrical characteristics of the junction must be arranged in a certain way. Figure 12.1 represents a junction between two cells, each of which is electrically excitable. The local circuit currents produced by inward movement of sodium ions in the presynaptic cell at A will be completed by currents flowing out of the synaptic cleft at B and out of the postsynaptic cell at C. Electrotonic transmission can occur only if the current density at C is large enough to cause a depolarization that is sufficient to initiate an action potential there. In normal chemical transmission there is effectively no current at C since the membrane resistance of the postsynaptic cell is too great.

Gap junctions act as a conduit for electric current to be carried directly from the presynaptic to the postsynaptic cell, so all the current can depolarize the membrane at C, with none leaving via the synaptic cleft at B. The size of the postsynaptic cell must not be much greater than that of the presynaptic cell, otherwise the postsynaptic current density

will be insufficient to cause excitation. Hence if transmission is to occur between a small axon and a large postsynaptic cell (as at the neuromuscular junction, for example), then it must be chemical in nature.

The advantages of electrotonic transmission, it would seem, are that it does not involve the complex apparatus of the chemical transmitter mechanism and that the synaptic delay can be almost negligible. In many cases it also acts as a synchronizing or *coupling* device. Here flow of current between adjacent cells enables them all to fire together at the same time.

Septal synapses in invertebrate giant nerve fibres

Many annelids and crustaceans possess giant fibres in the ventral nerve cord which are multicellular units, each being divided by a transverse septum in each segment. The giant fibres of the earthworm, mentioned in chapter 4, are of this type. The physiology of the transmission process across the septa was investigated by Watanabe & Grundfest (1961) in the lateral giant fibres of the crayfish. They found that currents passed through a microelectrode into one of the component cells of a fibre would readily cross the septum to cause a potential change in its neighbour. Nevertheless, the

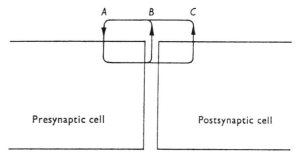

Figure 12.1. Current flow at a synapse produced by a presynaptic action potential. In chemical transmission all the current flowing into the presynaptic axon at A flows out via the synaptic cleft at B. In electrical transmission most of the current flowing in at A must enter the postsynaptic cell and depolarize its membrane by flowing out at C.

resistance of the septum is not negligible, so that its presence must slightly reduce the conduction velocity in the whole fibre.

Electrical interconnections between neurons

A number of cases are known in which there is some electrical interaction between neurons that frequently fire in synchrony. One example of this is in the two giant cells which occur in the segmental ganglia of leeches (Hagiwara & Morita, 1962). In fig. 12.2*a*, the upper trace shows the response of one of the cells to depolarizing current applied via an intracellular microelectrode, and the lower trace shows the electrical changes which occur simultaneously in the other cell. Figure 12.2*b* shows the very similar results obtained from a similar experiment on two adjacent pyramidal cells in the mammalian brain.

Similar electrotonic interconnections occur between pacemaker cells in the lobster heart ganglion (Hagiwara *et*

al., 1959), between spinal electromotoneurons in mormyrid electric fish (Bennett *et al.*, 1963), the motoneurons of toadfish sound-producing muscles (Bennett, 1966) and elsewhere (Bennett, 1977). Bennett points out that such electrotonic coupling is much used in synchronizing the activity of neurons innervating effector organs which show marked pulsatile activity. It is essential, for example, that all the cells of the lobster heart, the mormyrid electric organ or the toadfish sound-producing muscle should fire at the same time. In many cases, the impedance of the junction increases at higher frequencies, so that brief action potentials have less effect on an adjacent cell than do slow subthreshold potentials.

Structural studies may suggest the existence of electrotonic transmission between neurons by demonstrating the presence of gap junctions. In the mammalian spinal cord, small areas of gap junction occur in a proportion of excitatory synaptic boutons (Rash *et al.*, 1996). They are located near to the active zones of neurotransmitter release, and the synapses are consequently known as 'mixed synapses' (fig. 12.3). Their precise mode of functioning is not too clear as yet. It may be that they are concerned with enhancing the depolarization of the postsynaptic cell, or perhaps they serve to allow the passage of second messengers or other substances between the two cells.

The giant motor synapses of the crayfish

The flexor muscles in the abdomen of the crayfish are innervated by motor fibres of large diameter which can be excited by impulses in the lateral giant fibres of the nerve cord. This system provides the physiological basis of the 'tail-flick' escape response of the animal. Furshpan & Potter (1959) investigated the transmission process at this synapse using microelectrodes inserted into the pre- and postsynaptic axons, as is shown in fig. 12.4. Figure 12.5 shows the presynaptic and postsynaptic responses to ortho-

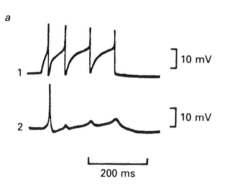

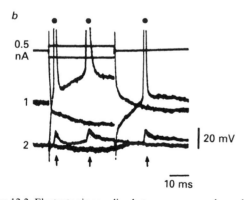

Figure 12.2. Electrotonic coupling between neurons, shown by passing current through one of a pair (trace 1 in each case) and recording responses from the other (trace 2). *a* shows responses in two leech neurons to depolarization of one of them. *b* is from two pyramidal cells in a rat brain slice and shows responses to depolarization (upper traces) and hyperpolarization (lower traces) of cell 1. (*a* from Hagiwara & Morita, 1962; *b* from MacVicar & Dudek, 1981.)

Figure 12.3. Mixed synapses in the rat spinal cord, with gap junctions located next to the active zone of neurotransmitter release. Presynaptic active zones here are either invaginated (left) or flat (right). (From Rash *et al.*, 1996. Reproduced from *Proceedings of the National Academy of Sciences USA* **93**, 4238. Copyright 1996 National Academy of Sciences, USA.)

dromic stimulation. Notice particularly that the postsynaptic response begins almost simultaneously with the presynaptic action potential; this suggests that the transmission process is electrical in nature. However, antidromic stimulation of the motor fibre produces only a very

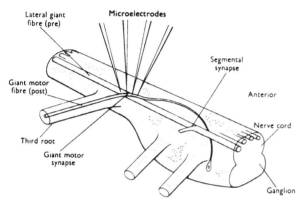

Figure 12.4. Part of the abdominal nerve cord of a crayfish, to show the position of the electrically transmitting synapse between the lateral giant fibre and the large motor axon in the third root. (From Furshpan & Potter, 1959.)

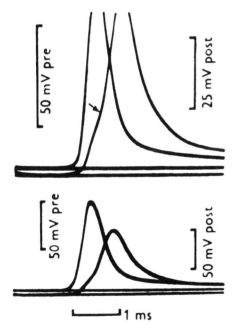

Figure 12.5. Pre- and postsynaptic action potentials at the crayfish giant motor synapse. Two pairs of records are shown, at different amplifications. The upper trace of each pair is the presynaptic response. The kink in the postsynaptic response (at the arrow) shows the level at which it crosses the threshold. Notice the negligible latency between the times of onset of the two responses. (From Furshpan & Potter, 1959.)

small response (less than 1 mV) in the lateral giant fibre, and so there can be no transmission of an impulse across the synapse in the reverse direction.

In further experiments, depolarizing and hyperpolarizing currents were applied through microelectrodes inserted into the axons on either side of the synapse. The results were very interesting. Depolarization of the presynaptic axon produced depolarization in the postsynaptic axon, whereas hyperpolarization had no effect; but depolarization of the postsynaptic axon did not produce changes in the presynaptic axon, whereas hyperpolarization did (fig. 12.6). This indicates that the synaptic membrane is a rectifier: current can flow only from the pre-fibre to the post-fibre, and not in the opposite direction.

The rectifying properties of the giant motor synapse are essential to the animal, since the motor fibre can also be excited (by chemical transmission) by other neurons. If there were no rectification at the junction, such excitation would always produce excitation of the lateral giant fibres and the crayfish would be continually subject to irrelevant escape responses.

Gap junction channels

Gap junctions are regions of fairly close apposition between the cell membranes of adjacent cells; the intercellular space is narrowed to about 20 to 30 Å instead of being in the region of 200 Å. They are called 'gap junctions' to distinguish them from 'tight junctions', where the cell membranes make contact so that there is no extracellular space between the cells. The distinction was made by Revel & Karnovsky (1967), who used electron microscopy to show that lanthanum salts could permeate between the two membranes at gap junctions but not at tight junctions. They also found that there was some material crossing the gap and that sections in the plane of the membranes showed particles with subunits in hexagonal arrays. Gap junctions occur at regions where cells are known to be in electrical connection with each other, such as in liver cells and heart muscle and at electrotonic synapses (see Bennett & Spray, 1985). Tight junctions are found where sheets of cells separate distinct fluid compartments, as in kidney tubules and intestinal epithelium.

Gap junction channels are often found in close arrays, as for example in liver cells and in the intercalated discs of heart muscle. This makes structural studies possible, using X-ray diffraction or quantitative electron microscopy. Such studies show that each gap junction channel is composed of a pair of hexamers, one in each of the two apposed membranes, as is shown in fig. 12.7 (see Bennett *et al.*, 1991; Yeager & Gilula, 1992). The individual subunits are called

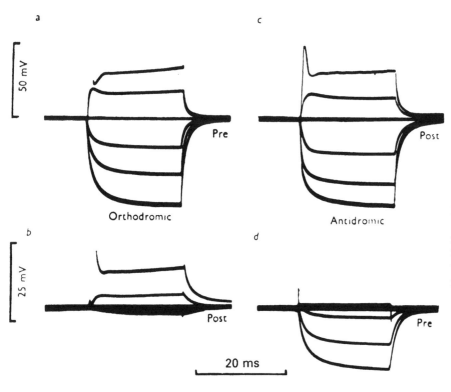

Figure 12.6. The rectifying properties of the crayfish giant motor synapse. Current is passed through the membrane of one of the fibres, and the membrane potential in both fibres is recorded. *a* and *b* show responses in the pre- and post-fibres to current through the pre-fibre membrane; notice that only depolarizing currents cross the synapse from pre- to post-fibre. *c* and *d* show responses in the pre- and post-fibres to current through the post-fibre membrane; notice that only hyperpolarizing currents cross the synapse from post- to pre-fibre. (From Furshpan & Potter, 1959.)

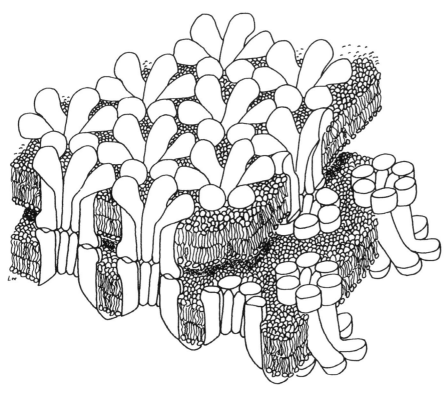

Figure 12.7. The structure of gap junctions isolated from mouse liver cells, as deduced from X-ray diffraction studies. (From Makowski *et al.*, 1984.)

connexins and the groups of six forming half a channel are called connexons. A complete gap junction channel is formed of two connexons from adjacent membranes, so that the cytoplasmic compartments of the two cells are connected by the channel pore passing through the middle.

At least sixteen different connexins have been cloned from vertebrate tissues. They range in size from 225 to 510 amino acid residues long and show considerable sequence identity. They are commonly named by reference to their relative molecular mass (M_r) in thousands as predicted from their amino acid sequence. For example, rat connexin 32 (Cx32) has an M_r of 32 007 (Paul, 1986) and rat connexin 43 (Cx43) has one of 43 036 (Beyer *et al.*, 1987). Alternatively, they may be named in accordance with their sequence similarities and presumed evolutionary relationships, which give two groups, α1 to α8 and β1 to β5 (Kumar & Gilula, 1996).

The amino acid sequences of connexins suggest that there are always four transmembrane segments, and that these are probably α-helical in structure. The N- and C-termini are cytoplasmic, and so also is the inner loop between the second and third transmembrane segments, and there are two outer loops on the extracellular side of the membrane (fig. 12.8). The connexin outer loops in one

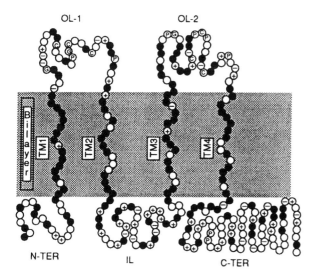

Figure 12.8. The membrane topology of connexins, based on rat liver connexin 32. There are four transmembrane segments, TM1 to TM4, two outer loops OL-1 and OL-2 on the extracellular side of the membrane, and the N- and C-termini and the inner loop IL on the cytoplasmic side. Black circles show hydrophobic residues. For the hydrophilic residues, + and − show charged residues, P shows proline, and in the outer loops the conserved cysteine (C) and glycine (G) residues are shown. (From Peracchia *et al.*, 1994.)

connexon must engage with those in the connexon forming the other half of the channel. Peracchia *et al.* (1994) suggested that one of the two connexons in a channel is rotated by 30° (half a connexin) with respect to the other so that each connexin in one half of the channel is connected via its outer loops to two in the other half. This arrangement would produce a relatively rigid structure.

There is no evidence that different connexins can be combined in the same connexon, but functional channels can sometimes be formed from two different types of connexon. Connexins can be expressed in *Xenopus* oocytes, so that two oocytes can be put side by side in contact and gap junction channels then form between them. White *et al.* (1994) found that heterotypic channels would form between Cx46 and either Cx43 or Cx50 from rat lens, but not between Cx43 and Cx50.

Some gap junction channels are closed by alterations in transmembrane voltage, some by pH changes and some by increases in cytoplasmic calcium ion concentration. The sensitivity of channel gating to these factors differs considerably in different tissues and in different animals. One model of gating suggests that it is produced by an alteration in the inclination of the six subunits in the membrane (Unwin & Zampighi, 1980; Bennett *et al.*, 1991).

In electrotonic synapses, the gap junctions clearly serve as routes for current flow between one cell and another. Undue voltage sensitivity would prevent such current flow, hence it is not surprising to find that it is absent in symmetrical junctions such as the septal synapses of crayfish abdominal lateral giant fibres (Johnston & Ramon, 1982). The rectifying synapses, however, may well have the connexons on one side of the junction which are voltage-sensitive while those on the other are not, as seems to occur in the giant motor axon synapse of the crayfish (Jaslove & Brink, 1986).

Gap junctions allow the passage of much larger molecules than will pass through most ion channels. Thus Kanno & Loewenstein (1964) found that fluorescein (M_r 376) diffused fairly freely from one cell to another in *Drosophila* salivary gland, but would not diffuse out of the cell into the external medium. Spray *et al.* (1979) investigated the electrotonic coupling between pairs of blastomeres from early amphibian embryos. They found that the junctional conductance was much reduced if the voltage across the junction was more than a few millivolts in either direction. Permeability to the dye lucifer yellow was altered in the same way by such voltage changes, implying that current flow and dye movement take place via the same channels.

By injecting fluorescent compounds of various sizes into

cells and looking for their appearance in neighbouring cells, it has been found in most cases that molecules up to about M_r 1000 in size can pass through them. This implies that the gap junction pores are about 16 Å in diameter (Schwarzmann *et al.*, 1981). Structural studies suggest that the pore is cylindrical in form and about 15 Å in diameter (Unwin & Ennis, 1984). Substances with M_r values up to about 1000 should be able to pass through a pore of this size. Gap junction channels will thus be permeable to most ions of physiological importance, and also to many intracellular messengers and metabolites. It seems likely that they form channels of communication by exchange of cytoplasmic small molecules in many cells, including glial cells and neurons (Dermietzel & Spray, 1993).

An inhibitory synapse operating by electrical transmission

Many fishes possess a large pair of interneurons, the Mauthner fibres, whose cell bodies lie in the brain and whose axons pass down the spinal cord. The space around the axon hillock is bounded by a dense layer of glial cells and extracellular material called the axon cap; this contains the terminals of a number of interneurons (fig. 12.9). Furukawa & Furshpan (1963) recorded an *external hyperpolarizing potential* (EHP) with a microelectrode placed inside the axon cap but outside the Mauthner cell, especially in response to stimulation of the contralateral Mauthner axon (fig. 12.10*a*). Firing of the Mauthner cell is inhibited during the course of the EHP, and it is therefore inhibitory in function. Similar hyperpolarizations can be detected in adjacent neurons when the Mauthner axon fires (Korn & Faber, 1975).

The EHP is unaffected by injection of tetrodotoxin into

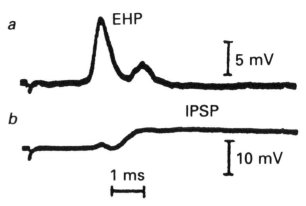

Figure 12.10. The external hyperpolarizing potential (EHP) produced in a Mauthner cell in response to stimulation of the axon of the other cell. *a* is an extracellular record with the tip of the microelectrode within the axon cap, *b* is an intracellular record from near the axon hillock. Since the reference electrode was distant in each case, the potential across the membrane is the difference between traces *a* and *b*. The EHP is followed by an IPSP, which is depolarizing in this case, probably as a result of chloride leakage from the microelectrode. (From Furukawa & Furshpan, 1963.)

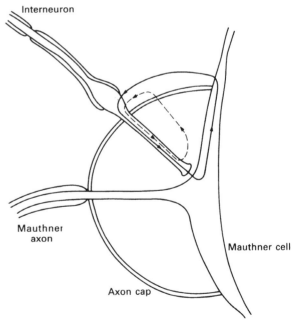

Figure 12.11. The origin of the EHP in Mauthner cells. The action potential in the interneuron travels only as far as the outside of the axon cap; inward flow of current at this point produces the local circuit currents shown in the diagram.

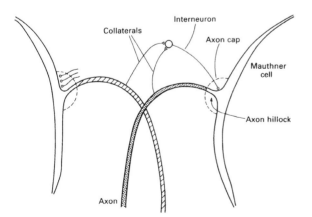

Figure 12.9. Mauthner cells in the brain of fishes, showing how the interneurons causing electrical inhibition are innervated by collateral branches from their axons.

the axon cap (Faber & Korn, 1978). The terminals of the interneurons ending within the axon cap are unmyelinated and so it seems likely that the action potential does not propagate into them. Local circuit currents in the terminals within the cap will therefore produce a brief positive-going potential change, seen as the EHP. This will increase the potential across the Mauthner cell membrane, given by the difference between the traces *a* and *b* in fig. 12.10, and so result in inhibition. Figure 12.11 shows the probable currents involved.

The function of Mauthner fibres is to elicit the startle response of the fish. This consists of a very rapid flexion of the body, brought about by a massive synchronous contraction of the muscles on that side, producing a tail-flick which drives the fish to one side or the other (Eaton *et al.*, 1977). Clearly it is most important that only one of the two Mauthner fibres should fire at one time, otherwise the fish's muscles would contract on both sides at once. Hence the need for effective and immediate inhibition of one cell by the other when it fires. The EHP is followed by a conventional chemically transmitted IPSP (fig. 12.10*b*), hence the functional value of the electrical inhibition may well lie in the rapidity with which inhibition becomes effective.

Part D
Sensory cells

13
The organization of sensory receptors

All animals are sensitive to some extent to changes in their environment. Special parts of the body are responsive to some of these changes and feed information concerning them into the central nervous system. Information about the workings of the animal's own body, or about communication signals from other members of its species, may be acquired in a similar fashion. These specially sensitive structures are known as sense organs or sensory receptors. They are crucially important in the lives of animals.

Since there is a very great variety of different types of sense organ, the following pages must necessarily be no more than a selective introduction to their physiology. This chapter attempts to give a general outline of the properties of sensory receptors, the next four chapters survey some selected different types, with particular emphasis on their sensory cells and transduction processes. The books by Barlow & Mollon (1982), Dawson & Enoch (1984), Darian-Smith (1984*b*), Corey & Roper (1992) and Dusenbery (1992) are some of the many useful sources of further information.

The methods used for investigating receptors fall into two main categories. Firstly, there is the behavioural, or psychophysical, approach, where the receptor is investigated indirectly by observation of the response of the animal to a sensory stimulus. For example, suppose we are interested in the ability of an animal to discriminate colours. In the case of humans, the experiments are not too difficult, since we can ask the subject if two colours look different. With other animals, however, we have to devise experiments in which the behaviour of the animal depends upon the ability to discriminate colour; for example, we can train a fish to feed from a red container and then see if it always chooses the red container from among a series of greys.

The second approach uses electrophysiological or similar methods to investigate the properties of the cells of the sensory receptor, such as by means of electrodes placed in its vicinity or on the sensory nerves leading from it. This type of study gives precise, quantitative information about the properties of the receptor, but it does not always enable us to decide which of these properties are relevant to its normal function. For example, we may be able to show by electrophysiological means that a receptor can be excited both by mechanical stimuli and by changes in temperature, but we cannot from these experiments alone determine how the animal interprets this information. Hence a thorough investigation of a sensory system must involve both behavioural and electrophysiological studies.

The variety of sense organs

Now let us consider how sensory systems can be classified. A moment's thought will show that Aristotle's list of 'the five senses' – sight, hearing, touch, taste and smell – is incomplete. We are also sensitive to changes in temperature, to our position with respect to gravity, to the movements of our bodies, and so on. These various categories of sensation are called sensory *modalities*. Some of these modalities can further be divided into *qualities*: for example, we can determine the pitch as well as the loudness of a sound, the colour as well as the brightness of a light source.

The sense organs themselves can be classified according to the type of stimulus which normally excites them. Thus *mechanoreceptors* are excited by mechanical stimuli, *photoreceptors* are sensitive to light, *thermoreceptors* are temperature sensitive, *chemoreceptors* are sensitive to the chemical composition of the surrounding medium, and *electroreceptors* are sensitive to very weak electric currents.

It is important to realize that the specificity of receptors is not absolute. Most receptors, for instance, can be excited by strong electric currents as well as by the stimulus to which they normally respond. However, as was first pointed out in 1826 by Müller in his 'law of specific nerve energies', the *sensation* produced by stimulation of sensory receptors for a particular modality is the same whatever the nature of the stimulus. One can easily demonstrate this by closing an eye and pressing it gently: the mechanical stimulation of the retina produces a visual sensation.

An alternative classification is based on the position of the receptors and the stimulus. Thus *exteroceptors* are sensitive to stimuli originating outside the body, and *interoceptors* are

Table 13.1 *Types of sensory receptor, with some examples*

Type of receptor	Exteroceptors		Interoceptors
	Contact	Distance	
Mechanoreceptors	Touch	Hearing Lateral line organs of fishes	Proprioceptors Equilibrium receptors Baroreceptors of carotid body
Photoreceptors	—	Sight. May or may not be image-forming and/or colour sensitive	—
Chemoreceptors	Gustatory (taste)	Olfactory (smell)	e.g. chemoreceptors of carotid body
Thermoreceptors	Most	Sensitivity to radiant heat, e.g. facial pit of crotaline snakes	e.g. hypothalamic thermoreceptors
Electroreceptors	—	Found in electric and some other fishes	—

excited by stimuli inside the body. Exteroceptors can be divided into *distance receptors* (such as those involved in sight and hearing), which are able to detect phenomena at a distance, and *contact receptors* (such as those involved in touch and taste), which can be stimulated only by contact with the stimulus. Interoceptors are divided into *equilibrium receptors*, which give information about the movement and position of the whole body, *proprioceptors*, which give information about the relative positions and movements of the muscles and skeleton, and *visceroceptors*, which monitor the conditions in the rest of the body.

Examples of these various categories are shown in table 13.1. In addition to these, there is a rather ill-defined sensation known as *pain*. Stimuli that produce reactions which appear to involve the sensation of pain in animals are sometimes called *nociceptive*, and the sensory endings that mediate these reactions (usually defined as those having a high threshold for mechanical or thermal stimuli) are known as *nociceptors*.

We should also mention here the question of sensitivity to

magnetism. There is excellent evidence that the dances of bees are affected by magnetic fields (Lindauer & Martin, 1972) and good evidence that migratory birds and homing pigeons can use the Earth's magnetic field for orientation (Keeton, 1981; Wiltschko & Wiltschko, 1996). But how this magnetic sense works is still a mystery; it is too soon for us to add a further category, of 'magnetoreceptors', to table 13.1.

The anatomy of sense organs shows a great deal of variety. All sense organs are innervated by sensory neurons. In the simplest type, there is but a single sensory neuron whose peripheral end is excited by the sensory stimulus. In the Pacinian corpuscle (fig. 13.1) for example, there is a single axon whose terminal is unmyelinated and surrounded by a series of lamellae. The mechanoreceptive hairs of arthropods (fig. 13.2) are also innervated by a single sensory hair, although in this case the neuron soma is placed distally instead of in the central nervous system. Notice that, in each case, the neuron ending is associated with *accessory structures* (the lamellae in the Pacinian corpuscle and the cuticular hair in the arthropod touch recep-

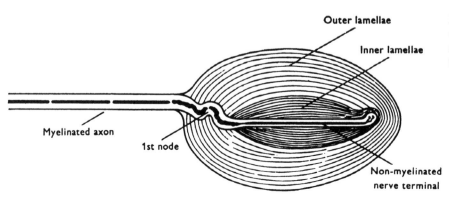

Figure 13.1. Diagrammatic section through a Pacinian corpuscle. (From Quilliam & Sato, 1955.)

Outer lamellae

Inner lamellae

Myelinated axon

1st node

Non-myelinated nerve terminal

tor). In some cases, no specialized accessory structures can be seen, and the nerve endings are 'free' in the tissues.

A more complex type of organization occurs when the sensory neuron is connected to a specialized *receptor cell* (or 'secondary sense cell'), so that excitation of the sensory neuron is preceded by excitation of the receptor cell. In the taste buds of the mammalian tongue, for example, there are a number of receptor cells, each connected, either alone or with others, to a sensory neuron. In very complex sense organs, such as the vertebrate eye and ear, there are large numbers of receptor cells and sensory neurons, and the accessory structures (such as the lens and the iris in the eye) may reach a high degree of sophistication.

Coding of sensory information

The first records of electrical activity in sensory nerves in response to sensory stimulation were made by Adrian in 1926. Using frog skin and muscle nerves, he showed that sensory stimulation was followed by action potentials in the sensory nerves. A more delicate analysis followed the recording of the electrical activity in a single sensory nerve fibre by Adrian & Zotterman (1926); let us have a look at their experiments.

The skeletal muscles of tetrapod vertebrates contain sense organs known as muscle spindles, which are responsive to stretch. Each spindle (fig. 13.3) consists of a bundle of modified muscle fibres whose central region is innervated by a sensory nerve fibre. The sternocutaneous muscle of the frog, a small muscle which is attached at one end to the sternum and at the other to the skin, contains three or four of these muscle spindles. Adrian & Zotterman cut the nerve supplying the sternocutaneous muscle so that it was connected to the muscle only, and placed it across their recording electrodes. The skin was cut so as to leave a small piece attached to the muscle, which could then be stretched by means of small weights hung from a thread tied to this piece of skin. They found that stretching the muscle produced a series of action potentials in the nerve. By progressively removing strips of muscle, they were able to reach a situation in which only one muscle spindle remained connected to the sensory nerve. Under these conditions the action potentials produced by stretch appeared at fairly regular intervals and were always the same size; this showed that the all-or-nothing law applied to sensory nerve impulses.

The next step was to see how this rhythmical discharge was affected by the intensity of the stimulus. Adrian &

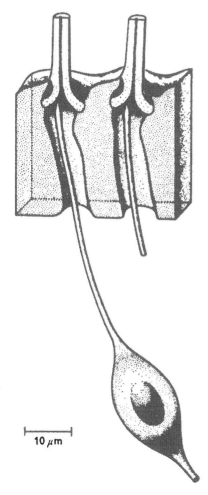

Figure 13.2. Diagram showing how the sense cell is attached to the base of a mechanoreceptive hair in the hair-plate organ in a honey-bee's neck. (From Thurm, 1965.)

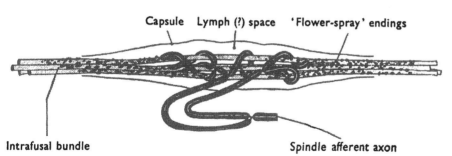

Capsule   Lymph (?) space   'Flower-spray' endings

Intrafusal bundle   Spindle afferent axon

Figure 13.3. The central capsular portion of a frog muscle spindle and its sensory nerve ending. (From Gray, 1957, by permission of the Royal Society.)

Zotterman discovered that the frequency of impulse discharge increased with increasing stretch of the muscle (fig. 13.4). This was a most important result, which has since been verified for a wide variety of different receptors. We can therefore say that, in a single sensory nerve fibre, information as to the intensity of the sensory stimulus is carried as a *frequency code*: the greater the intensity of the stimulus, the higher the frequency of the all-or-nothing action potentials in the fibre.

With many sensory stimuli, a stimulus of constant intensity produces a sensation which declines with time. This phenomenon, which is known as *sensory adaptation*, can readily be experienced by looking at a bright light or by resting light weights on the back of the hand. It is obviously of some interest to see whether adaptation is accompanied by any change in the nervous output of the sensory organ. Using their frog muscle spindle preparation, Adrian & Zotterman found that the impulse frequency in the sensory nerve fibre declined during a constant stretch, as is shown in fig. 13.5. This observation provides a partial explanation for the existence of sensory adaptation, and is further evidence for the frequency code hypothesis.

Adrian & Zotterman summarized their observations on the frog muscle spindle in the following words: 'The impulses set up by a single end-organ occur with a regular rhythm at a frequency which increases with the load on the muscle and decreases with the length of time for which the load has been applied.' It is a remarkable tribute to the quality of their investigation that much of the work done in

sensory physiology since that time has been concerned with expanding this statement for a wide variety of different sense organs.

Initiation of sensory nerve impulses

Adrian & Zotterman's experiments demonstrated that sensory information is conveyed to the central nervous system by means of impulses in the sensory nerve fibres. The next question was: how do these impulses arise? Katz (1950) placed an external electrode as near the sensory nerve ending as possible in the frog muscle spindle. He found that a local depolarization of the terminals appeared when the muscle was stretched, as well as the impulses in the sensory nerve axon (fig. 13.6). Treatment of the preparation with procaine prevented the production of impulses but did not affect the maintained depolarization. Finally, the degree of depolarization increased with increasing stretch, and the impulse frequency in the sensory nerve was linearly proportional to the extent of the local depolarization. Katz concluded from these results that the local depolarization is an intermediate link between the stimulus and the action potentials in the sensory nerve fibre. This local depolarization is called the *receptor potential* (Davis, 1961).

In the receptor neurons of crustacean stretch receptors (fig. 13.7), Eyzaguirre & Kuffler (1955a) were able to record receptor potentials with intracellular electrodes. Their results, shown in fig. 13.8, were in full agreement with those of Katz. Receptor potentials, defined as a potential change induced by the action of the sensory stimulus, have since been recorded from a wide variety of receptor cells or sensory nerve endings. The size of the receptor potential is usually roughly proportional to the intensity of the stimu-

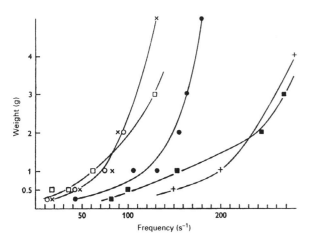

Figure 13.4. The relation between the load applied to a frog muscle and the impulse frequency in the sensory nerve fibre of one of its muscle spindles. Results from six different experiments, denoted by different symbols; in each case the weight had been applied to the muscle for 10 s before the measurement of nerve impulse frequency was made. (From Adrian & Zotterman, 1926.)

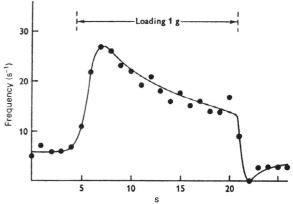

Figure 13.5. Adaptation in a frog muscle spindle: the action potential frequency declines during the application of a constant weight. (From Adrian & Zotterman, 1926.)

lus. It is thus a graded potential change, not an all-or-nothing one.

Receptor potentials usually arise when the sensory stimulus induces ion channels in the membrane of the receptor cell or sensory nerve terminal to open. In the crayfish stretch receptors, for example, stretch opens mechanosensitive channels in the fine terminals of the sensory nerves, and ion flow through these channels produces the receptor potential (Erxleben, 1989). The channels are generally cation selective, with sodium as the main but not the only ion involved (Diamond *et al.*, 1958; Edwards *et al.*, 1981). The greater the stretch, the more channels open (fig. 13.9), and so the larger the receptor potential.

The mechanism whereby the energy of the stimulus produces excitation of the receptor cell or sensory neuron terminal is known as *transduction*. Transducer mechanisms are necessarily different for different types of sensory stimulus, although as we shall see there are some features in common between them (Block, 1992; Torre *et al.*, 1995). In photoreceptors, for example, the primary stage in transduction is the capture of light quanta by the visual pigment molecules and the consequent photochemical changes in them. These changes then trigger a cascade of second-messenger signalling in the photoreceptor cell, leading ultimately to opening or closing of ion channels in the photoreceptor cell membrane. In olfactory receptors, transduction begins with the binding of odorant molecules to receptors in the plasma cell membrane and proceeds again via a second-messenger cascade. In mechanoreceptors, the mechanical stimulus probably opens ion channels more or less directly. In many cases, the transduction process acts as an amplification stage, in that the energy change comprising the receptor potential is much greater than the incident energy which produces it. A system whereby the incident energy alters the ionic permeability of the receptor cell membrane is, of course, ideally suited to providing such amplification.

Action potentials in the sensory nerve fibres are produced by a *generator potential* in the nerve terminal. For the crayfish stretch receptor or the Pacinian corpuscle, the receptor potential is also the generator potential. Here we describe the potential as a receptor potential if we are thinking mainly about the immediate response to the stimulus, and as a generator potential if we are thinking mainly about the initiation of action potentials. In other sense organs, however, where there is a separate receptor cell (as in taste buds, hair cells of the ear and retinal photoreceptors, for example), the receptor potential is quite separate from the generator potential: the receptor potential occurs in the receptor cell while the generator potential occurs in the sensory nerve terminal.

Where there are separate receptor cells, there must be some mechanism whereby the receptor potential in the receptor cell can produce the generator potential in the sensory neuron terminal. In most cases, there is a synapse between the receptor cell and the neuron terminal and a chemical transmission process is involved. Synaptic vesicles can be seen in the presynaptic regions of the rods and cones

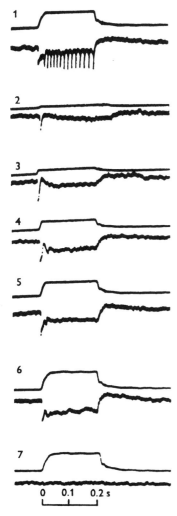

Figure 13.6. The receptor/generator potential recorded extracellularly from a frog muscle spindle in response to various degrees of stretch. The upper trace in each record indicates the change in length of the muscle and the lower trace records the activity of the sensory axon. Record 1 is from an intact preparation, records 2 to 6 from the same preparation after treatment with procaine, and record 7, obtained after crushing the sensory axon, shows that the potential changes in 1 to 6 are not artefacts. (From Katz, 1950.)

of the vertebrate eye, in the mechanoreceptive hair cells of the vertebrate labyrinth, and in the receptor cells of mammalian taste buds. The depolarization constituting the receptor potential causes release of neurotransmitter from the receptor cell, and the consequent postsynaptic depolarization of the sensory nerve terminal acts as the generator potential.

The general theme of this section – that there is a definite control sequence of events in the excitation of receptors – is illustrated diagrammatically in fig. 13.10. We must now take a closer look at the final stages in this sequence, the generator potential and the initiation of action potentials.

From generator potential to frequency code
We have seen that the generator potential carries sensory information in the form of an 'amplitude code' in which the degree of depolarization is related to the intensity of the stimulus. Let us see how this amplitude code is converted into the frequency code of action potentials in the sensory nerve fibres.

Consider a sensory nerve terminal which is functionally divided into two regions (fig. 13.11). There is a receptor region, which is particularly sensitive to the stimulus and responds to it by means of a graded receptor/generator potential, and there is a conductile region whose activity consists of all-or-nothing action potentials. The part of the conductile region next to the receptor region is the impulse initiation site. The membrane in the receptor region contains ion channels that are opened by the sensory stimulus so as to allow a non-selective cation current to flow inwards

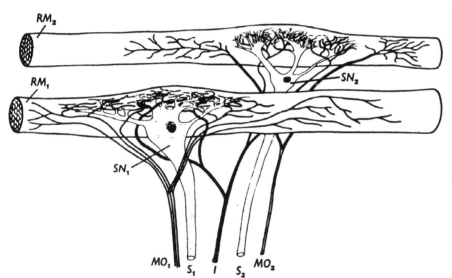

Figure 13.7. Diagram to show the structure of the abdominal muscle receptor organs in the crayfish. RM_1 is the muscle bundle of the tonic receptor neuron SN_1. RM_2 is the muscle bundle of the phasic receptor neuron SN_2. MO_1 and MO_2 are motor fibres of the tonic and phasic muscle bundles, respectively. I is an inhibitory axon. (From Burkhardt, 1958.)

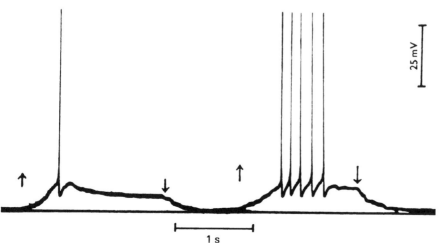

Figure 13.8. Intracellular records from a crayfish stretch receptor cell. The receptor muscle was stretched between the times marked by the arrows. The ensuing depolarization is called the receptor potential (since it arises in the sensory receptor) or the generator potential (since it generates action potentials in the sensory axon). (From Eyzaguirre & Kuffler, 1955*a*.)

25 mV

1 s

through them. The conductile region is well supplied with voltage-gated sodium and potassium channels so that it can support the propagated action potentials.

The stimulus in our model opens the ion channels in the receptor region so that there is an inward flow of current across the cell membrane; this will produce an outward flow

mmHg

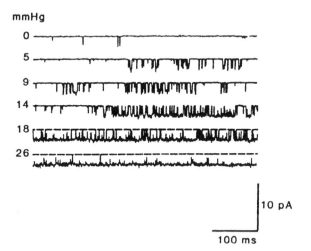

Figure 13.9. Unitary currents through stretch-activated channels in the crayfish stretch receptor. A patch clamp pipette was applied to the sensory neuron terminal membrane in the cell-attached mode, and the patch of membrane was stretched by applying suction (measured in milimetres of mercury, mmHg). Channel openings are seen as downward deflections of the trace; their probability increases as the suction increases. (From Erxleben, 1989. Reproduced from the *Journal of General Physiology*, by copyright permission of The Rockefeller University Press.)

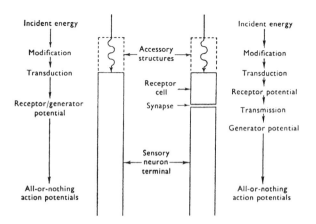

Figure 13.10. Sequences of events in the excitation of sensory nerve endings. The sequence on the left applies where there is no specialized receptor cell; that on the right applies where there is such a cell.

of current at the impulse initiation site. We assume that the stimulus is maintained and that the current it produces is constant. The electrical response at the impulse initiation site to this constant generator current is shown in fig. 13.12. The initial depolarization, from a to b, is a purely electrotonic potential. Then, from b to c, the depolarization is sufficient to cause a local response of the membrane, i.e. a further depolarization caused by the opening of some voltage-gated sodium channels. At c, the membrane potential crosses the threshold, the sodium conductance increases regeneratively as more and more sodium channels open, and an action potential results. After the action potential, there is a positive phase (d) as potassium ions flow out through the open voltage-gated potassium channels. From d to e, the combined effects of the potassium channel closure and the generator current produce depolarization again; at e the sodium channels begin to open again, and the process becomes regenerative, producing another action potential, at f. After this action potential, the sequence repeats itself for as long as the generator current persists.

If the intensity of the generator current is increased, the depolarization between d and f will occur more rapidly. This means that the time interval between successive impulses will be less, so that the impulse frequency is higher. At relatively high impulse frequencies the threshold for production of an action potential, which is crossed at f, will be raised, since the second action potential will fall within the refractory period of the first. This effect sets an upper limit to the discharge frequency.

Further aspects of sensory coding
Adaptation
Different sensory receptors show different adaptation rates. The rate and extent of adaptation is related to the function of the receptor, depending upon whether the animal needs information about the steady level of the stimulus or about changes that occur in the stimulus intensity. We find that the response of many receptors consists of two parts, a tonic response, which is proportional to the intensity of the stimulus, and a phasic response, which is produced by changes in stimulus intensity. An example of a case in which the response is predominantly tonic is in receptors indicating the position of the animal with respect to gravity. In touch receptors, on the other hand, the tonic response is very small and the phasic response is rapidly adapting; a circumstance which probably accounts in part for our not feeling the presence of the clothes we are wearing. Muscle spindles provide an example of a receptor which has both tonic and phasic responses.

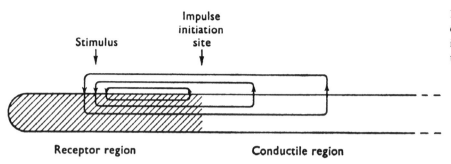

Figure 13.11. Local circuit currents producing impulse initiation in a sensory nerve terminal.

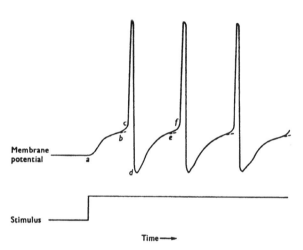

Figure 13.12. Generation of the frequency code in a sensory nerve terminal. For explanation, see the text.

The presence of adaptation probably has a number of different advantages. In many cases, it is much more important for an animal to be able to detect changes in its environment than to measure accurately the static properties of the environment. For example, objects moving in the visual field are likely to be associated with danger or (when the animal is a carnivore) food. Adaptation thus serves to reduce the amount of unimportant information which reaches the central nervous system. Adaptation also enables a receptor to supply information about the rate of change of a stimulus, and can thus make a receptor much more sensitive to small changes in stimulus intensity without sacrificing the ability to respond over a wide range of stimulus intensities.

Adaptation can occur at a number of stages in the control sequence of sensory excitation. In the Pacinian corpuscle, the accessory structures (the lamellae) are involved in the process; when pressure is applied to the outside of the corpuscle, the lamellae are temporarily deformed, so deforming the nerve terminal, but the inner lamellae rapidly move back to their original positions, so that excitation of the nerve terminal is not maintained (Hubbard, 1958). In other cases, such as in many photoreceptors, the receptor potential shows adaptation.

Adaptation at the impulse initiation site occurs in the tactile spine on the femur of cockroaches (French, 1984) and probably also in Pacinian corpuscles (Gray & Matthews, 1951). In lobster stretch receptors it involves mainly slow inactivation of the voltage-gated sodium channels (Edman *et al.*, 1987). Finally, the generator potentials in the afferent nerve terminals innervating the hair cells of the goldfish inner ear show adaptation as a result of exhaustion of the hair cell chemical transmitter (Furukawa & Matsuura, 1978).

The sensory threshold

Suppose we perform an experiment in which a male human subject listens to a series of sound pulses whose intensity is varied from quite loud to very soft, and is asked to state whether or not he can hear each pulse. At very low sound intensities, he will not be able to detect the pulses, whereas at higher intensities he will. In an intermediate range, he will be able to detect the pulses in a fraction of the times each pulse is presented, and this fraction will increase with increasing intensity. We may call the mid-point of this range, i.e. the sound intensity at which 50% of the pulses are detected, the sensory threshold. Obviously, this is a rather arbitrary concept: we could equally well call the sensory threshold the intensity at which there is a 10% or 75% probability of detection. But the important point here is that, with a stimulus whose intensity is variable over a continuous range (this proviso excludes vision and olfaction, where the stimulus is made up of discrete quanta or molecules), the probability of detection is a continuous function of the stimulus intensity; there are no discontinuities in the relation (fig. 13.13).

Some sense organs are amazingly sensitive. The rods of the human eye can respond to the absorption of a single photon, the human ear can detect sounds which produce movements of the basilar membrane only 3 Å in amplitude, and even the poor old human nose can detect as few as fifty odorant molecules under optimum conditions.

What limits the absolute sensitivity of sensory receptor cells? This question has been discussed by Bialek (1987), Block (1992) and Torre *et al.* (1995). They argue that, since animals are at nearly the same temperature as their surroundings, sensory stimuli have to be detected against a background of thermal energy. They distinguish two types of limit to the absolute sensitivity of sensory receptor cells, 'quantum' and 'classical'

Quantum stimuli are those which arrive in packets of energy such as individual photons for light stimuli or individual molecules for olfactory stimuli, and which the energy of the packets is sufficiently large to stand well clear of the background thermal energy. The thermal energy is given by kT, where T is the absolute temperature and k is the Boltzmann constant. The value of k is 1.38×10^{-23} J K^{-1}, so at 25 °C (298 K) kT is about 4×10^{-21} J. The energy of a photon is given by $h\nu$, where h is Planck's constant (6.62×10^{-34} J s) and ν (nu) is the frequency. For a photon of blue-green light (wavelength 500 nm, so frequency 6×10^{14} s^{-1}) $h\nu$ is 4×10^{-19} J, which is about 100 times larger than the thermal energy. This means that it should be possible in principle for a photoreceptor cell to detect a single photon against the thermal energy background (we shall see in chapter 15 just how they do so). And since no stimulus can be less than a single photon (half photons do not exist), we cannot expect receptor cells to be more sensitive than this.

Could we expect olfactory sense cells to respond to single molecules? We really need to know the energy released when an odorant molecule binds to a receptor molecule, but such figures are not available. The lowest values for binding energies of bacterial chemoreceptor molecules with their substrates are about 10 kJ mol^{-1}, i.e. about 1.7×10^{-20} J per molecule. This is about four times the background thermal energy. So it seems reasonable to conclude that in principle the sensitivity limit for olfactory sense cells is the single molecule.

'Classical' limits to sense cell sensitivity occur when the quantal energy is lower than the background thermal energy. This occurs in hearing, when the quantal energy in sound waves (the phonon) is much lower than kT. For a 1000 Hz sound wave (nearly two octaves above middle C), for example, the quantal energy $h\nu$ is only 6.6×10^{-31} J, vastly below the thermal energy background.

Stimulus–sensation relations

Let us now consider the question 'what is the relation between the intensity of the stimulus and the intensity of the sensation produced by it?' Weber, in 1846, found that human subjects could distinguish two weights as long as the difference between them was about 2.5%, irrespective of the absolute magnitude of the weights. This value – the minimum difference that can be detected – is known as the *increment threshold* or difference threshold. Weber's conclusion was later confirmed by Fechner (1862), who described the results in the form of an equation:

$$\Delta I / I = k\, \Delta S \qquad (13.1)$$

where I is the stimulus intensity, ΔI is the increment threshold, k is a constant, and ΔS is a 'unit of sensation'. Fechner then suggested that this equation could be converted into a differential equation by making $\Delta I = \mathrm{d}I$ and $\Delta S = \mathrm{d}S$, so that

$$\frac{\mathrm{d}S}{\mathrm{d}I} = \frac{1}{kI}$$

which, when integrated, gives

$$S = a \ln I + b \qquad (13.2)$$

where a and b are constants. Equation 13.2 is known as the 'Weber–Fechner law'; it states that the intensity of the sensation is proportional to the logarithm of the intensity of the stimulus.

In fact, it is found that the Weber–Fechner law is not universally applicable. In the human eye, for example, while it is approximately true for the medium range of light intensities, it does not apply at very low or very high light intensities.

An alternative formulation of the stimulus–sensation relation was produced by Stevens (1961*a,b*). For a number

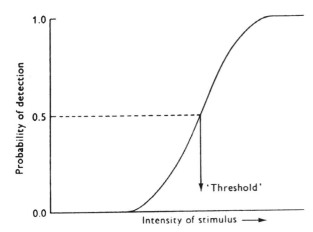

Figure 13.13. Diagram to show how the probability of detection of a stimulus rises with increasing stimulus intensity. The threshold is here arbitrarily defined as that stimulus intensity which is detected 50% of the times it is presented.

of different modalities, he found that the magnitude of the sensation, ψ (psi), equivalent to S in equation 13.2, was given by

$$\psi = k\phi^n \tag{13.3}$$

where ϕ (phi) is the intensity of the stimulus and k and n are constants. This relation has been called 'the power law' or 'the psychophysical law'.

Taking logarithms, equation 13.3 becomes

$$\log \psi = n \log \phi + \log k \tag{13.4}$$

so a plot of $\log \psi$ against $\log \phi$ will be straight line with a slope equal to the exponent n. Figure 13.14 shows three examples.

Stevens later produced evidence that the power law can

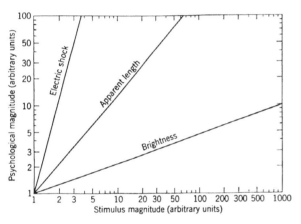

Figure 13.14. Scales of apparent magnitude for three different sensory qualities, plotted on log–log coordinates, to illustrate Stevens' power law. The lines follow equation 13.4, with different values of the exponent n. (From Stevens, 1961b.)

be demonstrated in physiological events as well as in psychological measurements. Thus the frequency of action potentials in a *Limulus* optic nerve fibre was proportional to the light intensity to the power 0.3, and the mammalian cochlear nerve compound action potential was proportional to sound pressure to the power 0.4 (Stevens, 1970).

Spontaneous activity

A number of receptors are spontaneously active: they produce nerve impulses in the sensory nerve in the absence of stimulation. An example of this is in the lateral line organs of fishes (Sand, 1937). In this case, it is clear that the spontaneous activity enables the mechanoreceptors in the lateral line canals to be sensitive to movements of the canal fluid in both directions. In fig. 13.15, headward perfusion of the canal causes an increase in the discharge frequency, and tailward perfusion causes a decrease; thus the receptor is analogous to a centre-reading voltmeter, which can register both positive and negative voltages, as opposed to a voltmeter in which the zero position is at one end of the scale.

An alternative function of spontaneous activity may be to increase the sensitivity of a receptor. If the receptor is spontaneously active, then there is no such thing as a subthreshold stimulus, since any increase in stimulus intensity will change the frequency of impulse discharge. The problem for the animal is then the detection of this change in sensory nerve impulse frequency. The facial pit receptors of crotaline snakes show spontaneous activity, which is modified by very small changes in the radiant heat impinging upon them; it seems probable that the existence of this spontaneous activity contributes to the very great sensitivity of these organs (Bullock & Dieke, 1956).

Finally, there is considerable evidence that the activity of

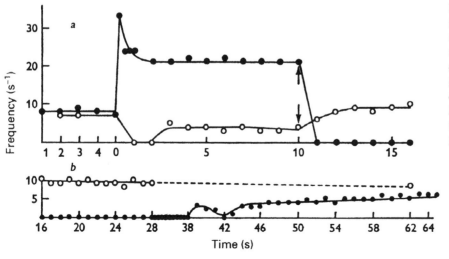

Figure 13.15. Impulse frequency in a single lateral line fibre of a ray in response to perfusion of the hyomandibular canal. Black circles, headward perfusion (between 0 and 10 s); white circles, tailward perfusion. Graph *b* is a continuation of *a*. (From Sand, 1937, by permission of the Royal Society.)

sense organs, spontaneous or induced, plays an important part in determining levels of activity in the central nervous system.

The use of multiple transmission lines

A large number of sense organs send sensory information into the central nervous system via a number of nerve fibres. Such an increase in the number of transmission lines will obviously increase the amount of information that can be carried by the system. It is useful at this point to introduce the concept of a *receptor unit* (J. A. B. Gray, 1962); this consists of a single sensory nerve fibre together with its branches and any receptor cells which may be connected with it. A multiple transmission line system is thus a sense organ, or a group of closely associated receptors, consisting of more than one receptor unit.

A multiple transmission line system can increase the sensitivity of a sense organ, as the following theoretical argument shows. Consider a sense organ consisting of n identical receptor units excited independently of one another, and assume that each unit has a probability x of firing in response to a small constant stimulus. Then the probability of any unit not firing is $(1 - x)$, and the probability of no unit firing is $(1 - x)^n$. Hence the probability p of one or more units firing (which, we shall assume, constitutes detection of the stimulus) is given by

$$p = 1 - (1 - x)^n \qquad (13.5)$$

Figure 13.16 shows how p varies with x for various values of n. It is evident that an increase in the number of trans-

mission lines can produce a considerable increase in sensitivity (just how much will depend on how x varies with the size of the stimulus). It will also produce an increase in the sharpness of the threshold, since the slopes of the curves increase with increasing values of n. For the particular case where $n = 2$, equation 13.5 becomes

$$p = 2x - x^2 \qquad (13.6)$$

This equation is fairly easy to test by psychophysical methods, since a number of sense organs are paired. Thus Pirenne (1943) found that the threshold for binocular vision in humans is lower than that for monocular vision, and Dethier (1953) showed that the threshold of the tarsal taste receptors in flies followed a similar relation (fig. 13.17).

Most information transmission lines are 'noisy' to some extent, i.e. there is some random activity (noise) in addition to the signal. In a sensory nerve fibre, this noise will be seen as random fluctuations in the interval between successive nerve impulses. How, then, is the animal to detect a small change in nerve impulse frequency against this noisy background? One way of doing this would be to measure the average frequency over a sufficiently long time, both before and during the stimulus. Use of this system must necessarily mean that the latency of detection will be appreciable, and that there will be no detection of very brief stimuli.

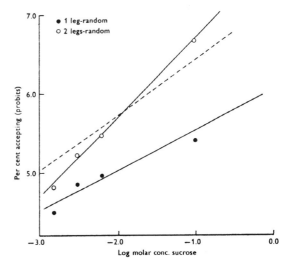

Figure 13.17. The greater sensitivity of flies with two legs in contact with a sucrose solution as compared with those having only one leg so applied. The dashed line shows the expected relation between sucrose concentration and the probability of response for two-legged flies, as calculated from application of equation 13.6 to the results for one-legged flies. The units (probits) on the ordinate are equal to one standard deviation; 5 probits correspond to 50% acceptance. (From Dethier, 1953.)

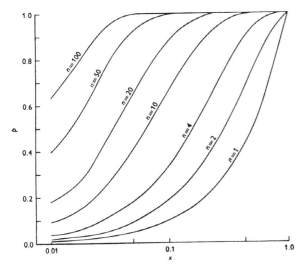

Figure 13.16. Graphical solutions of equation 13.5 for various values of n.

However, by using a number of transmission lines the same information can be obtained in a fraction of the time. Suppose that, in a single noisy sensory transmission line, the time needed to determine whether or not a new average frequency is significantly higher than the original one is t; then the equivalent time for n identical transmission lines is t/n. Hence an increase in the number of sensory transmission lines can reduce the response time.

Another way of looking at this problem of detection is to consider the signal-to-noise ratio of the system. If the signal-to-noise ratio of a single transmission line is q, then the signal-to-noise ratio of n identical transmission lines is $q\sqrt{n}$. Hence increase in the number of transmission lines increases the sensitivity of a receptor to changes in stimulus intensity. It seems very probable that considerations of this type apply to the large number of spontaneously active sensory transmission lines found in the facial pit sense organs of crotaline snakes.

A further use of multiple transmission lines is to ensure that the sense organ is sensitive to the direction or location of the stimulus. The most obvious example of this occurs in image-forming eyes, in which different photoreceptors respond when the stimulus (a light source, for example) is in different parts of the visual field.

We have so far considered the advantages of multiple transmission lines only for cases in which the transmission lines have identical receptor properties. A considerable increase in the information derived from a sense organ can be obtained by using a multiple transmission line system whose receptor units differ, quantitatively or qualitatively, in their sensitivity. One type of organization is what may be called *range fractionation* (Cohen, 1964), in which different receptor units respond to different parts of the range of stimulus intensities or qualities covered by the whole sense organ. This leads to an increase in the sensitivity of the sense organ, since a small change in the stimulus can produce a large change in the response of a particular receptor unit, whereas if that receptor unit had to be responsive over the whole of the stimulus range, its sensitivity over a fraction of the range would be correspondingly reduced. Examples of this type of organization occur in the vertebrate eye, where the rods and cones deal with low and high light intensities respectively, and different cone types may respond to different parts of the visual spectrum.

Receptor units in a system which utilizes range fractionation frequently show bell-shaped response curves. It follows that there must be some ambiguity in the information supplied by any one receptor unit. In the unit whose response is shown in curve *a* of fig. 13.18, for example, an impulse frequency of 7 s^{-1} is produced at both 24 and 35 °C. This

ambiguity can be resolved by comparison with another receptor unit with a different, but overlapping, response range; thus the unit whose response is shown by curve *b* in fig. 13.18 has an impulse frequency of 6 s^{-1} at 24 °C and 9 s^{-1} at 35 °C.

A similar type of argument can be applied to the fractionation of receptor units with respect to the qualitative aspects of the stimulus. Cohen and his colleagues (1955) measured the responsiveness of different gustatory receptor units in the afferent nerves of the cat's tongue. They found that many fibres were responsive to more than one of the modalities which psychophysical measurements suggest can be distinguished, and concluded that the different sensations arise from comparisons of the activity in a number of different receptor units. For example, they suggested that the 'sour' sensation is produced by simultaneous activity in the 'water', 'salt' and 'acid' fibres; if the 'water' and 'acid' fibres cease their activity, the sensation is then one of 'salt'. Thus the specificity of the response is sharpened by utilizing the multiple transmission line system. The trichromatic colour vision system in humans is another example of such sharpening of the specificity of sensation. Extreme instances of this type of system are found in some olfactory receptors; in the antennae of the silkmoth, for example, in over fifty cells whose response to various organic compounds was tested, no two cells showed the same response spectrum (fig. 16.15).

Interaction between receptors: lateral inhibition

A very interesting phenomenon occurs in the compound eye *of Limulus*, known as lateral inhibition (Hartline *et al.*, 1956, 1961). Hartline and his colleagues discovered that the discharge rate of a single eccentric cell decreases if the adja-

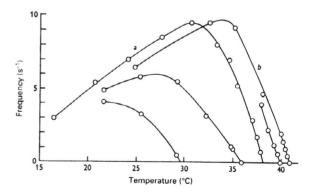

Figure 13.18. Steady impulse discharge frequencies of five 'cold' fibres in a cat's tongue at different temperatures. Curves *a* and *b* are referred to in the text. (From Hensel & Zotterman, 1951.)

cent ommatidium is illuminated, indicating that adjacent ommatidia are mutually inhibitory (fig. 13.19). This phenomenon leads to two interesting features in the pattern of information produced by the eye.

Firstly, the eye must become much more sensitive to edges. Consider the model system shown in fig. 13.20, where we have a row of ommatidia *A* to *J* of which *A* to *E* are exposed to a relatively high light intensity, and *F* to *J* are partially shaded. The extent of the inhibition produced by each cell on its immediate neighbours is then indicated by the diagonal arrows in the diagram, full arrows indicating strong inhibition and broken arrows light inhibition. Cells *A* to *D* are mutually inhibited to some extent, and hence their discharge frequencies are considerably less than they would be if there were no lateral inhibition. But cell *E* is only slightly inhibited by cell *F*, hence its discharge rate is higher than those of cells *A* to *D*. Similarly, cell *F* is more strongly inhibited than are cells *G* to *J*, hence its discharge rate is lower than theirs are. The net effect of these interactions is to emphasize the presence of the edge. A similar process appears to form the basis of edge perception in our own eyes. It may also account for the well-known optical illusion shown in fig. 13.21 – we may suggest that the grey spots seen at the junctions of the white lines are produced by lateral inhibition from the surrounding receptors produced by the image of the white lines.

The second feature of lateral inhibition is that it can increase the resolving power of the eye. The 'fields of view' of adjacent ommatidia show some considerable overlap. Lateral inhibition has the effect of cutting down this overlap, so that each optic nerve fibre responds only to the illumination of a small part of the visual field (Reichardt, 1962).

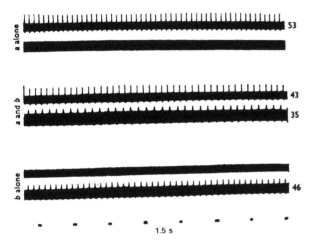

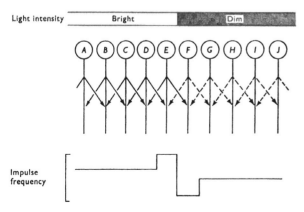

Figure 13.19. Simultaneous records of the sensory discharges in two *Limulus* optic nerve fibres serving adjacent ommatidia (*a* and *b*), in response to illumination of either *a* or *b* alone or both together. The records last 1.5 s in each case, and the figures to the right of them show the number of action potentials occurring in that time. (From Hartline & Ratliff, 1957.)

Figure 13.20. How lateral inhibition emphasizes edges. Each receptor in the field produces some inhibition of its neighbour in proportion to the extent to which it is illuminated. Receptors in bright light (*A* to *E*) produce stronger inhibition than do those in dim light (*F* to *J*). So *E* receives less inhibition than do *A* to *D*, and *F* receives more inhibition than do *G* to *J*.

Figure 13.21. An optical illusion that may be caused by lateral inhibition. Receptors imaging the intersections are surrounded by more active receptors than are those imaging other parts of the white lines, hence the appearance of grey spots at the intersections.

Transmitter–receiver systems

The energy for the activation of most distance exteroceptors is derived from the environment; a crotaline viper, for example, perceives the radiant heat emitted by its prey, and the prey (if it is lucky enough) hears the sounds produced by the approaching viper. However, there exist some specialized sensory systems in which the energy that stimulates the receptors is produced by the animal itself, being reflected back from or modified by objects in the environment.

One of the most well-known of such 'transmitter–receiver systems' is the echolocation system used by bats (see Griffin, 1958; Nachtigall & Moore, 1988). It has long been known that blinded bats can avoid obstacles and catch insects in flight just as well as normal bats. Pierce & Griffin (1938) discovered that flying bats emit a continual series of ultrasonic cries, and Griffin later showed that the rate at which these cries are made is much increased when a bat approaches an obstacle. Griffin & Galambos (1941) measured the ability of bats to fly through a barrier of metal wires spaced just wide enough apart to let them through. They found that deaf or dumb bats frequently hit the wires whereas normal or blind bats rarely did so. The conclusion from these experiments was that bats detect obstacles by emitting ultrasonic sound pulses and hearing the sound waves reflected back to them; the position of an object is determined by localizing the source of the reflected sound. Echolocation systems which are similar in principle to those used by bats are also found in porpoises and other toothed whales, certain cave-dwelling birds and possibly some other animals.

A different type of transmitter–receiver system is seen in the object location system used by some electric fish. Here the fish produces an electric field in the water by means of its electric organs, and can detect changes in this field with its electoreceptors. We shall examine these electroreceptors in chapter 17.

Finally, there are a number of deep-sea fishes which are bioluminescent. Since they have well-developed eyes, it seems probable that one of the functions of their luminescence is to produce light by which the fish can see.

Central control of receptors

The efferent nerve fibres innervating mammalian limb muscles are of two types: the large, rapidly conducting α fibres, and the small, more slowly conducting γ fibres. The γ fibres innervate the muscle fibres (intrafusal fibres) in the muscle spindles, whereas the α fibres innervate the extrafusal fibres and so are responsible for the contraction of the muscle (Leksell, 1945; Kuffler *et al.*, 1951). Stimulation of the γ fibres causes contraction of the intrafusal fibres so that the middle regions (which are not contractile) are stretched; this excites the sensory ending (fig. 13.22). Thus the output of the sensory fibres can be modified by the action of the γ motor fibres; in other words, it is to some extent under the control of the central nervous system.

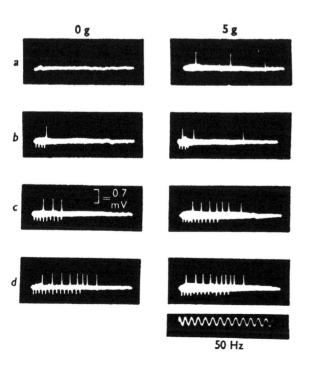

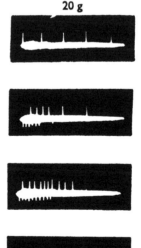

13.22. The effect of γ motor stimulation on the afferent discharges from a cat muscle spindle. Upward deflections on the records are sensory nerve impulses; downward deflections are artefacts produced by stimulating a single γ fibre at 200 s^{-1}. Columns show different loads on the muscle. Rows show the effects of γ stimulation: no stimulation for row *a*, increasing numbers of stimuli for rows *b* to *d*. (From Kuffler *et al.*, 1951.)

Such central nervous control of receptor output via efferent nerve fibres is not uncommon. The efferent fibres may act at various points in the sensory excitation control sequence. In the muscle spindles they act on the accessory structures; other examples of this type of action are the reflex control of pupil diameter in the vertebrate eye, and the action of the stapedius and tensor tympani muscles in the mammalian middle ear, which decrease the sensitivity of the ear to loud sounds. Direct efferent control of the size of the receptor/generator potential occurs in crustacean muscle receptor organs, where the sensory neuron is inhibited by an inhibitor axon (fig. 13.7) which, when stimulated, depresses the generator potential and prevents the initiation of sensory impulses (Eyzaguirre & Kuffler, 1955*b*).

Efferent nerve fibres occur in the auditory nerves of mammals (Rasmussen, 1953), and stimulation of them suppresses the afferent response in the auditory nerve to sound clicks. Electron microscopy shows that the receptor cells (hair cells) of the cochlea are innervated by two types of nerve ending, one of which contains synaptic vesicles and is therefore probably efferent (Engström & Wersäll, 1958). The extensive efferent innervation of the cochlear outer hair cells (p. 253) is probably concerned with reducing the sharp tuning of the cochlear response (Ashmore, 1994). Similar apparently efferent endings are found on the homologous receptor cells in other parts of the ear and in the lateral line organs of fishes and amphibians. Russell (1971) found that the lateral line organs of *Xenopus* (which are stimulated by water currents) are 'switched off' by the efferent nerves during voluntary movements of the animal.

Perception

Many questions about sensory processes can be considered at a number of different levels. How do we see things, for example? At one level we can look at the optics of the eye or investigate what happens in a photoreceptor cell when its visual pigment absorbs photons of light. At a higher level we can look at how the photoreceptor cells interact with each other and with other cells in the retina. Or we can proceed up the visual pathway into the brain and look at the properties of the nerve cells there that receive inputs from the retina. We would find much complexity there, and an increasing abstraction of the particular properties of the image falling on the retina. Just what these abstractions imply is still a matter for discussion.

Such investigations, all of them falling within the general field of sensory physiology, still do not answer the question 'how do we see things?' Our conscious perceptions are not the sensory messages themselves, they are interpretations of them (see, for example, Hochberg, 1984; Gregory, 1970, 1986). In vision, for example, we are able to understand and make sense of what we are looking at; it is very difficult to make a machine that will do this. There is a strong element of recognition in many perceptual processes, hence they are much affected by our previous experience of the world in which we live. But such questions transgress the cellular level of animal organization, and therefore would take us outside the bounds of this book.

14
Mechanoreceptors

Mechanoreceptors are sense organs that respond to mechanical stimuli, impinging upon them as forces or movements. They include receptors responsible for the senses of touch, hearing, acceleration, gravity, proprioception (sensitivity to movements and forces within the body) and pressure. It is now generally assumed that the transduction process in mechanoreceptors involves a direct action of the mechanical stimulus on ion channels in the membrane of the sensory cell or nerve terminal, although in most cases we know very little as yet about the nature of the channels involved.

The acoustico-lateralis system of vertebrates

In vertebrates, the sensory receptors involved in hearing, equilibrium reception and the detection of water movements are all of one basic type and are clearly of common evolutionary origin. In each case the receptor cells, or *hair cells* (fig. 14.1) possess fine processes, stereocilia or 'hairs', whose movement leads to modification of the sensory output. The receptor cells are all connected to sensory neurons which the enter the brain via the cranial nerves; neurons concerned with hearing and balance enter via the eighth nerve and those concerned with the detection of water currents enter via the various lateralis branches.

It is a fascinating feature of the acoustico-lateralis system that we can see how a particular type of receptor cell can be used to respond to a whole variety of different mechanical stimuli, largely by evolutionary elaboration of the accessory structures in the various sense organs (see fig. 14.2). It seems likely that the hair cells were developed in the first vertebrates and used by them for detection of water movements, then internalized in the vestibular apparatus of the inner ear so as to respond to linear and angular accelerations and the direction of gravity, and then utilized for the detection of sound waves.

The lateral line organs of fishes

The sensory cells of the lateral line system are grouped together in organs called *neuromasts*. Each neuromast (fig. 14.2*a*) consists of a number of receptor cells and supporting cells situated in the epidermis, and a gelatinous projec-

tion, the *cupula*, into which the sensory hairs of the receptor cells project. Neuromasts may be situated either freely on the outer surface of the animal or, more usually, in canals which open to the surface at intervals through small pores. The distribution of these canals in a typical fish is shown in fig. 14.3.

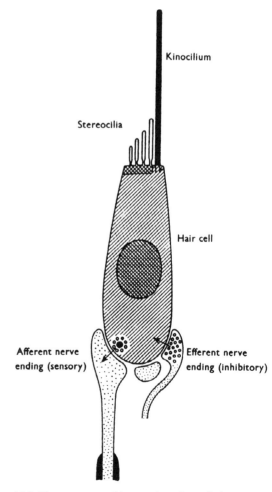

Figure 14.1. The structure and innervation of a typical receptor cell (a hair cell) in the vertebrate acoustico-lateralis system. (From Flock, 1971.)

If a water current impinges on the cupula of a neuromast it is moved to one side, and it is this movement, detected by the sensory hairs of the receptor cells, which acts as the effective stimulus. A fish whose lateral line system has been partially denervated no longer responds to water currents applied to the denervated region. This sensitivity to water movements enables the fish to detect moving objects in its immediate vicinity by means of the water currents which they produce (Dijkgraaf, 1934, 1963).

The electrical responses of the lateral line nerves to stimulation of the neuromasts by water currents of controlled velocity were investigated by Sand (1937), using a perfusion technique on the hyomandibular canal of the ray. He found that there was a considerable discharge in the nerve in the resting state, which was much increased by water movements in either direction. In a single unit the spontaneous discharge was increased by perfusion in one direction but decreased by perfusion in the other direction (fig. 13.15). Some units were excited by head-ward perfusion of water in the canal, others by tailward perfusion.

The semicircular canals

The inner ear, or labyrinth, of vertebrates is a complex organ concerned with equilibrium reception and usually also with hearing. The sensory portion is called the membranous labyrinth; it is a series of interconnected sacs and tubes containing the receptor cells and filled with an aqueous fluid called the endolymph. The osseous labyrinth is the close-fitting cavity in which the membranous labyrinth lies in the bone of the skull. The inner wall of the osseous labyrinth and the outer surface of the membranous labyrinth are separated by a thin layer of another fluid, the perilymph. The membranous labyrinth arises embryologically by an invagination of the ectoderm, and its receptor cells are very similar to those of the lateral line neuromasts.

The structure of the membranous labyrinth is shown in fig. 14.4. There are three semicircular canals, set in planes approximately at right angles to one another. The horizontal canals of the two labyrinths in the head lie in the same plane; the anterior vertical canal on one side is in approximately the same plane as the posterior vertical canal on the other, and *vice versa*. The canals are connected at each end to a sac called the utriculus, so that endolymph can move between canal and utriculus. At one end of each canal is a swelling, the ampulla, in which a group of receptor cells and supporting cells, the crista, is found. Sensory nerve fibres leave the crista to join the auditory nerve.

The sensory hairs of the crista are inserted into a gelatinous cupula. In sections of preserved material the cupula appears as a rather small structure, but in the living animal it extends right across the ampulla to form a fairly close-fitting seal with the opposite wall. This fact, which has important connotations for the mode of action of the semicircular canals, was discovered by Steinhausen (1931) when observing the appearance of the horizontal semicircular canal in a pike after indian ink had been injected into the endolymph.

It has been known for many years that the semicircular canals are concerned with the perception of rotational movements of the head. Such movements produce compensatory reflexes, especially in the extraocular muscles of the eyes, which are abolished by bilateral extirpation or denervation of the canals.

When a vertebrate is rotated at a constant angular velocity in the horizontal plane, its eyes move so as to compensate for this movement. If the head is passively turned to the left, for example, the eyes move to the right, so keeping the visual image fixed in position on the retina. As the rotation

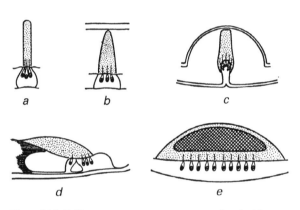

Figure 14.2. Different accessory structures associated with receptor cells of the vertebrate acoustico-lateralis system. *a*, a free neuromast of the lateral line system. *b*, a neuromast in a lateral line canal; *a* and *b* are both responsive to water currents. *c*, the ampullary sense organ of a semicircular canal, responsive to angular accelerations. *d*, the organ of Corti in the mammalian cochlea, responsive to sound. *e*, an otolith organ, responsive to linear accelerations and the direction of gravity. (From Dijkgraaf, 1963.)

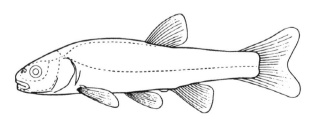

Figure 14.3. Distribution of the lateral line canals on the body surface of a fish. (From Dijkgraaf, 1952.)

continues, the eyes flick rapidly to the left, and then revert to moving to the right again. This pattern of movements, which is known as *nystagmus*, may be repeated for a time. With prolonged rotation however, the nystagmus dies away and the eyes remain in their normal position with respect to the head. On cessation of a prolonged rotation, the eyes show *after-nystagmus* for several seconds, i.e. they behave as though the head were being rotated in the opposite direction.

Can we relate these eye movements to the mechanics of the semicircular canals? Steinhausen (1933) found that the cupula takes several seconds to return to its resting position when it is displaced by a sudden change in the velocity of rotation. This fits well with his earlier observation that the cupula extends across the whole width of the ampulla, and suggests that the displacement of the cupula is strictly linked to the displacement of the endolymph. It would seem, then, that the essential stimulus for the hair cells of the crista is the angular deviation of the cupula from its normal position in the canal. This deviation is produced by the combination of three forces: (1) an inertial force, proportional to the moment of inertia of the endolymph and the angular acceleration of the canal; (2) a viscous (damping) force, proportional to the viscosity of the endolymph, the dimensions of the canal and the velocity of movement of the endolymph through it; and (3) an elastic restoring force in the cupula, proportional to its angular deviation from the resting position.

The simplest situation occurs when the canal is subjected to a period of rotation at a constant angular acceleration. Here the inertial force is constant and the viscous and elastic forces decrease and increase respectively as the cupula moves smoothly to a position determined by the acceleration. The behaviour of the system is more complex when it is subject to a long period of rotation at a constant angular velocity. At the start of this rotation the acceleration is very high and so the inertial force produces a rapid deflection of the cupula. Then, since there is no more acceleration, the inertial force falls and the cupula begins to return to its resting position under the influence of the elastic force. In order to do this it has to act against the viscous force by pushing the endolymph back through the canal, so that it is some time (about 30 s) before the cupula reaches its resting position. If the rotation is now stopped,

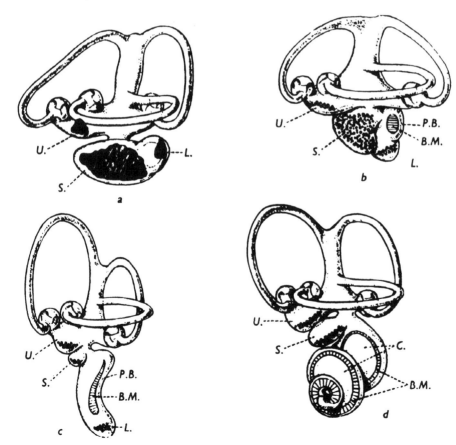

Figure 14.4. Membranous labyrinths of (*a*) a fish, (*b*) a reptile, (*c*) a bird, and (*d*) a mammal. U, utriculus; S, sacculus; L, lagena; P.B., papilla basilaris; B.M., basilar membrane; C, cochlea. (From von Frisch, 1936.)

the endolymph tends to continue on its way, so that there is for a short time a large inertial force which pushes the cupula in the opposite direction to that in which it was previously deflected; it is this 'overswing' that produces afternystagmus. Finally, over the next 20–30 s, the cupula returns to its resting position, again by pushing the endolymph through the canal against the viscous force.

This analysis of the cupula movements in terms of three forces can be expressed in terms of a differential equation as follows:

$$A\frac{\mathrm{d}^2x}{\mathrm{d}t^2} + B\frac{\mathrm{d}x}{\mathrm{d}t} + Cx = 0$$

where x is angular displacement, and A, B and C are constants. van Egmond and his colleagues (1949) made a series of measurements on the sensation of rotation in human subjects for different values of x and its derivatives, and showed that the results were consistent with this equation.

Lowenstein & Sand (1936, 1940*a*) measured the nervous discharge from the horizontal semicircular canals of the dogfish and the ray in response to rotations in the horizontal plane. They found that the changes in the nerve impulse frequency in the sensory fibres followed the same time course as the displacement of the cupula as measured by Steinhausen. These sensory fibres showed a steady spontaneous discharge when the cupula was in its resting position, which increased when the cupula was deflected towards the utriculus and decreased when it was deflected away from the utriculus. Similar features were seen in the responses of the vertical canals (Lowenstein & Sand, 1940*b*), except that here excitation occurred when the head was rotated so as to deflect the cupula away from the utriculus.

These experiments confirmed the view that the effective stimulus for the canals is angular acceleration produced by a rotation in the plane of the canal. Linear accelerations will have no effect since the inertial forces produced in all parts of the endolymph will be of equal magnitude and in the same direction; hence there will be no tendency for the endolymph to rotate.

So far we have considered the responses of a semicircular canal to rotation in the plane in which it lies, i.e. about an axis perpendicular to that plane. What happens if the axis of rotation is not perpendicular to the plane of the canal? Semicircular canals act as vectorial analysers of angular acceleration (Summers *et al.*, 1943). An angular acceleration can be represented by a vector parallel to the axis of rotation whose length represents the magnitude of the acceleration and whose direction indicates the direction of rotation. It is conventional to draw this so that the rotation

is clockwise when seen along the direction of the arrow. Now consider fig. 14.5, in which the line B is perpendicular to the plane in which the semicircular canal lies. The canal is subjected to an angular acceleration about an axis parallel to the line A, which lies at an angle θ with B. This angular acceleration is represented by the vector $\boldsymbol{\alpha}$. The canal is then effectively subjected to an angular acceleration about B represented by the vector $\boldsymbol{\beta}$, which is obtained by dropping a perpendicular from the tip of the vector $\boldsymbol{\alpha}$ onto B. It is clear that

$$\boldsymbol{\beta} = \boldsymbol{\alpha} \cos \theta$$

Hence if θ is 90°, i.e. if A lies in the plane of the semicircular canal, there can be no effective stimulation of the canal. This is the only set of positions of A for which this will occur; for all other positions there will be some response, and this will be maximal for a given angular acceleration when A coincides with B. Thus a single canal is subjected to the same effective stimulus when rotated about A with an angular acceleration $\boldsymbol{\alpha}$ as it is when rotated about B with an angular acceleration $\boldsymbol{\alpha} \cos \theta$. A single canal is therefore incapable of providing precise information as to the nature of the angular acceleration to which it is subjected.

In order to provide such precise information, it is necessary to have three canals set in three planes which intersect in lines that are not parallel to one another, since there are three degrees of freedom in the orientation of an axis in space. Any particular angular acceleration can then be resolved as is shown in fig. 14.6, by dropping perpendiculars from the tip of its vector onto lines per-

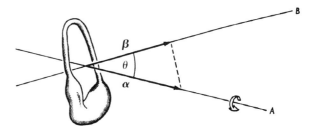

Figure 14.5. Rotation of a semicircular canal about an axis A which is not perpendicular to the plane in which the canal lies. An angular acceleration can be represented by a vector parallel to the axis of rotation whose length represents the magnitude of the acceleration; the rotation is clockwise when seen along the direction of the arrow. Here the vector $\boldsymbol{\alpha}$ represents the magnitude and direction of the angular acceleration. $\boldsymbol{\beta}$ is the projection of $\boldsymbol{\alpha}$ onto the line B, which is perpendicular to the plane of the canal. $\boldsymbol{\beta}$ is then the effective angular acceleration to which the canal responds.

pendicular to the planes in which the semicircular canals lie. It is in accordance with this that there are three semicircular canals in the labyrinths of all jawed vertebrates. It is not essential that the planes of these canals should be mutually at right angles to each other, but the sensitivity of measurement is maximal if they are.

The otolith organs

In addition to the semicircular canals, the labyrinth contains three sacs known as the utriculus, the sacculus and the lagena (fig. 14.4). The receptor cells of these sacs are arranged in groups to produce areas of sensory epithelia called maculae. The gelatinous material covering each macula is filled with calcareous granules (otoconia); in bony fishes the otoconia are fused to form otoliths (fig. 14.2*e*). There are typically three otolith organs, the macula utriculi, the macula sacculi and the macula lagenae, but the latter is absent in most mammals.

It seems that the essential stimulus for the receptor cells is a displacement of their sensory hairs, just as in the neuromasts of the lateral line and the cristae of the semicircular canals. In this case such displacement is brought about by movements of the otoconial mass under the influence of gravity or linear accelerations. Consider a flat macula which is horizontal when the head is in its normal position. The otoconial mass exerts a downward force on the cilia of the receptor cells; in the normal position these are upright and there is therefore no lateral force applied to them. If now the head is tilted through an angle α, then the cilia will be subject to a lateral force proportional to sin α (fig. 14.7). A system such as this will give a sinusoidal relationship between the angle of tilt and the degree of sensory excitation whatever the initial orientation of the receptor cell is. Now let us see how this model system corresponds with actuality.

Lowenstein & Roberts (1950) investigated the equilibrium function of the otolith organs electrophysiologically, using single-fibre preparations from the ray *Raia*. In almost all cases the sensory fibres showed a resting discharge in the normal position, which was increased or decreased by tilting the head. Many units appeared to act as true 'position receptors', with a more-or-less sinusoidal relation between the angle of rotation about one of the horizontal axes and the frequency of the sensory nerve impulses (fig. 14.8). In general, these units were of two types, those which had their positions of maximum activity in the 'side-up' and 'nose-up' positions, and those in which these positions were 'side-up' and 'nose-down'. In addition to these 'static' receptors, some units responded to changes in position irrespective of the direction of the change.

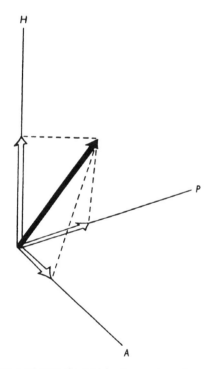

Figure 14.6. How the semicircular canals resolve an angular acceleration about any axis in space into three components. The lines *H*, *P* and *A* represent perpendiculars to the planes of the horizontal, posterior vertical and anterior vertical canals, respectively, and are therefore very approximately at right angles to one another. The black arrow is a vector indicating the imposed angular acceleration, equivalent to α in fig. 14.5. Perpendiculars dropped from the tip of this arrow onto *H*, *P* and *A* give the white arrows, which are vectors showing the components of angular acceleration about these axes, each equivalent to β in fig. 14.5.

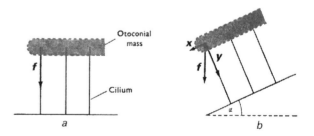

Figure 14.7. The forces on the sensory cilia of a flat macula: (*a*) when it is horizontal, and (*b*) when it is tilted through an angle α. The vertical force *f* represents the weight of the otoconial mass that is supported by one of the cilia. In *b* this force can be resolved into a component *y* acting along the axis of the cilium and a force *x* at right angles to this. Clearly $x = f \sin \alpha$.

Lowenstein & Roberts (1951) later showed that fibres from certain parts of the otolith maculae in the ray are very sensitive to vibration. The otoliths are thus potentially hearing organs, and the sacculus has been so developed in bony fishes and amphibians. This function is taken over by the cochlea in birds and mammals.

Orientation of the hair cells

The receptor cells of the different components of the acoustico-lateralis system are remarkably uniform in structure. They are all set in an epithelial layer, have ciliary processes projecting from their distal surfaces, and are innervated proximally (fig. 14.1). Each cell possesses two types of ciliary process: a group of *stereocilia*, which are densely packed with actin filaments inside, adjacent to a *kinocilium*, which contains nine peripheral microtubules and two central ones along its length just as in motile cilia. The stereocilia are shortest on the side away from the kinocilium, and their height increases the nearer they are to it. Each stereocilium is attached to its neighbour by a fine 'tip link' (see fig. 14.12); we shall return to the role of these later. The kinocilium is absent from the mature hair cells of the organ of Corti in the mammalian cochlea, but a basal region (the centriole) is still present.

The kinocilium is always placed on one side of the group of stereocilia, and possesses a 'basal foot' which projects from its basal body on the side away from the stereocilia; the sensory hair bundle is thus polarized in a particular direction. From electron microscopy of the sensory cells in fish labyrinths, Lowenstein and his colleagues were able to show that this polarization deter-mines the direction in which the sensory hair bundle must be bent in order to produce excitation. Figure 14.9 shows the orientation of the hair cells in the labyrinth of the burbot (Wersäll *et al.*, 1965); similar observations were made on the ray (Lowenstein & Wersäll, 1959; Lowenstein *et al.*, 1964).

The cells in the crista of any particular semicircular canal are all oriented in the same direction. As mentioned earlier, Lowenstein & Roberts had found that the receptor cells of the vertical canals are excited by movements of the cupula away from the utriculus whereas those of the horizontal canal are excited by movements towards it. In the vertical canals, the kinocilia are positioned on the side of the sensory hair bundle away from the utriculus, but those of the horizontal canal are on the side near to it. This implies that the cells are excited when the sensory hair bundle is bent towards the kinocilium and inhibited when it is bent away from it (fig. 14.10).

In the lateral line, the discharge frequency in any single sensory nerve fibre is increased by movement of the cupula in one direction and decreased by such movement in the opposite direction; each neuromast contains some fibres responsive to movements in one direction and some to movements in the other. Flock (1965) demonstrated the structural counterpart of these observations: adjacent hair cells in the neuromast are oriented with their kinocilia pointed in opposite directions, either up or down the canal. A similar situation occurs in free neuromasts (Dijkgraaf, 1963).

The centriole in the hair cells of the organ of Corti is always on the outer side of the ciliary bundle.

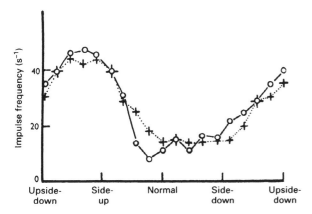

Figure 14.8. Discharge frequencies in a single sensory fibre from the utriculus of *Raja* during slow rotation about the longitudinal axis. For the time sequence of the observations, the continuous curve should be read from left to right and the dotted curve from right to left. (From Lowenstein & Roberts, 1950.)

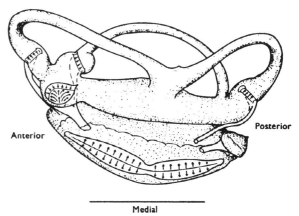

Figure 14.9. The orientation of the hair cells in the right labyrinth of the fish *Lota vulgaris*. The labyrinth is viewed from above. The arrows point towards the side of the ciliary bundle on which the kinocilium is situated. (From Wersäll *et al.*, 1965.)

How do the hair cells work?

We have seen that deflection of the sensory hair bundle towards its kinocilium side increases the frequency of action potentials in the afferent nerve fibres which contact the hair cell, and movement in the opposite direction reduces it. Corresponding excitatory and inhibitory

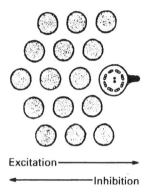

Figure 14.10. Schematic section through the ciliary bundle of a hair cell showing an array of stereocilia and a single kinocilium. The position of the basal foot of the kinocilium is also shown. Deflection towards the kinocilium excites the hair cell, deflection in the opposite direction inhibits it. (From Lowenstein *et al.*, 1964.)

changes are seen on intracellular recording from hair cells. Thus movement of the hair bundle towards the kinocilium produces depolarization and movement away from it produces hyperpolarization, both in lateral line cells (Flock, 1971) and in bullfrog sacculus (Hudspeth & Corey, 1977), as is shown in fig. 14.11. These receptor potentials control the release of chemical transmitter from the hair cells at their synapses with the afferent nerve fibres (Furukawa *et al.*, 1978; Fuchs, 1996).

A crucial question about hair cells is the nature of the transducer mechanism (Hudspeth, 1985, 1989): what is the causal link between deflection of the sensory hair bundle and the production of the receptor potential? Using microprobe stimulation of bullfrog saccular hair cells, Hudspeth & Jacobs (1979) found that movement of the kinocilium alone produced no receptor potential, whereas movement of the stereocilia alone (the kinocilium was held down flat by a microneedle) produced normal receptor potentials. So it seemed likely that the transducer mechanism is associated in some way with the stereocilia.

The stereocilia are rooted in an electron-dense structure called the cuticular plate, from which they emerge in a hexagonal array. They are more or less constant in diameter for most of their length but taper sharply at the base. They are packed with longitudinal actin filaments, all oriented in

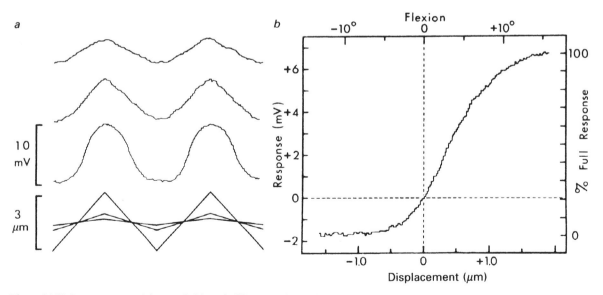

Figure 14.11. Receptor potentials recorded from bullfrog saccular hair cells with an intracellular microelectrode. The hair bundles were exposed by removal of the otolith and were stimulated by a vibrating probe. The three upper traces in *a* show the responses to the 10 Hz triangular wave stimulus monitored in the lower trace. The topmost trace, produced by a low amplitude stimulus, follows the waveform well, but the higher-amplitude stimuli produce receptor potentials which show saturation effects, as seen in the next two traces. Graph *b* shows how the amplitude of the response varies with that of the stimulus. (From Hudspeth & Corey, 1977.)

the same direction and cross-linked (Flock & Cheung, 1977; Tilney *et al.*, 1980). Consequently they are stiff structures which move essentially as rigid rods with basal pivots.

The stereocilia are grouped in ranks or rows across the hair bundle, with each rank containing stereocilia of a different length. The shortest are in the rank on the side away from the kinocilium (or the centriole in the mammalian cochlea), those in the next row are longer, and so on until we reach the tallest row next to the kinocilium. There are filamentous cross-links between adjacent stereocilia just above the basal tapered region; perhaps these account for the fact that the hair bundle moves as a whole when it is pushed by a microprobe (Flock, 1977).

There is some variation in the numbers and dimensions of the stereocilia in different hair cells. In the chick cochlea, for example, the hair cells at the distal end have about fifty stereocilia, the longest of which are 5.5 μm long and 0.12 μm thick, whereas those at the proximal end have about 300 stereocilia, with the longest 1.5 μm by 0.2 μm (Tilney & Saunders, 1983). Corresponding differences occur between hair cells in different organs and different species. In bullfrog saccular hair cells, much investigated by Hudspeth and others, there are about fifty to sixty stereocilia per cell, with the shortest about 5 μm long and the longest about 8 μm.

A striking discovery was made by Pickles and his colleagues (1984) using scanning electron microscopy of glutaraldehyde-fixed cochlear hair cells: there is a fine filament rising upwards from the tip of each stereocilium to the side of its taller neighbour. The ranked arrangement of the stereocilia means that these *tip links* are aligned with the axis of directional sensitivity of the cell, hence they will be stretched when the bundle is moved in the excitatory direction (fig. 14.12).

We might expect that the receptor potential would be produced by the opening of ion channels somewhere in the hair cell membrane as a result of movement of the stereocilia. Hudspeth (1982) used an extracellular electrode to map the electric field around bullfrog sacculus hair cells. He found that current flow is strongest near the tip of the hair bundle, suggesting that the transducer channels are located at or near the tips of the stereocilia. This, together with the discovery of the tip links, led to a model for transduction in hair cells that is illustrated in fig. 14.13. The idea is that the transducer channels in the tips of the shorter stereocilia are more likely to be open if the tip link is under tension, which will occur when the hair bundle is moved towards the taller stereocilia (Corey & Hudspeth, 1983; Hudspeth, 1985, 1989; Pickles & Corey, 1992).

Some neat evidence in favour of the importance of the tip links in transduction has been provided by Assad and

his colleagues (1991). They used the whole-cell patch clamp method to measure receptor currents produced by hair bundle displacement, and electron microscopy to look at the stereocilia. They found that treatment of the hair cell with a low calcium saline (10^{-9} M) buffered with the chelating agent BAPTA produced two effects: the transduction process was rapidly and irreversibly stopped, and the tip links disappeared.

Lumpkin & Hudspeth (1995) have produce further evidence in favour of the location of the transduction channels near the tips of the stereocilia rather than near their bases. They used the calcium-sensitive fluorescent dye fluo-3 and confocal microscopy to look at calcium ion concentrations inside hair cells, and found that stimulated hair cells show calcium entry at the top of the hair bundle, within 1 μm of the tips of the stereocilia.

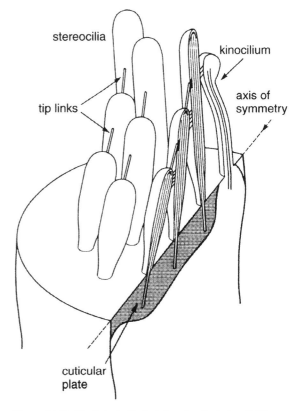

Figure 14.12. Tip links in the hair bundle of a hair cell. They run parallel to the axis of bilateral symmetry of the hair bundle, which is also the direction of maximum sensitivity. It is thought that they are elastic and attached at one or both ends to ion channels which open when the tip links are stretched. (From Pickles & Corey, 1992. Reproduced from: Mechanoelectrical transduction by hair cells. *Trends in Neurosciences* **15**, 254, with permission from Elsevier Trends Journals.)

Are the transducer channels sited at the upper end of the tip link, or at the lower end, or is there one at each end of the link? Calcium imaging using the high resolution provided by two-photon laser scanning microscopy seems to answer this question (Denk *et al.*, 1995; Denk & Svoboda, 1997). With this technique it is possible to observe calcium entry into individual stereocilia when a bundle is deflected. If the transducer channels were all at the top end of the tip link, the shortest stereocilia would show no calcium entry. If they were all at the bottom end, the tallest ones would show no entry. In fact Denk and his colleagues found that both the longest and the shortest stereocilia showed calcium entry, implying that there are transducer channels at both ends of the tip links.

We know rather little about the nature of the transducer channels at present. They are permeable to most cations

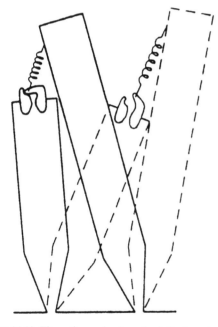

Figure 14.13. The gating spring hypothesis for transduction in hair cells. Two adjacent stereocilia are shown. The tip link between them is shown as a spring attached to an ion channel that is more likely to be open when the tip link is stretched. Deflection to the right causes the stereocilia to pivot about their bases and slide relative to one another, so stretching the tip link and opening the channel. Positive ions flow into the cell via the open channel, so producing the depolarization seen as the receptor potential. There may be a transduction channel at the upper end of the tip link, as well as or instead of that shown here at the lower end. (From Corey & Assad, 1992. Reproduced from *Sensory Transduction*, 45th Annual Symposium of The Society of General Physiologists, p. 328, 1992, by copyright permission of The Rockefeller University Press.)

(Corey & Hudspeth, 1979). Since the endolymph has a high potassium concentration, potassium is the major ion carrying the inward transducer current, but calcium is also important. Crawford *et al.* (1991) measured some single-channel events in hair cells that may have been partly damaged by low calcium ion concentrations so as to leave only one or two tip links intact. The single-channel conductance was quite high, at about 100 pS.

Hair cells usually show adaptation, i.e. the receptor potential declines during a maintained deflection, indicating that the probability of the transducer channels remaining open declines with time. An attractive model for this assumes that there is a molecular motor attached to the insertion plaque at the upper end of the tip link, as is shown in fig. 14.14 (Assad & Corey, 1992; Hudspeth & Gillespie, 1994). When the hair bundle is deflected in the positive direction and held there, the insertion site slips down the stereocilium so that the tension in the tip link is reduced and the channel closes. On returning to the resting condition the slack tip link becomes extended again as the motor climbs back up the stereocilium. If the hair bundle is deflected in the negative direction, the adaptation motor takes up some of the slack in the tip link so that the channel may open when the bundle returns to its resting position

It may be that calcium entry into the stereocilium via the transduction channel inhibits or alters the state of the adaptation motor. The insertion plaque would then slip down the taller stereocilium when there is tension in the tip link, but it would be pulled up by the motor when the calcium concentration fell in the presence of a closed channel. If this is so it may imply that the transduction channel is at the upper end of the tip link, next to the adaptation motor and part of the insertion plaque. A form of myosin, myosin Iβ, has been detected near the tips of the stereocilia and has been cloned from saccular tissue (Gillespie *et al.*, 1993; Metcalf *et al.*, 1994). Perhaps this forms the adaptation motor, with several myosin Iβ molecules attached to the insertion plaque and climbing up the actin filaments that pack the stereocilium.

What is the kinocilium for? It may well be vitally important in the morphogenesis of the hair cell, since its position and orientation indicates precisely the directional sensitivity of the cell. In some cells the kinocilium may help to couple the hair bundle to the surrounding accessory structures.

The mammalian cochlea

The cochlea contains the receptor cells responsible for the perception of sound in mammals. It consists of a tubular outgrowth of the membranous labyrinth, the scala media, surrounded by two tubular outgrowths of the osseous labyrinth, the scala vestibuli and the scala tympani. The structure of this complex, as seen in cross-section, is shown in fig. 14.15. The hair cells are arranged in a single row of inner hair cells and three rows of outer hair cells. In the human ear there are about 3500 inner hair cells and 12 000 outer hair cells.

The hair cells are mounted on the basilar membrane and the tips of their longest stereocilia are inserted into the tectorial membrane, although there is some evidence that inner hair cell stereocilia may be free from the tectorial membrane in some species (Slepecky, 1996). We would therefore expect the hair cells to be excited by movements

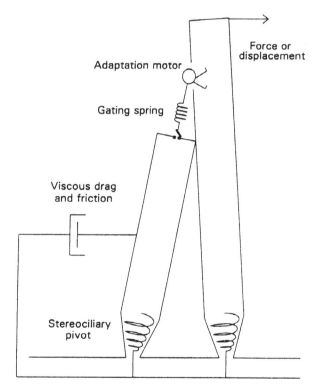

Figure 14.14. The active motor model of adaptation in hair cells. A positive deflection of the hair cell bundle opens the transduction channel, but then slippage of the insertion plaque down the taller stereocilium reduces the tension in the tip link and so allows the channel to close. After return to the resting condition, the adaptation motor climbs up the taller stereocilium again to reset the tension in the tip link. The adaptation motor may be switched off by entry of calcium ions through the transduction channel, which may therefore be located next to it at the upper end of the tip link, rather than at the lower end, as shown here. (From Corey & Assad, 1992. Reproduced from *Sensory Transduction*, 45th Annual Symposium of The Society of General Physiologists, p. 335, 1992, by copyright permission of The Rockefeller University Press.)

of the basilar and tectorial membranes relative to each other. The complex of hair cells, basilar membrane and tectorial membrane forms the *organ of Corti*.

Figure 14.16 shows how the displacements produced by sound waves pass through the ear. Sound waves entering the external ear cause vibrations of the ear drum, which are transmitted through the air-filled middle ear by a chain of small bones (the auditory ossicles) called the malleus, incus and stapes. These ossicles form an 'impedance matching' device whereby the forces over the relatively large area of the ear drum (50 to 90 mm² in humans) are concentrated on the relatively small area (3.2 mm²) of the foot-plate of the stapes. This concentration of forces produces an increase in pressure sufficient to ensure that the sound vibrations can be successfully transmitted from the air outside to the liquid perilymph of the inner ear. If there were no such mechanism, and the ear drum were directly backed by liquid, only a very small proportion of the sound energy would cross the air/liquid interface – the rest would be reflected.

The foot-plate of the stapes rests on a small membrane, the oval window, which covers a gap in the bone in which the inner ear is set. Behind this is the perilymph of the scala vestibuli, so that an inward movement of the stapes causes a displacement of fluid along the scala vestibuli. This displacement is accompanied by movement of the basilar membrane and a corresponding displacement of fluid in the scala tympani, which results in a bulging of the round window, a membrane between the scala tympani and the middle ear. The tectorial membrane moves up and down with the basilar membrane, but there is a lateral shearing action between them which displaces the ciliary processes of the hair cells and so provides the immediate stimulus for their excitation. The mechanical properties of this system are complex and difficult to investigate.

The general problem in cochlear function is how frequency discrimination arises. At moderate sound intensities, for example, we can tell the difference between pure tones of 1000 Hz and those of 1003 Hz (Moore, 1997). In a total frequency range of 20 to 17 000 Hz this is a high degree of sensitivity. Is such sensitivity in frequency discrimination built into the 32 mm length of the human cochlea? And if so how is it done? Or is frequency discrimination partly or entirely carried out in the higher centres of the auditory processing system in the brain?

These questions began to be asked in the nineteenth century, and the most influential answer was that suggested by Helmholtz in 1863. He assumed that the basilar membrane consisted of transverse fibres under tension, rather like the strings of a musical instrument. Since the width of the basilar membrane varies along its length these fibres would resonate at different frequencies. The cochlea is supplied along its length by a large number of nerve fibres, so it seemed reasonable to suppose that each nerve fibre was excited by the vibration of a particular part of the basilar membrane.

The first measurements of how the basilar membrane actually behaves were made by von Békésy, in an extensive and beautiful series of experiments on the physics of hearing, for which he was awarded the Nobel prize in 1961 (see von Békésy, 1960, 1964). He used cochleas derived from human cadavers and applied vibration at the oval window. He was then able to measure the displacement of different parts of the basilar membrane by placing silver grains on it and observing their movement with the aid of a microscope and a stroboflash. He found that a travelling wave of the same frequency as the sounds moves along the basilar membrane from base to apex, its amplitude varying

with the distance along the membrane (fig. 14.17). The position at which the vibration of the basilar membrane reached its maximum varied with sound frequency: it moved nearer to the stapes as the frequency was raised.

von Békésy's results emphasized that the basilar membrane is suspended in an incompressible salt solution within the bony shell of the cochlea. This means that the movement of any part of the cochlear partition must be accompanied by some compensating movement elsewhere. There is more recent evidence that the basilar membrane is not under tension, as Helmholtz had supposed, so that, as Patuzzi (1996) put it, 'it is more like a xylophone played under water than Helmholtz's piano played in air'.

von Békésy's results showed clearly that the basilar membrane responds differentially to different frequencies, and so it must be at least partially responsible for the process of frequency discrimination. However, they also seemed to show that the basilar membrane was not sharply tuned in

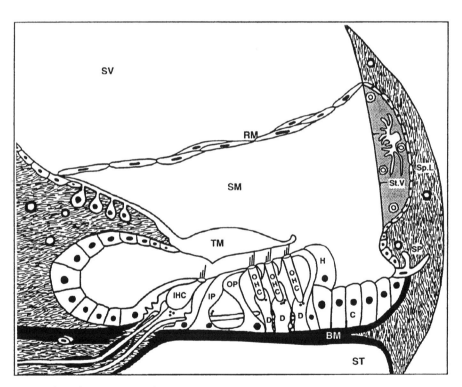

Figure 14.15. Schematic transverse section of part of the cochlea to show the sensory cells and other structures in the organ of Corti. The inner hair cells (IHC) and the outer hair cells (OHC) are supported on the basilar membrane (BM) by the inner and outer pillar cells (IP) and (OP), and the Dieters, Hensen and Claudius cells (D, H and C). The stereocilia of the hair cells are covered by the tectorial membrane (TM). The three fluid-filled compartments are the scala vestibuli (SV), the scala media (SM), and the scala tympani (ST). Reisner's membrane (RM) separates the SM and SV. The SM contains endolymph, a fluid with a high potassium ion concentration, secreted by the stria vascularis (StV). Movements of the basilar membrane produce deflections of the sterocilia of the hair cells. Also shown are the spiral ligament (Sp.L.) and the spiral prominence (SP). (From Slepecky, 1996, © Springer-Verlag 1996.)

the way that Helmholtz had suggested, since any one point would vibrate in response to a number of different sound frequencies. This implied that frequency discrimination must involve central analysis of the differential excitation of relatively large numbers of receptors.

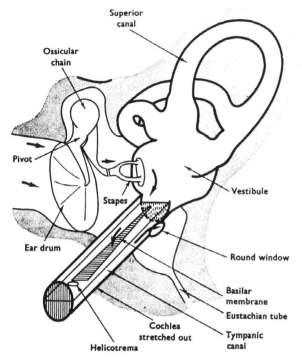

Figure 14.16. Schematic diagram of the middle and inner ear in a mammal. The cochlea is shown uncoiled. The arrows show the displacements produced by an inward movement of the ear drum. (From von Békésy, 1962.)

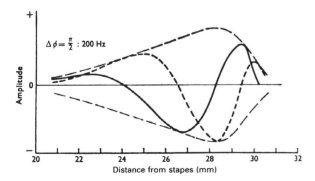

Figure 14.17. von Békésy's demonstration of the travelling wave in the cochlea. The continuous line shows the position of the basilar membrane at one instant during the response to a loud 200 Hz tone, and the middle dashed line shows its position one-quarter of a cycle later. The outer dashed curves indicate the envelope of the movement. (From von Békésy, 1960.)

The first results on the frequency responses of primary auditory nerve fibres were obtained by Tasaki (1954) from the guinea pig cochlea. He found that individual fibres were most sensitive at a particular frequency but that they would respond over a considerable frequency range, in accordance with von Békésy's observations on cochlear mechanics. Much sharper tuning curves were later obtained by Kiang (1965) in cats and Evans (1972) in guinea pigs. They found that any particular fibre is most sensitive at one particular frequency, called the characteristic frequency. The threshold rises very rapidly at higher frequencies and rather less rapidly at lower frequencies. Different fibres have different characteristic frequencies (fig. 14.18 shows a selection), and they are connected to inner hair cells at different places on the basilar membrane (Liberman, 1982).

The difficult business of making intracellular measurements of receptor potentials from inner hair cells was eventually successful in the hands of Russell & Sellick (1978). Particular cells showed sharp tuning curves just like those of the nerve fibres when stimulated with low intensity sounds.

The discrepancy between these sharp neural and cellular tuning curves and the broad curves produced by von Békésy, which is illustrated in fig. 14.18, was a considerable puzzle. For a time it seemed likely that there was a 'second

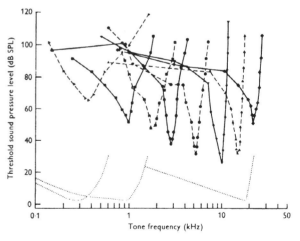

Figure 14.18. Sharp tuning of neural responses from the cochlea. The upper set of curves show how the threshold varies with sound frequency for eight different primary afferent fibres in the cochlear nerves of guinea pigs. There is a characteristic frequency for each fibre at which it is most sensitive, with the threshold intensity rising sharply at lower and higher frequencies. The dotted curves show tuning curves for the basilar membrane as they appeared to be in 1972, when this diagram was first published. More recent mechanical response curves are shown in fig. 14.19. (From Evans, 1972.)

filter' that in some way could sharpen up the tuning between the basilar membrane and the hair cells (Evans & Wilson, 1975). But further investigations on cochlear mechanics have since provided much of the answer to this problem.

von Békésy had to use stimuli equivalent to very high sound pressures in order to produce visible movements of the basilar membrane, and could not make measurements at high frequencies. More sensitive techniques have since been developed, and there has been much more emphasis on the use of living cochleas in good condition. Some very fruitful experiments were performed by Johnstone and his colleagues, utilizing the Mössbauer effect (Johnstone & Boyle, 1967; Johnstone *et al.*, 1970; Sellick *et al.*, 1982). More recently, Doppler-shift velocimetry systems have been used, giving even greater sensitivity and improved linearity (Nuttall *et al.*, 1991; Ruggero & Rich, 1991; Ruggero, 1992).

Both the Mössbauer technique and laser velocimetry utilize the Doppler effect, whereby the frequency of electromagnetic waves arising in or reflected from a moving object appears to rise or fall as the source approaches or recedes from the detector. In the Mössbauer technique the waves are gamma rays produced by radioactive cobalt in a small piece of foil placed on the basilar membrane. For the laser method the waves are monochromatic coherent light reflected from glass microspheres on the basilar membrane.

First results with the Mössbauer technique were largely compatible with von Békésy's measurements. But then doubts arose with the discovery of non-linearities in the behaviour of the basilar membrane at the point of maximum displacement: its movement was relatively greater, in relation to the movement of the malleus, at lower sound pressure levels (Rhode & Robles, 1974). These non-linearities disappeared after death, an effect paralleled in the neural tuning curves, which become much less sharp and lose their sensitive 'tips' when the oxygen supply is reduced (Evans, 1974).

Definitive results with the Mössbauer technique were produced by Sellick *et al.* (1982). They showed that the basilar membrane is indeed finely tuned in the living cochlea, and the mechanical tuning curve is similar in shape to that of individual auditory nerve fibres (fig. 14.19). This implies that, for soft sounds, the travelling wave on the basilar membrane shows a sharp peak in amplitude at a particular place determined by the frequency of the sound. Similar results have been obtained with the laser technique (Ruggero & Rich, 1991; Ruggero, 1992). So the fine tuning of the neural responses reflects the fine tuning of the basilar membrane movements.

How sensitive is the cochlea? The first calculations on this used linear extrapolation from von Békésy's measurements. At a sound intensity of 1000 dyn cm^{-2} (a very loud sound, 134 dB above threshold) the amplitude of the basilar membrane vibration was about 10^{-4} cm or 10 000 Å. At the standard threshold level of 0.0002 dyn cm^{-2}, therefore, one could calculate that the amplitude would be 10 000 × 0.0002/1000 Å, i.e. 0.002 Å. If this is correct, the ear should be able to detect sound when the basilar membrane moves by as little as 1/300th of the diameter of a hydrogen atom. This 'astonishing and baffling conclusion', as von Békésy & Rosenblith (1951) described it, was current for some years and is still occasionally quoted.

We now know, however, thanks to the Mössbauer technique, that the amplitude of the basilar membrane movement is not a linear function of the sound pressure level, a conclusion which invalidates the calculation given above. These later results suggest that the absolute threshold corresponds to a basilar membrane movement of about 3 Å or a little less (Sellick *et al.*, 1982; Bialek, 1987). This is still a

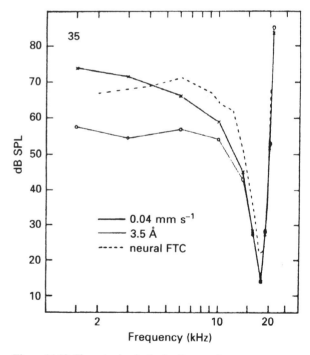

Figure 14.19. Sharp tuning in the basilar membrane as determined by the Mössbauer technique. The continuous lines show the sound intensities required to produce standard movements (a displacement of 3.5 Å or a velocity of μ40 m s^{-1}) of a particular point on the basilar membrane of a guinea pig at various frequencies. The dotted line shows a similar fine tuning curve for the response of a nerve fibre with a similar characteristic frequency. (From Sellick *et al.*, 1982.)

very small distance, but one which is in the atomic rather than the subatomic range.

The peak motion of the basilar membrane is thus about 100 times larger than is implied by applying a linear extrapolation to von Békésy's measurements, and this enhancement is sharply localized to within about 500 μm (Ashmore, 1994). This finely tuned enhancement of the cochlear mechanics is sometimes known as the 'cochlear amplifier'. How does it arise, and why does it disappear after death?

Let us first look at the results of an electron microscope study of the cat cochlea by Spoendlin (1975, 1984). He was able to distinguish the afferent and efferent nerve fibres by sectioning part of the auditory nerve; the efferent fibres in the cochlea, cut off from their cell nuclei, would degenerate and disappear after two weeks or so, leaving the afferent innervation intact. His conclusions, illustrated in fig. 14.20, were surprising and very interesting. Although there are three times as many outer hair cells as there are inner hair cells, 95% of the afferent nerve fibres are connected to inner hair cells. Each inner hair cell is connected to about twenty unbranched afferent neurons. Outer hair cell afferents branch to contact about ten separate outer hair cells, and each outer hair cell is contacted by about four afferents.

This disparity in the afferent innervation of the inner and outer hair cells is reflected in recordings of single fibre responses to sound. Thus Liberman (1982) recorded from fifty-six different primary afferents with an intracellular electrode, and in each case he then injected horseradish peroxidase through the electrode to label the fibre for microscopic analysis. All fifty-six fibres innervated inner hair cells. This suggests that all the tuning curves obtained by Kiang and by Evans (fig. 14.18) were similarly from afferent fibres connected to inner hair cells.

Spoendlin's results led to some hard thinking about cochlear function. It seemed particularly odd that the outer hair cells should be connected to such a small proportion of the afferents and yet should be so well supplied with efferents. Stimulation of the crossed olivocochlear bundle (the efferent pathway running from the brain stem to the cochlea, whose neurons end on the outer hair cells) causes tuning curves to become less sensitive at their tips (Kiang *et al.*, 1970; Brown *et al.*, 1983). This led to the idea that the outer hair cells might be important primarily in connection with the fine tuning of the basilar membrane, perhaps by some form of active response to sound.

This idea fitted in with a remarkable property of the cochlea that was discovered by Kemp (1978). He found that stimulation with a brief click is followed by sound output from the cochlea itself for several milliseconds afterwards. The frequency spectrum of the emitted sounds is very similar to that of the stimulus, suggesting that the sounds are generated at sites along the cochlea which correspond to their frequencies (Norton & Neely, 1987). It seemed possible that the phenomenon was connected with the sharpness of tuning of the basilar membrane; perhaps these active responses were in some way produced by the outer hair cells.

Direct evidence for active responses was provided by injecting current into isolated outer hair cells (Brownell *et al.*, 1985; Ashmore, 1987). The cells change their length by up to 4% under these conditions, as is shown in fig. 14.21.

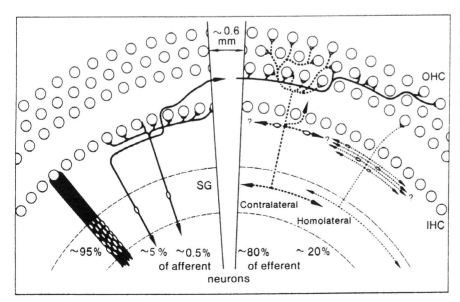

Figure 14.20. Innervation of the cat organ of Corti. The afferent neurons are shown by full lines, the efferent ones by broken ones. Notice that the great majority of afferent neurons come from the inner hair cells (IHC), whereas most of the efferent neurons supply the outer hair cells (OHC). (From Spoendlin, 1975.)

The speed of the response is remarkable: it begins within 70 μs of the change in membrane potential. This means that its mechanism is unlikely to involve chemical reactions such as occur in actin–myosin sliding.

There is good evidence that the active responses arise in the outer hair cell plasma membrane. Changes in membrane potential produce membrane charge movements analogous to the gating currents of nerve axons (Santos-Sacchi, 1991; Ashmore, 1992, 1994). Charge densities are estimated at 2000 to 4000 e_0 μm^{-2}. Electron microscopy shows the presence of particles in the membrane, densely packed at about 2500 μm^{-2}. Video-enhanced microscopy of patch-clamped membrane patches showed that they increased in area (seen as bowing or buckling of the patch under the patch electrode tip) when the membrane potential was hyperpolarized. Such changes did not occur in patches taken from the basal, synaptic membrane of the outer hair cells, or in membrane patches from inner hair cells. It looks as though the membrane particles are the sensor molecules and that they undergo some conformational change when charge movement occurs within them, brought about by changes in membrane potential, so acting as motor molecules as well. Inner hair cells do not possess these densely packed membrane particles, nor do membrane patches from them show changes in area when subject to membrane potential changes (Kalinec *et al.*, 1992).

There is a dense cytoskeletal network just below the basolateral membrane in outer hair cells (Holley &

Ashmore, 1988; Holley *et al.*, 1992). Fibres, probably of actin, run circumferentially at an angle of about 15° to the transverse axis, cross-connected by thinner filaments. They probably act as a coiled spring that can be compressed or extended, so converting the changes in membrane area into changes in the length of the hair cell.

It seems, then, that the active responses of the outer hair cells act as the cochlear amplifier, producing the sharp tuning of basilar membrane displacement. Just how their length changes act on the basilar membrane is not yet clear, but at least we now have something for the cochlear modellers to get their teeth into (see for example Geisler & Sang, 1995; Markin & Hudspeth, 1995; de Boer, 1996; Kolston & Ashmore, 1996; Patuzzi, 1996).

Mammalian muscle spindles

The afferent nerve fibres from mammalian limb muscles are of a number of different types (table 4.1), distinguishable according to their diameters and the nature of their sensory endings. There are two main types of receptor in the muscles: (1) the muscle spindles, which consist of small modified muscle fibres and are innervated by group Ia (primary) and group II (secondary) fibres and by motor fibres of the γ system; and (2) the Golgi organs in the tendons, which are innervated by group Ib fibres. In addition there are a number of small diameter fibres (groups III and IV) which have free or encapsulated endings and may be responsive to pressure or pain.

This pattern of innervation can be illustrated by refer-

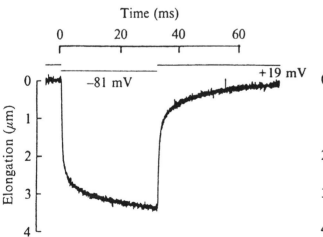

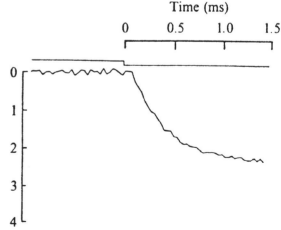

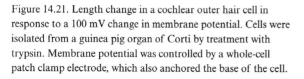

Figure 14.21. Length change in a cochlear outer hair cell in response to a 100 mV change in membrane potential. Cells were isolated from a guinea pig organ of Corti by treatment with trypsin. Membrane potential was controlled by a whole-cell patch clamp electrode, which also anchored the base of the cell. Changes in cell length were monitored by a photodiode at the hair bundle end. The records show averages of 100 events, that on the right shows the initial elongation at higher time resolution. (From Ashmore, 1994.)

ence to the soleus muscle of the cat (Matthews, 1964). There are about fifty spindles, with the same number of primary endings and fifty to seventy secondary endings. There are also about forty-five tendon organs, but individual group Ib fibres may innervate more than one tendon organ. A few group III and group IV afferents occur. The spindles contain a total of about 300 *intrafusal* muscle fibres, supplied by about 100 fusimotor nerve fibres of the γ system. Finally, there are about 25 000 *extrafusal* muscle fibres innervated by about 150 α motor nerve fibres.

Before examining the details of muscle spindle action, it would be as well to establish some general ideas as to how we might expect them to act under various conditions. Figure 14.22 shows, in a schematic and much simplified fashion, the mechanical relations between the extrafusal muscle fibres, the muscle spindles and the Golgi tendon organs. Notice that the tendon organs are in series with the extrafusal muscle fibres, whereas the muscle spindles are in parallel with them: these positions are just right for the tendon organs to act as tension receptors and the spindles to act as length receptors.

Our knowledge of the workings of mammalian muscle receptors begins with the work of B. H. C. Matthews (1933), and it is instructive to relate his results to the scheme

shown in fig. 14.22. If the resting muscle is passively stretched, then both the muscle spindles and the tendon organs will be stimulated (figs. 14.23a and b), the former by the stretch itself and the latter by the consequent increase in passive tension. If the α fibres are stimulated so that the muscle contracts, the tendon organs will be stimulated by the increase in tension (fig. 14.23c) and the muscle spindle will shorten (even in an 'isometric' contraction there will be some shortening because of the compliance of the tendons) and so cease to fire (fig. 14.23d).

The role of the γ fibres became evident later, when Leksell (1945) and Kuffler and his colleagues (1951) showed that their stimulation increases the sensory output of the spindle, as is shown in fig. 13.22. Stimulation of the γ fibres in an otherwise resting muscle will make the contractile regions of the intrafusal muscle fibres shorten and thus stretch their equatorial regions, so producing excitation of the muscle spindle afferents.

So much for the basic outlines of muscle spindle physiology. Detailed investigation has shown that muscle spindles are remarkably complex organs, whose study is no easy matter (Matthews, 1981; Hunt, 1990; Barker & Banks, 1994; Proske, 1997). In the 1960s there were divergent views about the structure of spindles and how it related to spindle physiology. These conflicts have since been largely resolved and the participants have been able to look back and see that each was partly right (Boyd, 1981). Let us look at some of the details.

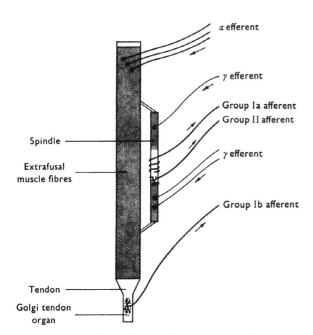

Figure 14.22. Simplified diagram of the innervation of a mammalian muscle and muscle spindle. Contractile regions are shaded. The muscle spindles contain intrafusal muscle fibres, whereas the fibres in the rest of the muscle are known as extrafusal.

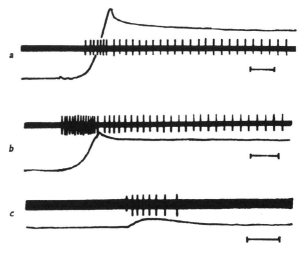

Figure 14.23. Responses of sensory fibres in cat leg muscles to passive stretch (*a* and *b*) and to twitch contractions of the muscle (*c* and *d*). *a* and *c* show responses of tendon organ endings; *b* and *d* show responses of spindle primary endings. The thick lines show nervous activity; the thin traces show tension changes in the muscle. Time scales 0.1 s. (From Matthews, 1933, redrawn.)

Structure

The most comprehensive studies on muscle spindles are of those in cat hind-limb muscles. What follows relates principally to these, but the spindles of other mammals, including humans, are not very different.

Each spindle contains five to ten intrafusal muscle fibres, held together centrally in a fluid-filled capsule. Two different types of muscle fibre were distinguished in the 1950s (Cooper & Daniel, 1956; Boyd, 1962). The *nuclear bag* fibres are about 8 mm long and their equatorial regions are swollen and contain numbers of nuclei clustered together. The *nuclear chain* fibres are narrower than the nuclear bag fibres and about half their length; their equatorial regions are not swollen but contain a single string of nuclei. Myofibrils are almost absent from the equatorial regions in both cases, specially in the nuclear bag fibres. Later it became clear that there are two types of nuclear bag fibre: bag$_1$ fibres are shorter and thinner than bag$_2$ fibres and there are histochemical differences between them

(Ovalle & Smith, 1972; Banks *et al.*, 1977). Most spindles contain two nuclear bag fibres, one of each type, and four or five nuclear chain fibres (fig. 14.24).

Can we relate these intrafusal fibre types to the different types of extrafusal fibre? Rowlerson *et al.* (1985) used immunohistochemistry to investigate the myosin isoforms of intrafusal fibres. They found that bag$_1$ fibres possessed myosin very like that of tonic fibres, and that the myosin of chain fibres was similar to that of embryonic and neonatal fast-twitch muscles. Myosin in bag$_2$ fibres seemed to be a distinct type, probably with some similarities to that of slow-twitch muscles. Bag$_1$ fibres show other similarities to extrafusal tonic fibres: the M line is absent and the sarcoplasmic reticulum is relatively sparse.

It was shown by Ruffini in 1898 that muscle spindles possess two types of sensory endings; later work has provided more detail but no cause to revise this conclusion (Banks *et al.*, 1982). Each spindle receives one group Ia afferent nerve fibre and from none to five group II fibres.

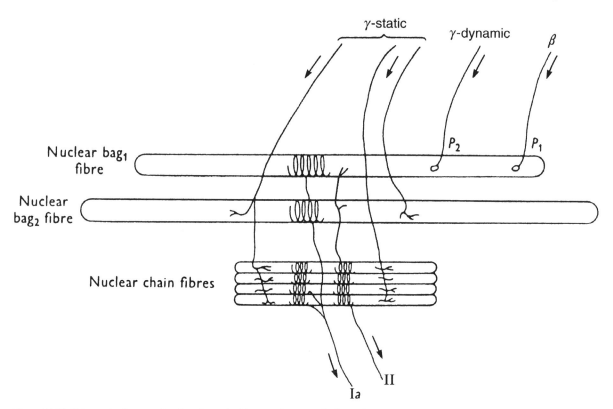

Figure 14.24. Schematic diagram showing the typical innervation pattern of a mammalian muscle spindle. There are three types of intrafusal muscle fibre: nuclear bag$_1$, nuclear bag$_2$ and nuclear chain. Sensory nerve axons (a group Ia fibre from the primary ending and a group II fibre from the secondary ending) are shown emerging below; the number of secondary endings per spindle varies from zero to five. Fusimotor axons are shown above; γ-dynamic axons have plate endings on bag$_1$ fibres and γ-static axons have trail endings on bag$_2$ and chain fibres. The β axon as shown may be called a β-dynamic axon; sometimes β-static axons occur, innervating the nuclear chain fibres. (Based largely on Boyd *et al.*, 1977, and Boyd & Gladden, 1986.)

The group Ia fibres end at the *primary endings*, where they wind round the equatorial regions of the intrafusal fibres to form 'annulospiral' terminations. All nuclear bag fibres have primary endings on them, and almost all nuclear chain fibres do. *Secondary endings*, derived from group II fibres, are found on either side of the primary endings on nuclear chain fibres; occasionally they also occur in a similar position on nuclear bag fibres. Most of the secondary endings in the cat are of the 'annulospiral' type, but there are some 'flower-spray' endings, in which the fibres split up into a number of branches which have slight thickenings at their ends.

The efferent nerve fibres supplying the muscle spindle are called fusimotor fibres. A number of these innervate extrafusal as well as intrafusal muscle fibres; they are known as β fibres and have terminals described as 'plate (p_1)' endings on the intrafusal muscle fibres. The γ fibres, which are exclusively fusimotor, have two types of terminal: 'plate (p_2)' and 'trail' endings. Plate endings form distinct endplates, the p_2 endings being larger than the p_1. Trail endings are more diffuse endings in which the fibre splits up into a number of terminals. γ-Plate axons innervate bag_1 fibres and γ-trail axons innervate bag_2 or chain fibres, or both. β axons replace γ axons on either bag_1 or long chain fibres in some spindles. Figure 14.24 summarizes the situation.

Sensory output

The properties of the primary and secondary endings were investigated by Cooper (1961) and Bessou & Laporte (1962), among others, by recording the responses of single afferent fibres from the limb muscles of anaesthetized cats. The group Ia (primary ending) and group II (secondary ending) afferents could be distinguished by their different conduction velocities. Both types of ending produced more action potentials at longer lengths, but only the primary

ending gave a high frequency burst of impulses when the muscle was being stretched. The primary ending is very sensitive at the start of a movement, less so as it proceeds; this means that even quite small movements are readily detected.

Hunt & Ottoson (1975) recorded receptor potentials of primary and secondary endings in isolated cat muscle spindles. They used extracellular electrodes and blocked the action potentials with tetrodotoxin; thus their experiments were similar in principle to those by Katz on frog muscle spindles, which we examined in chapter 13. The receptor potentials of primary endings showed a marked dynamic component: depolarization was greater during a stretch than after its completion. This effect was especially evident at the beginning of a stretch, in line with the high sensitivity. In secondary endings the dynamic component of the receptor potential was much smaller (fig. 14.25). Thus the nervous output of the two types of sensory ending is largely predictable from their receptor potentials.

Why should the receptor potentials of the two endings differ in their dynamic responsiveness? We might look for an answer in the mechanical properties of the intrafusal muscle fibres. This is likely to be more complicated than in extrafusal fibres, since the quantities of myofibrils and of elastic material vary along their length. Furthermore, there is evidence for some unexpected effects of stretching the bag fibre. Poppele & Quick (1981), using cinematography to look at isolated spindles, found that the sarcomere length in the region next to the sensory endings would actually decrease during a stretch. This stretch activation has been evoked to explain some of the after-effects of stretching and fusimotor stimulation (Emonet-Dénand *et al.*, 1985). The mechanism of stretch activation in spindles is probably rather different from that in insect asynchronous muscles (chapter 21) since it occurs in apparently resting muscle fibres.

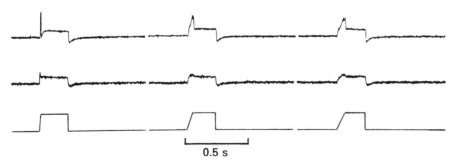

0.5 s

Figure 14.25. Receptor potentials from an isolated cat muscle spindle in response to stretching at three different velocities. Records were made with extracellular electrodes from the afferent fibres close to their sensory endings, with tetrodotoxin to prevent action potential production. The upper trace shows the response of the primary ending, the middle trace that of a secondary ending, and the lower trace monitors the length change applied to the spindle. (From Hunt & Ottoson, 1975.)

We know little about the transduction process in the sensory endings. It seems reasonable to suppose that there are mechanosensitive channels in the nerve terminals, perhaps attached to cytoskeletal elements on the inner side of the plasma membrane, and that these will open to let cations flow through when the membrane is stretched. The resulting receptor potential (fig. 14.25) then acts as the generator potential for the sensory nerve impulses. There may be more than one impulse initiation site in the primary terminal (Carr *et al.*, 1996; Proske, 1997).

Gamma efferent action
P. B. C. Matthews and his colleagues (Matthews, 1962; Crowe & Matthews, 1964; Brown *et al.*, 1965) showed that the γ efferent fibres in the cat are of two functionally distinct types, distinguished by their effects on the velocity-sensitive responses of the primary endings during extension. They defined the 'dynamic index' of a response to stretching at a constant velocity as the difference between the frequency of firing of an afferent fibre just before the end of the period during which the muscle is being extended and that occurring at the final length half a second later. The two types of fusimotor fibre are then *dynamic fibres*, which increase the dynamic index of the primary afferent fibres, and the *static fibres*, which reduce it. In other words, dynamic fibres make the primary endings relatively more sensitive to the velocity of stretching, and the static fibres make them less sensitive. These features are shown in fig. 14.26.

Is it possible to find any anatomical basis for this physiological division of fusimotor fibres into two types? An ingenious technique was applied to the problem (Brown & Butler, 1973; Barker *et al.*, 1976). If a fusimotor axon is stimulated repetitively for some time, all the intrafusal fibres which it innervates will contract and so deplete their glycogen reserves, and this depletion can be detected by suitable histological staining. In this way it has been found that static γ axons innervate both bag_2 and chain muscle fibres, whereas dynamic γ axons innervate predominantly bag_1 fibres. It looks as though static γ axons correspond to γ-trail axons and dynamic γ axons correspond to γ-plate axons. To put it another way, fusimotor axons innervating bag_1 fibres are dynamic and those which innervate bag_2 or chain fibres are static. In accordance with this, β axons have dynamic actions if they end on bag_1 fibres, static actions if they do not (Barker *et al.*, 1977; Jami *et al.*, 1982; Banks *et al.*, 1985).

This idea that the differences between the effects of static and dynamic γ axons arise from differences in the mechanical properties of the intrafusal muscle fibres which they innervate has an interesting history. Matthews (1964) suggested that dynamic γ axons innervated bag fibres while static γ axons innervated chain fibres; this suggestion fitted well with Boyd's anatomical views at the time, but not with Barker's. The ensuing controversy was not resolved until the existence of the two types of nuclear bag fibre was established.

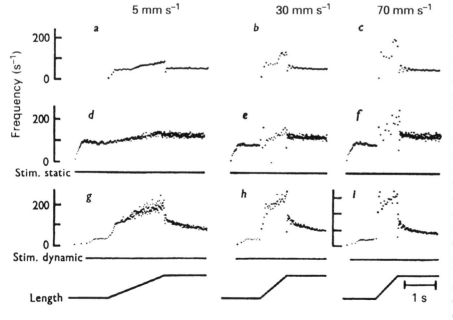

Figure 14.26. Effect of stimulating static and dynamic γ fibres on the response of a single primary ending in a cat soleus muscle to stretching the muscle at different velocities. Each action potential is shown as a dot whose vertical position is proportional to the instantaneous frequency (i.e. the reciprocal of the time since the preceding action potential). Traces *a* to *c* show the response to stretching at three different speeds with no stimulation of the fusimotor fibres. Traces *d* to *f* show corresponding responses during stimulation of a γ-static fibre at 70 s⁻¹. Traces *g* to *i* show responses to stimulation of a γ-dynamic fibre at 70 s⁻¹. (From Crowe & Matthews, 1964.)

There is still much to be discovered about muscle spindles. It would be good to understand how the contractions of the intrafusal muscle fibres produce their static or dynamic effects on the primary afferent discharge, and it would be good to have some more definite information about the nature of the transduction process.

Touch

Tactile sensations occur when an animal receives mechanosensory information about the objects it has contacted in its environment. All animals respond to such contact and there are a great variety of different sense organs involved. Here, however, we consider just two touch systems at very different levels of complexity.

Mechanoreceptors in the human hand

The sensitivity of the fingertips is remarkable. Think of the capabilities that we have for distinguishing different types of fabric by their feel, for example, or for picking out the right coin from a purse or pocket in the dark. This sensitivity depends on the presence of a large number of sensory nerve terminals in the skin.

Histologists have long been able to distinguish different forms of these sensory terminals, many of which are named after those who first described them in the nineteenth century (see Iggo & Andres, 1982; Darian-Smith, 1984*a*; Greenspan & Bolanowski, 1996). In the glabrous (hairless) skin of the fingertips, there are four main types of skin mechanoreceptor involved in tactile sensation. They are the Merkel receptors, the Meissner corpuscles, the Ruffini endings and the Pacinian corpuscles. Figure 14.27 shows their locations in the skin. There are other afferent nerve fibres without morphologically specialized endings, serving the sensations of temperature and pain.

The *Merkel cells* occur at the base of the epidermis; they make close contact with the expanded ends of the nerve terminals, known as Merkel discs. The Merkel cells are firmly attached to adjacent keratinocytes (epidermal skin cells) by desmosomes, and send finger-like projections into them. It seems likely that they are specialized receptor cells and that

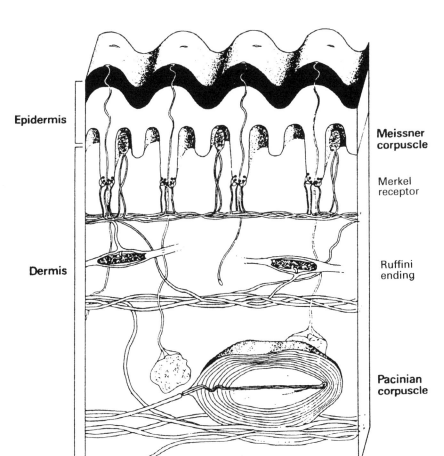

Figure 14.27. Skin of the primate fingerpad showing the four main sensory endings serving the sense of touch. (From Darian-Smith, 1984*a*.)

Epidermis

Dermis

Meissner corpuscle

Merkel receptor

Ruffini ending

Pacinian corpuscle

Table 14.1 *Mechanoreceptor afferents of the glabrous skin of the human hand*

Physiological type	FA I (= RA)	FA II (= PC)	SA I	SA II
Putative receptor	Meissner	Pacinian	Merkel	Ruffini
Adaptation	Rapid	Rapid	Slow	Slow
Receptive field size	Small	Large	Small	Large
Endings per cm²				
at the fingertip	140	21	70	49
on the palm	37	10	30	14
Sensations produced by intraneural microstimulation	Tap, flutter	Vibration	Pressure	None

Note:

FA, fast adapting; RA, rapidly adapting; PC, Pacinian corpuscle; SA, slowly adapting. (Simplified after Greenspan & Bolanowski, 1996.)

they excite the nerve terminals (the Merkel discs) by synaptic transmission (Ogawa, 1996).

The other three mechanoreceptor types we consider are all located in the dermis and have nerve terminals that are encapsulated in non-excitable cells. *Meissner's corpuscles* occupy the outermost part of the dermis, in between the epidermal ridges. The nerve terminal is surrounded by a rather loose capsule of perineural cells and collagen fibres. *Ruffini endings* have spindle-shaped capsules embedded in the dermis. They are connected to collagen fibres in the dermis and so are readily affected by mechanical tension in the skin. *Pacinian corpuscles* have a large capsule of twenty to seventy layers of perineural cells surrounding the nerve terminal (fig. 13.1).

There has been much controversy over the years as to how these anatomical types are related to the physiological responses of nerve fibres and the psychological aspects of tactile perception (Stevens & Green, 1996). For a time Müller's doctrine of 'specific nerve energies' (p. 225) was discarded in favour of the idea that sensations were determined by the pattern of activity in a group of nerve fibres rather than particular fibres responding to particular stimuli (Nafe, 1929; Bishop, 1946; Weddell, 1955). More recently, however, it has become evident that particular types of mechanoreceptor do respond to particular stimuli and thus that their nerve fibres do carry specific information. Let us look at some of the evidence.

Electrophysiological studies have distinguished four different types of response in the mechanoreceptor afferents from the hands of humans and monkeys (Lindblom, 1965); Knibestöl & Vallbo, 1970; Johansson, 1978. They are either rapidly or slowly adapting, and have either large or small receptive fields. The receptive field of a cutaneous sensory nerve fibre is the area of the skin in

which a tactile stimulus (touch by a fine point, for example) will excite the fibre. The four categories are: (1) rapidly adapting with small receptive fields, called RA, QA (quickly adapting) or FA I (fast adapting) units; (2) rapidly adapting with large receptive fields, FA II or PC (Pacinian corpuscle) units; (3) slowly adapting with small receptive field units, SA I; and (4) slowly adapting with large receptive fields, SA II. Table 14.1 summarizes the characteristics of these four types and fig. 14.28 shows examples of their receptive fields.

Which receptor endings are associated with these four functional types of mechanoreceptor afferent? Talbot and his colleagues (1968) anaesthetized the skin of human subjects and found that the threshold for the flutter-vibration sensation was raised at low frequencies but was unaffected at higher frequencies. In monkeys they showed that the low frequency response was produced by FA I units while the high frequency response was produced by FA II units. All this suggests that the low frequency FA I response is attributable to the Meissner corpuscles just under the epidermis, while the high frequency FA II response is produced by the Pacinian corpuscles at a much deeper level. This fits well with the extensive information that we have on the sensitivity of Pacinian corpuscles isolated from more accessible sites such as the mesenteries (see Bell *et al.*, 1994). The small receptive fields of the FA I receptors fits well with the position and number of the Meissner corpuscles close to the skin surface. We would expect the Pacinian corpuscles, much deeper in the skin, to respond to stimuli applied over a larger area at the surface; this fits well with the large receptive fields of the FA II fibres.

Iggo & Muir (1969) studied the elevated touch spots on the hairy skin of cats and monkeys, which contain numbers of Merkel receptors. Maintained pressure produced maintained output from the afferent nerve fibres, indicating that they were of the SA I type. Further evidence comes from an ingenious experiment by Ikeda *et al.* (1994), utilizing a method used for the destruction of cancerous cells. They loaded rat Merkel cells with the fluorescent dye quinacrine, then irradiated them with intense blue light to make the dye phototoxic. This treatment destroyed the Merkel cells and abolished the SA I responses, suggesting that the Merkel receptors produce the SA I responses. The experiment also implies that the Merkel cells are essential to the transduction process, and not simply part of the accessory structures surrounding a sensory nerve terminal.

Chambers and her colleagues (1972) investigated SA II responses in cats. By plotting the receptive field for a fibre they were able to localize the sensory terminal and then make a histological preparation of that piece of skin. They found that the SA II fibres terminated in the Ruffini endings.

The sensations corresponding to excitation of single sensory fibres in human subjects were determined by Ochoa & Torebjörk (1983). They used needle recording electrodes to measure the characteristics of single sensory units, and then applied electrical stimuli via the same electrodes; their volunteers said what sensations these stimuli aroused and where they felt them. The sensations following electrical stimulation seemed to come from particular areas of the skin, and the positions of these projected fields agreed closely with the receptive fields as determined by tactile stimulation. Stimulation of FA I units, presumably from Meissner corpuscles, produced sensations of tapping or (at higher frequencies) flutter or vibration. Repetitive stimulation of FA II units, from Pacinian corpuscles, produced sensations of vibration or tickling; no sensations were felt if the stimulation frequency was too low. Stimulation of SA I units, putatively connected to Merkel receptors, produced sensations of sustained pressure, in marked contrast to the intermittent sensations produced from FA I and FA II units. Stimulation of SA II units, probably from Ruffini endings, produced no conscious sensations.

Nematode mechanoreceptor genes

There are well over 10 000 species of nematodes or roundworms, found in all sorts of different habitats. They are cylindrical in shape and possess an unusually thick outer cuticle. They are remarkable for having fixed numbers of cells, so that they have an unusually determinate pattern of development.

The small soil nematode *Caenorhabditis elegans* has been increasingly used in recent years as a genetic model organism (Brenner, 1974; Hodgkin *et al.*, 1995). The animal is about a millimetre long and can readily be cultured in Petri dishes on a diet of bacteria. Its genome is about one-thirtieth

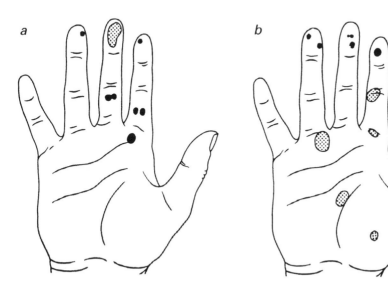

Figure 14.28. Some receptive fields of touch receptor afferents in the human hand. Tungsten needle electrodes were used to record from individual nerve fibres in the median nerve in the upper arm. Receptive fields were mapped by stimulation with a bristle (a von Frey hair) that delivered a stimulus four to five times the force required to elicit a single action potential. Adaptation rates were determined in response to pressure with a blunt probe or skin stretching with forceps. *a* shows rapidly adapting units: FA I (or RA) units (black) with small receptive fields, and an FA II (PC) unit with a large receptive field (dotted). *b* shows slowly adapting units: SA I (black) and SA II (dotted) units with small and large receptive fields, respectively. (From Johansson, 1978.)

the size of the human genome, probably contains about 13 100 genes, and is likely to be fully sequenced by the year 1999. Most populations consist of self-fertilizing hermaphrodites, but cross-fertilization can occur with the male sexual form. The adult hermaphrodites have 959 somatic nuclei and less than 2000 germ cell nuclei. Serial electron micrograph sections show that there are 302 neurons with 7600 synaptic junctions between them (White *et al.*, 1986).

The neurogenetics of mechanoreception in *C. elegans* was investigated by Chalfie & Sulston (1981). A light touch from an eyelash hair in the rear half of the animal would make it move forward, whereas a touch in the front half would produce a backward movement. The responses were mediated by six neurons, three in each half of the animal (fig. 14.29). Destruction of these neurons by laser microsurgery of their nuclei removed the touch sensitivity. The touch neurons have processes running longitudinally and close to the cuticle. They contain microtubules that are unusual in being formed from fifteen protofilaments; most microtubules have only thirteen protofilaments, but those in all other cells of *Caenorhabditis* have eleven. The processes are in contact on their outer sides with a patch of extracellular material called the mantle. The touch neurons seem to be firmly attached to the cuticle by fibrous connections passing through the thin epidermis; this presumably ensures that mechanical stimuli applied to the epidermis are transferred to the touch neurons.

Mutations can be induced in *C. elegans* by appropriate chemical or radiation treatment, and the worms can readily be screened for insensitivity to touch by their lack of response to stroking with an eyelash hair (a sharp prod with a fine wire enables non-motile mutants to be eliminated). Over 400 such mutants have been identified, involving eighteen different genes (Chalfie & Au, 1989). The functions of many of these *mec* (mechanosensory abnormal)

genes are now becoming evident, giving us a remarkable view into how a simple mechanoreceptor may be put together (Chalfie, 1993, 1995; Hamill & McBride, 1996; Herman, 1996; Tavernarakis & Driscoll, 1997).

It is clear that some of the *mec* genes are concerned with the formation of an ion channel. Chalfie & Wolinsky (1990) found that recessive mutations in *mec-4* produced touch insensitivity, but dominant mutations resulted in swelling of the touch neurons, followed by degeneration and death; one interpretation of this is that the recessive mutations produce channel block whereas the dominant mutations produce permanently open channels. Cloning of *mec-4* and *mec-10* showed that they were homologous with *deg-1*, one of the degenerin genes, mutants of which code for proteins that cause cell swelling-induced degeneration of other neurons (Driscoll & Chalfie, 1991; Hong & Driscoll, 1994; Huang & Chalfie, 1994). They also showed sequence similarities with the amiloride-sensitive sodium channel from rat epithelial cells (Cannessa *et al.*, 1994). The molecules appear to have two membrane-crossing α-helical segments, cytoplasmic N- and C-termini, and a possible pore-lining segment, and it seems highly likely that they form subunits of a transmembrane ion channel.

Figure 14.30 shows one way in which the various *mec* gene products may form a mechanosensory channel. *mec-6* also seems to be concerned in channel activity, perhaps as a subunit or in a regulatory role. *mec-7* and *mec-12* code for β-tubulin and α-tubulin, respectively, and are therefore involved in the production of the fifteen-protofilament microtubules. The *mec-2* product shows similarities to stomatin (a red cell transmembrane protein that binds to the cytoskeleton) and may act as a link between the channel and the microtubule.

On the extracellular side of the membrane, *mec-5* (coding for a unique isoform of collagen) and *mec-1* seem to be

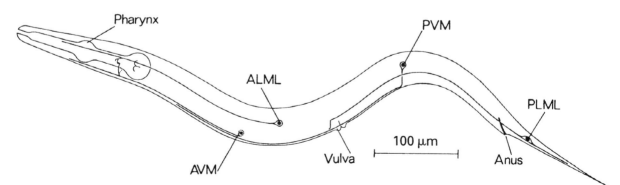

Figure 14.29. Touch sensitive neurons in *Caenorhabditis elegans*. AVM, anterior ventral microtubule cell; ALML, anterior lateral microtubule cell (left); PVM, posterior ventral microtubule cell; PLML, posterior lateral microtubule cell (left). The ALMR and PLMR, anterior and posterior lateral microtubule cells (right) are not shown in the diagram. (From Chalfie & Sulston, 1981.)

involved in the production of the mantle. The *mec-9* product is secreted by the touch neurons, and may perhaps act as a link between the channel and the mantle (Du *et al.*, 1996). The hypothetical model in fig. 14.30 suggests that it is part of the gating mechanism for the channel: movement of the mantle leads to channel opening as gating springs pull blocking particles away from the channel mouth.

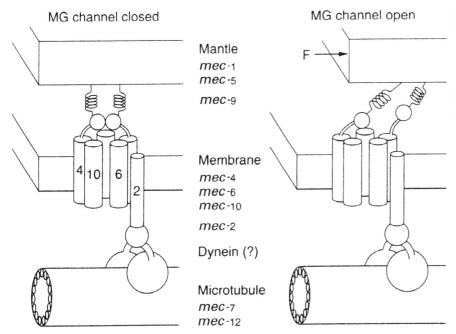

Figure 14.30. The molecular basis of touch sensitivity in *Caenorhabditis elegans*. The diagram shows one way in which the various products of *mec* genes may be put together to form a mechanically gated (MG) channel. Adaptation might be brought about by movement of the channel along the microtubule by means of dynein molecules. Alternative models exist. (From Hamill & McBride, 1996. Reproduced from: A supramolecular complex underlying touch sensitivity. *Trends in Neurosciences* **19**, 258, with permission from Elsevier Trends Journals.)

15
Photoreceptors

In this chapter we examine some aspects of the sensory receptor cells in our most complex sense organ, the eye. The optical apparatus of the eye focuses an image of the visual field on the retina. The retina contains, in humans, about 100 million photoreceptor cells, which are connected in a rather complicated fashion to about a million fibres in the optic nerve. When light falls upon the photoreceptor cells they are excited, and their excitation eventually leads to the production of action potentials in the optic nerve fibres.

The light sensitivity of the eye is caused primarily by the existence in the receptor cells of a *visual pigment*, whose function is to absorb light and, in so doing, to change in some way so as to start the chain of events leading to excitation of the optic nerve fibres. As a reflection of this photochemical change in the pigment, we find that the pigment molecules are bleached by illumination, and have to be regenerated before they regain their photosensitivity.

The range of sensitivity of the eye is enormous: the intensity of the brightest light which we can see is about 10^{10} times that of the dimmest. There are a number of mechanisms which enable this wide range to be perceived, which constitute the phenomena of visual adaptation. *Dark adaptation* is the increase in sensitivity which occurs when we pass from brightly lit to dim surroundings, and *light adaptation* is the reverse of this.

Visible light consists of electromagnetic radiation within a limited range of wavelengths. The visible spectrum, i.e. that range of electromagnetic radiation that can pass through the eye and cause a photochemical change in the visual pigments of the retina, covers wavelengths from about 400 nm (violet) to about 700 nm (deep red); within this range a number of different colours can be seen, as is shown in table 15.1. The colours, of course, are essentially properties of the retina, not of the light (Young, 1802; Wright, 1967).

Light energy is emitted and absorbed in discrete packets known as quanta or photons; there is no such thing as half a quantum. In a photochemical reaction, one molecule of pigment absorbs one quantum of light. The energy of a quantum varies with the wavelength of the light, and is

Table 15.1 *The colours of the visible spectrum. The wavelength ranges given are only an approximation, since the naming of colours is a somewhat subjective procedure and may differ for different observers*

Colour	Wavelength (nm)
Red	Above 620
Orange	590–620
Yellow	570–590
Green	500–570
Blue	440–500
Violet	Below 440

given by $h\nu$, where ν (nu) is the frequency (the velocity of light divided by the wavelength) and h is the Planck constant, 6.63×10^{-34} J s.

The eye
The photoreceptor cells of the eye, and the nerve cells that they innervate, are found in the retina, a thin layer of cells on the inner surface of the eye. The rest of the eye consists of accessory structures concerned, directly or indirectly, with assisting the retina in the perception of the visual field (see Davson, 1990).

The accessory structures
The gross structure of the human eye is shown in fig. 15.1. The outer coat of the eye is protective in function; it is called the sclera. The cornea is the transparent region of the sclera. Inside the sclera is a vascular layer, the choroid, which is usually pigmented. In some animals the choroid contains a reflecting layer, the tapetum lucidum, which serves to increase the effectiveness of the retina in catching light, but must necessarily lead to some loss of definition in the visual image. It is this layer which accounts for the reflecting properties of cats' eyes at night.

The interior of the eye contains the lens, with the

aqueous humour in front of it and the vitreous humour behind it. The curved surface of the cornea acts as a lens, so that a parallel beam of light entering a human lensless eye is brought to a focus at about 31.2 mm behind the front face of the cornea. Since the length of the eye is about 24.4 mm, the cornea, aqueous and vitreous alone cannot produce a sharp image on the retina. The necessary extra focusing power is provided by the lens, which is of higher refractive index than the aqueous and vitreous humours. The power of the lens can be increased by contraction of the ciliary muscle; this process of accommodation allows the images of near objects to be focused on the retina. In many of the lower vertebrates (fishes, amphibia and snakes – see Walls, 1942), accommodation is produced not by altering the power of the lens but by moving it backwards and forwards.

The lens is a remarkable structure: 30% of its weight is protein yet it is highly transparent. It consists of a large number of flattened cells or fibres that are electrically coupled together. The cells have surprisingly sophisticated signalling systems involving the control of intracellular calcium ion concentration. Cataract – the development of opacities in the lens – may occur when these go wrong (Duncan *et al.*, 1994).

In front of the lens is a diaphragm, the iris, the diameter of whose aperture (the pupil) can be varied by contraction of the iris muscles. Thus the activity of the iris can regulate the amount of light entering the eye; it is, in fact, reflexly controlled by the light intensity. The direction in which the eye looks is controlled by the extraocular muscles, of which there are six.

The retina

The nervous structure of the retina is much more complex than that of most other peripheral sense organs. The developmental reason for this is that it is produced by an outgrowth of the brain in the embryo. The photoreceptors themselves are probably derived from flagellated cells forming the ependymal lining of the cavities of the brain, so that they are on the internal side of the mature retina, and light must first pass through the rest of the retina before reaching them.

The structure of the retina, as determined by light microscopy, is shown in fig. 15.2. Using a conventional non-selective stain, a number of layers can be distinguished. The interrelations of the cells in these layers can be observed by using a selective stain such as the Golgi silver impregnation method (see Polyak, 1941). The photoreceptor cells are of two types, described, from their shapes, as *rods* and *cones*. These synapse with small interneurons, the *bipolar cells*,

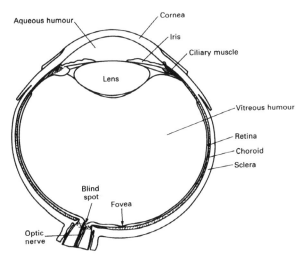

Figure 15.1. Diagrammatic horizontal section of the human right eye.

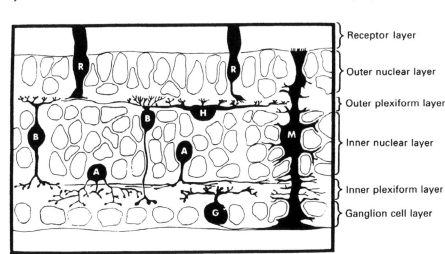

Figure 15.2. Cells of the vertebrate retina. based on observations on *Necturus*. The top of the section in this diagram is next to the pigment cell layer on the outer surface of the retina; light arrives from the bottom. A, amacrine cell; B, bipolar cell: G, ganglion cell; H, horizontal cell; M, glial cell or Müller fibre; R, receptor cell. (From Dowling, 1970.)

Receptor layer
Outer nuclear layer
Outer plexiform layer
Inner nuclear layer
Inner plexiform layer
Ganglion cell layer

which themselves synapse with the *ganglion cells.* The axons of the ganglion cells form the optic nerve, and carry visual information from the retina into the brain. In addition to this sequential information transfer system, there are two lateral systems of neurons: the *horizontal cells,* which form interconnections between the receptor cells, and the *amacrine cells,* which synapse with each other, with the ganglion cells, and with the proximal ends of the bipolars. Filling the spaces between these various neurons are the *Müller fibres,* which are elongated glial cells.

These retinal cells are arranged in the retinal layers as follows. Inside the pigment epithelium is the receptor layer, which consists of the inner and outer segments of the rods and cones. Beneath the receptor layer is the outer nuclear layer, which contains the nuclei of the receptor cells. Between the receptor layer and the outer nuclear layer is the 'external limiting membrane'; electron micrographs show that this is not a true membrane but a level at which the receptor cells are closely attached to each other via thickenings of the cell membrane. The next layer is the outer plexiform layer, which contains the dendrites and synapses of the rods and cones, the bipolars and the horizontal cells. The inner nuclear layer contains the nuclei of the bipolars, horizontal cells and Müller fibres. The inner plexiform layer contains the synapses and dendritic processes of the bipolar cells, amacrine cells and ganglion cells. The ganglion cell layer contains the nuclei of the ganglion cells and, finally, the nervous layer contains the axons of the ganglion cells.

The rods and cones are not distributed evenly over the surface of the retina. In humans (and in a number of other vertebrates, especially birds) there is a more or less central region which is specially modified for high visual acuity, known as the *fovea.* The human fovea contains cones only; it is surrounded by a region in which some rods occur, the *parafovea.* In the human extrafoveal retina the proportion of cones is very small.

There are about 100 million rods and about 6 million cones in the human eye. Since there are only about 1 million ganglion cells and optic nerve fibres, it follows that there must be some considerable convergence of the photoreceptors onto the ganglion cells. The anatomical basis of this convergence can be found in the synaptic contacts of the retinal cells. Figure 15.3 summarizes the results of a number of investigations by electron microscopy (see, for example, Dowling & Dubin, 1984; Dowling, 1987).

The presynaptic membrane of each receptor cell terminal is invaginated to form a pocket into which processes from bipolar and horizontal cells fit. These are called *ribbon synapses* since there is a dense ribbon or bar in the presynaptic cytoplasm, surrounded by an array of synaptic vesicles. Synapses of a more conventional structure are found between horizontal cells and bipolar cells. Some of the terminals of the bipolar cells also show ribbon synapses, again where there are two types of postsynaptic cell: amacrine and ganglion cells in this case. Many of the synapses between bipolar cells and amacrines are *reciprocal:* presynaptic and postsynaptic areas may occur at different places on each of the membranes bounding the synaptic cleft between the same two cells. Amacrine cells also form conventional synapses with ganglion cells and with other amacrine cells. Some of the latter form *serial synapses* in which a terminal may be postsynaptic to one cell and presynaptic to another.

All this structural complexity indicates that there must be

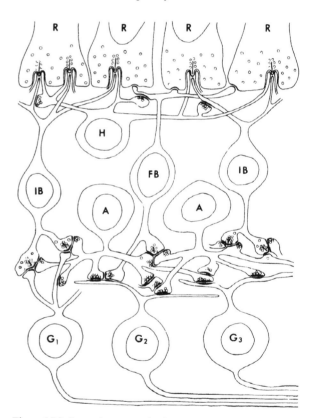

Figure 15.3. Synaptic contacts in the vertebrate retina, based largely on the frog. R, synaptic terminals of the photoreceptor cells. Notice their synaptic ribbons and the invaginations into which processes of the horizontal cells (H) and invaginating bipolars (IB) fit, and their simpler synapses with flat bipolars (FB). A, amacrine cells; notice their reciprocal synapses with bipolars and serial synapses with other amacrines and ganglion cells (G). The ganglion cells may receive input mainly from bipolars (G_1), from both bipolars and amacrines (G_2) or entirely from amacrines (G_3). (From Dowling & Dubin, 1984.)

an enormous amount of interaction between the various cells of the retina. Thus the information that passes up the optic nerve into the brain is already a highly processed version of the visual image.

The structure of rods and cones is shown schematically in fig. 15.4. The outer segment contains a stack of membranous discs (in rods) or infoldings of the cell membrane (in cones) in which the visual pigment molecules are embedded. The inner segment contains numerous mitochondria. It is connected to the outer segment via a thin neck whose structure is very like that of a cilium. Below the inner segment is a region which contains the nucleus, and the cell finally ends at the synaptic terminal.

The duplicity theory

The rods and cones have different functions as photoreceptors. The rods are used for vision at low light intensities and are not involved in colour vision. The cones are used at higher light intensities, and for colour vision. Visual acuity is higher for cone vision than for rod vision. These statements constitute the *duplicity theory*.

The duplicity theory was first propounded by Schultze in 1866. It is well known that, in very low light intensities such as occur on a dark night, the fovea is practically blind, and vision depends upon the extrafoveal regions of the retina; colour vision is absent under these conditions. Schultze pointed out that these features tie in with the distribution of the rods: there are no rods in the fovea. He went on to examine the retinae of a variety of different vertebrates, and showed that nocturnal animals tend to have a great preponderance of rods, and diurnal animals have a corresponding preponderance of cones.

The spectral sensitivity of rod (scotopic) vision differs from that of cone (photopic) vision. Scotopic vision is most sensitive to blue-green light and insensitive to red light, whereas photopic vision covers the whole of the visible spectrum, with the greatest sensitivity being in the yellow region. This movement of the region of maximum sensitivity to longer wavelengths, which accompanies the change from scotopic to photopic vision, is known as the 'Purkinje shift'.

A number of psychophysical experiments provide further evidence for the duplicity theory (see Hecht, 1937). A useful experimental method is that known as the 'fixation and flash' technique for determining visual thresholds. Since the threshold varies somewhat over different parts of the visual field, it is usually necessary in such experiments to ensure that a visual stimulus is always applied to the same part of the retina. This is done as follows. The subject 'fixates' on a small red light, so that its image falls on the fovea. The visual stimulus is then presented at some known angle to the line of fixation (fig. 15.5). In determining the threshold, the light intensity of the test field is adjusted so that the subject can just see it (or, when flashes of short duration are used, so that the subject sees the flash in 50% of the times that it is presented).

Using this technique, Pirenne (1944) investigated the threshold of the dark-adapted eye to deep-red and deep-blue

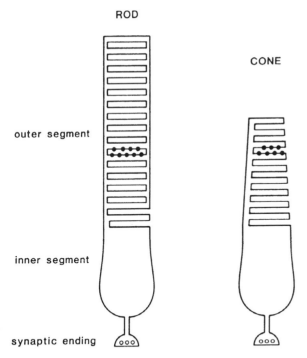

Figure 15.4. Schematic diagrams of vertebrate photoreceptors. The visual pigment molecules are embedded in the membranes of the outer segment. In the rod these membranes are in flattened discs with no contact with the surface membrane except in the most basal and recently formed discs. In the cone they are continuous with the surface membrane. The inner segment contains numerous mitochondria and the cell nucleus (not shown) and terminates at the synaptic ending, from which neurotransmitter vesicles release their contents. (From Baylor, 1987.)

Figure 15.5. The fixation and flash technique. The subject looks at (fixates) a small red light. The test flash is then presented at the desired position in the visual field. (After Pirenne, 1962.)

test fields of small diameter presented at small angles to the fixation line. His results are shown in fig. 15.6. The threshold for red light increases slightly as the field is presented further away from the fovea. The interpretation of this is that the cone density in the retina decreases correspondingly, and the rods are insensitive to red light. With blue light, however, to which the rods *are* sensitive, the threshold falls markedly with increasing eccentricities. This corresponds very neatly with the distribution of the rods, which are absent in the fovea and appear in increasing numbers at eccentricities greater than 0.5°. Furthermore, the blue flash appears to the observer to be blue when seen by the fovea (cone vision), but appears colourless at the threshold in the extrafoveal region (rod vision), as is indicated by the black and white circles in fig. 15.6.

During dark adaptation, the threshold falls progressively with time. Figure 15.7 shows the result of one experiment on this phenomenon, in which the test field was a blue flash placed at an eccentricity of 7° to the fixation line. The curve consists of two branches with a definite 'kink' at their intersection, and the flash appears blue above this kink and colourless below it. The interpretation of these results in terms of the duplicity theory is that the initial section of the curve is due to cone vision, and the later section to rod vision. In accordance with this, the later section is absent if the test field is small in size and viewed by the fovea, or if it is deep-red in colour; in each case we would not expect the rods to be excited.

Rhodopsin

Visual pigments were discovered in the nineteenth century, initially from observations on the retinas of frogs. Boll (1877) found that the reddish colour of the receptor outer segment layer fades on exposure to light. Kühne (1878) called the pigment sehpurpur or visual purple; it later became known as *rhodopsin*. He extracted it from the retina with bile salts, measured the spectral sensitivity of its bleaching by light, and concluded that the absorption of light by the pigment is the primary event in photoreception. Wald (1933, 1934) showed that vitamin A and a carotenoid which he called retinene were involved in visual function, and concluded that the visual pigment is a conjugated protein with retinene as a prosthetic group (see Wald, 1968).

The properties of the visual pigments can be examined in two situations, either *in vitro*, after extraction from the eye or production by gene cloning methods, or *in vivo*, while still in the intact eye. As a half-way house between these two extremes, they can also be examined in isolated retinae or in fragments of the photoreceptor cells. The two methods are complementary to each other: *in vivo* studies must be interpreted in the light of the properties of the pigments, which can be determined much more precisely *in vitro*, and the relevance of *in vitro* studies needs to be tested by reference to what happens in the intact eye.

Methods of investigating extracted pigments
In extracting a visual pigment from the retina of an animal, it is essential that the pigment should not be bleached by exposure to bright light. Hence it is usual to use dark-adapted animals, to dissect out the retina under darkroom conditions, using a dim deep-red light or infrared light and an infrared imaging system, and to carry out the extraction procedures from then on in the dark. Visual pigments are

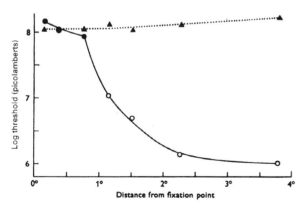

Figure 15.6. Thresholds for deep blue (circles) and deep red (triangles) flashes of 10′ diameter at points near to the fovea. Black symbols show points at which the flash appeared coloured, white symbols where it did not. The lambert is a unit of brightness. (From Pirenne, 1944.)

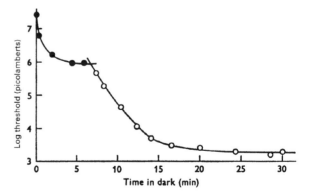

Figure 15.7. The fall in threshold in the dark following a 3 min exposure to bright light, as tested by deep blue light flashes seen at an eccentricity of 7°. For the first five points (black circles) the flash appeared blue or violet in colour; thereafter (white circles) it was apparently colourless. (From Hecht & Schlaer, 1938.)

not normally soluble in water, and have to be extracted from the retina by aqueous solutions of various detergents. Digitonin is the most commonly used extractant; it is not very efficient but gives solutions that are relatively stable (Fong *et al.*, 1982). Much purer solutions are obtained if the pigment is extracted from the outer segments of the rods and cones than if the whole retina is used; in particular, this method avoids contamination by haemoglobin from the retinal blood vessels. The outer segments are obtained by shaking the retina vigorously in 35% to 45% w/v sucrose solution followed by low-speed centrifugation, which leaves the outer segments suspended in the supernatant fluid.

Visual pigments can be characterized by measurement of the amount of light that they absorb at different wavelengths (see Dartnall, 1957; Knowles & Dartnall, 1977). Light from a monochromator is passed through a cell or cuvette containing an extract of the visual pigment, and measured by means of a photocell. The fraction of light absorbed is called the *absorptance* of the solution, J. Thus if I_i is the intensity of light incident upon the solution in the cell and I_t is the intensity of light transmitted by it, then

$$J = (I_i - I_t)/I_i \tag{15.1}$$

The absorptance of the solution varies with the wavelength of the light. If we plot the absorptance against the wavelength, the curve obtained is called the *absorptance spectrum* of the solution. Figure 15.8a shows the absorptance spectra of frog visual pigment solutions at different pigment concentrations, and in fig. 15.8b these curves have been replotted with their maxima made equal to unity.

It is evident in fig. 15.8 that the absorptance spectra become broader as the pigment concentration rises. The reason for this can be explained by an example. Suppose that the amount of light transmitted is halved if we double the concentration of the pigment. Then, at a wavelength where the absorptance is 90% of the maximum, doubling the pigment concentration will raise the absorptance to 95%. But, at a wavelength where the absorptance is only 40% of the maximum, doubling the concentration will raise it to 70%.

This dependence of the absorptance spectrum on the pigment concentration makes it difficult to compare results from solutions either of different strengths or in cells of different thicknesses. This difficulty is obviated by the use of *absorbance spectra*, constructed as follows.

Consider a very thin plane of pigment solution, of thickness dl. Of the light I incident on this plane, a portion dI will be absorbed. The fraction dI/I will be proportional to the thickness of the plane and the concentration of the pigment, c. Thus

$$\frac{dI}{I} = \alpha_\lambda \times c \, dl$$

where α_λ is a constant, the *extinction coefficient*, for any particular pigment at any particular wavelength λ. Integrating this equation between the limits I_i and I_t, we get

$$\alpha_\lambda \, c \, l = \ln I_i - \ln I_t$$

where l is the width of the cell. That is

$$\ln (I_i/I_t) = \alpha_\lambda \, c \, l. \tag{15.2}$$

This relation is known as Beer's law. When converted into decadic logarithms, equation 15.2 gives

$$A_\lambda = \alpha_\lambda \, c \, l / 2.303 \tag{15.3}$$

This quantity A_λ is called the *absorbance* or *optical density* of the solution. If we plot A_λ against λ, we obtain the

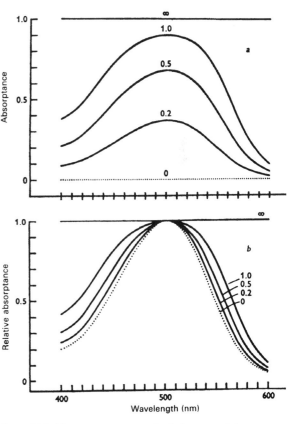

Figure 15.8. Absorptance spectra of solutions containing different concentrations of frog rhodopsin. In *a* the ordinate is the proportion of light absorbed by the pigment (*J* in equation 15.1). In *b* these curves are replotted so that their maxima coincide; notice that the curves become broader as the concentration of pigment increases. (From Dartnall, 1957, and Knowles & Dartnall, 1977.)

absorbance spectrum or *density spectrum* of the solution, as is shown in fig. 15.9*a*. The term 'extinction spectrum' has also been used for the absorbance spectrum.

When absorbance spectra for different concentrations of the same pigment are plotted as fractions of their maxima, all the curves coincide (fig. 15.9*b*). If there are two or more pigments in a solution, then the total absorbance is the sum of the absorbances of the individual pigments. These two properties make the absorbance spectrum a most useful tool for the characterization and comparison of different pigments. The absorptance spectrum, on the other hand, is normally used when questions involving the concentration of a pigment arise. To avoid confusion between absorptance and absorbance, the student may like to remember that the longer word refers to the broader spectrum.

One of the difficulties of measuring the absorbance

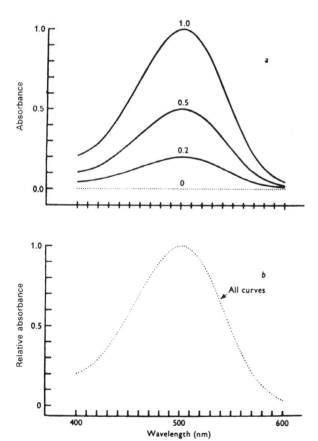

Figure 15.9. Absorbance spectra of solutions containing different concentrations of frog rhodopsin. In *a* the ordinate is the absorbance of the solution (*A* in equation 15.2). In *b* these curves are replotted so that their maxima coincide, with the result that all concentrations now give identical curves. (From Dartnall, 1957, and Knowles & Dartnall, 1977.)

spectra of visual pigments is that it is frequently very difficult to get rid of impurities in the extract, the absorbance spectra of which add to those of the visual pigment. This difficulty can be sidestepped by the measurement of the *difference spectrum*. It is a characteristic feature of visual pigments that they are bleached by exposure to bright lights. So the absorbance spectrum of the products of bleaching will differ from that of the unbleached pigment. The difference between the absorbance spectra before and after bleaching constitutes the difference spectrum. Since any impurities which are not visual pigments will have the same absorbance spectrum after the bleaching as they did before, they do not contribute to the difference spectrum.

Finally, how are we to distinguish between an extract containing only one visual pigment and one containing two or more? The technique for dealing with this problem is known as *partial bleaching*, and was developed by Dartnall (1952). The method consists in successively bleaching the extract with light of different wavelengths and measuring the difference spectrum after each bleach. If these difference spectra are different (when plotted on a percentage scale with their maxima made equal to 100%), then more than one pigment is present.

How this procedure works can best be explained by an example. Suppose we have a retinal extract containing two visual pigments, with maximal absorbances (λ_{max}) at wavelengths of 500 nm and 540 nm. If we illuminate this with red light (λ equal to, say, 640 nm), the 540 pigment will absorb much more than the 500 pigment and will therefore be preferentially bleached (the extinction coefficients at this wavelength are likely to be, respectively, about 8% and less than 1% of their maxima). Hence the difference spectrum after this first bleach will have its λ_{max} at about 540 nm. With a long enough exposure, nearly all the 540 pigment can be bleached. If we now illuminate the pigment with white light, or with light at any wavelength between about 560 and 420 nm, the 500 pigment will be bleached, and the difference spectrum for this second bleach will have its λ_{max} at about 500 nm. This shift in the difference spectrum indicates the presence of at least two visual pigments in the extract.

The visual pigment of human rods: rhodopsin

Rhodopsin (visual purple) is the principal photosensitive pigment of many vertebrate retinas. It is found throughout the human retina except at the fovea, since it is characteristic of the rods and absent from the cones. Thus we would expect rhodopsin to be the visual pigment involved in scotopic vision, or, to put it more precisely, the primary photoreceptor event in human scotopic vision should be the capture of light by rhodopsin. The correspondence between these two phenomena has been investigated many times, beginning with Kühne in 1878. Very clear evidence was produced by Crescitelli & Dartnall (1953); let us look at their results.

Crescitelli & Dartnall extracted human rhodopsin from an eye which had to be removed because of the presence of a ciliary body melanoma. The patient was fully dark-adapted before the operation, which was then performed under deep-red light. The absorbance spectrum of the rhodopsin extract was measured before and after bleaching, with the results shown in fig. 15.10a.

If the extract did not contain any light-absorbing impurities, then the 'unbleached' curve in fig. 15.10a would be the absorbance spectrum of human rhodopsin. But this is unlikely to be so, and indeed the 'unbleached' curve, by comparison with pure extracts of frog and cattle

rhodopsin, looks as though it is derived from a mixture of rhodopsin and some impurities which absorb light increasingly at wavelengths below about 480 nm. Is it possible, then, to make a reasonable calculation as to what the absorbance spectrum of pure human rhodopsin should be? Figure 15.10b shows the difference spectrum of the retinal extract. The curve is effectively identical with that of rat rhodopsin, and very similar in shape to that of frog rhodopsin, but shifted along the wavelength axis so that its λ_{max} is about 5 nm less. This indicates that the absorbance spectrum of human rhodopsin should also be similar to that of frog rhodopsin but, again, with its λ_{max} about 5 nm less.

Now the absorbance spectrum of frog rhodopsin is well established; its λ_{max} is 502 nm, and the full spectrum is shown by the broken line in fig. 15.11. Dartnall (1953) produced a nomogram which could be used to predict the form of the absorbance spectrum of any visual pigment, given the value of λ_{max}: since then this nomogram has been tested against a variety of visual pigments, and its use seems to be

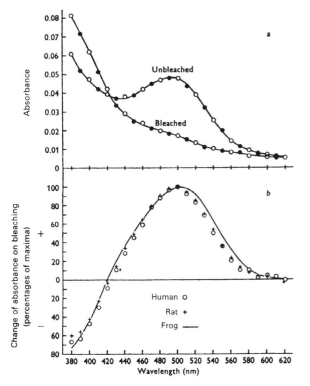

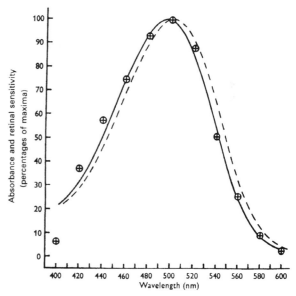

Figure 15.10. Absorbance and difference spectra of an extract containing human rhodopsin. Absorbance spectra of the extract before and after bleaching are shown in a; readings were taken in sequence from short to long wavelengths (white circles) and back again (black circles). The differences between the two curves in a are plotted in b (circles) on a percentage scale to give the difference spectrum; similar spectra for rat and frog rhodopsin are also shown. (From Crescitelli & Dartnall, 1953.)

Figure 15.11. Absorbance spectrum of human rhodopsin and the scotopic visibility function. The dashed curve is the absorbance spectrum of frog rhodopsin, as determined experimentally from very pure solutions. The continuous curve is the absorbance spectrum of human rhodopsin, calculated from the peak wavelength of the difference spectrum seen in fig. 15.10 using Dartnall's nomogram. The points represent the sensitivity of the eye at different wavelengths in very dim light, as determined by Crawford, corrected for the absorption of short wavelengths between the cornea and the retina and expressed as an equal quantal content spectrum. (From Crescitelli & Dartnall, 1953.)

fully justified in most cases. Using the nomogram to determine the absorbance spectrum of a pigment whose λ_{max} was equal to 497 nm, Crescitelli & Dartnall obtained the curve shown by the full line of fig. 15.11. This curve, they suggested, represents the true absorbance spectrum of human rhodopsin.

The next step was to compare the absorbance spectrum with the scotopic sensitivity to light of different wavelengths. The scotopic visibility function was determined by Crawford (1949) from observations on fifty subjects under the age of 30 (this matter of the age of the subjects is rather important, since sensitivity to short wavelengths declines after age 30 because of accumulation of yellow pigment in the lens). The subjects were dark adapted, and looked at a field 20° in diameter which was divided vertically into two halves. One half of the field was illuminated with white light of constant low intensity, and the other half with monochromatic light of different wavelengths, whose intensity could be altered so as to make both halves of the field appear equally bright. The actual intensities were about fifteen times threshold, well within the scotopic range.

Crawford's results were expressed as relative sensitivities, derived from the reciprocals of the light intensities needed to produce a certain sensation of brightness. Light intensity is usually measured in units equivalent to a certain amount of energy per unit time, hence Crawford's results were expressed as an 'equal energy spectrum'. But the absorption of light by a pigment takes place in quanta, whose energy is inversely proportional to the wavelength. Hence, in order for Crawford's results to be compared with the rhodopsin absorbance spectrum, they must be recalculated in terms of an 'equal quantal content spectrum'. A second correction is necessary, to allow for the absorption of light between the cornea and the retina, which is more pronounced at short wavelengths. Values for this preretinal absorption were obtained from the results of Luvigh & McCarthy (1938). The points shown in fig. 15.11 are Crawford's results as plotted by Crescitelli & Dartnall after these two corrections had been made. It is evident that the agreement between the scotopic sensitivity function and the absorption of light by rhodopsin is very precise, and we can therefore conclude that rhodopsin is the visual pigment used in human scotopic vision.

(There is a further correction that is not taken into account in fig. 15.11. The absorption of light by the retinal rhodopsin is given by its absorptance spectrum, not by its absorbance spectrum; the two are only equal when the net percentage of light absorbed is very low. Crescitelli & Dartnall estimated that the retinal absorption was only 3.5%, in which case the correction would be minute. Other estimates, however, suggest that this figure is too low – about 55% absorption seems more likely according to Alpern & Pugh (1974) – so that the curve in fig. 15.11 should be broadened somewhat.)

Rhodopsin photochemistry

It is evident that light must produce a chemical change in rhodopsin and that eventually the rhodopsin must be restored to its former state so as to be able to respond to light again. The processes that are involved in these changes constitute *the visual cycle*, a term introduced by Wald in 1934. It soon became clear that the cycle involved a splitting of rhodopsin into its protein portion and the substance detected by Wald, which he called retinene, and that reconstitution probably proceeded via vitamin A. Figure 15.12 shows the cycle as given by Morton in 1944.

Morton suggested that retinene might be the aldehyde of vitamin A, *retinal*. Evidence that this was so was soon forthcoming (Morton & Goodwin, 1944; Ball *et al.*, 1948). The protein chain in rhodopsin is called *opsin*. Retinal is attached to opsin via the ε-amino group of a lysine residue, by means of a protonated Schiff base link (Morton & Pitt, 1955; Bownds, 1967), i.e.

$$C_{19}H_{27}CH{=}O + H_3N^+.opsin \rightleftharpoons$$
$$C_{19}H_{27}CH{=}NH^+.opsin + H_2O.$$

Retinal can exist in a number of different stereoisomeric forms, some of which are shown in fig. 15.13. The most stable form is *all-trans*, with a straight chain. The 9-*cis* and 13-*cis* forms are common, but it was expected that the 11-*cis* form would be less stable, since steric hindrance between the methyl group on C-13 and the hydrogen atom on C-10 should twist the molecule at the *cis* linkage.

Rhodopsin can be reconstituted by mixing opsin with 11-*cis* retinal (Hubbard & Wald, 1952). Of the other isomers, only 9-*cis* retinal forms a photosensitive pigment (which

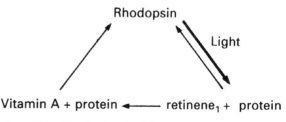

Figure 15.12. The visual cycle of rhodopsin in mammals, as understood in 1944. Since then further details of intermediate stages have been discovered (fig. 15.14 shows some of them), but the essence of the cycle remains the same. Retinene₁ is now known as retinal. (From Morton, 1944.)

has been named 'isorhodopsin'), and its difference spectrum is different from that of natural rhodopsin. The retinal released after exposure of rhodopsin to light, on the other hand, is in the all-*trans* configuration. Rhodopsin can also be broken down in the dark by heating so as to denature the protein part of the molecule, but in this case the retinal liberated is largely in the 11-*cis* form (Hubbard, 1959). From these various results we can conclude that the retinal in rhodopsin is in the 11-*cis* form, but that it is converted into the all-*trans* form by the action of light.

The breakdown of rhodopsin to all-*trans* retinal and opsin after exposure to light takes place in a series of stages (fig. 15.14). The direct action of light appears to be simply to change the retinal from the 11-*cis* to the all-*trans* form (Hubbard & Kropf, 1958); as a result of this a transient intermediate is produced, bathorhodopsin, which can be observed only at temperatures below -140 °C. Use of intense very brief laser pulses shows that bathorhodopsin is itself preceded by another intermediate, photorhodopsin, formed within 200 fs (femtoseconds) (i.e. 2×10^{-13} s) of the light flash (Yoshizawa *et al.*, 1984; Schoenlein *et al.*, 1991).

At temperatures above -140 °C there follows a series of changes that are apparently alterations in the internal structure of the opsin molecule. These conformational changes appear to be fairly minor in nature during the lumirhodopsin and metarhodopsin I stages; the evidence for this view is that metarhodopsin I is produced within microseconds of illumination at room temperatures, that totally dry rhodopsin is converted to metarhodopsin I by the action of light, and the overall change in the absorption spectrum up to this stage is not very great. The next stage, to metarhodopsin II, appears to involve a more drastic alteration in the structure of the molecule, since it takes longer (the half-time for the change is about 1 ms at room temperature), it requires hydrogen ions and it involves a large change in the absorption spectrum which is readily visible as 'bleaching' (see Birge, 1990; Hofmann *et al.*, 1995).

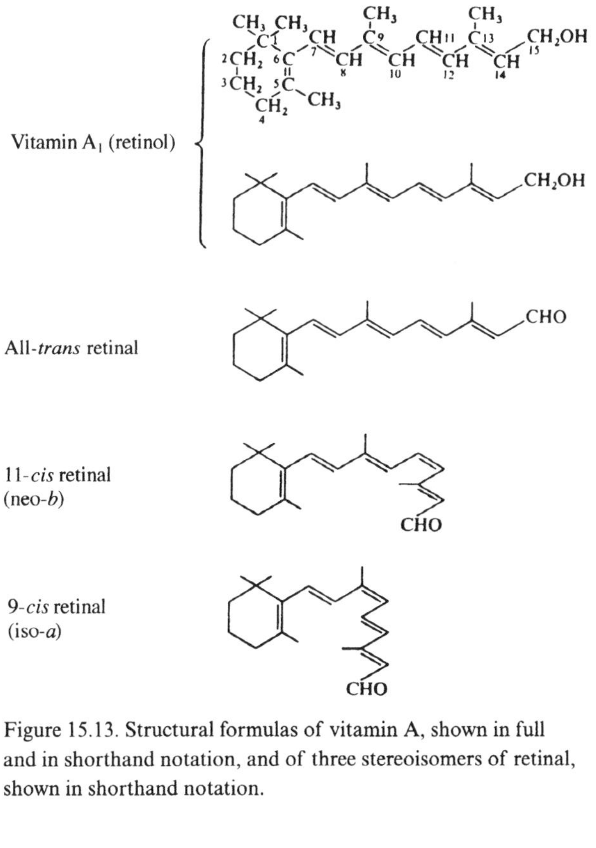

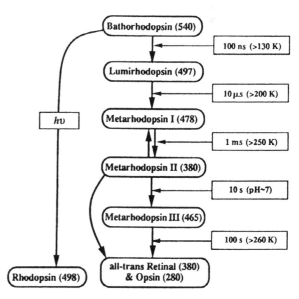

Figure 15.14. The sequence of intermediates in the splitting of retinal from rhodopsin by the action of light. Bathorhodopsin and lumirhodopsin can be detected by using very low temperatures. Absorption of a photon by rhodopsin raises its free energy by about 130 kJ mol^{-1}; thereafter the reactions in the sequence are thermal (they will take place in the dark) and spontaneous, and the relative free energies of the intermediates are suggested by their vertical positions. Figures in brackets after the names of the various intermediates show their $\lambda_{\max}$ values. Approximate conversion times at room temperature are shown, together with the temperatures required for the reactions to take place. Metarhodopsin III is not seen under all conditions. Bathorhodopsin is preceded by a very brief intermediate called photorhodopsin. (From Birge, 1990. Reprinted from *Biochemica et Biophysica Acta* **1016**, Birge, R. R., Nature of the primary photochemical events in rhodopsin and bacteriorhodopsin, p. 298, 1990, with kind permission of Elsevier Science-NL, Sara Burgerhaartstreet 25, 1055 KV Amsterdam, the Netherlands.)

Figure 15.13. Structural formulas of vitamin A, shown in full and in shorthand notation, and of three stereoisomers of retinal, shown in shorthand notation.

Hubbard & Kropf (1958) discovered that metarhodopsin II and the intermediates preceding it in the visual cycle can all absorb light to produce rhodopsin. This process is called photoregeneration; it seems to be more important in invertebrate visual cycles than in vertebrates. The normal regeneration process in vertebrates is enzymic, and takes place largely in the pigment epithelium. A number of specialized binding proteins are concerned with the transport of retinol (vitamin A) and retinal within and between the pigment epithelium and the receptor cells (Bridges, 1976; Bridges *et al.*, 1984).

Different visual pigments

We have already seen that the density spectrum of human rhodopsin is slightly different from that of frog rhodopsin, a circumstance which indicates that the two rhodopsins are different pigments. In fact, there are a large number of different pigments, produced by combination of retinal with different opsins. These various opsins are similar proteins with relatively minor differences in their amino acid sequences.

Different authors have used different nomenclatures for these pigments. Wald described all retinal pigments extracted from rods as 'rhodopsins' and those extracted from cones as 'iodopsins'. In Dartnall's system the pigment is named from its λ_{max} value, with a subscript to indicate that it contains retinal; thus frog rhodopsin is 'visual pigment 502_1', and human rhodopsin is 'visual pigment 497_1'. In human cone pigments, as we shall see, the pigments tend to be described by the colour to which they are most sensitive.

There is another set of pigments in which the chromophore is not retinal but 3-dehydroretinal, which used to be known as retinene$_2$. This is the aldehyde of vitamin A$_2$, which differs from vitamin A$_1$ in possessing a double bond between carbon atoms 3 and 4 (fig. 15.15). Wald described these pigments as 'porphyropsins' when they are extracted from rods; when 3-dehydroretinal is combined with cone

opsin (from iodopsin) an artificial pigment, 'cyanopsin', is produced. In Dartnall's nomenclature, 3-dehydroretinal pigments are indicated by the subscript; thus the porphyropsin of tadpoles is 'visual pigment 523_2'.

The organization of rhodopsin in rods

If the molecules of a light-absorbing substance are not arranged at random, but are oriented in one particular plane or one particular direction, then light waves vibrating in a particular plane are preferentially absorbed. Thus we can determine whether the visual pigment molecules in a photoreceptor cell are so organized by observing the absorption of light polarized in different directions. This was first done by Schmidt (1938), using the outer segments of frog rods. He found that when light was shone along their axis the rods appeared to be red (indicating that light was absorbed) whatever the plane of polarization. However, with illumination from the side, the rods appeared to be red when the plane of polarization was parallel to the transverse plane of the outer segment, but colourless (indicating that no light was absorbed) when it was parallel to the longitudinal axis (fig. 15.16). These results were confirmed and extended by Denton (1959) and by Wald and his colleagues (1962). They imply that the rhodopsin molecules (in particular, the retinal moieties) are oriented in transverse planes across the outer segment, but that there is no preferential direction of orientation within these planes. The most obvious way in which this could be done is by incorporation of the rhodopsin molecules into or onto the membranes of the sacs in the outer segment.

Such an arrangement must increase the efficiency of light absorption by the rod pigment, since light arrives 'end-on' at the rods in the intact eye; any molecules oriented so as to

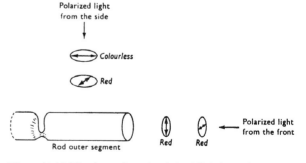

Figure 15.16. The absorption of polarized light by an intact retina. No absorption occurs (so that the rods appear colourless, instead of red) when the transverse components of the light waves are parallel to the longitudinal axes of the rods. This reflects the orientation of the rhodopsin molecules and their retinal chromophores in the disc membranes.

Vitamin A$_2$

All-*trans* 3-dehydroretinal

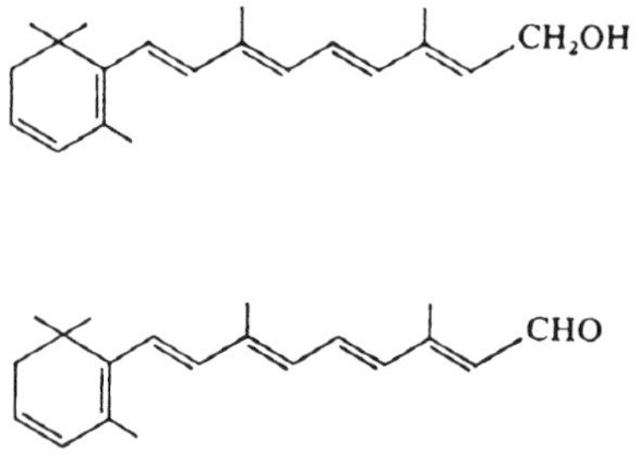

Figure 15.15. Vitamin A$_2$ and all-*trans* 3-dehydroretinal.

absorb light vibrating in a plane parallel to the longitudinal axis would be ineffective in vision. The absorption of unidirectional light by rhodopsin solutions (in which the molecules are, of course, oriented at random) is therefore less than that of rhodopsin in the rods, by a factor of two-thirds.

The molecular structure of rhodopsin

The amino acid sequence of cattle rhodopsin was determined by a number of laboratories, both directly from the protein itself and by molecular cloning techniques (Hargrave *et al.*, 1983; Nathans & Hogness, 1983; Ovchinnikov *et al.*, 1983). Rhodopsin was the first of the G-protein-linked receptor family to be sequenced. It consists of a single chain of 348 amino acid residues. Hydrophobicity analysis suggests that there are seven stretches of hydrophobic residues which form α-helices and cross the membrane from one side to the other (figs. 15.17 and 15.18); they are conventionally labelled A to G. A whole variety of physical techniques have provided confirmatory

evidence for this model (see Nathans, 1992). Cryo-electron microscopy with image reconstruction suggests that four of the helices are almost perpendicular to the plane of the membrane, while the other three are somewhat tilted, as we have seen in fig. 9.8 (Unger & Schertler, 1995).

Proteolytic enzymes can be used to attack the hydrophilic sections of the rhodopsin molecule, and can be used to tell which side of the membrane these sections are on. Thus intact outer segment discs will expose only the cytoplasmic side of the disc membrane to proteolysis, whereas those which have been frozen and thawed will break so as to expose the intradiscal or luminal side as well. The results show that that there are cytoplasmic sections between helices A and B, C and D, and E and F, and that there is an intradiscal section between D and E (Martynov *et al.*, 1983; Mullen & Akhtar, 1983). This means that the N-terminal end of the protein chain is on the intradiscal side of the membrane (this is equivalent to the extracellular side) and the C-terminus remains on the cytoplasmic side.

The retinal chromophore is attached to the lysine residue at position 296, which is in the middle of the seventh transmembrane section, helix G. Thus it sits in a largely hydrophobic environment, which is indeed to be expected from its structure and from our knowledge that retinol (vitamin A) is

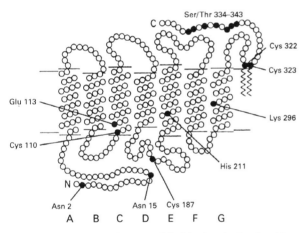

Figure 15.17. Transmembrane model of bovine rhodopsin with some functionally important amino acid residues pointed out. The N-terminal segment at the bottom is on the luminal side of the sac membrane, the C-terminal section above is on the cytoplasmic side; horizontal lines at the left show the thickness of the lipid membrane. The seven transmembrane α-helices are evident. Asparagine-2 and 15 are sites of glycosylation. Lysine-296 forms a Schiff base with the 11-*cis*-retinal, and glutamate-113 provides the negative charge to act as its counterion. Cysteine-110 and 187 form a disulphide bond that helps fix the form of the molecule. Histidine-211 modulates the equilibrium between metarhodopsins I and II. Cysteine-322 and 323 are sites of palmitoylation. Serine-334, 338 and 343, and threonine-335, 336, 340 and 342 can be phosphorylated by rhodopsin kinase. (From Nathans, 1992. Reprinted with permission from *Biochemistry* **31**, 4923–31. Copyright 1992 American Chemical Society.)

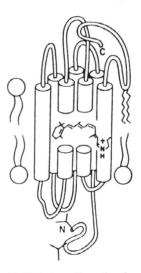

Figure 15.18. A three-dimensional model of the rhodopsin molecule. Two of the seven transmembrane α-helices have been partly cut away to show how the 11-*cis*-retinal chromophore sits in the molecule. The N terminal segment at the bottom is on the luminal side of the sac membrane, the C terminal section above is on the cytoplasmic side. Two lipid molecules at the left indicate the position of the lipid bilayer. (From Nathans, 1992. Reprinted with permission from *Biochemistry* **31**, 4923–31. Copyright 1992, American Chemical Society.)

one of the fat-soluble vitamins. It is surrounded by the seven α-helices, as is shown in fig. 15.18. The glutamate at position 113 on the helix C acts as the negative counterion for the positive charge on the Schiff base link at lysine-296. A disulphide link connects the cysteines at 110 and 187, connecting the intradiscal end of helix C to the loop between D and E. The cysteines at 322 and 323 are attached to palmitoyl fatty acid chains that enter the lipid bilayer. A group of serine and threonine residues near the C-terminus can be phosphorylated by the enzyme rhodopsin kinase.

The absolute threshold

What is the minimum amount of light that can be detected by the fully dark-adapted eye? Because of the quantal nature of light, this question can be rephrased as: what is the minimum number of quanta required for visual excitation? The problem was investigated, in a most elegant set of experiments, by Hecht and his colleagues (1942), whose work we shall now examine.

The principle of this work was to ensure that conditions were optimal for the perception of light by human subjects, to measure the intensity of light at the visual threshold, and finally to calculate the number of quanta absorbed by the visual pigment in the retina under these conditions.

In order to secure optimum conditions for perception, a number of precautions were taken. Subjects were fully dark adapted by remaining in complete darkness for at least half an hour. Using the 'fixation and flash' technique the test flash was positioned so that its image fell 20° to the left (in the left eye) of the fovea, a position which is in the region of maximum sensitivity of the retina. The diameter of the test flash was 10', which is the angle at which the product of area and threshold intensity is least. The product of threshold intensity times length of flash is constant at times up to about 10 ms, but increases with longer flashes; hence flashes of 1 ms duration were used. The wavelength of the light used was 510 nm, which is near the peak of the scotopic sensitivity curve.

The arrangement of the apparatus is shown in fig. 15.19. The eye looks through the artificial pupil P, fixates the red-light point FP, and sees the test field formed by the lens FL and the diaphragm D. The light for this field comes from the lamp L through the neutral filter F and the wedge W (which together control the intensity), and through the double monochromator $M_1 M_2$ (which selects the wavelength). The shutter S opens for 1 ms, and is controlled by the subject. Careful calibrations of the light intensity at P were made, so that the light energy incident on the cornea during a flash could be calculated for different neutral filters and different settings of the wedge W.

For each subject, a series of flashes of different intensities was presented many times, and the frequency of seeing the flash was determined for each intensity. Graphs of frequency of seeing against intensity were then drawn, and the threshold was regarded as that intensity at which the flash could be seen in 60% of the trials. Seven subjects were tested, some of them more than once. Their thresholds, measured as the light energy incident on the cornea, were in the range 2.1×10^{-17} to 5.7×10^{-17} J.

As we have seen, the energy content $h\nu$ of a quantum of light is dependent upon its wavelength. For light with a wavelength of 510 nm, $h\nu$ is 3.89×10^{-19} J. Hence the number of quanta incident on the cornea at the absolute threshold was in the range 54 to 148.

These values do not represent the number of quanta absorbed by the visual pigment, since some light is lost between the cornea and the retina, and not all of the light incident upon the retina is absorbed by it. Measurements by other workers had shown that about 4% of the light incident upon the cornea is reflected from its surface, and that (at a wavelength of 510 nm) about 50% of the light entering the eye is absorbed by the lens and ocular media before reaching the cornea. The main problem here, then, is the measurement of the proportion of the light incident upon the retina which is absorbed by the visual pigment. Hecht and his colleagues solved this problem in a rather ingenious manner. We have seen that the shape of the absorption spectrum of a pigment depends on its concentration (and therefore on the proportion of light absorbed), being broader at higher concentrations (see fig. 15.8*b*). It follows that the proportion of light absorbed by the rhodopsin in the retina can be obtained by comparing the shape of the scotopic luminosity curve with the shapes of the absorption spectra of different concentrations of rhodopsin in solution. The results suggested that between 5% and 20% of the light incident on the retina is absorbed by the visual pigment; for the purposes of calculation, the 20% value was

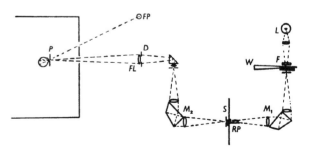

Figure 15.19. Optical system for determining the absolute threshold of human vision. For explanation, see the text. (From Hecht *et al.*, 1942.)

used. (More recent measurements suggest that this figure was still too low; but we shall return to this point later.)

Putting these figures together, we find that, of the light incident at the cornea at a wavelength of 510 nm, 96% enters the eye, 48% reaches the retina, and about 9.6% is absorbed by the visual pigment. Hence, at the threshold of vision, when 54 to 148 quanta are incident on the cornea, 5 to 14 quanta are absorbed by the visual pigment.

This figure is of such importance that it is most desirable to have an independent check on its accuracy. Such a check was obtained by Hecht and his colleagues from measurements of 'frequency-of-seeing' curves and consideration of the statistical properties of light flashes containing small numbers of quanta. Any flash of light of constant *average* quantal content (measured at the retina) will in fact produce fluctuating numbers of quanta. The actual number of quanta absorbed by the retina will also fluctuate.

The relative frequencies of the quantal contents of the flashes will be given by the terms of a Poisson distribution. Thus, if n is the number of quanta which it is necessary for the retina to absorb in order to be able to see a flash, and a is the average quantal content per flash in a series of flashes of 'constant' intensity, then the probability P_n that any one flash will yield exactly the necessary number of quanta is given by

$$P_n = e^{-a} \frac{a^n}{n!} \tag{15.4}$$

From this equation, it is possible to calculate the probabilities of n or more quanta occurring for different values of a and n; the results of this calculation are shown in fig. 15.20.

The next step was to determine 'frequency- of-seeing' curves at flash intensities in the region of the absolute threshold. The method used was identical with that for

determining the absolute threshold in the first instance. Three observers were used, and the results are shown in fig. 15.21. The theoretical curves fit the experimental curves well if the value of n is in the range 5 to 8. (This argument assumes that the variation in frequency-of-seeing is entirely caused by the fluctuations in the stimulus; if there is also some fluctuation in the stimulus:response ratio, then the frequency-of seeing curves would be stretched over a greater range of average quantal contents, so that their slopes would be less, and hence the value of n obtained would be too low. These experiments, therefore, show that the true value of n cannot be less than 5 to 8; it may be a little more.)

The general conclusion of this investigation is that, under optimal conditions, a minimum of about six quanta must be absorbed by the retina if a light source is to be detected. What does this mean in terms of the excitation of

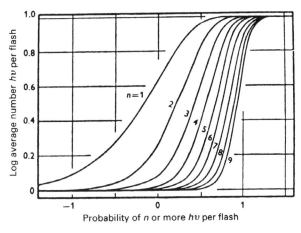

Figure 15.20. Solutions of equation 15.4 for different values of n. (From Hecht *et al.*, 1942.)

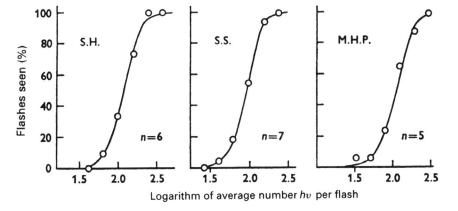

Figure 15.21. The relation between average quantal content (at the cornea) of a flash of light and the frequency with which it was seen, for three different observers (S. H., S. S., M. H. P.). The curves are drawn as in fig. 15.20 for $n = 5$, 6 and 7, but are moved along the abscissa in order to fit the points (this procedure is permissible, since the abscissa measures the number of quanta incident at the cornea, whereas n is the number absorbed by the retinal rods). (From Hecht *et al.*, 1942.)

the retinal cells? We can answer this question by indulging in a little probability theory.

Let us first consider the requirements for excitation of the rods. We can suppose either (1) that the absorption of one quantum is sufficient to excite a rod, or (2) that two or more quanta need to be absorbed by a rod before it is excited. We would eliminate the second hypothesis if we could show that the probability of any of the illuminated rods absorbing the requisite number of quanta from a flash of light at the visual threshold is less than the probability of seeing the flash. Our problem can therefore be stated as follows: if n quanta are randomly distributed over N rods, what is the chance of one or more of these rods receiving two or more quanta? The mean number of quanta absorbed per rod is n/N. If the actual numbers of quanta absorbed by each rod are distributed according to a Poisson distribution, then the probability P_x that any particular rod will absorb x quanta is

$$P_x = e^{-n/N}\frac{(n/N)^x}{x!}$$

Thus, when $x = 0$,

$$P_0 = e^{-n/N}$$

and, when $x = 1$,

$$P_1 = e^{-n/N}\frac{n}{N}$$

Hence the probability that any particular rod will absorb either one or no quanta is

$$P_0 + P_1 = e^{-n/N}\left(1 + \frac{n}{N}\right)$$

Therefore the probability that *every* rod in the group of N rods will absorb either one or no quanta is

$$(P_0 + P_1)^N = e^{-n}\left(1 + \frac{n}{N}\right)^N$$

Hence the probability, Q, that one or more rods will absorb more than one quantum is given by

$$Q = 1 - e^{-n}\left(1 + \frac{n}{N}\right)^N \qquad (15.5)$$

We have now to decide what values of n and N are to be inserted in equation 15.5 for the conditions obtaining at the absolute threshold. The power of the average human eye is about 60 dioptres, which is equivalent to a focal length of 16.67 mm. Thus the image of a 10' circular field is $(16.67 \times \sin 10')$ mm in diameter, i.e. 48.5 μm. Its area is therefore 1.84×10^{-3} mm^2. According to Østerberg (1935), there are

about 150 000 rods per square millimetre in the region of the retina on which the experiments of Hecht *et al.* were performed, so that a 10' field will contain about 273 rods. Let us take a conservative value of N: 250. Using a similarly conservative value of 10 for n, equation 15.5 becomes

$$Q = 1 - e^{-10}(1.04)^{250}$$
$$= 0.178$$

Thus, if the two-quanta-per-rod hypothesis were correct, we would expect only 17.8% of the flashes to be seen when the flash intensity is at the visual threshold. But, in fact, 60% of them are seen, since this is how the visual threshold has been defined. We must conclude that a visual sensation arises if a small number of rods absorb a single quantum each.

A complication

Hecht and his colleagues assumed that 20% of the light incident on the retina is absorbed by the photopigment. More direct measurements have since been made which suggest that a much higher figure should be used. Thus Alpern & Pugh (1974), using retinal densitometry concluded that about 55% of the light incident on the retina is absorbed, and Dobelle *et al.* (1969) using microspectrophotometry concluded that the figure is at least 50% (we examine these techniques later in the chapter). How does this affect Hecht and his colleagues' conclusions? Of the light incident at the cornea, about 24% is absorbed by the visual pigment. So at the threshold of vision, when 54 to 148 quanta are incident on the cornea, 13 to 35 are absorbed by the visual pigment. This makes the conclusion that one quantum suffices to excite a rod less well founded. Thus, in equation 15.5, when n is 13, Q is 0.28, when n is 20, Q is 0.53 and when n is 25, Q is 0.69.

We can get over this problem by looking at larger fields, where the measurements on thresholds are quite conclusive. Thus Stiles (1939), again using the fixation-and-flash method, measured the threshold for detection of a square of side 1.04° at 5° from the fovea, with a light of 510 nm exposed for 63 ms. His results indicate that such a stimulus had to contain 122 quanta (at the cornea) in order for 50% of the flashes to be seen. So about 59 quanta will reach the retina and about 33 of them will be absorbed by the visual pigment. At 5° from the fovea there are about 90 000 rods per square millimetre, and so the number of rods in the illuminated field is just over 8000. So, applying equation 15.5 with $n = 33$ and $N = 8000$, we get

$$Q = 1 - e^{-33}\left(1 + \frac{33}{8000}\right)^{8000}$$

$$= 0.066$$

which is clearly very much less than 0.50, the probability of seeing the flash. Even with $n = 59$ (corresponding to absorption of all the incident quanta by the retina), $Q = 0.195$, which is still a satisfactory figure for the one-quantum-per-rod hypothesis.

An alternative approach

The design of most threshold experiments implies that the subject either sees or does not see a flash of light. He or she has to make a yes or no decision, and 'false positives' (saying 'yes' when there was no flash) are discouraged. One can argue that this may lead to a cautious approach by the subject, so that an ability to respond to very low numbers of quanta would be missed.

Sakitt (1972) approached the problem by employing signal detection theory (Tanner & Swets, 1954; McNicol, 1972). This assumes that the observers can choose their own response criterion; the lower this is, the higher their frequency of seeing and also the higher their false positive rate. Sakitt's experimental design required the observer to rate the flash of light on a 0 to 6 scale. A rating of 0 meant that definitely no flash was seen, one of 1 meant that it was very doubtful that a flash was seen, 2 meant slightly doubtful, 3 a dim light, and so on. The experiments were done with three different subjects.

The results of Sakitt's experiments were very interesting. With a rating of 1, the proportion of correct answers was much higher than would be obtained by random guessing, so clearly a rating of 1 must count as detection of the signal. Frequency of seeing curves for ratings of 1 should then give the number of quanta required for detection. The answers were 1, 2 and 3 rod signals for the three subjects. This means that different subjects may choose different criteria for detection, and that these are low and may be as low as a single rod signal.

These results are completely in agreement with the major conclusion of the experiments by Hecht and his colleagues: one quantum suffices to excite a rod. They are largely in agreement over the second conclusion, that a small number of rods needs to be excited in order to produce a visual sensation, but it is a matter for debate as to just what that small number is.

Phototransduction

Phototransduction is the process whereby the capture of light energy results in an electrical output from the photoreceptor cell. We have seen that the absorption of one photon serves to excite a rod. This means that conformational change in one rhodopsin molecule can produce a response in the whole cell which is sufficient to affect the

activity of succeeding cells in the sequence from photoreceptor to brain. What is the mechanism of this remarkably sensitive process? Let us begin by looking at the electrical activity of the photoreceptors.

The electroretinogram

The electroretinogram is a mass electrical response of the retina recorded with fairly large external electrodes, one placed on the cornea and the other at some indifferent point in the body or (when using excised eyes) behind the optic bulb. An enormous amount of work has been done on this response since its discovery by Holmgren in 1865 (see Granit, 1947, 1962), but we shall here just outline the main features of the phenomenon.

Figure 15.22 shows an electroretinogram recorded by Granit (1933) from a decerebrate cat. The first part of the response is the small, cornea-negative a-wave. This followed by a rapid cornea-positive response, the b-wave, which is itself followed by a much slower response, the c-wave. On the cessation of illumination, there is a small cornea-positive deflection, the d-wave. In vertebrates other than mammals, the d-wave is usually much larger, and is comparable in size with the b-wave.

The components of the electroretinogram can also be detected when records of local activity in the retina are made with a glass micropipette electrode. Using this technique, Brown & Wiesel (1961) determined the amplitude of these various components at different levels in the retina, in an attempt to localize their origin. They found that the a- and c-waves were associated with the receptor layer; there was some evidence that the a-wave arises from activity in the rods and cones, whereas the c-wave arises from the cells of the pigment epithelium. The b-wave was maximal when the electrode was in the inner nuclear region; it appears to be caused by activity in the glial cells (Müller fibre, fig. 15.2) which occur at that level (Miller & Dowling, 1970).

Photoreceptor potentials

Since the early stages of the electroretinogram arise in the receptor cells they can be described as receptor potentials. Using intense brief flashes as the light stimulus, Brown &

Figure 15.22. The electroretinogram of a dark-adapted decerebrate cat, to show the a, b, c and d-waves. The eye was illuminated for the duration of the thick line below the record, on which time intervals of 0.5 s are marked. (From Granit, 1933.)

Murakami (1964) found that the early stages of the locally recorded electroretinogram consisted of two parts (fig. 15.23), an initial *early receptor potential* with very brief latency, followed by a larger *late receptor potential*. The rising phase of the late receptor potential corresponds to the *a*-wave of the electroretinogram.

Cone (1964) showed that the early receptor potential in the rat could be seen in electroretinograms recorded by conventional methods, that its action spectrum was similar to that of the visual pigment, and that its size was linearly proportional to the amount of pigment bleached by the flash. Many different observations suggest that the early receptor potential is not brought about by changes in membrane conductance (see Pak, 1968; Arden, 1969). It is resistant to anoxia and to treatment with a wide variety of different chemical agents that affect most other bioelectric potentials, such as potassium chloride solutions, acids and alkalis, fixatives and so on. It seems probable that the early receptor potential reflects charge displacement in the visual pigment molecules during the conformational changes that follow the absorption of photons. In other words, it is a consequence of light absorption, and not part of the causal sequence of events that leads to visual sensation.

The late receptor potential, on the other hand, appears to be produced by currents flowing through the receptor cell membranes, and is therefore an essential component of retinal signalling. Hence the next step was to record from individual photoreceptors with intracellular electrodes.

Rods and cones are rather small cells and it is technically not easy to make intracellular recordings from them. Tomita (1984) described his frustration at being unable to make successful penetrations of carp cones even with superfine high resistance microelectrodes. Eventually,

however, he met success by jolting the retina upwards towards the electrode in short steps of about 1 μm (Tomita, 1965, 1970; Tomita *et al.*, 1967). The microelectrode in such experiments records the membrane potential of the inner segment of the cone; outer segments are too small to give satisfactory records, but of course electrical changes in the outer segment cell membrane will be recorded, with some attenuation, in the inner segment.

The startling result of Tomita's experiments was that illumination produces hyperpolarization of the cone cell membrane, as is shown in fig. 15.24. Investigations on other

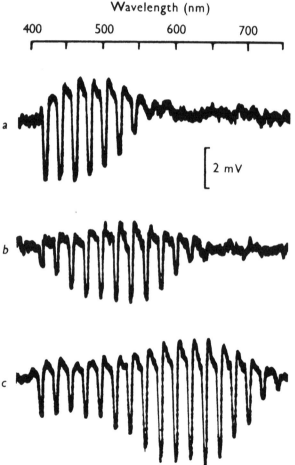

Figure 15.24. Intracellular recordings from cones of the carp, showing the hyperpolarizations produced by light. In each case the cone was illuminated with a series of flashes of monochromatic light in steps of 20 nm across the spectrum. The scale at the top gives the light wavelength. These records show the three types of spectral sensitivity found, with the maximal responses in the blue (*a*), green (*b*) and red (*c*) regions of the spectrum. (From Tomita *et al.*, 1967.)

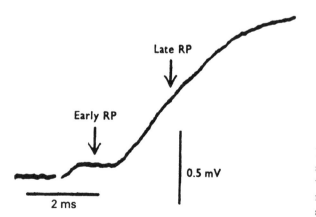

Figure 15.23. The early and late receptor potentials (RP), as recorded by a tungsten microelectrode in the extrafoveal retina of a monkey. (From Brown & Murakami, 1964.)

preparations (gecko and frog rods, mud-puppy and turtle cones, for example) have produced similar results. These hyperpolarizations produced by vertebrate photoreceptors on illumination are in marked contrast to the depolarizing responses of most other sensory cells. Invertebrate photoreceptors depolarize in response to light (Lisman *et al.*, 1992; Yarfitz & Hurley, 1994; Zuker, 1996).

The photoreceptor membrane resistance increases during the hyperpolarization (Toyoda *et al.*, 1969; Baylor & Fuortes 1970), suggesting that illumination reduces the conductance of the plasma membrane to some ion or ions. Figure 15.25 shows some of the evidence for this conclusion: the voltage change produced by brief current pulses increases on illumination.

Receptor coupling

So far we have assumed that single rods and cones are independent units, each one responding only when light is absorbed in its own outer segment. However, it has become clear that the receptors may be to some extent electrically coupled to each other, so that a change in potential in one cell can spread into neighbouring cells. Baylor and his colleagues (1971) measured the membrane potentials of turtle cones illuminated with small or large spots of light of the same intensity. The small spot (4 μm radius) was just sufficient in size to illuminate the whole of the cone, whereas the large one (70 μm radius) would also illuminate many neighbouring cones. They found that the resulting hyperpolarizations were larger with the large spots than with the small ones. A single cone would produce detectable hyperpolarizations in response to illumination of other cones up to about 40 μm away. Furthermore, when current was passed through one cone, potential changes could be recorded in a neighbouring one.

Further evidence was provided by some very interesting experiments by Fain (1975; Fain *et al.*, 1976) on the responses of toad rods to diffuse flashes of dim light. He found that the response of an individual rod to a series of flashes was much less variable than would be predicted by the Poisson distribution. For example in one series, when the flashes were of an intensity to bleach an average of 1.4 rhodopsin molecules per receptor, the Poisson distribution (equation 15.4) predicts that about 25% of the flashes would have bleached no rhodopsin molecules in that rod, 35% would have bleached one molecule, 24% two and 16% three or more. And yet the hyperpolarizations recorded from the rods all fell in the narrow range 440 to 660 μV. Fain concluded that a rod is able to respond even when no rhodopsin molecules are bleached in its own outer segment, because it receives signals from other rods, and that at least 85% of the response recorded from a single rod is generated by rhodopsin molecules activated in other rods.

The structural basis of coupling lies in the existence of gap junctions between the terminals of neighbouring receptors. The functional reasons for the phenomenon of receptor coupling are discussed by Detwiler *et al.* (1980), McNaughton (1990) and Hodgkin (1996). Perhaps it serves to reduce the noise in the system, but this must be at the expense of some spatial resolution.

Current flow in rods

The current flow associated with the late receptor potential and also in the dark was investigated by Hagins and his colleagues (Penn & Hagins, 1969; Hagins *et al.*, 1970; Hagins, 1972). They used slices of rat retina which they were able to keep alive while mounting them on a microscope stage. In order to examine these slices without bleaching their rhodopsin, the preparations were illuminated by infrared

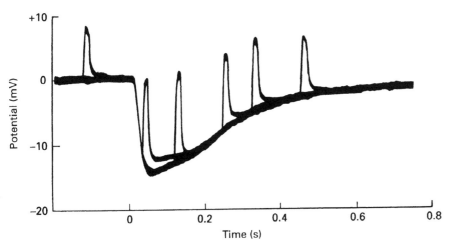

Figure 15.25. Evidence that light increases the membrane resistance in turtle cones. Superimposed tracings show six responses to brief flashes of light at zero time. For each trace a brief pulse of depolarizing current, of the same strength but occurring at different times, was applied through the microelectrode. The voltage changes produced by this pulse are larger during the hyperpolarization following the flash of light. (From Baylor & Fuortes, 1970.)

light and visualized by means of image converters. Microelectrodes could be accurately placed at different depths in the retina so that the extracellular potentials at these depths could be measured, and from these, currents could be calculated.

The results showed that in the dark there is a steady flow of current in the extracellular spaces between the rods towards their tips. It increased steadily as the electrodes traversed the layer of rod outer segments and then began to fall as they entered the layer beneath them (fig. 15.26). This means that there is a steady inflow of current into the rod outer segments, supplied by a corresponding outflow from the inner segments and nuclear regions.

Hagins and his colleagues found that illumination of the retina caused a reduction in the magnitude of the dark current. It is completely eliminated by a flash of light which allows 200 or more photons to be absorbed per rod, and is reduced to half by thirty to fifty photons per rod (Penn & Hagins, 1972). The relation between light absorption and response is linear at levels up to about thirty photons per rod, so we can see that absorption of one photon reduces the dark current by about 0.5 pA per rod. The response to a very brief flash reaches a peak after about 0.1 s and decays to zero within a second, and hence the reduction in dark

current corresponds to about 0.1 to 0.2 pC of electric charge or 1×10^{-18} to 2×10^{-18} moles of univalent ions. Since each mole contains 6×10^{23} ions, we can say that the absorption of one photon by a rat rod causes about a million ions not to flow through the rod membrane.

What is the nature of the dark current? The electroretinogram and the receptor potential with which it begins are dependent upon the presence of sodium ions (Furukawa & Hanawa, 1955; Sillman *et al.*, 1969), and frog rods hyperpolarize and become unresponsive to light in sodium-free solution (Brown & Pinto, 1974). It seems reasonable to suggest, then, that the dark current is caused by an inflow of sodium ions into the outer segments, which is reduced by the action of light. This would account for the hyperpolarization and increase in resistance of rods and cones in the light.

Some nice evidence in favour of this idea was provided by Korenbrot & Cone (1972) from measurements on the osmotic behaviour of rod outer segments. Suspensions of these can be obtained by shaking the retina gently in Ringer solution, and their volume can easily be measured from photomicrographs. When the outer segments were put into a strong solution of potassium chloride (3.5 times the isotonic concentration) in the dark, their volume shrank to about 70% of its initial value within about 3 s, and then remained constant. This suggests that there is a rapid exit of water from the outer segments in the first 3 s, after which they are in osmotic equilibrium; hence their plasma membranes must be impermeable to potassium ions. Illumination had no effect on this response. When they were put from Ringer into a similar strong solution of sodium chloride in the dark, however, their volume fell equally rapidly to 70% and then began to rise again, reaching 80% after 10 s. This suggests that after the initial rapid exit of water there is a slower entry of sodium ions which allows water to flow back into the outer segments. Illumination prevents the later increase in volume, so it must make the plasma membrane impermeable to sodium ions. All these results are in delightful agreement with expectation.

If sodium ions are continually flowing into the outer segment in the dark, there must be some region in the rod cell where they are extruded. Ouabain, an inhibitor of the sodium pump in many cells, greatly reduces the dark current within a minute or so of its application to the retina (Yoshikami & Hagins, 1973; Zuckerman, 1973), but Korenbrot & Cone found that it had no effect on the osmotic behaviour of isolated outer segments. So it looks as though there is a metabolically driven sodium extrusion pump in the plasma membrane of the inner segment. There are many mitochondria in the inner segment, presumably to

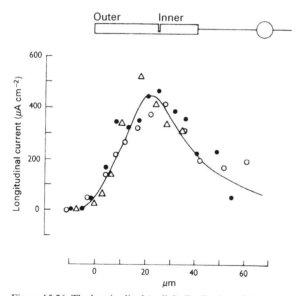

Figure 15.26. The longitudinal (radial) distribution of the dark current in the rat retina. The three symbols show results from three separate measurements. The curve is calculated on the assumption that the whole of the rod outer segment acts as a sink for the current, whose source is the inner segment and nuclear region. The positions of the outer and inner segments of the rods are shown. (From Hagins *et al.*, 1970; redrawn.)

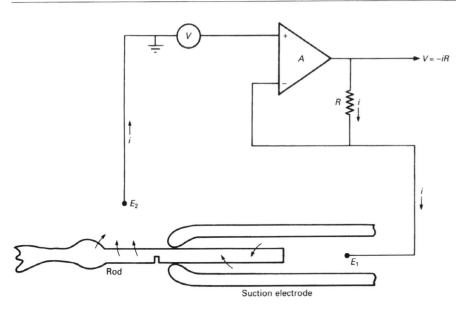

Figure 15.27. Measurement of the current through a rod outer segment using the suction pipette electrode technique. (Based on Baylor *et al.*, 1979*a*.)

supply the ATP to drive the pump. The pump serves to maintain the ionic gradient of sodium ions that provides the driving force for the flow of current into the outer segment through the open channels in its plasma membrane.

The extracellular currents measured by Hagins and his colleagues were produced by the massed activity of a number of rods. Intracellular measurements from individual rods also give responses with some degree of averaging because of the existence of electrical coupling between the cells. It would be highly desirable to look at individual outer segments, with the aim of examining the immediate electrical response to light.

A technique for doing this was developed by Baylor and his colleagues (1979*a*). They used a pipette electrode with a rounded tip just large enough for a toad rod outer segment to be sucked into it. The current flowing through the outer segment could then be measured with a feedback circuit similar to that used in patch clamp experiments, as is shown in fig. 15.27. The retina and the electrode system were mounted on a microscope stage, and manipulations were carried out using infrared illumination with an image converter.

Figure 15.28 shows the response of the outer segment to a brief flash of bright light. There is a current change in the outward direction of about 20 pA; this is maintained for a few seconds, with recovery after about 10 s or so. For technical reasons it was not possible to determine the baseline of zero current without damaging the rod, so the 20 pA change could represent an increase in an outward current, a decrease in an inward current, or perhaps a mixture of

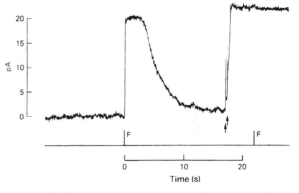

Figure 15.28. Toad rod outer segment currents measured with the suction electrode method. For technical reasons the dark current could not be measured directly, and the current scale therefore shows the difference between current in the dark at the beginning of the trace and that after subsequent treatments. A light flash (F) at time zero produces a saturating response. After recovery from this the rod was broken at the ciliary segment by giving the apparatus a sharp tap, at the arrows. This caused a change in current to a new level after which a second flash of light had no effect. The difference between old and new baselines gives the magnitude of the dark current. The difference between the original baseline and the response to light is sometimes called the photocurrent. (From Baylor *et al.*, 1979*a*.)

both. However, when the outer segment was broken off from the rest of the rod at the ciliary segment by means of a sharp tap (as indicated by the arrows in fig. 15.28), there was abrupt change in the current baseline by about 22 pA, and the response to illumination was abolished. This means that the baseline represents the dark current of the rod, an inward current of about 22 pA. The 'photocurrent' (the change in current on illumination) is thus a reduction in the dark current. Clearly these results agree well with the conclusions drawn by Hagins and his colleagues from extracellular measurements.

The size of the photocurrent is dependent upon the intensity of the flash (fig. 15.29). Bright flashes caused saturation at currents up to 27 pA. The relation between intensity and response was well fitted by the Michaelis equation:

$$r/r_{max} = i/(i + i_0)$$

where r is the amplitude of the response, r_{max} the value of r at saturation, i is the intensity of the flash and i_0 the intensity at half-saturation. The flash intensity at half-saturation was about 1.5 photons per square micrometre on average; this corresponds to a few hundred photons per outer segment. It is clear that, with about 3 billion rhodopsin molecules in the outer segment, saturation occurs when only a tiny fraction of them are photoactivated.

These experiments also provided information about the response of the outer segment to the absorption of single photons (Baylor *et al.*, 1979*b*). Figure 15.30 shows the current flow through a rod outer segment in response to a series of flashes of light at low intensity. Clearly some flashes fail to produce a response while others are followed by small discrete current changes of about 1 pA, with occasional larger responses. This suggests that the responses reflect the absorption of zero, one or more photons. The overall distribution of response sizes was well fitted by a Poisson distribution with some added variance in the amplitude of the quantal responses.

The suction pipette method was used by Hodgkin *et al.* (1984) to investigate the ionic basis of the dark current. They used isolated rods with either the outer or the inner segment sucked into the pipette, and looked at the effects of changing the external solution surrounding the part of the

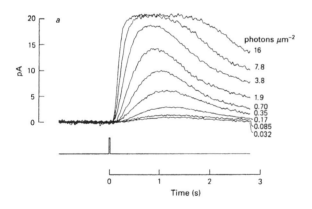

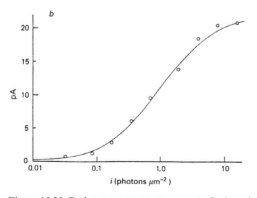

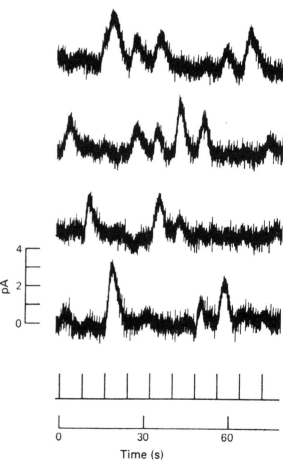

Figure 15.29. Rod outer segment responses to flashes of increasing intensity, measured with the suction electrode method. (From Baylor *et al.*, 1979*a*.)

Figure 15.30. Quantal responses in the toad rod outer segment. The record shows responses to a series of forty consecutive dim flashes of light at 8 s intervals. (From Baylor *et al.*, 1979*b*.)

rod protruding from it. Figure 15.31 shows the results of one of their experiments. Reduction of the sodium concentration surrounding the outer segment produced a corresponding reduction in the dark current, but a similar change for the inner segment was without effect. This confirms the view that the dark current is essentially a flow of sodium ions into the outer segment.

Some other interesting features emerged from these experiments. The selectivity of the light-sensitive channels is not very great: the permeability coefficient for potassium is about 0.8 that for sodium, and calcium and magnesium ions can also pass through them. The external calcium concentration had a marked effect on the dark current. A typical rod had a dark current of 18 pA in Ringer solution with a calcium ion concentration of 1 mM. This fell to 3 pA when the concentration was 10 mM and increased to about 300 pA when it was 1 μM or less. Substitution of lithium for sodium did not affect the dark current, but it made the recovery from a saturating flash much slower. The explanation suggested for this effect is that there is a sodium–calcium exchange system in the outer segment plasma membrane, which acts to remove calcium entering through the light-sensitive channels; after a flash this reduces the internal calcium concentration, and this in some way promotes the normal recovery process. But the exchange process will not work with external lithium in place of sodium, and so the recovery process is delayed (Yau & Nakatani, 1984*a*,*b*; Hodgkin *et al.*, 1985).

The sodium–calcium exchanger is electrogenic: positive charge is transferred into the cell as calcium ions are extruded, so that the exchanger activity can be measured as a transmembrane current (Yau & Nakatani, 1984*b*). Initially it was thought that three sodium ions were transported inwards for each calcium ion transported out, but further work showed that potassium ions were necessary for the calcium movement, so that four sodium ions pass inwards while one calcium ion and one potassium ion move outwards (Cervetto *et al.*, 1989). The sodium and potassium ions moving down their electrochemical gradients provide the energy to move the calcium ions up theirs.

Internal messenger hypotheses

The results in the previous section show that photoisomerization of one rhodopsin molecule causes a reduction in the dark current of about 0.5 pA in rat rods or about 1 pA in toad rods. In other words, change in one rhodopsin molecule

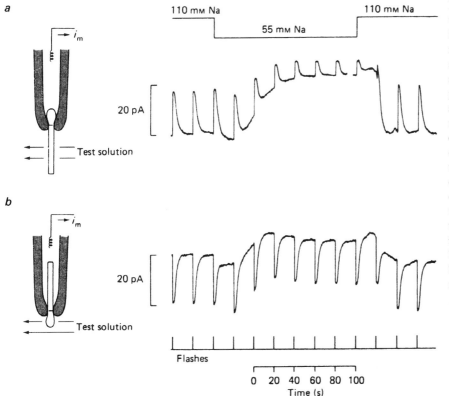

Figure 15.31. Demonstration that there is an inward flow of sodium ions across the outer segment membrane in the dark, using the suction electrode method. A test solution flows past the end of the pipette allowing rapid change of solutions. *a* shows the effect of reducing the external sodium ion concentration by replacing half of it with choline. The response to saturating flashes of light decreases, indicating that the dark current has decreased. In *b* the rod inner segment is exposed to reduced sodium, but here there is no reduction in dark current. In each record there is some baseline drift for technical reasons. (From Hodgkin *et al.*, 1984.)

can stop the flow of several million ions across the rod outer segment plasma membrane. The rhodopsin molecules are embedded in the disc membranes of the rod, which are not continuous with the plasma membrane. Consideration of this situation led to the suggestion that some internal messenger must carry information between the rhodopsin molecules and the plasma membrane. Such a system would permit amplification of the signal: one rhodopsin molecule might trigger the release of many messenger molecules, resulting in the closure of numbers of light-sensitive channels. (We adopt here the common usage of the term 'light-sensitive channels' to indicate channels in the outer segment plasma membrane which close on illumination. They are not themselves directly sensitive to light, of course: it is the rhodopsin molecules which absorb the light energy.)

The internal messenger hypothesis was stated in rather general terms by Fuortes & Hodgkin (1964), as part of a model system to describe the response of *Limulus* photoreceptors to light. Applications to vertebrate photoreceptors were made by Baylor & Fuortes (1970), Hagins (1972) and Cone (1973), among others.

Two alternative hypotheses about the nature of the internal messenger flourished in the 1970s and early 1980s. The first of these was the calcium hypothesis, which postulated that light produced an internal release of calcium ions and that these acted to close the light-sensitive channels in the outer segment plasma membrane. The second hypothesis invoked cyclic GMP (cyclic guanosine 3′,5′-monophosphate) as the internal messenger. In the mid 1980s the conflict between these two alternatives was resolved in favour of the cyclic GMP hypothesis (see, for example, Lamb, 1986; Pugh & Cobbs, 1986; Stryer, 1986, 1993; McNaughton, 1990; Miller, 1990; Yau, 1994*b*). It is instructive to see how this came about.

The calcium hypothesis was first proposed by Hagins & Yoshikami (Hagins, 1972; Yoshikami & Hagins, 1973; Hagins & Yoshikami, 1974). They found that the dark current is reduced as the external calcium ion concentration is raised, and suggested that this would raise the internal calcium ion concentration and so block the light-sensitive channels. Gold & Korenbrot (1980) used calcium-sensitive electrodes in the extracellular spaces of the retinal photoreceptor layer to show that light produced a rapid efflux of calcium ions from rods; they attributed this to a light-determined increase in intracellular calcium ion concentration.

Hagins (1972) suggested that calcium ions were released from a rod disc when one of its rhodopsin molecules absorbed a photon; at that time it seemed reasonable to suggest that the rhodopsin molecule might actually act as a calcium channel. In cones, he suggested, the photon-induced calcium entry occurred across the infolded plasma membrane where the visual pigment molecules are located.

Later experiments produced results that were not in agreement with the calcium hypothesis. Light causes a decrease, not an increase, in the calcium ion concentration in the outer segment. Evidence for this comes from the measurement of membrane currents in isolated rods (Yau & Nakatani, 1985) and, more directly, from experiments in which the calcium-sensitive protein aequorin is injected into rods (McNaughton *et al.*, 1986). Furthermore, when the internal calcium ion concentration in rod outer segments was reduced by the incorporation of the calcium buffer compound BAPTA, the dark current and the response to light increased (Matthews *et al.*, 1985). And finally, calcium was ineffective in increasing the conductance of inside-out patches of rod outer segment membrane when applied to the cytoplasmic side (Fesenko *et al.*, 1985). This brings us to the alternative hypothesis, that cyclic GMP is the internal messenger (fig. 15.32).

The cyclic GMP cascade

The story begins with the discovery by Bitensky *et al.* (1971) that rod outer segments would produce cyclic AMP from ATP in the dark but not in the light. Then Pannbacker and his colleagues (1972) found that cattle rods contain a phosphodiesterase (PDE) which will hydrolyse cyclic AMP and cyclic GMP. Since the enzyme was rather more active with cyclic GMP, they suggested that the latter could be an intermediate in the phototransduction process. An enzyme for synthesizing cyclic GMP, guanylate cyclase, was found in the outer segments (Pannbacker, 1973). Gorodis *et al.* (1974) showed that light produced a rapid drop in cyclic GMP levels in the retinas of cattle and frogs. In cattle, for example, cyclic GMP was present at the rather high concentration of 64 pmol (mg protein)$^{-1}$ in retinas which had been kept in the dark. This level was reduced by 70% after 15 s of illumination. Injection of cyclic GMP into toad rod outer segments produced depolarization and delayed the response to light (Miller & Nicol, 1979).

These observations served to establish cyclic GMP as a worthy candidate for the internal messenger role. There followed a period of biochemical detective work in which the proteins of the outer segment and their roles in controlling cyclic GMP levels were investigated. The essential conclusion of this work is that there is a cascade in which the photoactivation of one rhodopsin molecule leads to the breakdown of perhaps half a million or so cyclic GMP molecules. Let us look at some of the evidence for this conclusion.

Yee & Liebman (1978) used the pH change which accompanies cyclic GMP hydrolysis to measure its rate. They found that each photoexcited rhodopsin molecule (R*, probably metarhodopsin II) stimulated the hydrolysis of 4 × 10^5 cyclic GMP molecules. But since each PDE molecule would hydrolyse only 800 molecules of cyclic GMP per second, they concluded that there was a multiplier mechanism such that photoactivation of one rhodopsin molecule would lead to activation of about 500 PDE molecules. Woodruff & Bownds (1979) and Cote *et al.* (1984) confirmed the rapidity of the reaction by using rapid-quench

techniques in which the reaction mixture is plunged into perchloric acid a fraction of a second after illumination. Figure 15.33 illustrates one of these experiments, and shows that the light-induced fall in cyclic GMP concentration precedes the reduction of the dark current.

The next step was to show that R* does not activate PDE directly. The presence of a light-activated membrane-dependent GTPase (an enzyme which hydrolyses guanosine triphosphate, GTP, to guanosine diphosphate, GDP) in the rod outer segments was detected by a number of workers (Wheeler *et al.*, 1977; Godchaux & Zimmerman, 1979; Fung & Stryer, 1980). Fung and his colleagues (1981) isolated this protein and named it transducin. They showed that it consists of three subunits, that the α subunit binds GDP, and that photoexcited rhodopsin catalyses the exchange of GTP for the bound GDP, whereupon the α subunit (abbreviated T_α or $G_{T\alpha}$) separates from the βγ subunit. T_α-GTP activates the phosphodiesterase, whereas $T_{\beta\gamma}$ does not. Transducin is the best-known of the family of G proteins that we met previously in chapter 9.

Inactive PDE consists of four subunits, $\alpha\beta\gamma_2$. The α and β subunits are the ones that catalyse the hydrolysis of cyclic GMP, and the γ subunits inhibit this (Hurley & Stryer, 1982; Deterre *et al.*, 1988). T_α-GTP probably inactivates each inhibitory PDE γ-subunit by forming a complex with it (Yamazaki *et al.*, 1983).

There are thus two amplification stages in the cascade shown in fig. 15.32. Each R* molecule activates a large

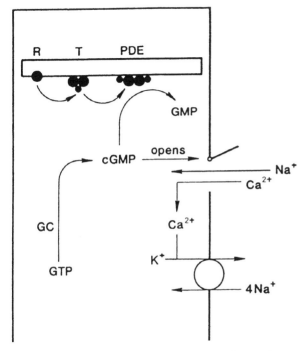

Figure 15.32. The light-sensitive cascade in a vertebrate rod photoreceptor. Levels of cyclic GMP are high in the dark, as GTP is converted to cyclic GMP by the enzyme guanylate cyclase (GC). Hence numbers of cyclic-GMP-gated channels are open, so that there is an inflow of (mainly) sodium ions through them, constituting the dark current. Some calcium ions also enter via the cyclic-GMP-gated channels, and these are removed via a sodium–calcium exchanger. Absorption of a photon by rhodopsin (R) in the disc membrane converts it to the photoactivated form, which then activates many molecules of the G protein transducin (T). Each transducin molecule then activates a phosphodiesterase molecule (PDE), which will then hydrolyse numbers of cyclic GMP molecules. Hence the cyclic GMP concentration falls and many of the cyclic-GMP-gated channels shut, so reducing the dark current. Further details are given in fig. 15.34. (From Baylor, 1996. Reproduced from *Proceedings of the National Academy of Sciences USA* **93**, 561. Copyright 1996 National Academy of Sciences, USA.)

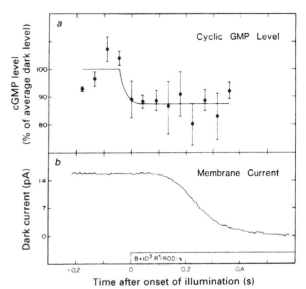

Figure 15.33. Results of a rapid-quench experiment, showing that the fall in cyclic GMP content in the rods (*a*) precedes the fall in membrane current (*b*). (From Cote *et al.*, 1984.)

number of transducin molecules, perhaps 1000 to 2000 s^{-1} at room temperature. Each activated transducin molecule itself activates a PDE molecule. The second amplification stage is then the breakdown of large numbers of cyclic GMP molecules by each activated PDE subunit: this is probably in the region of 2000 s^{-1}. Quantitative modelling of the kinetics of the cascade has been done by Lamb & Pugh (1992), Pugh & Lamb (1993) and Lamb (1994, 1996), with results that fit well with the initial stages of the photoreceptor response.

The idea that rhodopsin might act as an enzyme at the beginning of a multistage amplification process was suggested many years ago by Wald, and he was encouraged in this view by the discovery of the cascade of enzymic actions in blood clotting (Wald, 1965). But a clot of blood is an end-product, whereas a visual response is a transient phenomenon: the photoreceptor cell must be reset so that it can respond again to light a few moments later. So it is necessary that the individual steps in the cyclic GMP cascade should be cyclic rather than one-way, so that the various components can be restored to their original states.

How this is done can be outlined as follows (Stryer, 1991). The activity of activated transducin and PDE molecules is brought to an end as a result of the GTPase activity of transducin. T_α-GTP, bound to PDE_γ, splits GTP and converts to T_α-GDP, and then dissociates from the PDE_γ and recombines with $T_{\beta\gamma}$. The PDE_γ combines with $PDE_{\alpha\beta}$ and inactivates it. Cyclic GMP is resynthesized from GTP by the enzyme guanylate cyclase. There is thus a cycle of reactions involving transducin as it shuttles back and forth between rhodopsin and phosphodiesterase on the cytoplasmic face of the disc cell membrane. This is summarized in fig. 15.34.

Photoactivated rhodopsin molecules have a limited lifetime. Rhodopsin kinase is a 65 kDa protein which catalyses the phosphorylation of R* by ATP. The phosphorylated R* will then combine with another protein, called the 48 kDa protein or arrestin; it is then unable to combine with transducin, so its catalytic activity is brought to an end (Wilden

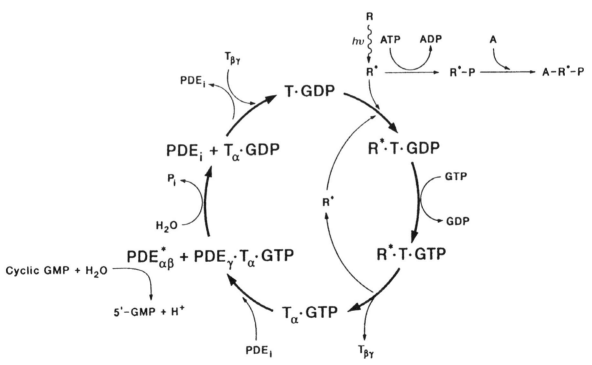

Figure 15.34. The light-activated transducin cycle in vertebrate photoreceptors. Absorption of a photon $h\nu$ by rhodopsin (R) converts it to the photoactivated form R*, which then activates transducin (T) by catalysing the exchange of GDP for GTP. The α subunit of transducin combines with the γ subunit of phosphodiesterase (PDE) so as to allow the PDE α and β subunits to hydrolyse cyclic GMP. The GTPase activity of T_α converts it to T_α-GDP, allowing PDE to revert to the inactive form PDE_i. Finally the T_α-GDP recombines with $T_{\beta\gamma}$. Amplification occurs at two stages: each R* molecule can activate many T molecules, and each $PDE_{\alpha\beta}$ can hydrolyse many cyclic GMP molecules. R* is inactivated by phosphorylation followed by combination with the protein arrestin (A). (From Stryer, 1991.)

et al., 1986). The importance of phosphorylation has been emphasized by studies on transgenic mice in which a proportion of the rhodopsin molecules were truncated by removal of the last fifteen amino acid residues of the C-terminal section they were thus lacking the phosphorylation sites (J. Chen *et al.*, 1995). Rod outer segments from these animals would produce two forms of quantal response in dim light: most were of the normal type (cf. fig. 15.30) which lasted less than a second, but others were of larger amplitude and lasted for up to 30 s. Each of these long responses, absent from normal animals, must have come from the activation of a truncated rhodopsin molecule that could not be inactivated because it had no phosphorylation sites.

Channels opened by cyclic GMP
The existence of the cyclic GMP cascade implies rather strongly that cyclic GMP must play an important role in visual transduction, but it does not tell us just what this role is. Miller & Nicol (1979) were among the first to get to grips with this question; they found that a rapid depolarization followed injection of cyclic GMP into the rod outer segments of intact retinas. Matthews *et al.* (1985) and Cobbs & Pugh (1985) used the suction electrode technique with isolated salamander rods and introduced cyclic GMP via a patch electrode applied to the inner segment. They found that this produced dramatic increases in the dark current and that recovery from light flashes was much delayed.

The situation was greatly clarified following some elegant experiments by Fesenko *et al.* (1985), using the patch clamp technique. Their recording method is shown in fig. 15.35a. A frog retina was treated with trypsin and shaken gently to release isolated rod cells and outer segments. The patch clamp electrode was brought into contact with the rod, brief suction produced a gigaohm seal, and a sharp tap excised a patch of rod surface membrane. The cytoplasmic side of the excised patch was thus exposed to the bathing solution in the chamber. They measured the conductance of the patch by applying voltage changes and observing the currents that these produced.

Figure 15.35b shows the effect of applying cyclic GMP to the inside of the membrane. The currents produced by 10 mV pulses increased from about 1 pA to nearly 4 pA, indicating a four-fold increase in membrane conductance. Fesenko and his colleagues were surprised to find that this effect was independent of the presence of ATP or GTP; this indicates that the cyclic GMP acts directly on the membrane channels, not indirectly by activating a protein kinase. Changing the calcium concentrations on the cytoplasmic side had no effect on the conductance in the absence of cyclic GMP and only a small effect in its pres-

ence. This provided strong evidence against the idea that calcium ions are the internal messenger in phototransduction.

Similar experiments on the effects of cyclic GMP on the conductance of isolated patches were performed by Haynes & Yau (1985) on the outer segment membranes of catfish cones. The results were very similar to those on rod membranes, indicating that cones are also activated by a cyclic GMP cascade. Baylor & Hodgkin (1973) calculated that the mean hyperpolarization produced by a single photon was 25 μV in turtle red- and green-sensitive cones, but 130 μV in the rods. This smaller quantal response in the cones could be due to a smaller gain in the cyclic GMP cascade, i.e. each photoactivated visual pigment molecule may cause the breakdown of fewer cyclic GMP molecules.

Is the membrane conductance which is increased by cyclic GMP in excised patches the same as that which is reduced by light in intact rods? Some pleasing experiments

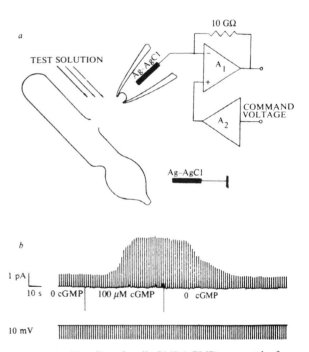

Figure 15.35. The effect of cyclic GMP (cGMP) on a patch of rod outer segment membrane. The experimental arrangement is shown in *a* and the response of the patch in *b*. Constant voltage pulses were applied to the patch (lower trace), and the currents produced by them were measured (upper trace); the currents are thus proportional to the membrane conductance. When cyclic GMP was applied to the cytoplasmic side of the patch the conductance increased four-fold, indicating the presence of channels that are held open by cyclic GMP. (From Fesenko *et al.*, 1985.)

on this question were performed by Yau & Nakatani (1985). They recorded the currents flowing through a rod outer segment drawn partly into a suction electrode; the response to a flash of light is shown in fig. 15.36a. Then the inner segment and the basal part of the outer segment of the rod were knocked off with a probe, leaving an open-ended outer segment whose interior was now accessible to the bath solution. This arrangement allowed the concentrations of ions and small soluble molecules in the outer segment to be controlled, while leaving most of its proteins and membrane systems intact.

Figure 15.36b shows that a large inward current could be induced by cyclic GMP. This current was not affected by light unless GTP was also present in the bath solution, when it was greatly reduced (fig. 15.36c and d). GTP is necessary for the activation of transducin by photoactivated rhodopsin, so there will be no activation of PDE in its absence. Thus the conductance which is reduced by light is the same as that which is increased by cyclic GMP.

Fesenko and his colleagues did not observe individual channel openings and closings in their patch clamp records, indicating that the single-channel conductance was much smaller than in many of the other channels which have been investigated. Bodoia & Detwiler (1985) measured the current noise in patch clamp records from the outer segments of isolated rods. Analysis of the high frequency component of the noise suggested that it was composed of single-channel currents of only 3 to 5 fA. Thus a response

to a single photon, producing a dark current reduction of about 1 pA, would be brought about by the closure of 200 to 300 channels. A 20 pA dark current in a whole outer segment would require about 5000 open channels. Excised patches of rod membrane may show currents much larger than this, suggesting that only 1% to 2% of the total number of cyclic-GMP-gated channels present are open at any time (Yau & Baylor, 1989).

Direct observation of channel opening has been made by using patch clamp methods in the absence of divalent cations (Haynes *et al.*, 1986; Zimmerman & Baylor, 1986). Under these conditions brief single-channel currents in the region of 1.5 pA could be observed, implying a conductance of about 25 pS. These conductance levels are high enough to rule out the possibility that the light-sensitive conductance involves carriers rather than channels. Calcium and magnesium ions may reduce the channel conductance by their greater affinity for a binding site in the channel pore; other ions might not then be able to pass through while this site is occupied, so the average current will be reduced (Zimmerman & Baylor, 1992).

The cyclic-GMP-gated channel has been cloned by the usual molecular techniques. Its subunits show appreciable similarity in sequence and deduced structure with voltage-gated channels such as the *Shaker* potassium channel: in each subunit there appear to be six transmembrane helices including a charged S4 segment, together with an H5 pore-lining region (p. 99), and a C-terminal region that contains

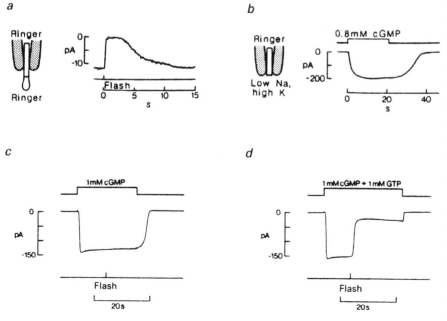

Figure 15.36. Membrane currents in dialysed rod outer segments. In *a* an intact isolated toad rod is held in a suction pipette electrode; the record shows the dark current and the response to a flash of light. The inner segment and basal part of the outer segment were then knocked off, leaving the inside of the rest of the outer segment membrane open to the bath solution. Raising the cyclic GMP (cGMP) concentration from zero to 0.2 mm then produced a large inward current (*b*). The current induced by cyclic GMP was not sensitive to light (*c*) unless GTP (which is necessary for the activation of transducin) was also present (*d*). (From Yau & Nakatani, 1985.)

the cyclic-GMP-binding site (Kaupp *et al.*, 1989; Jan & Jan, 1990, 1992). Indeed, *Shaker* channels can be made to have an ionic selectivity similar to that of the cyclic-GMP-gated channel by removal of just two amino acids, tyrosine-445 and glycine-446, from the H5 region (Heginbotham *et al.*, 1992).

The subunits are of two types, α and β (Körschen *et al.*, 1995; Molday, 1996). The α subunit, the first to be discovered, will form channels on its own when expressed in oocytes. β subunits form channels only when coexpressed with α subunits; they probably then form channels with $\alpha_2\beta_2$ stoichiometry. The N-terminal region of the β subunit is extensive and rich in glutamate residues. The transmembrane region is very similar to that of the α subunit. The C-terminal section of the β subunit contains the cyclic-GMP-binding site as it does in the α subunit, but also, it seems, a calcium–calmodulin binding site.

The steepness of the relation between cyclic GMP concentration and the number of channels that are open suggests that a number of cyclic GMP molecules have to bind to the channel to make it open (Yau & Nakatani, 1985). Goulding *et al.* (1994) concluded that up to four cyclic GMP molecules bind to the channel and that the probability of the channel opening increases markedly as more of them are bound.

Calcium ions and sensitivity regulation

We have seen that there is a sodium–calcium exchange system in the rod plasma membrane, such that four sodium ions are transported into the rod for each calcium and potassium ion extruded. Under normal conditions in the dark, therefore, there must be a balance between the inward leakage of calcium ions through the light-sensitive channels and their extrusion through the sodium–calcium exchange system. On exposure to light, the light-sensitive channels close but the exchange system continues for a time, so there should be a net extrusion of calcium from the rods. This would explain the observation by Gold & Korenbrot (1980) of an increase in extracellular calcium under these conditions, which was wrongly interpreted at the time as suggesting an increase in internal calcium ion concentration. It is also in accordance with the later evidence that the internal free calcium ion concentration decreases in the light (McNaughton *et al.*, 1986; Ratto *et al.*, 1988).

Does this reduction in internal calcium ion concentration in the light have any effect upon the phototransduction process? Torre *et al.* (1986) approached this question by introducing a calcium buffer (BAPTA) into the cell via a patch electrode so as to prevent or reduce the change in calcium concentration. Using the suction electrode technique to measure the photocurrent, they found that rods

with BAPTA produce longer and larger responses, but with an initial rate of rise that is the same as in normal rods. This suggests that the reduction in internal calcium that normally occurs serves to speed up the time course of the response and reduce its sensitivity. These features are characteristic of light adaptation, the process whereby the visual process responds to increased light intensities.

It is now evident that calcium affects a number of the components of visual transduction (fig. 15.37). Guanylyl cyclase, the enzyme that converts GTP to cyclic GMP, is inhibited by calcium ions. Koch & Stryer (1988) showed that the cyclase activity in rod outer segments increases five- to twenty-fold when the calcium ion concentration is lowered from 200 nM to 50 nM. This action is mediated by a calcium-binding protein called guanylyl cyclase-activating protein, CGAP (Dizhoor *et al.*, 1994, 1995).

The activity of phosphodiesterase is increased by increased calcium ion concentrations; this is probably a secondary consequence of an increase in the catalytic activity of rhodopsin (Lagnado & Baylor, 1994), and a reduction in the activity of rhodopsin kinase. The action on rhodopsin kinase is mediated by a calcium-binding protein called recoverin (Klenchin *et al.*, 1995; C. K. Chen *et al.*, 1995).

The cyclic-GMP-gated channels are also affected by the calcium ion concentration. Binding of calcium-calmodulin to the β subunit decreases the affinity of the channels for cyclic GMP, so the channels are more likely to open again when the calcium concentration falls in light adaptation (Hsu & Molday, 1993, 1994).

Photoreceptors in which much of the visual pigment has been bleached are much less sensitive to light, a phenomenon known as bleaching adaptation. The loss of sensitivity is much more than one would expect simply from the reduction in concentration of unbleached pigment. During recovery (dark adaptation) in rods, for example, the sensitivity is proportional to the logarithm of the concentration of regenerated rhodopsin (Dowling, 1960; Rushton, 1965). It is as if the bleached pigment acts like a continuous background light. Bleaching adaptation can be explained in terms of the reduced calcium ion concentration that occurs until the pigment is regenerated. The bleached pigment can apparently continue to activate the phototransduction cascade, although at a much lower gain: the bleached rhodopsin might activate one or two transducin molecules instead of the 500 to 1000 activated by metarhodopsin II (Fain *et al.*, 1996; Matthews *et al.*, 1996).

Colour vision

By 'colour vision', we mean the ability to distinguish lights of different wavelengths. A glance at a paint specification chart will immediately show that we are able to recognize an enormous variety of different colours. Although the spectrum is conventionally divided into six or seven different colours (table 15.1), lights of wavelengths as little as 1 or 2 nm apart can be clearly distinguished from each other. What is the physiological basis of this ability?

Human colour vision is trichromatic. The meaning of this statement, and the main evidence for it, has been explained by Brindley (1970) as follows:

> Given any four lights, whether spectroscopically pure or not, it is always possible to place two of them in one half of a photometric field and two in the other half, or else three in one half and one in the other, and by adjusting the intensities of three of the four lights to make the two halves of the field appear indistinguishable to the eye. This property of human colour discrimination, whereby the adjustment of three independent continuous controls makes possible an exact match, though two are generally insufficient, is known as trichromacy.

Trichromacy was first hinted at by Newton, and became generally accepted during the eighteenth century. But at this time it was thought to be a property of light, and it

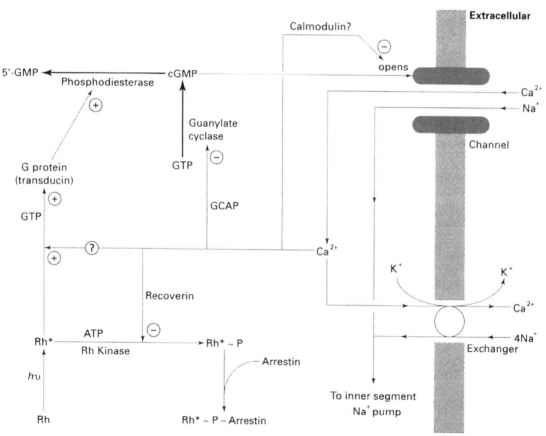

Figure 15.37. Regulatory actions of calcium ions on phototransduction in rods. The rhodopsin–transducin–phosphodiesterase cascade on the disc membranes is shown on the left, the ion movements across the plasma membrane are shown on the right. Illumination causes closure of the cyclic-GMP-gated channels and hence a fall in calcium ion concentration in the outer segment. This in turn produces an increase in the rate of inactivation of photoexcited rhodopsin (Rh*), an increase in the activity of guanylate cyclase, a reduction in the catalytic activity of Rh* (all of which tend to increase the cyclic GMP concentration) and an increase in the affinity of cyclic-GMP-gated channels for cyclic GMP. All these effects reduce the size of the photocurrent and increase its rate of fall, as occurs in light adaptation. GCAP, guanylyl cyclase activating protein. (From Koutalos & Yau, 1996. Reproduced from: Regulation of sensitivity in vertebrate rod photoreceptors by calcium. *Trends in Neurosciences* **19**, 73, with permission from Elsevier Trends Journals.)

needed the genius of Thomas Young (1802) to suggest that trichromacy arose from the properties of the eye: that, to put it in modern terms, colour vision is mediated by three sensory channels which have different spectral sensitivities. Young's suggestion was adopted by Helmholtz (1866) in his great work on physiological optics, so that the idea that colour vision is brought about by three mechanisms with differing spectral sensitivities is frequently known as 'the Young–Helmholtz theory'. There are still many problems to be solved in colour vision (see Mollon & Sharpe, 1983; Dickinson *et al.*, 1997). However, it is quite clear that, at least at the photoreceptor level with which we are concerned here, Thomas Young got the answer right.

We can state the Young–Helmholtz theory in modern terms as follows: colour vision is mediated by three types of cone, each containing a different visual pigment. Two colours can be distinguished if they stimulate at least two of the different cone mechanisms to different extents. Pure yellow, for example, wavelength 580 nm, is absorbed by the pigments in the red-sensitive and the green-sensitive cones. The same sensation is produced by a mixture of green light at 540 nm and red light at 620 nm, provided the intensities of these two colours are adjusted appropriately (see Baylor, 1995, for example). Under these conditions the total number of photoisomerizations per second in each pigment is the same for the red–green mixture as it is for the pure yellow light.

Abnormalities of colour vision, the various types of 'colour blindness', are surprisingly common. People missing one cone pigment are called dichromats: the commoner types, both described as red–green colour-blind, are protanopes (with no red-sensitive pigment) and deuteranopes (no green-sensitive pigment). People with only one cone pigment cannot distinguish colours at all and are called monochromats. People with three cone pigments of which at least one has an unusual absorbance spectrum are known as anomalous trichromats.

Microspectrophotometry

Excellent evidence for the existence of three different cone pigments has been provided by the technique of microspectrophotometry. This is a method for measuring the light-absorbing properties of individual rods and cones. It is not an easy technique: a very small beam of light has to be focused on the receptor cells of isolated retinas. Difficulties arise in eliminating light scattering and in measuring the low light intensities that have to be used in order to avoid excessive bleaching of the pigments (see Liebman, 1972; Knowles & Dartnall, 1977; Bowmaker, 1984).

First results with the technique established a number of

crucial facts. Liebman (1962) identified rhodopsin and some of its photoproducts in frog rods. Marks (1965) found that the difference spectra of goldfish cones fell into three separate groups, suggesting that there is one of three different pigments in each cone. Similar results were obtained for human and monkey retinas by Marks *et al.* (1964) and Brown & Wald (1964). These findings were very exciting, since they fitted precisely the requirements of the Young–Helmholtz theory of colour vision.

Many more measurements have since been made, with improved apparatus. Let us look at some results obtained by Bowmaker & Dartnall (1980) on human rods and cones. A retina was obtained from a patient whose eye had to be removed because of a malignant choroid tumour. Pieces of it were teased apart on a microscope slide and the photoreceptor cells were immobilized under a coverslip. The microspectrophotometer compared the intensity of two monochromatic light beams, one passing through the outer segment of a rod or cone, and the other not. The wavelength of the light was scanned from 700 to 350 nm and back. This double scan checks that no appreciable bleaching has taken place during the determination of the absorbance spectrum.

A typical record, giving the absorbance spectrum of a rod, is shown in fig. 15.38. Absorbance spectra from cones fell into three groups, corresponding to blue-sensitive, green-sensitive and red-sensitive cones. The mean absorbance curves for these four types are shown in fig. 15.39. Similar results were obtained from a further six

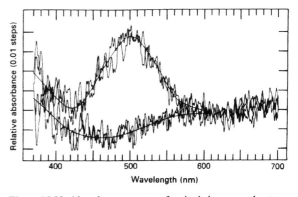

Figure 15.38. Absorbance spectra of a single human rod outer segment, determined by microspectrophotometry. The microspectrophotometer beam was scanned from blue (short wavelengths) to deep red and back again, giving the noisy traces shown. Continuous curves are drawn through the middle of the noise. The upper curve was determined for the dark-adapted rod exposed only to dim deep red light after surgery, and the lower curve shows a similar record after bleaching with white light. (From Bowmaker & Dartnall, 1980.)

subjects (Dartnall *et al.*, 1983). The mean λ_{max} values were 496 nm for the rods, and 419, 531 and 558 nm for the cones. For the red- and green-sensitive cones, there was some evidence that different cones may have slightly different pigments, with λ_{max} values separated by up to 9 nm. The three distinct curves for cones are just what is required to account for the trichromacy of colour vision.

(We digress here briefly to consider nomenclature. The three human cone pigments can be called blue-sensitive, green-sensitive and red-sensitive, or – commonly but much less precisely – blue, green and red. The green-sensitive pigment, however, would look magenta in colour if we could actually see it in quantity, just as rhodopsin, with its peak sensitivity in the blue-green, looks purple in colour. The absorption spectrum of the red-sensitive pigment actually peaks in the yellow-green. An alternative is to describe the pigments as being sensitive to short, medium and long wavelengths: so SW, MW, and LW refer to the blue-, green- and red-sensitive pigments, respectively. Following the imprecise custom, however, we shall hereafter refer to the three pigments and the cones they are found in as blue, green and red.)

Some other techniques
There are other methods of investigating the spectral sensitivity of the cone colour mechanisms. Psychophysical methods, such as colour-matching or increment threshold measurements, have been used for many years (Maxwell, 1860; Stiles, 1978): the conclusions, however, have to be reached somewhat indirectly.

Retinal densitometry is an ingenious method used by Rushton and his colleagues, among others (see Rushton, 1972). The principle of the method is to shine lights of two wavelengths, of which one is absorbed by the pigment and the other not, into the eye and measure the intensity of the light reflected back from the choroid layer behind the retina. Since this light passes through the retina twice, some of it will be absorbed by the visual pigment, and so the absorptive properties of the pigment can be determined from comparison of the intensities of the reflected beams of the two lights.

Campbell & Rushton (1955) measured the concentration of rhodopsin in the human eye under various conditions. They used blue-green light for absorption by rhodopsin, with orange light for the reference beam. They were able to follow changes in rhodopsin concentration during bleaching and dark adaptation, and to show that the density of rhodopsin in different parts of the retina is proportional to the numbers of rods there.

Cone pigments can be investigated from retinal densitometry observations on the fovea, since rods are absent from there. Using a subject with protanopia (a colour vision defect in which there is no red-sensitive pigment), Rushton (1963) detected a foveal green visual pigment with a λ_{max} near to 540 nm. By comparison with the fovea of a subject with normal colour vision, the properties of a second, red, pigment could be inferred.

The electroretinogram has been used as a means of measuring cone pigment spectral sensitivities, using flickering monochromatic light as the stimulus (Jacobs *et al.*, 1985; Neitz *et al.*, 1991). Green and red curves peaking at 530 nm and 561 nm, respectively, were obtained from humans, and observations were also made on various monkeys. Again, however, this is a somewhat indirect method.

The suction electrode technique (fig. 15.27) allows the response of individual photoreceptors to be measured, and

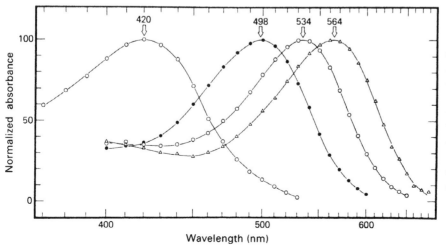

Figure 15.39. Mean absorbance spectra of the four classes of human photoreceptors. The 498 curve is that for rods, the others are for cones: blue-sensitive (420), green-sensitive (534) and red-sensitive (564). Notice that the ratios of the absorbances for the three types of cone are different at different wavelengths; this is the physical basis for colour vision. (From Bowmaker & Dartnall, 1980.)

their sensitivity to colour can determined by adjusting the intensity of monochromatic light at different wavelengths until a standard electrical response is achieved. Figure 15.40 shows the spectral sensitivity of macaque monkey cones, obtained by Baylor and his colleagues (1987) using this method. The curves show peaks at 430 nm, 531 nm and 561 nm, corresponding to the blue, green and red pigments, respectively. Measurements on human cones gave similar results (Schnapf et al., 1987).

The suction electrode technique has an advantage over microspectroscopy in that spectral sensitivities can be measured to much lower levels, so that it is possible to determine where the low edges of the curves lie with greater accuracy. This has an interesting consequence with regard to receptor purity. Referring again to fig. 15.40, at 600 nm the blue cone response is 5 log units less than that of the green and red cones. This means that the blue-sensitive cones must

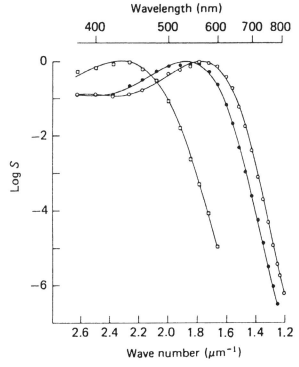

Figure 15.40. Spectral sensitivities of monkey cones, derived from photocurrent measurements on single cones using the suction electrode technique. The sensitivity S (plotted here on a logarithmic scale) is the reciprocal of the flash intensity required to produce a photocurrent of the same size at different wavelengths, scaled so that $\log S = 0$ at λ_{max}. The curves are drawn with $\lambda_{max} = 430$ nm (squares, average of 5 blue cones), 531 nm (black circles, average of twenty green cones) and 561 nm (white circles, sixteen red cones). (From Baylor et al., 1987.)

contain only blue pigment, or, to put it more precisely, less than one molecule in 10^5 is of the green or red type.

Gene cloning has been put to work in determining human cone pigment properties (Merbs & Nathans, 1992a,b). The cDNAs coding for cone pigment genes were expressed after transfection into a kidney cell culture. Cells expressing a particular pigment could be isolated and further propagated, and from them the cone pigment could be extracted and its absorption spectrum determined. We look at some of the results of this method below.

The molecular basis of colour vision

By using DNA encoding for cattle rhodopsin as a hybridization probe, Nathans and his colleagues isolated the genes coding for rhodopsin and the cone pigments from human DNA (Nathans & Hogness, 1984; Nathans et al., 1986a,b, 1992; Nathans, 1994). Genes for the red and green pigments are carried on the X chromosome, hence they show sex-linked inheritance. The blue pigment gene is on chromosome 7 and the rhodopsin gene on chromosome 3.

The deduced amino acid sequence for human rhodopsin is identical with the cattle sequence at 94% of its residues. Sequences for the cone pigments show that the green and red pigments are very similar to each other, with 96% of their residues identical. Comparisons between the blue and green pigments, and between each of them and rhodopsin, show greater differences, with 40% to 44% identity. This suggests that the green and red pigments are derived from a common ancestor relatively recently in evolutionary terms.

The red gene shows a polymorphism at position 180 in the amino acid sequence, and this underlies some variation in the colour-matching of normal men. In a group of young white males, 62% had a red pigment with serine at this position, 38% had alanine in its place. Those with the serine form needed less red light in a red–green mixture to match yellow (Winderickx et al., 1992). Expression of the two pigments in cultured cells allowed their absorption spectra to be measured: λ_{max} for the alanine-180 form was 552 nm whereas that for the serine-180 form was 557 nm (Merbs & Nathans, 1992a).

The red and green genes lie next to each other on the X chromosome, and there may be one, two or even three copies of the green gene. This may have arisen, it is thought, as a result of the high sequence similarity between the two genes: crossing over may have occurred at different points on the two DNA strands, as suggested in fig. 15.41. This could lead to multiple copies of the green gene, and also to the loss of a green gene, as occurs in deuteranopes. Crossing over within misplaced red and green genes can produce hybrid genes with anomalous absorption spectra (Merbs & Nathans, 1992*b*).

One of the first precise descriptions of colour blindness was by John Dalton, the chemist who deduced the atomic nature of matter from the laws of chemical combination, in 1794. He suggested that the deficiencies of his own colour vision arose because the vitreous humour of his eyes was tinted blue, and left instructions that this should be investi-

gated after his death. His doctor found that this was not the case, but the remains of the eyes were preserved as museum specimens by the Manchester Literary and Philosophical Society. One hundred and fifty years later, Hunt and his colleagues (1995) used the techniques of molecular biology to extract and sequence the opsin genes from them. They found that Dalton possessed the red cone pigment but not the green pigment: he was a deuteranope.

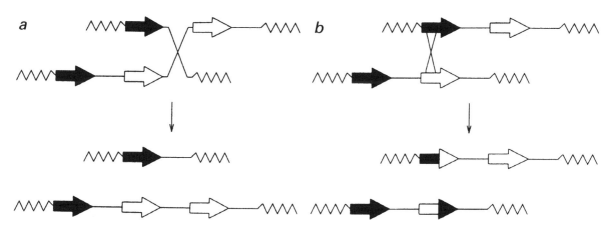

Figure 15.41. Recombination of human red and green genes by crossing over, either between (*a*) or within (*b*) misplaced genes. Red genes are shown as black arrows, green genes as white arrows. The event in *a* leads to deuteranopia (no green pigment) if the person is male and also to someone with two copies of the green gene. The event in *b* leads to two hybrid genes and (in

males) various types of anomalous trichomacy. (From Tovée, 1994, based on Nathans *et al.*, 1986*a,b*. Reproduced from: The molecular genetics and evolution of primate colour vision. *Trends in Neurosciences* **17**, 30, with permission from Elsevier Trends Journals.)

16
Chemoreceptors

It is likely that all living organisms respond to some of the chemicals in their environment. Animals need to detect food sources at a distance, and to test the nature of those sources when they have reached them. Often they use chemical methods for communication between different individuals, and especially for sexual interactions.

The chemical senses are conventionally divided into smell, or olfaction, and taste, or gustation. For humans and other mammals the distinction seems fairly clear: we use our sense of smell to detect air-borne chemicals arising from a distant source, and we use our sense of taste to sample solid or liquid material in our mouths. For fish and other aquatic animals, where odorant substances are necessarily carried in water, the logic of distinguishing between olfaction and taste is less clear. Even in ourselves, the olfactory receptors are essential components of the sensations we receive from food or drink in the mouth, as anyone whose nasal passages are blocked during a cold will know.

In this chapter we concentrate on the transduction mechanisms in the receptors mediating taste and olfaction in mammals, and by way of contrast we also have a quick look at how insects do it.

Taste mechanisms in mammals
Soluble tastant molecules are detected by taste receptor cells on the tongue (see Roper, 1992; McLaughlin & Margolskee, 1994; Lindemann, 1996). The sensitive cells are grouped in *taste buds* (fig. 16.1). Taste receptor cells are long and thin; their apical surfaces have numbers of microvilli that project through the taste pore to contact the tastant molecules in the mouth. Their basal surfaces make synaptic contact with the terminals of the gustatory afferent neurons.

There are about 50 to 100 receptor cells in each taste bud. Basal cells, that do not extend to the taste pore, also occur. There are usually 2000 to 5000 taste buds in the human mouth, on the tongue, palate and epiglottis. On the tongue they are grouped in papillae of various forms.

The epidermal cells of the skin and the intestine have a lifetime of only a few days and are continually being replaced. Does the same apply to taste receptor cells? Beidler & Smallman (1965) used tritiated thymidine as a radioactive marker for newly formed DNA and found that cell division occurs in the epithelial cells surrounding the taste bud. The daughter cells enter the bud as basal stem cells, differentiate and live for ten to fifteen days.

Tastes are usually divided into four basic types: sweet, bitter, sour and salty. This classification derives ultimately from Aristotle (who had some further categories – astringent, pungent and harsh – that involve olfactory or tactile components) and was settled during the nineteenth century. Whether or not these four categories can be regarded as 'primary' tastes comparable to the three primary colours, and how they relate to the activity of individual gustatory nerve fibres is still not entirely clear (see Bartoshuk, 1988). Two further categories, *umami* (the pleasant taste of certain amino acids and peptides, a Japanese word) and the taste of water, should perhaps be included (Lindemann, 1996).

The first electrical recordings from taste-sensitive neurons were made by Pfaffmann (1941, 1984). He found

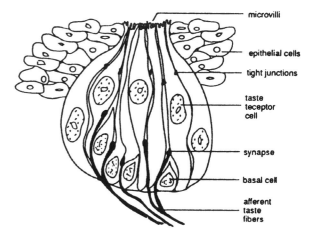

Figure 16.1. Diagram to show the structure of a mammalian taste bud. The group of cells sits in the skin epithelium of the tongue, with the microvilli of the receptor cells in contact with the contents of the mouth. (From Margolskee, 1993. With permission from the Company of Biologists Ltd.)

that individual fibres in the cat chorda tympani nerve usually responded to more than one of the four taste categories. He distinguished three types of fibre: those responsive to acid ('sour') alone, those responsive to acid and also to sodium chloride ('salt'), and those responsive to acid and also to quinine ('bitter'). Later work has tended to produce neural response patterns that are either wider (Erickson, 1963) or narrower (Frank, 1973) than those described by Pfaffmann. The ambiguities in the relations between taste cells and the nerve fibres with which they synapse, and the difficulties caused by sensory adaptation, may account for this somewhat unsatisfactory situation.

Roper (1983) used fine glass microelectrodes to record from the large taste cells of the mud-puppy *Necturus*. He found that they had resting potentials up to −90 mV and would produce all-or-nothing action potentials if stimulated by depolarizing current. The action potentials could be blocked by tetrodotoxin, suggesting that they were dependent on voltage-gated sodium channels. Blockage of potassium channels with tetraethylammonium restored the electrical excitability of the tetrodotoxin-treated cells, but now the action potentials were slower and longer, suggesting that voltage-gated calcium channels were involved.

The patch clamp technique has produced confirmatory evidence for the presence of voltage-gated sodium and calcium channels in taste cells of the rat (Béhé *et al.*, 1990; Avenet & Lindemann, 1991). It seems that the taste stimulus produces a depolarization of the apical membrane, and that this results in an action potential that carries depolarization into the basal region of the cell. This allows calcium ions to enter the basal regions of the cell via the voltage-gated calcium channels, thereby triggering the release of neurotransmitter.

How does the transduction process in the apical membrane of the taste cells work? There are a number of different mechanisms, of two main types. Some tastants open or close ion channels in the apical membrane. Others combine with G-protein-coupled receptors. Let us have a look at them.

Transduction via ion channels

Amiloride is a diuretic that blocks the sodium channels of the distal parts of kidney tubules. It also reduces the sensation of *saltiness* when sodium salts are applied to the tongue, and reduces the flow of sodium ions across the tongue epithelium (Schiffman *et al.*, 1983; Heck *et al.*, 1984). Patch clamp experiments show that the response to salt involves the inflow of sodium ions into the taste cell via amiloride-sensitive sodium channels in the apical cell membrane (Avenet & Lindemann, 1988, 1991). Amiloride-sen-

sitive sodium channels from the kidney and elsewhere have subunits with two membrane-crossing segments; there may be four subunits per channel, with $\alpha_2\beta\gamma$ stoichiometry (Cannessa *et al.*, 1994). It seems likely that the entry of sodium ions into the taste cell via these channels produces sufficient depolarization to trigger an action potential and hence release of neurotransmitter (fig. 16.2). A sodium pump in the basolateral membrane extrudes sodium ions from the cell (DeSimone *et al.*, 1981).

Some of the neural response to sodium chloride is unaffected by amiloride (Formaker & Hill, 1988). Currents across the tongue epithelium are dependent on the anion present: thus sodium chloride solutions produce larger current flows than do sodium acetate or gluconate solutions (Ye *et al.*, 1991). These results suggest that there is an alternative (paracellular) pathway for sodium ions, via the tight junctions between taste cells, so that the ions can act directly on the basolateral membranes of the cell.

Acids, which ionize to produce hydrogen ions (protons), taste *sour*. In amphibians hydrogen ions are detected in part by blocking of potassium channels in the apical membrane (see fig. 16.3). This stops potassium ion efflux from the cell,

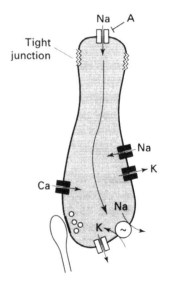

Figure 16.2. Taste receptor cell response to salt. Sodium ions enter through amiloride-sensitive sodium channels in the apical membrane, shown at the top of the diagram. The consequent depolarization activates some of the voltage-gated sodium, potassium and calcium channels in the basolateral membrane leading to further depolarization. This allows calcium ions to enter via the voltage-gated calcium channels, and these act as a trigger to release neurotransmitter, which excites the afferent nerve fibres with which the receptor cell synapses. The sodium is eventually removed by the action of the Na,K-ATPase (~) in the basolateral membrane. (From Lindemann, 1996.)

so producing depolarization. There is also a proton-gated cation channel, and hydrogen ion action via the paracellular pathway (Kinnamon *et al.*, 1988; Okada *et al.*, 1994; DeSimone *et al.*, 1995).

A number of compounds that taste *bitter* may also block potassium channels in the apical membrane; examples are quinine and 4-aminopyridine. Figure 16.3 shows an example of this. Much of the bitter response also occurs via specific receptors.

The barbels of the catfish *Ictalurus* contain receptors sensitive to low concentrations of L-arginine. These seem to be closely associated with cation channels, or may them-

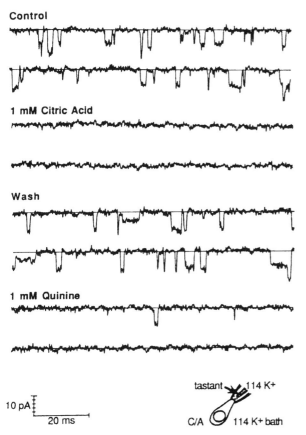

Control

1 mM Citric Acid

Wash

1 mM Quinine

10 pA

20 ms

tastant 114 K+

C/A 114 K+ bath

Figure 16.3. Potassium channels blocked by sour and bitter stimuli. The records show channel activity in a cell-attached patch from the apical membrane of the amphibian *Necturus*, and its block by citric acid and quinine; they were obtained in sequence from top to bottom. Both patch pipette and bath contained 114 mM KCl, holding potential was −60 mV for the upper four traces, then −50 mV for the lower four. (From Kinnamon, 1992. Reproduced from *Sensory Transduction*, 45th Annual Symposium of the Society of General Physiologists, p. 265, 1992, by copyright permission of The Rockefeller University Press.)

selves possess intrinsic ion channels: when barbel membrane vesicles are inserted into lipid bilayers the conductance of the bilayer is readily increased by treatment with L-arginine (Caprio *et al.*, 1993).

Transduction via G-protein-linked receptors

Various lines of evidence imply that some taste mechanisms involve specific membrane receptors that use G proteins to activate second-messenger systems. Cell membranes from tongue epithelium containing taste buds convert ATP to cyclic AMP if stimulated with sucrose or other sweet substances (Striem *et al.*, 1989). Injection of cyclic AMP into taste cells activates a cyclic-AMP-dependent protein kinase, which phosphorylates potassium channels and closes them, thus producing depolarization (Avenet *et al.*, 1988). A specific G protein, called gustducin, is found in taste cells; it is similar in sequence to the transducin of photoreceptor cells (McLaughlin *et al.*, 1992). Mice with no gustducin (produced by genetically engineered 'knockout') showed reduced behavioural and neurophysiological responses to both bitter and sweet compounds (Wong *et al.*, 1996). Other G proteins, including transducin itself, are also found in taste cells (Ruiz-Avila *et al.*, 1995; Kinnamon & Margolskee, 1996).

Transduction of *sweet* taste involves two separate second-messenger pathways, as has been demonstrated in some elegant experiments by Bernhardt and his colleagues (1996). The non-sugar sweeteners saccharin and SC-45647 produced rapid increases in inositol trisphosphate (IP$_3$) concentration in rat tongue epithelium containing taste buds, whereas sucrose had only a slight effect. Imaging with the calcium-sensitive fluorescent dye fura-2 showed that SC-45627 produced marked increases in calcium ion concentration in particular cells of isolated taste buds. Sucrose also induced calcium increases in these cells, but not if the external calcium ion concentration was lowered, suggesting that in this case the calcium enters from the outside. These cells did not respond to the bitter substance denatonium, suggesting that bitter and sweet responses are from different cells.

The probable system for this dual response to sweet tastants is shown in fig. 16.4. Sugars combine with G-protein-linked receptors to activate a cyclic AMP pathway, this closes potassium channels and so depolarizes the basolateral membrane, perhaps with production of action potentials, and thereby opens voltage-gated calcium channels. The inflow of calcium ions then acts as a trigger for the release of transmitter, and this excites the sensory nerve fibre with which the cell synapses. Non-sugar sweeteners also combine with G-protein-linked receptors, but these activate phospholipase C so as to produce IP₃, which releases calcium ions from the endoplasmic reticulum. Both pathways can occur in the same cell.

The response to *bitter* substances involves a number of different transduction mechanisms, probably acting in parallel (fig. 16.5). Some compounds may block potassium channels directly, as mentioned above, but the major action is probably via G-protein-coupled receptors and second-messenger pathways. Some taste receptor cells respond to the bitter substance denatonium by an increase in IP₃ content and calcium ion concentration, brought about by activation of phospholipase C (Spielman *et al.*, 1994; Bernhardt *et al.*, 1996). An alternative pathway involves the activation of phosphodiesterase, which breaks down cyclic AMP to AMP and so opens cation channels that are normally held closed by combination with cyclic nucleotides (Kolesnikov & Margolskee, 1995; Ruiz-Avila *et al.*, 1995; Wong *et al.*, 1996).

The pleasant *umami* taste, as in chicken broth and meat extracts, is largely attributable to L-glutamate, other amino acids, and small peptides. The receptors are probably metabotropic glutamate receptors; mGluR4 has been located in the taste buds of rats (Chaudhari *et al.*, 1996).

There are still many questions to be answered in relation to taste transduction mechanisms. At the time of writing none of the G-protein-linked receptors for sweet and bitter tastes have been cloned, and it is still no more than a reasonable expectation that they are members of the 7TM superfamily. To what extent different receptors occur in the same taste receptor cell is still not clear, and whether individual nerve fibres are linked synaptically to taste receptor cells with different sensitivities has still to be worked out. But we know much more than we did a few years ago, and

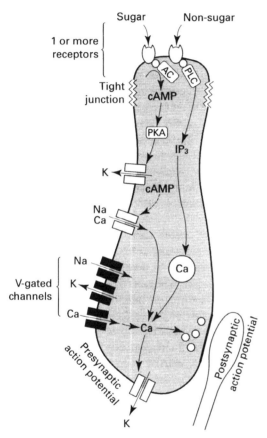

Figure 16.4. Taste receptor cell response to sweet substances. Sugars combine with a G-protein-linked receptor, which in turn activates G protein molecules to activate adenylyl cyclase (AC), so producing the second-messenger cyclic AMP (cAMP). This closes potassium channels in the basolateral membrane, probably by phosphorylation with cyclic-AMP-dependent protein kinase A (PKA), leading to depolarization, opening of voltage-gated calcium channels, calcium inflow and release of neurotransmitter. Non-sugar sweet substances combine with a receptor (possibly the same one as for sugars, possibly a different one) to activate phospholipase C (PLC), leading to production of inositol trisphosphate (IP₃) and hence release of calcium ions from the endoplasmic reticulum. (From Lindemann, 1996.)

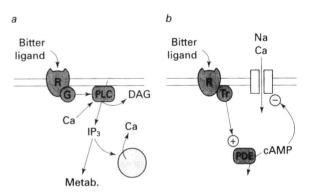

Figure 16.5. Probable mechanisms of bitter taste transduction via G-protein-linked receptors. *a* shows the inositol trisphosphate (IP₃) pathway. The receptor (R) activates a G protein (probably G₁) which activates phopholipase C (PLC), so producing inositol trisphosphate (IP₃) and diacylglycerol (DAG). The IP₃ then combines with intracellular IP₃ receptor channels to release calcium ions from the endoplasmic reticulum. In *b* the receptor activates the G protein transducin (Tr) and/or gustducin so as to activate the enzyme phosphodiesterase (PDE). This then breaks down cyclic AMP (cAMP), so bringing to an end its inhibitory action on cation channels in the plasma membrane. Hence calcium ions enter the cell. In both *a* and *b* the calcium ions act as the trigger for neurotransmitter release. (From Lindemann, 1996.)

we can expect some of the answers to emerge in the near future.

Olfactory transduction in mammals

Most mammals have a good sense of smell: they can detect and discriminate between thousands of odours made up of different air-borne chemicals. The main olfactory organ is the olfactory epithelium in the nasal passages, an area of 2 to 4 cm^2 on each side in humans, containing about 10 million olfactory receptor cells in all. In dogs the patches are 20 to 200 cm^2 in area, with proportionately more receptor cells (Cain, 1988). The epithelium is covered by a thin layer of mucus that protects it from desiccation by the air stream.

Each olfactory receptor cell is a neuron with a peripheral cell body and nucleus, sending its axon to the olfactory bulb at the front of the brain. It sits in the olfactory epithelium surrounded by supporting cells (fig. 16.6). A single dendrite extends to the epithelial surface and into the mucus layer.

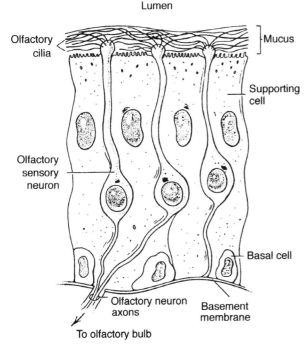

Lumen

Olfactory cilia

Mucus

Supporting cell

Olfactory sensory neuron

Basal cell

Olfactory neuron axons

Basement membrane

To olfactory bulb

Figure 16.6. The olfactory epithelium in mammals. Each olfactory receptor cell is a neuron with a peripheral cell body and nucleus, whose axon runs from the cell body to the terminals in the olfactory bulb. The receptor cell dendrites extend to the surface of the epithelium, and numbers of cilia spread out from them through the mucus layer. Odorants combine with specific receptor molecules on the surface of the cilia. (From Buck & Axel, 1991. Reproduced with permission from *Cell*. Copyright Cell Press.)

The tip of the dendrite is swollen slightly, forming the olfactory knob. Attached to the knob are five to twenty cilia that spread out laterally through the mucus layer. The cilia have the usual $9 + 2$ microtubular structure near the base, but lose the central pair of microtubules more distally; they are not motile.

The olfactory receptor cells are unlike most other neurons in that they have a lifetime of just a few weeks. Radioactive thymidine is taken up by the basal cells in the olfactory epithelium, indicating DNA synthesis and therefore cell division. About a week later the label is found in mature receptor cells, but after a few weeks there are none still labelled (Graziadei, 1973; Graziadei & Graziadei, 1979). The receptor cells may become progressively damaged by injurious chemicals in the air stream, so their continuous replacement may be a mechanism to deal with this (Shepherd, 1994).

Olfactory receptor molecules

The idea that olfactory transduction begins with the binding of odorant molecules to particular receptor molecules on the olfactory epithelium has been held for some time. Perhaps the differences between different smells, it was thought, arose because their constituent odorants combined with different types of receptor. Amoore (1967) produced some evidence in favour of this view when he discovered that a proportion of people with otherwise normal smell had a much higher threshold for the detection of isobutyric acid than do most of us. This, he suggested, implies that they are lacking in one particular receptor which normally binds isobutyric acid. Other examples of specific anosmia, as the condition is known, occur; this implies that the people involved are lacking in other olfactory receptors.

Freeze-etching techniques show that the plasma membranes of the olfactory cilia contain numbers of intramembranous particles at a density of about 1000 μm^{-2}; such particles are absent from the cilia of respiratory epithelia (Kerjaschki & Hörander, 1976). It seems reasonable to suggest that many of these particles are receptor molecules, and that olfactant molecules must bind to them as the first step in olfactory transduction.

By 1990, as we shall see, it seemed very likely that olfactory transduction involves a G-protein-dependent cascade which begins when a stereospecific receptor molecule binds a particular odorant molecule. But very little was known about these putative olfactory receptor proteins. What sort of proteins were they? Were there just a few different types, as implied by Amoore (1967), or many hundreds, as predicted by Lancet (1986)? Answers to these questions were provided by Buck & Axel (1991), in a most elegant and informative example of the use of molecular gene cloning methods to solve physiological problems. Let us have a look at this work.

Buck & Axel began by assuming that the putative receptors were members of the 7TM superfamily of G-protein-linked receptors such as rhodopsin and the β-adrenergic receptors. They expected the receptors to show considerable diversity, and therefore to be encoded by a family of similar but not identical genes. They also expected them to be expressed in the olfactory epithelium but not in other tissues of the body. They began by using the polymerase chain reaction (PCR) to isolate members of the 7TM superfamily from cDNA prepared from the mRNA of rat olfactory epithelium. To do this they used PCR primers – oligonucleotides about thirty bases long – corresponding to parts of the amino acid sequences of transmembrane segments 2 or 7 of a number of different 7TM superfamily members. This led to the isolation of a number of PCR products.

Buck & Axel were particularly interested in looking for multigene products, i.e. for PCR products that consisted of a number of similar DNA species rather than just a single one. They did this in an ingenious fashion by digesting the products with restriction enzymes (enzymes that cut DNA molecules at particular sequence patterns) and measuring the size of the fragments so produced. If there were just one DNA species present, then the molecular weights of all the different fragments should add up to that of the original DNA from which they were derived. But the chains of different DNA species would be cut in different places, so that, with a mixture of DNA species present, the sum of the weights of the fragments would be greater than the average weight of the DNA in the PCR product. In this way they were able to find members of a multigene 7TM family expressed in olfactory epithelium. They could then use these sequences to look for further homologous sequences in other DNA libraries.

The results of this investigation were the sequences of eighteen different members of a large multigene family of 7TM receptors whose expression is restricted to the olfactory epithelium. There is appreciable variation in the amino acid sequences of transmembrane segments 3, 4 and 5, as is shown in fig. 16.7. Perhaps it is these segments that are involved in binding to the different odorants.

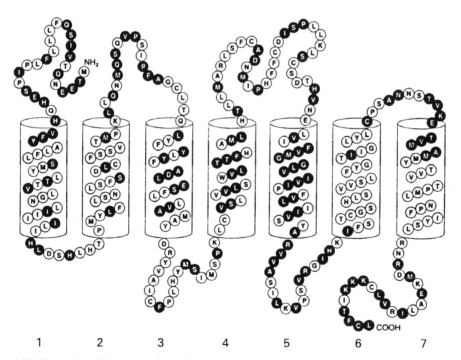

1   2   3   4   5   6   7

Figure 16.7. The amino acid sequence of an olfactory receptor protein, obtained by cDNA cloning. The cylinders show the seven putative α-helices that span the membrane. The N-terminus (at the left) is situated extracellularly and the C-terminus is in the cytoplasm. The sequence (single-letter amino acid code) is that of one particular clone in a set of ten. Black circles show residues that show appreciable variation in different clones of the set, whereas white circles show residues that are shared by six or more of them. The variability of the residues in the transmembrane helices 3, 4 and 5 may reflect the positions of stereospecific binding sites for different odorants. (From Buck & Axel, 1991. Reproduced with permission from *Cell*, copyright Cell Press.)

Clearly there are many more than eighteen members of the odorant receptor family. Buck & Axel concluded that their methods had disclosed the presence of at least 100 to 200 different odorant receptor genes. Taking account of those that had been missed by the screening method, it seems likely that there are about 1000 different genes altogether (Chess *et al.*, 1992, 1994; Axel, 1995). This represents about 1% of the mammalian genome, and demonstrates the biological importance of the sense of smell.

The second-messenger cascade

What happens after the odorants bind to the receptors? It is possible to remove the cilia from the olfactory epithelium by treatment with high calcium ion concentrations, and then to separate them from the other cellular components by differential centrifugation. Pace and his colleagues (1985) used such preparations to show that the cilia contained a high concentration of adenylyl cyclase (the enzyme that produces cyclic AMP from ATP) and that this was dependent on guanosine triphosphate (GTP), suggesting G protein involvement. The cyclase activity was further stimulated by exposure to a mixture of odorants containing citral, L-carvone, 1,8-cineol and *n*-amylacetate. Adenylyl cyclase activity in membranes from other sources (deciliated olfactory epithelium or brain cells, for example) was not affected by the odorant mixture.

This striking set of experiments suggested that olfactory transduction has much in common with visual transduction and other G-protein-dependent cascades: combination of an effector molecule with the membrane receptor activates a G protein which then activates an enzyme leading to the production of a cyclic nucleotide second messenger. Good evidence for this idea appeared when molecular techniques led to the discovery of relevant proteins particular to the olfactory epithelium, a G protein known as G_{olf} (Jones & Reed, 1989) and a cyclase known as type III adenylyl cyclase (Bakalyar & Reed, 1990). Immunocytochemistry shows that these are found in the distal segments of the olfactory cilia (Menco *et al.*, 1992).

What does the cyclic AMP second messenger do? Nakamura & Gold (1987) applied patch clamp electrodes to the cilia of isolated toad olfactory receptor cells. They found that the conductance of excised patches was much increased by application of cyclic AMP or cyclic GMP to the cytoplasmic surface, indicating the presence of cyclic-nucleotide-gated channels. cDNA coding for the channel protein was cloned by Dhallan *et al.* (1990). The channel was similar in sequence to the cyclic-GMP-gated channel of retinal photoreceptors, and showed more a distant relationship to the voltage-gated channels of nerve cells. Patch

clamp studies showed that the density of channels in the cilia of toad receptor neurons was about 2000 μm^{-2}, and very much less than this elsewhere in the cell; the open channel conductance was about 30 pS (Kurahashi & Kaneko, 1993). So odorant stimulation leads to opening of channels in the cilia, and the depolarization consequent on this, illustrated in fig. 16.8, will produce action potentials in the olfactory receptor neuron axon.

The main sequence of events in olfactory transduction, then, involves a G-protein-dependent cascade. It begins when a stereospecific receptor molecule binds a particular odorant molecule. This then activates the G protein G_{olf}, which itself then activates the enzyme adenylyl cyclase III, so producing cyclic AMP. The cyclic AMP then triggers the opening of cyclic-nucleotide-gated channels (fig. 16.9). There are two amplification stages in this sequence: each activated receptor will activate many G protein molecules, and each activated adenylyl cyclase molecule will produce many molecules of cyclic AMP. Patch clamp measurements on isolated salamander olfactory receptor cells suggest that each odorant molecule produces a current flow of 0.3 to 1 pA, which is probably sufficient, in these small cells, to produce an action potential (Menini *et al.*, 1995).

There may be other second-messenger pathways involved in olfactory transduction. Some odours lead to IP_3 production (Breer & Boekhoff, 1990), but the precise significance of this is not clear.

How is the excitation of the olfactory receptor neuron switched off? There is no doubt that olfactory adaptation is rapid. At the behavioural level, we can very rapidly become accustomed to the presence of some unpleasant smell. At the cellular level, the depolarization produced by an odorant soon returns to the resting level after a short exposure or during a long one (Firestein *et al.*, 1993; Menini *et al.*, 1995).

There may be many ways in which this adaptation is effected. Olfactory binding proteins may remove odorants from the mucus (Pevsner & Snyder, 1990). Calcium ions, entering via the cyclic-AMP-gated channels, may inactivate those channels (see Zufall *et al.*, 1994). The receptors may be inactivated by phosphorylation produced by a number of protein kinases, and by the protein β-arrestin-2 (Dawson *et al.*, 1993).

Information coding

The expression of individual olfactory receptor genes in the olfactory epithelium can be detected by *in situ* hybridization of their messenger RNA with specific radioactive antisense probes (Raming *et al.*, 1993; Ressler *et al.*, 1993; Vassar *et al.*, 1993). On average, each gene is expressed in

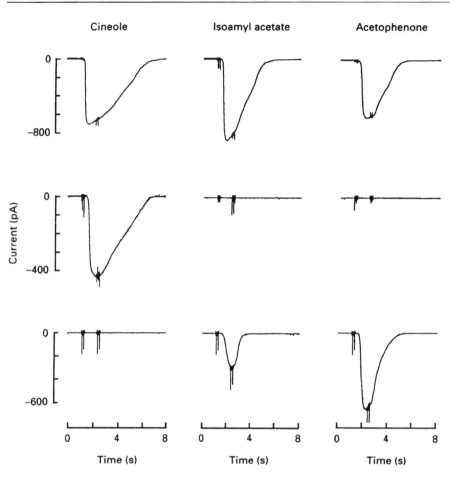

Figure 16.8. Responses of three isolated salamander olfactory receptor cells to three different odorants; each row shows the responses of one cell. The currents were measured using the whole-cell patch clamp system. The three odorants were each applied at a concentration of 5×10^{-4} M for 1.2 s, with the beginning and end of the stimulus shown by artefact marks. Notice how each cell has a different pattern of responses. (From Firestein *et al.*, 1993.)

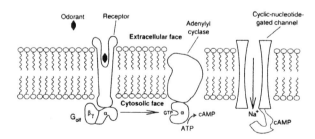

Figure 16.9. The cyclic AMP cascade for olfactory signal transduction in olfactory receptor neuron cilia. Stereospecific receptor molecules bind particular odorants, activating the specific G protein G_{olf}. The α subunit of G_{olf} then activates adenylyl cyclase III, so producing cyclic AMP which then opens cyclic-nucleotide-gated channels. Inflow of sodium and calcium ions produces depolarization which leads to action potentials in the neuron axon. (From Buck & Axel, 1991. Reproduced with permission from *Cell*. Copyright Cell Press.)

only 0.1% to 0.2 % of the olfactory receptor cells. Since there are about 1000 different receptor types, therefore, it seems likely that any particular cell expresses just one of them. In catfish, where there are probably only about 100 different receptor types, individual genes are expressed in about 1% of the total number of cells (Ngai *et al.*, 1993).

How do 1000 different receptor types encode about 10 000 different odours? Sicard & Holley (1984) measured the responses of sixty different frog olfactory neurons to twenty different odorants. They found that the cells all responded to more than one of the odorants, but the pattern of response was different for each cell (fig. 16.10). Thus one cell might respond to odorants *a*, *b* and *c*, say, whereas another responded to *a*, *c* and *d*, while a third responded to *b*, *d* and *e*. There is enough information in this set of responses to work out what any odorant in the range *a* to *e* is: if the second and third cells respond to the odorant but the first does not, for example, then the odorant must be *d*.

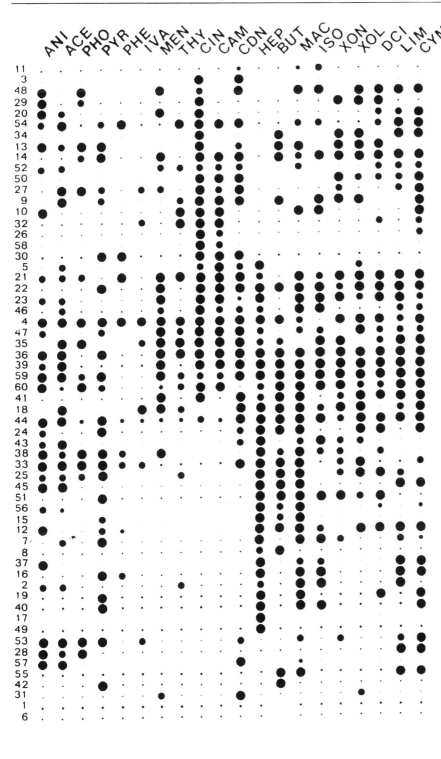

Figure 16.10. The range of responsiveness to twenty different odorants in sixty different receptor neurons from frog olfactory epithelium. The size of the spots indicates the action potential frequency (from 1 to over 500 min^{-1}) measured by extracellular recording. The odorants, from left to right in the diagram, were anisole, acetophenone, thiophenol, pyridine, phenol, isovaleric acid, l-menthol, thymol, 1,8-cineole, camphor, cyclodecanone, *n*-heptanol, *n*-butanol, methylamylketone, isoamylacetate, cyclohexanone, cyclohexanol, (+)-citronellol, (+)-limonene and *p*-cymene. Notice that each neuron responds to some odorants but not others, but no two neurons show the same pattern of responses. (From Sicard & Holley, 1984. Reprinted from *Brain Research* **292**, Sicard, G. & Holley, A., Receptor cell responses to odorants: similarities and differences among odorants, p. 289, 1984, with kind permisssion of Elsevier Science-NL, Sara Burgerhartsstraat 25, 1055 KV Amsterdam, The Netherlands.)

Extracting the information present in the array of responses from the olfactory receptor cells is a matter for the brain, beginning with the olfactory bulb. This contains a number of glomeruli, where the terminals of the olfactory receptor cells synapse with mitral and tufted cells, the next neurons in the ascending olfactory pathway. There are about 1800 glomeruli in the mouse olfactory bulb. Mombaerts and his colleagues (1996; see also Axel, 1995) used an elegant genetic engineering method to investigate the connections between the olfactory receptor cells and the glomeruli. They produced mice in which one of the receptor genes was linked to an enzyme that would produce a blue stain when treated appropriately, so that the axons of the cells expressing this particular gene could be seen in histological preparations. They found that the olfactory receptor cells expressing this one particular molecular receptor species all send their axons to the same two glomeruli in the olfactory bulb. The pattern of connections is bilaterally symmetrical, and the patterns in different individuals are similar.

So each glomerulus responds to the activation of just one molecular receptor type, and the positions of the glomeruli appear to map the nature of the odorants in some way. Then there is some evidence that a lateral inhibition process sharpens the specificity of the mitral and tufted cells; fig. 16.11 shows how this may be done (Yokoi *et al.*, 1995). Clearly the information for the analysis of odorants is there to be sorted out, although just how the brain does it is not yet entirely clear (see Shepherd, 1994; Buck, 1996; Mombaerts, 1996).

The vomeronasal organ

The vomeronasal organ is a chemoreceptive structure found in most terrestrial vertebrates. In mammals it is usually tubular in form and found at the base of the nasal cavity, separate from the olfactory epithelium. It seems to be specialized for the detection of pheromones, chemicals produced by other individuals of the same species that may affect the behaviour of the recipient (Halpern, 1987; Liman, 1996). Vomeronasal receptor cells differ from those of the olfactory epithelium in that they do not possess cilia.

What sort of molecular receptors are involved in pheromone detection? Dulac & Axel (1995) used gene cloning methods to answer this question. They made cDNA libraries from single vomeronasal receptor neurons, and then looked for differences between them, arguing that different cells might express different pheromone receptors. They found that any two neurons would commonly differ

in just one cDNA species, implying that a protein expressed in one neuron was replaced by another in the other, and suggesting the both proteins were pheromone receptors. Sequencing showed that the putative receptors were all of the 7TM G-protein-linked receptor superfamily, but of a different group from those of the olfactory epithelium.

Dulac & Axel concluded that there are about 100 different members of the vomeronasal molecular receptor family. Each neuron apparently expresses one receptor, and the neurons expressing a particular receptor are distributed at random through the vomeronasal epithelium. It may be that the receptors are specific to particular pheromones rather than to a range of related ones (Buck, 1995).

The transduction mechanism in vomeronasal neurons must be different from that of neurons in the olfactory

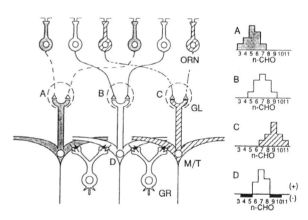

Figure 16.11. Model of neural connections for information processing in the olfactory bulb. The array of olfactory receptor neurons (ORN) in the olfactory epithelium is shown at the top. Each ORN expresses just one olfactory receptor species, which responds to a group of similar odorant compounds. Axons from all the ORNs expressing one particular receptor converge on a particular glomerulus (GL), where they synapse with mitral or tufted cells (M/T). Lateral dendrites of the M/T cells synapse with granule cells (GR) so as to produce inhibitory interactions between adjacent M/T cells. The result of this arrangement for lateral inhibition is to enhance the specificity of the responses of individual M/T cells, as is shown in the graphs at the right. These show responses of ORN cells converging on glomeruli A, B and C to aldehydes of different carbon chain lengths (n-CHO). The output of the middle M/T cell D has a sharper response spectrum, with inhibitory responses (−) at very short and long chain lengths. (From Yokoi *et al.*, 1995. Reproduced from *Proceedings of the National Academy of Sciences USA* **92**, 3374. Copyright 1995 National Academy of Sciences, USA.)

epithelium, since many of the proteins of the second-messenger cascade there, including the G proteins and the adenylyl cyclase, are different, and cyclic-nucleotide-gated channels appear to be absent (Berghard *et al.*, 1996; Berghard & Buck, 1996; Liman & Corey, 1996). A hamster mounting pheromone, aphrodisin, affects IP_3 production in male hamster vomeronasal organ membranes (Kroner *et al.*, 1996).

Chemoreception in insects

Chemical stimuli play an important part in the lives of insects, especially in the identification of food substances and the location of mating partners. One of the great advantages of insects as material for the study of chemoreception is that single sense organs (sensilla) can be stimulated and recorded from individually in isolation from their neighbours.

Each chemoreceptive sensillum consists of a modified region of cuticle through which the dendrites of one or more bipolar nerve cells make contact with the exterior. The cuticular part of the sensillum may be in the form of a hair (sensilla trichodea), a projecting peg (sensilla basiconica), a flat plate (sensilla placodea), or a short peg set in a small pit (sensilla coelonconica). A typical olfactory sensillum trichodeum will have many thousands of fine pores along its length, whereas a similar gustatory hair will have a single pore at its tip. Olfactory sensilla sunk in pits are adapted for use with high concentrations of odorant substances, those in the form of protruding pegs or hairs are for lower ones.

Stocker (1994) has summarized the distribution of chemoreceptive sensilla in the fruit fly *Drosophila*. There are about 400 olfactory sensilla on the third segment of each antenna, and a further 60 on each maxillary palp. There are over 500 gustatory sensilla in all; they occur mainly on the labellum (the end of the proboscis, through which food is tasted, dissolved and ingested), the legs, and the wing margin, with some on the pharynx and the female genitalia.

Figure 16.12 shows the structure of a typical olfactory sensillum. The sense cell is a neuron whose cell body sits among some accessory cells in the epidermis. On its inner

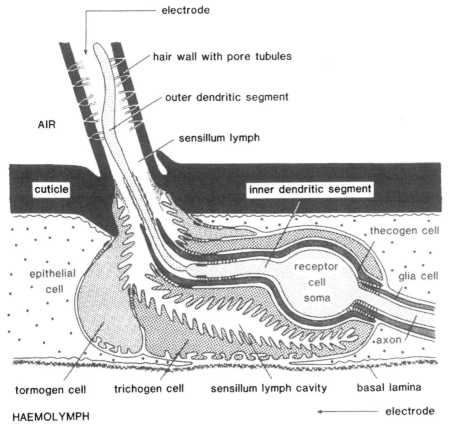

Figure 16.12. Diagram of an insect olfactory sensillum trichodeum with one receptor cell and three auxiliary (thecogen, trichogen and tormogen) cells. The ciliary portion of the dendrite occurs at the narrow connection between its inner and outer segments. (From Kaissling, 1986.)

side its axon connects it to the central nervous system. On its outer side is the sensory dendrite, divided into inner and outer segments by a short ciliary section. The outer segment contains longitudinal microtubules; it is in contact with a fluid called the sensillum liquor or lymph. The ciliary section contains a ring of nine microtubular doublets, as in motile cilia, but there is usually no central pair. The inner segment contains mitochondria and other components of cytoplasm as in the cell body with which it connects.

There are usually four sheath cells associated with each sensillum. They are connected to each other, to the sense cells and to the surrounding epidermal cells by tight junctions. This ensures that the sensillum liquor is kept separate from the haemolymph, which is the main body fluid. The sensillum liquor is probably secreted by one of the sheath cells; it has a much higher potassium concentration than the haemolymph.

Behavioural experiments on taste

A remarkably large amount of information has been obtained from experiments involving measurement of the concentration of a substance required to elicit or inhibit a feeding response (Dethier, 1955, 1976). Suppose, for example, that we wish to measure the minimum concentration of sucrose which is detectable by a blowfly such as *Phormia* or *Calliphora*. The fly is suspended from a glass rod, starved, and then allowed to drink as much water as it wants. A drop of sucrose solution of known concentration

is then brought into contact with one of the tarsi; if the sugar concentration is sufficient, the fly extends its proboscis, but there is no response if it is not. In this way it has been shown that the tarsal receptors of blowflies are most sensitive to sucrose and maltose, less sensitive to glucose, and insensitive to lactose.

The relative sensitivity to compounds which do not evoke the feeding response can be found by measuring the 'rejection thresholds' of the various compounds. In this technique, the concentration of the compound which is just necessary to prevent the feeding response to a sucrose solution is determined. Dethier and his colleagues measured the chemoreceptor sensitivity of blowflies to over 2000 aliphatic organic compounds by this technique. A number of correlations between activity and molecular structure appeared, of which perhaps the most clear-cut was that the rejection thresholds were inversely related to the water solubility of the compounds involved. The simplest explanation of this phenomenon is that a response should depend on the stimulating substance becoming incorporated into the lipid membrane of the receptor cell.

Electrophysiological experiments on taste

Owing to the small diameter of the axons of the chemoreceptor sense cells, their responses to stimulation cannot readily be recorded in the usual manner; two novel techniques had to be developed before electrophysiological investigations could be carried out. In the first of these,

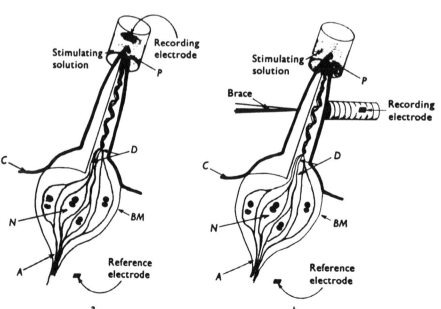

Figure 16.13. Methods of stimulating and recording from individual taste receptor hairs in insects. *a* shows the technique in which the stimulating solution is held in the recording electrode, *b* shows the 'side-wall cracking' technique. A, axons; BM, basement membrane; C, cuticle; D, dendrites; N, sensory nerve cell; P, pore in cuticle at tip of hair. (From Wolbarsht, 1965.)

shown in fig. 16.13*a*, a pipette containing the stimulating solution is placed over the tip of the sensory hair, and also acts as the recording electrode (Hodgson *et al.*, 1955). This method can only be used with electrolytes in the stimulating solution. The second method, sometimes known as 'side wall cracking' (fig. 16.13*b*), is not subject to this limitation, since the recording electrode is applied to the base of the hair and gains contact with its interior by fracturing the cuticle; stimulating solutions are applied, as before, by means of a pipette applied to the tip of the hair (Morita & Yamashita, 1959).

Some records obtained with the side-wall cracking technique, with glucose as the stimulant, are shown in fig. 16.14. The response consists of a negative receptor potential upon which are superimposed a number of small, largely positive-going action potentials. Notice that the size of the receptor potential and the frequency of the action potentials both increase with increasing glucose concentration, that the response shows adaptation, and that at least two sizes of action potential are evident, indicating that at least two sense cells are stimulated.

How do these responses arise? We may assume that ion channels in the outer segment membrane are opened as a

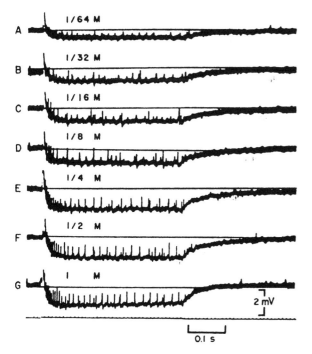

Figure 16.14. Records obtained from a single sensillum on the labellum of a fleshfly by the side-wall cracking method. The stimulus was glucose solutions of different concentrations. (From Morita & Shiraishi, 1985.)

result of contact with the stimulant solution, either directly or via combination with particular receptors. Movement of ions through the channels would then depolarize the membrane to produce the receptor potential. This depolarization, we assume, spreads down the dendrite to reach an impulse initiation site near the cell body. Action potentials then propagate down the sensory axon. The local circuits associated with these action potentials spread back up the dendrite and are recorded by the extracellular electrode next to the outer segment (Morita & Shiraishi, 1985; Morita, 1992).

Electron micrographs show that there are four, or sometimes five, receptor cells associated with each sensillum trichodeum in *Phormia*. In accordance with this, four physiologically distinct receptors have been found by electrophysiological methods. These are the 'salt' and 'sugar' receptors (Hodgson & Roeder, 1956), which respond to monovalent salts and certain sugars respectively, the 'water' receptor (Evans & Mellon, 1962), and a mechanoreceptor whose dendrite is attached to the base of the hair (Wolbarsht & Dethier, 1958). The function of the fifth fibre may be connected with sensitivity to anions (Dethier & Hanson, 1968).

Some very specific contact chemoreceptors exist. The beetle *Chrysolina brunsvicensis* feeds only on plants of the genus *Hypericum* (Saint John's wort). These plants contain the compound hypericin, which is found nowhere else. Rees (1969) found that the tarsal chemoreceptors of the beetle are stimulated by hypericin, whereas those of related insects are not. Another example is provided by the larvae of the cabbage white butterfly (*Pieris brassicae*). Ma (1972) found that sensilla on the mouthparts are stimulated by the mustard oil glucosides, and that this leads to feeding; thus compounds which act as a feeding deterrent for most other insects are here used by the larva to keep it eating the right food plant. Further examples are given by Städler (1984).

Olfaction in insects

Insects use the sense of smell to detect odours from such sources as plants, animals, carrion or dung, so as to find their food, oviposition sites and so on. Many insects also communicate by chemical means: one individual releases a substance which is detected by other members of the species. Such substances are called pheromones. They include sex attractants, aggregation pheromones and alarm substances.

Much of our knowledge of insect olfaction is derived from work by Schneider and his colleagues at the Max

Planck Institute at Seewiesen (Schneider *et al.*, 1964; Boeckh *et al.*, 1965; Schneider, 1965, 1984; Kaissling, 1971, 1986). They used two methods of measuring the receptor responses: (1) the mass response from the whole antenna (the electroantennogram), obtained with recording electrodes placed at the tip and base of an antenna, and (2) single unit responses recorded by means of a microelectrode inserted through the thin cuticle of a sensillum. They placed solutions of odoriferous substances on small pieces of filter paper in the orifice of an air jet, and stimulated the antenna by directing the air stream at it.

By measuring the responsiveness or otherwise of a large number of sensilla on the antennae of moths and honeybees, Schneider and his colleagues found that the receptor cells fall into two main classes, called 'generalists' and 'specialists'. Odour generalists are cells which respond to a number of different substances, some causing an increase in discharge frequency and others inhibiting it. The responses of twenty-seven different sensilla basiconica of a moth to ten different odorants are shown in fig. 16.15. What is particularly interesting about these results is that no two cells have the same reaction spectrum. It seems probable that odours are identified by the patterns of responsiveness which they elicit in a large number of receptors.

We have seen that results showing a similar organization have also been obtained from frog olfactory receptor cells (fig. 16.10). Indeed, it is becoming clear that there are common principles of organization for olfactory transduction and neural processing in different animal groups (Hildebrand & Shepherd, 1997).

Odour specialists are olfactory receptor cells which respond to a very limited number of compounds and which are usually present in large numbers, each cell having the same action spectrum. The antennae of male silkmoths (*Bombyx mori*), for example, contain a large number of sen-

silla trichodea which are specifically responsive to the sex attractant pheromone which is produced by the females (fig. 16.16). The main component of the *Bombyx* pheromone was eventually isolated (from half a million female moths!) and synthesized by Butenandt and his colleagues in 1961 (see Hecker & Butenandt, 1984). It is a long-chain unsaturated alcohol (hexadeca-dien-10-*trans*-12-*cis*-1-ol), which was given the name bombykol.

The sensitivity of the bombykol response was investigated by Kaissling & Priesner (1970; Kaissling, 1971). For the absolute threshold the relevant questions are how many molecules are required (1) to excite one sensory cell, and (2) to produce a behavioural response in the male. The response of an individual sensillum to puffs of air containing different concentrations of bombykol is shown in fig. 16.17. With an odour source of 10^{-4} μg, the stimulus sometimes elicits an action potential, sometimes not; the mean response was 0.34 action potentials per puff. The average number of molecules reaching the sensillum at this concentration is also less than one; hence we can say that one molecule of bombykol is sufficient to excite one sensory cell.

Males respond to a puff of air containing bombykol by fluttering their wings. This reaction occurred in some moths with an odour source of 3×10^{-6} μg. Kaissling & Priesner calculated that under these conditions about 310 molecules would contact sensilla, out of a total number of about 30 000 sensilla in each antenna. The moth thus responds when about 1% of its pheromone detectors are activated.

The cuticular wall of the hair sensillum is perforated by large numbers of pores (fig. 16.16*d* and *e*). Each one is connected to a number of tubules which open into the sensillum liquor space surrounding the sensory cell dendrite. Experiments with radioactive pheromone in *Antheraea polyphemus* show that diffusion occurs rapidly from the outer surface of the sensillum through the pores to the sensillum liquor; 40% of the radioactivity appeared in sensil-

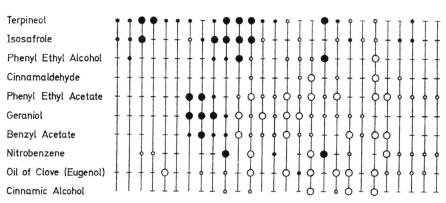

Terpineol
Isosafrole
Phenyl Ethyl Alcohol
Cinnamaldehyde
Phenyl Ethyl Acetate
Geraniol
Benzyl Acetate
Nitrobenzene
Oil of Clove (Eugenol)
Cinnamic Alcohol

Figure 16.15. Olfactory responses of twenty-seven sensilla basiconica in the moth *Antherea pernei* to ten different compounds. Black circles indicate excitation, white circles indicate inhibition, and the sizes of these circles indicate the extent of the change in nerve impulse frequency. Small horizontal lines indicate no effect. (From Kaissling, 1971, after Boeckh *et al.*, 1965.)

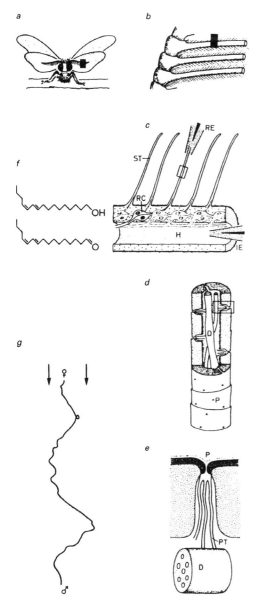

Figure 16.16. Pheromone reception in *Bombyx mori*. The male moth is shown in *a*, and parts of its antennae are shown at increasing magnifications in *b* to *e*. In *c* the tip of a sensillum (ST) is cut off and the recording electrode (RE) placed over it; the indifferent electrode (IE) is placed in the haemolymph (H). Each sensillum contains the dendrites from two sensory receptor cells (RC). *d* shows part of the two dendrites (D); the outer surface of the sensillum is covered with pores (P). A section through one of the pores is shown in *e*; pore tubules (PT) connect the pore to the sensory liquor and possibly to the dendrite surface (D). The formulae of the two components of the pheromone, bombykol (above) and bombykal, are shown in *f*. A male moth reaches a female by flying along the odour plume containing her pheromone as is indicated in *g*. (From Steinbrecht & Schneider, 1980.)

lum liquor extruded 2 min after exposure (Kanaujia & Kaissling, 1985).

What happens to the pheromone after it has contacted the sensory dendrite? The moth needs to respond rapidly to a fall in pheromone concentration, otherwise it would not know when it had left the odour plume downwind of the female. Prolonged after-effects of pheromone exposure would interfere with this. Vogt & Riddiford (1981) found that the sensillum liquor *in Antheraea* contains a very active esterase which will hydrolyse the pheromone (which in this case is an acetate ester) and so inactivate it. There is also a pheromone-binding protein present, and Vogt *et al.* (1985) suggested that this serves to protect the pheromone from the enzyme during its passage from the pore tubules to the receptors on the dendrite membrane. It also possible that the pheromone-binding protein inactivates the pheromone after it has contacted its first receptor molecule, so that it is not able to stimulate further ones (Ziegelberger, 1995).

It seems reasonable to assume that the molecular receptors for olfactory stimuli may be G-protein-linked receptors of the 7TM superfamily, although at the time of writing there is no direct evidence for this. G proteins have been detected in cockroach antennae, and phospholipase C activity (producing IP_3) is stimulated by pheromones and blocked by pertussis toxin (Boekhoff *et al.*, 1990). Some of the *Drosophila* mutants with defective olfaction have defects in their IP_3 metabolism (Carlson, 1996; Smith, 1996). Cyclic GMP and calcium ions may also act as second messengers here (Ziegelberger *et al.*, 1990; Kaissling, 1996).

In the early years of pheromone research it was assumed that each sex attractant pheromone consisted of a single substance only. This indeed seems to be the case for cockroaches (Roelofs, 1995), but the situation is more complicated in moths. Meijer and his colleagues (1972) found that the pheromone of the summer fruit tortrix moth (*Adoxophyes orana*) contains two components, *cis*-9- and *cis*-11-tetradecen-1-ol acetate. Traps containing either component alone were ineffective in capturing males, but those containing both components, in the ratio 3:1 of *cis*-9:*cis*-11, were highly effective. den Otter (1977) showed that each antennal sensillum contains two sensory cells, one responsive to the *cis*-9 isomer only and the other much more responsive to the *cis*-11. Thus the sensitivity to the ratios of the two components resides in the central nervous system of the moth, not in its individual sensory cells.

Further work showed that it is the normal situation for moth sex-attractant pheromones to have two or more components. In accordance with this, the aldehyde of bombykol, called bombykal, was discovered as a minor component in the pheromone of *Bombyx mori* (Kaissling *et al.*, 1978). The sensitivity to the component ratio means that male moths can readily distinguish their own species pheromone from that of a related species which uses the same components in a different ratio (Priesner, 1979; Steinbrecht & Schneider, 1980).

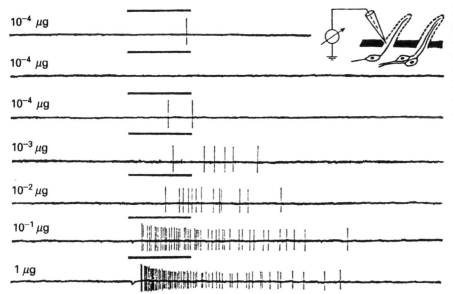

Figure 16.17. Action potentials from a bombykol receptor cell in a male *Bombyx mori*. The quantities of bombykol on the odour source are shown at the left, and the bar indicates the duration of the puff of air containing it. The experiment shows that one bombykol molecule is sufficient to excite a receptor cell. (From Kaissling & Priesner, 1970.)

17
Some other sensory receptors

In previous chapters we examined the physiology of sensory receptors responsive to mechanical, visual and chemical stimuli, and we saw that in some cases our knowledge of the transduction processes and related cellular mechanisms is quite extensive. In this chapter we look briefly at some further senses about whose mechanisms we have rather less information.

Thermoreceptors

Sensations of warm and cold are important to ourselves and other animals in finding a suitable environment in which to live. Conditions are suboptimal if this is too cold or too hot, and extremes of temperature are fatal. The temperature of particular objects may be important, as in the detection of warm-blooded prey by snakes or fleas, or the control of the incubation mound by the mallee fowl.

Temperature-sensitive sensory nerve fibres with endings in mammalian skin are of two types, cold fibres and warm fibres, as is shown in fig. 17.1. Both groups have a low level of activity at the body heat level of 37 °C. Cold fibres discharge increasingly as the skin temperature falls below this, reaching a maximum in the region of 27 °C; sometimes they also discharge at very high temperatures, above 50 °C.

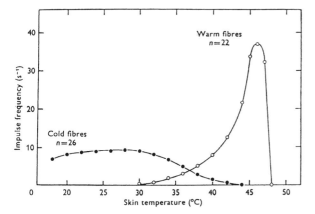

Figure 17.1. Average discharge frequencies at different steady temperatures for cold and warm fibres from the skin of the nose region in cats. (From Hensel & Kenshallo, 1969.)

Warm fibres discharge increasingly at higher temperatures, with a maximum at about 46 °C and a very rapid fall-off above this (Hensel & Zotterman, 1951; Hensel & Kenshallo, 1969; Pierau & Wurster, 1981).

These fibres also show marked responses to *change* in temperature. Cold fibres increase their discharge rate for a few seconds on rapid cooling and decrease it on rapid warming. The reverse happens with warm fibres. These effects are referred to as dynamic responses, as opposed to the static responses to constant temperatures. In one warm fibre, for example, a sudden rise from 38 to 42 °C produced a change in discharge frequency from 9 to 50 impulses s^{-1}, but within half a minute it had fallen back to a new static response level at 18 s^{-1}. On returning to 38 °C the discharge frequency fell briefly to zero before recovering to its former level at 9 s^{-1} (Hensel & Kenshallo, 1969).

The two types of physiological response correspond on the whole to two different anatomical types of fibre. Cold receptors are mainly small myelinated fibres of the Aδ group, with diameters of 1 to 3 μm and conduction velocities of 5 to 15 m s^{-1}, whereas warm fibres are mainly unmyelinated C fibres, with diameters less than 1 μm and conduction velocities of 0.5 to 2 m s^{-1} (Iggo, 1964, 1969, 1990; Spray, 1986). The cold fibres divide near their endings to form a group of terminals just under the epidermis. The terminals are swollen and contain numbers of mitochondria (Hensel et al., 1974). Warm fibres apparently have free nerve endings.

What is the transduction mechanism for thermoreceptors? There is not much direct evidence on this question. Many biological processes are affected by temperature. Spray (1974, 1986) suggested that cold receptors are dependent on the activity of the sodium–potassium pump in the nerve endings: a drop in temperature would reduce the activity of the pump, depolarize the ending and so lead to more action potentials. In accordance with this he found that ouabain, an inhibitor of the pump, produces a brief discharge of frog cold receptors when added to the skin, after which the endings become insensitive to temperature changes. Mammalian cold endings also become unresponsive to cold in the presence of ouabain (Pierau *et al.*, 1975).

Nociceptors and pain

Nociceptors are sensory neurons that respond to tissue-damaging stimuli or the existence of tissue damage, and whose activity is believed to be associated with the sensation of pain (see Belmonte & Cervero, 1996). There are two main types of afferent fibre involved, small myelinated fibres of the Aδ (delta) group, and unmyelinated fibres of the C group. The conduction velocity of the Aδ fibres (5 to 25 m s^{-1}) is much higher than that of the C fibres (0.5 to 2 m s^{-1}); they have a lower threshold to electrical stimulation and are more sensitive to ischaemic block.

Bishop & Landau (1958) used these differences to demonstrate the existence of two separate types of pain in humans. Electrical stimulation of a peripheral skin nerve by single shocks or at a low repetition rate produces a tactile sensation of tapping at threshold intensities. At higher shock intensities, three to five times that required to elicit the tapping sensation, the stimulus feels painful, with a sharp pricking sensation produced by each shock at frequencies up to 30 s^{-1}. The tactile tapping sensation is attributed to larger Aβ myelinated fibres, the painful pricking sensation to smaller Aδ fibres. The Aδ response is known as 'fast' pain.

Activity in the Aδ fibres of the lower arm can be prevented by cutting off the blood supply with a pressure cuff. After half an hour or so conduction in the myelinated fibres ceases and all sensations of touch and of pricking pain beyond the cuff are lost. Strong electrical stimuli or firm pressure will now produce an excruciating burning pain. This is attributed to C fibre action and is known as 'slow' pain.

Studies of electrical activity in Aδ and C fibres fit well with these conclusions, although there were initial difficulties in making the recordings (see Perl, 1996). Burgess & Perl (1967) used fine glass microelectrodes to record from single myelinated fibres in the skin nerves of anaesthetized cats. They found numbers of fine fibres of the Aδ group that were responsive only to damaging mechanical stimuli, such as pinching the skin with serrated forceps or cutting it. Records from C fibres were more difficult: Bessou & Perl (1969) solved the problem by dividing the peripheral nerve into finer and finer bundles until single-fibre responses of low conduction velocity could be obtained, and then finding what sensory stimuli would excite them. About half their C fibres responded only to intense mechanical stimuli, painfully high temperatures or irritant acids; some of these were responsive to high thresh-

old mechanical stimuli only, but most were polymodal, responding to the range of noxious stimuli.

Ochoa & Torebjörk (1989) used their microstimulation technique (p. 261) to stimulate individual C fibres in human volunteers, and provided excellent evidence of the link between electrical activity and sensation. They found that each C fibre was associated with a particular receptive field on the skin of the hand, as determined by mechanical stimulation, and that this corresponded with the perceived area of induced dull or burning pain when the fibre was stimulated electrically in the nerve trunk via the recording electrode (fig. 17.2).

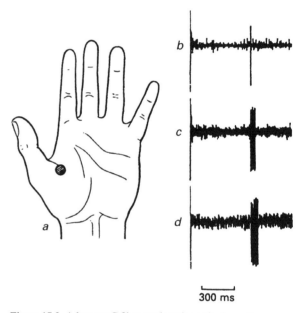

300 ms

Figure 17.2. A human C fibre nociceptive unit. A needle microelectrode inserted into the nerve trunk at the elbow recorded the activity of a single C fibre unit in a conscious volunteer. Then electrical stimulation via this electrode produced a sensation of dull pain that was localized to the shaded area of the skin of the hand (*a*). Painful stimuli evoked action potentials in this unit when applied to the area of skin shown by the black dot inside the shaded area. Trace *b* shows the action potential elicited by intradermal electrical stimulation of the receptive field (the black dot). The conduction velocity determined from such records was 0.9 m s^{-1}, showing that the unit was a C fibre. Traces *c* and *d* show that this fibre really was the one made active by painful stimulation. Application of a painful heat stimulus to the receptive field produced a transient fall in the conduction velocity (presumably as a result of the burst of activity produced by this stimulus); trace *c* shows twelve superimposed records in the recovery period. Trace *d* shows a similar recovery of conduction velocity after a period of electrical stimulation of the unit at the elbow. (From Ochoa & Torebjörk, 1989.)

Hyperalgesia

It is a well-known feature of life that tissue injury can lead to an increased sensitivity to painful stimuli in the region of the affected area. Such increased sensitivity in the injured tissue is called primary hyperalgesia. When an increased sensitivity occurs in the surrounding uninjured tissue it is called secondary hyperalgesia.

Primary hyperalgesia seems to be produced by the release of a variety of substances from damaged and inflamed cells, and from leucocytes, blood platelets, mast cells and various other cells including sympathetic neurons. These substances include prostaglandins E_2 and I_2, bradykinin, leukotrienes, serotonin, adenosine, histamine and others (Levine & Taiwo, 1994). Patch clamp experiments on cells in culture suggest that the prostaglandins act via cyclic AMP as a second messenger, perhaps by closing potassium channels in nociceptor nerve endings and so reducing their thresholds to noxious stimuli (Pitchford & Levine, 1991). It may be that other hyperalgesic substances act in a similar way, presumably via G-protein-coupled 7TM receptors.

Capsaicin, the active component producing the 'hot' taste of chilli peppers, is a potent algesic substance. When injected under the skin it produces a strong pain sensation and increased activity in C fibre nociceptors (LaMotte *et al.*, 1992). It also has a striking effect in producing secondary hyperalgesia and a conversion of mechanical stimuli from simply tactile to painful. This is illustrated in the experiment by Torebjörk and his colleagues (1992) that is shown in fig. 17.3. Here the volunteer experienced a sensation of touch in the area shown black when a particular A fibre was stimulated by an intraneural microelectrode. Then some capsaicin was injected under the skin a short distance away, and an area of hyperalgesia spread out from the injection site. Intraneural stimulation of the mechanoreceptor afferent now produced a painful sensation. As the effect of the capsaicin wore off, the sensation following intraneural stimulation returned to normal.

This experiment is illustrative of a well-known feature of pain, that the sensation is not simply a consequence of the excitation of nociceptor afferents alone. The Aβ mechanoreceptor input produces a different sensation after C fibres from a nearby area have been stimulated by the capsaicin. A related observation was made by Cook and her colleagues (1987) on second-order neurons in the rat spinal cord. These were initially responsive only to noxious stimulation of a particular skin area, but repetitive

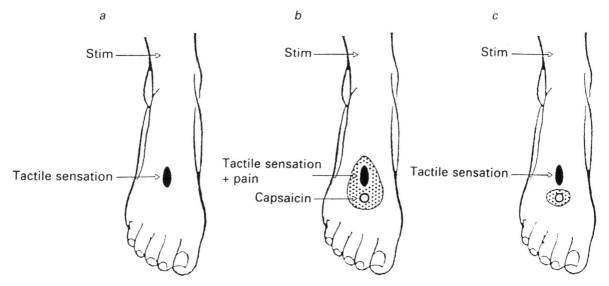

Figure 17.3. A constant mechanoreceptor input becomes painful during a period of secondary hyperalgesia. *a* shows how intraneural microstimulation of a myelinated afferent fibre (probably an Aβ fibre) in the peroneal nerve produced a non-painful tactile sensation projected to a small area of skin on the upper surface of the foot. *b* shows the situation 14 min after injection of capsaicin under the skin nearby (white circle); this produced an area of hyperalgesia (dotted), i.e. light touch in this area was perceived as painful. The hyperalgesic area overlapped the projected receptive field of the tactile afferent, and intraneural stimulation of the afferent now produced a painful sensation. The hyperalgesic area later contracted, so that at 39 min after the injection (*C*), it no longer overlapped the receptive field of the mechanoreceptor afferent, and intraneural stimulation of the afferent no longer produced a sensation of pain. The experiment shows that painful sensory information is not limited to nociceptor afferents. (From Torebjörk *et al.*, 1992.)

stimulation of peripheral C fibres would extend their receptive fields and make them responsive to light mechanical stimuli. These two experiments each suggest that there is some interaction between the Aβ and C fibre inputs at some point in the central nervous system so that higher interneurons leading to the conscious sensation of pain are now activated by the previously innocuous Aβ input. The phenomenon is known as central sensitization (Devor, 1996)

Thus the subject of pain is much more complex than simply the question of how nociceptor afferent fibres are excited. Accounts of these wider issues can be found in Wall & Melzack (1994) and Melzack & Wall (1996), among others.

Electroreceptors

Most sense cells are responsive to electric currents of sufficient magnitude, but this does not justify description of them as 'electroreceptors'. As Machin (1962) pointed out, a sense organ can only be described as an electroreceptor if it responds to electrical stimuli present in the environment and if the organism responds in a way appropriate to the detection of these electrical stimuli. Many fish possess such electroreceptors, sometimes in association with electric organs producing weak discharges, as in gymnotids and mormyrids, sometimes in their absence, as in cartilaginous fish, catfish and others.

The weak electric discharges produced by certain freshwater fish were discovered by Lissmann (1951), and he suggested that they might use them in conjunction with an electric sense as an object location system. According to this idea, the activity of the electric organ sets up an electric field in the water, and the form of this field will be distorted if objects of different conductivity are placed in it, as is shown in fig. 17.4. The disturbance could be detected by the fish if it is able to measure the currents flowing across the skin at different points on its body surface, for which the presence of an array of electroreceptors must be necessary.

Lissmann (1958) described a number of experiments as evidence for sensitivity to weak electric fields in electric fish. Specimens of *Gymnarchus niloticus* responded to the closing of a switch between two wires dipped into the aquarium tank, and to movements of magnets or electrostatic charges outside the tank. He was able to show that *Gymnotus carapo* can be trained to differentiate between metal and plastic discs in total darkness, or between presence and absence of a magnet placed outside the tank and out of sight of the fish.

These experiments clearly establish that electric fish are sensitive to electrical changes in the environment, but they do not necessarily imply that the fish can detect objects by means of their distortion of the electric field produced by the electric organs, since metals immersed in water can give rise to small voltages and it might be these that the fish is responding to. The question was further investigated by Lissmann & Machin (1958) in some elegant experiments on *Gymnarchus*. They used as test objects porous pots which were filled with materials of different conductivities; a porous pot is opaque and is an effective diffusion barrier for short-term experiments, but does not offer any considerable resistance to electric current when it is immersed in a conducting solution. The fish could easily be trained to distinguish between pots filled with aquarium water and pots filled with distilled water. Such a fish would then respond differently to a pot filled with aquarium water according to whether or not it contained a glass rod more than 1.5 mm in diameter; the presence of the glass rod would, of course, alter the overall conductance of the pot.

In discussing the evolution of electric fishes, Lissmann (1958) argued that the possession of an electric sense must

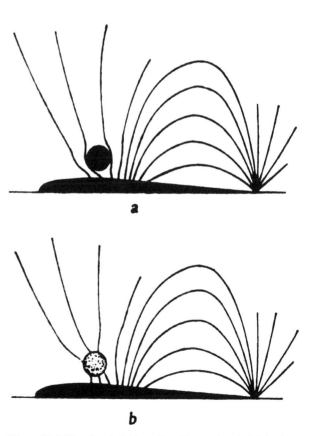

Figure 17.4. The electric field produced by an electric fish in the presence of objects of (*a*) low and (*b*) high conductivity. (From Lissmann & Machin, 1958.)

be a prerequisite for the evolution of an electrolocation system involving electric organs, and thus that we might expect to find electroreceptors in some non-electric fish. He suggested that certain sense organs known to exist in the skin of fishes, often connected to the surface by jelly-filled pores, would in fact be electroreceptors. Such structures included the ampullae of Lorenzini in elasmobranchs, the pit organs of catfish, and the mormyromasts of mormyrids.

Confirmation of Lissmann's predictions was rapidly forthcoming. Murray (1962) showed that the ampullae of Lorenzini were extremely sensitive to electric currents. Dijkgraaf & Kalmijn (1963) found that dogfish and skate show behavioural responses to weak electric currents, and that these are abolished by cutting the nerves from the ampullae. Action potentials in the nerves supplying the electroreceptors of weak electric fish were recorded by Bullock *et al.* (1961) in gymnotiforms and by Fessard & Szabo (1961) in mormyrids. Since then electroreceptors have been described in catfish, African knifefish, most non-teleost fish and some amphibians (see Bullock & Heiligenberg, 1986), and even in the duck-billed platypus (Scheich *et al.*, 1986).

The anatomy of the electroreceptors of various fish was investigated by Szabo (1965, 1974). He distinguished two general types, called the ampullary organs and the tuberous organs (fig. 17.5). An ampullary organ consists of a group of sensory cells at the bottom of a jelly-filled canal which opens to the exterior. A tuberous organ consists of a small group of sensory cells in a capsule which projects into the epidermis from below. The capsule is connected to the outer surface of the fish by a canal which contains loosely packed epidermal cells with large extracellular spaces. The receptor cells form part of an epithelial sheet with supporting cells, and with tight junctions between adjacent cells. The recep-

tor cells in each organ synapse with the branches of a single afferent axon, and sometimes a number of organs are connected to the same axon.

Ampullary organs are found in most electrosensitive fish. Tuberous organs are found only in the Gymnotiformes and Mormyriformes, both of them teleost orders whose members possess electric organs. In line with this distribution we find that the two types have different functions: ampullary organs are used primarily for detecting electric fields arising in the environment; tuberous organs are used by electric fish for electrolocation and communication.

Transduction of the electric signal appears to be done by voltage-gated calcium channels (Clusin & Bennett, 1977*a,b*; Sugawara & Obara, 1989). In elasmobranchs (skates and sharks), these are in the apical membrane of the cell (that part of the plasma membrane on the exterior side of the tight junctions), whereas in teleosts they are in the basal membrane (fig. 17.6). In teleosts, lumen-positive

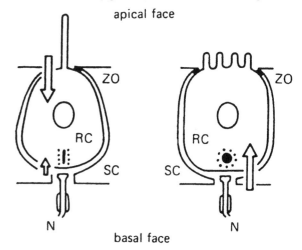

SKATE   *PLOTOSUS*

Figure 17.6. How electroreceptor cells work in non-teleost (skate) and teleost (the marine catfish *Plotosus*) fish. The receptor cells (RC) are polarised, with an apical face towards the outside of the fish and a basal face synapsing with the sensory nerve terminal. They sit between supporting cells (SC) in an epithelial layer. Current flow across the epithelium is forced to pass through the cells by the tight junctions (ZO; zona occludens) between them. Voltage-gated calcium channels are found in the apical membrane of the skate and in the basal membrane of *Plotosus*; the thick white arrows show calcium entry when these channels are activated by depolarization. Calcium entry through these channels triggers neurotransmitter (N) release in *Plotosus*. In the skate it seems that neurotransmitter release follows the opening of further calcium channels (small arrow) as a result of cell depolarization. (From Obara & Sugawara, 1984.)

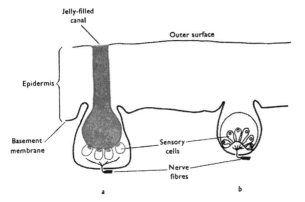

Figure 17.5. Gymnotid electroreceptors. A diagrammatic section through the skin of a gymnotid fish showing (*a*) an ampullary organ and (*b*) a tuberous organ. (Redrawn after Szabo, 1965.)

currents depolarize the basal membrane and so open the calcium channels, hence there is an inflow of calcium ions, leading to neurotransmitter release. In elasmobranchs the calcium channels in the apical membrane open with lumen-negative currents; the further depolarization that this produces opens another set of calcium channels in the basal membrane, again leading to transmitter release. In both cases the depolarization may be ended by the action of calcium-activated potassium channels.

Electric fish can be divided roughly into pulse species, which produce brief discharges at various intervals, and wave species, which produce high frequency discharges more or less continuously. The coding methods of their tuberous electroreceptors, in their responses to the electric organ discharge are affected by these differences. Figure 17.7*a*, for example, shows responses of two types of unit in

the pulse species *Hypopomus*: M units respond with a single action potential; B units give a burst of impulses whose number and frequency are proportional to the strength of the stimulating current. In *Eigenmannia*, a wave species, there are again two types of unit (fig. 17.7*b*): T units always respond once per cycle when the electric organ discharge intensity is above threshold, while P units increase their probability of firing in any one cycle (and therefore their overall firing frequency) as the intensity of the stimulating current increases.

Units analogous to the M and B units of gymnotiform pulse species occur in mormyrids; they correspond to two different morphological types, known as knollenorgans and mormyromasts (Bennett, 1965). In both gymnotiforms and mormyrids, the units that simply respond to the presence or absence of the electric organ discharge (M and T units, knollenorgans) may serve to detect the presence of other discharging individuals. On the other hand, the units which code for the intensity of the electric organ discharge (B and P units, mormyromasts) are probably concerned with electrolocation (Heiligenberg, 1977).

How do the electroreceptors signal the changes in the electric field produced by the electric organ? Figure 17.8 shows how a single P unit in the gymnotiform wave species *Sternarchus* can respond to the position of objects of

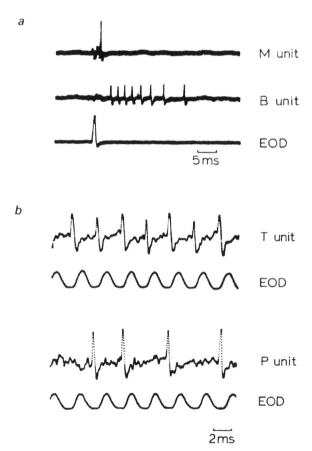

Figure 17.7. Coding in tuberous electroreceptors of gymnotiforms. *a* shows responses of an M unit and a B unit in response to a single discharge of the electric organ (EOD) in the pulse species *Hypopomus*. *b* shows corresponding responses of a T unit and a P unit in the wave species *Eigenmannia*. (From Heiligenberg, 1977, data of J. Bastian.)

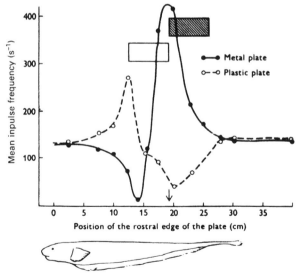

Figure 17.8. The relation between nerve impulse frequency in an electroreceptor nerve fibre of *Sternarchus* and the position of metal and plastic plates placed close to the fish. The arrow indicates the location of the electroreceptor ending. The plates (the metal plate is shown cross-hatched and the plastic plate plain) are drawn to scale in the positions where they elicited the maximal response. (From Hagiwara *et al.*, 1965.)

different conductivity in its immediate vicinity, as determined by Hagiwara and his colleagues (1965). The normal firing rate of the unit is about 130 s^{-1}; at an electric organ discharge frequency of 800 s^{-1} this corresponds to a probability of firing of 0.16 cycle^{-1}. When a metal plate is brought towards the electroreceptor from the tail, the firing rate increases to reach a maximum of over 400 s^{-1}, corresponding to a firing probability of about 0.52 cycle^{-1}. As the plate passes over the electroreceptor, the firing rate drops markedly to a very low value (probability about 0.01) and then recovers as the plate is taken away towards the head. An inverse set of responses occurs when a plastic plate is moved over the same track.

Clearly an array of a large number of electroreceptors of the type studied by Hagiwara and his colleagues will provide the information for a highly sensitive location system. Szabo found that a small specimen of *Sternarchus* possessed about 130 ampullary organs and over 2000 tuberous organs.

The electric sense of elasmobranchs was investigated by Kalmijn (1971). He found that plaice (flatfish upon which dogfish and skates feed) produce an appreciable electric field, up to 1000 μV cm^{-1}, from the action potentials of muscles used in respiration or movement. This compares with Murray's value of 1 to 10 μV cm^{-1} for the threshold field detectable by a single ampulla of Lorenzini. A dogfish can find a live plaice even when it is held in an agar chamber under sand, whereas it cannot find pieces of fish under the same conditions. So it seems as though these elasmobranchs are using their electroreceptors as highly sensitive detectors of live food buried in the sand; they are detecting the electric currents produced by the muscles of their prey.

It is also possible that elasmobranchs are sensitive to the induced currents set up in moving water currents or a moving fish by the earth's magnetic field (Kalmijn, 1982, 1984). Kalmijn was able to train dogfish and rays to respond to fields as low as 5 nV cm^{-1}. Fields of higher strengths than this occur in ocean currents, and would be induced in a fish swimming at the low speed of just a few centimetres per second. The fact that the behavioural threshold is below the apparent threshold for a single ampulla (which is detected by the experimenter as an increase or decrease in the spontaneous level of neural firing) can probably be explained by the multiple channel effects on the signal-to-noise ratio that were discussed in chapter 16.

Part E
Muscle cells

18
Mechanics and energetics of muscle

The function of muscle cells is to contract: to shorten and develop tension. This means that the end-product of cellular activity can be measured with considerable precision, by mechanical measurement of the change in length or tension or both. Such activity must obviously involve the consumption of energy, some of which may appear as heat.

In this chapter and the next two we shall be concerned mainly with the properties of rapidly contracting vertebrate skeletal muscles, such as frog sartorius and the rabbit psoas. Skeletal muscles are activated by motoneurons, as we have seen in previous chapters. Their cells are elongate and multinuclear and the contractile material within them shows cross-striations; hence skeletal muscle is a form of striated muscle. Some of the special properties of other muscles are examined in chapter 21.

The normal stimulus for the contraction of a muscle fibre in a living animal is an impulse in the motor nerve by which it is innervated. The sequence of events following the nerve impulse is shown schematically in fig. 18.1. We have examined stages 1 to 4 of this sequence (the excitation processes) in previous chapters. This chapter is concerned with some of the overall consequences of contraction (stage 6); details of the cellular mechanisms involved in stages 5 and 6 are considered in the following chapters.

Anatomy

Skeletal muscle fibres (fig. 18.2) are multinucleate cells formed by the fusion of numbers of elongated uninucleate cells called myoblasts. Mature fibres may be as long as the muscle of which they form part, and 10 to 100 μm in diameter. The nuclei are arranged around the edge of the fibre. Most of the interior of the fibre consists of the protein filaments which constitute the contractile apparatus, grouped together in bundles called myofibrils. The myofibrils are surrounded by cytoplasm (or sarcoplasm), which also con-

tains mitochondria, the internal membrane systems of the sarcoplasmic reticulum and the T system, and a fuel store in the form of glycogen granules and sometimes a few fat droplets. We shall examine the structure of the contractile apparatus and the internal membrane systems in more detail in the following chapters.

The muscle fibre is bounded by its cell membrane, sometimes called the sarcolemma, to which a thin layer of connective tissue (the endomysium, formed of collagen fibres) is attached. Bundles of muscle fibres are surrounded by a further sheet of connective tissue (the perimysium) and the whole muscle is contained within an outer sheet of tough connective tissue, the epimysium. These connective tissue sheets are continuous with the insertions and tendons which serve to attach the muscles to the skeleton.

Mammalian muscles have an excellent blood supply, with blood capillaries forming a network between the individual fibres. Sensory and motor nerve fibres enter the muscle in one or two nerve branches. The sensory nerve endings include those on the muscle spindles (sensitive to length), in the Golgi tendon organs (sensitive to tension), and a variety of free nerve endings in the muscle tissue, some of which are involved in sensations of pain.

In mammals the γ motoneurons provide a separate motor nerve supply for the muscle fibres of the muscle spindles, while the bulk of the muscle fibres are supplied by the α motoneurons. Each α motoneuron innervates a number of muscle fibres, from fewer than ten in the extraocular muscles (those which move the eyeball in its socket) to over a thousand in a large limb muscle. The complex of one motoneuron plus the muscle fibres which it innervates is called a *motor unit*. Since they are all activated by the same nerve cell, all the muscle fibres in a single motor unit contract at the same time. Muscle fibres belonging to different motor units, however, may well contract at different times.

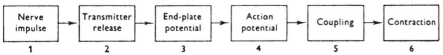

Figure 18.1. The control sequence leading to contraction in a vertebrate 'twitch' muscle fibre.

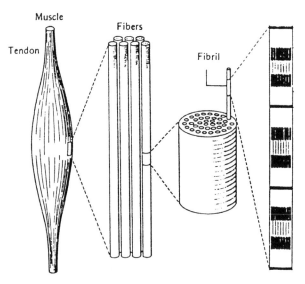

Figure 18.2. Diagram to show the arrangement of fibres in a vertebrate striated muscle. The cross-striations on the myofibrils can be seen with light microscopy. (After Schmidt-Nielsen, 1983.)

Most mammalian muscle fibres are contacted by a single nerve terminal, although sometimes there may be two terminals originating from the same nerve axon. Muscle fibres of this type are known as *twitch fibres*, since they respond to nervous stimulation with a rapid twitch. In the frog and other lower vertebrates, another type of muscle fibre is also commonly found, in which there are a large number of nerve terminals on each muscle fibre. These are known as *tonic fibres*, since their contractions are slow and maintained. There are some tonic fibres in the extraocular muscles of mammals, and also in the muscles of the larynx and the middle ear.

Mechanical properties
The mechanical properties of whole muscles can be investigated with isolated muscle or nerve–muscle preparations such as the gastrocnemius or sartorius in the frog. The sartorius muscle is attached at one end to the pelvic girdle, and at the other end to the tibia at the knee joint. Contraction of the muscle moves the leg forward and flexes it at the knee. For physiological experiments, the muscle can be removed intact from the animal by cutting the tendons, or the bones to which they are attached, which can then be connected to the recording apparatus.

We have seen in chapter 7 that the fibres of frog twitch muscles are electrically excitable. Thus a muscle can be simulated by direct application of electric shocks, as well as via its motor nerve. For many experiments it is more convenient to use such direct stimulation, often via a number of electrodes spaced along the length of the muscle so that all parts are activated simultaneously.

When muscles contract they exert a force on whatever they are attached to (this force is equal to the *tension* in the muscle) and they shorten if they are permitted to do so. Hence we can measure two different variables during the contraction of a muscle: its length and its tension. Most often one of these two is maintained constant during the contraction. In *isometric* contractions the muscle is not allowed to shorten (its length is held constant) and the tension it produces is measured. In an *isotonic* contraction the load on the muscle (which is equal to the tension in the muscle) is maintained constant and its shortening is measured.

Sometimes the tension is measured while length changes, linear or sinusoidal, are imposed on the muscle. Contractions during constant velocity length changes may be called isovelocity responses. Various other types of load can be used. An *elastic load* causes the shortening of a muscle to be proportional to its tension. An *inertial load* consists of a mass which is accelerated by contraction of the muscle. Loads of these types, although very commonly those which the muscle contracts against in nature, are not much used in experiments on the mechanics of muscle, since they do not allow either the length or the tension to be extrinsically determined during the course of the contraction.

Isometric contractions
In isometric contractions both ends of the muscle are fixed so that it cannot shorten, and we measure the contraction as a change in tension. This means that one end of the muscle has to be attached to a suitable tension-recording device. A common device in early experiments was a lever pivoted on a torsion wire, so that its deflection was proportional to the force applied to it. The deflection of the lever could be recorded by a revolving smoked drum on which the tip of the lever wrote, or, later, by a beam of light and a photoelectric cell. Unfortunately such a system is not very stiff (the muscle must always shorten to some extent in order to move the lever) and has appreciable inertia.

Better isometric recording devices are much stiffer and have negligible inertia. One method is to use a special triode valve whose anode projects through a stiff spring diaphragm. Movement of the anode then alters the current flowing through the valve, so the force applied to the anode peg can be displayed as a voltage change on an oscilloscope. Another useful method is to attach the muscle to a steel bar of suitable stiffness which has semiconductor strain gauges bonded onto it (fig. 18.3). The resistance of the strain gauges then varies with muscle tension. For recording the tension from single muscle fibres, more sensitive devices are required, such as a silicon beam strain gauge system or a variable capacitance (see Woledge *et al.*, 1985).

The tension produced by a muscle is a force, and is therefore measured in newtons or in grams weight (1 kg weight is equivalent to 9.80 N). Figure 18.4 shows the time course of the tension development in isometric contractions. A single stimulus produces a rapid increase in tension which then decays; this is known as a *twitch*. The duration of the twitch varies from muscle to muscle, and decreases with increasing temperature. For a frog sartorius at 0 °C, a typical value for the time between the beginning of the contraction and its peak value is about 200 ms, tension falling to zero again within 800 ms.

If a second stimulus is applied before the tension in the first twitch has fallen to zero, the peak tension in the second twitch is higher than that in the first; this effect is known as

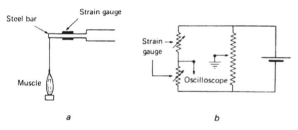

Figure 18.3. An isometric lever system for measuring the force exerted by a muscle without allowing it to shorten. Semiconductor strain gauges are bonded to a steel bar (*a*), and form two arms of a resistance bridge connected to a battery (*b*).

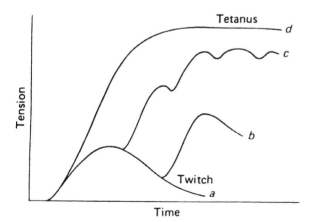

Figure 18.4. Isometric contractions. *a* is the response to a single stimulus, producing a twitch; *b* is the response to two stimuli, showing mechanical summation; *c* is the response to a train of stimuli, showing an 'unfused tetanus'; *d* is the response to a train of stimuli at a higher repetition rate, showing a maximal fused tetanus.

mechanical summation. Repetitive stimulation at a low frequency thus results in a 'bumpy' tension trace. As the frequency of stimulation is increased, a point is reached at which the bumpiness is lost and the tension rises smoothly to reach a steady level. The muscle is then in *tetanus*, and the minimum frequency at which this occurs is known as the *fusion frequency*.

A resting muscle is resistant to stretching beyond a certain length, so that it is possible to determine a passive length–tension curve, as is shown in fig. 18.5. The full isometric tetanus tension of the stimulated muscle is also dependent on length, shown as the 'total active tension' curve in fig. 18.5. The difference between the two curves is known as the 'active increment' curve; notice that this passes through a maximum at a length near to the maximum length in the body, falling away at longer or shorter lengths.

When a muscle in isometric tetanus is suddenly shortened, the tension falls abruptly, and then rises to a new maximum level which is determined by the isometric length–tension relation (fig. 18.6). This type of experiment is known as a *quick release* (Gasser & Hill, 1924). The amount by which the tension falls is roughly proportional to the length change during the quick release, up to the point at which the tension falls to zero. With quick releases of greater extents, there is an increasing delay between the time of release and the redevelopment of tension. The time course of the redevelopment of tension after the quick release is similar to, but not usually identical with, that of the development of tension at the beginning of the period of stimulation. The converse of this type of experiment occurs when the muscle is stretched during contraction.

Isotonic contractions

In the measurement of isotonic contractions, the tension exerted by the muscle is maintained constant (usually by

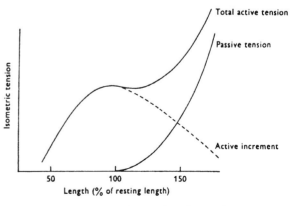

Figure 18.5. The length–tension relation of a muscle.

allowing it to lift a constant load) and its length change is measured. A *free-loaded* contraction is one in which the muscle is loaded in the resting state and then stimulated. This type of measurement is not much used, since the initial

length of the muscle will vary with different loads. In an *after-loaded* contraction, the muscle is not loaded at rest, but must lift a load in order to shorten.

The apparatus used for measuring an after-loaded contraction is shown schematically in fig. 18.7. The muscle is attached to a light, freely pivoted isotonic lever so that it must lift a weight when it shortens. The movement of the lever can be recorded by the interruption of a beam of light focused on a photocell. When the muscle is relaxed, the lever rests against a stop, so that the resting muscle does not have to support the load. If the stop were not there the muscle would take up longer initial lengths with heavier loads, which would make it more difficult to interpret the results of experiments with different loads.

Figure 18.8*a* shows what happens when the muscle has to lift a moderate load while being stimulated tetanically. The tension in the muscle starts to rise soon after the first stimulus, but it takes some time to reach a value sufficient to lift the load, so that there is no shortening at first and the muscle is contracting isometrically. Eventually the tension becomes equal to the load and so the muscle begins to shorten; the tension remains constant during this time and the muscle contracts isotonically. It is noticeable that initially the velocity of shortening during the isotonic phase is constant, provided that the muscle was initially at a length near to its maximum length in the body. As the muscle shortens further, however, its velocity of shortening falls until eventually it can shorten no further and shortening ceases. When the period of stimulation ends, the muscle is extended by the load as it relaxes until the lever meets the afterload stop, after which relaxation becomes isometric

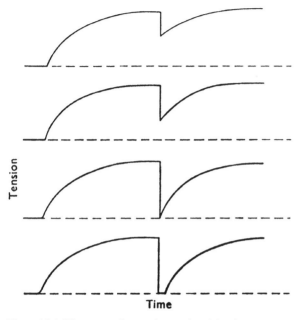

Figure 18.6. Diagram to show an isometric quick release experiment. Tetanic stimulation starts shortly after the beginning of each trace, producing a rise in isometric tension to plateau at its maximum level. The release to a shorter length produces an immediate fall in tension, followed by a recovery to reach the plateau level again. The releases are of increasing extents for the lower traces.

Figure 18.7. Isotonic lever arranged for use with after-loaded contractions. The insert shows the photocell system used to record the position of the lever.

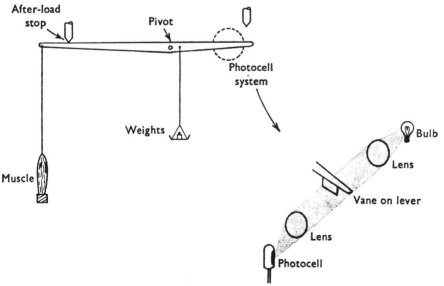

and the tension in the muscle falls to its resting level. If we repeat this procedure with different loads (as in fig. 18.8*b*), we find that the contractions are affected in three ways:

1 The delay between the stimulus and the onset of shortening is longer with heavier loads. This is because the muscle takes longer to reach the tension required to lift the load.
2 The total amount of shortening decreases with increasing load. This is because the isometric tension falls at shorter lengths (fig. 18.5) and so the more heavily loaded muscle can only shorten by a smaller amount before its isometric tension becomes equal to the load. Figure 18.9 illustrates this point.
3 During the constant velocity section of the isotonic contraction, the velocity of shortening decreases with increasing load. It becomes zero when the load equals the maximum tension which can be reached during an isometric contraction of the muscle at that length.

Notice that the first two of these observations are essentially predictable from what we already know about isometric contractions.

An alternative way of measuring isotonic contractions is by means of the *isotonic release* method. The apparatus used for this is shown in fig. 18.10. The muscle is attached to an isotonic lever as in fig. 18.7, but the lever can here be held up against the contraction of the muscle by means of a stop which can be immediately withdrawn at any desired time during the contraction. The results of a typical experiment are shown in fig. 18.11. At the beginning of the stimulation period the muscle contracts isometrically and the tension rises to its maximum value. When the release relay stop is withdrawn, the tension falls abruptly to a lower level which is equal to the value of the load. The change in length seems to consist of two components: firstly, an abrupt shortening which is coincident with the change in tension; and, secondly, a steady isotonic shortening at more or less constant velocity. Since the mass of the lever and the load can never be negligible, the lever system possesses inertia, and so there is usually some oscillation at the change-over between these two phases. With increasing loads (fig. 18.11*b*), the initial length change is less, and the velocity of shortening during the isotonic phase is less. The initial phase of length change thus behaves as an elasticity, since its extent is proportional to the change of tension. The

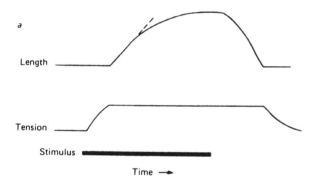

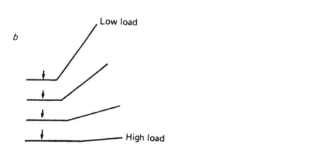

Figure 18.8. After-loaded isotonic tetanic contractions. *a* shows the length and tension changes during a single contraction, with shortening as an upward deflection of the length trace. Shortening is initially at a constant velocity, as is suggested by the dashed line. *b* shows the initial length changes in contractions against different loads.

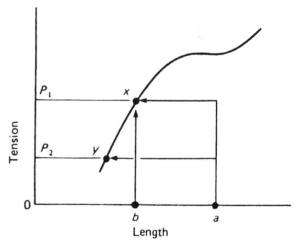

Figure 18.9. Diagram showing why it is that a lightly loaded muscle can shorten further than a heavily loaded one. Starting from point *a* on the length axis, the muscle contracts isometrically until its tension is equal to the load it has to lift, and then it shortens until it meets the isometric length–tension curve. With a heavy load P_1 this occurs at *x*, with a lighter load P_2 it occurs at *y*. Notice that point *x* can also be reached by an isometric contraction from point *b*. (When starting from a much extended length, a muscle may in practice stop short of point *x* when lifting load P_2; this is probably caused by inequalities in the muscle, so that some sarcomeres shorten more than others.)

isotonic phase shows the same relation between force and velocity as is seen in afterloaded contractions.

If we plot the velocity of shortening against the load in isotonic contractions we get a *force–velocity curve*, as in fig. 18.12. Similar results are obtained whether the velocity is measured in an after-loaded contraction or in an isotonic release, provided all the measurements are made at the same length. The form of the curve is well described by an equation produced by A. V. Hill (1938):

$$(P + a)(V + b) = \text{constant} \qquad (18.1)$$

where P is the force or load on the muscle, V is the velocity of shortening, and a and b are constants with the dimensions of force and velocity, respectively. The constant on the right-hand side of the equation is equal to $b(P_0 + a)$, where P_0 is the isometric tension.

Careful experiments with single muscle fibres show that the velocity of shortening in the isotonic phase of an isotonic release does not immediately settle down to its final value, but is transiently higher and then lower than the steady value (Podolsky, 1960; A. F. Huxley & Simmons, 1971). We shall look at some investigations on these transient effects in chapter 20. Their existence means that the force–velocity relation is essentially a steady-state relation, and that it applies only approximately to conditions when the tension is changing.

Power output

Power is the rate of doing work, so since work is force × distance, power must be force × velocity. Since work is mechanical energy, power is the rate of mechanical energy expenditure. Power output during isotonic shortening can

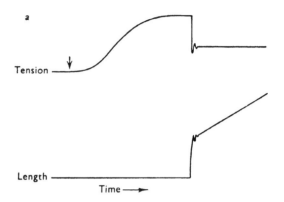

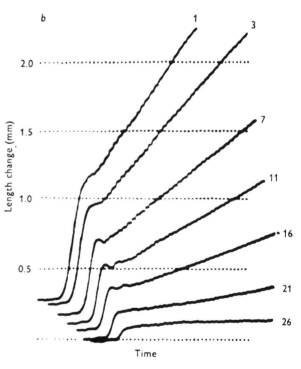

Figure 18.11. Isotonic releases. *a* is a diagram of the length and tension changes during an isotonic release; shortening is shown upwards on the length trace. *b* shows length changes during a series of isotonic releases against different loads. Shortening is again shown upwards, the dots on the grid lines are at 1 ms intervals, and the figures opposite each trace indicate the load (in grams) on the muscle after release. From a frog sartorius muscle, with maximum isometric tension 32 g. (*b* from Jewell & Wilkie, 1958.)

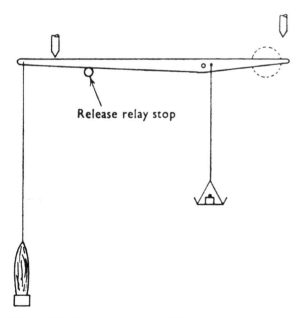

Figure 18.10. Isotonic lever arranged for isotonic releases. The system is as in fig. 18.7 except that the lever is held up by a stop mounted on a relay. When the relay is activated, the stop is immediately withdrawn so that the muscle can now shorten.

be calculated from the force–velocity curve. It is zero when $P = 0$ or $P = P_0$ and positive at intermediate values. It is maximal at a point near to 0.3 V_{max} or 0.3 P_0 in muscles where a/P_0 is about 0.25 to 0.3. In muscles where a/P_0 is lower, giving a more concave force–velocity curve, the maximum power occurs at lower values of P and correspondingly higher values of V (fig. 18.13). Values of maximum power obtained from force–velocity curves in this way range from 7 to 500 W kg^{-1}.

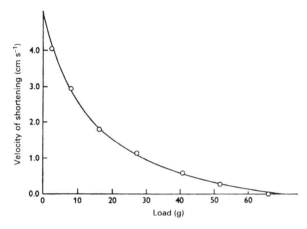

Figure 18.12. The force–velocity relation of a frog sartorius muscle at 0 °C. The experimental points were determined from after-loaded contractions as in fig. 18.8. The curve is drawn according to equation 18.1 (Hill's equation) with $a = 14.35$ g, $b = 1.03$ cm s^{-1}, and $P_0 = 65.3$ g. (From Hill, 1938, by permission of the Royal Society.)

The shape of the power–load curve has consequences for the design of bicycles, amongst other things. A cyclist's leg muscles will not produce their best power output if they are contracting against a load that is too heavy or too light, so cyclists need a suitable set of gears to arrange that their legs are moving at much the same speed whether the bicycle is going slowly uphill or rapidly on the flat. Equally, animals need muscles with force–velocity curves appropriate to the speeds at which they move (Hill, 1950b).

The maximum power measured from a force–velocity relation refers only to power output during the isotonic contraction. For sustained work the muscle usually has to contract and relax again and again, as in running, swimming, or flying, for example. During these activities it is likely to spend at least half the time relaxed and being extended, when the power output will be zero or (if there is some resistance to extension) negative. Power under these conditions has been measured by subjecting the muscle to sinusoidal length changes synchronized with the stimulus frequency, a procedure known as the work-loop method. Measured values range from less than 5 to around 150 W kg^{-1} (Josephson, 1993).

Heat production

Almost all chemical reactions, and a number of physical changes, are accompanied by the evolution or absorption of heat. Contracting muscles produce heat, and the measurement of this heat production is of considerable interest in

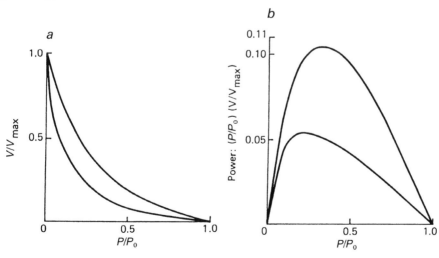

Figure 18.13. Diagram to show power output during isotonic contractions, calculated from the force–velocity curve. *a* shows two force–velocity curves drawn according to equation 18.1, one with $a/P_0 = 0.3$ and the other (the lower, more concave curve) with $a/P_0 = 0.1$. The curves are scaled to units of V_{max} and P_0. *b* shows the two power–load curves calculated from the two curves in *a* by multiplying V and P together at individual points. Notice that the more curved force–velocity curve has a relatively lower maximum power output that is produced at a relatively lower force. (From Woledge *et al.*, 1985.)

that we might expect it to be related to the chemical reactions involved in the contraction process. Furthermore, measurement of the heat production is an essential prerequisite in any attempt to measure the energy expended by a contracting muscle. This energy, $-\Delta E$, is the sum of the work w done by the muscle on the load and the heat h produced during the contraction, or

$$-\Delta E = h + w \qquad (18.2)$$

The minus sign on the left-hand side of equation 18.2 is a matter of thermodynamic convention.

The technical problems involved in the measuring the heat production of muscles are considerable, since the temperature changes involved may be extremely small. If an electric circuit is made up of two different metals, an electromotive force is produced if the temperature of one of the junctions is different from that of the other; such a device is called a thermocouple. Hill and his colleagues used a thermopile to measure heat production; this consists of a large number of thermocouples in series, and is therefore more sensitive than a single thermocouple.

The main features of the heat production during isometric contractions were determined by Hill & Hartree in 1920. Later work has been much concerned with increasing the accuracy and time resolution of the measurements involved, and with the heat changes associated with shortening or lengthening of the muscle.

In resting muscle, various metabolic processes occur throughout the life of the muscle, and these processes are associated with the liberation of heat. This *resting heat* is about 10 mJ g^{-1} min^{-1} in frog sartorius muscles at 20 °C (Hill & Howarth, 1957). The nomenclature of the heat changes produced by contracting muscles has varied somewhat since the original formulation by Hill & Hartree; we shall adopt the scheme given by Kushmerick (1983), in which the heat produced during contraction and relaxation is called the *initial heat*, and the heat produced after the muscle has relaxed is called the *recovery heat.*

In an isometric tetanus, the initial heat consists of three phases (fig. 18.14). The *activation heat* is a burst of heat production produced soon after the onset of stimulation. It is thought to accompany the processes whereby the force-producing reactions are switched on; we shall see later that these processes involve calcium ion movements within the muscle cells. The *maintenance heat* is produced steadily throughout a tetanus and is thought to reflect largely the force-producing reactions themselves. The *relaxation heat* is produced while the muscle is relaxing, and is probably largely due to the degradation of mechanical work within the muscle.

Elastic bodies usually undergo heat changes when they are stretched or allowed to shorten; they are known as *thermoelastic* changes. Most substances, such as steel or wood, absorb heat when stretched and release heat when released; active muscle shows this type of behaviour (Woledge, 1961). Rubber, on the other hand, releases heat when stretched and cools on release; this type of behaviour is shown by resting muscle.

When an active muscle shortens, an extra amount of heat is released, in addition to the activation heat. This is called the *heat of shortening*, a term introduced by Hill in 1938. One of Hill's experiments is shown in fig. 18.15; he concluded that the extra heat released on shortening is proportional to the distance shortened.

When a contracting muscle is stretched, its heat production is greatly reduced, so that it is actually absorbing energy during the period of stretch. There has been much discussion over the years about the precise relations between length changes and heat production; they are not as simple as was first thought (Hill, 1938, 1964; Homsher & Rall, 1973; Woledge *et al.*, 1985; Homsher, 1987).

The high energy compounds which have been broken down during contraction have to be resynthesized. For a long period after the end of a contraction, the heat production of a muscle is higher than the normal resting level. This is known as the *recovery heat*; it corresponds to the respiratory processes involved in resynthesis of the high energy phosphate compounds. It is greatly reduced in the absence of oxygen.

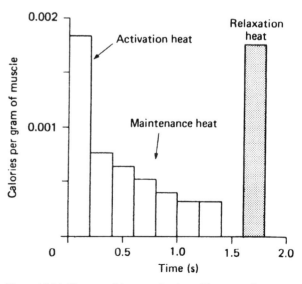

Figure 18.14. The rate of heat production of frog sartorius muscle during an isometric tetanus. The muscle was stimulated for a period of 1.2 s. (From Hill & Hartree, 1920.)

The Fenn effect

Most early theories of the mechanism of muscular contraction suggested that the muscle became effectively a stretched elastic body when it was stimulated. But the force exerted by a purely elastic body when stretched is determined by its length, and is independent of the velocity of shortening. Thus the observation that, during isotonic shortening, force is determined by velocity indicates that the 'elastic' theory cannot be correct. It was then suggested (Gasser & Hill, 1924; Levin & Wyman, 1927) that some of the elasticity of the muscle is damped by an internal viscosity. The idea was that some of the tension in the elasticity was used up internally in working against the viscosity. More tension would be so absorbed at high velocities, leaving less to appear at the ends of the muscle.

Evidence incompatible with the viscoelastic theory was provided by Fenn (1923, 1924). He measured the shortening of a muscle under various loads and also measured its heat production at the same time. He was then able to calculate, from equation 18.2, the total energy release and found that more energy is released when the muscle is allowed to shorten during the contraction. This is not what one would expect from the elastic and viscoelastic theories, which postulate that the muscle contains a fixed amount of potential energy at the beginning of a contraction, which must be converted into heat and work during the contraction. Fenn concluded that the energy released during a contraction 'can be modified by the nature of the load which the muscle discovers it must lift after the stimulus'.

Confirmation and extension of Fenn's results was obtained by Hill in 1938. The extra release of energy on shortening is frequently known as the 'Fenn effect'.

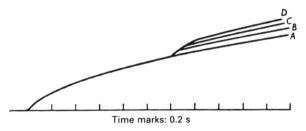

Time marks: 0.2 s

Figure 18.15. Heat production in a frog sartorius muscle when it was allowed to shorten during a tetanus at 0 °C, to show the heat of shortening. The curves are galvanometer deflections proportional to the amount of heat produced by the muscle. Curve *A* shows the isometric response, with no shortening. For curves *B* to *D* the muscle was released at 1.2 s after the onset of stimulation and allowed to shorten various distances (1.9, 3.6 and 5.2 mm) against a constant load (3 g) before being stopped. (From Hill, 1938, by permission of the Royal Society.)

Chemical change in muscle

Hill's investigations in 1938 and later provided an approximate but influential description of the mechanical properties of contracting muscle, and his estimates of total energy release gave a quantitative description of the Fenn effect. But a description of muscle solely in terms of its gross output is necessarily incomplete, since it does not answer one of the crucial questions about muscle: how is chemical energy converted into work?

Two major lines of investigation have approached this question. Firstly, there has been a search for the fuel for muscular contraction and attempts to relate energy output to chemical breakdown. This became increasingly fruitful from 1927 onwards. Secondly, beginning in the 1950s, structural studies have led to considerable advances in our understanding of the mechanisms involved in the contractile process. Both these paths have drawn upon the mechanical and thermal studies we have described so far. For the remainder of this chapter we consider chemical change; the nature of the contractile process forms the subject of chapter 19.

The nature of the energy source

The energy for muscular contraction is derived ultimately from the chemical energy released by the oxidation of food substances. For our purposes it is more pertinent to enquire what the immediate source of the energy for contraction is. Quantitative studies on this question began with the work of Fletcher & Hopkins in 1907. They showed that muscles can continue to contract in the absence of oxygen, and that they produce lactic acid under these conditions. It was later shown by Meyerhof and others that this lactic acid is formed by the breakdown of glycogen, which itself is formed by polymerization of glucose derived from carbohydrates in the food. Until about 1930, then, it was generally believed that the immediate source of energy for contraction was the reaction which results in the formation of lactic acid; this was known as 'the lactic acid theory'.

The next step was the discovery of 'phosphagen', soon shown to be creatine phosphate, and the demonstration that it was broken down during contraction (Eggleton & Eggleton, 1927; Fiske & Subbarow, 1927). In a crucial series of experiments, Lundsgaard (1930a,b) determined the concentrations of lactic acid and creatine phosphate (also called phosphorylcreatine) in muscles poisoned with iodoacetate and held in anaerobic conditions. He found that stimulation and contraction did not cause the production of lactic acid under these conditions, whereas creatine phosphate was broken down in proportion to the tension produced by the muscle. These results clearly disproved the lactic acid theory.

The final stage in this 'revolution in muscle physiology', as it was called by Hill in 1932 (see Hill, 1965), was the discovery of adenosine triphosphate (ATP) and its role as a carrier of energy within the cell (Lohmann, 1929, 1934; Meyerhof & Lohmann, 1932; Lipmann, 1941). This led to an understanding of the basic scheme of energy production and utilization in cells (fig. 18.16) which is now familiar to all students of cell biology. A fascinating account of these discoveries, and of much else in the history of muscle research, is given by Needham (1971).

If the scheme in fig. 18.16 is correct, then, as Hill (1950*a*) pointed out in his 'challenge to biochemists', it should be possible to show that ATP is broken down during the contraction of living muscle cells. The general method used by biochemists responding to this challenge has been to stimulate a muscle, then immediately and very rapidly freeze it so as to prevent any further biochemical changes, and then determine how much of the substance one is interested in is present. A similar determination is performed on an unstimulated muscle, usually the equivalent muscle on the opposite leg of the same animal. Using these techniques, it was shown by a number of workers that contraction leads to the breakdown of creatine phosphate in muscles poisoned with iodoacetic acid (which prevents glycolysis, and hence prevents resynthesis of creatine phosphate). This could mean either that creatine phosphate is the immediate source of energy for contraction, or that creatine phosphate reacts with adenosine diphosphate (ADP) to give ATP and creatine, ATP being the immediate energy source.

The substance 1-fluoro-2,4-dinitrobenzene (FDNB) blocks the action of the enzyme creatine phosphotransferase, so that ADP cannot be rephosphorylated to ATP by the breakdown of creatine phosphate. Using this substance, Davies and his colleagues were able to show that break-down of ATP occurs during contractions of frog rectus abdominis and sartorius muscles (Cain *et al.*, 1962; Infante & Davies, 1962).

The average amount of ATP lost during the rising phase of each twitch in the sartorius muscle experiments was 0.22 μmol per gram of muscle. Is this sufficient to account for the energy used in the contraction? The muscle shortened against a light load in these experiments, and the average value for the work done in each contraction was 17.4 g cm g^{-1}; this is equivalent to 1.71×10^{-3} J g^{-1}. The heat of hydrolysis of the terminal phosphate bond in the conversion of ATP to ADP is probably about 47 kJ mol^{-1} (Homsher, 1987). Hence the dephosphorylation of 0.22 μmoles of ATP should make available about 10^{-2} J. The amount of ATP broken down is therefore more than enough to account for the work done in a twitch; the excess energy appears as heat.

Energy balance during contraction

The first law of thermodynamics states the principle of the conservation of energy: energy is neither created nor destroyed in a chemical reaction. In contracting muscle, chemical energy is converted into heat and work, and hence it should be possible to show that the energy available from chemical breakdown is equal to the heat released plus the work done. If we cannot account for the energy release (heat plus work) by estimating the energy available from the known chemical reactions in the muscle, then there must be some unknown reaction or reactions present. Energy balance experiments therefore provide a clear test of our understanding of the overall chemical changes that take place in contracting muscle. The subject has been well reviewed by Curtin & Woledge (1978), Kushmerick (1983), Woledge *et al.* (1985) and Homsher (1987).

In thermodynamic terms, it is useful to refer to the enthalpy change of a reaction, ΔH. At constant pressure P, this is related to the energy change ΔE by

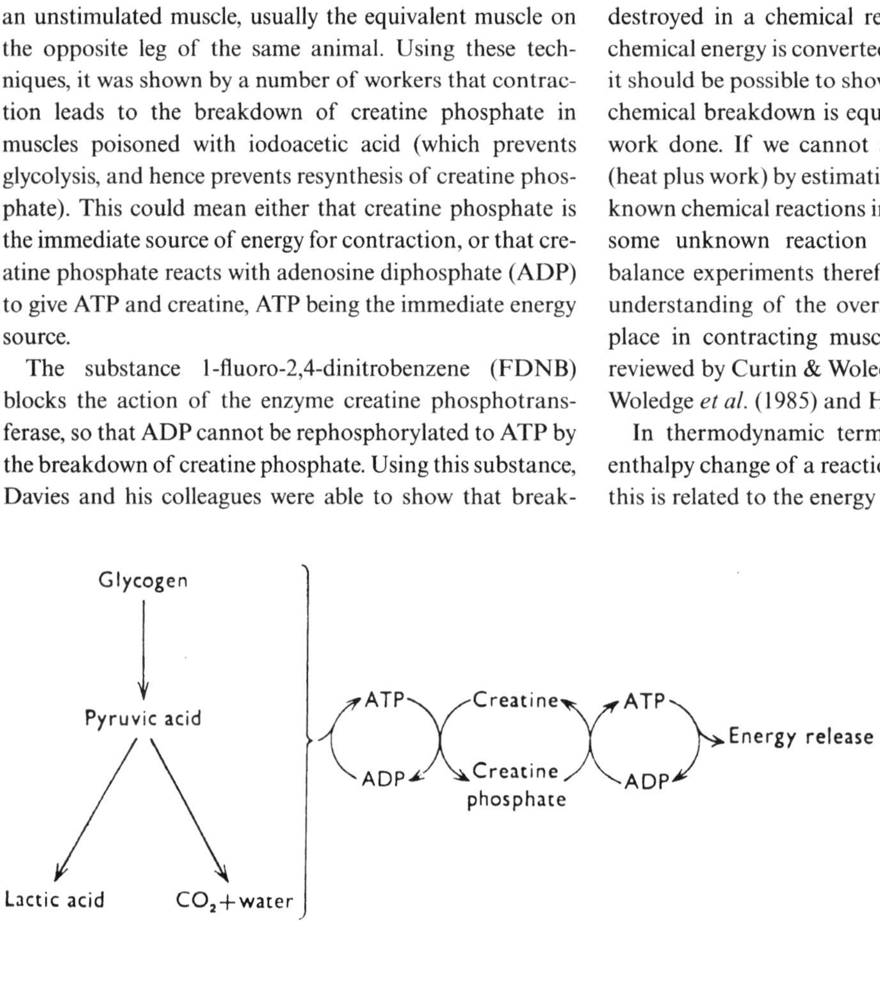

Figure 18.16. Diagram to show, very schematically, the respiratory energy production in an animal cell. Oxygen is needed to convert pyruvic acid to carbon dioxide and water. Many invertebrates use arginine phosphate instead of creatine phosphate for storage of high energy phosphate.

$$\Delta H = \Delta E + P\Delta V$$

where ΔV is the change in volume. If ΔV is zero (as is very nearly the case in contracting muscle) the enthalpy change is effectively the same as the energy change. We may thus rewrite equation 18.2 more precisely as

$$-\Delta H = h + w$$

If there are n reactions occurring in the contracting muscle, then we have to add the enthalpy changes for all n of them together to get the total enthalpy change. Thus if ΔH_i is the molar enthalpy change for the ith reaction (i.e. the heat of reaction per mole of one of the reactants) and ξ_i (xi$_i$) is the extent of the reaction (i.e. the number of moles of that reactant which are converted), then the total enthalpy change is given by

$$h + w = -\sum_{i=1}^{n} \xi_i \Delta H_i$$

Let us first examine a piece of work by Wilkie (1968) that at first sight provided a clear-cut correspondence between energy release and chemical breakdown. Wilkie used frog sartorius muscles which were poisoned with iodoacetate to prevent glycolysis and nitrogen to prevent oxidative phosphorylation. Under these conditions we would expect any ADP formed from ATP breakdown to be immediately rephosphorylated by the breakdown of creatine phosphate, so that change in creatine phosphate content should represent the net chemical energy consumption of the muscle. Wilkie therefore measured the amount of creatine phosphate in individual muscles after they had performed a contraction in which he had also measured the work done by the muscle and the heat it produced. He also measured the creatine phosphate content of the unstimulated muscle from the other leg of the frog so as to estimate the change in creatine phosphate content due to activity. He repeated this experiment for a variety of different types of contraction: isometric and isotonic, twitches and tetani. The results are shown in fig. 18.17. Clearly the breakdown of creatine phosphate is directly proportional to the sum of the heat and work produced by the muscle, 46.4 kJ of energy being released per mole of creatine phosphate broken down.

Does all of this energy come from the breakdown of creatine phosphate? Although the figure of 46.4 kJ mol^{-1} (or 11.1 kcal mol^{-1}) did at first appear to be in reasonable agreement with the expected values for the heat of hydrolysis of creatine phosphate, doubts soon arose. Woledge (1973), as a result of careful calorimetric measurements, concluded that the expected value should be about 34 kJ mol^{-1}.

In another investigation, Kushmerick & Davies (1969) measured the ATP breakdown during the contraction phase of muscles poisoned with DTNB. They found that here also the chemical changes could not account for all the heat produced by the muscle. The difference between the energy that could be explained in terms of creatine phosphate and ATP breakdown and the total observed energy release became known as 'unexplained enthalpy'.

It was desirable to determine the time course of the unexplained energy release. In order to do this it was necessary to freeze muscles at different times during the contraction cycle, and hence to use a very rapid freezing technique. Kretschmar & Wilkie (1969) designed a rather dramatic apparatus to do this: the muscle was hit simultaneously from each side by two aluminium hammers which until that moment had been sitting in liquid nitrogen. Thus the creatine phosphate content of muscles could be determined at different times during a period of contraction, and compared with the heat production of other muscles. Using this method Gilbert *et al.* (1971) found that the initial rate of heat production was notably higher than could be accounted for by chemical breakdown. Curtin & Woledge

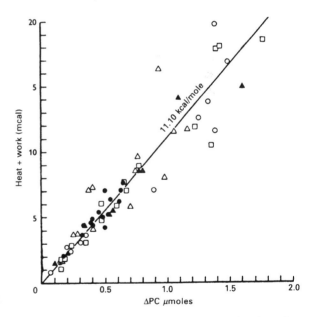

Figure 18.17. The relation between energy production (heat plus work) and creatine phosphate breakdown (ΔPC) in frog sartorius muscles poisoned with iodoacetate and nitrogen. Each point represents a determination on one muscle after the end of a series of contractions. The symbols indicate different types of contraction: isometric switches and tetani, isotonic tetani, and tetani in which the muscle was stretched by the load. The slope of the diagonal is 46.4 kJ mol^{-1} (= 11.1 kcal mol^{-1}). (From Wilkie, 1968.)

(1979) reached similar conclusions. In a 5 s isometric tetanus, the unexplained energy accounted for about 28% of the total energy release, but thereafter little extra unexplained energy was produced (fig. 18.18).

At this point the faint-hearted might begin to doubt whether the first law of thermodynamics applies to frog muscles, but Paul (1983) produced clear evidence that it does. He stimulated muscles so as to produce an isometric tetanus for 3 s every 256 s over a period of some minutes. During this time they reached a steady-state condition in which the heat produced per cycle was constant; this implies that the recovery processes had proceeded to the same extent after each tetanus. Oxygen consumption measurements provided an estimate of the total chemical change during both contraction and recovery phases. Measurements of creatine phosphate and ATP levels (by using the cooled hammer method) in experimental and control muscles allowed the chemical change during a tetanus to be determined.

Paul found that that total observed enthalpy change, measured as heat production (averaging 181.5 mJ g^{-1} per cycle of tetanus plus recovery) was precisely explained by the total chemical change as measured by oxygen consump-

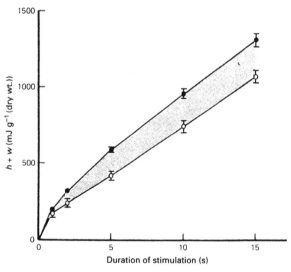

Figure 18.18. Enthalpy production during isometric tetani of different durations in frog muscle. The black symbols show the observed enthalpy (heat plus work), as measured at different times during a 15 s tetanus; the heat was measured with a thermopile and the work was calculated as that done against the series elasticity. The white symbols show the explained enthalpy, obtained from measurements of the breakdown of creatine phosphate and ATP. The difference between the two curves is the unexplained enthalpy. (From Curtin & Woledge, 1979.)

tion (180.6 mJ g^{-1}). The excess unexplained enthalpy during the contractions (the enthalpy difference between the heat production and the creatine phosphate breakdown) was thus balanced by a corresponding deficiency during the recovery periods. Apart from restoring our confidence in the laws of nature, this experiment provides further evidence for the reality of the reactions producing the unexplained enthalpy and shows that they are reversed during the recovery period.

What is the nature of these reactions? One way of approaching this question is to try to see whether or not they are associated with the contractile mechanism itself or with the coupling process (the intracellular control processes which switch contraction on and off). We shall see in the next chapter that when muscles are greatly stretched they produce little or no tension, but the intracellular calcium movements which form part of the coupling process still occur. Unexplained enthalpy is still produced at these long lengths, so it is unlikely to be associated with the contractile mechanism (Curtin & Woledge, 1981; Homsher & Kean, 1982).

We shall also see later (chapter 20) that the coupling process involves release of calcium ions from the sarcoplasmic reticulum into the sarcoplasm. Some of these calcium ions combine with the protein troponin, which forms part of the contractile mechanism. Frog muscles also contain appreciable quantities of the calcium-binding protein parvalbumin (see Wnuk *et al.*, 1982). Calcium is therefore likely to bind to both these proteins soon after the onset of stimulation.

Homsher *et al.* (1987) discuss the amount of heat that would be released by these reactions. The concentration of parvalbumin in frog muscle is about 0.4 μmol g^{-1}, and each molecule has two binding sites for calcium. There are 0.2 μmol g^{-1} of similar binding sites on troponin so in all there is about 1 μmol of calcium-binding sites per gram of muscle. Displacement of magnesium by calcium from these proteins results in a heat release of about 31 kJ per mole of calcium bound, so calcium binding in the muscle could account for a heat release of about 31 mJ g^{-1}, in good agreement with the size of the unexplained enthalpy. Thus it seems likely that the isometric unexplained enthalpy is largely a product of calcium movements associated with the coupling process.

Another form of unexplained enthalpy occurs during rapid shortening. Homsher and his colleagues (1981) found that the extra energy release on shortening (Hill's heat of shortening plus work) was not accounted for by the chemical breakdown of high energy phosphate (creatine phosphate plus ATP) during the period of shortening. The unexplained enthalpy production under the conditions used (a 30% shortening in 0.3 s) was about 6 mJ per gram of muscle. This was balanced by a corresponding deficiency of heat production during the isometric contraction, which continued immediately after the end of the shortening period. It looks as though there is some exothermic change in the contractile mechanism during rapid shortening and that this change to some extent precedes the breakdown of high energy phosphate.

19
The contractile mechanism of muscle

Microscopic examination reveals one of the most characteristic features of vertebrate skeletal muscles: they are striated. Suitable optical techniques or staining methods show that bands of light and dark material alternate along the length of the myofibrils, and that these bands are aligned across the breadth of the fibre.

Detailed descriptions of these striation patterns, and sometimes observations on how they altered with changes in fibre length, were made by a number of nineteenth-century microscopists. But, as Andrew Huxley (1980) has pointed out, this knowledge was disregarded and further structural studies were largely neglected in the first half of the twentieth century. The advances in understanding of muscle during this time arose largely from biochemical and physiological studies, and the nature of the striation pattern seemed to have no relevance to these approaches.

All this changed with the advent of the sliding filament theory in 1954. Quite suddenly muscle fine structure made sense in terms of function. The search for structural detail as the means of interpreting physiological and biochemical observations began afresh, with new and increasingly sophisticated methods. As a result we now have some exciting glimpses of the molecular activity that underlies muscular contraction. But let us first put the sliding filament theory in its context by taking a look at the biochemical and structural background from which it emerged.

The myofibril in 1953
The contractile machinery of striated muscle cells consists of a small number of different proteins which are aggregated together in filaments. The two major components of this system are the proteins *actin* and *myosin*, which interact with each other to produce the contraction.

Biochemistry
The name myosin was first used by Kühne in 1864 to describe the protein mixture which he isolated by saline extraction from frozen frog muscle. But pure myosin was not obtained until the 1940s, when Straub (1943), working with Szent-Györgyi in Hungary separated it from the

second major protein, which he called actin. A number of other constituents of the myofibril have been discovered since then; they serve either to modify the interaction between actin and myosin or to determine the structural organization of the system.

Myosin is a rather complex protein with an M_r of about 520 000. One of its most important properties is that it is an ATPase, i.e. it will enzymically hydrolyse ATP to form ADP and inorganic phosphate (Engelhardt & Ljubimowa, 1939). This reaction is activated by calcium ions, but inhibited by magnesium ions. Treatment with the proteolytic enzyme trypsin splits the myosin molecule into two sections known as light meromyosin and heavy meromyosin; of these, only heavy meromyosin acts as an ATPase (Mihalyi & Szent-Györgyi, 1953).

Isolated *actin* exists in two forms: G-actin, a more or less globular molecule of about 42 000 M_r, and F-actin, a fibrous protein which is a polymer of G-actin. Neither form has any ATPase activity.

If solutions of actin and myosin are mixed, a great increase in viscosity occurs, due to the formation of a complex called *actomyosin*. Actomyosin is an ATPase but, unlike myosin ATPase, it is activated by magnesium ions. 'Pure' actomyosin (a mixture of purified actin and purified myosin) will split ATP in the absence of calcium ions, but 'natural' actomyosin (an actomyosin-like complex which can be extracted from minced muscle with strong salt solutions) can only split ATP if there is a low concentration of calcium ions present. In the absence of calcium ions, addition of ATP to a solution of natural actomyosin results in a decrease in viscosity, suggesting that the actin–myosin complex becomes dissociated.

These properties of actomyosin solutions can be paralleled by the properties of glycerol extracted muscle fibres. Szent-Györgyi (1949) prepared these by soaking a muscle in cold 50% glycerol for a period of weeks. Such treatment removes most of the sarcoplasmic material from the fibres, leaving only the contractile structures. In the absence of ATP the fibres do not contract, but cannot be extended without exerting considerable force; they are said to be in *rigor.* This stiffness is caused by the formation of cross-linkages between the actin and myosin, and corresponds to the formation of actomyosin when solutions of actin and myosin are mixed. On addition of ATP in the presence of magnesium ions the fibres become readily extensible. This relaxing or 'plasticizing' action of ATP is a result of the breakage of the cross-links, and corresponds to the dissociation of actomyosin in solution by ATP. Finally, if calcium ions are added to the glycerinated fibres in the presence of ATP and magnesium ions, ATP is split and the fibre contracts; this corresponds to the calcium-activated splitting of ATP by 'natural' actomyosin.

Structure

The striation pattern of myofibrils, as seen by conventional light microscopy after fixing and staining, or by phase-contrast, polarized light or interference microscopy, is shown in fig. 19.1*a*. The two main bands are the dark, strongly birefringent A band and the lighter, less birefringent I band. These bands alternate along the length of the myofibril. In the middle of each I band is a dark line, the Z line. In the middle of the A band is a lighter region, the H zone, which is sometimes bisected by a darker line, the M line. (The origins of these letters used in the description of the striation pattern are seen in the original names: *I*sotropisch, *A*nisotropisch *Z*wischenscheibe, *H*ensen's disc, and *M*ittelmembran. These names are no longer used.) The unit of length between two Z lines is called the *sarcomere.*

The structural basis of the myofibrillar striation pattern

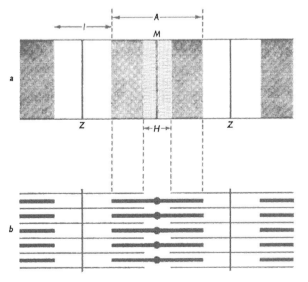

Figure 19.1. The striation pattern of a myofibril as seen by light microscopy (*a*), and its structural basis, seen by electron microscopy: interdigitating arrays of thick and thin filaments (*b*).

was obscure until the advent of the techniques of electron microscopy and thin sectioning, and their application to muscle by Hugh Huxley and Jean Hanson (Hanson & Huxley, 1953, 1955; H. E. Huxley, 1953*a*, 1957). They found that the myofibrils are composed of two interdigitating sets of filaments, about 50 and 110 Å in diameter. The thin filaments are attached to the Z lines and extend through the I bands into the A bands. The position of the thick filaments is coincident with that of the A band. The H zone is that region of the A band between the ends of the two sets of thin filaments, and the M line is caused by cross-links between the thick filaments in the middle of the sarcomere. Because of their positions, the thick filaments are sometimes called A filaments and the thin filaments are called I filaments. This arrangement can be seen in the electron micrographs shown in figs. 19.2 and 19.3, and, diagrammatically, in fig. 19.1*b*.

As we have seen, the major part of the myofibrillar material consists of two proteins, *actin* and *myosin,* and the interaction between these two seems to be the chemical basis of muscular contraction. Consequently, it was desirable to determine the localization of actin and myosin in the myofibrils, and to see whether their distribution was related to that of the thick and thin filaments discovered by electron microscopy. The question was investigated by Hanson & Huxley (1953, 1955) using isolated myofibrils from a muscle which had been previously extracted with glycerol; glycerol extraction removes most of the sarcoplasmic material from rabbit psoas muscles, leaving only the contractile structures. The appearance of such a myofibril, viewed by phase-contrast microscopy, is shown in fig. 19.4*a*. After treatment with a 0.6 M solution of potassium chloride containing some pyrophosphate and a little magnesium chloride, the dark material of the A bands disappeared, as is shown in fig. 19.4*b*. This solution had previously been used to extract myosin from minced muscle, and it was therefore concluded that the A filaments are composed largely of myosin. The myosin-extracted fibre was then treated with a 0.6 M potassium iodide solution, which was known to extract actin from muscle. This removed the substance of the I bands (fig. 19.4*c*), showing that the I filaments are composed mainly of actin. It would seem that the Z lines, which were not affected by the extraction of myosin and actin, are composed of some other substance.

Transverse sections through the A band in the region of overlap show that the filaments are arranged in a hexagonal array so that each myosin filament is surrounded by six actin filaments, and each actin filament is surrounded by three myosin filaments (fig. 19.5). Hence there are twice as many actin filaments as there are myosin filaments. Cross-sections through the H zone show only myosin filaments, and cross-sections through the I band show only actin filaments. Previous to these electron microscope studies, Hugh Huxley (1953*b*) had obtained evidence for a hexagonal array of two types of filament from equatorial measurements of low angle X-ray diffraction patterns. Taking his results in conjunction with the electron micrographs, it was deduced that the centre-to-centre distance between each myosin filament in the array was about 450 Å at rest length. This dis-

tance increases at shorter lengths and becomes less if the muscle is stretched, indicating that the myofibril is effectively a constant volume system. In electron micrographs the apparent distance between filaments may appear to be less than 450 Å, as a result of the shrinkage which occurs during fixation and embedding.

High magnification electron micrographs of glycerol-extracted muscles (fig. 19.3) show that the thick filaments are covered with projections (H. E. Huxley, 1957). In the overlap region, these projections may be joined to the thin filaments, and so they are known as *cross-bridges*. There are no projections in the very middle of the filaments; this produces a light region, variously known as the L zone, pseudo H zone, or M region, about 0.15 μm long, in the middle of the sarcomere (fig. 19.2).

The sliding filament theory

Prior to 1954, most suggestions as to the mechanism of muscular contraction involved the coiling and contraction

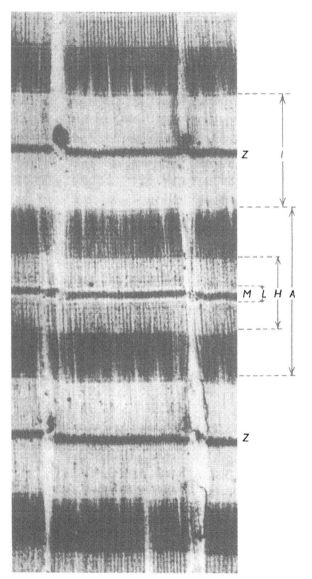

Figure 19.2. Thick longitudinal section of a frog sartorius muscle fibre, showing the striation pattern as seen by electron microscopy at relatively low magnification. Magnification 27 000×. (Photograph kindly supplied by Dr H. E. Huxley.)

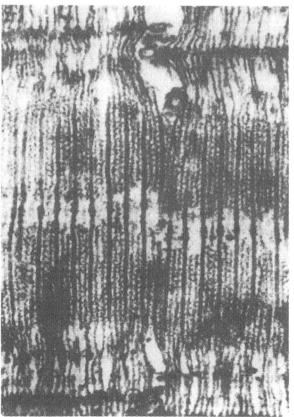

Figure 19.3. Thin longitudinal section of a glycerol-extracted rabbit psoas muscle fibre. Notice, particularly, the cross-bridges between the thick and thin filaments. (Photograph kindly supplied by Dr H. E. Huxley.)

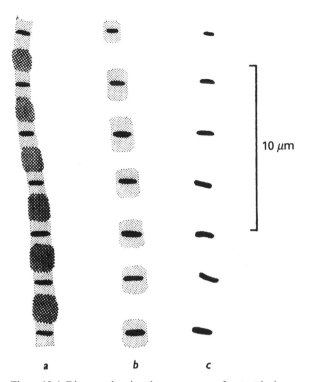

10 μm

a   b   c

Figure 19.4. Diagram showing the appearance of a stretched myofibril from glycerol-extracted rabbit muscle, as viewed by phase-contrast microscopy: (*a*), before treatment; (*b*) after extraction of myosin; (*c*) after extraction of actin. (Drawn from a photograph in Hanson & Huxley, 1955.)

of long protein molecules, rather like the shortening of a helical spring. In that year, the *sliding filament theory* was independently formulated by Hugh Huxley and Jean Hanson (from phase-contrast observations on glycerinated myofibrils) and by Andrew Huxley and Rolf Niedergerke (using interference microscopy of living muscle fibres). In each case the authors showed that the A band does not change in length either when the muscle is stretched or when it shortens actively or passively. This observation, interpreted in terms of the interdigitating filament structure described in the previous section, suggests that contraction involves sliding of the I filaments between the A filaments, with the lengths of both sets of filaments remaining unchanged, as is indicated in fig. 19.6.

The formulation of the sliding filament theory is a fine example of the rewards of hard work in the hands of those with a lively imagination. It is interesting that all four authors of the 1954 papers were in their thirties, had only recently come to the muscle field, and had utilized relatively new techniques to make their observations. Fascinating accounts of their work at this time are available (Randall, 1975; Simmons, 1992; Maruyama, 1995; H. E. Huxley, 1996).

What makes the filaments slide past each other? The generally accepted view is that sliding is caused by a series of cyclic reactions between the projections on the myosin filaments and the active sites on the actin filaments. Each projection is first attached to the actin filament to form a cross-bridge, then it moves or pulls on the actin filament,

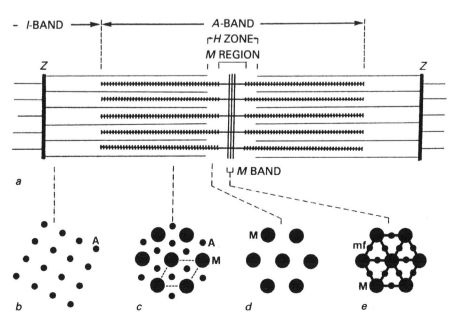

Figure 19.5. The filament array in vertebrate striated muscle. *a* shows a longitudinal view, *b* to *e* show transverse sections at various levels: *b*, in the I band near to the Z line; *c*, in the overlap region of the A band; *d*, in the H zone; and *e*, at the M line. The transverse sections show the thick (myosin, M) and thin (actin, A) filaments and, in *e*, the M filaments (mf). (From Squire, 1981.)

and finally it lets go, moving back to attach to another site further along the actin filament. The cross-bridges thus act as *independent force generators*, to use Andrew Huxley's useful term. This view is embodied in an alternative name for the mainstream version of the sliding filament theory: the 'cross-bridge theory' (see e.g. Tregear & Marston, 1979).

Let us have a look at some of the evidence for the sliding filament theory. There are two points to be established: (1) that the filaments remain constant in length during shortening or contraction of the muscle, and (2) that the cross-bridges, acting as independent force generators, provide the motive force for the sliding movement.

The lengths of the A and I filaments

We have met a number of examples of the use of electron microscopy in previous chapters, but it may be as well to consider at this point some of the difficulties encountered in making accurate measurements with the technique. The study of the structure of cells and tissues usually involves the preparation of thin sections of the material. The tissue is first fixed in a suitable fixative such as a solution of osmium tetroxide or glutaraldehyde. Then it is dehydrated in alcohol and embedded in a substance such as methacrylate (Perspex) or various epoxy resins (such as Araldite) so as to form a relatively hard block. This block is then sectioned on a special microtome to provide sections of the order of 0.02 to 0.05 μm thick.

The electron microscope can produce an image of an

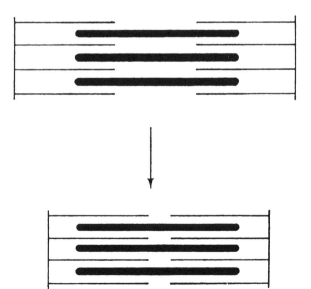

Figure 19.6. The structural changes in a sarcomere on shortening, according to the sliding filament theory.

object only if parts of the object are 'electron dense', i.e. if they are opaque to electrons. In practice, this means that the tissue has to be 'stained' by depositing heavy metal atoms upon it; some suitable stains are osmium tetroxide, potassium permanganate, lead hydroxide and uranyl acetate. In the final electron micrograph, therefore, one observes the distribution of heavy metal atoms on a thin section of tissue which has been fixed, dehydrated and embedded. Hence it is as well to check that the structures seen after any particular treatment are compatible with those seen after other procedures (use of a different fixative, for example) or deduced from the use of other techniques, such as light microscopy or X-ray diffraction.

The preparatory procedures used in electron microscopy frequently result in shrinkage of the tissue involved. So it is no easy matter to determine the lengths of the A and I filaments in living or glycerinated muscles. However, Page & Huxley (1963) investigated this problem in a piece of work that is a most beautiful example of mensurative electron microscopy. They began by showing that the sarcomere length of unrestrained muscles (measured by an optical diffraction technique) decreased during the fixation and dehydration processes, but was unaffected by the embedding process. These effects could be eliminated if the muscle was fixed at each end, so this procedure was followed in the rest of their experiments.

By homogenizing glycerinated fibres in a suitable medium, Page & Huxley obtained 'I segments' consisting of a Z line and a set of I filaments. They examined these segments by negative staining and shadow-casting. Both of these techniques can be used without fixation or dehydration in alcohol, and so shrinkage from these causes could be eliminated. Shadowed segments and those stained with sodium phosphotungstate were 2.05 μm long; those stained with uranyl acetate were slightly shorter.

Page & Huxley used particular care in the electron microscopy of longitudinal sections of muscles: the blocks were cut at right angles to the longitudinal axis so as not to shorten the filaments by the pressure of the microtome knife, and the effectiveness of this procedure was checked by measuring the width of the sections and the width of the block, the magnification of the electron microscope was calibrated at frequent intervals, and allowances were made for the shrinkage of the photographic prints.

Muscles were fixed in osmium tetroxide when contracting isometrically under the influence of electrical stimulation or solutions containing a high potassium ion concentration. Measurements for the I filament lengths (i.e. the total length of I filament on each side of a Z line, including the thickness of the Z line) were 2.01 to 2.05 μm for the

I filaments, except at a sarcomere length of 3.7 μm, when they were somewhat less. Measurements for the A filaments were 1.56 to 1.61 μm. These results indicate that very little shrinkage of the filaments occurs during osmium fixation of active muscles, in spite of the fact that appreciable shrinkage occurs when resting muscles are fixed. Page & Huxley suggested that the reason for this is that the A and I filaments are cross-linked during activity, whereas in the resting condition they are able to shrink by sliding past each other. At a sarcomere length of 3.7 μm, the filaments will not be overlapping, therefore no cross-bridges can be formed, which accounts for the shrinkage of the I filaments at this length.

The conclusion from this set of experiments, illustrated diagrammatically in fig. 19.7, was that the lengths of the filaments do not change in the resting or active muscles, whatever the length of the sarcomere. In frog striated muscles the I filaments are 2.05 μm long in total (1.0 μm long in each half-sarcomere, attached to the Z line which is 0.05 μm thick) and the A filaments are 1.6 μm long.

This conclusion has been contested from time to time (e.g. Pollack, 1990), but a careful reinvestigation using quick freezing methods has clearly confirmed it for mammalian muscles (Sosa *et al.*, 1994). Glycerinated rabbit psoas fibres were plunged into liquid ethane while relaxed, in rigor, or while contracting isometrically or shortening. Electron microscopy showed that the myofilaments were always the same length to within 3.5%: at 1.63 μm for the A filaments and 1.13 μm for the I filaments. Rabbit I filaments are clearly a little longer than frog ones. We shall see in the next section that there are some minor elastic changes

in filament lengths when they are subjected to tension, but these are too small to require any major revision of the sliding filament theory.

X-ray diffraction measurements

X-rays are a type of electromagnetic radiation; they differ from visible light in that their wavelength is extremely short, being about 1 Å. (The ångström unit, Å, is 0.1 nm; it is frequently used by X-ray crystallographers as a more convenient unit of length than the nanometre.) The resolution of an optical microscope is ultimately limited by the wavelength of the light which it uses. Thus if we could make a microscope which used X-rays instead of visible light, we should be able to examine very small structures indeed. This cannot be done, however, since it is not possible to focus X-rays so as to form an image. But the light in an optical microscope passes through two stages between the object and its image: first it is scattered by the object; then it is focused by the lenses of the microscope. X-rays are also scattered by objects. The essence of the X-ray diffraction technique is that the pattern produced by this scattering is interpreted by mathematical means so as to indicate what a focused image would look like. In practice, it is only possible to do this if we possess certain information about the X-rays and if the scattering pattern shows a sufficient degree of regularity. A regular pattern is produced only if there are corresponding regularities in the object, such as are produced by the regular packing of atoms and molecules in crystals or fibres.

The arrangement for an examination of a muscle fibre by the X-ray diffraction technique is shown schematically in

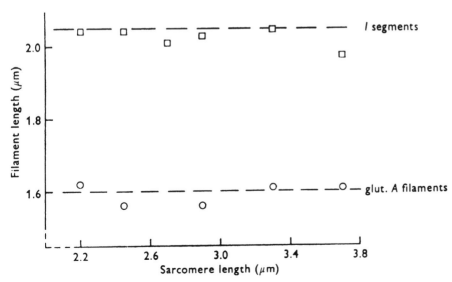

Figure 19.7. The lengths of the A filaments (circles) and I filaments (squares) in frog muscle fixed in osmium tetroxide during isometric tetani at various sarcomere lengths. The dashed lines show the lengths of isolated I segments and the lengths of the A filaments in glutaraldehyde-fixed muscles. (From Page, 1964, by permission of the Royal Society.)

fig. 19.8. If we were to draw a line on the photographic plate parallel to the fibre axis and passing through the point at which the undiffracted X-ray beam hits the plate, such a line would lie along the *meridian* of the pattern. A similar line passing through the undiffracted beam but at right angles to the fibre axis would lie along the *equator* of the pattern. Spots or lines produced on the plate by diffracted X-rays are called 'reflections'. Reflections lying on the meridian are produced by regularly repeating structures spaced along the fibre axis. Reflections lying on the equator are produced by regularly repeating structures in the transverse plane of the fibre. Structures which repeat in directions other than axially or radially (as, for example, in helically arranged units) produce off-meridional reflections, which may be arranged in a series of lines parallel to the equator and known as layer lines.

Each set of repeating structures produces a series of reflections on the photographic plate at regular distances from the line of the original X-ray beam, and their intensity usually decreases at increasing distances from this line. The first member of a series of this type (i.e. that nearest to the centre of the pattern) is known as a first-order reflection, the second is known as a second-order reflection, and so on. The distance between reflections of successive orders (which is equal to the distance between the first-order reflection and the centre of the pattern) is inversely related to the distance between the repeating units in the fibre. Hence low angle patterns (patterns in which the emerging rays do not diverge much from their original direction) give information about structures repeating at relatively long distances, such as might be seen with the electron microscope, and wide angle patterns give information about repeating structures which are closer together. Figure 19.9 shows the main reflections seen in low angle diffraction patterns from muscle.

Wide angle X-ray diffraction measurements on muscle show the α-helix pattern, irrespective of the length of the muscle. This α-helix pattern appears to be derived mainly from myosin, and its constancy at different lengths implies that there is little change in the lengths of the myosin molecules during stretching or contraction. This observation is in accordance with the sliding filament theory, although it does not provide any strong evidence in its favour.

Much more conclusive evidence is provided by low angle X-ray diffraction measurements of meridional reflections, which indicate the distances between repeating units along the axis of the myofibrillar filaments. Hugh Huxley (1953b) showed that such measurements on resting muscles were independent of the length of the muscle. This is just what one would expect from the sliding filament theory. If, on the other hand, the filaments did shorten during shortening of the muscle, we would expect the distance between their repeating units to decrease.

It is obviously desirable that this conclusion should be confirmed for actively contracting muscles. This was done by two groups of workers, in London (Elliott *et al.*, 1965) and in Cambridge (H. E. Huxley *et al.*, 1965). The chief technical difficulty of this type of experiment is that the exposure times for X-ray diffraction experiments may have to be a matter of hours, whereas it is impossible to excite a vertebrate striated muscle for more than a few seconds at a time without fatiguing it. The problem was overcome by stimulating the muscle at intervals and, using refined X-ray diffraction methods, passing the X-ray beam through the muscle only when the isometric tension was above a certain level. In each case it was found that the axial spacings during isometric contractions were not significantly different from those in the resting muscle and were unaffected by the length of the muscle.

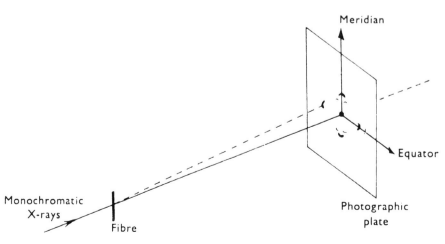

Figure 19.8. Schematic diagram to show the arrangement for making low angle X-ray diffraction observations on fibres. The dashed line shows one of the many diffracted rays, producing a reflection where it hits the photographic plate.

More precise measurements by Hugh Huxley & Brown (1967) showed that there is in fact a slight *increase* in the axial spacings derived from the myosin filaments during an isometric contraction. Their measurements gave the value of the principal myosin subunit spacing as 143 ± 0.1 Å in the resting muscle and 144.6 ± 0.1 Å during contraction, an increase of just over 1%. More recent measurements using synchroton radiation as an intense X-ray source have shown that the actin filaments are extended by about 0.3% during an isometric contraction (H. E. Huxley *et al.*, 1994; Wakabayashi *et al.*, 1994).

Do the cross-bridges move?

The sliding filament theory postulates that the contractile force is mediated via the cross-bridges originating on the myosin filaments. This suggests that the cross-bridges should move during the contraction, and there is some evidence that this is so. Reedy *et al.* (1965) found that glycerol-extracted insect flight muscles produced different X-ray diffraction and electron micrograph patterns according to whether they were in rigor or were relaxed. They concluded that the projections from the myosin filaments in resting muscle stick out at right angles from the myosin filaments, but that in rigor

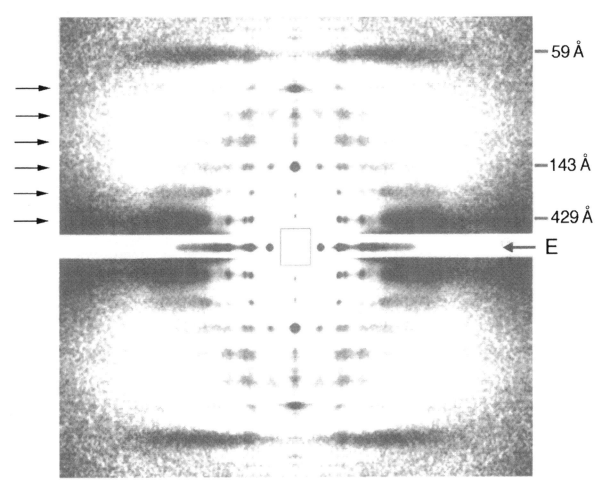

Figure 19.9. The main reflections seen in a typical low angle X-ray diffraction pattern from relaxed vertebrate skeletal muscle. E shows the equator, at right angles to the fibre axis and passing through the position of the undiffracted beam. Parallel to the equator is a set of layer lines based on 429 Å, indicated by the arrows. These arise from the helical arrangement of the cross-bridges emerging from the myosin filaments (see fig. 19.24). The meridional reflection at 143 Å represents the axial distance between adjacent 'crowns' of myosin cross-bridges. The 59 Å off-meridional reflection represents the longer-pitch primitive helix of the actin filament; the shorter-pitch primitive helix produces a similar reflection at 51 Å (see fig. 19.31). The muscle was from a fish (plaice), and the X-ray source was the Daresbury synchroton radiation facility. The diffraction pattern was recorded with a two-dimensional electronic detector; the intensity at each point has been multiplied by the square of its distance from the centre of the pattern to enhance the weaker outer reflections. (Courtesy of Dr J. Harford and Prof J. M. Squire.)

they become attached to the actin filaments and move through an angle of about 45° so as to pull the actin filaments towards the centre of the sarcomere (fig. 19.10).

Changes in the orientation of the cross-bridges in living muscles during isometric contractions are apparent from the X-ray diffraction measurements of Huxley & Brown. This conclusion is based on the observation that, during contraction, the intensity of the off-meridional reflection at 429 Å is only about 30% of that in resting muscle, whereas the intensity of the meridional reflection at 143 Å is about 66% of that in resting muscle. This implies that the helically repeating structures on the myosin filament become much less regularly arranged than do the axially repeating structures, and therefore that the outer portions of the myosin projections move with respect to their bases.

Remarkably precise measurements on the timing of these changes have since been made by Hugh Huxley and his colleagues. They used the intense X-ray sources available from the synchroton radiation emitted by high energy electron accelerators such as the electron–positron storage ring DORIS near Hamburg, coupled with the development of sophisticated electronic position-sensitive X-ray detectors (see H. E. Huxley & Faruqi, 1983). Such techniques have made it possible to measure X-ray reflection intensities during the course of a contraction for time periods as short as 1 ms in some cases.

Figures 19.11 and 19.12 show time-resolved observations on frog sartorius muscle twitches (H. E. Huxley *et al.*, 1982). There is a marked fall in the intensity of the 429 Å off-meridional layer line during the twitch. The time course of the change closely parallels that of the rise and fall in tension, preceding it by a few milliseconds. This is interpreted as a disordering of the helical arrangement of the cross-bridges as they become attached to the actin filaments and pull on them.

An alternative approach has been adopted by Hirose and his colleagues (1993, 1994). They looked for changes in cross-bridge positions using electron microscopy in combination with rapid activation and rapid freezing techniques; the procedure is engagingly known as 'flash and smash'. The experiments were done on chemically skinned fibres from rabbit psoas; such fibres have their cell membranes disrupted and made permeable by treatment with solutions containing the calcium-chelating agent EGTA, so that the interior of the fibre is accessible to substances which the experimenter may wish to apply, such as ATP or calcium ions (Eastwood *et al.*, 1979). The specimens were mounted on a plunger which could fall towards a copper block cooled with liquid helium; contact with the block would produce almost instanteous freezing, and the

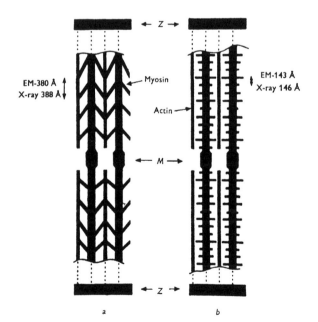

Figure 19.10. Diagram, derived from X-ray diffraction and electron microscopy measurements, to show the positions of the cross-bridges of glycerinated insect flight muscle in the relaxed state (*b*) and in rigor (*a*). (From Reedy *et al.*, 1965.)

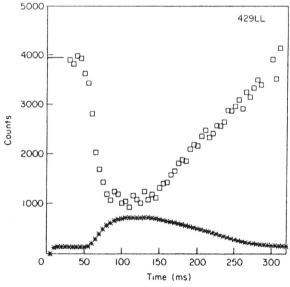

Figure 19.11. Intensity of the 429 Å off-meridional layer line during an isometric contraction as measured by time-resolved X-ray diffraction using synchrotron radiation. The upper curve (squares) shows the intensity of the layer line, measured as counts per 5 ms in single twitches at 10 °C, the lower curve (asterisks) shows the corresponding isometric tension. The stimulus was at 42.5 ms. (From H. E. Huxley *et al.*, 1982.)

frozen muscle fibres could then be prepared for electron microscopy.

Rapid and synchronous activation of these muscle fibres was brought about by flash photolysis of 'caged' ATP. This is an inert precursor of ATP, P^3-1-(2-nitro)phenylethyl-adenosine-5'-triphosphate, which can be instantly broken down to release ATP by an intense light flash, such as is produced by a frequency-doubled ruby laser. In the absence of calcium ions the cross-bridges detach and the muscle fibres relax when the ATP is released, but if they are present the cross-bridges enter their active cycle and contraction occurs (Goldman *et al.*, 1984*a*,*b*). So by arranging that the flash illuminates the muscle fibres as they fall towards the freezing block, the cross-bridges can be frozen in position at various times after they have begun to move from the rigor position.

Using this system, Hirose *et al.* (1994) found that the shapes of the cross-bridges were markedly different according to the condition that they were in at the instant of freezing. Figure 19.13 summarizes their results. The majority of cross-bridges were detached in relaxed fibres, whereas all were attached in rigor. Immediately after ATP release the cross-bridges became much more diverse in shape, and these shape changes continued over the next few hundred

milliseconds as the muscle fibres developed tension. These experiments thus provide remarkably direct evidence that the cross-bridges move during contraction.

The relation between sarcomere length and isometric tension

If the sliding filament theory is correct, and the cross-bridges act as independent force generators during contraction, then the isometric tension should be proportional to the degree of overlap of the filaments. In terms of fig. 18.5, the decline in the active increment curve at longer lengths should be caused by the reduced degree of overlap (and therefore smaller number of cross-bridges) between the two types of filaments as the muscle is extended. This conclusion was first pointed out by Andrew Huxley & Niedergerke in 1954.

The obvious way to test this suggestion is to measure the active increment length–tension curve, with the abscissa as sarcomere length rather than the length of the whole muscle, and compare it with the lengths of the A and I filaments. However, this turns out to be not so easy as one might think. Andrew Huxley & Peachey (1961) found that the sarcomere length in single frog muscle fibres was shorter at the ends than in the middle. This means that at long

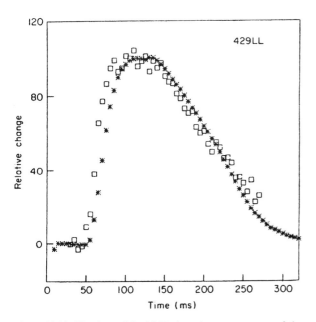

Figure 19.12. The data of fig. 19.11 plotted as a percentage of the maximum change. Notice that the change in layer line intensity (squares) precedes the tension change (asterisks) by about 10 ms during the rising phase of the contraction. (From H. E. Huxley *et al.*, 1982.)

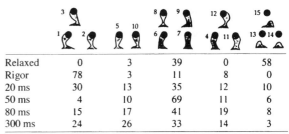

Relaxed	0	3	39	0	58
Rigor	78	3	11	8	0
20 ms	30	13	35	12	10
50 ms	4	10	69	11	6
80 ms	15	17	41	19	8
300 ms	24	26	33	14	3

Figure 19.13. Distribution of cross-bridge shapes in electron micrographs of muscle fibres subjected to the flash and smash technique. The fibres were rapidly frozen while relaxed, or in rigor, or at various times after the release of caged ATP. Over 70 000 cross-bridge images from transverse sections were digitized and subjected to correspondence analysis, a statistical and computational procedure for objective classification of images. This led to the fifteen different types indicated at the head of the table. In each image we are looking along the axis of the filaments; the black circle at the top is the actin filament, and the edge of the myosin filament backbone is at the bottom. Notice particularly that the cross-bridges of fibres frozen in rigor are mainly of the asymmetrical types 1 to 3, and that a much greater variety of cross-bridge form is found after release of ATP when the fibres start to contract. (From Hirose *et al.*, 1994. Reproduced from *The Journal of Cell Biology*, 1988, **127**, 2598, by copyright permission of The Rockefeller University Press.)

lengths there may be overlap of the filaments in the sarcomeres at the ends of the fibre, but not in those in the middle, so that the ends contract at the expense of the middle section if the fibre is fixed at each end. In order to overcome this difficulty, it is necessary to hold a small portion of the fibre (in which the sarcomere length *is* constant) at a constant length while its tension is measured. This tricky problem was solved by Andrew Huxley and his colleagues (Gordon *et al.*, 1966*a,b*); let us see how they did it.

We first need to consider a useful electronic feedback device known as a 'spot follower' (fig. 19.14). This consists of a cathode ray tube, a photocell, two lenses, and a movable vane placed between the lenses. The object of the system is to ensure that the movement of the spot follows exactly the vertical movement of the edge of the vane, however irregular this may be. This is achieved by feeding the output of the photocell into an amplifier connected to the Y plates of the cathode ray tube. The system is arranged so that, if the photocell can 'see' more than half the spot, the spot is raised; if it can see less than half the spot, it is lowered (remember that the spot on the screen moves in the direction opposite to its image on the vane). Thus the movement of the spot follows (inversely) the movement of the edge of the vane. It follows that the output of the photocell, since it is this that directly determines the position of the spot, must be proportional to the position of the edge of the vane.

Now let us look at some of the details of the apparatus used by Gordon *et al.* (fig. 19.15). A single muscle fibre is used, mounted on a microscope stage, and stimulated electrically. It is connected via its tendons to a tension transducer valve at one end and to the arm of a galvanometer at

the other. The galvanometer arm moves according to the force exerted on it by the muscle fibre and the current in its coil. Two small pieces of gold leaf are attached to the fibre with grease, to act as markers. The position of these markers is chosen so that the sarcomere length (observed through the microscope) does not vary along the length of the fibre between them. Below the microscope is a double-beam cathode ray tube, mounted vertically. The substage lens is positioned so that images of the two spots on the cathode ray tube screen can be focused on an edge of each marker. Light from these images is then collected separately onto two photocells.

The output of the right-hand photocell is fed to the amplifiers controlling the position of both spots. Thus if both markers move to the right or left by the same amount, the follower system works so that both spots move correspondingly, and there is therefore no change in the output of the left-hand photocell. But if the distance between the markers alters, the left-hand photocell (and its spot-follower loop) is activated, and the right-hand spot (which is focused on the left-hand marker) moves independently of the left-hand spot. The output of the left-hand photocell then indicates the distance between the markers, shown as the box marked 'length' in fig. 19.15.

Besides the spot-follower feedback loops, there is another feedback loop, indicated as the 'length-regulator loop' in the diagram. The output of the left photocell ('length') is fed to the main amplifier, which drives the galvanometer. This circuit is arranged so that the galvanometer moves so as to bring the output of the photocell back to its mid-position, i.e. so as to keep the 'length' (the distance between the markers on the muscle fibre) constant. This, of course, is just what the apparatus is required to do. The

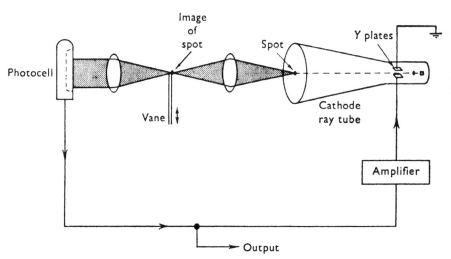

Figure 19.14. A simple 'spot-follower' circuit. Feedback from the photocell ensures that the spot on the cathode ray tube moves so as to keep its image fixed on the edge of the moving vane.

length can be set at different values by altering the 'length input signal', since the input to the main amplifier is the difference between this and the 'length' signal.

A single experiment with this apparatus consists of a series of isometric tetani obtained at different sarcomere lengths. The results of a number of experiments are summarized in fig. 19.16. It is evident that the length-tension diagram consists of a series of straight lines connected by short curved regions. There is a 'plateau' of constant tension at sarcomere lengths between 2.05 and 2.2 μm. Above this range, tension falls linearly with increasing length; the projected straight line through most of the points in this region passes through zero at 3.65 μm, but there is in fact a very slight development of tension at this point. Below the plateau, tension falls gradually with decreasing length down to about 1.65 μm, then much more steeply, reaching zero at about 1.3 μm.

According to the sliding filament theory, the isometric tension is directly proportional to the number of cross-bridges that can be formed between the A and I filaments, less any internal resistance tending to extend the sarcomeres. At long lengths, therefore, the tension ought to be

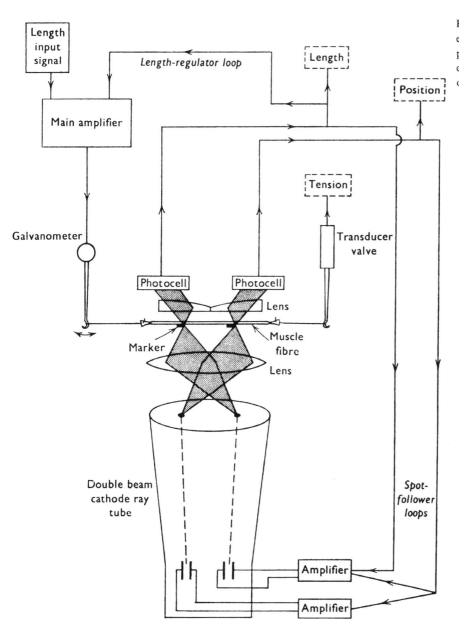

Figure 19.15. Apparatus used for experiments on the mechanics of portions of a muscle fibre. For explanation, see the text. (Based on Gordon *et al.*, 1966*a*.)

proportional to the degree of overlap of the A and I filaments. In order to see whether the length–tension diagram fits with this prediction we need to know the dimensions of the filaments and the position of the cross-bridge-forming projections on the myosin filaments. From the measurements of Page & Huxley (1963), we know that the A filaments are 1.6 μm long (symbol *a* in fig. 19.17) and the I filaments, including the Z line, are 2.05 μm long (*b*). The middle region of the A filaments (*c*), which is bare of projections and therefore cannot form cross-bridges, is 0.15 to 0.2 μm long and the thickness of the Z line (*z*) is about 0.05 μm.

Now let us see if the length–tension diagram shown in fig. 19.16 can be related to these dimensions, starting at long sarcomere lengths and working through to short ones. Figure 19.17*B* shows the points at which qualitative changes in the relations between the elements of the sarcomere occur. These stages are numbered 1 to 6, and the corresponding lengths are shown by the arrows in fig. 19.16. Above 3.65 μm (1) there should be no cross-bridges, and therefore no tension development. In fact there is some very small tension development up to about 3.8 μm, which may be caused by some lack of register between the A or I filaments so that some residual overlap occurs in places.

Between 3.65 μm and 2.25 to 2.2 μm (stages 1 to 2 in figs. 19.16 and 19.17) the number of cross-bridges increases linearly with decrease in length, and therefore the isometric tension should show a similar increase. It does. With further shortening (2 to 3) the number of cross-bridges remains constant and therefore there should be a plateau of constant tension in this region; there is. So over the sarcomere length range 3.65 μm down to about 2.0 μm the tension is precisely proportional to the number of cross-bridges that can be formed.

After stage 3, we might expect there to be some increase in the internal resistance to shortening, as the I filaments must now overlap; after stage 4 the I filaments from one half of the sarcomere might interfere with the cross-bridge formation between the A and I filaments in the other half of the sarcomere. These effects would be expected to reduce the isometric tension. This does in fact occur, although no extra reduction corresponding to stage 4 is detectable. At 1.65 μm (5), the A filaments hit the Z lines, and therefore there should be a considerable increase in the resistance to shortening; it is found that there is a distinct kink in the length–tension curve at almost exactly this point, after which the tension falls much more sharply. The curve reaches zero tension at about 1.3 μm, before stage 6 (when the I filaments would have hit the Z line) is reached.

It would be difficult to find a more precise test of the sliding filament theory than is given by this experiment, and the theory obviously passes the test with flying colours. Similar experiments using laser diffraction methods for the determination of sarcomere lengths and for length clamping have produced results wholly in agreement (Bagni *et al.*, 1988).

The results of these experiments tell us something extra about the mode of action of the cross-bridges. The filament array in the myofibrils is a constant volume system, so the transverse distance between the actin and myosin filaments must decrease as the muscle is stretched. But the decrease in isometric tension with increasing sarcomere length above 2.2 μm is linear; this means that the tension per cross-bridge is independent of length. Therefore the tension produced per cross-bridge is independent of the transverse distance between the filaments. We shall return to this point later (p. 363).

Structure of the contractile machinery

The success of the sliding filament theory led to further questions. What makes the filaments slide and how is the force produced? The search for answers has involved a great

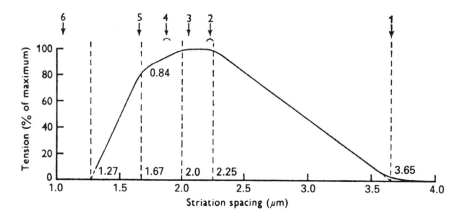

Figure 19.16. The isometric tension (active increment) of a frog muscle fibre at different sarcomere lengths. The numbers 1 to 6 refer to the myofilament positions shown in fig. 19.17*b*. (From Gordon *et al.*, 1966*b*.)

deal of work on the fine detail of myofibrillar structure, including especially studies of the protein molecules and how they fit together to form the thick and thin filaments. Let us have a selective look at some of this work; more detailed accounts are given by Craig (1994), Haselgrove (1983) and Squire (1981, 1986, 1997). Table 19.1 gives a summary of the main proteins of the myofibril.

Myosin

The myosin superfamily is a group of related motor molecules that convert energy from ATP hydrolysis into mechanical force or movement by pulling on actin filaments (Moosecker & Cheney, 1995). Other molecular motors are the kinesins and dyneins, which produce motion in opposite directions when interacting with microtubules (Goldstein, 1993; Holzbauer & Vallee, 1994). The myosin found in muscle is sometimes called myosin II to distinguish it from the other members of the superfamily, and it is this form that we refer to here.

Myosin molecules can be seen by electron microscopy by using the shadow-casting technique (H. E. Huxley, 1963;

Slayter & Lowey, 1967; Elliott & Offer, 1978). They are tadpole-like structures with a 'tail' about 1500 Å long and two 'heads' at one end. Under suitable conditions they are able to aggregate to form filaments, as we shall see later.

The myosin molecule consists of six separate protein chains (Lowey *et al.*, 1969; Weeds & Lowey, 1971; see also Warrick & Spudick, 1987; Bagshaw, 1993; Lowey, 1994). There are two heavy chains, each 220 kDa, and two pairs of light chains of about 20 kDa each (fig. 19.18). The light chains are of two types: essential (ELC, also called alkali light chains) and regulatory (RLC, also called phosphorylatable or DTNB light chains). Each heavy chain has a globular 'head' and a long α-helical 'tail'. Each heavy chain head has an ELC and an RLC associated with it. The tails of the two heavy chains are wound round each other in a coiled-coil structure. Each head can apparently swing freely about its junction with the rod and there is also a 'hinge' in the rod about 430 Å from the head–rod junction (Elliott & Offer, 1978).

The myosin molecule can be split by proteolytic enzymes into various subfragments (Mihaly & Szent-Györgyi, 1953;

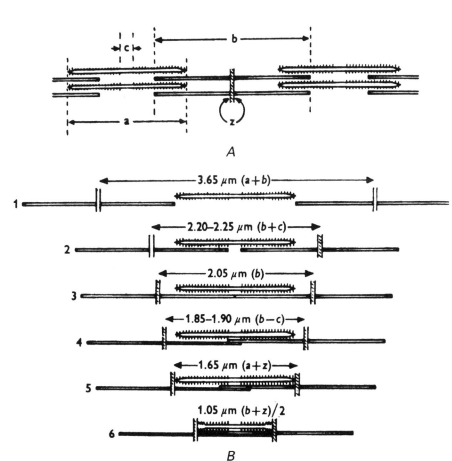

Figure 19.17. Myofilament dimensions in frog muscle. The lower diagram *B* shows the myofilament arrangements at different lengths; the letters *a*, *b*, *c* and *z* refer to the dimensions given in the upper diagram *A*. The sarcomere lengths corresponding to the positions labelled 1 to 6 are indicated by the arrows in fig. 19.16. (From Gordon *et al.*, 1966*b*.)

Table 19.1 *The main myofibrillar proteins of vertebrate skeletal muscle*

Protein	M_r $(\times 10^{-3})$	Content %	Localization
Contractile			
Myosin	520	43	A band
Actin	42	22	I band
Regulatory			
Troponin	70	5	I band
Tropomyosin	66	5	I band
M-protein	165	2	M line
C-protein	135	2	A band
α-Actinin	95×2	2	Z line
Structural			
Titin	2800	10	A and I bands
Nebulin	750	5	I band

Note:

Proteins amounting to less than 1% of the total are omitted. (Data from Maruyama, 1986.)

Mueller & Perry, 1962; Lowey *et al.*, 1969). Mild digestion by trypsin produces two sections known as light meromyosin (LMM) and heavy meromyosin (HMM); HMM acts as an ATPase and binds to actin, but LMM does not. Examination of the meromyosins by electron microscopy shows that HMM has two more or less globular 'heads' and a short 'tail', whereas LMM is a rod-like molecule. HMM can be further split into subfragments, two globular S1 portions and a rod-like S2 portion, by digestion with papain. The S2 rod ('long S2') can be split again to remove a 'hinge' region, leaving a section called 'short S2'. LMM molecules will aggregate to form filaments under suitable conditions, but neither HMM nor its constituent subfragments will (H. E. Huxley, 1963).

The amino acid sequence of the myosin heavy chain from the nematode worm *Caenorhabditis* was determined by gene cloning techniques (Karn *et al.*, 1983). It shows considerable sequence identity with the corresponding sequences later determined for vertebrate skeletal and smooth muscle and for various non-muscle myosins (Strehler *et al.*, 1986; Yanagisawa *et al.*, 1987; Warrick & Spudich, 1987). Hence the myosin molecule is rather conservative in evolutionary terms. The mammalian skeletal muscle heavy chain has 1939 amino acid residues, with an M_r of 223 900. The N-terminal sequences are appropriate to a globular head region, and the C-terminal rod shows the extended α-helix structure which we would expect from a coiled-coil dimer.

The charge distribution in the rod section, evident from the amino acid sequence, shows some very interesting regularities (McLachlan & Karn, 1982; McLachlan, 1984). Firstly, there are those connected with the coiled-coil structure. Many α-helical coiled-coil proteins have a characteristic pattern whereby hydrophobic residues occur

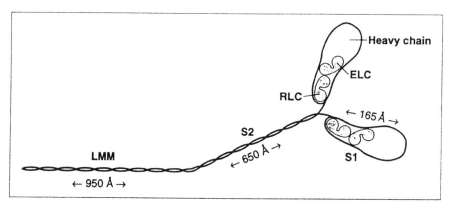

Figure 19.18. The six polypeptide chains that form the myosin molecule. The whole molecule consists of two globular heads attached to a long tail. The tail or rod is a coiled-coil formed from the α-helical regions of the two heavy chains; it is divided into an LMM (light meromyosin) section, and an S2 (subfragment 2) section. Each heavy chain has a globular region that combines with two light chains to form an S1 head. The light chains are of two types, called essential (ELC) and regulatory (RLC). Heavy meromyosin (HMM) consists of the S1 and S2 subfragments. Enzymic activity and the molecular motor are found in the S1 heads, LMM will aggregate with others to form the backbone of the myosin filament, and the S2 part of the rod connects the two. (From Rayment & Holden, 1994. Reproduced from: The three-dimensional structure of a molecular motor. *Trends in Biochemical Sciences* **19**, 129, with permission from Elsevier Trends Journals.)

alternately every three or four residues, so producing a seven-residue repeat (Crick, 1953). We can thus designate the residues in this repeating sequence as *a*, *b*, *c*, *d*, *e*, *f* and *g*, with *a* and *d* as the hydrophobic residues. Successive *a* and *d* residues then lie on a zigzag line down the length of the α-helix, so that the two helices can lie with their lines of hydrophobic residues in mutual contact (fig. 19.19).

Acidic and basic residues are clustered on the outside of the coiled-coil, and so arranged that there is a regular repeating pattern along its entire length (fig. 19.20). The regular bands of charged residues, alternating positive and negative, suggest that strong interactions will arise if two molecules (i.e. two coiled-coils) are placed side by side. If they are aligned precisely with each other, or if they are displaced by steps of 28 residues, there will be a strong repulsion between them since similar charges will be opposite to each other. If one molecule is displaced by 14 residues, however, then the two molecules will attract each other since the negative charges in one will be opposite the positive charges in the other, and vice versa.

The same thing will happen at the similar displacements of 42, 70, 98 etc., i.e. at displacements of $14n$ residues, where n is an odd number. Successive 28 residue sections are not precisely identical in their amino acid sequence, so there may be some variation in the degree of attraction between adjacent molecules at different values of n; McLachlan & Karn calculated that the attraction is particularly strong at $n = 7$, and also at $n = 21$ and $n = 35$, i.e. at displacements of 98, 294 and 490 residues. We might therefore expect myosin molecules to pack alongside each other with displacements of 98 residues in their chain positions. The helical rise in an α-helix is 1.485 Å per residue, so 98 residues occupy a

length of 145.5 Å. This is in most attractive agreement with the 143 Å X-ray meridional reflection attributed to myosin which, as we shall see, probably represents the stagger of successive triplets of myosin molecules along the myosin filament (fig. 19.25).

The regular 28 residue cycle is interrupted at four places in the chain by the insertion of an extra residue; McLachlan & Karn called these 'skip residues'. Their presence probably makes the coiled-coil slightly unwound, so that it has a longer pitch than usual, near these locations. We would expect this to have consequences for the packing of the molecules in the thick filaments.

The amino acid sequence shows that the S1 head is a much less regular structure than the rod; it contains only short sections of α-helix. The S1 heavy chain can be readily split in two places by proteolytic enzymes, to give 25 kDa,

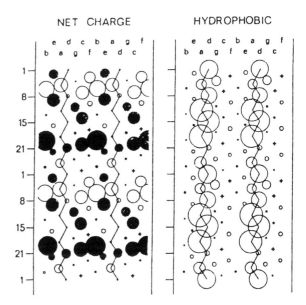

Figure 19.20. Distribution of different types of amino acid residue on the surface of the supercoil of the two α-helices in the myosin rod. A section of coil fifty-six residues long, containing two repeats of the averaged twenty-eight residue repeating sequence, has been projected onto a cylinder and unrolled so that each section of the sequence appears twice side-by-side, first from one helix and then from the other. The sequence reads across from right to left with a slight downward slant and a pitch of 3.5 residues per turn of each α-helix. The zigzag lines connect the residues *a* and *d* on the side of the supercoil. On the left the radii of the circles represent the average net charge of the residues in that position (white circles for positive charge, black for negative), on the right they represent the average hydrophobicity of the residues. Notice the repeating bands of positive and negative charge alternating along the length of the molecule. (From McLachlan & Karn, 1982.)

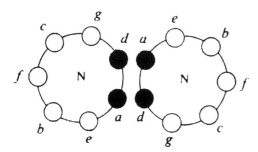

Figure 19.19. Cross-section of the rod section of the myosin molecule. Successive amino acid residues in the two α-helices are labelled *a* to *g*, and of these normally only *a* and *d* are hydrophobic. Consequently the *a* and *d* residues of the two chains associate with each other to form the coiled-coil structure. The diagram is drawn as if there were 3.5 residues per turn in each α-helix; in fact there are about 3.6, so that the two helices coil around each other. (From McLachlan & Karn, 1982.)

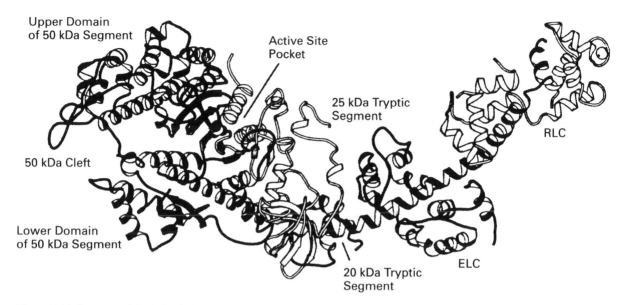

Figure 19.21. Structure of the S1 head of chicken skeletal muscle myosin, derived from high resolution X-ray diffraction studies. The protein chains are shown in the form of a ribbon diagram, with coiled sections indicated α-helices and arrows indicating β-strands. In the heavy chain, the N-terminal 25 kDa segment is shown in light grey, the 50 kDa segment in medium grey, and the 20 kDa segment in dark grey. The light chains ELC and RLC are associated with the long α-helix forming the 20 kDa segment. (From Rayment *et al.*, 1996. Reproduced with permission from the *Annual Review of Physiology* Volume 58, © 1996 by Annual Reviews Inc.)

50 kDa and 20 kDa fragments, reading in that order from the N-terminus.

High resolution X-ray diffraction is the best method of determining how the polypeptide chain of a protein is folded in three dimensions. For this it is essential to have crystals of the protein. For many years attempts to crystallize myosin S1 heads failed, until Rayment and his colleagues used an unconventional method involving methylation of lysine side-chain amino groups. The results that flowed from this breakthrough have considerably increased our understanding of structure of the myosin motor and how it works (Rayment *et al.*, 1993a,b; Fisher *et al.*, 1995a, b; Rayment *et al.*, 1996).

The structure of the S1 head as determined by Rayment and his colleagues is shown in fig. 19.21. The ATPase active site is in a pocket between the 25 kDa and 50 kDa segments. The actin binding site appears to be associated with the 50 kDa domain at the left hand end of the diagram. The 20 kDa segment is in the form of a long α-helix, surrounded by the two light chains, ELC and RLC, and perhaps stiffened by them.

The thick filaments

Hugh Huxley (1963) isolated myosin filaments by homogenizing portions of glycerinated fibres, and examined them by electron microscopy using the negative-staining technique. The most noticeable feature of these filaments was the presence of fairly regularly spaced projections on them, clearly corresponding to the projections and cross-bridges seen in thin sections of glycerinated muscle fibres. In the middle of each filament was a section, 0.15 to 0.2 μm long, from which these projections were absent, and which must correspond to the 'pseudo H zone' of intact muscle fibres.

We have seen that the myosin molecule is a tadpole-like structure with two 'heads' attached to a rod-like 'tail' section. A most interesting feature of these molecules, discovered by Hugh Huxley (1963), is that they are able to aggregate, under suitable conditions, to form filaments. The 'artificial filaments' so formed are of varying lengths, but all otherwise show the same general structure as the isolated natural filaments, including the projection-free region in the middle. Huxley suggested that the 'tails' of the myosin molecules become attached to each other to form a filament as is shown in fig. 19.22, with the 'heads' projecting from the body of the filament. Notice particularly that this type of arrangement accounts for the bare region in the middle. It also has two crucial implications for the design of the contractile apparatus: (1) the polarity of the myosin molecules is reversed in the two halves of the filament, and (2) the particularly reactive regions of the myosin molecule, the

ATPase site and the actin-binding site, are placed on the outside of the filament at the ends of the projections from it.

The analysis of the amino acid sequence of the myosin rod by McLachlan & Karn (1982), referred to in the previous section, fits very well with Hugh Huxley's model of filament structure. Figure 19.23 shows how we would expect adjacent molecules to be staggered by 143 Å with respect to one another.

In his pioneer investigation on the low angle X-ray diffraction pattern of living muscle, Hugh Huxley (1953*b*) observed a series of reflections corresponding to an axial repeat distance of about 420 Å. Using more accurate methods, Worthington (1959) observed a strong reflection at 145 Å with a fainter one at 72 Å, and suggested that these corresponded to the third and sixth orders of a 435 Å repeat distance. These reflections did not correspond to those obtained from actin (which could also be seen as a separate series), and Worthington therefore suggested that they were produced by myosin.

Huxley & Brown (1967) made a detailed investigation of the X-ray diffraction pattern of living muscle. The strong meridional reflection at 143 Å (equivalent to Worthington's 145 Å), together with the off-meridional layer lines based on 429 Å (Worthington's 435 Å) led them to suggest that the cross-bridges emerge from the filament in pairs every 143 Å, there being a rotation of 60° between each successive pair, so that projections oriented in the same direction occur every 429 Å. An alternative model was proposed by Squire (1974), in which a 'crown' of three cross-bridges emerged every 143 Å and there was a rotation of 40° between each successive crown, as is shown in fig. 19.24.

In principle it is possible to distinguish between these different models by measuring the number of molecules in each filament. This demands very careful extraction techniques so that all the myosin, and nothing but the myosin, is measured. Application of such techniques was initially inconclusive, with different investigators producing answers of two, three or even four myosin molecules per crown.

A fresh approach was provided by experiments involving scanning transmission electron microscopy (Lamvik, 1978;

Figure 19.22. Hugh Huxley's suggestion as to how the myosin molecules aggregate to form an *A* filament with a projection-free shaft in the middle and reversed polarity of the molecules in each half of the sarcomere. (From H. E. Huxley, 1971.)

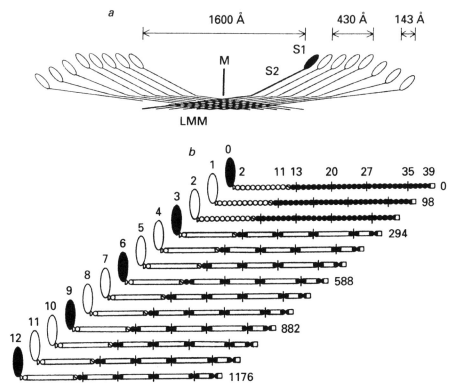

Figure 19.23. Longitudinal packing of myosin molecules in the thick filament, as suggested by the amino acid sequence of the rod section of the molecule. *a* shows the bipolar arrangement in the middle of the filament. *b* shows the parallel array of staggered myosin rods. Twenty-eight residue repeats are shown as circles in rods 0 to 2, with the first twelve repeats (white circles) being in the short S2 section. Skip residues (see text) are shown by vertical lines, and the half-stagger between them is emphasized in rods 3 to 12. Successive rods are shown displaced by ninety-eight residues. Vertebrate thick filaments probably have three such arrays in each cross-section. (From McLachlan & Karn, 1982.)

Reedy *et al.*, 1981; Knight *et al.*, 1986). Measurements of electron scattering by individual filaments provide an estimate of filament mass. Thus Knight and his colleagues found that the mass per half filament of rabbit thick filaments was 91.7 MDa, spread along a length of 8170 Å. If the myosin crowns occur every 143 Å along the filament from the edge of the bare zone to the tip, then there will be about fifty-two crowns per half filament. If we further assume that 6% of the mass is proteins other than myosin (see later), then the mass per crown is 91.7 × 0.94/52, i.e. 1.66 MDa. Since the mass of the myosin molecule is 0.52 MDa there are 1.66/0.52 i.e. 3.19 myosins per crown. This suggests that there are indeed three myosins per crown, as indicated in fig. 19.24.

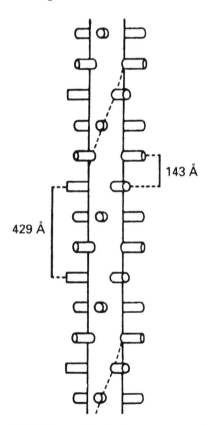

Figure 19.24. Diagram showing the arrangement of cross-bridges on the myosin filament in vertebrate striated muscle. (From Offer, 1974.)

Do cross-bridges occur regularly at all the 143 Å levels along the length of the filament, or are there some gaps in the sequence? Craig & Offer (1976) investigated this question by labelling glycerinated rabbit fibres with antibodies to the S1 subfragment. They found that there was dense labelling throughout the A band except at the centre and along a narrow stripe near its ends. They concluded that the cross-bridges extend in a regular sequence from the end of the bare zone to the end of the filament except for a single gap between the second and third cross-bridges from the end (fig. 19.25). Notice that in this model there are only forty-nine crowns per half filament; adoption of this value would produce a slight revision in the calculations in the previous paragraph.

Titin and other thick filament accessory proteins

The protein *titin* (also called *connectin*) was discovered by Maruyama and Wang and their colleagues (Maruyama *et al.*, 1977*b*; Wang *et al.*, 1979). Titin is a very large molecule, with an M_r of about 3 000 000 to 3 700 000 (different isoforms are found in different muscles), which occupies about 9% of the myofibrillar mass in vertebrates. Similar but smaller molecules (minititins, 700 000–1 200 000 M_r) occur in invertebrates. Titin can be separated by using SDS-polyacrylamide gels with very large pores. Isolated molecules are filaments over 1 μm long and about 4 nm thick, with a globular head at one end (Nave *et al.*, 1989). In the myofibril each titin molecule extends from the Z line to the M line as is indicated in fig. 19.26. The section of the titin molecule in the A band seems to be bound to the thick filaments, whereas the section in the I band is in part extensible (Fürst *et al.*, 1988; K. Wang *et al.*, 1993).

Cloning studies show that titin contains large numbers of repeated domains 100 residues long (Labeit & Kolmerer, 1995*a*; Linke *et al.*, 1996). In human cardiac titin there are 112 C-2 immunoglobulin-like (Ig) and 132 fibronectin-3-like (Fn) repeats; in skeletal muscle titin there are further Ig repeats in the I band. The Fn repeats are confined to the section of the molecule in the A band. The molecule is anchored at its N-terminus in the Z line and at its C-terminus in the M line, and there is a phosphorylation site near each end. There is also a kinase domain near where the molecule enters the M line.

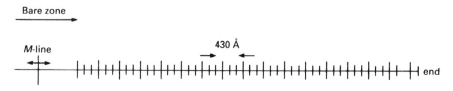

Figure 19.25. Distribution of cross-bridges in half a myosin filament, as deduced from antibody staining experiments. (From Craig & Offer, 1976.)

The A band section of titin seems to act as a 'ruler' or template for the assembly of myosin molecules in the thick filament (Trinick, 1994; Houmeida *et al.*, 1995). Electron micrographs show that single molecules of titin and myosin will bind to each other; the binding sites are near the tail end of the mysosin molecule and distributed in the *A* band section of the titin molecule. Titin also binds to C-protein and X-protein, accessory proteins in the thick filaments, and to the M line proteins to be discussed below.

The I band section of the titin molecule contains numbers of immunoglobulin repeats, interrupted in skeletal muscle by a non-modular section that is rich in proline, glutamate, valine and lysine residues, and is hence known as the PEVK region. Careful immunological electron microscopy, using antibodies to different parts of the titin molecule, shows that the titin does not stretch homogeneously when the myofibril is stretched (Gautel & Goulding, 1996; Linke *et al.*, 1996). For small stretches, it seems that the string of immunoglobulin domains is simply straightened out, with little increase in resting tension. Further

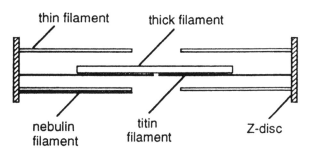

Figure 19.26. The positioning of titin and nebulin in the sarcomere. (From Schiaffino & Reggiani, 1996.)

stretch has to straighten the rubber-like PEVK region, with an associated rise in resting tension. Once this is fully straight the myofibril cannot be further stretched without damage to the system. Different muscles have PEVK regions of different lengths, and this correlates well with their different passive tension properties. Figure 19.27 indicates how this is thought to happen.

Vertebrate thick filaments contain some accessory proteins, called C-, X- and H-proteins, which bind to myosin filaments at regular intervals (Offer *et al.*, 1973; Starr & Offer, 1983). Examination of isolated A bands using negative staining shows a series of eleven stripes about 430 Å apart, with the first one at the edge of the bare zone (Craig, 1977). Antibody staining reveals that some of these stripes are sites of attachment of the accessory proteins (Bennett *et al.*, 1986). Cloning studies show that the amino acid sequences of these proteins contain domains homologous with immunoglobulin and fibronectin (Vaughan *et al.*, 1993; Fürst & Gautel, 1995). Since these proteins bind to titin as well as to myosin, it seems very likely that they are in some way an important component in the integrity of the thick filaments. The titin-binding site of C-protein is in its C-terminal region. This is deleted from the cardiac version of C-protein in patients suffering from an inherited heart disease known as the chromosome-11-associated form of familial hypertrophic cardiomyopathy, suggesting that the disorder is a result of thick filament misassembly (Freiburg & Gautel, 1996).

At the M line, each thick filament is connected to its neighbours by a series of 'M bridges'. There are also 'M filaments', which lie halfway between pairs of thick filaments and parallel to them and are contacted by the M bridges as

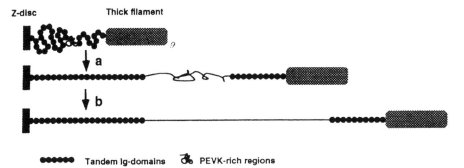

Tandem Ig-domains   PEVK-rich regions

Figure 19.27. Sequential extension of titin I band structures. In extending from short to medium sarcomere lengths (*a*), the main change is a straightening of the string of Ig domains. Further extension (*b*) involves a straightening of the rubber-like PEVK region, involving an appreciable increase in resting tension. (From Gautel & Goulding, 1996. Reprinted from *FEBS Letters*

385, Gautel, M. & Goulding, D. A molecular map of titin/connectin elasticity reveals two different mechanisms acting in series, p. 13, 1996, with kind permission of Elsevier Science-NL, Sara Burgerhartstraat 25, 1055 KV Amsterdam, The Netherlands.)

is shown in fig. 19.28 (Knappeis & Carlsen, 1968; Luther & Squire, 1978). These structures must serve to maintain the thick filaments in their hexagonal array. The M line contains two similar proteins known as M-protein and myomesin (Strehler *et al.*, 1983; Noguchi *et al.*, 1992). Creatine kinase (creatine phosphotransferase), the enzyme that catalyses the transfer of phosphate from creatine phosphate to ATP, is also localized at the M line (Turner *et al.*, 1973).

The molecular organization of the M line has recently been investigated by Obermann and his colleagues (1996, 1997), using electron microscopy of muscle fibres treated with antibodies to particular regions of the titin, myomesin and M-protein molecules. Their conclusions are shown in fig. 19.29. Myomesin molecules bind to myosin at their N-termini and to titin, titin molecules bind to each other. M-protein molecules bind at their ends to adjacent myosin filaments. So the M filaments are probably titin and myomesin, and the M bridges are probably M-protein and perhaps myomesin.

Actin

G-actin is a globular protein of about 42 000 M_r. Each molecule normally contains one molecule of bound ATP or ADP and one bound calcium ion. In salt solutions of physiological ionic strength, G-actin containing bound ATP will polymerize to form F-actin, during which the

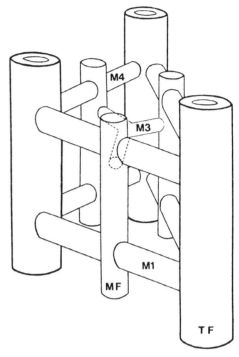

Figure 19.28. Part of the *M* line showing thick filaments (TF), *M* filaments (MF), *M* bridges (M1 and M4) and the postulated secondary *M* bridges (M3). The whole structure is symmetrical about the M1 plane. (From Luther & Squire, 1978.)

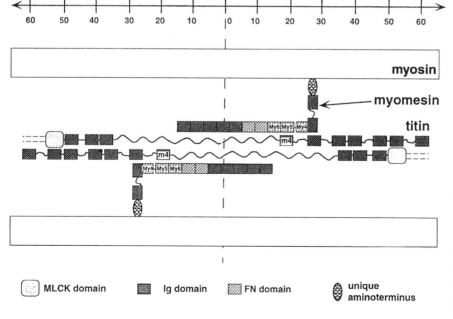

Figure 19.29. Arrangement of titin and myomesin in the *M* line, as determined by electron microscopy of muscle fibres treated with antibodies to particular domains of the two molecules. The scale shows distances in nanometres from the centre of the sarcomere. Myomesin binds to myosin at its N-terminus and to titin at the My4 to My6 domains. M protein molecules (not shown in the diagram) probably bind to myosin at each end and run transversely between adjacent thick filaments, interlacing at right angles with the myomesin and titin molecules. M4 represents the fourth Ig-like domain in the titin molecule. (From Obermann *et al.*, 1997, by permission of Oxford University Press.)

ATP is split to leave bound ADP in its place (Szent-Györgyi, 1951).

As well as forming the main protein of the thin filaments in striated and smooth muscles, actin occurs in a wide variety of other animal and plant cells, where it plays an important role in cell motility and the cytoskeleton. Actin from rabbit skeletal muscle contains a sequence of 375 amino acid residues (Elzinga *et al.*, 1973; Vandekerckhove & Weber, 1978*a,b*). Actins from heart muscle, smooth muscle and non-muscle tissues show only very small differences in the sequence, indicating that the actin molecule is highly conserved in evolution (Sheterline & Sparrow, 1994). Perhaps this is because actin has to bind to a large number of different proteins (including, in skeletal muscle, other actin molecules, myosin, tropomyosin, troponin, Z line proteins and perhaps others), so that the possibilities for change may be strictly limited.

Pure G-actin cannot be crystallized because of its readiness to polymerize to form F-actin. However it forms a one-to-one complex with pancreatic DNase I, and this can be crystallized, and so it is possible to investigate the structure of the molecule by X-ray diffraction (Kabsch *et al.*, 1990). The molecule consists of two domains named 'large' and 'small'. There is a cleft between them, into which the ATP molecule and its bound calcium ion fit (fig. 19.30). The 'small' domain contains the N- and C-termini and a number of amino acid residues concerned with binding to myosin.

The thin filaments of all muscles consist principally of the polymerized form of actin, F-actin. X-ray diffraction patterns of F-actin were first obtained by Selby & Bear (1956), using dried muscles from the clam *Venus*. They concluded that the actin monomers were arranged either in a net-like structure or in helices (topologically a helix is similar to a cylindrical net); the pitch of the helix model was 350 or 406 Å, corresponding to 13 or 15 G-actin monomers, respectively. Using negative staining and other techniques, Hanson & Lowy (1963) concluded that F-actin consists of

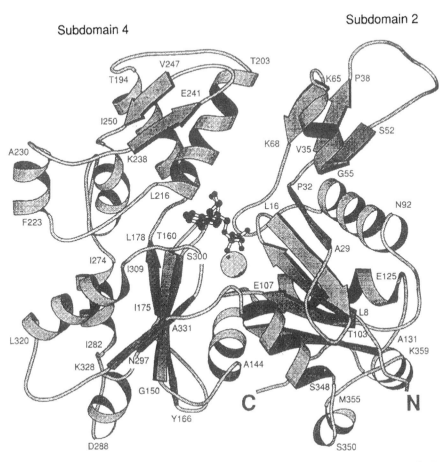

Figure 19.30. The structure of G-actin as determined by X-ray crystallography of the actin–DNase complex, with later refinements. The 'small' domain is on the right. The cleft in the middle contains an ATP molecule and a calcium ion. (From Lorenz *et al.*, 1993.)

two chains of monomers connected together in a double helical form, as is shown in fig 19.31. X-ray diffraction measurements by Hugh Huxley & Brown (1967) suggested that the pitch of the long helix was probably 2 × 370 Å, with about 13.5 monomers per turn.

The monomers in the Hanson & Lowy model are arranged on two primitive helices with pitches of 51 and 59 Å, corresponding to the off-meridional X-ray reflections at 51 and 59 Å. (A primitive or genetic helix is one that includes all the monomers, as opposed to the two long helices, which each include half of them.) We can see how these arise by using a simple thought experiment. Imagine a double string of beads with no twist to it, and with the beads in the two strings offset by 27.3 Å, half of the distance between the centres of adjacent beads in each of the strings. The beads are now on two primitive helices, one left-handed and the other right-handed, each with a pitch of 56.6 Å. Now we apply a twist to the top of the double string so that the two strings become twisted round each other in two right-handed helices. This action will wind up the right-handed primitive helix, making its pitch shorter,

and will unwind the left-handed primitive helix, making its pitch longer. When the pitches of these two reach 51 and 59 Å, respectively, the pitch of the long double helix will be about 2 × 375 Å, as shown in fig. 19.31.

Holmes and his colleagues (1990; Lorenz *et al.*, 1993) have produced a model of the actin filament based on the atomic model of G-actin deduced by Kabsch *et al.* (1990). They calculated the X-ray reflections that would be expected from G-actin monomers placed in various orientations in a model filament, and compared them with the actual X-ray diagram obtained from F-actin in gels oriented in capillary tubes. One particular orientation produced good agreement with actuality, and so was used to model the filament. Individual monomers show strong hydrophobic interactions with their neighbours on their own strand of the double helix. A hydrophobic loop from the opposite strand plugs into a pocket between adjacent monomers on the same strand so that all the components of the double helix are held tightly together.

Other thin filament proteins

Tropomyosin was first isolated by Bailey (1948) and its amino acid sequence was determined by Stone *et al.* (1974). The molecule consists of two α-helical chains which intertwine to form a coiled-coil structure of the type that we have seen in the myosin rod. It has a molecular weight of 2 × 33 000 and length of about 400 Å.

The role of tropomyosin began to emerge with the discovery of a second accessory protein, troponin, by Ebashi and his colleagues (see Ebashi *et al.*, 1968, 1969). The responsiveness of 'natural' actomyosin, and of glycerinated muscle fibres, to calcium ions depends on the presence together of both tropomyosin and *troponin*. Addition of tropomyosin and troponin to purified actomyosin systems inhibits their ATPase activity in the absence of calcium, but does not do this if calcium ions are present. Thus troponin and tropomyosin serve to sensitize the actomyosin ATPase to the presence of calcium ions. We shall see in the following chapter (where we shall examine the properties of troponin in more detail) that this effect is crucial to the control of contractile activity in the living muscle cell.

Electron micrographs show a periodicity of about 400 Å in the I band, which is greatly enhanced after staining with ferritin-labelled antibody to troponin (Ohtsuki *et al.*, 1967). In their study of the X-ray diffraction patterns of living muscle, Huxley & Brown (1967) found a marked meridional reflection at 385 Å, and they concluded that this arose from the thin filament. Thus it seems likely that this 385 Å reflection represents the distance between the points at which troponin is attached via tropomyosin molecules to the actin

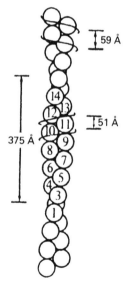

Figure 19.31. Helical geometry of the actin filament based on the 'double string of beads' model originally proposed by Hanson & Lowy. Successive monomers can be arranged on two primitive helices, a right-handed one whose pitch is about 51 Å and a left-handed one at about 59 Å. This gives the appearance of two intertwining helices whose pitch is 20 × 375 Å. The axial distance between the centres of adjacent monomers (such as 5 and 6) is 27.3 Å, hence that between successive monomers on one of the long-pitch helices (such as 5 and 7) is 54.6 Å. X-ray diffraction diagrams show relatively strong off-meridional reflections at 51 and 59 Å. (From Egelman *et al.*, 1982.)

backbone of the filament. Ebashi and his colleagues therefore suggested that the tropomyosin molecules lie end-to-end along the grooves between the two strings of actin monomers, each tropomyosin with a troponin molecule attached to it, as is shown in fig. 19.32.

The ratio of actin to tropomyosin to troponin molecules is 7:1:1 (Greaser *et al.*, 1973). If each tropomyosin molecule binds to seven actin monomers on one of the long-pitch actin helices, then the distance between successive tropomyosin molecules will be $(7 \times 55) = 385$ Å. The amino acid sequence of tropomyosin fits very well with this idea: there are fourteen pseudo-repeats containing similar sequences of about 200 amino acid residues (Parry, 1975; McLachlan & Stewart, 1976). McLachlan & Stewart suggested that the molecule could lie alongside the F-actin filament and bind at 55 Å intervals to seven successive actin monomers on one of the actin long-pitch helices. Studies on the structure of F-actin are in accordance with this view (Milligan *et al.*, 1990). The fourteen pseudo-repeats in tropomyosin perhaps correspond to two sets of alternative binding sites, which can be locked on or off by rotating the tropomyosin molecule by about a quarter of a turn about its long axis. We shall see in the next chapter that this idea that the tropomyosin molecule can adopt two alternative positions in the thin filament is crucial to modern views of how the interaction between actin and myosin is controlled in the living muscle.

Nebulin is a protein of large M_r value that is associated with the thin filaments, as indicated in fig. 19.26 (Wang, 1984). Its molecular size varies from 600 to 900 kDa in different skeletal muscles, in accordance with the lengths of their thin filaments. It is not, contrary to some early suggestions, an elastic molecule, since bands seen by electron microscopy of antibody-treated myofibrils occur at the same distance from the Z line whether or not the myofibril is stretched. Hence it seems likely that nebulin in some way regulates the length of the thin filament in development (Kruger *et al.*, 1991; Trinick, 1994). Perhaps there are two nebulin molecules for each thin filament, occupying the two grooves of F-actin (Pfuhl *et al.*, 1994). The amino acid sequence shows 185 repeats of a 35 residue module, and the central 154 copies of this are grouped into twenty-two seven-module super repeats, which seem to correspond to the 385 Å thin filament periodicity (Labeit & Kolmerer, 1995*b*).

Tropomodulin is a tropomyosin-binding protein. One molecule caps the free (pointed) end of each thin filaments, and so it also may be important in determining thin filament length (Weber *et al.*, 1994; Gregorio *et al.*, 1995).

Decorated thin filaments

In his study of the structure of muscle molecules and myofibril filaments with the negative staining technique, Hugh Huxley (1963) discovered a fruitful example of actin–myosin interaction. Under suitable conditions individual myosin molecules or their HMM or S1 subunits will attach to thin filaments or F-actin to produce 'decorated' thin filaments (fig. 19.33). It was immediately obvious that such filaments are polarized; the overall appearance shows a series of 'arrowheads' which always point in the same direction on any one filament, away from the Z line.

The detailed structure of decorated filaments has been

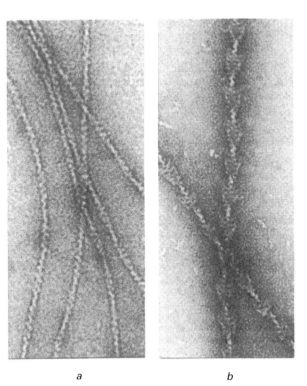

a   *b*

Figure 19.33. Negatively stained actin filaments from vertebrate striated muscle. In *a* the filaments are as isolated from the muscle, in *b* they have been 'decorated' with molecules of vertebrate myosin subfragment 1. Magnification 270 000× for *a* and 260 000× for *b*. (Photographs kindly supplied by Dr R. Craig.)

Figure 19.32. A model for the structure of the thin filament, showing the probable position of the tropomyosin and troponin components. (From Ebashi *et al.*, 1969.)

investigated by computational analysis of electron micrographs using a three-dimensional reconstruction technique developed by DeRosier & Klug (1968) and used by them to determine the structure of the tail of the T4 bacteriophage. The results show that the 'arrowheads' are caused by the binding of one S1 head to each actin monomer, with each S1 head being tilted relative to the filament axis (Moore *et al.*, 1970; Milligan & Flicker, 1987; Milligan *et al.*, 1990). Each arrowhead is about 370 Å long, reflecting the periodicity of the actin long-pitch helices.

The Z line

Longitudinal sections through vertebrate muscle show the Z line (Z disc or Z band) as a zigzag structure, thicker in some muscles than in others. Transverse sections give the appearance of either a square lattice or a 'basket weave' structure. Serial sections show that these are produced by the ends of a series of tetragons formed by links ('Z filaments') connecting the thin filaments on each side of the Z line. Each thin filament is connected by four 'Z filaments' to four thin filaments on the other side of the Z line, but the details of just how this is done are still under discussion (Knappeis & Carlsen, 1962; Goldstein *et al.*, 1986; Luther, 1991; Vigoreaux, 1994). Figure 19.34 shows a likely arrangement.

The major constituent of the Z line is α-*actinin*, a 180 kDa protein which is not found elsewhere in the myofibril

Figure 19.34. Possible structure of the Z line. Four Z filaments (white and shaded rods) are connected to each thin filament (black rod pairs). (From Squire, 1981.)

(Masaki *et al.*, 1967; Blanchard *et al.*, 1989). It is an actin-binding protein with two subunits arranged back-to-back with each other. Each subunit has a structure similar to the membrane-associated actin-binding proteins spectrin and dystrophin. It may well form the Z filaments seen in structural studies.

The actin-binding protein CapZ, also known as β-actinin or capping protein, binds to the barbed ends of actin filaments and so is found at the Z line (Maruyama *et al.*, 1977*a*, 1990). A number of other proteins whose status is rather uncertain have also been localized at the Z line (see Vigoreaux, 1994).

Cytoskeletal proteins

There is a striking feature about the experiment by Huxley & Hanson (1954) shown in fig. 19.4, which was for a long time unexplained. After the myosin and actin filaments were extracted from the isolated myofibril, the Z discs remained in position. If the thick and thin filaments were the only longitudinal components in the myofibril, as implied in diagrams of the contractile apparatus such fig. 19.5, then we would expect its longitudinal integrity to disappear so that the individual Z discs should float away from each other. Since they did not, it looked as though there should be some longitudinal structures in the myofibril in addition to the thick and thin filaments.

It is now evident that these longitudinal elements are part of the cytoskeleton of the muscle fibre. Titin and nebulin are sometimes described as the endosarcomeric cytoskeleton; titin is responsible for much of the resting tension at sarcomere lengths between about 2.5 and 4.5 μm (K. Wang *et al.*, 1993).

The exosarcomeric network is composed of intermediate filaments that surround the myofibril. Intermediate filaments are 7 to 11 nm thick, and found in the cytoplasm of most cells (see Amos & Amos, 1991). Those in muscle are largely made of the protein *desmin*. The transverse components are attached to the myofibrils at the Z line and the M band; their strong association with the Z line serves to bind adjacent myofibrils together (Granger & Lazarides, 1978; Wang & Ramirez-Mitchell, 1983). The longitudinal components become fully stretched and bear resting tension at sarcomere lengths above 4.5 μm (K. Wang *et al.*, 1993).

Dystrophin is a muscle cytoskeletal protein discovered as a result of investigations on Duchenne muscular dystrophy (Hoffman *et al.*, 1987; Koenig *et al.*, 1988; see Worton & Brooke, 1995). This is wasting disease of genetic origin. It affects about 1 in 3500 boys, most of whom do not survive to adulthood. The gene is carried on the X chromosome and the defect is recessive, so its inheritance is sex-linked. About

a third of the cases appear to arise by spontaneous mutation. This unusually high mutation rate is partly related to
the very large size of the gene: it is over 2.5 megabases long
(the largest human gene to date) and contains 79 exons.

Dystrophin is a 427 kDa protein associated with the
inner face of the muscle plasma membrane (Ervasti &
Campbell, 1991; Tinsley *et al.*, 1994; Campbell, 1995). The
N-terminal region shows some similarity to α-actinin and
binds to actin, whereas the C-terminal region binds to β-
dystroglycan, part of a glycoprotein complex sitting in the
membrane, and to a group of cytoplasmic proteins called
syntrophins. The central region shows some similarity with
the erythrocyte cytoskeletal protein spectrin; it is largely α-
helical in form, and pairs of molecules probably form antiparallel coiled-coil dimers. The glycoprotein complex
connects on the outer side of the membrane with laminin,
one of the proteins of the extracellular matrix (fig. 19.35).
The function of this complex of dystrophin-associated
proteins seems to be to link the contractile material to the
extracellular matrix so that the sarcolemma does not get
damaged when the muscle contracts.

The nature of cross-bridge action
We have seen that the sliding filament theory provides an
excellent explanation for the structural changes that occur

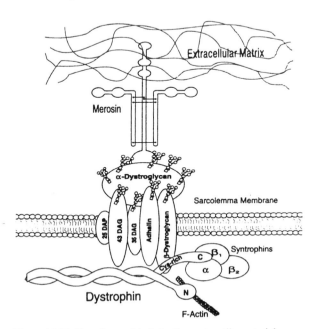

Figure 19.35. How dystrophin links the contractile material
(actin) via various dystrophin-associated proteins to the
extracellular matrix. (From Campbell, 1995. Reproduced with
permission from *Cell*, copyright Cell Press.)

when striated muscles contract. When Hugh Huxley (1957)
discovered the cross-bridges linking the thick and thin filaments (fig. 19.3), he realized immediately that if the cross-
bridges do indeed provide the active force for the sliding of
the filaments, then cross-bridge action must be a cyclic
process, since the cross-bridges are simply too short to
account for the extensive length changes that can occur in
contracting muscle. We would therefore expect each cross-
bridge to undergo a series of cycles of attachment, pulling
and detachment. The energy for each cycle, we would
expect, should come from the splitting of ATP.

Can such a concept of cyclic cross-bridge activity
account for the properties of contracting muscles as
described in the previous chapter? And what actually
happens to each cross-bridge as it undergoes a cycle of
activity? Partial answers to these questions are now available. Let us begin by looking at a theoretical approach produced by Andrew Huxley soon after the sliding filament
theory had been formulated.

Andrew Huxley's 1957 theory
The theory produced by Andrew Huxley in 1957 was a
first attempt to devise a precise model based firmly on the
sliding filament theory. It has proved to be most successful in the sense that it has suggested a number of experiments and provided the initial inspiration for several
daughter theories. Although Huxley's suggestions regarding the particular mechanism whereby force is generated
between the myosin and actin filaments may not be
correct, some of the conclusions which he reached are of
general application.

He began by assuming that shortening and the development of tension are produced by independent force generators (the cross-bridges) which can be effective only in the
region of overlap of the thick and thin filaments. Within
each contraction, each force generator undergoes cycles of
activity: attachment, pulling, and detachment, followed by
reattachment (perhaps at a different site) and a new cycle.
An essential point in what follows is the postulate that the
probabilities of attachment and detachment of the cross-
bridges are determined by their position.

The hypothetical force-generating mechanism is shown
in fig. 19.36. An active site M on the myosin filament oscil-
lates by thermal agitation backwards and forwards along
the length of the filament, but is restrained by elastic elements on each side (the 'springs' in the diagram); its equilibrium position is denoted by O. On the actin filament is
an active site A, at a longitudinal distance x from O. The A
and M sites can become attached to each other (forming a
cross-bridge), and this reaction has a rate constant f:

$$A + M \xrightarrow{f} AM \qquad (1)$$

This AM link can be broken by combination of the A site with a high-energy phosphate compound XP, the reaction proceeding with a rate constant g:

$$AM + XP \xrightarrow{g} AXP + M \qquad (2)$$

(We now know that high energy phosphate in the form of ATP combines with myosin rather than with actin, but this detail does not affect the development and conclusions of Huxley's theory.) Finally the system is reset so that reaction (1) can occur again by dissociation of AXP and splitting of the high energy phosphate bond (which supplies the energy needed to bring about this dissociation):

$$AXP \rightarrow A + X + phosphate \qquad (3)$$

The rate constants f and g are assumed to be dependent upon x, the position of the A site with respect to O (fig. 19.37). f is zero when x is negative (A to the left of O), and

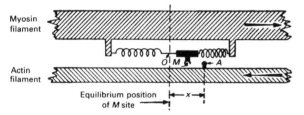

Figure 19.36. The tension-generating mechanism in Andrew Huxley's 1957 model. The part of a fibril which is shown is in the right-hand half of a sarcomere. Details in the text. (From A. F. Huxley, 1957, by permission of Pergamon Press, Ltd.)

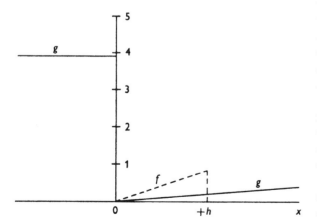

Figure 19.37. The dependence of the rate constants f and g on x in Andrew Huxley's 1957 model. The unit of the ordinate scale is the value of $(f + g)$ when $x = h$. (From A. F. Huxley, 1957, by permission of Pergamon Press, Ltd.)

increases linearly with increasing x up to the point h, beyond which it is again zero. g is very high (and constant) when x is negative, but zero at O, and small when x is positive, increasing slowly with increasing x.

During shortening, a single cycle of reactions (1) to (3) occurs as follows. An A site approaches an oscillating M site from the right; reaction (1) cannot occur until it passes the point h. When x is less than h, there is a fairly high probability that reaction (1) will occur; assume that, in this case, it does. M is now drawn towards O by the elastic force in the left-hand 'spring'. During this time there is a low probability that reaction (2) will occur, but as soon as the link passes O (i.e. as soon as x becomes negative) g becomes very high, and thus the probability of reaction (2) occurring increases enormously, and so the link is broken. Then reaction (3) occurs, and the A site is therefore ready to interact with the next M site that it meets. It is assumed that similar reactions occur asynchronously along the length of the sarcomere. This means that there will always be some links formed at any one time, so that the filaments will slide smoothly past each other.

The tension generated at any one contraction site is the tension in the left-hand 'spring' when the AM link is in existence. The tension produced by the whole muscle is the sum of the tensions at all contraction sites in a length of muscle equal to half a sarcomere (the forces generated in the two halves of a sarcomere, and in other sarcomeres along the length of the muscle, are in series with each other, and are therefore not additive). In an isometric contraction, some sliding occurs until the full isometric tension is reached and the series elastic component is fully stretched. All the AM links formed will then be to the right of O. There will be continual breakage of these links (reaction 2) since g, though small, is not zero. Thus each M site will be continually going through the cycle of reactions from (1) to (3), and hence the high-energy phosphate compound XP will be continually broken down, with the release of energy. This accounts for the activation heat and the consumption of ATP in an isometric contraction.

When the muscle shortens, the average value of g will rise, since the A sites are continually moving to the left of O. Thus the rate at which the cycle of reactions occurs will also increase. This accounts for the extra energy release during shortening.

Another consequence of shortening of the muscle is that any one A site will only be in a position to react with the corresponding M site for a limited time; this time will decrease as the velocity of shortening increases. Hence the probability of a link being formed between any one pair of sites will decrease as the velocity of shortening increases. This means that the total number of links formed at any one time will be lower at higher shortening velocities, and therefore the tension of the whole muscle will also be lower. At higher velocities, moreover, more links will remain in existence when x is negative; when this happens the tension in the right-hand 'spring' will pull against the shortening generated by other links where x is positive, so further reducing the tension developed by the whole muscle. These two effects account for the force–velocity relation.

Andrew Huxley's 1957 model is concerned mainly with the mechanism of contraction. However, it is desirable that it should be related to the properties of muscle under other conditions. In the resting muscle, we must assume that either reaction (1) or reaction (3) cannot occur; Huxley did not specify the activation mechanism which is necessary for contraction (we may assume that calcium ions are involved), but this is not an essential point in relation to the characteristics of contraction. At the end of a contraction, the reaction sequence ceases and the tension falls as reaction (2) proceeds to completion. When the muscle is in rigor, reaction (2) cannot take place, so that the AM links cannot be broken, which accounts for the inextensibility of the muscle in this condition.

Hugh Huxley's 1969 model

Early ideas of cross-bridge movement often invoked binding to actin at the tip of the cross-bridge and motive power supplied by bending or pulling at its base. This view was supplanted by a rather convincing picture produced by Hugh Huxley in 1969 (fig. 19.38). He suggested that that the myosin molecule possesses flexible linkages at two points in its structure: at the link between LMM and S2, the rod-like portion of HMM, and within HMM at the link between S2 and the globular S1 head. The primary source of movement in this model is a rotation of the S1 head about its attachment to the actin filament. The S2 rod serves to transmit the force produced by this rotation to the backbone of the myosin filament.

This model explains the appearance of insect flight muscles in rigor, agrees well with what we know of the structure of the myosin molecule, and is compatible with the results of X-ray diffraction studies. It also enables us to understand how the force per cross-bridge can be unaffected by the change in the transverse distance between the thick and thin filaments at different lengths, since the S2 link allows the S1 head to move out from the backbone of the myosin filament just as far as is necessary to contact the actin filament.

Andrew Huxley and Simmons's theory

This theory differs from Andrew Huxley's 1957 theory in its model of the tension-generating site. It is based on careful measurements of the mechanical transient responses of single muscle fibres and takes into account Hugh Huxley's 1969 model of cross-bridge structure. The apparatus was developed from the spot-follower system used by Gordon *et al.* (1966*a,b*) but had significant improvements in its time resolution. The results and the theory which arose from them were first reported by Andrew Huxley & Simmons (1971, 1973) and are discussed by Huxley (1974, 1980). More detailed descriptions and further experiments are given in a series of papers by Ford, Huxley & Simmons (1977, 1981, 1985, 1986).

Figure 19.39 shows the two types of transient obtainable from single muscle fibres: the length changes which follow a step change in tension (the velocity transient) and the tension changes which follow a step change in length (the tension transient). The events in the tension transient fall into four fairly distinct phases. First there is a sudden drop in tension during the length change (phase 1). Next, immediately after the length change, there is a rapid rise in

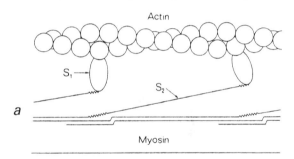

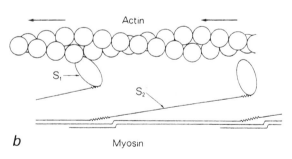

Figure 19.38. Hugh Huxley's 1969 model of how sliding is brought about. The S_1 heads rotate about their points of attachment to the thin filament. In *a* the left-hand cross-bridge has just attached, whereas the S_1 head of the right-hand one has nearly completed its rotation. *b* shows the situation a short time later: the S_1 head of the left-hand cross-bridge has now rotated, so pulling the actin filament to the left, and the right-hand cross-bridge is now detached. (From H. E. Huxley, 1976. Reproduced with permission from *Molecular Basis of Motility*, ed. L. M. G. Heilmeyer *et al.*, p. 12, fig. 2, Copyright by Springer-Verlag Berlin. Heidelberg 1976.)

Table 19.2 *Phases of the transient responses to sudden reduction of length (the tension transient) or of load (the velocity transient)*

Phase	Time of occurrence	Tension transient	Velocity transient
1	During applied step change	Simultaneous drop of tension	Simultaneous shortening
2	Next 1–2 ms	Rapid early tension recovery	Rapid early shortening
3	Next 5–20 ms	Extreme reduction or even reversal of rate of tension recovery	Extreme reduction or even reversal of shortening speed
4	Remainder of the response	Gradual recovery of tension, with asymptotic approach to isometric tension	Shortening at steady speed sometimes with superposed damped oscillation

Note:
The stated times are appropriate for frog muscle at about 5 °C. (From A. F. Huxley, 1974.)

tension (phase 2). This is followed by a period of a few milliseconds during which the recovery of tension is greatly slowed or even reversed (phase 3). Finally the tension gradually climbs back to the isometric tension appropriate to its length (phase 4). A related set of four phases can be seen in velocity transients; table 19.2 summarizes these events.

Huxley & Simmons investigated the nature of the first two phases by making a series of length steps of different

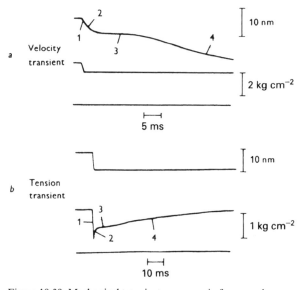

Figure 19.39. Mechanical transient responses in frog muscle during tetanic stimulation. *a* shows the length change following a sudden change in tension; *b* is the tension change following a sudden change in length. In each case the upper trace shows length (shortening downwards), the middle trace tension and the bottom trace the tension zero baseline. The numbers 1 to 4 indicate corresponding phases in the two types of transient response, described in table 19.2. (From A. F. Huxley, 1974.)

sizes and observing the height of the consequent tension changes. Their results are shown in fig. 19.40, where T_1 represents the tension at the end of the first phase and T_2 that at the end of the second. The T_1 curve is very nearly a straight line and this represents the behaviour of a passive elastic element. But the T_2 curve is quite different from this: tension falls very little for small length changes, after which the curve is roughly parallel to the T_1 curve with a displacement of about 6 nm per half sarcomere. This suggests that T_2 represents the properties of the force generators in the cross-bridges, each of which is capable of moving through about 6 nm while still exerting nearly maximum tension. To put it another way, each of the force generators seems capable of taking up about 6 nm of slack.

This analytical model of a cross-bridge can be compared with the structure of the myosin molecule. The tension generator is the S1 subfragment, which can attach to a site on the actin filament and rotate about it. In doing so it pulls on an elastic element which is probably the S2 subfragment connecting the S1 head to the backbone of the myosin filament.

Since these experiments were done, X-ray diffraction evidence has shown that the compliance of the actin and myosin filaments is larger than was expected, and hence that the cross-bridges must be less elastic than was thought (H. E. Huxley *et al.*, 1994; Wakabayashi *et al.*, 1994). At normal lengths, it turns out, the elasticity is divided roughly equally between the cross-bridges and the thin filaments. This means that we cannot measure the proportion of attached bridges simply by measuring the stiffness of a muscle fibre. The implications of this discovery are still under discussion, but it seems that the original Huxley & Simmons ideas are still compatible with the new data (Goldman & Huxley, 1994; A. F. Huxley & Tideswell, 1996).

Biochemical events in the cross-bridge cycle

The cycle of cross-bridge activity – attachment, movement, detachment and resetting ready for reattachment – must somehow be related to the splitting of ATP, which provides the energy for the whole process. We may ask how the reactions between the contractile proteins and ATP in test-tubes relate to the situation in the intact myofibril, and, particularly, at what stage in the cross-bridge cycle is the ATP split?

Lymn & Taylor (1970, 1971) used rapid reaction techniques to investigate the reaction kinetics of ATP hydrolysis by a mixture of actin and heavy meromyosin (HMM). Assay of the products of ATP splitting (ADP and inorganic phosphate) by enzymic methods measures their presence in free solution, but assay using strong acid can detect ATP splitting even if the products are still bound to myosin. In this way it was found that the splitting of ATP was much faster than release of its products. The initial rate of hydrolysis of ATP (measured with strong acid quenching) was much the same for both HMM alone and acto-HMM, although the steady-state rate for HMM alone was much reduced. This suggests that the first molecule of ATP which an HMM molecule meets is readily split, whereas later ones are not unless actin is present. This situation could arise if actin promotes the dissociation of the products of splitting from myosin.

Further work led to modification of the Lymn & Taylor scheme (see Eisenberg & Hill, 1985; Hibberd & Trentham, 1986; Goldman, 1987; Geeves, 1991; Bagshaw, 1993). Thus Stein *et al.* (1979) found that the actin concentration required for 50% maximal ATPase activity was only a quarter of that required for 50% binding of S1. This implies that ATP splitting and actomyosin dissociation are not absolutely linked to one another. From this conclusion arose the concept that there might be two types of actomyosin states, weakly bound and strongly bound, the weakly bound state being converted to the strongly bound state by the release of inorganic phosphate. This idea is illustrated in fig. 19.41.

A number of schemes have been produced that are reasonably successful in combining biochemical reaction schemes with models of cross-bridge movement in a plausible manner (e.g. Tregear & Marston, 1979; Eisenberg & Hill, 1985; Smith & Geeves, 1995). It is quite clear from such models that a cross-bridge working in the filament array is in a situation very different from that of HMM or S1 molecules in solution. In particular, there is no analogue of muscle force in solution and so no reason why cross-bridge states should be held away from their equilibrium positions. Is it possible, therefore, to investigate the biochemistry of cross-bridges *in situ*?

One useful technique that has been used to approach this problem enables the concentration of ATP in the fibre to be suddenly raised from zero to a useful level (Goldman *et al.*,

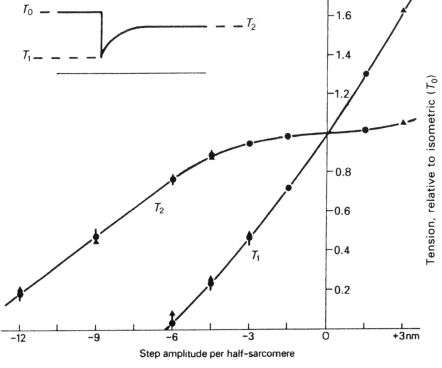

Figure 19.40. Tension levels in the early stages of tension transients in a frog muscle fibre. The points show the tension at the end of phase 1 (the T_1 curve) and phase 2 (T_2), in tension transients produced by different length changes. The T_1 curve represents an elastic component of the cross-bridges, and the horizontal difference between the T_1 and T_2 curves represents their capacity to shorten actively. (From A. F. Huxley, 1980, after Ford *et al.*, 1977.)

Step amplitude per half-sarcomere

1984*a,b*). The compound P³-1-(2-nitro)phenylethyl-ATP is an ATP complex which is itself biologically inert but which can be broken down by an intense flash of light to release ATP; it is known as 'caged ATP'. Such a rapid release of ATP produced very rapid relaxation of muscle fibres in rigor, followed (in the presence of calcium ions) by a redevelopment of tension as the cross-bridges proceeded through the tension-generating steps of their cycle (fig. 19.42). Both these events are rapid, suggesting that the rate-limiting steps in the cycle occur when the cross-bridges are attached and exerting tension. We can compare this conclusion with Andrew Huxley's 1957 model, in which breakage of the actin–myosin link will not normally occur until it has moved beyond the position $x = 0$.

Molecular motility assays

In recent years we have learned much more about the myosin molecular motor and its interaction with actin from some remarkable experiments involving *in vitro* motility assays. These use purified actin and myosin in systems which allow the movement of single filaments or small numbers of molecules to be observed by light microscopy. Much of the work has been done by Spudich and his colleagues at Stanford University and by Yanagida and his colleagues at Osaka University (H. E. Huxley, 1990; Spudich, 1994; Yanagida & Ishijima, 1995; Warshaw, 1996).

It may be useful to settle some terminology at this point. We shall use the term *working stroke* or *power stroke* to designate the distance that a single myosin molecule will move an actin filament while it is bound to one particular actin monomer in the filament. We shall use the term *step size* to represent the distance that a single myosin molecule can move an actin filament when one ATP molecule is split. So far we have implicitly assumed that these two distances are the same, but we shall see that 'it ain't necessarily so'.

Yanagida and his colleagues (1984) labelled individual actin filaments with phalloidin (a cyclic peptide found in the mushroom *Amanita phalloides*) to which the fluorescent dye tetramethylrhodamine was attached, and viewed them using fluorescence light microscopy with video image enhancement. They found that the filaments would move in the presence of myosin molecules and ATP. Kron & Spudich (1986) made a 'lawn' of myosin molecules on a flat surface and found that fluorescent actin filaments could be

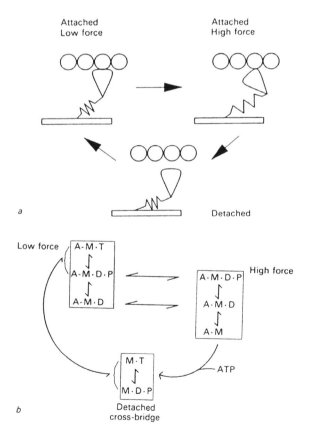

Figure 19.41. Three mechanical states in the cross-bridge cycle (*a*) and the corresponding chemical states in solution according to the model produced by Geeves and his colleagues (*b*). A, actin; M, myosin; T, ATP; D, ADP; P, inorganic phosphate. (From Geeves, 1991.)

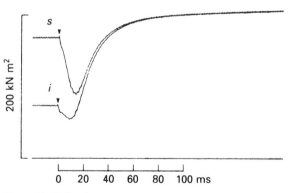

Figure 19.42. Tension transients initiated by flash photolysis of caged ATP within a single glycerinated muscle fibre. At the start of each trace the fibre was in rigor with 10 mM caged ATP and 100 μM Ca²⁺. At the time zero a 50 ns flash from a frequency-doubled ruby laser released ATP at a concentration of 700 μm. For trace *i* the fibre was held isometrically at a sarcomere length of 2.6 μm. For trace *s* the fibre was stretched slightly just before the laser pulse so as to increase the strain on the cross-bridges. The traces show the fall in tension as the rigor cross-bridges are dissociated by ATP, followed by the rise in tension as the active cycle proceeds. (From Goldman, 1987.)

seen moving over it (fig. 19.43). Toyoshima *et al.* (1987) found that the S1 heads of myosin were sufficient to produce movement on their own. The speed of the filaments was about 7 μm s^{-1} with HMM, whereas with S1 it was 1–2 μm s^{-1}. Perhaps the higher speed of movement with HMM arises because the extra flexibility introduced by the S2 links allows more of the S1 heads to be correctly oriented for binding to actin.

The myosin step size, the distance that an actin filament slides when a myosin molecule splits one ATP molecule, has been estimated by measuring the ATPase activity of motility assays together with data about the movements of the actin filaments in them. Different researchers, however, have reached different conclusions. Thus the Stanford group (Toyoshima *et al.*, 1990) calculated a step size of not much more than 8 nm, whereas the Osaka group (Harada *et al.*, 1990) estimated that it is over 100 nm.

Let us see how the Stanford group reached their conclusion about the length of the step size. Suppose we apply a low concentration of actin filaments of various lengths to a lawn of myosin heads; all of them will be bound by rigor linkages to at least one myosin head. If we now irrigate the lawn with ATP, the myosin heads will start their cycles of detachment, reattachment and movement, so moving the actin filaments. Long actin filaments will interact with a number of myosin molecules, and so will always be held in contact with the lawn by a proportion of them. But very short actin filaments, held by only one myosin head, for example, will float away from the lawn when the head reaches the detachment part of its cycle. Hence there will be a minimum filament length, long enough for there always to be at least one myosin head attached to it. Above this length, the sliding velocity will be independent of filament length and determined by the movement of the individual myosin head; since extra heads will act in parallel, their velocities do not sum.

An actin filament of minimum length *l* will have at least one myosin head bound to it and producing movement at

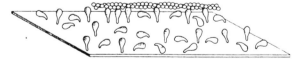

Figure 19.43. Diagram to show an actin filament moving on a lawn of myosin S1 heads in an *in vitro* motility assay. The direction of sliding is determined by the polarity of the actin filament. It seems that the S1 heads are sufficiently flexible for many of them to bind to the actin filament and move so as to make it slide past them. ATP is split in the process. (From H. E. Huxley, 1990.)

any moment. Let *n* be the average number of heads bound to a minimum length filament at any instant: this will be a little more than 1 in order to ensure that the actual number of heads bound never falls below 1. So when the minimum length filament moves a distance equal to the step size *d* m, *n* ATP molecules will be split. So if *na* ATP molecules are split in one second in producing the movement of our actin filament of minimum length, then it will move a distance *ad* metres in that second, i.e. its velocity *v* will be *ad* m s^{-1}. Its velocity will be the same as the average velocity of all the filaments, irrespective of their lengths.

The ratio of the minimum length of an actin filament to the total length *L* of actin filaments in the assay system must be equal to the ratio of the numbers of ATP molecules split per second by one filament of minimum length to the total numbers *A* of ATP molecules split per second, i.e.

$$\frac{l}{L} = \frac{na}{A}$$

Then, multiplying both sides by *d* and rearranging, we get

$$\frac{d}{n} = \frac{vL}{Al}$$

The actual values measured by the Stanford group were $v = 4.6$ μm s^{-1}, $L = 86$ m, $A = 3.4 \times 10^{11}$ s^{-1}, and $l = 150$ nm. So

$$\frac{d}{n} = \frac{4.6 \times 10^{-6} \times 86}{3.4 \times 10^{11} \times 150 \times 10^{-9}} \text{ m}$$

$$= 7.8 \times 10^{-9} \text{ m}$$

So the step size is 8 nm multiplied by the factor *n*. Since *n* is most unlikely to be more than 5, the authors conclude, the step size is most unlikely to be more than 40 nm, and is probably less. And since the value adopted for *l* was itself a lower limit (the length was below the resolving power of the microscope and had to be estimated indirectly), it may well be much less than this.

As mentioned above, the Osaka group have consistently obtained step size estimates much larger than this, using similar but not identical techniques (e.g. Harada *et al.*, 1990; Yanagida, 1990). Possible reasons for the discrepancies have been discussed by Hugh Huxley (1990) and Burton (1992).

Further advances have been made in the techniques of measuring forces and movements produced by single motor molecules or filaments, and these have been used to estimate the size of the working stroke. Yanagida and his colleagues have measured the forces exerted on a single actin filament by attaching it to a fine glass needle and observing the deflection of the needle tip (Yanagida *et al.*, 1993;

Yanagida & Ishijima, 1995; Ishijima *et al.*, 1996). They found that the maximum force exerted by single myosin molecules was 5 to 6 pN, and that the displacement produced by single molecules at low loads was about 20 nm.

An alternative approach, adopted by the Stanford group and others, is to use a laser optical trapping technique (sometimes called 'optical tweezers') to control the position of single actin filaments and to measure the forces and displacements produced by single myosin molecules (Finer *et al.*, 1994; Miyata *et al.*, 1995; Molloy *et al.*, 1995; Simmons *et al.*, 1996). If a laser beam is focused on a small particle, the radiation pressure of the light keeps the particle at the point of focus (Ashkin & Dziedzic, 1977; Kuo & Sheetz, 1992). With a suitable feedback system, movements of the particle produced by molecular forces can be followed by movements of the trapping laser beam. The use of this ingenious device for investigating the myosin motor is illustrated schematically in fig. 19.44.

Figure 19.45 shows some of the results obtained by the Stanford group with this method (Finer *et al.*, 1994). With a light load and in the presence of ATP, small movements of the actin filament averaging 11 nm in size were recorded. These were always in the same direction, as would be

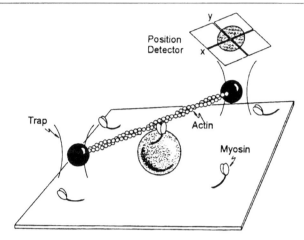

Figure 19.44. The use of optical trapping to measure the forces and movements produced by a single myosin molecule. The two laser optical traps ('optical tweezers') are focused on polystyrene beads attached at each end of a single actin filament. Silica beads are attached rigidly to a coverslip and the whole covered sparsely with myosin molecules. One silica bead acts as a platform to lift a myosin molecule off the coverslip. The design of the optical trap enables the movements of the polystyrene beads to be followed and recorded, and a feedback system allows the forces required to do this to be measured. (From Warshaw, 1996.)

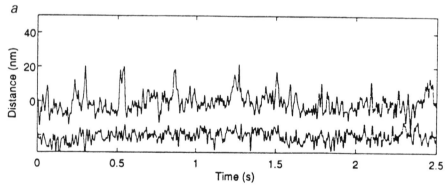

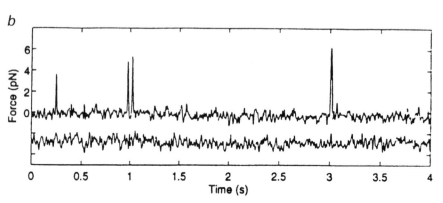

Figure 19.45. Forces and movements produced by single myosin molecules pulling on an actin filament in the presence of 2 mM ATP, measured using the optical tweezers system shown in fig. 19.44. *a* shows displacements of a bead attached to an actin filament in contact with an HMM molecule with a light load. The upper trace shows the movement of the bead along the direction of the actin filament, the lower trace shows its movement at right angles to this. *b* shows the force transients produced with a much stiffer system approximating to isometric conditions. Again, the upper trace shows the forces on the bead along the direction of the actin filament, whereas the lower trace shows the forces at right angles to this. (From Finer *et al.*, 1994. Reprinted with permission from *Nature* **368**, 115. Copyright 1994 Macmillan Magazines Limited.)

expected from the polarity of the actin filament, whereas Brownian or other random movements would not show this characteristic, and hence it seems that they provide a direct measure of the size of the myosin working stroke. When the system was adjusted to make the tweezers much stiffer, brief force transients of 3 to 4 pN were measured. Molloy and his colleagues (1995) obtained somewhat smaller transients, and concluded that the working stroke is about 4 nm.

Is the working stroke the same as the step size? The conventional view, supported by the results of the Stanford group, is that they are, and that each working stroke is necessarily accompanied by the breakdown of a single ATP molecule. Yanagida and his colleagues, however, suggest that the coupling between the working stroke and the breakdown of ATP is much looser than this, so that, when the load is light, several working strokes can occur for every ATP molecule broken down (Yanagida, 1990; Yanagida *et al.*, 1993; Yanagida & Ishijima, 1995). This idea is illustrated in fig. 19.46.

Some evidence in favour of the loose coupling model comes from experiments on mechanical transients. Lombardi *et al.* (1992) showed that the ability to redevelop tension after a quick release (the T_2 component in fig. 19.40) is largely recovered within a few milliseconds of a previous quick release. In one experiment they imposed four quick releases on a frog muscle fibre at 8 ms intervals, and found that the T_2 recovery in tension, thought to represent the working stroke of the cross-bridges, was still occur-

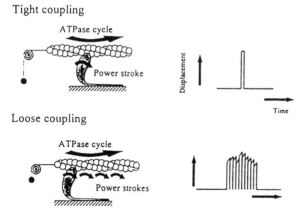

Figure 19.46. Tight coupling and loose coupling models of myosin action. In the tight coupling model each working stroke (called the power stroke here) is associated with the breakdown of one ATP molecule, so the step size is the same as the working stroke. In the loose coupling model, one ATPase cycle causes one working stroke at high loads but many at low loads, so the step size can be longer than the working stroke. (From Yanagida & Ishijima, 1995.)

ring at the fourth release. They calculated that this implied a working rate of about 77 strokes s^{-1} for each cross-bridge. However, the maximum ATPase rate per myosin head under steady-state conditions is only 6 s^{-1} (Kushmerick & Davies, 1969), from which Lombardi and his colleagues concluded that there are many working strokes for each ATP molecule hydrolysed.

How do the cross-bridges move?

Hugh Huxley's 1969 model suggests that the S1 heads rotate about their points of attachment to the actin filament. Attempts to find evidence for this view have been made with spectroscopic probes for the detection of molecular motion (see Thomas, 1987). One method which has proved useful is electron paramagnetic resonance (EPR) spectroscopy. This involves measuring the absorption of energy by paramagnetic substances in a radio-frequency oscillating magnetic field. The paramagnetism arises from the presence of unpaired electrons, so a synthetic probe which can be bound to the muscle protein has to be used. Suitable probes include nitroxide derivatives of iodoacetamide or maleimide, which will bind to one of the sulphydryl groups in the S1 head.

The results of EPR spectroscopy showed that probes attached to the S1 heads in glycerinated muscle fibres in rigor are uniformly and rigidly oriented (Thomas & Cooke, 1980). Addition of ATP in the absence of calcium (which produces relaxation from the rigor state as the actin–myosin links dissociate) gave rise to a random distribution of probe orientations with rotational movements in the 10 μs range. Similar rotational mobility was seen in isolated myosin filaments and in stretched fibres in rigor; in the latter the proportion of fixed myosin heads was proportional to the degree of filament overlap (Barnett & Thomas, 1984). Thus the rotational mobility of free myosin heads seemed to be very high, but heads bound to actin filaments in rigor were held rigidly at a constant orientation.

In the presence of both ATP and calcium ions, myofibrils split ATP and contract. The EPR spectrum of glycerinated fibres contracting in the presence of ATP and calcium ions was a combination of the spectra obtained from fibres in rigor and in relaxation (Cooke *et al.*, 1984). During an isometric contraction, about 20% of the myosin heads had probes oriented at the same angle as in rigor, whereas the remainder of the heads were disoriented. These results implied that the myosin heads do not rotate about their point of attachment to the actin filament during the working stroke, contrary to the model shown in fig. 19.38. Perhaps, it was suggested, the S1 head bends in some way while its anchorage to the actin filament remains firm

(Cooke, 1986). An alternative idea, that the S2 rod might shorten near its junction with the LMM rod (Harrington, 1979), seems to have been ruled out by the successful use of the S1 head alone in motility assays.

The determination of the molecular structure of the myosin head by Rayment and his colleagues (fig. 19.21) has greatly cleared the air about how the myosin motor might work. These studies have led to the 'swinging neck lever model' (Rayment *et al.*, 1993*b*; Spudich, 1994; Uyeda, 1994; Spudich *et al.*, 1995; Uyeda *et al.*, 1996; Block, 1996; Holmes, 1997), although they have not provided direct evidence for it. The 20 kDa C-terminal domain of the S1 head forms a long α-helix to which the two light chains are bound, as is shown on the right of fig. 19.21. This region is sometimes called the 'neck' of the molecule. The swinging lever model suggests that it acts as a lever arm, converting a small rearrangement at a 'hinge' region into a large movement at its C-terminal end, as is shown in fig. 19.47.

Some excellent evidence for the swinging lever model comes from motility assays with some mutant myosins (Spudich *et al.*, 1995; Uyeda *et al.*, 1996). The object of the experiments was to alter the length of the lever and see how this affected the velocity of actin filaments sliding on a lawn of the modified myosins. The longer the lever, the faster the sliding, we would expect. The mutant myosin heavy chains were obtained from the cellular slime mould *Dictyostelium*. They were genetically engineered to have a lever arm with either both light chain binding sections removed, or just

one of them removed, or with an extra light chain binding section added. Analysis showed that the resultant proteins had zero, one, two (for the wild type) or three light chains in each S1 head. The results of the motility assay were dramatically in line with expectation, as can be seen in fig. 19.48: sliding velocity was linearly proportional to the calculated length of the lever arm.

An alternative approach uses a fluorescent probe attached to the lever arm. Irving and his colleagues (1995) replaced the regulatory light chain of glycerol-extracted muscle fibres with one labelled with a rhodamine probe. When the probes are excited by polarized light, the fluorescence intensity varies with the position of the molecular probe. Irving and his colleagues measured the fluorescence intensity during synchronized myosin head movements produced by mechanical transients. They found that the fluorescence signals changed during these stretches and

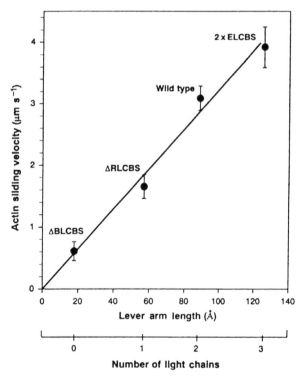

Figure 19.48. Sliding velocities of actin filaments in motility assays using mutant and normal (wild type) *Dictyostelium* myosins with neck lever arms of different lengths. Mutants had sections binding the light chains removed or added: ΔBLCBS with both chains removed, ΔRLCBS with just the RLC chain removed, 2 × ELCBS with an extra ELC-binding section. The velocities of at least forty-three actin filaments were measured for each point; bars indicate standard deviations. (From Spudich *et al.*, 1995.)

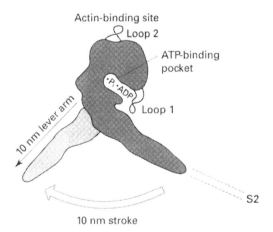

10 nm stroke

Figure 19.47. The swinging level arm model for S1 action. A change in shape of the molecule near to the ATP-binding pocket produces a movement of about 10 nm at the end of the lever arm. This pulls on the S2 link which is attached to the myosin filament backbone. (After Spudich, 1994, redrawn. Reprinted with permission from *Nature* **372**, 517. Copyright 1994 Macmillan Magazines Limited.)

releases, and in the phase 2 periods following them (table 19.2), showing that the cross-bridges do indeed tilt during active contraction.

So it looks as though the swinging lever model is the correct one. What is not so clear is precisely how far the lever swings, and whether this agrees with the estimates of working stroke distance obtained from motility assays.

Where does the motive power for the movement of the lever come from? Rayment and his colleagues (1993*b*) thought at first that the opening and closing of the cleft in which the ATP and its products were bound was the originator of this movement. Further experiments with stable analogues of the ADP-P$_i$ complex, however, led to a revision of this view (Fisher *et al.*, 1995*a*,*b*; Rayment *et al.*,

1996). Rayment and his colleagues prepared catalytic portions of the S1 head (i.e. the S1 head without most of the lever arm) of *Dictyostelium* myosin complexed to either MgADP-beryllium fluoride or MgADP-aluminium fluoride, and determined their structures by crystallization and X-ray analysis. The structure of the beryllium fluoride complex was very similar to that of the S1 head of chicken myosin (fig. 19.21). The aluminium fluoride complex, however, showed some significant differences in the position of the upper and lower parts of the 50 kDa domain. It is therefore suggested that the motive power for the movement involves a partial closure of the cleft between the two parts of the 50 kDa domain and that this results in a rotation of the lever arm.

20
Activation of muscular contraction

The normal stimulus for the contraction of a skeletal muscle fibre in a living animal is an impulse in the motor nerve that innervates it. In the twitch muscles of vertebrates with which we are concerned in this chapter, this nerve impulse leads to a propagated action potential in the muscle fibre, which is then followed by a twitch contraction. The time relations of the action potential and twitch tension in a single muscle fibre are shown in fig. 20.1.

The sequence of these events is shown schematically in fig. 18.1. We have examined stages 1 to 4 of this sequence (the excitation processes) in earlier chapters of this book, and stage 6 (contraction) in chapters 19 and 20. Here we consider how excitation of the muscle fibre membrane initiates contraction of the myofibrils in the interior of the fibre. This constitutes stage 5 of fig. 18.1, the excitation–contraction coupling process. In terms of fig. 20.1, then, how does the action potential produce the contraction?

Excitation–contraction coupling
The importance of depolarization of the cell membrane

When muscle fibres are immersed in solutions containing a high concentration of potassium ions, they undergo a relatively prolonged contraction known as a *contracture*. Contractures can also be produced by various drugs, such as acetylcholine, veratridine and others. Kuffler (1946) showed that many of these substances produce depolarization of the cell membrane; furthermore, if the substance was applied locally the resulting contracture was limited to that part of the muscle fibre where depolarization occurred. The relation between membrane potential and tension in potassium contracture was determined quantitatively by Hodgkin & Horowicz (1960). They found that no contracture occurred when the membrane potential was more negative than about −55 to −50 mV; above this threshold value, tension increased rapidly with increasing depolarization, reaching a maximum above about −40 mV (fig. 20.2). These results show quite clearly that contracture tension is related to the degree of depolarization of the cell membrane.

What happens when the period of depolarization is very brief, as occurs during the action potential that elicits a twitch contraction? Adrian and his colleagues (1969), and later Costantin (1974) and Gilly & Hui (1980) investigated this question. They used voltage clamp techniques to produce short depolarizations, tetrodotoxin to prevent action potential production, and visual observation to detect the mechanical threshold. The results showed that the mechanical threshold is commonly around −50 mV at pulse lengths of 20 ms or more, but is progressively higher at shorter pulse lengths. Adrian and his colleagues interpreted this in terms of the release of an internal activator, with rate constants for the process dependent upon membrane potential.

Potassium contractures may differ in different types of muscle. When a frog rectus abdominis muscle (a 'tonic' muscle) is depolarized by placing it in a solution containing a high potassium ion concentration, the resulting contracture lasts for several minutes. On the other hand, a frog sartorius muscle (a 'twitch' muscle) subjected to the same

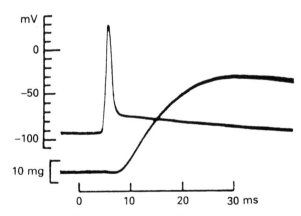

Figure 20.1. Electrical and mechanical responses of a single frog twitch muscle fibre to an electrical stimulus. The upper trace shows the action potential, recorded with an intracellular microelectrode, and the lower trace the isometric tension, recorded with a sensitive force transducer. Temperature 20 °C. (From Hodgkin & Horowicz, 1957.)

treatment contracts to maintain a steady tension for a few seconds and then relaxes; this relaxation is not accompanied by any change in membrane potential. A second contracture can be obtained only if the potassium ion concentration is sufficiently reduced for a short time. There is thus some kind of restitution or 'priming' process occurring. Hodgkin & Horowicz (1960) showed that the extent of this restitution is inversely related to the potassium ion concentration (and therefore to the membrane potential) in the intervening period between two exposures to a high potassium ion concentration. This 'priming' process does not occur in those muscles which show maintained potassium contractures.

The importance of calcium ions

It is clear that depolarization of the cell membrane cannot itself be the ultimate trigger for the contraction process. Apart from the difficulty of seeing how this could work, we know that glycerinated fibres will contract even though there is no membrane present. Now glycerol-extracted muscle fibres bathed in a solution containing ATP and magnesium ions are extremely sensitive to the calcium ion concentration; concentrations as low as 10^{-6} M are sufficient to cause some contraction and ATP splitting. Thus we might expect that depolarization of the cell membrane of an intact muscle fibre would cause an increase in the internal calcium ion concentration, so leading to contraction. Calcium ions, in other words, are the means whereby the contraction of the muscle is switched on and off. This suggestion was first made by Heilbrunn & Wiercinski in 1947, and it is now very well established (see Ashley *et al.*, 1991; Rüegg, 1992; Melzer *et al.*, 1995). Let us look at some of the evidence.

A number of investigations have been made on the effect of calcium ions on the potassium contractures of various muscles. Contractures eventually fail in the absence of calcium, and low calcium ion concentrations reduce the contracture tension (Niedergerke, 1956; Frank, 1960; Edman & Schild, 1962; Aidley, 1965).

Caldwell and his colleagues developed a technique for injecting solutions into the interior of the large muscle fibres of the spider crab *Maia*. Injection of potassium, sodium or magnesium ions or ATP did not produce any contraction, but the injection of calcium ions did (Caldwell & Walster, 1963). In further experiments they injected calcium ion buffer solutions, prepared by mixing known quantities of calcium chloride and the chelating agent EGTA (ethylene glycol bis(β-aminoethyl ether)-*N*,*N*'-tetraacetate), so that the free calcium ion concentration inside the fibre could be stabilized at a predetermined value (Portzehl *et al.*, 1964). They found that the threshold calcium ion concentration necessary for contraction was about 10^{-6} M. This is significantly close to the calcium ion concentration necessary to activate the ATPase system of isolated myofibrils (Weber & Herz, 1963).

The relation between calcium ion concentration and tension has been further investigated with experiments on 'skinned' muscle fibres (Hellam & Podolsky, 1969; Ashley & Moisescu, 1977; Stephenson & Williams, 1981, 1982). A muscle fibre is held under paraffin or silicone oil and its sarcolemma and surrounding connective tissues are removed with a fine needle, so leaving the interior of the fibre directly accessible to aqueous solutions. Alternatively, the fibre can be 'chemically skinned' by removing its sarcolemma with a suitable detergent solution. A sensitive force-transducer measures the tension produced by the fibre when solutions

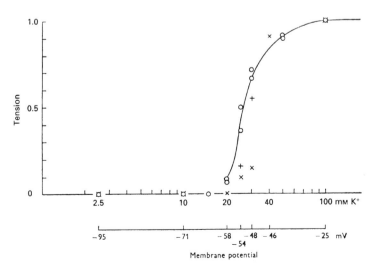

Figure 20.2. The relation between peak contracture tension and potassium ion concentration or membrane potential in single frog muscle fibres. (From Hodgkin & Horowicz, 1960.)

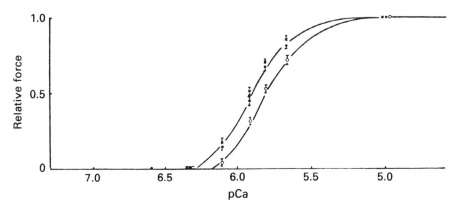

Figure 20.3. The relation between calcium ion concentration (pCa) and force in skinned muscle fibres. Measurements were made at two different sarcomere lengths; the left-hand curve shows results from the longer length. pCa is given by $-\log_{10}[Ca^{2+}]$, where $[Ca^{2+}]$ is the molar concentration of calcium ions. (From Stephenson & Williams, 1982.)

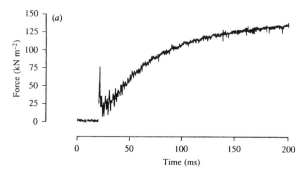

Figure 20.4. Contraction of a frog muscle fibre following release of caged calcium ions. The fibre was chemically skinned and contained 5 mM nitr-5. An intense brief laser flash caused photolysis of the caged compound and the calcium ions were set free inside the fibre. (From Ashley *et al.*, 1991.)

with different concentrations of calcium ions (using calcium-EGTA buffers) are applied.

Figure 20.3 shows the results of an experiment of this type done with a rat fast-twitch muscle fibre. The threshold for contraction is just below 10^{-6} M calcium ions and full contraction is reached by 10^{-5} M. Various factors affect the position of such curves. Increasing magnesium ion concentration and higher acidity (lower pH) push them to the right, i.e. the calcium sensitivity of the system is reduced. Stretching the muscle makes the system slightly more sensitive to calcium, as is evident in the diagram.

The calcium ion concentration inside a muscle fibre can be almost instantaneously increased by the used of 'caged' calcium. The calcium is chelated by a photolabile compound such as nitr-5, from which it is released by an intense flash of light. The results show that tension rises rapidly after the release of the calcium ions in the vicinity of the myofibrils, as is indicated in fig. 20.4 (Ashley *et al.*, 1991).

A very direct demonstration that the calcium ion concentration in the sarcoplasm rises immediately after

stimulation was made by Ashley & Ridgeway (1968, 1970). Their experiments made use of the protein aequorin, which is isolated from a bioluminescent jellyfish and emits light in the presence of calcium ions. They injected solutions of aequorin into the large muscle fibres of the barnacle *Balanus nubilus*. When such a fibre was stimulated electrically it produced a faint glow of light, indicating the presence of calcium ions in its interior. Relative measurements of the internal calcium ion concentration at different times and under different conditions can be made by measuring the light emission with a photomultiplier tube. The time course of light emission after a stimulus is called a 'calcium transient'.

One of these experiments is shown in fig. 20.5 and is worth taking a look at. The size of the calcium transient increases with the degree of depolarization, beyond a certain threshold value. The degree of tension development rises with increasing size of the calcium transient. The calcium transient starts after a short latent period and begins to fall soon after the end of the stimulus pulse. But the tension keeps rising until the calcium transient is long past its peak and almost finished; the time course of the calcium transient is in fact very similar to the rate of change of tension during the rising phase of the tension change.

Calcium transients following stimulation can also be demonstrated by means indicator dyes such as murexide, arsenazo III or antipyralazo III (Jöbsis & O'Connor, 1966; Miledi *et al.*, 1982). Fluorescent dyes such as fura-2 or mag-fura-2 have been much used for cellular imaging in recent years, and have sometimes been used in muscle cells (Klein *et al.*, 1988; Delbono & Stefani, 1993). Garcia & Schneider (1993) used antipyralazo III to measure the large rapid responses at the beginning of a transient and the higher affinity fura-2 to measure the slower low level changes during the return to the resting condition.

In general, it seems clear that the free calcium ion concentrations in skeletal muscle cells rise from a low level in the region of 0.1 μM at rest to up to 10 μM during activity.

The T system

The experiments so far described imply that depolarization of the cell membrane causes an increase in the calcium ion concentration in the interior of the muscle fibre. An early suggestion as to how this is done was that depolarization allows calcium ions to enter the muscle fibre from outside; these calcium ions would then diffuse into the interior of the fibre and activate the contractile system. However, the evidence is against this idea. The amount of calcium entering the fibre during a twitch is too little to account for the necessary increase in calcium ion concentration inside the fibre (Sandow, 1965). And Hill (1949) calculated that the time taken for a substance released from the membrane to diffuse into the interior of the fibre would be much too long to account for the speed with which activation becomes maximal after the stimulus.

How then does excitation at the cell surface cause release of calcium inside the fibre? The first step in the solution of this problem was provided by the demonstration by Andrew Huxley & Taylor (1955, 1958) that there is a specific inward-conducting mechanism located (in frog sartorius muscles) at the Z line. In these experiments, the fibres

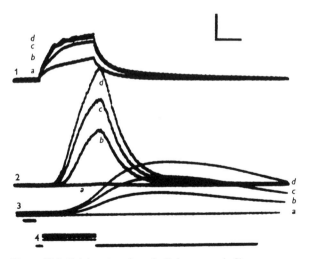

Figure 20.5. Calcium transients in *Balanus* muscle fibres as measured by the aequorin technique described in the text. Trace 4 shows the stimulus pulse, which is applied via an intracellular silver wire electrode. Trace 1 shows the resulting depolarization, trace 2 the photomultiplier output (the calcium transient) and trace 3 the tension. *a*, *b*, *c* and *d* are the results of applying four different current pulses of increasing intensity. The calibration marks are 100 ms (horizontal), 20 mV (for trace 1) and 5 g (for trace 3). (From Ashley & Ridgeway, 1968.)

were viewed by polarized light microscopy (so as to make the striation pattern visible) and stimulated by passing current through an external microelectrode, of diameter 1 to 2 μm, applied to the fibre surface. Hyperpolarizing currents did not produce contraction. Depolarizing currents did produce contraction, but only when the electrode was positioned at certain 'active spots' located at intervals along the Z line (fig. 20.6). In these cases the A bands adjacent to the I bands opposite the electrode were drawn together, and the extent to which this contraction passed into the interior of the fibre was proportional to the strength of the applied current.

At first it was thought that the inward-conducting mechanism was the Z line itself, but on repeating the experiments with crab muscle fibres, Andrew Huxley & Straub (1958) found that the 'active spots' were localized not at the Z line but near the boundary between the A and I bands. This suggests that there is some transverse structure located at the Z lines in frog muscles and at the A–I boundary in crab muscles.

Such a structure was found by Porter & Palade (1957) in their electron microscope study of the endoplasmic (or 'sarcoplasmic') reticulum in various vertebrate skeletal muscles. The sarcoplasmic reticulum consists of a network of vesicular elements surrounding the myofibrils (fig. 20.7). At the Z lines in frog muscle, and at the A–I boundaries in most other striated muscles (including crab muscle), are structures known as 'triads', in which a central tubular element is situated between two vesicular elements. These central elements of the triads are in fact tubules which run transversely across the fibre; they form the transverse tubular system or *T system*. There is no communication between the lumina of the T system and those of the sarcoplasmic reticulum vesicles, although their respective membranes are in close contact at the triads.

Clearly the T system tubules are in just the right position to account for the 'inward-conducting mechanism' suggested by the experiments of Huxley & Taylor. Hence it is important to know whether or not the T system is connected to the cell membrane. In the earlier studies the material was fixed with osmium tetroxide and the T system appeared as a row of elongated vesicles. In material fixed with glutaraldehyde, however, it is evident that the system really is tubular, and in favourable preparations it is possible to see that the tubules are invaginations of the cell membrane (Franzini-Armstrong & Porter, 1964). Further convincing evidence that this is so was provided by Hugh Huxley (1964), who soaked muscle fibres in solutions containing ferritin. Ferritin is an iron-storage protein found in the spleen; because of its high iron content it is electron dense, and the individual molecules can be seen by electron microscopy. In Huxley's experiments, ferritin appeared in the T system tubules, indicating that they must be in contact with the external medium. No ferritin appeared in the sarcoplasmic reticulum or in the rest of the sarcoplasm.

How does the electrical signal at the cell surface membrane travel down the T tubules into the interior of the fibre? There are two possibilities to be considered: either it is a passive process of electrotonic spread, or the T tubule membranes are electrically excitable and can conduct action potentials. The T tubules are too small for their membrane potentials to be measured directly, and hence it is necessary to use contraction of the myofibrils in different parts of the muscle fibre cross-section as an indicator of the inward spread of activity. Using voltage-clamped fibres, Costantin (1970) found that the inward spread of activity was reduced in fibres treated with tetrodotoxin or a low sodium solution. This suggests that inward propagation is an electrically excited process with a propagated action potential involving sodium ion flow, just as at the cell surface membrane.

The speed of inward propagation was measured in a most ingenious way by Gonzalez-Serratos (1971). He placed a single muscle fibre in a Ringer solution containing gelatin, and allowed it to cool and set. He then compressed the gelatin block so as to shorten the muscle fibre without bending it. This made the myofibrils wavy. But when the myofibrils were activated after stimulation of the fibre, they

shortened and so became straight. Using a high-speed movie camera, it was possible to measure the time at which straightening of the myofibrils occurred at different points in the fibre cross-section, and so to determine the speed of propagation of the activating signal. This was about 7 cm s^{-1} at 20 °C, with a Q_{10} of 2. Thus in a fibre 100 μm in diameter the wave of activation would take about 0.7 ms to propagate from the surface to the centre.

The sarcoplasmic reticulum

If a skeletal muscle is homogenized, the myofibrils in the homogenate do not contract on the addition of ATP, and the rate of ATP splitting is very low. However, if the myofibrils are isolated from the rest of the homogenate by low speed centrifugation and subsequent washing, then they do contract in the presence of ATP and have a high ATPase activity. Marsh (1952) suggested that the ATPase activity of the myofibrils was inhibited by a 'relaxing factor' present in the muscle homogenate. Bendall (1953) showed that this factor prevented the contraction of glycerol-extracted muscle fibres. A crucial observation on the nature of relaxing factor was made by Portzehl (1957). She found that, if a relaxing factor extract is centrifuged for some time at a high speed, the relaxing activity is confined to the resulting precipitate, which is of course particulate; the supernatant is without relaxing effect. Electron microscopy of this precipitate by Nagai *et al.* (1960) and others showed that it consists of small vesicles, often ellipsoidal or tubular

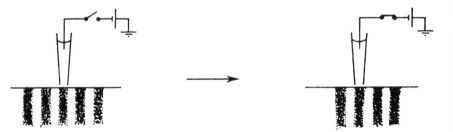

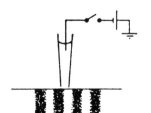

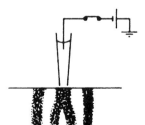

Figure 20.6. The effect of local depolarizations on an isolated frog muscle fibre. Diagrams on the left show the resting condition, those on the right show the condition during passage of a depolarizing current. Contractions occur when the electrode is opposite the I band (*b*), but not when it is opposite the A band (*a*). (Based on A. F. Huxley & Taylor, 1958.)

in shape. It would seem that these vesicles are derived from the sarcoplasmic reticulum.

The vesicular fraction of homogenized muscle is able to accumulate calcium from solutions containing ATP, magnesium ions and a small amount of calcium ions (Hasselbach, 1964; Martonosi & Beeler, 1983; Läuger, 1991). This calcium uptake is associated with ATP splitting; Hasselbach & Makinose (1963) calculated that two calcium ions are taken up for each ATP molecule split at calcium ion concentrations above 10^{-7} M, this ratio falling to one in the range 10^{-7} to 10^{-8} M. When the calcium ion concentration is below 10^{-8} M, the rate of ATP splitting is very low, and no further accumulation of calcium occurs. These experiments show that the vesicles of the sarcoplasmic reticulum can reduce the calcium ion concentration to below that necessary for contraction, by means of an ATP-driven 'calcium pump' in the vesicular membrane. The free calcium ion concentration inside the sarcoplasmic reticulum is in the range 0.5 to 1 mM, so the pump can work against a concentration gradient of more than 10^4 to 1.

The pump itself is a calcium–magnesium-activated ATPase of about 110 000 M_r, which is firmly bound in the sarcoplasmic reticulum membrane. Freeze-fracture electron micrographs show numbers of particles densely packed on the cytoplasmic side (see Franzini-Armstrong, 1994). The amino acid sequence of the calcium-ATPase molecule has been determined by cDNA sequencing (MacLennan *et al.*, 1985). The molecule has a globular region containing the ATPase site in the cytoplasm, connected via a stalk to a series of ten hydrophobic α-helices anchored in the membrane. Similar calcium-ATPases are widespread in non-muscle cells and in the plasma membrane (Song & Fambrough, 1994).

The action of the pump is aided by binding of calcium inside the sarcoplasmic reticulum by the protein calsequestrin (MacLennan & Wong, 1971). This is concentrated

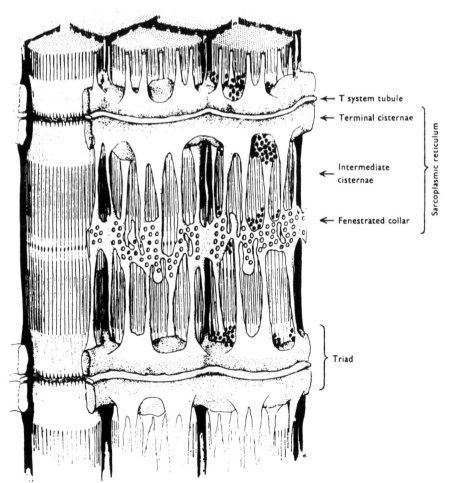

Figure 20.7. The internal membrane systems of a frog sartorius muscle fibre. (From Peachey, 1965.)

← T system tubule

← Terminal cisternae

Intermediate cisternae

← Fenestrated collar

Sarcoplasmic reticulum

Triad

in the terminal cisternae of the sarcoplasmic reticulum where it is attached to the inner side of the membrane (Jorgensen *et al.*, 1983; Franzini-Armstrong *et al.*, 1987).

Relaxation, then, is brought about by pumping of calcium ions from the myofibrils into the sarcoplasmic reticulum. We have seen that activation of contraction is effected by a sudden increase in myofibrillar calcium ion concentration, and so it seems very likely that this calcium is released from the sarcoplasmic reticulum.

How does the T *system signal to the sarcoplasmic reticulum?*

The nature of the link between depolarization of the T system tubule and release of calcium ions from the sarcoplasmic reticulum has long been a major problem. Clues to its solution have emerged from investigations on charge transfer in the T tubules, on the fine structure of the triad, and especially on the nature of the signalling molecules in the two membranes.

Since the activation of contraction is voltage dependent, we might expect that activation should depend upon the movement of charged particles in the potential gradient of the T tubule membrane. The problem was investigated by Schneider & Chandler (1973), using techniques similar to those used in investigations of the gating currents in nerve axons. They found that asymmetrical displacement currents that appear to be caused by the movement of membrane charge can indeed be detected. Charge flow was not evident at membrane potentials more negative than -80 mV and saturated above about -10 mV. The time course of the currents was much slower than that of nerve gating currents, by 20 to 100 times. The charge has been divided into two components, q_β and q_γ, thought to be derived from the cell surface membrane and the T tubule membrane, respectively (Adrian & Huang, 1984; Huang, 1988). Chandler and his colleagues (1976) suggested that each charge forms part of a long molecule which extends from the T tubule membrane to the sarcoplasmic reticulum membrane, and that movement of the charge might unplug channels in the sarcoplasmic reticulum membrane, so allowing calcium ions to escape.

Whether or not the 'long molecule' idea is correct, it seems very likely that the charge movements are related to calcium release (Schneider, 1994; Huang, 1993). Melzer and his colleagues (1986) measured charge movements and calcium transients in cut segments of single frog muscle fibres. They found that charge movement and calcium release were both dependent on membrane potential, as is shown in fig. 20.8. The close relation between the three quantities suggests strongly that there is a causal relation

between them: depolarization determines charge movement, which itself produces calcium release.

The structure of the triad has been much studied by Franzini-Armstrong and her colleagues (Franzini-Armstrong, 1970, 1994; Block *et al.*, 1988; Franzini-Armstrong & Jorgensen, 1994). She found that the T tubule and its adjacent sarcoplasmic reticulum sac are connected by an array of electron-dense structures called 'feet' (fig. 20.9). These are relatively large structures anchored in the sarcoplasmic reticulum membrane and each consisting of four subunits. The T tubule membrane contains particles grouped in fours ('tetrads'), and positioned opposite the feet, as is indicated in fig. 20.10. About half of the feet have no corresponding tetrad.

Our understanding of these structures has been much advanced by the isolation, cloning and sequencing of two major signalling molecules that occur in the two membranes. The T tubules contain numbers of the dihydropyridine (DHP) receptor, a voltage-gated calcium channel, so it is these that are grouped in fours to form the tetrads. The sarcoplasmic reticulum contains numbers of the ryanodine receptor, a calcium-release channel, and it is these that form the feet.

The DHP receptor is an L-type voltage-gated calcium channel with an amino acid sequence showing considerable

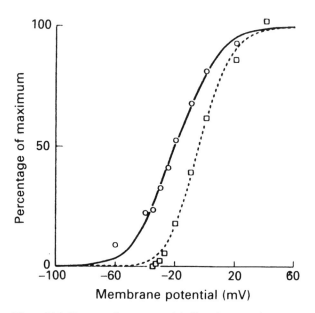

Figure 20.8. How membrane potential affects intramembrane charge movement (circles) and peak rate of calcium release (squares) in frog twitch muscle fibres. Measurements were made on voltage-clamped fibres using antipyralazo III to measure calcium ion concentrations. (From Melzer *et al.*, 1986.)

similarities to the voltage-gated sodium channel, as we have seen in chapter 6 (Tanabe *et al.*, 1987). Its crucial role in T tubule function has been most convincingly demonstrated: when cDNA coding for the molecule was introduced into dysgenic muscle fibres it restored their excitation–contraction coupling (Tanabe *et al.*, 1988).

Ryanodine is a plant alkaloid which locks the calcium-release channels of the sarcoplasmic reticulum open; hence radioactive ryanodine can act as a marker for these channels during their isolation. The receptor forms tetrameric particles with a characteristic clover-leaf appearance very like that of the feet seen at the triad. It can be incorporated into lipid bilayers, when it forms channels with relatively high conductances, up to 595 pS or more in some conditions (Inui *et al.*, 1987; Lai *et al.*, 1988; Meissner, 1994).

Cloning and sequencing of the skeletal muscle ryanodine receptor shows that it is a very large molecule: each protein chain has 5032 or 5037 residues, so the whole tetrameric complex must have a molecular weight over 2 million. One model of the chain has four membrane-crossing α-helices in the C-terminal region, another has ten or twelve (Takeshima *et al.*, 1989; Zorzato *et al.*, 1990). Presumably the C-terminal regions from the four subunits come together to make a transmembrane pore in the sarcoplasmic

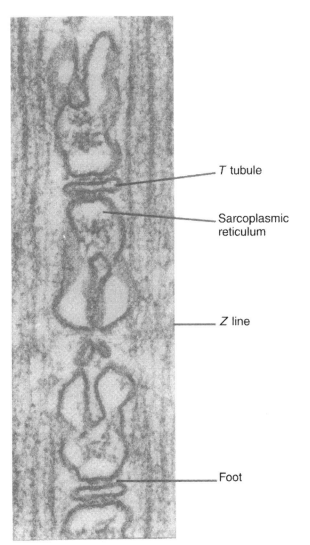

Figure 20.9. Electron micrograph of a longitudinal section of toadfish swimbladder muscle, showing the connections ('feet') between the T tubules and the sarcoplasmic reticulum. Magnification 180 000×. (Photograph kindly supplied by Dr C. Franzini-Armstrong.)

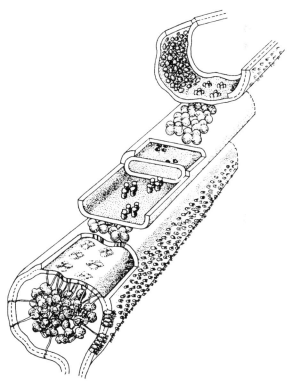

Figure 20.10. Membranes of the triad, the junction between T tubule and sarcoplasmic reticulum, showing the positions of feet and tetrads. Each tetrad is a group of four DHP receptors or voltage-gated calcium channels in the T tubule membrane. Each foot is a ryanodine receptor or calcium-release channel, rooted in the sarcoplasmic reticulum membrane and spanning the space between it and the T tubule. Each tetrad is associated with a foot, but note that half of the feet do not have corresponding tetrads. Also shown are the calcium pump (CaATPase) molecules in the membrane of the sarcoplasmic reticulum and calsequestrin molecules inside it. (From Block *et al.*, 1988. Reproduced from *The Journal of Cell Biology*, 1988, **107**, 387, by copyright permission of The Rockefeller University Press.)

reticulum membrane. The great mass of the molecule is in the foot region; cryo-electron microscopy shows that this is a rather open structure 'suggestive of scaffolding' that may serve to hold the triadic membranes together while allowing free diffusion of calcium ions away from the sarcoplasmic reticulum (Radermacher *et al.*, 1994).

The close association of the DHP receptors and the ryanodine receptor is indicated in fig. 20.11. It seems likely that the DHP receptors in skeletal muscle do not function as channels for the entry of calcium ions into the muscle fibre; rather, they act as voltage sensors to detect the change in T tubule membrane potential (Ríos *et al.*, 1991; Schneider, 1994). We may assume that the S4 segments (p. 102) are the voltage sensors, but just how the change in the DHP receptors opens the calcium release channel of the ryanodine receptor is yet to be determined.

Image analysis of calcium ion concentration changes in muscle, using fluorescent calcium-sensitive dyes and confocal microscopy, shows spontaneous 'sparks' of calcium ions at the triads, whose frequency is greatly enhanced by depolarization (Tsugorka *et al.*, 1995; Klein *et al.*, 1996). These may well be produced by the opening of one or a few calcium release channels activated by a single tetrad of DHP receptors; perhaps some of the uncoupled calcium release channels are opened by calcium ions released by the coupled one. This idea is illustrated in fig. 20.12.

The molecular basis of activation

As mentioned in the previous chapter, a mixture of purified F-actin plus purified myosin shows ATPase activity in the absence of calcium ions, but 'natural' actomyosin (the complex extracted from minced muscle with strong salt solutions) will split ATP only if there is a low concentration of calcium ions present. Ebashi and his colleagues found that this responsiveness to calcium ions is dependent upon

the presence of troponin and tropomyosin (Ebashi *et al.*, 1969). So it looked as though an understanding of the mechanism of activation must depend upon knowledge of the behaviour of these two proteins. We have examined some of the properties of tropomyosin already (p. 358), but troponin deserves a further look.

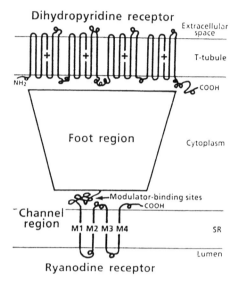

Figure 20.11. Schematic diagram showing the membrane topology of the skeletal muscle ryanodine receptor and its relation to the DHP (dihydropyridine) receptor. Only one of the four identical subunits forming the calcium-release channel is shown. An alternative model proposes 10 or 12 transmembrane segments in the sarcoplasmic reticulum membrane (SR). Each subunit is closely associated with a voltage-gated calcium channel (the DHP receptor), which probably acts primarily as a voltage sensor. (From Takeshima *et al.*, 1989. Reprinted with permission from *Nature* **339**, 445. Copyright 1989 Macmillan Magazines Limited.)

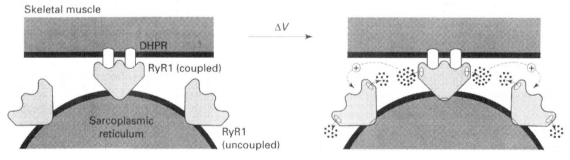

Figure 20.12. How an elementary unit of calcium signalling (a 'spark') may be produced in skeletal muscle. Depolarization (ΔV) produces a conformational change in a tetrad of dihydropyridine receptors (DHPR), opening the calcium-release channel (ryanodine receptor, RyR1) coupled to it and so releasing calcium ions (shown as dots) from the sarcoplasmic reticulum. These calcium ions may then activate adjacent uncoupled calcium-release channels. (From Berridge, 1997.)

Troponin

Troponin was first isolated by Ebashi & Kodama (1966*a,b*). Further work showed that it consists of three components, called troponin-C (18 000 M_r), troponin-I (21 000 M_r) and troponin-T (31 000 M_r) by Greaser & Gergely (1973; see also Perry, 1994, 1996).

Troponin-C has four calcium-binding sites in each molecule. It binds firmly to troponin-I and less firmly to troponin-T. Conformational changes occur when calcium is bound. The four sites show homology in their amino acid sequences; each consists of a loop between two α-helices (Collins *et al.*, 1973). There are considerable sequence homologies with other calcium-binding molecules such as parvalbumin, calmodulin and some of the myosin light chains. High resolution X-ray diffraction studies on troponin-C crystals show that the molecule has two distinct calcium-binding domains connected by a long section of α-helix (Herzberg & James, 1985).

The most notable property of troponin-I is that it inhibits actomyosin ATPase, an effect which is much enhanced if tropomyosin is present. This inhibition is relieved by troponin-C if it is binding calcium; thus the presence of calcium ions switches off the inhibition produced by the troponin-I–troponin-C complex.

Troponin-T is an elongated molecule which binds firmly to tropomyosin and troponin-I, and apparently less firmly to troponin-C. Thus it serves to link the whole troponin complex to tropomyosin.

Overall, then, the troponin system forms a complex which binds to tropomyosin and actin and which inhibits actomyosin ATPase unless calcium ions are present. It is not hard to believe that the structural arrangement of this system would be altered by conformational change in one of its members, such as occurs when calcium is bound by troponin-C. Such a rearrangement in the thin filament might well form the basis of the activation process.

Thin filament activation

Direct evidence that the structure of the thin filaments alters during activation has been provided by X-ray diffraction measurements. The details of the actin pattern alter on contraction; in particular the second layer line at about 180 Å becomes much more evident and the sixth layer line at 59 Å also increases in intensity somewhat (Vibert *et al.*, 1972; H. E. Huxley, 1973). Calculations by Parry & Squire (1973), Haselgrove (1973) and Hugh Huxley showed that these changes could have been produced by a movement of tropomyosin in the groove between the two long-pitch helices of actin monomers in the thin filament. Figure 20.13 shows Hugh Huxley's picture of this phenomenon in 1973.

A movement of this type might be brought about by conformational changes in the troponin complex following binding of calcium ions to troponin C. Hugh Huxley (1973) suggested that it would expose sites on the actin monomers which would allow the binding of the S1 myosin heads, and that such actin–myosin binding could not happen at rest because the tropomyosin was in the way. Each tropomyosin molecule would thus unblock seven actin monomers. This idea is known as the steric blocking hypothesis. It may not be the whole story (Chalovich & Eisenberg (1982) suggested that the troponin–tropomyosin system affects the kinetics of actomyosin ATPase, and Perry (1996) suggested that conformational changes in actin are involved) but it certainly seems to be a very persuasive suggestion.

Confirmation for the idea has been sought from electron micrographs of thin filaments, using the three-dimensional reconstruction technique described earlier. Thin filaments

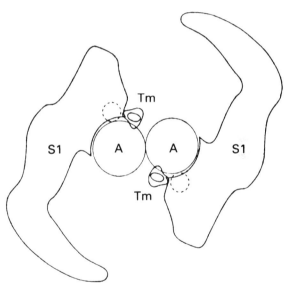

Figure 20.13. The steric blocking model of thin filament activation as proposed in 1972. The diagram is based on X-ray diffraction studies and three-dimensional reconstruction from electron micrographs. A thin filament is seen in cross-section with actin (A) and tropomyosin (Tm) molecules. Two myosin S1 heads are shown in the position they were thought to occupy in decorated thin filaments. Tropomyosin positions are shown for the muscle at rest (dotted circles) and when activated (continuous contours). The probable positions and shapes of the protein molecules have been revised since 1972, but the conclusion that the tropomyosin molecule moves into the 'groove' between the actin monomers remains. The steric blocking model suggests that until this movement takes place the S1 heads are physically unable to attach to the actin monomers. (From H. E. Huxley, 1973.)

from *Limulus* have their tropomyosin molecules positioned over the myosin-binding regions in the absence of calcium ions, but in their presence the tropomyosin is moved around the actin monomers closer to their inner regions. Studies on vertebrate thin filaments show that in the resting state the tropomyosin is positioned over clusters of amino acid residues known to be involved in cross-bridge docking (Lehman *et al.*, 1994, 1995).

Modelling studies based on X-ray diffraction data and knowledge of the molecular structure of the actin filament have provided results in good agreement with the steric blocking model. It may be that the tropomyosin does not prevent weak binding of myosin to actin in the resting condition, but that it does prevent strong binding and movement of the myosin head in the power stroke (Squire *et al.*, 1993).

The timing of thin filament activation

Hugh Huxley and his colleagues extended their time-resolved X-ray diffraction measurements using synchroton radiation to a study of the activation process (Kress *et al.*, 1986). They looked at the increase in the intensity of the 179 Å off-meridional reflection (the actin second layer line) after stimulation; we have seen that this change had been interpreted as a lateral movement of tropomyosin across the thin filament surface.

The results of these experiments were clear-cut (fig. 20.14). The increase in the 179 Å reflection intensity was the first indication of structural change in the contractile apparatus, reaching half its final value by about 17 ms after the stimulus. This was considerably quicker than the changes attributed to movement of the myosin heads: the half-times for change of the equatorial pattern and of the 429 Å myosin layer line were at least 25 ms (see fig. 19.12). Tension changes were slower still.

When muscles were greatly stretched so that the thick and thin filaments no longer overlapped, change in the intensity of the 179 Å reflection on stimulation still occurred and followed the same time course. This implies that tropomyosin movement is independent of and precedes cross-bridge attachment. At these lengths the changes on the equator and at 429 Å, attributed to cross-bridge movement, no longer occur.

On relaxation at normal lengths the fall in intensity of the 179 Å reflection is somewhat more rapid than the fall in tension. In greatly stretched muscles, however, the fall in intensity is quicker still. In one set of experiments, the average time for the intensity to fall to half its value after the end of stimulation was 407 ms at normal lengths, but only 128 ms in greatly stretched muscles. This suggests that the presence of attached cross-bridges may interfere with the return of the tropomyosin to its resting position.

These beautiful experiments provide excellent confirmation for the view that movement of tropomyosin further into the groove of the thin filament is the essential prerequisite for the attachment of the cross-bridges. They fit well with the steric blocking hypothesis, although they do not rule out the possibility that tropomyosin movement is associated with allosteric effects on actin structure.

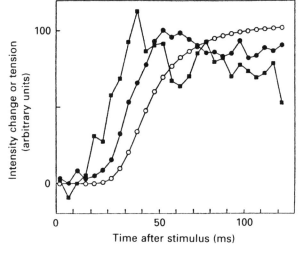

Figure 20.14. Time-resolved X-ray diffraction measurements on activation, from a frog twitch muscle at 5 °C. The curves show the intensity increases of the second actin layer line at 179 Å (squares) and of the equatorial (1,1) reflection (black circles), following an electrical stimulus. These curves indicate the movements of tropomyosin and of the myosin heads, respectively. White circles show the development of isometric tension. (From Kress *et al.*, 1986.)

21
Varieties of muscle design

In chapters 18 to 20 we have been very largely concerned with the properties of vertebrate twitch skeletal muscle, exemplified by the sartorius of the frog. But it must be realized that the frog sartorius represents only one of a considerable variety of types of muscle, and a rather specialized one at that.

Muscles vary greatly in the time characteristics of their contractions. Some have much higher velocities of shortening than others; the muscles of the hare contract faster than do those of the tortoise. Muscles involved in sound production often contract more rapidly than do the locomotor muscles. Even similar muscles from similar animals have higher contraction speeds in small animals than in large animals; compare the wing-beat frequencies of a sparrow and a pelican (Hill, 1950b). These differences are correlated with differences in the ATPase activities of the myosin extracted from the various muscles (Bárány, 1967), and these in turn with differences in the amino acid sequences of the myosin and other myofibrillar molecules (Schiaffino & Reggiani, 1996).

The way the contractile machinery is organized is not uniform, although all muscles work via actin–myosin interactions involving the splitting of ATP. Striated muscles from different animal phyla may have different arrangements of the thick and thin filaments. Not all muscles are striated. The arrangement of myosin molecules in the thick filaments varies. Some muscles contain extra proteins such as paramyosin in their filaments.

There are also differences in the excitation and regulatory processes of different types of muscle fibre, although all seem to use calcium ions as the primary trigger for contraction. Some have propagated action potentials; some do not. Some are electrotonically linked by gap junctions; some are not. Some have large amounts of sarcoplasmic reticulum; some do not. Some have troponin–tropomyosin regulatory systems; some do not. In this chapter we sample some of this variety.

Vertebrate skeletal muscles
Twitch and tonic fibres

The skeletal muscle fibres of frogs and toads fall into two distinct classes: twitch fibres and tonic fibres. These are frequently referred to as 'fast' and 'slow' fibres, respectively, but since this nomenclature can lead to some confusion we shall not adopt it here. The sartorius is composed entirely of twitch fibres. Tonic fibres are found to varying extents in other muscles; the rectus abdominis is particularly rich in them.

Most twitch fibres have only one motor end-plate, although a few may have two. The cell membrane is electrically excitable and, as we have seen in chapter 7, will produce an all-or-nothing propagated action potential if it is sufficiently depolarized. The end-plate potential is sufficiently large to cause such a depolarization, so that, just as in the nerve axon, excitation consists of all-or-nothing events.

The excitation process in tonic fibres is quite different (Kuffler & Vaughan Williams, 1953a,b). The fibres are electrically inexcitable, so that no propagated action potentials occur. There are a large number of neuromuscular junctions on each fibre; this condition is known as multiterminal innervation (fig. 21.1b). Each motor impulse causes the release of only a relatively small amount of transmitter substance, so that the end-plate potentials are small in size. These end-plate potentials show pronounced summation on repetitive stimulation; thus the amount by which the fibre membrane is depolarized is dependent upon the frequency of the nervous input. There is negligible contractile response to a single stimulus; with repetitive stimulation the tension rises with increasing stimulus frequency, reaching a maximum at about 50 impulses s^{-1}. It seems probable that the tonic fibres are used for maintained low tension contractions such as are involved in the maintenance of posture, whereas the twitch fibres are used for rapid movements.

Electron microscopy shows that frog tonic muscle fibres differ from twitch fibres in a number of ways. There are no M lines, the Z lines are thicker, there is much less sarcoplasmic

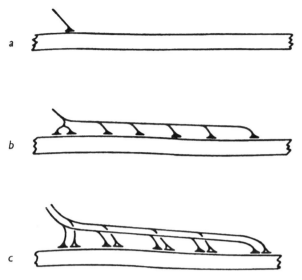

Figure 21.1. Varieties of innervation in muscle fibres. *a* shows uniterminal innervation; *b* shows multiterminal innervation; *c* shows multiterminal and polyneuronal innervation.

reticulum and there are fewer triads in tonic fibres (Peachey & Huxley, 1962; Page, 1965).

Lännergen (1979) found a third type of muscle fibre in *Xenopus*. These intermediate fibres had multiterminal innervation and showed maintained potassium contractures just like tonic fibres, but would produce a slow twitch on electrical stimulation.

The muscle fibres of fishes, with some exceptions, fall into two classes which seem to correspond to those of amphibians (Bone, 1978; Johnston, 1980). The tonic fibres are usually reddish since they contain the oxygen-storage pigment myoglobin, and are located peripherally. They differ from the tonic fibres of amphibians in having an M line, a well-developed sarcoplasmic reticulum and extensive triads; they are usually well supplied with mitochondria. White (twitch) fibres have rather more sarcoplasmic reticulum but very few mitochondria.

These cytological features are related to the functions of the fibres in the living fish. The red (tonic) fibres are used during continuous cruising at relatively low swimming speeds, and so they will require a continuous supply of ATP produced by oxidative metabolism in the mitochondria. The white (twitch) fibres are responsible for short bursts of high speed activity, which can be supported by an anaerobic energy supply. Nevertheless it is clear that in many species white muscle fibres are also involved in continuous swimming at faster speeds (Greer Walker & Pull, 1973; Johnston *et al.*, 1977).

Birds also have two types of fibre, with uniterminal and multiterminal innervation, respectively, but those with multiterminal innervation are electrically excitable and may show propagated action potentials (Ginsborg, 1960).

Mammalian twitch fibre types

Almost all the skeletal muscles of mammals are composed entirely of twitch fibres. Tonic muscle fibres are found in some of the muscles innervated by the cranial motor nerves, including the extraocular muscles which determine the direction of gaze, the tensor tympani of the middle ear, and some of the muscles of the larynx.

The twitch fibres can be divided into two types according to their myofibrillar ATPase histochemistry: type I fibres stain densely after preincubation in acid conditions, whereas type II fibres stain after preincubation in alkaline conditions (Engel, 1962). These two types correspond to differences in contractile properties; they are named slow twitch and fast twitch, respectively, according to their speeds and duration of contraction (Close, 1972). Sometimes whole muscles are composed of the same type of fibre. The soleus, for example, is largely a slow-twitch muscle while the gastrocnemius (which is larger and acts in parallel with it) contains a majority of fast-twitch fibres. Slow-twitch muscles are usually used in the control of posture, whereas fast-twitch muscles are used for locomotory and other movements.

Fast-twitch (type II) fibres have been further divided into two or three types. The criteria for this division include the number of mitochondria present and the relative importance of oxidative and glycolytic pathways in respiration (Peter *et al.*, 1972), the ATPase activity at acid or alkaline pH (Brooke & Kaiser, 1970), the contraction time of the fibres (i.e. the time between the onset of a twitch contraction and its peak) and especially their resistance to fatigue (Burke *et al.*, 1973).

To simplify the situation, then, there are three main types of twitch fibre in mammals:

1 red-slow, type I, SO (slow oxidative) or S (slow) fibres,
2 red-fast, type IIa, FOG (fast oxidative-glycolytic) or FR (fast fatigue-resistant) fibres, and
3 white, type IIb, FG (fast glycolytic) or FF (fast fatigue-sensitive) fibres.

This classification is imperfect in a number of respects (see Pette & Staron, 1990; Kelly & Rubinstein, 1994). Fibres intermediate between types IIa and IIb (called IIc) are sometimes found. The subdivisions of the fast-twitch fibres by one criterion may not always agree with those by another; it may not always be correct to equate IIa and IIb fibres precisely with FOG and FG fibres, for example

(Green, 1986). The characteristics of fibres may change during development or adaptation. Although the motor units in a muscle may show a considerable range of physiological and histochemical properties, the fibres of any one motor unit are always all of one type (Burke *et al.*, 1973; Nemeth *et al.*, 1981).

Vertebrate heart muscle

The hearts of animals can be divided into two general types according to how their rhythmic contractile activity is initiated. In *myogenic* hearts, found in vertebrates and molluscs, excitation arises spontaneously in the heart muscle fibres themselves. In *neurogenic* hearts, found in many arthropods, the heart muscle fibres contract only in response to nervous stimuli. Here we examine some aspects of the excitation and coupling processes in vertebrate cardiac muscle.

Cardiac electrophysiology

Intracellular recordings from heart muscle fibres were first made by Draper & Weidmann (1951), using isolated bundles of Purkinje fibres from dogs. In mammalian hearts, the Purkinje fibres form a specialized conducting system which serves to carry excitation through the ventricle. After being isolated for a short time, they begin to produce rhythmic spontaneous action potentials, as is shown in fig. 21.2. The form of the action potentials differs from those of nerve axons and vertebrate twitch muscle fibres in that there is a prolonged *plateau* between the peak of the spike and the repolarization phase. The action potential in spontaneously

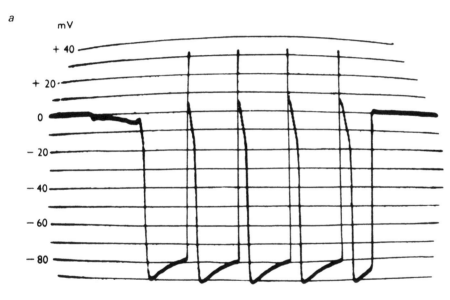

Figure 21.2. Cardiac action potentials. *a* shows an intracellular record from a Purkinje fibre of a dog heart. The microelectrode entered the fibre soon after the start of the trace, and was withdrawn before its end. The interval between successive action potentials was 1.4 s. *b* shows the nomenclature for different parts of such action potentials. (*a* from Draper & Weidmann, 1951.)

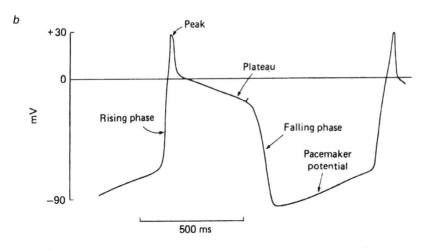

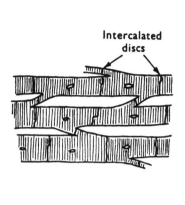

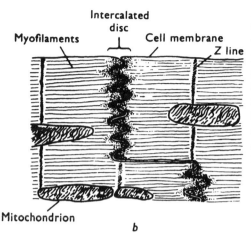

a

b

Figure 21.3. Schematic diagrams to show the structure of vertebrate heart muscle. Light microscopy (*a*) shows uninucleate muscle cells separated by intercalated discs. Under the electron microscope (*b*) the intercalated discs are seen to be concentrations of dense material on each side of an intercellular boundary.

active cells is initiated when the slowly rising *pacemaker potential* crosses a threshold level.

A heart would be ineffective if the rhythmic action potentials and their associated contractions in any one fibre were independent of those of its fellows: it is obviously essential that the contractions of the fibres should be synchronized. Vertebrate heart muscle consists of a network of branching fibres which are connected to each other by *intercalated discs* (fig. 21.3). Under the electron microscope these are seen to be concentrations of dense material on each side of two cell membranes. Desmosomes, providing mechanical attachment, are present, and there are gap junctions between the cells (Fawcett & McNutt, 1969).

We have seen in chapter 12 that gap junction channels consist of two connexons in the adjacent membranes, each consisting of six subunits known as connexins. Connexin 43 was the first connexin to be isolated from cardiac muscle, and at least four others have since been identified. Their expression varies somewhat in different parts of the heart (Beyer *et al.*, 1987, 1994).

The gap junctions connecting adjacent cells offer a low resistance to current flow, so that local circuits set up by an action potential in one cell can cross into the next cell and so excite it. The electrical activity of the whole of the heart will then be driven by those cells which have the most rapid spontaneous frequency. In the frog, these pacemaker fibres occur in the sinus venosus, and the excitation spreads from there into the atria and ventricle. In mammals the pacemaker region is the sinuatrial node; excitation sweeps across the atria to the atrioventricular node, and from there it is carried into the ventricles along the Purkinje fibres (fig. 21.4). The Purkinje fibres are specialized for conduction in that they are elongated, relatively large in diameter, and have little contractile material.

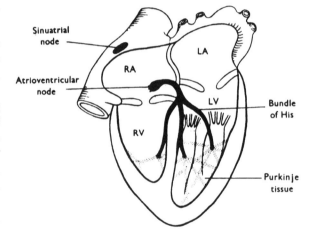

Figure 21.4. Pacemaker and specialized conductile regions of the mammalian heart. RA, LA, right and left auricles; RV, LV, right and left ventricles. (From Scher, 1965.)

It is possible to measure the electrical activity of the human heart simply by attaching leads to the wrists and ankles of the subject. The resulting record is known as an *electrocardiogram*, or ECG for short. ECGs were first measured by Einthoven, using the string galvanometer which he invented for the purpose (see Einthoven, 1924), and their measurement (now by more modern methods) has long been a standard procedure in medical practice. Figure 21.5 shows a typical ECG, recorded between the right arm and left leg. The different peaks in the electrical cycle of events were labelled the P, Q, R, S, and T waves by Einthoven.

The events in the heart cycle to which these electrical waves in the ECG correspond were worked out by recording with surface electrodes from exposed hearts in experimental animals or surgical patients. Figure 21.6 shows how the ECG relates to action potentials in different heart muscle cells. The heart beat is initiated by pacemaker activity in the cardiac cells of the sinuatrial node. This activity is not evident in the ECG, since a relatively small number of cells are involved and so the currents produced are too small to be seen by electrodes placed on the skin. The pacemaker activity then excites the adjacent cells of the atria and a wave of depolarization sweeps over the whole of the atria. The currents associated with this much larger number

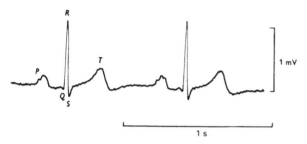

Figure 21.5. A human electrocardiogram, recorded between electrodes applied to the right wrist and left ankle.

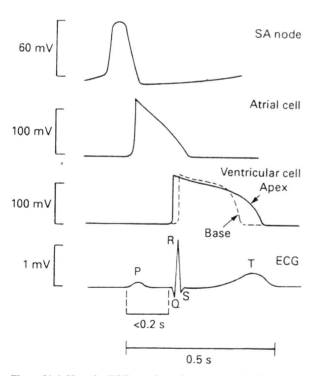

Figure 21.6. How the ECG waveform (lowest trace) is related to the timing of action potentials seen by intracellular recording at different sites in the heart. SA, sinuatrial. (From Levick, 1995. Reproduced with permission from *An Introduction to Cardiovascular Physiology*, 2nd edition, Copyright 1995 Butterworth-Heinemann Ltd.)

of cells are recorded in the ECG as the P wave. During the plateau of the atrial action potential there is little current flow and so the level of the ECG is not affected.

The atria and ventricles are separated by a sheet of connective tissue. The only electrically excitable pathway between the two is the atrioventricular bundle, which connects the atrioventricular node at the base of the atrial septum to the bundle of His in the ventricular septum. Conduction in the atrioventricular node is slow, so that there is some delay between the activity in the atria and that in the ventricles.

From the atrioventricular node the wave of depolarization spreads via the atrioventricular bundle to excite the specialized conducting tissue of the ventricles, the Purkinje fibres of the bundle of His. The action potential passes down the left and right branches of this system, but the activity is again not evident in the ECG because the number of cells involved is relatively small and hence the current flow is small also. The bundle of His brings excitation to the mass of the ventricular muscle cells, beginning in the septum and then spreading from the apex of the ventricle up to the base. The net currents involved in the depolarization of all these ventricular cells are very large and are seen in the ECG as the QRS complex.

After this the whole of the ventricle is depolarized and there is very little electric current flow; the ventricular muscle is contracting so as to pump blood out along the aorta and pulmonary artery. Then repolarization of the ventricular fibres occurs, at slightly different times in different places, and the current flow associated with this is seen as the T wave. After this the heart is electrically at rest except in the pacemaker regions, its muscles are relaxing and it is refilling with blood ready for the next cycle. The pacemaker potentials preceding the next P wave are not visible in the ECG.

Recording with intracellular electrodes shows that there is some variation in the form of the action potential in different regions of the heart (fig. 21.7). In the sinuatrial and atrioventricular nodes the action potential is slow to rise and relatively small in amplitude, with no plateau. Elsewhere the action potential rises rapidly and does have a plateau, which is somewhat shorter in the ventricles than in the Purkinje tissue and shorter still in the atria. Cells of the sinuatrial node have pronounced pacemaker potentials; these are not normally found elsewhere, although slow pacemaker potentials may arise after a time in isolated Purkinje fibres.

Action potentials and ionic currents
What is the ionic basis of heart muscle action potentials? Draper & Weidmann (1951) showed that the membrane

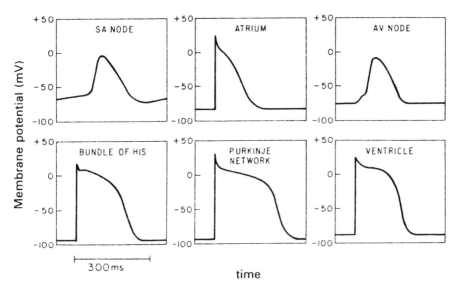

Figure 21.7. The shape of the action potential in different regions of the mammalian heart. SA, sinuatrial; AV, atrioventricular. (From Katz, 1977.)

potential at the peak of the early spike was reduced when the external sodium concentration was lowered. This suggests that, as in the action potential of nerve axons, the initial rapid depolarization is brought about by a regenerative increase in the sodium conductance of the membrane. But what happens during the plateau? Weidmann (1951) made the crucial observation that membrane resistance at this time is quite high, being much the same or even higher than it is during the pacemaker potential. So, in contrast to nerve axons, the membrane conductance to potassium cannot begin to rise soon after depolarization; there must be some long delay before this happens. Further evidence for this was produced by Hutter & Noble (1960), who showed that, in the absence of sodium ions, the cardiac cell membrane shows marked inward rectification: the membrane resistance is higher for depolarizing currents than it is for hyperpolarizing ones.

Direct measurements of membrane ionic currents need an effective voltage clamp system, but there are difficulties in producing one because of the complex geometry of the heart muscle and the small size of the cells. One way is to use the 'sucrose gap' method. Part of the muscle bundle is enclosed in an isotonic sucrose solution; this penetrates the intercellular spaces between the cells and so renders them incapable of carrying electric current. It is therefore possible to pass current between the two Ringer-filled compartments and know that all of it is passing through the intracellular material; and hence it must all pass out across the plasma membrane. Another method has been used for Purkinje fibres, which are larger than other cardiac cells. Here a short length of fibre is cut out (the injured ends heal over) and two microelectrodes are inserted. Neither method is very good at dealing with large currents.

Isolated cardiac cells have been used increasingly in recent years (see Noble & Powell, 1987). The cells are separated from one another by treatment of the heart with the enzyme collagenase, which breaks down the connective tissue holding them together. Brown *et al.* (1981) used suction electrodes to pass current into the cell and to record its membrane potential in a voltage clamp system, and others since then have used the whole-cell patch clamp technique (e.g. DiFrancesco & Tortora, 1991).

An early result of voltage clamp investigations was the discovery that depolarization results in an inward flow of calcium ions. Beeler & Reuter (1970) were able to demonstrate this by first depolarizing the fibres to -40 mV so as to inactivate the sodium current; further depolarization then resulted in a slow inward current (fig. 21.8). This slow current was unaffected by tetrodotoxin (which blocks the fast inward current) but was abolished in calcium-free solutions.

Sorting out the full nature of the cardiac action potential has proved to be a complicated and as yet unfinished task (see Noble, 1984, 1994, 1995; DiFrancesco, 1993; Irisawa *et al.*, 1993; Boyett *et al.*, 1996). The number of different ionic currents and channels involved is surprisingly large. The theoretical models developed by Noble and his colleagues have become progressively more complex over the years as they have had to accommodate more and more awkward facts.

The first of these models was produced by Noble in 1962. It was based on the Hodgkin–Huxley equations for the nerve action potential, with two changes. The potassium

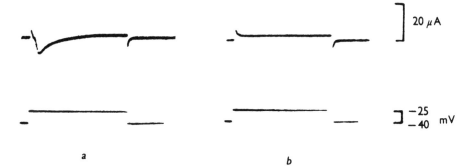

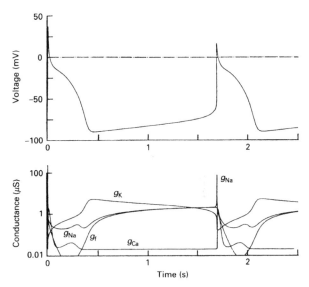

a **b**

Figure 21.8. Calcium current in voltage-clamped ventricular muscle from a dog heart. External calcium ion concentration was 1.8 mM in *a* and zero in *b*. The holding potential was −40 mV, which is sufficiently depolarized to inactivate the sodium current. Further depolarization to −25 mV for 570 ms resulted in the slow transient inward current shown in *a*. (From Beeler & Reuter, 1970.)

current was separated into two components, an inward rectifier current and a greatly delayed current that was elicited by depolarization. Computer simulations mimicked the form of the action potential in what appeared to be a very satisfactory manner.

The next model (McAllister *et al.*, 1975) was produced in response to new observations, especially the discovery of the calcium current. In this version the inward calcium current flowed during the whole of the plateau of the action potential. There was also a transient outward current and another outward current (called i_{K2} at this time) which was deactivated during the pacemaker potential.

It became clear that the 1975 model would have to be replaced when DiFrancesco (1981*a*,*b*) showed that the principal event during the pacemaker phase was not a reduction in outward current (i_{K2}) but an increase in an inward current. The new current was called i_f. It is time dependent and activated by hyperpolarization, and is carried largely by sodium ions. The i_f channels are, however, quite different from the fast sodium channels: they are activated by hyperpolarization instead of depolarization, they are also permeable to potassium ions, and they are insensitive to tetrodotoxin.

A new model incorporating i_f was produced by DiFrancesco & Noble (1985). Other changes in this model included currents activated by internal calcium, currents due to the sodium–potassium pump and the sodium–calcium exchange system, and consideration of the effects of changes in the intracellular and extracellular ionic concentrations as a result of activity.

The sequence of events in the cardiac action potential according to the 1985 model is shown in fig. 21.9. The curves show ionic conductance changes in a computer simulation appropriate to Purkinje fibres showing some

Figure 21.9. Computer simulation of the mammalian cardiac action potential and the conductance changes which are responsible for it, based on Purkinje fibres. The inward current which becomes apparent during the pacemaker potential flows through g_f. The sodium conductance g_{Na} includes both the conductance of the fast sodium channel and the sodium component of g_f. g_K, potassium conductance; g_{Ca}, calcium conductance. (From DiFrancesco & Noble, 1985.)

spontaneous activity. Let us begin at the point in the cycle where the membrane potential is at its most negative, at about 0.4 s on the time trace. It has reached this negative value because the potassium conductance g_K is high. However, the pacemaker conductance g_f has been switched on by the hyperpolarization, and it rises steadily for the next second or so. The slow sodium ion inflow that this permits results in a steady depolarization, the pacemaker potential.

After a time the pacemaker potential has depolarized the membrane sufficiently to open the fast-activating sodium channels; these are of the type found in nerve axons, i.e. they are voltage gated, rapidly inactivating, and sensitive to tetrodotoxin. Then follows the familiar runaway relation between membrane potential and sodium conductance just as in the nerve axon, so that there is a massive inflow of sodium ions and a rapid overshooting depolarization.

But now the model departs radically from the situation in nerve axons. The potassium conductance has fallen to a low level and there is an elevated calcium conductance attributable to voltage-gated calcium channels; hence the membrane potential remains near zero for some hundreds of milliseconds. The calcium conductance falls as the calcium channels inactivate, but the inward current is maintained by the electrogenic action of the sodium–calcium exchange transporter, which allows three sodium ions to enter for every calcium ion extruded. The plateau declines gradually, however, and is brought to an end as a result of the long-delayed increase in potassium conductance, and any calcium and fast sodium channels remaining open are finally closed during the repolarization phase. By the end of the action potential the pacemaker conductance g_f has already begun to rise and so the new cycle continues on its way.

The ionic basis of pacemaker and action potentials at the sinuatrial node is somewhat different. In particular, there is no fast sodium current, so the action potentials do not have the sharp leading edges seen in ventricular cells. In the pacemaker potential the i_f current may be the major and crucial component (DiFrancesco, 1993), or it may be that a slow inward calcium current and a time-independent ('background') slow sodium current are also at least as important (Campbell *et al.*, 1992; Irisawa *et al.*, 1993).

Excitation–contraction coupling

The importance of calcium ions for heart muscle contraction was first indicated by Ringer in 1883, in work that is memorialized in his famous saline solution for the maintenance of frog tissues in isolation from the body. We have seen in the previous chapter how the calcium-sensitive luminescent protein aequorin or fluorescent calcium-sensitive dyes can be used to measure the rise and fall of the intracellular calcium ion concentration in skeletal muscle fibres. Similar calcium transients have been observed in frog and mammalian cardiac fibres, using thin strips of muscle or, more recently, disaggregated cells (Allen & Blinks, 1978; Wier, 1990; Cannell *et al.*, 1994).

The intracellular calcium transient is probably derived from two sources. Firstly, there is entry of calcium via the voltage-gated calcium channels that open during the action potential. Secondly, there is release of calcium ions from the sarcoplasmic reticulum. There is good evidence that the second of these processes is triggered by the first, i.e. that calcium release from the sarcoplasmic reticulum is induced by a rise in calcium ion concentration within the cell (Fabiato, 1983, 1985; Beuckelmann & Wier, 1988; Valdeolmillos *et al.*, 1989). This phenomenon is known as calcium-induced calcium release, CICR. In one of Fabiato's experiments, he showed that contraction of skinned fibres could be induced by relatively low calcium ion concentrations if the sarcoplasmic reticulum was intact, but that this would not happen if the sarcoplasmic reticulum had been removed with detergent.

Cardiac muscle contains DHP receptors (voltage-gated calcium channels) and ryanodine receptors (calcium release channels) as does skeletal muscle, but the molecular nature of the receptors and their structural arrangement is somewhat different. Although the two types of channel occur in clusters near to each other in cardiac muscle, the DHP receptors in the T tubules or cell surface membrane do not form tetrads in contact with the ryanodine receptor feet (Sun *et al.*, 1995). There are three different isoforms of the mammalian ryanodine receptor, found in skeletal muscle, cardiac muscle, and brain; the cardiac isoform is also found in brain cells (see Meissner, 1994). Different isoforms of the DHP receptors occur in skeletal and cardiac muscle (Mikami *et al.*, 1989).

The relations between the DHP receptors and the ryanodine receptors in cardiac muscle are thus different from the situation in skeletal muscle. The DHP receptors act as functional calcium channels, so that they are opened by depolarization, and the calcium entering the cell then triggers the opening of the adjacent ryanodine receptor calcium-release channels (fig. 21.10).

The consequences of calcium-induced calcium release have been investigated theoretically by Stern (1992). An apparent difficulty of the system is that we might expect it to operate in an all-or-nothing manner: small increases in calcium concentration would be damped by the calcium uptake and calcium-binding systems in the cell, whereas larger ones would trigger release which would then trigger further release, and so on. Stern calculates that such all-or-nothing responses would indeed occur if all the calcium passed through a common cytoplasmic pool. However, if the effects of individual DHP receptor calcium channels are localized to just a few adjacent ryanodine receptors, then the amount of calcium released from the sarcoplasmic reticulum is proportional to the number of active DHP receptors, in accordance with what we know about the behaviour of cardiac muscle cells. Investigation of the calcium transient in time and space shows the existence of spatial non-uniformities which fit very well with this model (Cannell *et al.*, 1994; Cheng *et al.*, 1996).

The sarcoplasmic reticulum of heart muscle accumulates calcium by the action of a calcium-ATPase pump during relaxation, just as does that of skeletal muscle. Some calcium ions are also removed during relaxation by the action of the sodium–calcium exchange system; this accounts for 20% to 30% of the total, the rest being largely removed by uptake into the sarcoplasmic reticulum (Bers, 1991; Cannell, 1991).

Overall calcium movements have been measured by Hilgemann (1986), who used the calcium-sensitive dye tetramethylmurexide to measure the calcium concentrations in the extracellular space. He found that there is a net calcium inflow during the early part of the action potential, but this is replaced by a net outflow (presumably via the sodium–calcium exchanger) during the later part. The system has been modelled by Hilgemann & Noble (1987) using the DiFrancesco–Noble equations, as is shown in fig. 20.11.

Control of the heartbeat

Our common experience shows that the heart is a remarkably adaptable organ. If we take exercise, both the rate at which it beats and the volume of blood pumped per beat (the stroke volume) increase: we become conscious of the heart pounding away within us. These effects are controlled via the autonomic nervous system. The parasympathetic system, acting via muscarinic acetylcholine receptors, is inhibitory in that its activity reduces the heart rate and heart output. The sympathetic system and the hormone adrenaline act largely via β-adrenergic receptors to stimulate heart rate and output.

Stimulation of parasympathetic fibres in the vagus nerve, or application of acetylcholine, causes slowing of the heart rate, a decrease in the height and duration of the action potential, and a decrease in the tension during contraction. Strong inhibition may prevent spontaneous activity altogether. These effects are confined largely to the atria and the sinuatrial and atrioventricular nodes; the ventricles are not much affected. The effects on the action potentials are caused by an increased permeability to potassium ions, with the opening of more inward-rectifier potassium channels (Hutter & Trautwein, 1956; Sakmann *et al.*, 1983) and a reduction in the slow inward calcium current (Giles & Noble, 1976). The slowing of depolarization during the pacemaker potential, leading to a reduction in heart rate, seems to be due entirely to a reduction of the pacemaker current i_f (Zaza *et al.*, 1996).

Acetylcholine acts on the heart by combining with muscarinic receptors and so activating G proteins. This leads directly to opening of the inward rectifier potassium channels GIRK as described on p. 161 and indirectly to a reduction of the pacemaker current i_f by reduction in the cellular concentration of cAMP (Yatani *et al.*, 1990; DiFrancesco & Tortora, 1991).

Stimulation of the sympathetic nervous input to the heart produces acceleration of the heartbeat, shortening of the action potential, increased conduction velocity in the atrioventricular node, and increased contractility in the atria and ventricles. These effects are mediated through β-adrenergic receptors, so producing activation of G proteins and causing an increase in cyclic AMP levels in the cell. This in turn leads to increased calcium currents (Reuter, 1984) and increased i_f pacemaker currents (Yatani *et al.*, 1990; Zaza *et al.*, 1996).

The contractility of isolated cardiac muscle is affected by the timing of electrical stimuli. Up to a point higher

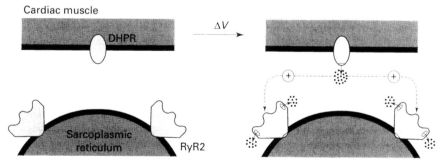

Figure 21.10. The elementary unit of calcium signalling (a 'spark') in cardiac muscle, showing calcium-induced calcium release. Depolarization (ΔV) opens a voltage-gated calcium channel (dihydropyridine receptor, DHPR) in the T tubule or cell surface membrane. The calcium entering the cell then opens adjacent calcium-release channels (ryanodine receptors, RyR2) so that more calcium is released into the cytoplasm. Compare with fig. 20.12. (From Berridge, 1997.)

frequencies lead to stronger contractions; this is called the staircase effect. Thus if the heart rate rises the stroke volume usually rises also.

A number of poisonous drugs called cardiac glycosides also increase the strength of the heart beat. Examples are ouabain and digitoxin; they inhibit the sodium pump of cell membranes. In heart muscle this may produce a number of indirect effects as a result of the consequent increase in internal sodium ion concentration; one such change may be a decrease in the activity of the sodium–calcium exchange system, so that cytoplasmic calcium levels remain higher. This would account for the therapeutic effect of foxglove (*Digitalis*) extracts in cases of heart failure, described by William Withering in 1785. One of the triumphs of the DiFrancesco–Noble model is that it can predict with remarkable precision the effects of blockage of the sodium pump (Earm & Noble, 1990; Noble, 1995).

Since each heart beat involves only one action potential in every cardiac muscle cell, the large changes in contractile force which can take place must involve changes in the relation between depolarization and contraction. It is now clear that there are two ways in which this can be brought about: either the amount of calcium released within the fibre is changed or the sensitivity of the myofibrils to calcium ion concentration alters. The size of the calcium transient in cardiac muscles fibres is much less than in skeletal muscle fibres, and would normally provide insufficient calcium to saturate all the troponin C sites in the cell. This allows changes in the amount of calcium released to alter the tension produced in the ensuing contraction.

The amplitude of the calcium transients (and of the consequent contractions) can be increased by a number of agents known to increase the force of the heartbeat. These include increased external calcium concentration, increased

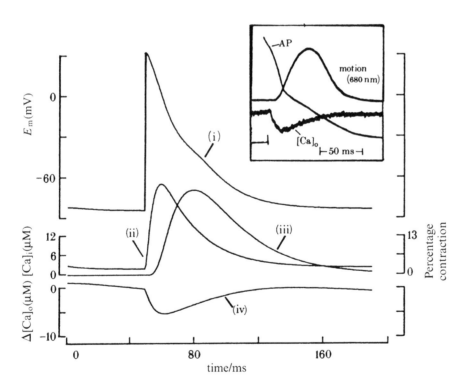

Figure 21.11. Model of the atrial action potential, and its success in reconstructing the net calcium movements, as shown by the extracellular calcium transient. Trace (i) shows the computed action potential, (ii) shows the internal calcium ion concentration, (iii) shows the contraction and (iv) predicts the external calcium transient. Notice that the model predicts a fall in external calcium as calcium ions enter via the voltage-gated calcium channels early in the action potential, followed by a recovery as calcium is extruded via the sodium–calcium exchanger. The inset shows the experimental measurements made on strips of rabbit left atrium, showing the action potential (AP), cell movement, and the change in external calcium ion concentration measured using a calcium-sensitive dye (tetramethylmurexide) in the extracellular space. (From Hilgemann & Noble, 1987, with permission of the Royal Society.)

stimulus frequency (the staircase effect), noradrenaline and other catecholamines acting via β-adrenergic receptors, and cardiac glycosides (fig. 21.12).

It is well known that increasing the volume of the heart immediately before a beat increases the force of the contraction. This relation is known as Starling's law of the heart or alternatively the Frank–Starling law (Frank, 1895; Patterson & Starling, 1914). The length–tension curve of intact cardiac muscle, at lengths below that at which maximum tension is obtained, is much steeper than the corresponding curve for skeletal muscle (Allen *et al.*, 1974; Allen & Kentish, 1985). But if we look at the length–tension curve for skinned cardiac fibres fully activated by relatively high calcium concentrations, this steepness disappears (Fabiato & Fabiato, 1975). This implies that the degree of activation of the contractile apparatus is affected by length. But since increasing the length is not accompanied by an increase in the size of the calcium tran-

sient (see fig. 21.12*b*), we must conclude that the contractile apparatus becomes more sensitive to calcium at longer lengths.

This conclusion was confirmed by Hibberd & Jewell (1982), who found that the steepness of the length–tension relation in skinned fibres was greater at lower calcium concentrations and that the force production was more sensitive to calcium ion concentration at longer lengths. Interestingly, this reduction in calcium sensitivity at shorter lengths is eliminated if the cells are compressed by immersion in hypertonic solutions (McDonald & Moss, 1995). This implies that the lateral distance between the contractile filaments plays an important part in the Frank–Starling law, but just how it does so is not yet clear.

Vertebrate smooth muscle

Smooth muscles – muscles in which there are no oblique or cross-striations – form the muscular walls of the viscera

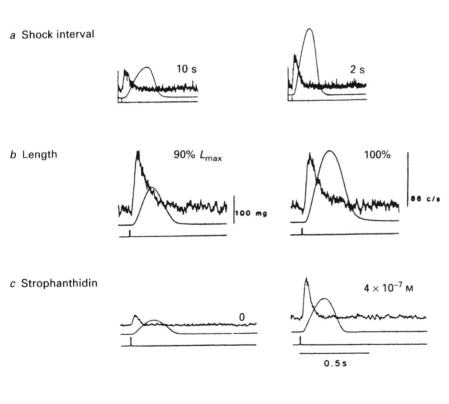

a Shock interval

10 s

2 s

b Length

90% L_{max}

100%

100 mg

88 c/s

c Strophanthidin

0

4×10^{-7} M

0.5 s

d Noradrenaline

0

10^{-7} M

0.5 s

Figure 21.12. Calcium transients in cat heart muscle in response to electrical stimulation. The noisy traces show the aequorin signal, averaged from a number of successive responses, the smooth traces show force. Pairs of records show the effects of: (*a*), stimulation frequency; (*b*), muscle length; (*c*), strophanthidin (a cardiac glycoside); and (*d*), noradrenaline. Calibrations of light output and force are the same for each member of a pair of records, but differ between pairs. (From Morgan & Blinks, 1982.)

Table 21.1 *Electrical activity in various mammalian smooth muscles*

Muscle	Species	Spontaneous activity	Action potential
Trachea	Dog	Absent	Absent
Gall bladder	Guinea pig	Absent	Graded
Artery	Rabbit	Absent	Absent
Vas deferens	Guinea pig	Absent	All-or-nothing
Uterus, non-pregnant	Rat	Bursts	Graded
Uterus, progesterone	Rat	Bursts	All-or-nothing
Bile duct	Guinea pig	Bursts	Graded
Taenia coli	Guinea pig	Continuous	All-or-nothing
Urinary bladder	Guinea pig	Bursts or continuous	All-or-nothing

Note:
Simplified after Creed, 1979.

and blood vessels in vertebrates. They also occur in the iris, ciliary body and nictitating membrane in the eye, in the trachea, bronchi and bronchioles of the lungs, and they erect the hairs in mammals.

Smooth muscles are usually concerned with relatively slow shortening, or with the maintenance of tension for long periods. This means that their power requirements (remember that power equals force times velocity) are relatively low. However, they do need to be economical, maintaining appreciable tension at low rates of ATP breakdown. This is exactly what is found in practice: smooth muscle has much lower rates of ATP utilization than does skeletal muscle (Paul, 1989).

Individual smooth muscle cells are uninucleate, and much smaller than the multinucleate fibres of skeletal muscles; they are usually about 4 μm in diameter and up to 400 μm long. The cells are held together in bundles, 20–200 μm in diameter, by thin sheets of connective tissue, and these bundles themselves are frequently interconnected. Adjacent cells are connected by small flask-shaped protrusions from one into pockets of the other, and here the opposing cell membranes are brought close together to form gap junctions (see Gabella, 1994).

Electrical activity

Smooth muscles vary widely in their patterns of activity. In some activity is more or less continuous, as in the iris, or the walls of blood vessels, in others there may be only occasional bursts of activity, as in the uterus. The walls of some hollow organs contract as a unit, as in the bladder, whereas in others waves of contraction and relaxation may pass from one end to another, as in parts of the intestine. This diversity of mechanical behaviour is associated with a cor-

responding variety in electrical activity and nervous control; table 21.1 gives some examples.

The electrical activity of a spontaneously active muscle consists of slow waves of variable amplitude and all-or-nothing action potentials. The fibres are depolarized and the frequency of the action potentials increases if the muscle is stretched (Bülbring, 1955). The spontaneous activity can be modified by the action of neurotransmitters and hormones.

Action potentials can be initiated in electrically excitable smooth muscles by electrical stimulation. Current can flow from one cell to another via the gap junctions, so the action potentials may propagate along the axes of the muscle cells and bundles and also, more slowly, across them. The action potentials of smooth muscles are slower than those of nerve axons and vertebrate skeletal muscles; their duration at half-maximum amplitude varies from 7 to 20 ms in different muscles. They are insensitive to tetrodotoxin and can be produced in the absence of sodium ions in some muscles, but are prevented by calcium channel blocking agents such as manganese, cobalt or lanthanum ions or the drugs verapamil or nifedipine. Hence the principal inward current during the action potential must be via calcium channels rather than sodium channels (see Holman & Neild, 1979; Tomita, 1981).

Neurotransmission

The terminals of the sympathetic nerves supplying smooth muscle can be prepared for light microscopy by exposing freeze-dried tissues to hot formaldehyde vapour; the catecholamines present are then converted to substances which fluoresce strongly under ultraviolet light. The relations of the terminals to the smooth muscle cells can be determined

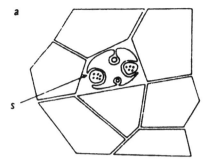

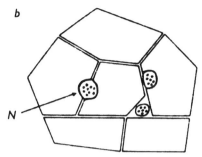

Figure. 21.13. Neuromuscular junctions in vertebrate smooth muscle. In *a* the axons occur in small bundles, *s*; they do not closely contact the muscle cells. In *b* the axons *N* occur singly and their varicosities make close synaptic contacts with the muscle cells, with a 20 nm synaptic cleft. (From Bennett, 1972.)

by electron microscopy. These methods show that the fine terminal branches have frequent swellings along their length, known as varicosities, which contain synaptic vesicles. In some muscles each varicosity is intimately apposed to an individual muscle cell (fig. 21.13*b*). In the spontaneously active muscles of the gut, the axon terminals remain in small bundles, associated with Schwann cells, and their varicosities do not form close contacts with individual muscle cells (fig. 21.13*a*). And in some muscles innervation is via both small axon bundles and close-contact varicosities.

Stimulation of the motor nerves innervating smooth muscle results in junction potentials (postsynaptic potentials) of various types (fig. 21.14). Excitatory nerves produce depolarizing potentials, whereas inhibitory nerves produce hyperpolarizing ones. In muscles with innervation via close-contact varicosities, these have a relatively rapid time course, whereas, in those with innervation via axon bundles only, the junction potentials may last for half a second or more.

The synaptic mechanisms involved in these various junction potentials are correspondingly diverse (Bolton, 1979; Bolton & Large, 1986). Their slowness, in comparison with the end-plate potentials of skeletal muscles, implies that the channels involved are activated indirectly via G proteins (fig. 9.3*b* and *c*) rather than directly as part of a unitary receptor–channel complex such as the nicotinic acetylcholine receptor. This conclusion is in part confirmed by what we know of the postsynaptic receptors involved: both the muscarinic acetylcholine receptor and the β-adrenergic receptor act via G proteins (see chapter 9).

Acetylcholine, released from parasympathetic nerve fibres, produces excitatory responses in a large variety of visceral muscles (Bolton, 1981). Since these responses are blocked by atropine the receptors involved are all muscarinic. Excitatory responses are also produced in many smooth muscles via α-adrenergic receptors, purinergic receptors activated by ATP, or both of these acting together. In the guinea pig vas deferens, for example, there is good evidence that cotransmission occurs in the sympathetic nerves, with both ATP and noradrenaline being

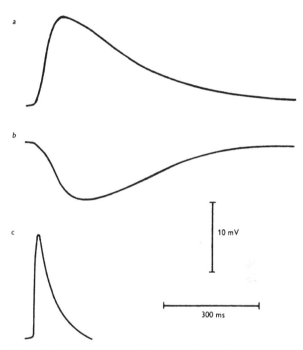

Figure 21.14. Junction potentials in smooth muscle. *a* shows an excitatory junction potential in guinea pig taenia coli muscle; *b* shows an inhibitory junction potential in guinea pig taenia coli; and *c* shows an excitatory junction potential in mouse vas deferens. Innervation is by small axon bundles in the taenia coli (*a* and *b*). and by close-contact varicosities in the vas deferens (*c*). Notice that *c* has a much faster time course than *a* or *b*. (From Bennett, 1972.)

released on stimulation; the ATP produces a relatively fast response and the noradrenaline a slower one (Burnstock 1995*b;* Sneddon & Westfall, 1984).

Many inhibitory and excitatory responses are produced by stimulation of sympathetic nerves acting via adrenergic receptors (Bülbring & Tomita, 1987). In the guinea pig taenia coli, for example, activation of α-adrenergic receptors produces hyperpolarization and relaxation. Activation of β-adrenergic receptors also produces relaxation but with only slight hyperpolarization, suggesting that they may act directly on the calcium stores of the cell (Bülbring *et al.*, 1981).

Both excitatory and inhibitory responses may also be elicited by neurotransmitters other than noradrenaline and acetylcholine. This non-adrenergic non-cholinergic (NANC) neurotransmission is mediated by different neurotransmitters at different sites: candidates include nitric oxide, ATP, and peptides such as substance P, VIP, CGRP, tachykinins and PACAP. Furthermore, smooth muscle is responsive to a wide variety of hormones and other endogenous substances, as well as a whole host of pharmacological agents (Gintzler & Hyde, 1983; Belai & Burnstock, 1994; Szekeres & Papp, 1994; Holzer *et al.*, 1995; Maggi & Giuliani, 1995; Zagorodnyuk *et al.*, 1996).

The contractile apparatus

The major contractile proteins of smooth muscle are actin and myosin, just as in striated muscle but with different isoforms. As mentioned in chapter 19, actin is a remarkably conservative protein in evolutionary terms. There are six main isoforms of human actin, found in skeletal muscle, cardiac muscle, vascular smooth muscle (these are sometimes called α-actins), enteric smooth muscle (a γ-actin), and two non-muscle or cytoplasmic β and γ isoforms. The muscle forms are more similar to each other than to the cytoplasmic forms (Sheterline & Sparrow, 1994).

The actin of vertebrate smooth muscle occurs in thin filaments that are readily seen by electron microscopy. They contain tropomyosin as in striated muscles, but troponin is absent and the protein caldesmon occurs instead. Many thin filaments appear to be attached at one end to structures in the cytoplasm called dense bodies, which contain α-actinin and are probably analogous to the Z lines of striated muscle. Others are attached to dense patches next to the cell membrane. The dense patches of adjacent cells appear to connect with each other so that there is mechanical continuity between the constituent cells of the muscle (see Gabella, 1981, 1984; Small *et al.*, 1992; Lehman *et al.*, 1996).

Smooth muscle myosin is somewhat different in sequence from the striated muscle myosins, and indeed shows similarities to certain non-muscle myosins (Goodson, 1994). Nevertheless the myosin molecule shows the familiar structure with two S1 heads attached to a coiled-coil rod. It consists of two heavy chains and four light chains, two essential and two regulatory, with M_r values around 17 000 and 20 000, respectively.

Early studies by electron microscopy were unable to demonstrate the presence of thick filaments. This was rather surprising, not least because of the existence of a 143 Å reflection in the X-ray diffraction pattern (Lowy *et al.*, 1970). Improvements in fixation technique showed that thick filaments are indeed present, but there was then some controversy as to whether the myosin occurs as filaments or as ribbons (Shoenberg & Needham, 1976). It now seems that the ribbons were artefacts, perhaps produced by low temperature, but the episode serves to demonstrate the surprising lability of organization of the myosin in smooth muscle.

The arrangement of myosin molecules in smooth muscle thick filaments is different from that in skeletal muscles. The myosin molecules are oriented in opposite directions on the two faces of the filament, as is indicated in fig. 21.15 (Small & Squire, 1972; Craig & Megerman, 1977; Cross *et al.*,

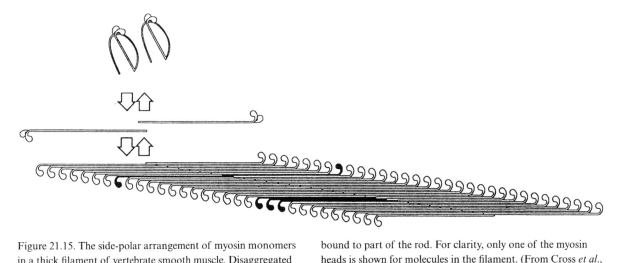

Figure 21.15. The side-polar arrangement of myosin monomers in a thick filament of vertebrate smooth muscle. Disaggregated monomers tend to adopt an inert conformation with the heads bound to part of the rod. For clarity, only one of the myosin heads is shown for molecules in the filament. (From Cross *et al.*, 1991, by permission of Oxford University Press.)

1991). Such filaments would have no bare zone in the middle but would have a bare region on one side at each end. Cross-sections of smooth muscle tend to show filaments that are square in section, and it may be that the side-polar filaments will themselves aggregate along their flat surfaces.

At the molecular level the motor mechanism of smooth muscle seems to be essentially similar to that of skeletal muscle. Smooth muscle myosin pulls on actin filaments in motility assays as does skeletal muscle myosin, with similar forces and displacements for each cycle of myosin head movement. The duration of the attachment part of the cycle is longer for smooth muscle myosin, however, suggesting that it spends a greater part of the cycle attached to the actin. This would lead to slower shortening velocities and higher average forces (VanBuren *et al.*, 1995; Warshaw, 1996).

All this implies that the sliding filament theory applies to smooth muscle as well as to striated muscle. It is possible that groups of filaments in smooth muscle can form 'contractile units' attached at each end to the sarcolemmal cytoskeleton; they would be analogous in function to a short series of sarcomeres in a striated muscle cell (Small & Squire, 1972). According to this idea, contraction occurs as thin filaments are pulled past thick filaments by cross-bridge action just as in striated muscle (fig. 21.16). But because of the side-polar arrangement of the cross-bridges, a long thin filament can completely overlap a short thick filament and can be effectively pulled over the whole of its length by that filament. This allows the muscle to operate at near maximum tension over a very wide range of lengths.

Activation

As in other muscle types, calcium is the trigger for the activation of contraction. Depolarization opens L-type voltage-gated calcium channels, allowing calcium ions into the cell. The sarcoplasmic reticulum is a calcium store;

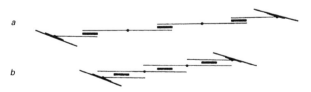

Figure 21.16. A contractile unit of vertebrate smooth muscle, showing how the filaments could slide past each other during contraction. The muscle is extended in *a* and shortened in *b*. Thin filaments are connected to the sarcolemma at the ends of the contractile unit. The model assumes a side-polar arrangement in the thick filaments. (From Squire, 1986, based on Small & Squire, 1972.)

calcium may be released from it by the activation of IP_3 receptors or ryanodine receptors. It may be that calcium can activate either or both of these receptors so as to produce calcium-induced calcium release, as in heart muscle. The normal activator for the IP_3 receptor is, of course, IP_3; this is released as a second messenger inside the cell by the phosphatidylinositol signalling system (fig. 9.6), which is itself activated by a G-protein-coupled receptor in response to some neurotransmitter or hormone (Somlyo & Somlyo, 1994; Horowitz *et al.*, 1996; Bárány, 1996).

Contraction following calcium influx triggered by depolarization is sometimes known as electromechanical coupling; contraction following calcium release triggered by IP_3 produced by activation of some G-protein-coupled receptor, which may occur without appreciable depolarization, is known as pharmacomechanical coupling (Somlyo & Somlyo, 1968, 1994; Rembold, 1996).

The primary activation process in smooth muscle is not via troponin on the thin filaments, as in skeletal muscle, but via myosin. A rise in intracellular calcium ion concentration leads to phosphorylation of the regulatory light chain MLC_{20} and this is essential to the actin–myosin interaction. Thus Sobieszek (1977) found that MLC_{20} was the only component of chicken gizzard actomyosin that would incorporate phosphate from ^{32}P-labelled ATP in the presence of calcium ions, and that the ATPase activity of actomyosin was directly proportional to the amount of this phosphorylation. The ATPase activity began to rise only after phosphorylation was well advanced. Similar results were obtained by others, and the conclusion that phosphorylation of MLC_{20} is the primary means whereby the actin–myosin interaction is controlled is now well established (Górecka *et al.*, 1976; Small & Sobieszek, 1977; Horowitz *et al.*, 1996).

MLC_{20} is phosphorylated by the action of the enzyme myosin light chain kinase (MLCK), which is itself activated by calcium–calmodulin. MLCK has been cloned and sequenced, and we know something about its mode of action (Stull *et al.*, 1996), although its precise location in the muscle cell remains to be determined. The MLC_{20} phosphorylation is reversed by the action of a phosphatase closely bound to the myosin (Shimizu *et al.*, 1994; Shirazi *et al.*, 1994).

MLCK is inactive until it combines with calcium–calmodulin. Calmodulin is a globular protein with an M_r of 17 000 and four calcium-binding sites per molecule (Babu *et al.*, 1985). It combines with calcium in the physiologically important concentration range 10^{-7} to 10^{-5} M, and will then combine with a number of target proteins, one of which is MLCK. Thus it looks as though calcium combines with calmodulin, which then activates myosin light chain kinase, which then phosphorylates the regulatory light chains, and this then activates the actomyosin ATPase and the cross-bridge cycle (see Marston, 1982; Kamm & Stull, 1985; Somlyo & Somlyo, 1994; Bárány, 1996). Figure 21.17 summarizes this sequence of events.

There is also a thin filament regulatory system present in smooth muscle. The thin filaments contain actin, tropomyosin and caldesmon, but no troponin. Caldesmon is a long protein (about twice as long as tropomyosin) that is tightly bound to the thin filaments. There is probably one caldesmon molecule for every two tropomyosin molecules and every fourteen actin monomers. It acts as an inhibitor of actin–myosin interactions in the presence of tropomyosin, perhaps by holding the tropomyosin molecules

over the myosin-binding sites on the actin. Combination with calcium–calmodulin removes this inhibition, perhaps by moving the tropomyosin molecules away from the myosin-binding sites (Sobue *et al.*, 1981; Marston & Redwood, 1991, 1992; Marston & Huber, 1996).

The precise role of the thin filament regulatory system and the function of caldesmon in the living cell is not entirely clear, since the myosin light chain phosphorylation system seems to be the main method of controlling contraction. One possibility is that it is used as a means of enhancing relaxation when calcium levels fall at the end of a period of excitation.

Another protein apparently associated with thin filaments, perhaps with filaments that do not contain caldesmon, is calponin. It will bind to actin and inhibit

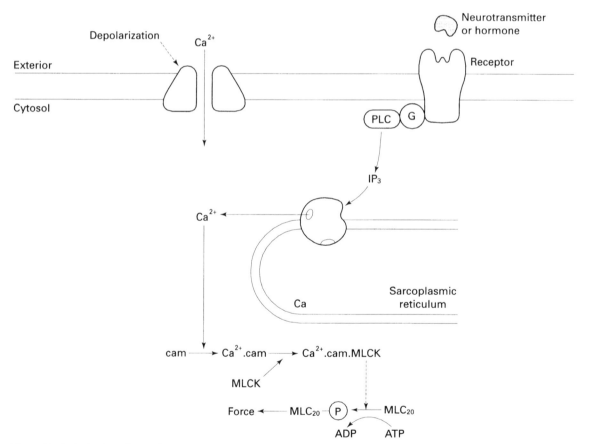

Figure 21.17. Activation of smooth muscle actomyosin. Electromechanical coupling occurs when depolarization opens voltage-gated calcium channels in the plasma membrane, so allowing calcium ions to flow into the fibre. Pharmacomechanical coupling occurs when a neurotransmitter or hormone combines with a G-protein-coupled receptor to activate the phosphatidylinositol signalling system; the IP_3 triggers the release of calcium ions from the sarcoplasmic reticulum. The calcium ions combine with calmodulin (cam) to activate myosin light chain kinase (MLCK), which then catalyses the phosphorylation of the myosin regulatory light chain MLC_{20}. This leads to cross-bridge formation and movement, the splitting of ATP and the production of force. PLC, phospholipase C; G, G protein.

actomyosin ATPase. This inhibition is partly removed in the presence of calcium ions by the calcium-binding proteins caltropin and calmodulin. We are still rather in the dark as to how calponin acts and what its role is (see Gimona & Small, 1996).

Latch

We have seen that smooth muscles are very economical in that they can maintain an appreciable force for some time while splitting very little ATP. Phosphorylation of the regulatory light chain MLC_{20}, however, is associated with ATPase activity and the onset of contraction. The timing of these features was investigated by Dillon and his colleagues (1981) in arterial smooth muscle depolarized by solutions containing a high concentration of potassium ions. They found that the isometric tension rose steadily over the first minute and then was maintained at a high level for 8 min or more; load-bearing capability followed a similar time course. But the ability to shorten peaked at about 30 s and then fell back to pre-stimulus levels over the next few minutes. This shortening capability curve was closely paralleled by the time course of MLC_{20} phosphorylation, which also peaked at around 30 s and then declined (fig. 21.18).

A likely explanation for these effects is that cross-bridge cycling, involving ATP splitting, filament sliding and shortening, is dependent on phosphorylation of the regulatory light chain, but that tension is dependent merely on cross-bridges having been formed. As the MLC_{20} molecules are dephosphorylated, the cross-bridges that they form part of stay in position for a long time so that isometric tension is maintained, although there is no more shortening or ATP breakdown. This condition is known as the 'latch' state. Latch tension will be proportional to the degree of phosphorylation in the early stages of a contraction, and so dependent on the amount of calcium released then. Model systems based on the idea have been successful in describing the properties of actual muscles (Hai & Murphy, 1989; Hartshorne & Kawamura, 1992; Murphy, 1994).

Arthropod muscles

All arthropod muscles are striated, so they have much in common with the skeletal muscles of vertebrates. The two groups have separate evolutionary origins, however, hence we can assume that the sarcomeric arrangement of the thick and thin filaments arose separately in each of them. This accounts for a number of differences in the detail of muscle structure and function.

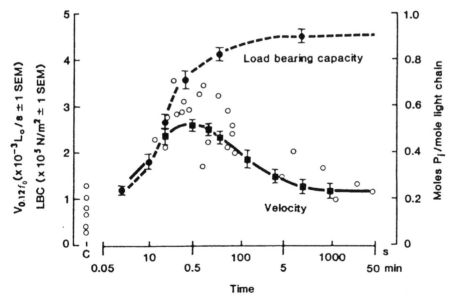

Figure 21.18. Evidence for non-cycling ('latch') cross-bridges in smooth muscle. Light chain phosphorylation (white circles) and mechanical properties of pig arterial muscle were measured during depolarization with a saline solution containing 109 mM potassium ions in place of the normal sodium ions. The MLC_{20} phosphorylation (measured after rapid freezing) peaked at about 65% after about half a minute but then declined, paralleling the shortening velocity $V_{0.12f_0}$ at a light load (0.12 of isometric tension). However the load bearing capacity (LBC) reached a high level at 8 min and (in further experiments) continued for much longer. (From Dillon *et al.*, 1981. Reprinted with permission from *Science* **211**, 495–7. Copyright 1981 American Association for the Advancement of Science.)

Innervation and excitation

All arthropod muscles are multiterminally innervated. In many of them the cell membrane is electrically excitable, so that responses to nervous stimulation may look very similar to those recorded at the neuromuscular junction of a vertebrate twitch fibre (fig. 21.19a). In other words, the neuromuscular transmitter (glutamate) opens cation-selective transmitter-gated ion channels in the subsynaptic membrane, and the depolarization that this produces then opens voltage-gated channels in the adjacent membrane. A chemically excited response, the junction potential or postsynaptic potential, is followed by an electrically excited response.

However, on closer investigation, it is found that these electrically excited responses are not 'all-or-nothing' propagated action potentials, but graded responses whose size is roughly proportional to the initial depolarization (fig. 21.20). In this respect, these graded responses are similar to the subthreshold local responses of nerve axons. It seems reasonable to suggest that they are produced by a similar mechanism, i.e. that the increase in inward current which follows depolarization is too small to counteract the effects of the increase in outward current.

The electrically excited responses of arthropod muscles differ from those of nerve axons and vertebrate twitch muscle fibres in that the inward current is carried very largely by calcium ions. This has been shown, for example, in the muscle fibres of barnacles (Hagiwara & Naka, 1964) and stick insects (Ashcroft, 1981). Thus Ashcroft found that the membrane potential at the peak of the action potential increased with increasing calcium ion concentration, and that the action potentials were unaffected by the absence of sodium ions or the presence of tetrodotoxin.

The motor innervation of arthropod muscles differs strikingly from that of vertebrate muscles in four respects: (1) many muscles are supplied by an inhibitor axon, stimulation of which causes relaxation if the muscle is excited; (2) each muscle is innervated by only a small number of motor axons; (3) polyneuronal innervation occurs, in which an individual muscle fibre may be innervated by two or more axons (fig. 21.1c); and (4) neuromodulation, involving neural modification of responses without direct synaptic action on the muscle fibres, occurs in some cases.

These points are well illustrated by the jumping muscle (the extensor tibiae) of the hind leg of locusts (Hoyle, 1955a,b, 1978; Usherwood & Grundfest, 1965). This is supplied by three conventional axons: a 'fast' excitor, a 'slow' excitor and an inhibitor. The majority of the fibres are innervated by the 'fast' excitor. These fibres respond to stimulation of the 'fast' axon by means of a postsynaptic potential and a large electrically excited response, similar to that shown in fig. 21.19a; the mechanical response of the muscle is a rapid twitch. About a quarter of the fibres in the muscle, including some of those also innervated by the 'fast' excitor, are innervated by the 'slow' excitor. Stimulation of the 'slow' excitor produces postsynaptic potentials which are not large enough to elicit electrically excited responses (fig. 21.19b); the mechanical response of the muscle is scarcely detectable after a single stimulus, but repetitive stimulation produces a slow, fairly smooth

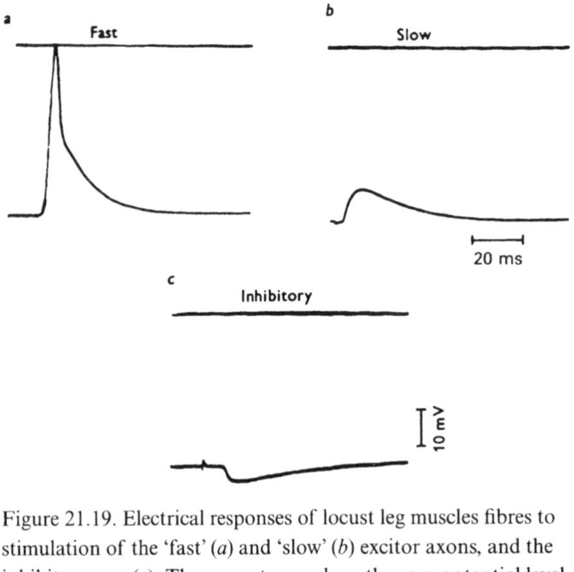

Figure 21.19. Electrical responses of locust leg muscles fibres to stimulation of the 'fast' (a) and 'slow' (b) excitor axons, and the inhibitor axon (c). The upper traces show the zero potential level. (From Usherwood, 1967.)

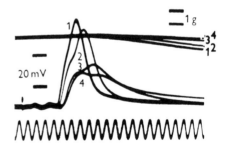

Figure 21.20. The effect of a neuromuscular blocking agent (tryptamine) on the 'fast' response in a locust leg muscle. The upper traces show the tension developed during a twitch and, initially, zero membrane potential. The lower traces show the electrical response of a muscle fibre. Time signal 500 Hz. Trace 1 shows the normal responses, traces 2 to 4 show responses at 5 s intervals after application of tryptamine. Notice that the electrically excited component of the response is not an all-or-nothing phenomenon. (From Hill & Usherwood, 1961.)

contraction whose intensity increases with increasing stimulus frequency. Some of the fibres innervated by the 'slow' axon are also innervated by the inhibitor axon. These fibres respond to inhibitory stimulation by hyperpolarizing potentials (IPSPs) similar to those of vertebrate motoneurons (fig. 21.19c); the tension produced by stimulation of the 'slow' excitor axon is reduced if the inhibitor is active at the same time.

The neuromodulator axon for the locust extensor tibiae arises from a dorsal unpaired median (DUM) neuron in the metathoracic ganglion, discovered by Hoyle and his colleagues (1974), and called by them the DUMETi neuron. Stimulation of the DUMETi neuron releases octopamine in the region of the muscle, although the neuron does not appear to form neuromuscular synapses of the normal closely apposed type. Such stimulation has no direct effect on the muscle fibre membrane potential, but it does slightly increase the response to slow axon activity and increases the rate of relaxation after both fast and slow axon stimulation (see Evans, 1985). This suggests that there are receptors for octopamine both on the slow axon terminals and on the muscle fibre surface.

Slight differences from this pattern are seen in other arthropod muscles. In the dorsal longitudinal flight muscles of locusts, the fibres are divided into five groups ('motor units'), each innervated by an axon which produces responses of the 'fast' type (Neville, 1963). The fibres of some muscles (such as locust spiracular muscles; Hoyle, 1959) are electrically inexcitable; the 'fast' response is then merely a larger version of the 'slow' response. The pentapeptide proctolin is an important or even essential neuromodulator for some muscles (Belanger & Orchard, 1993a,b). The EPSPs of crustacean muscles frequently show marked summation and facilitation (fig. 21.21).

The electrical effects of postsynaptic inhibition in arthropod muscle fibres are essentially similar to those in vertebrate motoneurons. The sign and magnitude of the IPSPs can be altered by changing the membrane potential. The reversal potential is usually fairly near to the resting potential; it is practically unaffected by changes in external potassium ion concentration, but very sensitive to changes in external chloride ion concentration (Boistel & Fatt, 1958; Usherwood & Grundfest, 1965). Thus the action of the inhibitory transmitter GABA is to increase the chloride conductance of the membrane by opening GABA-activated chloride channels.

Presynaptic inhibition occurs in certain crustacean muscles, as has been shown very elegantly by Dudel & Kuffler (1961). Using an extracellular microelectrode, they were able to record the currents associated with EPSPs at single nerve terminals on the muscle fibres (fig. 21.22). These fluctuated in a discontinuous manner, suggesting corresponding fluctuations in the number of quanta of transmitter released. When the inhibitory axon was also stimulated, the number of 'failures' at any one terminal increased, and the EPSP recorded intracellularly (which is caused by transmitter release from all the terminals on the fibre) decreased in size. A statistical analysis of the type

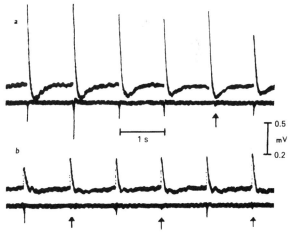

Figure 21.22. Experiment showing presynaptic inhibition in a crayfish claw muscle. Upper traces show intracellular responses to stimulation of the excitor axon at a rate of 1 s^{-1}. Lower traces are simultaneous extracellular records showing activity at a single functional region; the arrows indicate failure of transmission. *a* shows stimulation of the excitor axon alone. *b* shows stimulation of the inhibitor axon 2 ms before each excitatory stimulus. The decrease in the size of the intracellular responses in *b* shows that inhibition is occurring; the increase in the number of failures in transmission recorded by the extracellular electrode indicates that this inhibition must be at least partially presynaptic. (From Dudel & Kuffler, 1961.)

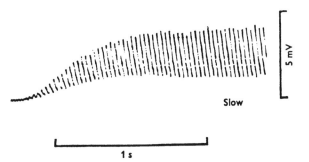

Figure 21.21. Membrane potential changes in a *Panulirus* (rock lobster) muscle fibre during repetitive stimulation of the 'slow' excitor axon. Note the facilitation and summation of the responses. (From Hoyle & Wiersma, 1958.)

used by del Castillo & Katz on magnesium-treated frog muscle fibres (see chapter 7) showed that the average number of quanta released per terminal per impulse was reduced from 2.4 to 0.6 during inhibition. This implies that the inhibitory process must be presynaptic.

The means whereby gradation of skeletal muscle contraction is achieved in arthropods is quite different from that in vertebrates. Whereas vertebrate skeletal muscles are composed of an appreciable number of motor units and the total force depends largely on the proportion of them that are active, arthropod muscles have very few motor units and other methods of gradation are required. The size of the electrical response in any fibre is dependent on: (1) which axons, 'fast' or 'slow', are active; (2) the frequency of arrival of excitatory nerve impulses, which affects the degree of membrane depolarization by summation and facilitation of the postsynaptic potentials; (3) the activity of the inhibitor axon; and (4) the activity of the neuro-modulator axon, if there is one.

Structural organization of arthropod muscle cells
We have seen that, in vertebrate skeletal muscles, the myofilament array is such that each A filament is sur-rounded by six I filaments and each I filament by three A filaments, so that there are two I filaments per A filament in each half-sarcomere. Different arrangements occur in arthropod muscles. In most insect flight muscles each A fila-ment is surrounded by six I filaments, but each I filament is placed between two A filaments so that the I to A ratio is 3:1 (fig. 21.23). In thoracic muscles of the cockroach there are eight or nine I filaments round each A filament and the I to A ratio is 4:1 (Hagopian & Spiro, 1968). Finally, in some insect leg and visceral muscles there are twelve I fila-ments round each A filament and the I to A ratio is 6:1 (Hagopian, 1966; Smith *et al.*, 1966).

What is the significance of these 'extra' I filaments? We have seen that, other things being equal, the maximum iso-metric tension produced by a muscle is proportional to the length of the A band. Hence the longer the A band, the greater the strain on the myofilaments during contraction. The presence of 'extra' I filaments is associated with long A bands, so it is possible that the greater total number of I fila-ments is needed to reduce the strain on each one of them.

What limits the size of the various internal components in muscle cells? The muscle cell can be regarded as an arena of competing subcellular interests: the principal contest-ants are the myofibrils, the mitochondria and the sarco-plasmic reticulum. To maximize the work output per contraction, the myofibrillar fraction should be maximal. To increase the rate of activation and relaxation, the sarco-plasmic reticulum should be increased in volume. And to increase the level of maintained power output, the mito-chondrial mass should be increased.

Figure 21.24 shows the relative proportions of these

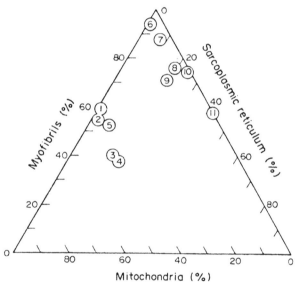

Figure 21.24. The relative proportions of myofibrils, mitochondria and sarcoplasmic reticulum in various insect muscle cells. The axes show the relative proportions by volume of each of the three components, expressed on a scale in which the volume of all three is 100%; the proportions of nuclei, sarcoplasm etc. are thus neglected. 1 and 2 are asynchronous (fibrillar) flight or singing muscles; 3 to 5 are synchronous flight or singing muscles; 6 to 11 are leg muscles. The muscles are: 1, blowfly flight muscle; 2, a cicada asynchronous tymbal muscle; 3, a cicada synchronous tymbal muscle; 4, katydid flight/singing muscle; 5, dragonfly flight muscle; 6, locust jumping muscle, tonic and, 7, phasic fibres; 8, locust toe muscle; 9, flea jumping muscle; 10, wasp extensor tibialis and, 11, toe muscle. (From Aidley, 1985, data from various authors.)

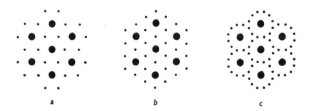

Figure 21.23. Myofilament arrays in vertebrate and insect striated muscles. The diagrams show the arrangement of myosin (large spots) and actin (small spots) filaments as seen in transverse sections through the overlap region. *a* shows vertebrate muscles, *b* insect flight muscles, and *c* most insect leg and visceral muscles. (From Smith *et al.*, 1966.)

three fractions of the cell in various insect muscles. Slowly contracting leg muscles (6 and 7 in the diagram) do not require or possess large quantities of the non-myofibrillar components. Leg muscles which need to be rapidly activated (such as 8 to 11) need to have much more sarcoplasmic reticulum, and hence must make do with a corresponding reduction in myofibril content; but they have no requirement for maintained power output, hence they have few mitochondria. Flight muscles and singing muscles (1 to 5) are repetitively active for long periods, so they need a large mitochondrial component to supply the energy for contraction as well as (in the conventional synchronous muscles, 3 to 5) an appreciable sarcoplasmic reticulum. The asynchronous muscles, 1 and 2 in the diagram, have very little sarcoplasmic reticulum; we shall see later that this is related to their ability to produce oscillatory power output while being fully active.

Insect asynchronous muscles

The wing-beat frequency of many insects is very high; the record is held by the midge *Forcipomyia*, in which is it about 1000 beats s^{-1} (Sotavalta, 1953). This means that the tension in the flight muscles in *Forcipomyia* must rise and fall within a millisecond, which seems almost impossibly fast. Pringle (1949) measured the electrical responses of the flight muscles of a flying blow-fly *Calliphora* and showed that their frequency was very much less than that of the wing-beat (fig. 21.25). This indicates that the individual contractions of the muscles, which produce the wing-beat, are not 'twitches' analogous to the twitches of other muscles. Pringle suggested that the function of excitation is to bring the contractile apparatus into a state of activity (analogous to the tetanus condition in other muscles) in which rhythmic contractions are possible.

Further work by Roeder (1951) and others (see Pringle, 1981) showed that this asynchronous relation between the neurally excited muscle potentials and the contraction frequency occurs in the flight muscles of flies, beetles, bees and wasps, and certain bugs; it is also found in the sound-producing tymbal muscles of some cicadas (Pringle, 1954). These muscles are described as 'fibrillar' (since their fibres contain closely packed myofibrils) or 'asynchronous'. The flight muscles of more primitive insects, such as locusts, dragonflies and moths, are not of the asynchronous type, and show the familiar one-to-one ratio between the electrical and mechanical responses to nervous stimulation.

A closer investigation of the mechanical properties of asynchronous muscle was carried out by Machin & Pringle (1959), using a flight muscle from the rhinoceros beetle *Oryctes*. The muscle was partly dissected from the insect

and connected to a piezoelectric force transducer at one end and to a moving-coil vibrator at the other. The position of the vibrator arm was measured by means of a light beam and a phototransistor, so as to give the length of the muscle. The load on the muscle (produced by the action of the vibrator) was electrically controlled from the output of the phototransistor, allowing regulation of its stiffness, damping and mass. The outputs of the length and tension transducers could be connected to the X and Y amplifiers of an oscilloscope, giving a direct display of the length–tension diagram.

In the resting condition, the muscle was very stiff, giving a steep passive length–tension curve. With an isometric load, stimulation of the motor nerve resulted in the development of an extra steady maintained tension, just as in other muscles. But with an inertial load, obtained by a suitable setting of the load parameters, the muscle would undergo an oscillatory contraction, giving a length–tension loop such as is shown in fig. 21.26. The frequency of this oscillatory contraction is the same as the mechanical resonant frequency of the muscle and its load. In a flying insect, the load is determined by the inertia of the wings, the aerodynamic forces on the wings, and the stiffness of the thoracic skeleton by whose movements the wings are moved.

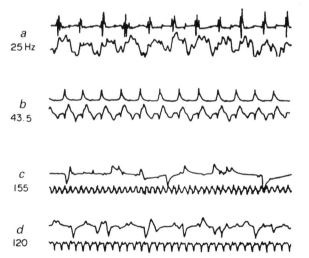

Figure 21.25. Records of the electrical activity of the flight muscles (upper traces) and mechanical movements of the thorax (lower traces) from various flying insects: (*a*) a cockroach; (*b*) a moth; (*c*) a fly; (*d*) a wasp. In *a* and *b* the flight muscles are of the synchronous type, where each contraction is preceded by an action potential. In *c* and *d* they are of the asynchronous type, where the oscillatory contraction frequency is higher than the frequency of the action potentials. The wing-beat frequency is shown for each insect. (From Pringle, 1981. With permission of the Company of Biologists Ltd.)

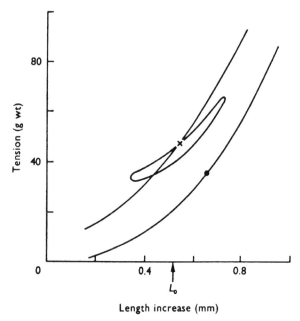

Tension (g wt)

80

40

0

0.4 L_0 0.8

Length increase (mm)

Figure 21.26. The oscillatory contraction of a rhinoceros beetle basalar muscle. The lower sloping line shows the length–tension relation of the unstimulated muscle, the upper sloping line shows the length–tension relation when the muscle is stimulated repetitively under isometric conditions. The loop shows the oscillatory contraction produced by such stimulation when the load possesses suitable inertial and damping components; it is traced out in an anticlockwise direction and at a frequency in the region of 25 Hz. (From Machin & Pringle, 1959, by permission of the Royal Society.)

Thus the wing-beat frequency of insects with asynchronous flight muscles can be altered by altering the resonant frequency of the wing–thorax system, such as by cutting off the tips of the wings (when the frequency increases) or by loading them with small amounts of wax (the frequency decreases).

Much information about the properties of asynchronous muscles has been obtained by using the method of sinusoidal analysis (Machin & Pringle, 1960; see Aidley, 1985). The muscle, or a bundle of glycerol-extracted fibres, is subjected to a sinusoidal length change of small amplitude, and the resulting sinusoidal change in tension is measured. In resting muscle the tension changes lead the length changes, i.e. the tension is higher on average when the muscle is being stretched than when it is shortening, so that the apparatus is doing work on the muscle. Similar behaviour is found in many passive materials and indeed in most muscles if they are fully active for the whole of the cycle. With stimulated asynchronous muscle, however, the tension changes lag behind the length changes at the lower frequencies. This

means that the tension is higher on average when the muscle is shortening than when it is being stretched, so that the muscle is doing work on the apparatus.

If we progressively increase the frequency of an oscillating asynchronous muscle we find that the work output per cycle rises to a maximum, then falls and eventually becomes negative. Maximal power output (work per second) occurs at a slightly higher frequency than maximum work per cycle. Since the wing-beat frequency of a flying insect is determined by the mechanical resonant frequency of the wing–thorax system, we would expect characteristics of the flight muscles to be such that maximum power output occurs at or near this frequency; there is some evidence that this is so (Molloy *et al.*, 1987).

Jewell & Rüegg (1966) found that glycerinated fibres from the wing muscles of giant water bugs (Belostomatidae) showed the typical oscillatory phenomena in the presence of suitable concentrations of ATP and calcium ions. This indicates that the oscillation really is a property of the contractile apparatus, not of the excitation or coupling processes, which merely serve to keep the muscle in the active condition. The ATPase activity of glycerol-extracted water bug flight muscle fibres in the presence of calcium ions was investigated by Rüegg & Tregear (1966). They found that more ATP is split when the fibres are subjected to oscillatory length changes, and that this 'extra' ATPase activity is greatest when the frequency is that at which maximum power output occurs.

The essential property of asynchronous muscle is called activation by stretch: an increase in length is followed, after a short delay, by a rise in tension. There is a corresponding delayed fall in tension after shortening, and the two effects together produce the oscillatory capabilities of the muscle. The question that arises is, therefore, what is the mechanism of this activation by stretch?

The I bands of asynchronous muscles are very short; hence the thick filaments extend to very near the Z line. This means that length changes are very small. The oscillations shown in fig. 21.26 are about 4% of the muscle length, for example, and high speed video measurements on bumblebee flight muscles seen through a window cut in the cuticle suggested that they shorten by just 1.9% during flight (Gilmour & Ellington, 1993).

One of the striking features of asynchronous muscle is its very great stiffness both at rest and during contraction. White (1983) demonstrated that the high stiffness is accounted for by the existence of 'C filaments', first seen in blowfly flight muscle by Auber & Couteaux (1963), which connect the ends of the thick filaments to the Z line. The C filaments are probably composed of the protein projectin (see Bullard & Leonard, 1996). The presence of the C filaments means that stretching the muscle as a whole will also stretch the individual thick filaments. Thorson & White (1969, 1983) suggested that such an alteration in thick filament length could alter the rate constants for cross-bridge action, and that this change could account for the property of activation by stretch.

Wray (1979) produced an attractive suggestion for the origin of stretch activation, arising from structural considerations. The thick filaments of insect flight muscles are different from those of vertebrate skeletal muscles; there are four myosin heads per crown, emerging every 145 Å along the filament, but the rotation of successive crowns gives an overall pitch of 4×385 Å (Reedy et al., 1981; see Tregear, 1983). Thus the pitch of the thick filaments matches almost precisely that of the thin filaments, a very unusual situation. Hence the number of cross-bridges that can be formed must depend critically on the relative axial alignment of the two sets of filaments. Wray suggested that stretch brings more myosin heads into positions where they can readily form links with actin molecules.

The high degree of geometrical regularity which is a characteristic of asynchronous muscles could be related to this mechanism. Their well-developed M bands and Z lines might be needed to maintain the lateral alignment of the thick and thin filament arrays. A prime function of the C filaments could be to ensure that the thick filaments are positioned precisely midway between the two ends of the sarcomere so that the relative positions of the two sets of filaments are the same in each half sarcomere and at different positions along the length of the muscle. However, a detailed look at filament geometry by J. M. Squire (1992) suggests that Wray's match–mismatch hypothesis does not in fact fit the facts.

From an evolutionary of view, it is interesting to speculate on whether the oscillatory characteristics of asynchronous muscle are dependent on some basically new properties of the muscle, or whether they arose by enhancement of properties already present to some extent in other muscles. Some evidence in favour of the second possibility is provided by the properties of the sound-producing muscles in the cicada *Fidicina*: although these muscles are neurogenically activated, with the conventional one-to-one relation between nerve impulse and muscle contraction, their glycerinated fibres show all the mechanical properties of glycerinated water bug flight muscles (Aidley & White, 1969). This suggests that the work on insect asynchronous muscles (reviewed by Pringle, 1979, 1981; Tregear, 1983) may help in understanding how more conventional muscles work.

Molluscan and other invertebrate muscles
Striated and smooth muscles
The skeletal and cardiac muscles of vertebrates and arthropods, and the visceral muscles of arthropods, are all striated. Striated muscles also occur locally in many other animals. Some examples are the swimming muscles of jellyfish, the pedicellariae muscles of echinoderms, and certain muscles in the heads of some annelids.

Many muscles in invertebrates show no signs of striation when examined by light microscopy. Electron micrograph sections show two types of filament, thick and thin, but these are not transversely aligned (Hanson & Lowy, 1960; Paniagua et al., 1996).

The 'catch' muscles of molluscs (see below) have been particularly investigated (Sobieseck, 1973; Elliott & Bennett, 1984; Bennett & Elliott, 1989). The thin filaments are similar to those of striated muscles and show the same F-actin double helix structure when isolated. But the thick filaments vary in thickness from 150 to 1500 Å, and appear to consist of a number of closely packed ribbon-like elements. The muscles contain a high proportion of paramyosin, and it is this that forms the major part of the thick filaments. Myosin can be extracted from the thick filaments, when an array of roughly triangular pits are seen on their surface (Szent-Györgyi et al., 1971). Neighbouring triangles all point in the same direction, but this reverses in the middle of the filament, so that the two halves have opposite polarities, rather as in the myosin filaments of vertebrate striated muscles. Many of the thin filaments are attached to dense bodies. They form 'arrowheads' when decorated with heavy meromyosin, and these are of opposite polarity on each side of a dense body, just as are those on each side of a Z band in striated muscles.

A number of invertebrate muscles possess striations which follow a helical or oblique course within the fibres. They are known as *oblique striated* muscles. They are found in annelids, molluscs, nematodes and some other invertebrates (Rosenbluth, 1968). The fibres contain two types of filament, corresponding to the A and I filaments of striated muscles. The oblique striations arise from a staggered arrangement of the thick filaments, as can be seen in electron micrographs of tangential longitudinal sections (fig. 21.27). During contraction, the thin filaments slide between the thick ones, so shortening the distance between successive striations. Since the muscle cell is a constant volume system, shortening is accompanied by widening, hence the angle of the striations alters: they become less oblique (Lanzavecchia, 1968; Mill & Knapp, 1970).

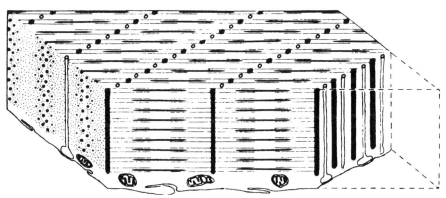

Figure 21.27. Diagram showing the arrangement of myofilaments in an oblique striated earthworm muscle fibre. The thin filaments are attached to Z rods, shown in black here. Alternating with the Z rods are tubules of the sarcoplasmic reticulum. The angle of the oblique striations is exaggerated. (From Mill & Knapp, 1970.)

Calcium-sensitive regulatory proteins

In vertebrate skeletal muscles, as we have seen in chapter 20, contraction is regulated by the calcium-sensitive troponin–tropomyosin system on the thin filaments. Similar thin filament linked regulatory systems are found in amphioxus and mysid crustaceans and in the fast muscles of decapod crustaceans (Lehman, 1976).

In molluscan muscles, however, Kendrick-Jones *et al.* (1970) found that there is no troponin present. Molluscan thin filaments will react with rabbit myosin so as to split ATP, in the absence of calcium ions. But there is a calcium-sensitive mechanism on the thick filaments: molluscan myosin will combine with pure actin and split ATP only in the presence of calcium ions. Similar myosin-linked regulatory systems are found in echinoderms and in a number of minor groups.

In insects and most other arthropods, and in annelids and nematodes, both types of regulation are present simultaneously.

The mechanism of molluscan myosin-linked regulation has been investigated in the fast striated adductor muscle of the scallop, *Pecten*. Each myosin molecule is similar in general structure to those of vertebrate skeletal muscles, with two heavy chains, two regulatory light chains which can be removed with EDTA, and two essential light chains which contain SH groups. After removal of the two regulatory light chains, ATPase activity and tension development will occur in the absence of calcium ions (Szent-Györgyi *et al.*, 1973; Chantler & Szent-Györgyi, 1980; Simmons & Szent-Györgyi, 1985). The regulation does not depend on the presence of actin; the ATPase activity of the myosin alone is 100 times greater in the presence of calcium than in its absence (Wells & Bagshaw, 1985).

Hardwicke and his colleagues used photo-cross-linking techniques to investigate myosin regulation in the scallop. They found that a substituted photoactivated agent would link the regulatory and SH light chains under conditions when rigor was produced (calcium ions but no ATP) but not under conditions of rest (ATP but no calcium ions). This implies that the distance between the two light chains is greater at rest and suggests that movement of at least one of them occurs during activation by calcium (Hardwicke *et al.*, 1983; Hardwicke & Szent-Györgyi, 1985).

'Catch' muscles

Certain muscles of molluscs, of which the most closely investigated is the anterior byssus retractor of the mussel *Mytilus*, are remarkable in that their rate of relaxation is dependent upon the nature of the excitatory stimulus. In 'phasic' responses, the relaxation following an isometric contraction is complete within a few seconds, whereas in 'tonic' or 'catch' responses, the relaxation phase lasts for several minutes or sometimes hours. Phasic responses can be produced by electrical stimulation using repetitive brief pulses or alternating current, whereas tonic responses can be produced by continuous direct current stimulation (Winton, 1937). Tonic responses are also elicited by treatment with acetylcholine; they are abolished by 5-hydroxytryptamine (Twarog, 1954).

The mechanical responses of the *Mytilus* anterior byssus retractor muscle were investigated by Jewell (1959) and Lowy & Millman (1963), among others. Jewell found that the behaviour of the muscle during stimulation was not greatly different from that of frog leg muscles, except that the time course of contraction was much slower. During the tonic response, however, the ability to shorten and to redevelop tension after a quick release was very much reduced. If the muscle was restretched after a quick release during this time, the full 'catch' tension was re-established.

The control mechanism for the catch system has been illuminated by Cornelius (1982), in experiments on *Mytilus* anterior byssus retractor muscles which had been chemically 'skinned' with saponin. He found that maintained contractions, with the ability to shorten, were produced in the presence of ATP by calcium ion concentrations over 10^{-7} M. Reduction of the calcium ion concentration to below this value produced a very slow relaxation during which the muscle was in the catch state, i.e. tension was maintained but shortening could not occur. The duration of the catch was greatly reduced in the presence of cyclic AMP or 5-hydroxytryptamine. Cyclic AMP had no effect on the response in the presence of calcium. Catch contractions are associated with large inflows of calcium ions (Pelc *et al.*, 1996).

5-Hydroxytryptamine activates adenylate cyclase and so leads to cyclic AMP production in catch muscles. There is some evidence that catch is associated either with the phosphorylation of paramyosin (Cooley *et al.*, 1979; Achazi, 1982), or with the dephosphorylation of myosin (Castellani & Cohen, 1987).

Myosin forms only 20% of the content of the thick filaments of catch muscles, the rest being paramyosin. It would not be surprising, therefore, if paramyosin were crucially involved in the mechanism of the catch. At one time it was thought that separate links between the paramyosin molecules, in parallel with the normal myosin–actin cross-bridges, might be involved in the catch system, but no such links have been seen in electron micrographs. It seems more likely that changes in the paramyosin might affect the rate of breakage of the myosin–actin links (Szent-Györgyi *et al.*, 1971). The structure of the thick filaments, in which myosin molecules form a layer just one molecule thick over the surface of the paramyosin core, would seem to permit some such interaction. Perhaps, as Cohen (1982) suggested, catch occurs when phosphorylation of the paramyosin somehow restricts the mobility of the myosin cross-bridges.

References

Where there are a number of papers with the same first author, those with two or more further authors, referred to in the text as Someone *et al.*, are listed in chronological order, after those with the first author and one other.

Abraham, W. C., Dragunow, M. & Tate, W. P. (1991). The role of immediate early genes in the stabilization of long-term potentiation. *Molec. Neurobiol.* **5**, 297–314.

Achazi, R. K. (1982). Catch muscle. In *Basic Biology of Muscles: A Comparative Approach*, ed. B. M. Twarog, R. J. C. Levine & M. M. Dewey, pp. 291–308. New York: Academic Press.

Adams, D. J., Dwyer, T. M. & Hille, B. (1980). The permeabillty of endplate channels to monovalent and divalent metal cations. *J. Gen. Physiol.* **75**, 493–510.

Adams, M. E. & Olivera, B. M. (1994). Neurotoxins: overview of an emerging research technology. *Trends Neurosci.* **17**, 151–5.

Adams, P. R. & Brown, D. A. (1982). Synaptic inhibition of the M-current: slow post-synaptic potential mechanism in bullfrog sympathetic neurones. *J. Physiol.* **332**, 263–72.

Adams, P. R., Jones, S. W., Pennefather, P., Brown, D. A., Koch, C. & Lancaster, B. (1986). Slow synaptic transmission in frog sympathetic ganglia. *J. Exp. Biol.* **124**, 259–85.

Adelman, J. P., Shen, K.-Z., Kavanaugh, M. P., Warren, R. A., Wu, Y.-N., Lagrutta, A., Bond, C. T. & North, R. A. (1992). Calcium-activated potassium channels expressed from cloned complementary DNAs. *Neuron* **9**, 209–16.

Adrian, E. D. (1914). The all-or-none principle in nerve. *J. Physiol.* **47**, 460–74.

Adrian, E. D. (1926). The impulses produced by sensory nerve endings. I. *J. Physiol.* **61**, 47–72.

Adrian, E. D. & Bronk, D. W. (1928). The discharge of impulses in motor nerve fibres. Part I. Impulses in single fibres of the phrenic nerve. *J. Physiol.* **66**, 81–101.

Adrian, E. D. & Huang, C. L.-H. (1984). Experimental analysis of the relationship between charge movement components in skeletal muscle of *Rana temporaria*. *J. Physiol.* **353**, 419–34.

Adrian, E. D. & Zotterman, Y. (1926). The impulses produced by sensory nerve endings. II. The response of a single end-organ. *J. Physiol.* **61**, 151–71.

Adrian, R. H., Chandler, W. K. & Hodgkin, A. L. (1969). The kinetics of mechanical activation in frog muscle. *J. Physiol.* **204**, 207–30.

Agnew, W. S., Levinson, S. R., Brabson, J. S. & Raftery, M. A. (1978). Purification of the terodotoxin-binding component associated with the voltage-sensitive channel from *Electrophorus electricus* electroplax membranes. *Proc. Natl. Acad. Sci. USA* **75**, 2606–11.

Agnew, W. S., Miller, J. A., Ellisman, M. H., Rosenberg, R. L., Tomiko, S. A. & Levinson, S. R. (1983). The voltage-regulated sodium channel from the electroplax of *Electrophorus electricus. Cold Spr. Harb. Symp. Quant. Biol.* **48**, 165–179.

Ahlquist, R. P. (1948). A study of the adrenotropic receptors. *Amer. J. Physiol.* **153**, 586–600.

Aidley, D. J. (1965). The effect of calcium ions on potassium contracture in a locust leg muscle. *J. Physiol.* **177**, 94–102.

Aidley, D. J. (1985). Muscular contraction. In *Comprehensive Insect Physiology, Biochemistry and Pharmacology*, ed. G. A. Kerkut & L. I. Gilbert, vol. 5, pp. 407–37. Oxford: Pergamon Press.

Aidley, D. J. & Stanfield, P. R. (1996). *Ion Channels: Molecules in Action*. Cambridge: Cambridge University Press.

Aidley, D. J. & White, D. C. S. (1969). Mechanical properties of glycerinated fibres from the tymbal muscles of a Brazilian cicada. *J. Physiol.* **205**, 179–92.

Alberini, C. M., Ghirardi, M., Metz, R. & Kandel, E. R. (1994). C/EBP is an immediate-early gene required for the consolidation of long-term facilitation in *Aplysia. Cell* **76**, 1099–114.

Albers, R. W., Fahn, S. & Koval, G. J. (1963). The role of sodium ions in the activation of *Electrophorus* electric organ adenosine triphosphatase. *Proc. Natl. Acad. Sci. USA* **50**, 474–81.

Alberts, B., Bray, D., Lewis, J., Raff, M., Roberts, K. & Watson, J. D. (1994). *Molecular Biology of the Cell*, third edition. New York: Garland.

Allen, D. G. & Blinks, J. R. (1978). Calcium transients in aequorin-injected frog cardiac muscle. *Nature* **273**, 509–13.

Allen, D. G. & Kentish, J. C. (1985). The cellular basis of the length–tension relation in cardiac muscle. *J. Molec. Cell. Cardiol.* **17**, 821–40.

Allen, D. G., Jewell, B. R. & Murray, J. W. (1974). The contribution of activation processes to the length–tension relation of cardiac muscle. *Nature* **248**, 606–7.

Almers, W. (1990). Exocytosis. *Ann. Rev. Physiol.* **52**, 607–24.

Almers, W. & McCleskey, E. W. (1984). Non-selective conductance in calcium channels of frog muscle: calcium selectivity in a single-file pore. *J. Physiol.* **353**, 585–608.

Almers, W., McCleskey, E. W. & Palade, P. T. (1984). A non-selective cation conductance in frog muscle membrane blocked by micromolar external calcium ions. *J. Physiol.* **353**, 565–83.

Almers, W., Breckenridge, L. & Spruce, A. (1988). Time course of fusion pore conductance during exocytosis of beige mouse mast cells. *J. Physiol.* **407**, 96P.

Alpern, M. & Pugh, E. N. (1974). The density and photosensitivity of human rhodopsin in the living retina. *J. Physiol.* **237**, 341–70.

Altamarino, M., Coates, C. W. & Grundfest, H. (1955). Mechanisms of direct and neural excitability in electroplaques of electric eel. *J. Gen. Physiol.* **38**, 319–60.

Alvarez-Leefmans, F. J. (1990). Intracellular Cl⁻ regulation and synaptic inhibition in vertebrate and invertebrate neurons. In *Chloride Channels and Carriers in Nerve, Muscle, and Glial Cells*, ed. F. J. Alvarez-Leefmans & J. M. Russell, pp. 109–58. New York: Plenum Press.

Amara, S. G. & Pacholczyk, T. (1991). Sodium-dependent neurotransmitter reuptake systems. *Curr. Opin. Neurobiol.* **1**, 84–90.

Amara, S. G., Jones, V., Rosenfeld, M. G., Ong, E. & Evans, R. M. (1982). Alternative RNA processing in calcitonin gene expression generates mRNAs encoding different polypeptide products. *Nature* **298**, 240–4.

Amoore, J. E. (1967). Specific anosmia: a clue to the olfactory code. *Nature* **214**, 1095–8.

Amos, L. A. & Amos, W. B. (1991). *Molecules of the Cytoskeleton*. Basingstoke: Macmillan.

Amos, L. A., Henderson, R. & Unwin, P. N. T. (1982). Three-dimensional structure determination by electron microscopy of two-dimensional crystals. *Prog. Biophys. Molec. Biol.* **39**, 183–231.

Anderson, C. R. & Stevens, C. F. (1973). Voltage clamp analysis of acetylcholine produced end-plate current fluctuations at frog neuromuscular junction. *J. Physiol.* **235**, 655–91.

Anegawa, N. J., Lynch, D. R., Verdoom, T. A. & Pritchett, D. B. (1995). Transfection of *N*-methyl-D-aspartate receptors in a nonneural cell line leads to cell death. *J. Neurochem.* **64**, 2002–12.

Araki, T., Eccles, J. C. & Ito, M. (1960). Correlation of the inhibitory postsynaptic potential of motoneurones with the latency and time course of inhibition of monosynaptic reflexes. *J. Physiol.* **154**, 354–77.

Arden, G. B. (1969). The excitation of photoreceptors. *Prog. Biophys. Molec. Biol.* **19**, 371–421.

Armstrong, C. M. (1981). Sodium channels and gating currents. *Physiol. Rev.* **61**, 644–83.

Armstrong, C. M. (1992). Voltage-dependent ion channels and their gating. *Physiol. Rev.* **72**, S5–S13.

Armstrong, C. M. & Bezanilla, F. M. (1973). Current related to the movement of the gating particles of the sodium channels. *Nature* **242**, 459–61.

Armstrong, C. M. & Bezanilla, F. (1974). Charge movement associated with the opening and closing of the activation gates of the Na channels. *J. Gen. Physiol.* **63**, 675–89.

Armstrong, C. M. & Bezanilla, F. (1977). Inactivation of the sodium channel. II. Gating current experiments. *J. Gen. Physiol.* **70**, 567–90.

Armstrong, C. M., Bezanilla, F. M. & Rojas, E. (1973). Destruction of sodium conductance inactivation in squid axon perfused with pronase. *J. Gen. Physiol.* **62**, 375–91.

Ashcroft, F. M. (1981). Calcium action potentials in the skeletal muscle fibres of the stick insect *Carausius morosus*. *J. Exp. Biol.* **93**, 257–67.

Ashkin, A. & Dziedzic, J. M. (1977). Feedback stabilization of optically levitated particles. *Appl. Physics Letters* **30**, 202–4.

Ashley, C. C. & Moisescu, D. G. (1977). Effect of changing the composition of the bathing solutions upon the isometric tension–pCa relationship in bundles of crustacean myofibrils. *J. Physiol.* **270**, 627–52.

Ashley, C. C. & Ridgeway, E. B. (1968). Simultaneous recording of membrane potential, calcium transient and tension in single muscle fibres. *Nature* **219**, 1168–9.

Ashley, C. C. & Ridgeway, E. B. (1970). On the relationships between membrane potential, calcium transient and tension in single barnacle muscle fibres. *J. Physiol.* **209**, 105–30.

Ashley, C. C., Mulligan, I. P & Lea, T. J. (1991). Ca^{2+} and activation mechanisms in skeletal muscle. *Quart. Rev. Biophys.* **24**, 1–73.

Ashmore, J. F. (1987). A fast motile response in guinea-pig outer hair cells: the cellular basis of the cochlear amplifier. *J. Physiol.* **388**, 323–47.

Ashmore, J. F. (1992). Mammalian hearing and the cellular mechanisms of the cochlear amplifier. In *Sensory Transduction* (*Society of General Physiologists Series*, vol. 47), ed. D. P. Corey & S. D. Roper, pp. 395–412. New York: Rockefeller University Press.

Ashmore, J. F. (1994). The cellular machinery of the cochlea. *Exp. Physiol.* **79**, 113–34.

Assad, J. A. & Corey, D. P. (1992). An active motor model for adaptation by vertebrate hair cells. *J. Neurosci.* **12**, 3291–309.

Assad, J. A., Shepherd, G. M. G. & Corey, D. P. (1991). Tip-link integrity and mechanical transduction in vertebrate hair cells. *Neuron* **7**, 985–94.

Atkinson, R. C. & Shiffrin, R. M. (1971). The control of short-term memory. *Sci. Amer.* **225**(2), 82–90.

Atwater, I., Bezanilla, F. & Rojas, E. (1969). Sodium influxes in internally perfused squid giant axon during voltage clamp. *J. Physiol.* **201**, 657–64.

Auber, J. & Couteaux, R. (1963). Ultrastructure de la strie Z dans des muscles de Diptères. *J. Microscopie* **2**, 309–24.

Auerbach, A. & Sachs, F. (1983). Flickering of a nicotinic ion channel to a subconductance state. *Biophys. J.* **42**, 1–10.

Avenet, P. & Lindemann, B. (1988). Patch-clamp study of isolated taste receptor cells of the frog. *J. Membrane Biol.* **105**, 245–55.

Avenet, P. & Lindemann, B. (1991). Noninvasive recording of receptor cell action potentials and sustained currents from single taste buds maintained in the tongue: the response to mucosal NaCl and amiloride. *J. Membrane Biol.* **124**, 33–41.

Avenet, P., Hofmann, F. & Lindemann, B. (1988). Transduction in taste receptor cells requires cAMP-dependent protein kinase. *Nature* **331**, 351–4.

Axel, R. (1995). The molecular logic of smell. *Sci. Amer.* **273**(4), 130–7.

Axelsson, J. & Thesleff, S. (1959). A study of supersensitivity in denervated mammalian skeletal muscle. *J. Physiol.* **147**, 178–93.

Babu, Y. S., Sack, J. S., Greenhough, T. J., Bugg, C. E., Means, A. R. & Cook, W. J. (1985). Three dimensional structure of calmodulin. *Nature* **315**, 37–40.

Bacskai, B. J., Hochner, B., Mahaut-Smith, M., Adams, S. R., Kaang, B.-K., Kandel, E. R. & Tsien, R. Y. (1993). Spatially resolved dynamics of cAMP and protein kinase A subunits in *Aplysia* sensory neurons. *Science* **260**, 222–6.

Baddeley, A. D. (1976). *The Psychology of Memory*. New York: Harper & Row.

Bagni, M. A., Cecchi, G., Colomo, F. & Tesi, C. (1988). Plateau and descending limb of the sarcomere length–tension relation in short length-clamped segments of frog muscle fibres. *J. Physiol.* **401**, 581–95.

Bagshaw, C. R. (1993). *Muscle Contraction*, second edition. London: Chapman & Hall.

Bailey, C. H. & Chen, M. (1983). Morphological basis of long-term habituation and sensitization in *Aplysia*. *Science* **220**, 91–3.

Bailey, C. H. & Chen, M. (1988*a*). Long-term memory in *Aplysia* modulates the total number of varicosities of single identified neurons. *Proc. Natl. Acad. Sci. USA* **85**, 2373–7.

Bailey, C. H. & Chen, M. (1988*b*). Long-term sensitization in *Aplysia* increases the number of presynaptic contacts onto the identified gill motor neuron L7. *Proc. Natl. Acad. Sci. USA* **85**, 9356–9.

Bailey, C. H., Chen, M., Keller, F. & Kandel, E. R. (1992). Serotonin-mediated endocytosis of apCAM: an early step of learning-related synaptic growth in *Aplysia*. *Science* **256**, 645–9.

Bailey, C. H., Alberini, C., Ghirardi, M. & Kandel, E. R. (1994). Molecular and structural changes underlying long-term memory storage in Aplysia. In *Molecular and Cellular Mechanisms of Neurotransmitter Release* (*Adv. Second Mess. Phosphoprotein Res.* **29**), ed. L. Stjärne, P. Greengard, S. E. Grillner, T. G. M. Hökfelt & D. R. Ottoson, pp. 529–44. New York: Raven Press.

Bailey, K. (1948). Tropomyosin: a new asymmetrical protein component of the muscle fibril. *Biochem. J.* **43**, 271–9.

Bakalyar, H. A. & Reed, R. R. (1990). Identification of a specialized adenylyl cyclase that may mediate odorant detection. *Science* **250**, 1403–6.

Baker, P. F. (1986). The sodium–calcium exchange system. In *Calcium and the Cell*, ed. D. Evered & J. Whelan, *Ciba Foundation Symposium* **122**, pp. 73–86. Chichester: John Wiley.

Baker, P. F. & Dipolo, R. (1984). Axonal calcium and magnesium homeostasis. In *The Squid Axon*, ed. P. Baker (*Current Topics in Membranes and Transport* **22**), pp. 195–247. Orlando, FL: Academic Press.

Baker, P. F. & Willis, J. S. (1972). Binding of the cardiac glycoside ouabain to intact cells. *J. Physiol.* **224**, 441–62.

Baker, P. F., Hodgkin, A. L. & Shaw, T. I. (1961). Replacement of the protoplasm of a giant nerve fibre with artificial solutions. *Nature* **190**, 885–7.

Baker, P. F., Hodgkin, A. L. & Shaw, T. I. (1962*a*). Replacement of the axoplasm of giant nerve fibres with artificial solutions. *J. Physiol.* **164**, 330–54.

Baker, P. F., Hodgkin, A. L. & Shaw, T. I. (1962*b*). The effects of changes in internal ionic concentrations on the electrical properties of perfused giant axons. *J. Physiol.* **164**, 355–74.

Baker, P. F., Blaustein, M. P., Keynes, R. D., Mansil, J., Shaw, T. I. & Steinhardt, R. A. (1969). The ouabain-sensitive fluxes of sodium and potassium in squid giant axons. *J. Physiol.* **200**, 459–96.

Baker, P. F., Hodgkin, A. L. & Ridgeway, E. B. (1971). Depolarization and calcium entry in squid giant axons. *J. Physiol.* **218**, 709–55.

Baldwin, J. M. (1993). The probable arrangement of the helices in G protein-coupled receptors. *EMBO J.* **12**, 1693–703.

Baldwin, J. M. (1994). Structure and function of receptors coupled to G proteins. *Curr. Opin. Cell Biol.* **6**, 180–90.

Ball, S., Goodwin, T. W. & Morton, R. A. (1948). Studies on vitamin A. 5. The preparation of retinene$_1$–vitamin A aldehyde. *Biochem. J.* **42**, 516–23.

Banks, R. W., Harker, D. W. & Stacey, M. J. (1977). A study of mammalian intrafusal muscle fibres using a combined histochemical and ultrastructural approach. *J. Anat.* **123**, 783–96.

Banks, R. W., Barker, D. & Stacey, M. J. (1982). Form and distribution of sensory terminals in cat hindlimb muscle spindles. *Phil. Trans. R. Soc. Lond.* B **299**, 329–64.

Banks, R. W., Barker, D. & Stacey, M. J. (1985). Form and classification of motor endings in mammalian muscle spindles. *Proc. R. Soc. Lond.* B **225**, 195–212.

Bar, R. S., Deamer, D. W. & Corwell, D. G. (1966). Surface area of human erythrocyte lipids: reinvestigation of experiments on plasma membrane. *Science* **153**, 1010–12.

Bárány, M. (1967). ATPase activity of myosin correlated with speed of muscle shortening. *J. Gen. Physiol.* **50**, 197–218.

Bárány, M. (ed.) (1996). *Biochemistry of Smooth Muscle Contraction.* San Diego, CA: Academic Press.

Barchi, R. L. (1995). Molecular pathology of the skeletal muscle sodium channel. *Ann. Rev. Physiol.* **57**, 355–85.

Bargmann, W. & Schairer, E. (1951). The site of origin of the hormones of the posterior pituitary. *Amer. Sci.* **29**, 255–9.

Barker, D. & Banks, R. W. (1994). The muscle spindle. In *Myology: Basic and Clinical*, second edition, ed. A. G. Engel & C. Franzini-Armstrong, vol. 1, pp. 333–60. New York: McGraw-Hill.

Barker, D., Emonet-Dénand, F., Harker, D. W., Jami, L. & Laporte, Y. (1976). Distribution of fusimotor axons to intrafusal muscle fibres in cat tenuissimus spindles as determined by the glycogen-depletion method. *J. Physiol.* **261**, 49–69.

Barker, D., Emonet Dénand, F., Harker, D. W., Jami, L. &

Laporte, Y. (1977). Types of intra- and extrafusal muscle fibre innervated by dynamic skeleto-fusimotor axons in cat peroneus brevis and tenuissimus muscles, as determined by the glycogen depletion method. *J. Physiol.* **266**, 713–26.

Barlow, H. B. & Mollon, J. D. (eds.) (1982). *The Senses.* Cambridge: Cambridge University Press.

Barnard, E. A. (1992). Subunits of GABA$_A$, glycine, and glutamate receptors. In *Receptor Subunits and Complexes*, ed. A. Burgen & E. A. Barnard, pp. 163–87. Cambridge: Cambridge University Press.

Barnard, E. A., Wieckowski, J. & Chiu, T. H. (1971). Cholinergic receptor molecules and cholinesterase molecules at mouse skeletal muscle junction. *Nature* **234**, 207–9.

Barnard, E. A., Miledi, R. & Sumikawa, K. (1982). Translation of exogenous messenger RNA coding for nicotinic acetylocholine receptors produces functional receptors in *Xenopus* oocytes. *Proc. R. Soc. Lond.* B **215**, 241–6.

Barnett, V. A. & Thomas, D. D. (1984). Saturation transfer electron paramagnetic resonance of spin-labelled muscle fibres. *J. Molec. Biol.* **179**, 83–102.

Barrett, J. N., Magleby, K. L. & Pallotta, B. S. (1982). Properties of single calcium-activated potassium channels in cultured rat muscle. *J. Physiol.* **331**, 211–30.

Bartoshuk, L. M. (1988). Taste. In *Stevens' Handbook of Experimental Psychology*, second edition, ed. R. C. Atkinson, R. J. Herrnstein, G. Lindzey & R. D. Luce, vol. 1, pp. 461–99. New York: John Wiley & Sons.

Bartsch, D., Ghirardi, M., Skehel, P. A., Karl, K. A., Herder, S. P., Chen, M., Bailey, C. H. & Kandel, E. R. (1995). *Aplysia* CREB2 represses long-term facilitation: relief of repression converts transient facilitation into long-term functional and structural change. *Cell* **83**, 979–92.

Barzilai, A., Kennedy, T. E., Sweatt, J. D. & Kandel, E. R. (1989). 5–HT modulates protein-synthesis and the expression of specific proteins during long-term facilitation in *Aplysia* sensory neurons. *Neuron* **2**, 1577–86.

Bass, A. H. (1986). Electric organs revisited. In *Electroreception*, ed. T. H. Bullock & W. Heiligenberg, pp. 13–70. New York: John Wiley.

Baylor, D. A. (1987). Photoreceptor signals and vision. *Invest. Ophthalmol.* **28**, 34–49.

Baylor, D. A. (1995). Colour mechanisms of the eye. In *Colour: Art & Science*, ed. T. Lamb & J. Bourriau, pp. 103–26. Cambridge: Cambridge University Press.

Baylor, D. A. (1996). How photons start vision. *Proc. Natl. Acad. Sci. USA* **93**, 560–5.

Baylor, D. A. & Fuortes, M. G. F. (1970). Electrical responses of single cones in the retina of the turtle. *J. Physiol.* **207**, 77–92.

Baylor, D. A. & Hodgkin, A. L. (1973). Detection and resolution of visual stimuli by turtle photoreceptors. *J. Physiol.* **234**, 163–98.

Baylor, D. A., Fuortes, M. G. F. & O'Bryan, P. (1971). Receptive fields of cones in the retina of the turtle. *J. Physiol.* **214**, 265–94.

Baylor, D. A., Lamb, T. D. & Yau, K.-W. (1979*a*). The membrane current of single rod outer segments. *J. Physiol.* **288**, 589–611.

Baylor, D. A., Lamb, T. D. & Yau, K.-W. (1979*b*). Responses of retinal rods to single photons. *J. Physiol.* **288**, 613–34.

Baylor, D. A., Nunn, B. J. & Schnapf, J. L. (1987). Spectral sensitivity of cones in the monkey *Macaca fascicularis*. *J. Physiol.* **390**, 145–60.

Bear, M. F. & Malenka, R. C. (1994). Synaptic plasticity: LTP and LTD. *Curr. Opin. Neurobiol.* **4**, 389–99.

Beeler, G. W. & Reuter, H. (1970). Membrane calcium current in ventricular myocardial fibres. *J. Physiol.* **207**, 191–209.

Begenisich, T. & Cahalan, M. D. (1980). Sodium channel permeation in squid axons. I. Reversal potential experiments. *J. Physiol.* **307**, 217–42.

Béhé, P., DeSimone, J. A., Avenet, P. & Lindemann, B. (1990). Membrane currents in taste cells of the rat fungiform papilla. *J. Gen. Physiol.* **96**, 1061–84.

Beidler, L. M. & Smallman, R. L. (1965). Renewal of cells within taste buds. *J. Cell Biol.* **27**, 263–72.

Beirão, P. S. L., Davies, N. W. & Stanfield, P. R. (1994). Inactivating 'ball' peptide from *Shaker* B blocks Ca²⁺-activated but not ATP-dependent K⁺ channels of rat skeletal muscle. *J. Physiol.* **474**, 269–74.

Bekkers, J. M. (1994). Quantal analysis of synaptic transmission in the central nervous system. *Curr. Opin. Neurobiol.* **4**, 360–5.

Bekkers, J. M., Greef, N. G. & Neumcke, B. (1983). The conductance of sodium channels in the squid giant axon. *J. Physiol.* **343**, 24P–25P.

Belai, A. & Burnstock, G. (1994). Evidence for coexistence of ATP and nitric oxide in nonadrenergic, noncholinergic (NANC) inhibitory neurons in the rat ileum, colon and anococcygeus muscle. *Cell Tissue Res.* **278**, 197–200.

Belanger, J. H. & Orchard, I. (1993*a*). The locust ovipositor opener muscle: properties of the neuromuscular system. *J. Exp. Biol.* **174**, 321–42.

Belanger, J. H. & Orchard, I. (1993*b*). The locust ovipositor opener muscle: proctolinergic central and peripheral neuromodulation in a centrally driven motor system. *J. Exp. Biol.* **174**, 343–62.

Bell, J., Bolanowski, S. J. & Holmes, M. H. (1994). The structure and function of Pacinian corpuscles: a review. *Prog. Neurobiol.* **42**, 79–128.

Belmonte, C. & Cervero, F. (eds.) (1996). *Neurobiology of Nociceptors*. Oxford: Oxford University Press.

Bendall, J. R. (1953). Further observations on a factor (the 'Marsh' factor) effecting relaxation of ATP-shortened muscle fibre models, and the effect of Ca and Mg ions on it. *J. Physiol.* **121**, 232–54

Bennett, J. A. & Dingledine, R. (1995). Topology profile for a glutamate receptor: three transmembrane domains and a channel-lining reentrant membrane loop. *Neuron* **14**, 373–84.

Bennett, M. R. (1972). *Autonomic Neuromuscular Transmission*. Cambridge: Cambridge University Press.

Bennett, M. R. (1994). The concept of neurotransmitter release. In *Molecular and Cellular Mechanisms of Neurotransmitter Release* (*Adv. Second Mess. Phosphoprotein Res.* **29**), ed. L. Stjärne, P. Greengard, S. E. Grillner, T. G. M. Hökfelt & D. R. Ottoson, pp. 1–29. New York: Raven Press.

Bennett, M. R., Fisher, C., Florin, T., Quine, M. & Robinson, J. (1977). The effect of calcium ions and temperature on the binomial parameters that control acetylcholine release by a nerve impulse at amphibian neuromuscular synapses. *J. Physiol.* **271**, 641–72.

Bennett, M. V. L. (1965). Electroreceptors in mormyrids. *Cold Spr. Harb. Symp. Quant. Biol.* **30**, 245–62.

Bennett, M. V. L. (1966). Physiology of electrotonic junctions. *Ann. N.Y. Acad. Sci.* **137**, 509–39.

Bennett, M. V. L. (1971). Electric Organs. In *Fish Physiology*, ed. W. S. Hoar & D. J. Randall, vol. 5, pp. 347–492. New York: Academic Press.

Bennett, M. V. L. (1977). Electrical transmission: a functional analysis and comparison to chemical transmission. In *Handbook of Physiology*, section 1, vol. 1, ed. E. R. Kandel, pp. 357–416. Bethesda, MD: American Physiological Society

Bennett, M. V. L. (1985). Nicked by Occam's razor: unitarianism in the investigation of synaptic transmission. *Biol. Bull.* **168**, suppl., 159–67.

Bennett, M. V. L. & Spray, D. C. (1985). *Gap Junctions*. New York: Cold Spring Harbor Laboratory.

Bennett, M. V. L., Wurzel, M. & Grundfest, H. (1961). The electrophysiology of electric organs of marine electric fishes. I. Properties of electroplaques of *Torpedo nobiliana*. *J. Gen. Physiol.* **44**, 757–804.

Bennett, M. V. L., Aljure, E., Nakajima, Y. & Pappas, G. D. (1963). Electrotonic junctions between teleost spinal neurones: electrophysiology and ultrastructure. *Science* **141**, 262–4.

Bennett, M. V. L., Barrio, L. C., Bargellio, T. A., Spray, D. C., Herzberg, E. & Sáez, J. C. (1991). Gap junctions: new tools, new answers, new questions. *Neuron* **6**, 305–20.

Bennett, P., Craig, R., Starr, R. & Offer, G. (1986). The ultrastructural location of C-protein, X-protein and H-protein in rabbit muscle. *J. Mus. Res. Cell Motil.* **7**, 550–67.

Bennett, P. M. & Elliott, A. (1989). The 'catch' mechanism in molluscan muscle: an electron microscopy study of freeze-substituted anterior byssus retractor muscle of *Mytilus edulis*. *J. Mus. Res. Cell Motil.* **10**, 297–311.

Berghard, A. & Buck, L. B. (1996). Sensory transduction in vomeronasal neurons: evidence for $G_{\alpha o}$, $G_{\alpha i2}$ and adenylyl cyclase II as major components of a pheromone signaling cascade. *J. Neurosci.* **16**, 909–18.

Berghard, A., Buck, L. B. & Liman, E. R. (1996). Evidence for distinct signaling mechanisms in 2 mammalian olfactory sense organs. *Proc. Natl. Acad. Sci. USA* **93**, 2365–9.

Bergold, P. J., Sweatt, J. D., Winicov, I., Weiss, K. R., Kandel, E.R. & Schwartz, J. H. (1990). Protein synthesis during acquisition of long-term facilitation is needed for the persistent loss of regulatory subunits of the *Aplysia* cAMP-dependent protein kinase. *Proc. Natl. Acad. Sci. USA* **87**, 3788–91.

Bernard, C. (1857). *Leçons sur les Effects des Substances Toxiques et Médicamenteuses.* Paris.

Bernhardt, S. J., Naim, M., Zehavi, U. & Lindemann, B. (1996). Changes in IP_3 and cytosolic Ca^{2+} in response to sugars and non-sugar sweeteners in transduction of sweet taste in the rat. *J. Physiol.* **490**, 325–36.

Berninger, B. & Poo, M. (1996). Fast actions of neurotrophic factors. *Curr. Opin. Neurobiol.* **6**, 324–30.

Bernstein, J. (1902). Untersuchungen zur Thermodynamik der bioelektrischen Ströme. *Pflügers Archiv* **92**, 521–62.

Berridge, M. J. (1993). Inositol trisphosphate and calcium signalling. *Nature* **361**, 315–25.

Berridge, M. J. (1997). Elementary and global aspects of calcium signalling. *J. Physiol.* **499**, 291–306.

Bers, D. M. (1991). Species differences and the role of sodium–calcium exchange in cardiac muscle relaxation. *Ann. N. Y. Acad. Sci.* **639**, 375–85.

Bessou, P. & Laporte, Y. (1962). Responses from primary and secondary endings of the same neuromuscular spindle of the tenuissimus muscle of the cat. In *Symposium on Muscle Receptors*, ed. D. Barker, pp. 105–19. Hong Kong: Hong Kong University Press.

Bessou, P. & Perl, E. R. (1969). Response of cutaneous sensory units with unmyelinated fibers to noxious stimuli. *J. Neurophysiol.* **32**, 1025–43.

Betz, W. J. & Bewick, G. S. (1992). Optical analysis of synaptic vesicle recycling at the frog neuromuscular junction. *Science* **255**, 200–3.

Beuckelmann, D. J. & Wier, W. G. (1988). Mechanism of release of calcium from sarcoplasmic reticulum of guinea-pig cardiac cells. *J. Physiol.* **405**, 233–55.

Beyer, E. C., Paul, D. L. & Goodenough, D. A. (1987). Connexin43: a protein from rat heart homologous to a gap junction protein from liver. *J. Cell. Biol.* **105**, 2621–9.

Beyer, E. C., Veenstra, R. D., Kanter, H. L. & Saffitz, J. E. (1994). Molecular structure and patterns of expression of cardiac gap junctions. In *Cardiac Electrophysiology: From Cell to Bedside*, 2nd edition, ed. D. P. Zipes & J. Jalife, pp. 31–8. Philadelphia: W. B. Saunders.

Bezanilla, F. (1986). Voltage-dependent gating: current measurement and interpretation. In *Ionic Channels in Cells and Model Systems*, ed. R. Latorre, pp. 37–72. New York: Plenum Press.

Bezanilla, F. & Stefani, E. (1994). Voltage-dependent gating of ionic channels. *Ann. Rev. Biophys.* **23**, 819–46.

Bialek, W. (1987). Physical limits to sensation and perception. *Ann. Rev. Biophys.* **16**, 455–78.

Binstock, L. & Goldman, L. (1969). Current- and voltage-clamped studies on *Myxicola* giant axons. Effect of tetrodotoxin. *J. Gen. Physiol.* **54**, 730–40.

Birge, R. R. (1990). Nature of the primary photochemical events in rhodopsin and bacteriorhodopsin. *Biochim. Biophys. Acta* **1016**, 293–327.

Birks, R., Huxley, H. E. & Katz, B. (1960). The fine structure of the neuromuscular junction of the frog. *J. Physiol.* **150**, 134–44.

Birnbaumer, L. & Rodbell, M. (1969). Adenyl cyclase in fat cells. II. Hormone receptors. *J. Biol. Chem.* **244**, 3477–82.

Bishop, G. H. (1946). Neural mechanisms of cutaneous sense. *Physiol. Rev.* **26**, 77–102.

Bishop, G. H. & Landau, W. M. (1958). Evidence for a double peripheral pathway for pain. *Science* **128**, 712–13.

Bitensky, M. W., Gorman, R. E. & Miller, W. H. (1971). Adenyl cyclase as a link between photon capture and changes in membrane permeability of frog photoreceptors. *Proc. Natl. Acad. Sci. USA* **75**, 5217–18.

Black, J. W., Duncan, W. A. M., Durrant, C. J., Ganellin, C. R. & Parsons, E. M. (1972). Definition and antagonism of histamine H_2-receptors. *Nature* **236**, 385–90.

Blanchard, A., Ohanian, V. & Critchley, D. (1989). The structure and function of α-actinin. *J. Mus. Res. Cell Motil.* **10**, 280–9.

Blaustein, M. P. & Hodgkin, A. L. (1969). The effect of cyanide on the efflux of calcium from squid axons. *J. Physiol.* **200**, 497–527.

Bliss, T. V. P. & Collingridge, G. L. (1993). A synaptic model of memory: long-term potentiation in the hippocampus. *Nature* **361**, 31–9.

Bliss, T. V. P. & Gardner-Medwin, A. R. (1973). Long-lasting potentiation of synaptic transmission in the dentate area of the unanaesthetised rabbit following stimulation of the perforant path. *J. Physiol.* **232**, 357–74.

Bliss, T. V. P. & Lømo, T. (1973). Long-lasting potentiation of synaptic transmission in the dentate area of the anaesthetised rabbit following stimulation of the perforant path. *J. Physiol.* **232**, 331–56.

Block, B. A., Imagawa, T., Leung, A., Campbell, P. & Franzini-Armstrong, C. (1988). Structural evidence for direct interaction between the molecular components of the transverse tubules/sarcoplasmic reticulum junction in skeletal muscle. *J. Cell Biol.* **107**, 2587–600.

Block, S. M. (1992). Biophysical principles of sensory transduction. In *Sensory Transduction (Society of General Physiologists Series*, vol. 47), ed. D. P. Corey & S. D. Roper, pp. 1–17. New York: Rockefeller University Press.

Block, S. M. (1996). Fifty ways to love your lever: myosin motors. *Cell* **87**, 151–7.

Bockaert, J., Fozard, J. R., Dumuis, A. & Clarke, D. E. (1992). The $5-HT_4$ receptor: a place in the sun. *Trends Pharmacol. Sci.* **13**, 141–5.

Bodoia, R. D. & Detwiler, P. B. (1985). Patch-clamp recordings of the light-sensitive dark noise in retinal rods from the lizard and frog. *J. Physiol.* **367**, 183–216.

Boeckh, J., Kaissling, K.-E. & Schneider, D. (1965). Insect olfactory receptors. *Cold Spr. Harb. Symp. Quant. Biol.* **30**, 263–80.

Boekhoff, I., Raming, K. & Breer, H. (1990). Pheromone-induced stimulation of inositol-trisphosphate formation in insect antennae is mediated by G-proteins. *J. Comp. Physiol.* B **160**, 99–103.

Boistel, J. & Fatt, P. (1958). Membrane permeability change during transmitter action in crustacean muscle. *J. Physiol.* **144**, 176–91.

Boll, F. (1877). Zur Anatomie und Physiologie der Retina. *Arch. Anat. Physiol.* 1877, 4–35.

Bolton, T. B. (1979). Mechanisms of action of transmitters and other substances on smooth muscle. *Physiol. Rev.* **59**, 606–718.

Bolton, T. B. (1981). Action of acetylcholine on the smooth muscle membrane. In *Smooth Muscle*, ed. E. Bülbring, A. F. Brading, A. W. Jones & T. Tomita, pp. 199–217. London: Edward Arnold.

Bolton, T. B. & Large, W. A. (1986). Are junction potentials essential? Dual mechanism of smooth muscle cell activation by transmitter released from autonomic nerves. *Quart. J. Exp. Physiol.* **71**, 1–28.

Bone, Q. (1978). Locomotor muscle. In *Fish Physiology*, ed. W. S. Hoar & D. J. Randall, vol. 7, pp. 361–424. New York: Academic Press.

Bormann, J., Hamill, O. P. & Sakmann, B. (1987). Mechanism of anion permeation through channels gated by glycine and γ-aminobutyric acid in mouse cultured spinal neurones. *J. Physiol.* **385**, 243–86.

Bortoletto, Z. A., Bashir, Z. I., Davies, C. H. & Collingridge, G. L. (1994). A molecular switch activated by metabotropic glutamate receptors regulates induction of long-term potentiation. *Nature* **368**, 740–3.

Bourtchuladze, R., Frenguelli, B., Blendy, J., Cioffi, D., Schutz, G. & Silva, A. J. (1994). Deficient long-term memory in mice with a targeted mutation of the cAMP-responsive element-binding protein. *Cell* **79**, 59–68.

Bowmaker, J. K. (1984). Microspectrophotometry of vertebrate photoreceptors. *Vision Res.* **24**, 1641–50.

Bowmaker, J. K. & Dartnall, H. J. A. (1980). Visual pigments of rods and cones in a human retina. *J. Physiol.* **298**, 501–11.

Bownds, D. (1967). Site of attachment of retinal in rhodopsin. *Nature* **216**, 1178–81.

Boyd, I. A. (1962). The structure and innervation of the nuclear bag muscle fibre system and nuclear chain muscle fibre system in mammalian muscle spindles. *Phil. Trans. R. Soc. Lond.* B **245**, 81–136.

Boyd, I. A. (1981). The muscle spindle controversy. *Sci. Prog. Oxf.* **67**, 205–21.

Boyd, I. A. & Gladden, M. (1986). Morphology of mammalian muscle spindles: review. In *The Muscle Spindle*, ed. I. A. Boyd & M. Gladden, pp. 3–22. Basingstoke: Macmillan.

Boyd, I. A., Gladden, M. H., McWilliam, P. N. & Ward, J. (1977). Control of dynamic and static nuclear bag fibres and nuclear chain fibres by gamma and beta axons in isolated cat muscle spindles. *J. Physiol.* **265**, 133–62.

Boyett, M. R., Harrison, S. M., Janvier, N. C., McMorn, S. O., Owen, J. M. & Shui, Z. (1996). A list of vertebrate cardiac ionic currents – nomenclature, properties, function and cloned equivalents. *Cardiovasc. Res.* **32**, 455–81.

Boyle, P. J. & Conway, E. J. (1941). Potassium accumulation in muscle and associated changes. *J. Physiol.* **100**, 1–63.

Brake, A. J., Wagenbach, M. J. & Julius, D. (1994). New structural motif for ligand-gated channels defined by an ionotropic ATP receptor. *Nature* **371**, 519–23.

Bredt, D. S. & Snyder, S. H. (1994). Nitric oxide: a physiologic messenger molecule. *Ann. Rev. Biochem.* **63**, 175–95.

Breer, H. & Boekhoff, I. (1992). Second messenger signalling in olfaction. *Curr. Opin. Neurobiol.* **2**, 439–44.

Brenner, S. (1974). The genetics of *Caenorhabditis elegans*. *Genetics* **77**, 71–94.

Bretscher, M. S. (1973). Membrane structure: some general principles. *Science* **181**, 622–9.

Bridges, C. D. B. (1976). Vitamin A and the role of the pigment epithelium during bleaching and regeneration of rhodopsin in the frog eye. *Exp. Eye Res.* **22**, 435–55.

Bridges, C. D. B., Alvarez, R. A., Fong, S.-L., Gonzalez-Fernanez, F., Lam, D. M. K. & Liou, G. I. (1984). Visual cycle in the mammalian eye: retinoid-binding proteins and the distribution of 11-cis retinoids. *Vision Res.* **24**, 1581–94.

Brindley, G. S. (1970). *Physiology of the Retina and the Visual Pathway*, second edition. London: Arnold.

Brisson, A. & Unwin, P. N. T. (1985). Quaternary structure of the acetylcholine receptor. *Nature* **315**, 474–7.

Brock, J. A. & Cunnane, T. C. (1993). Neurotransmitter release mechanisms at the sympathetic neuroeffector junction. *Exp. Physiol.* **78**, 591–614.

Brock, L. G., Coombs, J. S. & Eccles, J. C. (1952). The recording of potentials from motoneurones with an intracellular electrode. *J. Physiol.* **117**, 431–60.

Brock, L. G., Eccles, R. M. & Keynes, R. D. (1953). The discharge of individual electroplates in *Raia clavata*. *J. Physiol.* **122**, 4P–6P.

Brooke, M. H. & Kaiser, K. K. (1970). Muscle fibre types: how many and what kind? *Arch. Neurol.* **23**, 369–79.

Brown, A. M., Lee, K. S. & Powell, T. (1981). Sodium current in single rat heart muscle cells. *J. Physiol.* **318**, 479–500.

Brown, D. A. & Adams, P. R. (1980). Muscarinic suppression of a novel voltage-sensitive K^+-current in a vertebrate neurone. *Nature* **283**, 673–6.

Brown, G. L., Dale, H. H. & Feldburg, W. (1936). Reactions of the normal mammalian muscle to acetylcholine and to eserine. *J. Physiol.* **87**, 394–424.

Brown, J. E. & Pinto, L. H. (1974). Ionic mechanism for the photoreceptor potential of the retina of *Bufo marinus*. *J. Physiol.* **236**, 575–91.

Brown, J. H. & Brown, S. L. (1984). Agonists differentiate muscarinic receptors that inhibit cyclic AMP formation from those that stimulate phosphoinositide metabolism. *J. Biol. Chem.* **254**, 3777–81.

Brown, K. T. & Murakami, M. (1964). A new receptor potential of the monkey retina with no detectable latency. *Nature* **201**, 6268.

Brown, K. T. & Wiesel, T. N. (1961). Localization of origins of electroretinogram components by intraretinal recording in the intact cat eye. *J. Physiol.* **158**, 257–80.

Brown, M. C. & Butler, R. G. (1973). Studies on the site of termination of static and dynamic fusimotor fibres within spindles of the tenuissimus muscles of the cat. *J. Physiol.* **233**, 553–73.

Brown, M. C., Crowe, A. & Matthews, P. B. C. (1965). Observations on the fusimotor fibres of the tibialis posterior muscle of the cat. *J. Physiol.* **177**, 140–59.

Brown, M. C., Nuttall, A. L. & Masta, R. I. (1983). Intracellular recordings from cochlear inner hair cells: effects of stimulation of the crossed olivocochlear efferents. *Science* **222**, 69–72.

Brown, P. K. & Wald, G. (1964). Visual pigments in single rods and cones of the human retina. *Science* **144**, 45–52.

Brown, R. E. (1994). *An Introduction to Neuroendocrinology.* Cambridge: Cambridge University Press.

Brown, R. H. (1993). Ion channel mutations in periodic paralysis and related myotonic diseases. *Ann. N. Y. Acad. Sci.* **707**, 305–16.

Brown, T. H. & Zador, A. M. (1990). Hippocampus. In *The Synaptic Organization of the Brain*, third edition, ed. G. M. Shepherd, pp. 346–88. Oxford: Oxford University Press.

Brownell, W. E., Bader, C. R., Bertrand, D. & de Ribaupierre, Y. (1985). Evoked mechanical responses of isolated cochlear outer hair cells. *Science* **227**, 194–6.

Buck, L. B. (1995). Unraveling chemosensory diversity. *Cell* **83**, 349–52.

Buck, L. B. (1996). Information coding in the vertebrate olfactory system. *Ann. Rev. Neurosci.* **19**, 517–44.

Buck, L. & Axel, R. (1991). A novel multigene family may encode odorant receptors: a molecular basis for odor recognition. *Cell* **65**, 175–87.

Bülbring, E. (1955). Correlation between membrane potential, spike discharge and tension in smooth muscle. *J. Physiol.* **128**, 200–21.

Bülbring, E. & Tomita, T. (1987). Catecholamine action on smooth muscle. *Pharmacol. Rev.* **39**, 49–96.

Bülbring, E., Ohashi, H. & Tomita, T. (1981). Adrenergic mechanisms. In *Smooth Muscle*, ed. E. Bülbring, A. F. Brading, A. W. Jones & T. Tomita, pp. 219–48. London: Edward Arnold.

Bullard, B. & Leonard, K. (1996). Modular proteins of insect muscle. *Adv. Biophys.* **33**, 211–21.

Bullock, T. H. & Dieke, F. P. J. (1956). Properties of an infra-red receptor. *J. Physiol.* **134**, 47–87.

Bullock, T. H. & Heiligenberg, W. (eds.) (1986). *Electroreception.* New York: John Wiley.

Bullock, T. H., Hagiwara, S., Kusand, K. & Negishi, K. (1961). Evidence for a category of electroreceptors in the lateral line of gymnotid fishes. *Science* **134**, 1426–7.

Bult, H., Boeckxstaens, G. E., Pelckmans, P. A., Jordaens, F. H., Van Maercke, Y. M. & Herman, A. G. (1990). Nitric oxide as an inhibitory non-adrenergic non-cholinergic neurotransmitter. *Nature* **345**, 346–7.

Bunzow, J. R., van Tol, H. H. M., Grandy, D. K., Albert, P. A., Salon, J., Christie, M., Machida, C. A., Neve, K. A. & Civelli, O. (1988). Cloning and expression of a rat D2 dopamine receptor cDNA. *Nature* **336**, 783–7.

Burgess, P. R. & Perl, E. R. (1967). Myelinated afferent fibres responding specifically to noxious stimulation of the skin. *J. Physiol.* **190**, 541–562.

Burgoyne, R. D. & Morgan, A. (1995). Ca^{2+} and secretory-vesicle dynamics. *Trends Neurosci.* **18**, 191–6.

Burke, R. E., Levine, D. N., Tsairis, P. & Zasac, F. (1973). Physiological types and histochemical profiles in motor units of the cat gastrocnemius. *J. Physiol.* **234**, 723–48.

Burkhardt, D. (1958). Die Sinnesorgane des Skeletmuskels und die nervöse Steuerung der Muskeltätigkeit. *Ergebn. Biol.* **20**, 27–66.

Burnashev, N., Schoepfer, R., Monyer, H., Ruppersberg, J. P., Günther, W., Seeburg, P. H. & Sakmann, B. (1992). Control by asparagine residues of calcium permeability and magnesium blockade in the NMDA receptor. *Science* **257**, 1415–19.

Burnstock, G. (1995a). Current state of purinoceptor research. *Pharmacol. Acta Helv.* **69**, 231–42.

Burnstock, G. (1995b). Noradrenaline and ATP: cotransmitters and neuromodulators. *J. Physiol. Pharmacol.* **46**, 365–84.

Burnstock, G. (1996). A unifying purinergic hypothesis for the initiation of pain. *Lancet* **347**, 1604–5.

Burnstock, G., Campbell, G., Satchell, D. G. & Smyth, A. (1970). Evidence that adenosine triphosphate or a related nucleotide is the transmitter substance released by non-adrenergic inhibitory nerves in the gut. *Brit. J. Pharmacol.* **40**, 668–88.

Burton, K. (1992). Myosin step size: estimates from motility assays and shortening muscle. *J. Mus. Res. Cell Motil.* **13**, 590–607.

Butler, A., Tsunoda, S., McCobb, D. P., Wei, A. & Salkoff, L. (1993). *mSlo*, a complex mouse gene encoding "maxi" calcium-activated potassium channels. *Science* **261**, 221–4.

Byers, D., David, R. L. & Kiger, J. A. (1981). Defect in cyclic AMP phosphodiesterase due to *dunce* mutation of learning in *Drosophila melanogaster*. *Nature* **289**, 79–81.

Bylund, D. B. & U'Prichard, D. C. (1983). Characterization of α_1- and α_2-adrenergic receptors. *Int. Rev. Neurobiol.* **24**, 343–431.

Cahalan, M. & Neher, E. (1992). Patch clamp techniques: an overview. *Methods Enzymol.* **207**, 3–14.

Cain, D. F., Infante, A. A. & Davies, R. E. (1962). Chemistry of muscle contraction. Adenosine triphosphate and phosphoryl creatine as energy supplies for single contractions of working muscle. *Nature* **196**, 214–17.

Cain, W. S. (1988). Olfaction. In *Stevens' Handbook of Experimental Psychology*, second edition, ed. R. C. Atkinson, R. J. Herrnstein, G. Lindzey & R. D. Luce, vol. 1, pp. 409–59. New York: John Wiley & Sons.

Calakos, N. & Scheller, R. H. (1996). Synaptic vesicle biogenesis, docking, and fusion: a molecular description. *Physiol. Rev.* **76**, 1–29.

Caldwell, P. C. & Walster, G. E. (1963). Studies on the micro-injection of various substances into crab muscle fibres. *J. Physiol.* **169**, 353–72.

Caldwell, P. C., Hodgkin, A. L., Keynes, R. D. & Shaw, T. I. (1960). The effects of injecting 'energy-rich' phosphate compounds on the active transport of ions in the giant axons of *Loligo*. *J. Physiol.* **152**, 561–90.

Campbell, D. L., Rasmusson, R. L. & Strauss, H. C. (1992). Ionic current mechanisms generating vertebrate primary cardiac pacemaker activity at the single cell level: an integrative view. *Ann. Rev. Physiol.* **54**, 279–302.

Campbell, F. W. & Rushton, W. A. H. (1955). Measurement of the scotopic pigment in the living human eye. *J. Physiol.* **130**, 131–47.

Campbell, G. (1987). Cotransmission. *Ann. Rev. Pharmacol. Toxicol.* **27**, 51–70.

Campbell, K. P. (1995). Three muscular dystrophies: loss of cytoskeleton–extracellular matrix linkage. *Cell* **80**, 675–9.

Cannell, M. B. (1991). Contribution of sodium–calcium exchange to calcium regulation in cardiac muscle. *Ann. N. Y. Acad. Sci.* **639**, 428–43.

Cannell, M. B., Cheng, H. & Lederer, W. J. (1994). Spatial non-uniformities in [Ca²⁺] during excitation–contraction coupling in cardiac myocytes. *Biophys. J.* **67**, 1942–56.

Cannessa, C. M., Schild, L., Buell, G., Thorens, B., Gautschi, I., Horisberger, J.-D. & Rossier, B. C. (1994). Amiloride-sensitive epithelial Na⁺ channel is made of three homologous subunits. *Nature* **367**, 463–7.

Cannon, S. C. (1996). Sodium channel defects in myotonia and periodic paralysis. *Ann. Rev. Neurosci.* **19**, 141–64.

Caprio, J., Brand, J. G., Teeter, J. H., Valentincic, T., Kalinowski, D. L., Kohbara, J., Kumazawa, T. & Wegert, S. (1993). The taste system of the channel catfish: from biophysics to behaviour. *Trends Neurosci.* **16**, 192–7.

Carafoli, E. (1991). The calcium pumping ATPase of the plasma membrane. *Ann. Rev. Physiol.* **53**, 531–47.

Carafoli, E. & Zurini, M. (1982). The Ca²⁺ pumping ATPase of plasma membranes. *Biochim. Biophys. Acta* **683**, 279–301.

Carew, T. J. (1996). Molecular enhancement of memory formation. *Neuron* **16**, 5–8.

Carew, T. J., Pinsker, H. M. & Kandel, E. R. (1972). Long-term habituation of a defensive withdrawal reflex in *Aplysia*. *Science* **175**, 451–4.

Carew, T. J., Hawkins, R. D. & Kandel, E. R. (1983). Differential

classical conditioning of a defensive withdrawal reflex in *Aplysia californica*. *Science* **219**, 397–400.

Carlson, J. R. (1996). Olfaction in *Drosophila*: from odor to behaviour. *Trends Genetics* **12**, 175–80.

Carlsson, A. (1987). Perspectives on the discovery of central monoaminergic neurotransmission. *Ann. Rev. Neurosci.* **10**, 19–40.

Carlsson, A. (1993). Thirty years of dopamine research. *Adv. Neurol.* **60**, 1–10.

Carlsson, A., Falck, B. & Hillarp, N.-Å. (1962). Cellular localization of brain monoamines. *Acta Physiol. Scand.* **56** (suppl. 196), 1–27.

Carr, R. W., Morgan, D. L. & Proske, U. (1996). Impulse initiation in the mammalian muscle spindle during combined fusimotor stimulation and succinyl choline infusion. *J. Neurophysiol.* **75**, 1703–13.

Carstén, E., Pitkänen, M., Sirviö, J., Parsadanian, A., Lindholm, D., Thoenen, H. & Riekkinen, P. J. (1993). The induction of LTP increases BDNF and NGF mRNA but decreases NT-3 mRNA in the dentate gyrus. *Neuroreport* **4**, 895–8.

Cartaud, J., Benedetti, E. L., Cohen, J. B., Meunier, J.-C. & Changeux, J.-P. (1973). Presence of a lattice structure in membrane fragments rich in nicotinic receptor protein from the electric organ of *Torpedo marmorata*. *FEBS Lett.* **33**, 109–13.

Cartaud, J., Benedetti, E. L., Sobel, A. & Changeux, J.-P. (1978). A morphological study of the cholinergic receptor protein from *Torpedo marmorata* in its membrane environment and in its detergent-extracted purified form. *J. Cell Sci.* **29**, 313–37.

Castellani, L. & Cohen, C. (1987). Myosin rod phosphorylation and the catch state of molluscan muscles. *Science* **235**, 334–7.

Castellucci, V. F. & Kandel, E. R. (1976). Presynaptic facilitation as a mechanism for behavioural sensitization in *Aplysia*. *Science* **194**, 1176–8.

Castellucci, V. F., Kandel, E. R., Schwartz, J. H., Wilson, F. D., Nairn, A. C. & Greengard, P. (1980). Intracellular injection of the catalytic subunit of cyclic AMP-dependent protein kinase simulates facilitation of transmitter release underlying behavioral sensitization in *Aplysia*. *Proc. Natl. Acad. Sci. USA* **77**, 7492–6.

Catterall, W. A. (1986). Voltage-dependent gating of sodium channels: correlating structure and function. *Trends Neurosci.* **9**, 7–10.

Catterall, W. A. (1992). Cellular and molecular biology of voltage-gated sodium channels. *Physiol. Rev.* **72**, S15–S48.

Catterall, W. A. (1993). Structure and function of voltage-gated ion channels. *Trends Neurosci.* **16**, 500–6.

Caulfield, M. P., Jones, S., Vallis, Y., Buckley, N. J., Kim, G.-D., Milligan, G. & Brown, D. A. (1994). Muscarinic M-current inhibition via $G_{\alpha q/11}$ and α-adrenoceptor inhibition of Ca^{2+} current via $G_{\alpha o}$ in rat sympathetic neurones. *J. Physiol.* **477**, 415–22.

Cervetto, L., Lagnado, L., Perry, R. J., Robinson, D. W. & McNaughton, P. A. (1989). Extrusion of calcium from rod outer segments is driven by both sodium and potassium gradients. *Nature* **337**, 740–3.

Chalfie, M. (1993). Touch receptor development and function in *Caenorhabditis elegans. J. Neurobiol.* **24**, 1433–41.

Chalfie, M. (1995). The differentiation and function of the touch receptor neurons of *Caenorhabditis elegans. Prog. Brain Res.* **105**, 179–82.

Chalfie, M. & Au, M. (1989). Genetic control of differentiation of the *Caenorhabditis elegans* touch receptor neurons. *Science* **243**, 1027–33.

Chalfie, M. & Sulston, J. (1981). Developmental genetics of the mechanosensory neurons of *Caenorhabditis elegans. Dev. Biol.* **82**, 358–70.

Chalfie, M. & Wolinsky, E. (1990). The identification and suppression of neurodegeneration in *Caenorhabditis elegans. Nature* **345**, 410–16.

Chalovich, J. M. & Eisenberg, E. (1982). Inhibition of actomyosin ATPase activity by troponin-tropomyosin without blocking the binding of myosin to actin. *J. Biol. Chem.* **257**, 2432–7.

Chambers, M. R., Andres, K. H., von Duering, M. & Iggo, A. (1972). The structure and function of the slowly adapting type II mechanoreceptor in hairy skin. *Quart. J. Exp. Physiol.* **57**, 417–45.

Chandler, W. K. & Meves, H. (1965). Voltage clamp measurements on internally perfused giant axons. *J. Physiol.* **180**, 788–820.

Chandler, W. K., Hodgkin, A. L. & Meves, H. (1965). The effect of changing the internal solution on sodium inactivation and related phenomena in giant axons. *J. Physiol.* **180**, 821–36.

Chandler, W. K., Rakowski, R. F. & Schneider, M. (1976). Effects of glycerol treatment and maintained depolarization on charge movement in skeletal muscle. *J. Physiol.* **254**, 285–316.

Chang, F.-L. & Greenough, W. T. (1984). Transient and enduring morphological correlates of synaptic activity and efficacy change in the rat hippocampal slice. *Brain Res.* **309**, 35–46.

Chang, M. M., Leeman, S. E. & Niall, H. D. (1971). Amino acid sequence of substance P. *Nature New Biol.* **232**, 86–7.

Changeux, J.-P. (1995). The acetylcholine receptor: a model for allosteric membrane proteins. *Biochem. Soc. Trans.* **23**, 195–205.

Changeux, J.-P., Galzi, J.-L., Devillers-Thiéry, A. & Bertrand, D. (1992). The functional architecture of the acetylcholine nicotinic receptor explored by affinity labelling and site-directed mutagenesis. *Quart. Rev. Biophys.* **25**, 395–432.

Chantler, P. D. & Szent-Györgyi, A. G. (1980). Regulatory light chains and scallop myosin: full dissociation, reversibility and co-operative effects. *J. Molec. Biol.* **138**, 473–92.

Charest, R., Prpic, V., Exton, J. H. & Blackmore, P. (1985). Stimulation of inositol trisphosphate formation in hepatocytes by vasopressin, epinephrine and angiotensin II and its relationship to changes in cytosolic free Ca^{2+}. *Biochem. J.* **227**, 79–90.

Charnock, J. S. & Post, R. L. (1963). Evidence of the mechanism of ouabain inhibition of cation activated adenosine triphosphatase. *Nature* **199**, 910–11.

Chaudhari, N., Yang, H., Lamp, C., Delay, E., Cartford, C., Than, T. & Roper, S. (1996). The taste of MSG: membrane receptors in taste buds. *J. Neurosci.* **16**, 3817–26.

Chavez, R. A. & Hall, Z. W. (1992). Expression of fusion proteins of the nicotinic acetylcholine receptor from mammalian muscle identifies the membrane-spanning regions in the α and δ subunits. *J. Cell. Biol.* **116**, 385–93.

Chen, C.-C., Akopian, A. N., Sivilotti, L., Colquhoun, D., Burnstock, G. & Wood, J. N. (1995). A P_{2X} purinoceptor expressed by a subset of sensory neurons. *Nature* **377**, 428–31.

Chen, C.-K., Inglese, J., Lefkowitz, R. J. & Hurley, J. B. (1995). Ca^{2+}-dependent interaction of recoverin with rhodopsin kinase. *J. Biol. Chem.* **270**, 18060–6.

Chen, C.-N., Denome, S. & Davis, R. L. (1986). Molecular analysis of cDNA clones and the corresponding genomic coding sequences of the *Drosophila* dunce$^+$ gene, the structural gene for cAMP phosphodiesterase. *Proc. Natl. Acad. Sci. USA* **83**, 9313–17.

Chen, J., Makino, C. L., Peachey, N. S., Baylor, D. A. & Simon, M. I. (1995). Mechanisms of rhodopsin inactivation by a COOH-terminal truncation mutant. *Science* **267**, 374–7.

Cheng, H., Lederer, M. R., Xiao, R. P., Gomez, A. M., Zhou, Y. Y., Ziman, B., Spurgeon, H., Lakatta, E. G. & Lederer, W. J. (1966). Excitation–contraction coupling in heart: new insights from Ca^{2+} sparks. *Cell Calcium* **20**, 129–40.

Chess, A., Buck, L., Dowling, M. M., Axel, R. & Ngai, J. (1992). Molecular biology of smell: expression of the multigene family encoding putative olfactory receptors. *Cold Spr. Harb. Symp. Quant. Biol.* **57**, 505–16.

Chess, A., Simon, I., Cedar, H. & Axel, R. (1994). Allelic inactivation regulates olfactory receptor gene expression. *Cell* **78**, 823–34.

Chetkovich, D. M., Gray, R., Johnston, D. & Sweatt, J. D. (1991). *N*-methyl-D-aspartate receptor activation increases cAMP levels and voltage-gated Ca^{2+} channel activity in area CA1 of the hipocampus. *Proc. Natl. Acad. Sci. USA* **88**, 6467–71.

Chiu, S. Y. (1980). Asymmetry currents in the mammalian myelinated nerve. *J. Physiol.* **309**, 499–519.

Chiu, S. Y. & Ritchie, J. M. (1981). Evidence for the presence of potassium channels in the paranodal region of acutely demyelinated mammalian single nerve fibres. *J. Physiol.* **313**, 415–37.

Chiu, S. Y. & Ritchie, J. M. (1982). Evidence for the presence of potassium channels in the internode of frog myelinated nerve fibres. *J. Physiol.* **322**, 485–501.

Chiu, S. Y. & Ritchie, J. M. (1984). On the physiological role of internodal potassium channels and the security of conduction in myelinated nerve fibres. *Proc. R. Soc. Lond.* B **220**, 415–22.

Chiu, S. Y., Ritchie, J. M., Rogart, R. B. & Stagg, D. (1979). A quantitative description of membrane currents in rabbit myelinated nerve. *J. Physiol.* **292**, 149–66.

Choi, D. W. (1988). Calcium-mediated neurotoxity: relationship to specific channel types and role in ischemic damage. *Trends Neurosci.* **11**, 465–9.

Choi, D. W. (1992a). Bench to bedside: the glutamate connection. *Science* **258**, 241–3.

Choi, D. W. (1992b). Excitotoxic cell death. *J. Neurobiol.* **23**, 1261–76.

Choi, D. W. (1995). Calcium: still center-stage in hypoxic-ischemic neuronal death. *Trends Neurosci.* **18**, 58–60.

Clapham, D. E. (1994). Direct G-protein activation of ion channels. *Ann. Rev. Neurosci.* **17**, 441–64.

Clark, K. A. & Collingridge, G. L. (1995). Synaptic potentiation of dual-component excitatory postsynaptic currents in the rat hippocampus. *J. Physiol.* **482**, 39–52.

Claudio, T., Ballivet, M., Patrick, J. & Heinemann, S. (1983). Nucleotide and deduced amino acid sequences of *Torpedo californica* acetylcholine receptor α subunit. *Proc. Natl. Acad. Sci. USA* **80**, 1111–15.

Clay, J. R. & DeFelice, L. J. (1983). Relationship between membrane excitability and single channel open–close kinetics. *Biophys. J.* **42**, 151–7.

Close, R. I. (1972). Dynamic properties of mammalian skeletal muscle. *Physiol. Rev.* **52**, 129–97.

Clusin, W. T. & Bennett, M. V. L. (1977a). Calcium-activated conductance in skate electroreceptors. Current clamp experiments. *J. Gen. Physiol.* **69**, 121–43.

Clusin, W. T. & Bennett, M. V. L. (1977b). Calcium-activated conductance in skate electroreceptors. Voltage clamp experiments. *J. Gen. Physiol.* **69**, 145–82.

Cobbs, W. H. & Pugh, E. N. (1985). Cyclic GMP can increase rod outer-segment light-sensitive current 10-fold without delay of excitation. *Nature* **313**, 585–7.

Cohen, M. J. (1964). The peripheral organization of sensory systems. In *Neural Theory and Modelling*, ed. R. Reiss, pp. 273–92. Stanford, CA: Stanford University Press.

Cohen, M. J., Hagiwara, S. & Zotterman, Y. (1955). The response spectrum of taste fibres in the cat: a single fibre analysis. *Acta Physiol. Scand.* **33**, 316–32.

Cohen, S. (1982). Matching molecules in the catch mechanism. *Proc. Natl. Acad. Sci. USA* **79**, 3176–8.

Cole, A. J., Saffen, D. W., Baraban, J. M. & Worley, P. F. (1989). Rapid increase of an immediate early gene messenger RNA in hippocampal neurons by synaptic NMDA receptor activation. *Nature* **340**, 474–6.

Cole, K. S. (1949). Dynamic electrical characteristics of the squid axon membrane. *Arch. Sci. Physiol.* **3**, 253–8.

Cole, K. S. (1968). *Membranes, Ions and Impulses*. Berkeley, CA: University of California Press.

Cole, K. S. & Curtis, H. J. (1939). Electric impedance of the squid giant axon during activity. *J. Gen. Physiol.* **22**, 649–70.

Collingridge, G. L. & Bliss, T. V. P. (1987). NMDA receptors – their role in long-term potentiation. *Trends Neurosci.* **10**, 288–93.

Collingridge, G. L. & Bliss, T. V. P. (1995). Memories of NMDA receptors and LTP. *Trends Neurosci.* **18**, 54–6.

Collingridge, G. L., Kehl, S. J. & McLennan, H. (1983). Excitatory amino acids in synaptic transmission in the Schaffer collateral-commissural pathway of the rat hippocampus. *J. Physiol.* **334**, 33–46.

Collingridge, G. L., Herron, C. E. & Lester, R. A. J. (1988). Synaptic activation of *N*-methyl-D-aspartate receptors in the Schaffer collateral-commissural pathway of rat hippocampus. *J. Physiol.* **399**, 283–300.

Collins, C. A., Rojas, E. & Suarez-Isla, B. A. (1982a). Activation and inactivation characteristics of the sodium permeability in muscle fibres from *Rana temporaria*. *J. Physiol.* **324**, 297–318.

Collins, C. A., Rojas, E. & Suarez-Isla, B. A. (1982b). Fast charge movements in skeletal muscle fibres from *Rana temporaria*. *J. Physiol.* **324**, 319–45.

Collins, J. H., Potter, J. D., Horn, M. J., Wiltshire, G. & Jackman, N. (1973). The amino acid sequence of rabbit skeletal muscle troponin C: gene replication and homology with calcium-binding proteins from carp and hake muscle. *FEBS Lett.* **36**, 268–72.

Colquhoun, D. & Hawkes, A. G. (1977). Relaxation and fluctuations of membrane current that flow through drug-operated ion channels. *Proc. R. Soc. Lond.* B **199**, 231–62.

Colquhoun, D. & Hawkes, A. G. (1981). On the stochastic properties of single ion channels. *Proc. R. Soc. Lond.* B **211**, 205–34.

Colquhoun, D. & Hawkes, A. G. (1982). On the stochastic properties of bursts of single ion channel openings and of clusters of bursts. *Phil. Trans. R. Soc. Lond.* B **300**, 1–59.

Colquhoun, D. & Sakmann, B. (1981). Fluctuations in the microsecond time range of the current through single acetylcholine receptor ion channels. *Nature* **294**, 464–6.

Colquhoun, D. & Sakmann, B. (1983). Bursts of openings in transmitter-activated ion channels. In *Single Channel Recording*, ed. B. Sakmann & E. Neher, pp. 345–64. New York: Plenum Press.

Colquhoun, D. & Sakmann, B. (1985). Fast events in single-channel currents activated by acetylcholine and its analogues at the frog muscle end-plate. *J. Physiol.* **369**, 501–57.

Comb, M., Birnberg, N. C., Seasholtz, A., Herbert, E. & Goodman, H. M. (1986). A cyclic AMP and phorbol-ester inducible DNA element. *Nature* **323**, 353–6.

Comb, M., Hyman, S. E. & Goodman, H. M. (1987). Mechanisms of trans-synaptic regulation of gene expression. *Trends Neurosci.* **10**, 473–8.

Cone, R. A. (1964). The early receptor potential of the vertebrate retina. *Nature* **204**, 736–9.

Cone, R. A. (1973). The internal transmitter model for visual excitation: some quantitative implications. In *Biochemistry and Physiology of Visual Pigments*, ed. H. Langer, pp. 275–82. Berlin: Springer-Verlag.

Connor, J. A. & Stevens, C. F. (1971). Voltage clamp studies of a transient outward membrane current in gastropod neural somata. *J. Physiol.* **213**, 21–30.

Conti, F. & Neher, E. (1980). Single channel recordings of K currents in squid axons. *Nature* **285**, 140–3.

Conti, F., DeFelice, L. J. & Wanke, E. (1975). Potassium and sodium ion current noise in the membrane of the squid giant axon. *J. Physiol.* **248**, 45–82.

Conti, F., Hille, B., Neumcke, B., Nonner, W. & Stämpfli, R. (1976*a*). Measurement of the conductance of the sodium channel from current fluctuations at the node of Ranvier. *J. Physiol.* **262**, 699–727.

Conti, F., Hille, B., Neumcke, B., Nonner, W. & Stämpfli, R. (1976*b*). Conductance of the sodium channel in myelinated nerve fibres with modified sodium inactivation. *J. Physiol.* **262**, 729–42.

Conway, E. J. (1957). Nature and significance of concentration relations of potassium and sodium ions in skeletal muscle. *Physiol. Rev.* **37**, 84–132.

Cook, A. J., Woolf, C. J., Wall, P. D. & McMahon, S. B. (1987). Dynamic receptive field plasticity in rat spinal cord dorsal horn following C-primary afferent input. *Nature* **325**, 151–3.

Cooke, R. (1986). The mechanism of muscle contraction. *CRC Crit. Rev. Biochem.* **21**, 53–118.

Cooke, R., Crowder, M. S., Wendt, C. H., Barnett, V. A. & Thomas, D. D. (1984). Muscle cross-bridges: do they rotate? In *Contractile Mechanisms in Muscle*, ed. G. H. Pollack & H. Sugi, pp. 413–23. Tokyo: University of Tokyo Press.

Cooley, L. B., Johnson, W. H. & Krause, S. (1979). Phosphorylation of paramyosin and its possible role in the catch mechanism. *J. Biol. Chem.* **254**, 2195–8.

Coombs, J. S., Eccles, J. C. & Fatt, P. (1955*a*). The electrical properties of the motoneurone membrane. *J. Physiol.* **130**, 291–325.

Coombs, J. S., Eccles, J. C. & Fatt, P. (1955*b*). The specific ionic conductances and the ionic movements across the motoneuronal membrane that produce the inhibitory postsynaptic potential. *J. Physiol.* **130**, 326–73.

Coombs, J. S., Eccles, J. C. & Fatt, P. (1955*c*). Excitatory synaptic action in motoneurones. *J. Physiol.* **130**, 374–95.

Coombs, J. S., Eccles, J. C. & Fatt, P. (1955*d*). The inhibitory suppression of reflex discharges from motoneurones. *J. Physiol.* **130**, 396–413.

Coombs, J. S., Curtis, D. R. & Eccles, J. C. (1957*a*). The interpretation of spike potentials of motoneurones. *J. Physiol.* **139**, 198–231.

Coombs, J. S., Curtis, D. R. & Eccles, J. C. (1957*b*). The generation of impulses in motoneurones. *J. Physiol.* **139**, 232–49.

Cooper, E., Couturier, S. & Ballivet, M. (1991). Pentameric structure and subunit stoichiometry of a neuronal acetylcholine receptor. *Nature* **350**, 235–8.

Cooper, J. R., Bloom, F. E. & Roth, R. H. (1996). *The Biochemical Basis of Neuropharmacology*, seventh edition. New York: Oxford University Press.

Cooper, S. (1961). The responses of the primary and secondary endings of muscle spindles with intact motor innervation during applied stretch. *Quart. J. Exp. Physiol.* **46**, 398–98.

Cooper, S. & Daniel, P. M. (1956). Human muscle spindles. *J. Physiol.* **133**, 1P–3P.

Corey, D. P. & Assad, J. A. (1992). Transduction and adaptation in vertebrate hair cells: correlating structure with function. In *Sensory Transduction* (*Society of General Physiologists Series*, vol. 47), ed. D. P. Corey & S. D. Roper, pp. 325–42. New York: Rockefeller University Press.

Corey, D. P. & Hudspeth, A. J. (1979). Ionic basis of the receptor potential in a vertebrate hair cell. *Nature* **281**, 675–7.

Corey, D. P. & Hudspeth, A. J. (1983). Kinetics of the receptor current in bullfrog saccular hair cells. *J. Neurosci.* **3**, 962–76.

Corey, D. P. & Roper, S. D. (eds.) (1992). *Sensory Transduction* (*Society of General Physiologists Series*, vol. 47). New York: Rockefeller University Press.

Cornelius, F. (1982). Tonic contraction and the control of relaxation in a chemically skinned molluscan smooth muscle. *J. Gen. Physiol.* **79**, 821–34.

Costantin, L. L. (1970). The role of sodium current in the radial spread of contraction in frog muscle fibres. *J. Gen. Physiol.* **55**, 703–15.

Costantin, L. L. (1974). Contractile activation in frog skeletal muscle. *J. Gen. Physiol.* **63**, 657–74.

Cote, R. H., Biernbaum, M. S., Nicol, G. D. & Bownds, M. D. (1984). Light-induced decreases in cGMP concentration precede changes in membrane permeability in frog rod photoreceptors. *J. Biol. Chem.* **259**, 9635–41.

Couceiro, A. & de Almeida, D. F. (1961). The electrogenic tissue of some Gymnotidae. In *Bioelectrogenesis*, ed. C. Chagas & A. Paes de Carvalho, pp. 3–13. Amsterdam: Elsevier.

Couteaux, R. & Pécot-Dechavassine, M. (1970). Vésicules synaptique et poches au niveau des 'zones actives' de la jonction neuromusculaire. *Comptes Rendus Acad. Sci. D* **271**, 2346–9.

Couturier, S., Bertrand, D., Matter, J. M., Hernandez, M.-C., Bertrand, S., Millar, N., Valera, S., Barkas, T. & Ballivet, M. (1990). Neuronal nicotinic acetylcholine receptor subunit ($\alpha 7$) is developmentally regulated and forms a homo-oligomeric channel blocked by α-BTX. *Neuron* **5**, 847–56.

Covarrubias, M., Wei, A. & Salkoff, L. (1991). *Shaker, Shal, Shab* and *Shaw* express independent K^+ current systems. *Neuron* **7**, 763–73.

Craig, R. (1977). Structure of A-segments from frog and rabbit skeletal muscle. *J. Molec. Biol.* **109**, 69–81.

Craig, R. (1994). The structure of the contractile filaments. In

Myology, second edition, ed. A. G. Engel & C. Franzini-Armstrong, pp. 134–75. New York: McGraw-Hill.

Craig, R. & Megerman, J. (1977). Assembly of smooth muscle myosin into side-polar filaments. *J. Cell Biol.* **75**, 990–6.

Craig, R. & Offer, G. (1976). Axial arrangement of cross-bridges in thick filaments of vertebrate skeletal muscle. *J. Molec. Biol.* **102**, 325–32.

Crawford, A. C., Evans, M. G. & Fettiplace, R. (1989). Activation and adaptation of transducer currents in turtle hair cells. *J. Physiol.* **419**, 405–34.

Crawford, A. C., Evans, M. G. & Fettiplace, R. (1991). The actions of calcium on the mechano-electrical transducer current of turtle hair cells. *J. Physiol.* **434**, 369–98.

Crawford, B. H. (1949). The scotopic visibility function. *Proc. Phys. Soc.* B **62**, 321–34.

Creed, K. E. (1979). Functional diversity of smooth muscle. *Brit. Med. Bull.* **35**, 243–7.

Creighton, T. E. (1993). *Proteins: Structures and Molecular Properties.* New York: W. H. Freeman & Co.

Crescitelli, F. & Dartnall, H. J. A. (1953). Human visual purple. *Nature* **172**, 195–6.

Crick, F. H. C. (1953). The packing of α-helices: simple coiled-coils. *Acta Cryst.* **6**, 689–97.

Cross, M. & Dexter, T. M. (1991). Growth factors in development, transformation, and tumorigenesis. *Cell* **64**, 271–80.

Cross, R. A., Geeves, M. A. & Kendrick-Jones, J. (1991). A nucleation-elongation mechanism for the self-assembly of side-polar sheets of smooth muscle myosin. *EMBO J.* **10**, 747–56.

Crowe, A. & Matthews, P. B. C. (1964). The effects of stimulation of static and dynamic fusimotor fibres on the response to stretching of the primary endings of the muscle spindles. *J. Physiol.* **174**, 109–31.

Curtin, N. A. & Woledge, R. C. (1978). Energy changes and muscular contraction. *Physiol. Rev.* **58**, 690–761.

Curtin, N. A. & Woledge, R. C. (1979). Chemical change and energy production during contraction of frog muscle: how are their time-courses related? *J. Physiol.* **288**, 353–66.

Curtin, N. A. & Woledge, R. C. (1981). Effect of muscle length on energy balance in frog skeletal muscle. *J. Physiol.* **316**, 453–68.

Curtis, D. R., Phillis, J. W. & Watkins, J. C. (1960). The chemical excitation of spinal neurones by certain acidic amino acids. *J. Physiol.* **150**, 656–82.

Curtis, H. J. & Cole, K. S. (1942). Membrane resting and action potentials in giant fibres of squid nerve. *J. Cell. Comp. Physiol.* **19**, 135–44.

Dale, H. H. (1914). The action of certain esters and ethers of choline, and their relation to muscarine. *J. Pharmacol. Exp. Ther.* **6**, 147–90.

Dale, H. H. & Dudley, H. W. (1929). The presence of histamine and acetylcholine in the spleen of the ox and the horse. *J. Physiol.* **68**, 97–123.

Dale, H. H., Feldburg, W. & Vogt, M. (1936). Release of acetylcholine at voluntary motor nerve endings. *J. Physiol.* **86**, 353–80.

Darian-Smith, I. (1984a). The sense of touch: performance and peripheral neural processes. In *Handbook of Physiology*, section 1 *The Nervous System*, vol. 3 *Sensory Processes*, ed. I. Darian-Smith, part 2, pp. 739–88. Bethesda, MD: American Physiological Society.

Darian-Smith, I. (ed.) (1984b). *Handbook of Physiology*, section 1 *The Nervous System*, vol. 3 *Sensory Processes*. Bethesda, MD: American Physiological Society.

Darnell, J. E., Lodish, H. & Baltimore, D. (1986). *Molecular Cell Biology.* New York: Scientific American Books.

Dartnall, H. J. A. (1952). Visual pigment 467, a photosensitive pigment present in tench retinae. *J. Physiol.* **116**, 259–89.

Dartnall, H. J. A. (1953). The interpretation of spectral sensitivity curves. *Brit. Med. Bull.* **9**, 24–30.

Dartnall, H. J. A. (1957). *The Visual Pigments.* London: Methuen.

Dartnall, H. J. A., Bowmaker, J. K. & Mollon, J. (1983). Human visual pigments: microspectrophotometric results from the eyes of seven persons. *Proc. R. Soc. Lond.* B **220**, 115–30.

Dash, P. K., Hochner, B. & Kandel, E. R. (1990). Injection of the cAMP-responsive element into the nucleus of *Aplysia* sensory neurons blocks long-term facilitation. *Nature* **345**, 718–21.

Davis, H. (1961). Some principles of sensory receptor action. *Physiol. Rev.* **41**, 391–416.

Davis, H. P. & Squire, L. R. (1984). Protein synthesis and memory: a review. *Psychol. Bull.* **96**, 518–59.

Davis, R. L. & Dauwalder, B. (1991). The *Drosophila dunce* locus – learning and memory genes in the fly. *Trends Genetics* **7**, 224–9.

Davson, H. (1990). *Physiology of the Eye*, fifth edition. London: Macmillan.

Davson, H. & Danielli, J. F. (1943). *The Permeability of Natural Membranes.* Cambridge: Cambridge University Press.

Dawson, A. P (1990). Regulation of intracellular Ca^{2+}. *Essays in Biochemistry* **25**, 1–37.

Dawson, T. M. & Snyder, S. H. (1994). Gases as biological messengers: nitric oxide and carbon monoxide in the brain. *J. Neurosci.* **14**, 5147–59.

Dawson, T. M., Arriza, J. L., Jaworsky, D. E., Borisy, F. F., Attramadal, H., Lefkowitz, R. J. & Ronnett, G. V. (1993). Beta-adrenergic receptor kinase–2 and beta-arrestin–2 as mediators of odorant-induced densensitization. *Science* **259**, 825–9.

Dawson, V. L., Dawson, T. M., London, E. D., Bredt, D. S. & Snyder, S. H. (1991). Nitric-oxide mediates glutamate neurotoxicity in primary cortical cultures. *Proc. Natl. Acad. Sci. USA* **88**, 6368–71.

Dawson, W. W. & Enoch, J. M. (eds.) (1984). *Foundations of Sensory Science.* Berlin: Springer-Verlag.

de Boer, E. (1996). Mechanics of the cochlea: modeling efforts. In *The Cochlea*, ed. P. Dallos, A. N. Popper & R. R. Fay, pp. 258–317. New York: Springer-Verlag.

de Robertis, E. & Bennett, H. S. (1954). Submicroscopic vesicular component in the synapse. *Federation Proc.* **13**, 38.

DeFelice, L. J. (1981). *Introduction to Membrane Noise.* New York: Plenum Press.

del Castillo, J. & Engbaek, L. (1954). The nature of the neuromuscular block produced by magnesium. *J. Physiol.* **124**, 370–84.

del Castillo, J. & Katz, B. (1954a). Quantal components of the end-plate potential. *J. Physiol.* **124**, 560–73.

del Castillo, J. & Katz, B. (1954b). Statistical factors involved in neuromuscular facilitation and depression. *J. Physiol.* **124**, 574–85.

del Castillo, J. & Katz, B. (1955). On the localization of acetylcholine receptors. *J. Physiol.* **128**, 157–81.

del Castillo, J. & Katz, B. (1956). Biophysical aspects of neuromuscular transmission. *Prog. Biophys.* **6**, 121–70.

del Castillo, J. & Katz, B. (1957). Interaction at end-plate receptors between different choline derivatives. *Proc. R. Soc. Lond.* B **146**, 369–81.

del Castillo, J. & Moore, J. W. (1959). On increasing the velocity of a nerve impulse. *J. Physiol.* **148**, 665–70.

Delbono, O. & Stefani, E. (1993). Calcium transients in single mammalian skeletal muscle fibres. *J. Physiol.* **463**, 689–707.

den Otter, C. J. (1977). Single sensillum responses in the male moth *Adoxophyes orana* (F. v. R.) to female sex pheromone components and their geometrical isomers. *J. Comp. Physiol.* **121**, 205–22.

Denk, W. & Svoboda, K. (1997). Photon upmanship: why multiphoton imaging is more than a gimmick. *Neuron* **18**, 351–7.

Denk, W., Holt, J. R., Shepherd, G. M. G. & Corey, D. P. (1995). Calcium imaging of single stereocilia in hair cells: localization of transduction channels at both ends of tip links. *Neuron* **15**, 1131–21.

Denk, W., Yuste, R., Svoboda, K. & Tank, D. W. (1996). Imaging calcium dynamics in dendritic spines. *Current Opinion Neurobiol.* **6**, 372–8.

Denton, E. J. (1959). The contributions of the oriented photosensitive and other molecules to the absorption of whole retina. *Proc. R. Soc. Lond.* B **150**, 78–94.

Derkach, V., Surprenant, A. & North, R. A. (1989). 5-HT$_3$ receptors are membrane ion channels. *Nature* **339**, 706–9.

Dermietzel, R. & Spray, D. C. (1993). Gap junctions in the brain: where, what type, how many and why? *Trends Neurosci.* **16**, 186–92.

DeRosier, D. J. & Klug, A. (1968). Reconstruction of three dimensional structures from electron micrographs. *Nature* **217**, 130–4.

DeSimone, J. A., Heck, G. L. & DeSimone, S. K. (1981). Active ion transport in dog tongue: a possible role in taste. *Science* **214**, 1039–41.

DeSimone, J. A., Callaham, E. M. & Heck, G. L. (1995). Chorda tympani taste response of rat to hydrochloric acid subject to voltage-clamped lingual receptive field. *Amer. J. Physiol.* **37**, C1295–C1300.

Deterre, P., Bigay, J., Forquet, F., Robert, M. & Chabre, M. (1988). cGMP phosphodiesterase of retinal rods is regulated by 2 inhibitory subunits. *Proc. Natl. Acad. Sci. USA* **85**, 2424–8.

Dethier, V. G. (1953). Summation and inhibition following contralateral stimulation of the tarsal chemoreceptors of the blowfly. *Biol. Bull.* **105**, 257–68.

Dethier, V. G. (1955). The physiology and histology of the contact chemoreceptors of the blowfly. *Quart. Rev. Biol.* **30**, 348–71.

Dethier, V. G. (1976). *The Hungry Fly.* Cambridge, MA: Harvard University Press.

Dethier, V. G. & Hanson, F. E. (1968). Electrophysiological responses of the chemoreceptors of the blowfly to sodium salts of fatty acids. *Proc. Natl. Acad. Sci. USA* **60**, 1296–1303.

Detwiler, P. B., Hodgkin, A. L. & McNaughton, P. A. (1980). Temporal and spatial characteristics of the voltage response of rods in the retina of the snapping turtle. *J. Physiol.* **300**, 213–50.

Devane, W. A., Hansus, L., Breuer, A., Pertwee, R. G., Stevenson, L. A., Griffin, G., Mandelbaum, A., Etinger, A. & Mechoulam, R. (1992). Isolation and structure of a brain constituent that binds to the cannabinoid receptor. *Science* **258**, 1946–9.

Devillers-Thiéry, A., Giraudat, J., Bentaboulet, M. & Changeux, J.-P. (1983). Complete mRNA coding sequence of the acetylcholine binding α subunit of *Torpedo marmorata* acetylcholine receptor: a model for the transmembrane organization of the peptide chain. *Proc. Natl. Acad. Sci. USA* **80**, 2067–71.

Devor, M. (1996). Pain mechanisms. *Neuroscientist* **2**, 233–44.

Dhallan, R. S., Yau, K.-Y., Schrader, K. & Reed, R. R. (1990). Primary structure and functional expression of a cyclic nucleotide-activated channel from olfactory neurons. *Nature* **347**, 184–7.

Diamond, J., Gray, J. A. B. & Inman, D. R. (1958). The relation between receptor potentials and the concentration of sodium ions. *J. Physiol.* **142**, 382–94.

Dickinson, C., Murray, I. & Carden, D. (eds.) (1997). *John Dalton's Colour Vision Legacy.* London: Taylor & Francis.

DiFrancesco, D. (1981a). A new interpretation of the pace-maker current in calf Purkinje fibres. *J. Physiol.* **314**, 359–76.

DiFrancesco, D. (1981b). A study of the ionic nature of the pace-maker current in calf Purkinje fibres. *J. Physiol.* **314**, 377–93.

DiFrancesco, D. (1993). Pacemaker mechanisms in cardiac tissue. *Ann. Rev. Physiol.* **55**, 455–72.

DiFrancesco, D. & Noble, D. (1985). A model of cardiac electrical activity incorporating ionic pumps and concentration changes. *Phil. Trans. R. Soc. Lond.* B **307**, 353–98.

DiFrancesco, D. & Tortora, P. (1991). Direct activation of cardiac pacemeker channels by intracellular cyclic AMP. *Nature* **351**, 145–7.

Dijkgraaf, S. (1934). Untersuchungen über die Funktion der Seitenorgane an Fischen. *Z. Vergl. Physiol.* **20**, 162–214.

Dijkgraaf, S. (1952). Bau und Funktionen der Seitenorgane und des Ohrlabyrinths bei Fischen. *Experientia* **8**, 205–16.

Dijkgraaf, S. (1963). The functioning and significance of the lateral line organs. *Biol. Rev.* **38**, 51–105.

Dijkgraaf, S. & Kalmijn, A. J. (1963). Untersuchungen über die Funktion der Lorenzinischen Ampullen an Haifischen. *Z. Vergl. Physiol.* **47**, 438–56.

Dillon, P. F., Aksoy, M. O., Driska, S. P. & Murphy, R. A. (1981). Myosin phosphorylation and the cross-bridge cycle in arterial smooth muscle. *Science* **211**, 495–7.

DiPolo, R. (1989). The sodium–calcium exchange in intact cells. In *Sodium–Calcium Exchange*, ed. T. J. A. Allen, D. Noble & H. Reuter, pp. 5–26. Oxford: Oxford University Press.

Dixon, R. A. F., Kobilka, B. K., Strader, D. J., Benovic, J. L., Dohlman, H. G., Frielle, T., Bolanowski, M. A., Bennett, C. D., Rands, E., Diehl, R. E., Mumford, A., Slater, E. E., Signal, I. S., Caron, M. G., Lefkowitz, R. J. & Strader, C. D. (1986). Cloning of the gene and cDNA for mammalian β-adrenergic receptor and homology with rhodopsin. *Nature* **321**, 75–9.

Dixon, R. A. F., Sigal, I. S., Rands, E., Register, R. B., Candelore, M. R., Blake, A. D. & Strader, C. D. (1987). Ligand binding to the β-adrenergic receptor involves its rhodopsin-like core. *Nature* **326**, 73–7.

Dizhoor, A. M., Lowe, D. G., Olshevskaya, E. V., Laura, R. P. & Hurley, J. B. (1994). The human photoreceptor membrane guanylyl cyclase, RetGC, is present in outer segments and is regulated by calcium and a soluble activator. *Neuron* **12**, 1345–52.

Dizhoor, A. M., Olshevskaya, E. V., Henzel, W. J., Wong, S. C., Stults, J. T., Ankoudinova, I. & Hurley, J. B. (1995). Cloning, sequencing, and expression of a 24-kDa Ca^{2+}-binding protein activating photoreceptor guanylyl cyclase. *J Biol. Chem.* **270**, 25200–6.

Dobelle, W. H., Marks, W. B. & MacNichol, E. F. (1969). Visual pigment density in single primate foveal cones. *Science* **166**, 1508–10.

Dodd, J. & Horn, J. P. (1983). Muscarinic inhibition of sympathetic C neurones in the bullfrog. *J. Physiol.* **334**, 271–91.

Dodge, F. A. & Frankenhaeuser, B. (1958). Membrane currents in isolated frog nerve fibre under voltage clamp conditions. *J. Physiol.* **143**, 76–90.

Dodge, F. A. & Rahamimoff, R. (1967). On the relationship between calcium concentration and the amplitude of the end-plate potential. *J. Physiol.* **189**, 90P–92P.

Dohlman, H. G., Bouvier, M., Benovic, J. L., Caron, M. G. &

Lefkowitz, R. J. (1987). The multiple membrane spanning topography of the β_2-adrenergic receptor. Localization of the sites of binding, glycosylation and regulatory phosphorylation by limited proteolysis. *J. Biol. Chem.* **262**, 14282–8

Dohlman, H. G., Thorner, J., Caron, M. G. & Lefkowitz, R. J. (1991). Model systems for the study of seven-transmembrane-segment receptors. *Ann. Rev. Biochem.* **60**, 653–88.

Douglass, J., Civelli, O. & Herbert, E. (1984). Polyprotein gene expression: generation of diversity of neuroendocrine peptides. *Ann. Rev. Biochem.* **53**, 665–715.

Doupnik, C. A., Davidson, N. & Lester, H. A. (1995). The inward rectifier potassium channel family. *Curr. Opin. Neurobiol.* **5**, 268–77.

Dowdall, M. J., Boyne, A. F. & Whittaker, V. P. (1974). Adenosine triphosphate: a constitutent of cholinergic synaptic vesicles. *Biochem. J.* **140**, 1–12.

Dowling, J. E. (1960). Chemistry of visual adaptation in the rat. *Nature* **188**, 114–18.

Dowling, J. E. (1970). Organization of vertebrate retinas. *Invest. Ophthalmol.* **9**, 655–80.

Dowling, J. E. (1987). *The Retina, an Approachable part of the Brain.* Cambridge, MA: Harvard University Press.

Dowling, J. E. & Dubin, M. W. (1984). The vertebrate retina. In *Handbook of Physiology*, section 1, vol. 3 *Sensory Processes*, ed. I. Darian-Smith, pp. 317–39. Bethesda, MD: American Physiological Society.

Draper, M. H. & Weidmann, S. (1951). Cardiac resting and action potentials recorded with an intracellular electrode. *J. Physiol.* **115**, 74–94.

Driscoll, M. & Chalfie, M. (1991). The *mec–4* gene is a member of a family of *Caenorhabditis elegans* genes that can mutate to induce neuronal degeneration. *Nature* **349**, 588–93.

du Bois Reymond, E. (1877). *Gesammelte Abhandlungen zur allgemeinen Muskel und Nervenphysik.* Leipzig.

du Vigneaud, V. (1956). Hormones of the posterior pituitary gland: oxytocin and vasopressin. *Harvey Lect.* **50**, 1–26.

Du, H., Gu, G., William, C. M. & Chalfie, M. (1996). Extracellular proteins needed for *C. elegans* mechanosensation. *Neuron* **16**, 183–94.

Dudai, Y. (1988). Genetic dissection of learning and short-term memory in *Drosophila*. *Ann. Rev. Neurosci.* **11**, 537–63.

Dudai, Y., Jan, Y.-N., Byers, D., Quinn, W. G. & Benzer, S. (1976). *dunce*, a mutant of *Drosophila* deficient in learning. *Proc. Natl. Acad. Sci. USA* **73**, 1684–8.

Dudel, J. & Kuffler, S. W. (1961). Presynaptic inhibition at the crayfish neuromuscular junction. *J. Physiol.* **155**, 543–62.

Dulac, C. & Axel, R. (1995). A novel family of genes encoding putative pheromone receptors in mammals. *Cell* **83**, 195–206.

Duncan, G. (1990). *Physics in the Life Sciences*, second edition. Oxford: Blackwell Scientific Publications.

Duncan, G., Williams, M. R. & Riach, R. A. (1994). Calcium,

cell signalling and cataract. *Prog. Retinal Eye Res.* **13**, 623–52.

Dunlap, K. & Fischbach, G. D. (1981). Neurotransmitters decrease the calcium conductance activated by depolarization of embryonic chick sensory neurones. *J. Physiol.* **317**, 519–35.

Dura, J. M., Taillebourg, E. & Preat, T. (1995). The *Drosophila* learning and memory gene *linotte* encodes a putative receptor tyrosine kinase homologous to the human *RYK* gene product. *FEBS Letters* **370**, 250–4.

Dusenbery, D. B. (1992). *Sensory Ecology.* New York: Freeman.

Dwyer, T. M., Adams, D. J. & Hille, B. (1980). The permeability of the endplate channel to organic cations in frog muscle. *J. Gen. Physiol.* **75**, 469–92.

Earm, Y. E. & Noble, D. (1990). A model of the single atrial cell: between calcium current and calcium release. *Proc. R. Soc. Lond.* B **240**, 83–96.

Eastwood, A. B., Wood, D. S., Bock, K. L. & Sorenson, M. M. (1979). Chemically skinned mammalian skeletal muscle. 1. The structure of skinned rabbit psoas. *Tiss. Cell* **11**, 553–66.

Eaton, R. C., Bombardieri, R. A. & Meyer, D. (1977). The Mauthner-initiated startle response in teleost fish. *J. Exp. Biol.* **66**, 65–81.

Ebashi, S. & Kodama, A. (1966*a*). A new factor promoting aggregation of tropomyosin. *J. Biochem.* **58**, 107–8.

Ebashi, S. & Kodama, A. (1966*b*). Native tropomyosin-like action of troponin on trypsin-treated myosin B. *J. Biochem.* **60**, 733–4.

Ebashi, S., Kodama, A. & Ebashi, F. (1968). Troponin. 1. Preparation and physiological function. *J. Biochem.* **64**, 465–77.

Ebashi, S., Endo, M. & Ohtsuki, I. (1969). Control of muscle contraction. *Quart. Rev. Biophys.* **2**, 351–84.

Eccles, J. C. (1957). *The Physiology of Nerve Cells.* London: Oxford University Press.

Eccles, J. C. (1964). *The Physiology of Synapses.* Berlin: Springer-Verlag.

Eccles, J. C. (1976). From electrical to chemical transmission in the central nervous system. *Notes Records R. Soc. Lond.* **30**, 219–30.

Eccles, J. C., Eccles, R. M. & Magni, F. (1961). Central inhibitory action attributable to presynaptic depolarization produced by muscle afferent volleys. *J. Physiol.* **159**, 147–66.

Eccles, J. C., Magni, F. & Willis, W. D. (1962). Depolarization of central terminals of group I afferent fibres from muscle. *J. Physiol.* **160**, 62–93.

Edman, Å., Gestrelius, S. & Grampp, W. (1987). Analysis of gated membrane currents and mechanisms of firing control in the rapidly adapting lobster stretch receptor neurone. *J. Physiol.* **384**, 649–69.

Edman, K. A. P. & Schild, H. O. (1962). The need for calcium in the contractile responses induced by acetylcholine and potassium in the rat uterus. *J. Physiol.* **161**, 424–41.

Edmonds, B., Gibb, A. J. & Colquhoun, D. (1995). Mechanisms of activation of glutamate receptors and the time course of excitatory synaptic currents. *Ann. Rev. Physiol.* **57**, 495–519.

Edwards, C., Ottoson, D., Rydqvist, B. & Swerup, C. (1981). The permeability of the transducer membrane of the crayfish stretch receptor to calcium and to other divalent cations. *Neuroscience* **6**, 1455–60.

Edwards, C., Dolezal, V., Tucek, S., Zemkova, H. & Vyskocil, F. (1985). Is an acetylcholine transport system responsible for nonquantal release of acetylcholine at the rodent myoneural junction? *Proc. Natl. Acad. Sci. USA* **82**, 3514–18.

Edwards, F. A., Konnerth, A., Sakmann, B. & Busch, C. (1990). Quantal analysis of inhibitory synaptic transmission in the dentate gyrus of rat hippocampal slices: a patch-clamp study. *J. Physiol.* **430**, 213–49.

Edwards, F. A., Gibb, A. J. & Colquhoun, D. (1992). ATP receptor-mediated synaptic currents in the central nervous system. *Nature* **359**, 144–7.

Egelman, E. H., Francis, N. & DeRosier, D. J. (1982). F-actin is a helix with a random variable twist. *Nature* **298**, 131–5.

Eggleton, P. & Eggleton, G. P. (1927). The inorganic phosphate and a labile form of organic phosphate in the gastrocnemius of the frog. *Biochem. J.* **21**, 1–5.

Ehrenstein, G. & Gilbert, D. L. (1966). Slow changes of potassium permeability in the squid giant axon. *Biophys. J.* **6**, 553–66.

Einthoven, W. (1924). The string galvanometer and measurement of the action currents of the heart. Republished in 1965 in *Nobel lectures, Physiology or Medicine 1921–41.* Amsterdam: Elsevier.

Eisenberg, E. & Hill, T. L. (1985). Muscle contraction and free energy transduction in biological systems. *Science* **227**, 999–1006.

Elgoyhen, A. B., Johnson, D. S., Boulter, J., Vetter, D. E. & Heinemann, S. (1994). α9: an acetylcholine receptor with novel pharmacological properties expressed in rat cochlear hair cells. *Cell*, **79**, 705–15.

Elliott, A. & Bennett, P. M. (1984). Molecular organization of paramyosin in the core of molluscan thick filaments. *J. Molec. Biol.* **176**, 477–93.

Elliott, A. & Offer, G. (1978). Shape and flexibility of the myosin molecule. *J. Molec. Biol.* **123**, 505–19.

Elliott, G. F., Lowy, J. & Millman, B. M. (1965). X-ray diffraction from living striated muscle during contraction. *Nature* **206**, 1357–8.

Elliott, T. R. (1904). On the action of adrenaline. *J. Physiol.* **31**, 20P.

Ellisman, M. H. & Levinson, S. R. (1982). Inununocytochemical localization of sodium channel distributions in the excitable membranes of *Electrophorus electricus. Proc. Natl. Acad. Sci. USA* **79**, 6707–11.

Elzinga, M., Collins, J. H., Kuehl, W. M. & Adelstein, R. S. (1973). Complete aminoacid sequence of actin of rabbit skeletal muscle. *Proc. Natl. Acad. Sci. USA.* **70**, 2687–91.

Emonet-Dénand, F., Hunt, C. C. & Laporte, Y. (1985). Effects of stretch on dynamic fusimotor after-effects in cat muscle spindles. *J. Physiol.* **360**, 201–13.

Engberg, I., Flatman, J. A. & Lambert, J. D. C. (1979). The actions of excitatory amino acids on motoneurones in the feline spinal cord. *J. Physiol.* **288**, 227–61.

Engel, W. K. (1962). The essentiality of histo- and cytochemical studies of skeletal muscle in the investigation of neuromuscular disease. *Neurology* **12**, 778–84.

Engelhardt, W. A. & Ljubimowa, M. N. (1939). Myosine and adenosinetriphosphatase. *Nature* **144**, 668–9.

Engström, H. & Wersäll, J. (1958). The ultrastructural organization of the organ of Corti and the vestibular sensory epithelia. *Exp. Cell Res.* suppl. **5**, 460–92.

Erickson, J. D., Eiden, L. E. & Hoffman, B. J. (1992). Expression cloning of a reserpine-sensitive vesicular monoamine transporter. *Proc. Natl. Acad. Sci. USA* **89**, 10993–7.

Erickson, J. D., Varoqui, H., Schafer, M. K., Midi, W., Diebler, M. F., Weihe, E., Rand, J., Bonner, T. I. & Usdin, T. B. (1994). Functional identification of a vesicular acetylcholine transporter and its expression from a "cholinergic" gene. *J. Biol. Chem.* **269**, 21929–32.

Erickson, R. P. (1963). Sensory neural patterns and gustation. In *Olfaction and Taste I*, ed. Y. Zotterman, pp. 205–13. New York: Pergamon Press.

Erlanger, J. & Gasser, H. S. (1937). *Electrical Signs of Nervous Activity*. Philadelphia: University of Pennsylvania Press.

Ervasti, J. M. & Campbell, K. P. (1991). Membrane organization of the dystrophin–glycoprotein complex. *Cell* **66**, 1121–31.

Erxleben, C. (1989). Stretch-activated current through single channels in the abdominal stretch receptor organ of the crayfish. *J. Gen. Physiol.* **94**, 1071–83.

Evans, D. R. & Mellon, de F. (1962). Electrophysiological studies of a water receptor associated with the taste sensilla of the blowfly. *J. Gen. Physiol.* **45**, 487–500.

Evans, E. F. (1972). The frequency response and other properties of single fibres in the guinea-pig cochlear nerve. *J. Physiol.* **226**, 263–87.

Evans, E. F. (1974). The effects of hypoxia on the tuning of single cochlear nerve fibres. *J. Physiol.* **338**, 65P–67P.

Evans, E. F. & Wilson, J. P. (1975). Cochlear tuning properties: concurrent basilar membrane and single nerve fiber measurements. *Science* **190**, 1218–21.

Evans, P. D. (1985). Octopamine. In *Comprehensive Insect Physiology, Biochemistry and Pharmacology*, ed. G. A. Kerkut & L. I. Gilbert, vol. 11, pp. 499–530. Oxford: Pergamon Press.

Exton, J. H. (1985). Mechanisms involved in (-adrenergic phenomena. *Amer. J. Physiol.* **248**, E633–E647.

Exton, J. H. (1990). Signaling through phosphatidylcholine breakdown. *J. Biol. Chem.* **265**, 1–4.

Eyzaguirre, C. & Kuffler, S. W. (1955a). Processes of excitation in the dendrites and in the soma of single isolated sensory nerve cells of the lobster and crayfish. *J. Gen. Physiol.* **39**, 87–119.

Eyzaguirre, C. & Kuffler, S. W. (1955b). Synaptic inhibition in an isolated nerve cell. *J. Gen. Physiol.* **39**, 155–84.

Faber, D. S. & Korn, H. (1978). Electrophysiology of the Mauthner cell: basic properties, synaptic mechanisms, and associated networks. In *Neurobiology of the Mauthner Cell*, ed. D. Faber & H. Korn, pp. 47–131. New York: Raven Press.

Faber, D. S., Young, W. S., Legendre, P. & Korn, H. (1992). Intrinsic quantal variability due to stochastic properties of receptor-transmitter interactions. *Science* **258**, 1494–8.

Fabiato, A. (1983). Calcium-induced release of calcium from the cardiac sarcoplasmic reticulum. *Amer. J. Physiol.* **245**, C1–C14.

Fabiato, A. (1985). Time and calcium dependence of activation and inactivation of calcium-induced release of calcium from the sarcoplasmic reticulum of a skinned canine cardiac Purkinje cell. *J. Gen. Physiol.* **85**, 247–89.

Fabiato, A. & Fabiato, F. (1975). Dependence of the contractile activation of skinned cardiac cells on the sarcomere length. *Nature* **256**, 54–6.

Fain, G. L. (1975). Quantum sensitivity of rods in the toad retina. *Science* **187**, 838–41.

Fain, G. L., Gold, G. H. & Dowling, J. E. (1976). Receptor coupling in the toad retina. *Cold Spr. Harb. Symp. Quant. Biol.* **40**, 547–61.

Fain, G. L., Matthews, H. R. & Cornwall, M. C. (1996). Dark adaptation in vertebrate photoreceptors. *Trends Neurosci.* **19**, 502–7.

Fakler, B., Brändle, U., Glowatzki, S., Weidemann, S., Zenner, H.-P. & Ruppersberg, J. P. (1995). Strong voltage-dependent inward rectification of inward rectifier K$^+$ channels is caused by intracellular spermine. *Cell* **80**, 149–54.

Fatt, P. & Katz, B. (1951). An analysis of the end-plate potential recorded with an intracellular electrode. *J. Physiol.* **115**, 320–69.

Fatt, P. & Katz, B. (1952). Spontaneous subthreshold activity at motor nerve endings. *J. Physiol.* **117**, 109–28.

Fawcett, D. W. & McNutt, N. S. (1969). The ultrastructure of the cat myocardium. I. Ventricular papillary muscle. *J. Cell Biol.* **42**, 1–45.

Feany, M. B. & Quinn, W. G. (1995). A neuropeptide gene defined by the *Drosophila* memory mutant *amnesiac*. *Science* **268**, 869–73.

Fechner, G. T. (1862). *Elemente der Pscyhophysik*. Leipzig: Breitkopf und Härtel.

Feldburg, W. & Gaddum, J. H. (1934). The chemical transmitter at synapses in a sympathetic ganglion. *J. Physiol.* **81**, 305–19.

Fenn, W. O. (1923). A quantitative comparison between the energy liberated and the work performed by the isolated sartorius of the frog. *J. Physiol.* **58**, 175–203.

Fenn, W. O. (1924). The relation between the work performed and the energy liberated in muscular contraction. *J. Physiol.* **58**, 373–95.

Fertuck, H. C. & Salpeter, M. M. (1974). Localization of

acetylcholine receptor by ^{125}I-labelled α-bungarotoxin binding at mouse motor end-plates. *Proc. Natl. Acad. Sci. USA* **71**, 1376–8.

Fertuck, H. C. & Salpeter, M. M. (1976). Quantitation of junctional and extrajunctional acetylcholine receptors by electron microscope autoradiography after ^{125}I-α-bungarotoxin binding at mouse neuromuscular junctions. *J. Cell Biol.* **69**, 144–58.

Fesenko, E. E., Kolesnikov, S. S. & Lyubarsky, A. (1985). Induction by cyclic GMP of cationic conductance in plasma membrane of retinal rod outer segment. *Nature* **313**, 310–3.

Fessard, A. & Szabo, T. (1961). Mise en évidence d'un récepteur sensible à l'électricité dans la peau des Mormyres. *Comptes Rendus Acad. Sci.* **253**, 1859–60.

Fillenz, M. (1990). *Noradrenergic neurons*. Cambridge: Cambridge University Press.

Finer, J. T., Simmons, R. M. & Spudich, J. A. (1994). Single myosin molecule mechanics: piconewton forces and nanometre steps. *Nature* **368**, 113–19.

Firestein, S., Picco, C. & Menini, A. (1993). The relation between stimulus and response in olfactory receptor cells of the tiger salamander. *J. Physiol.* **468**, 1–10.

Fisher, A. J., Smith, C. A., Thoden, J., Smith, R., Sutoh, K., Holden, H. M., Rayment, I. (1995*a*) Structural studies of myosin-nucleotide complexes – a revised model for the molecular basis of muscle contraction. *Biophys. J.* **68**, S19-S28.

Fisher, A. J., Smith, C. A., Thoden, J., Smith, R., Sutoh, K., Holden, H. M., Rayment, I. (1995*b*). X-ray structures of the myosin motor domain of *Dictyostelium discoideum* complexed with MgADP·BeF$_x$ and MgADP·AlF$_4^-$. *Biochemistry* **34**, 8960–72.

Fiske, C. H. & Subbarow, Y. (1927). The nature of the 'inorganic phosphate' in voluntary muscle. *Science* **65**, 401–3.

Fletcher, P. & Forrester, T. (1975). The effect of curare on the release of acetylcholine from mammalian motor nerve terminals and an estimate of quantum content. *J. Physiol.* **251**, 131–44.

Fletcher, W. M. & Hopkins, F. G. (1907). Lactic acid in amphibian muscle. *J. Physiol.* **35**, 247–309.

Flexner, J. B., Flexner, L. B. & Stellar, E. (1963). Memory in mice as affected by intracerebral puromycin. *Science* **141**, 57–9.

Flock, Å. (1965). Transducing mechanisms in the lateral line canal organ receptors. *Cold Spr. Harb. Symp. Quant. Biol.* **30**, 133–44.

Flock, Å. (1971). Sensory transduction in hair cells. In *Handbook of Sensory Physiology*, vol. 1 *Principles of Receptor Physiology*, ed. W. R. Loewenstein, pp. 396–441. Berlin: Springer-Verlag.

Flock, Å. (1977). Physiological properties of sensory hairs in the ear. In *Psychophysics and Physiology of Hearing*, ed. E. F. Evans & J. P. Wilson, pp. 15–25. London: Academic Press.

Flock, Å. & Cheung, H. C. (1977). Actin filaments in sensory hairs of inner ear receptor cells. *J. Cell Biol.* **75**, 339–43.

Fong, S.-L., Tsin, A. T. C., Bridges, C. D. B & Liou, G. I. (1982). Detergents for extraction of visual pigments: types, solubilization, and stability. *Methods Enzymol.* **81**, 133–40.

Ford, L. E., Huxley, A. F. & Simmons, R. M. (1977). Tension responses to sudden length change in stimulated frog muscle fibres near slack length. *J. Physiol.* **269**, 441–515.

Ford, L. E., Huxley, A. F. & Simmons, R. M. (1981). The relation between stiffness and filament overlap in stimulated frog muscle fibres. *J. Physiol.* **311**, 219–49.

Ford, L. E., Huxley, A. F. & Simmons, R. M. (1985). Tension transients during steady shortening of frog muscle fibres. *J. Physiol.* **361**, 131–50.

Ford, L. E., Huxley, A. F. & Simmons, R. M. (1986). Tension transients during the rise of tetanic tension in frog muscle fibres. *J. Physiol.* **372**, 595–609.

Formaker, B. L. & Hill, D. L. (1988). An analysis of residual NaCl response after amiloride. *Amer. J. Physiol.* **255**, 1002–7.

Forsythe, I. D. & Westbrook, G. L. (1988). Slow excitatory postsynaptic currents mediated by *N*-methyl-D-aspartate receptors on cultured mouse central neurones. *J. Physiol.* **396**, 515–33.

Frank, D. A. & Greenberg, M. E. (1994). CREB: a mediator of long-term memory from mollusks to mammals. *Cell* **79**, 5–8.

Frank, G. B. (1960). Effects of changes in extracellular calcium concentration on the potassium-induced contracture of frog's skeletal muscle. *J. Physiol.* **151**, 518–38.

Frank, K. & Fuortes, M. G. F. (1957). Presynaptic and postsynaptic inhibition of monosynaptic reflexes. *Fedn Proc.* **16**, 39–40.

Frank, M. (1973). An analysis of hamster afferent taste nerve response functions. *J. Gen. Physiol.* **61**, 588–618.

Frank, O. (1895). Zur Dynamik des Herzmuskels. *Z. Biol.* **32**, 370–447.

Frankenhaeuser, B. (1965). Computed action potential in nerve from *Xenopus laevis*. *J. Physiol.* **180**, 780–7.

Frankenhaeuser, B. & Hodgkin, A. L. (1957). The action of calcium on the electrical properties of squid axons. *J. Physiol.* **137**, 218–44.

Franks, N. P. & Lieb, W. R. (1994). Molecular and cellular mechanisms of general anaesthesia. *Nature* **367**, 607–14.

Franzini-Armstrong, C. (1970). Studies of the triad. I. Structure of the junction in frog twitch fibers. *J. Cell Biol.* **47**, 488–99.

Franzini-Armstrong, C. (1994). The sarcoplasmic reticulum and transverse tubules. In *Myology: Basic and Clinical*, second edition, ed. A. G. Engel & C. Franzini-Armstrong, vol. 1, pp. 176–99. New York: McGraw-Hill.

Franzini-Armstrong, C. & Jorgensen, A. O. (1994). Structure and development of e-c coupling units in skeletal muscle. *Ann. Rev. Physiol.* **56**, 509–34.

Franzini-Armstrong, C. & Porter, K. R. (1964). Sarcolemmal invaginations constituting the T system in fish muscle fibers. *J. Cell Biol.* **22**, 675–96.

Franzini-Armstrong, C., Kenney, L. J. & Varriano-Marston, E.

(1987). The structure of calsequestrin in triads of vertebrate skeletal muscle: a deep etch study. *J. Cell Biol.* **105**, 49–56.

Freiburg, A. & Gautel, M. (1996). A molecular map of the interactions between titin and myosin-binding protein-C: implications for sarcomeric assembly in familial hypertrophic cardiomyopathy. *Eur. J. Biochem.* **235**, 317–23.

French, A. S. (1984). Action potential adaptation in the femoral tactile spine of the cockroach, *Periplaneta americana. J. Comp. Physiol.* A **155**, 803–12.

Frey, U., Krug, M., Reymann, K. G. & Matthies, H. (1988). Anisomycin, an inhibitor of protein-synthesis, blocks late phases of LTP phenomena in the hippocampal CA1 region *in vitro. Brain Res.* **452**, 57–65.

Frost, W. N., Castellucci, V. F., Hawkins, R. D. & Kandel, E. R. (1985). Monosynaptic connections made by the sensory neurons of the gill- and siphon-withdrawal reflex in *Aplysia* participate in the storage of long-term memory for sensitization. *Proc. Natl. Acad. Sci. USA* **82**, 8266–9.

Fuchs, P. A. (1996). Synaptic transmission at vertebrate hair cells. *Curr. Opin. Neurobiol.* **6**, 514–19.

Fung, B. K.-K. & Stryer, L. (1980). Photolyzed rhodopsin catalyzes the exchange of GTP for bound GDP in rod outer segments. *Proc. Natl. Acad. Sci. USA* **77**, 2500–4.

Fung, B. K.-K., Hurley, J. B. & Stryer, L. (1981). Flow of information in the light-triggered cyclic nucleotide cascade of vision. *Proc. Natl. Acad. Sci. USA* **78**, 152–6.

Fuortes, M. G. F. & Hodgkin, A. L. (1964). Changes in time scale and sensitivity in the ommatidia of *Limulus. J. Physiol.* **172**, 239–63.

Furchgott, R. F. & Zawadzki, J. V. (1980). The obligatory role of endothelial cells in the relaxation of arterial smooth muscle by acetylcholine. *Nature* **288**, 373–6.

Furshpan, E. J. & Potter, D. D. (1959). Transmission at the giant synapses of the crayfish. *J. Physiol.* **145**, 289–325.

Fürst, D. O. & Gautel, M. (1995). The anatomy of a molecular giant: how the sarcomere cytoskeleton is assembled from immunoglobulin superfamily molecules. *J. Molec. Cell. Cardiol.* **27**, 951–9.

Fürst, D. O., Osborn, M., Nave, R. & Weber, K. (1988). The organization of titin filaments in the half-sarcomere revealed by monoclonal antibodies in immunoelectron microscopy: a map of ten non-repetitive epitopes starting at the Z-line extends close to the M-line. *J. Cell Biol.* **120**, 711–24.

Furuichi, T., Kohda, K., Miyawaki, A. & Mikoshiba, K. (1994). Intracellular channels. *Curr. Opin. Neurobiol.* **4**, 294–303.

Furukawa, T. & Furshpan, E. J. (1963). Two inhibitory mechanisms in the Mauthner neurons of goldfish. *J. Neurophysiol.* **26**, 140–76.

Furukawa, T. & Hanawa, I. (1955). Effects of some common cations on electroretinogram of the toad. *Jap. J. Physiol.* **5**, 289–300.

Furukawa, T. & Matsuura, S. (1978). Adaptive rundown of excitatory post-synaptic potentials at synapses between hair cells and eighth nerve fibres in the goldfish. *J. Physiol.* **276**, 193–209.

Furukawa, T., Hayashida, Y. & Matsuura, S. (1978). Quantal analysis of the excitatory post-synaptic potentials at synapses between hair cells and afferent nerve fibres in goldfish. *J. Physiol.* **276**, 211–26.

Gabella, G. (1981). Structure of smooth muscles. In *Smooth Muscle*, ed. E. Bülbring, A. F. Brading, A. W. Jones & T. Tomita, pp. 1–46. London: Edward Arnold.

Gabella, G. (1984). Structural apparatus for force transmission in smooth muscles. *Physiol. Rev.* **64**, 455–77.

Gabella, G. (1994). Structure of smooth muscles. In *Pharmacology of Smooth Muscle*, ed. L. Szerekes & J. G. Papp. *Handb. Exp. Pharmacol.* **111**, 1–34. Berlin: Springer-Verlag.

Galzi, J-L., Devillers-Thiéry, A., Hussy, N., Bertrand, S., Changeux, J.-P. & Bertrand, D. (1992). Mutations in the channel domain of a neuronal nicotinic receptor convert ion selectivity from cationic to anionic. *Nature* **359**, 500–5.

Garcia, J. & Schneider, M. F. (1993). Calcium transients and calcium release in rat fast-twitch skeletal muscle fibres. *J. Physiol.* **463**, 709–28.

Garrahan, P. J. & Glynn, I. M. (1967). The stoichiometry of the sodium pump. *J. Physiol.* **192**, 217–35.

Garthwaite, J. (1991). Glutamate, nitric oxide and cell-cell signalling in the nervous system. *Trends Neurosci.* **14**, 60–7.

Garthwaite, J. (1995). Neural nitric oxide signalling. *Trends Neurosci.* **18**, 51–2.

Garthwaite, J. & Boulton, C. L. (1995). Nitric oxide signaling in the central nervous system. *Ann. Rev. Physiol.* **57**, 683–706.

Garthwaite, J., Charles, S. L. & Chess-Williams, R. (1988). Endothelium-derived relaxing factor release on activation of NMDA receptors suggests role as intercellular messenger in the brain. *Nature* **336**, 385–8.

Gasser, H. S. & Hill, A. V. (1924). The dynamics of muscular contraction. *Proc. R. Soc. Lond.* B **96**, 398–437.

Gautel, M. & Goulding, D. (1996). A molecular map of titin/connectin elasticity reveals two different mechanisms acting in series. *FEBS Lett.* **385**, 11–14.

Geeves, M. A. (1991). The dynamics of actin and myosin association and the cross-bridge model of muscular contraction. *Biochem. J.* **274**, 1–14.

Geinisman, Y., de Toledo-Morrell, L., Morrell, F., Heller, R. E., Rossi, M. & Parshall, R. F. (1993). Structural synaptic correlate of long-term potentiation: formation of axospinous synapses with multiple, completely partitioned transmission zones. *Hippocampus* **3**, 435–45.

Geisler, C. D. & Sang, C. N. (1995). A cochlear model using feed-forward outer hair cell forces. *Hearing Res.* **86**, 132–46.

Gerald, C., Adham, N., Kao, H. T., Olsen, M. A., Laz, T. M., Schechter, L. E., Bard, J. A., Vaysse, P. J., Hartig, P. R., Branchek, T. A. & Weinshank, R. L. (1995). The 5-HT$_4$ receptor: molecular cloning and pharmacological characterization of two splice variants. *EMBO J.* **14**, 2806–15.

Geren, B. B. (1954). The formation from the Schwann cell surface of myelin in the peripheral nerves of chick embryos. *Exp. Cell Res.* **7**, 558–62.

Gilbert, C., Kretscmar, K. M., Wilkie, D. R. & Woledge, R. C. (1971). Chemical change and energy output during muscular contraction. *J. Physiol.* **218**, 163–93.

Gilbert, D. L. & Ehrenstein, G. (1984). Membrane surface charge. In *The Squid Axon*, ed. P. F. Baker. *Current Topics in Membranes and Transport*, vol. 22. pp. 407–43. Orlando, FL: Academic Press.

Giles, W. & Noble, S. J. (1976). Changes in membrane currents in bullfrog atrium produced by acetylcholine. *J. Physiol.* **261**, 103–23.

Gillespie, P. G., Wagner, M. C. & Hudspeth, A J. (1993). Identification of a 120-kD hair-bundle myosin I located near stereociliary tips. *Neuron* **11**, 581–94.

Gilly, W. F. & Armstrong, C. M. (1980). Gating current and potassium channels in the giant axon of the squid. *Biophys. J.* **29**, 485–92.

Gilly, W. F. & Hui, C. S. (1980). Mechanical activation in slow and twitch skeletal muscle fibres of the frog. *J. Physiol.* **301**, 137–56.

Gilman, A. G. (1987). G proteins: transducers of receptor-generated signals. *Ann. Rev. Biochem.* **56**, 615–49.

Gilman, A. G. (1995). G proteins and regulation of adenylate cyclase (Nobel lecture). *Angew. Chem. Int. Ed. Eng.* **34**, 1406–19.

Gilmour, K. M. & Ellington, C. P. (1993). In vivo muscle length changes in bumblebees and the in vitro effects on work and power. *J. Exp. Biol.* **183**, 101–13.

Gimona, M. & Small, J. V. (1996). Calponin. In *Biochemistry of Smooth Muscle Contraction*, ed. M. Bárány, pp. 91–103. San Diego, CA: Academic Press.

Gingrich, J. A. & Caron, M. G. (1993). Recent advances in the molecular biology of dopamine receptors. *Ann. Rev. Neurosci.* **16**, 299–321.

Ginsborg, B. L. (1960). Some properties of avian skeletal muscle fibres with multiple neuromuscular junctions. *J. Physiol.* **154**, 581–98.

Gintzler, A. R. & Hyde, D. (1983). A specific substance P antagonist attenuates noncholinergic electrically induced contractures of the guinea-pig isolated ileum. *Neurosci. Lett.* **40**, 75–9.

Glanzman, D. L., Kandel, E. R. & Schacher, S. (1990). Target-dependent structural changes accompanying long-term synaptic facilitation in *Aplysia* neurons. *Science* **249**, 799–802.

Glynn, I. M. (1984). The electrogenic sodium pump. In *Electrogenic Transport*, ed. M. P. Blaustein & M. Lieberman, pp. 33–48. New York: Raven Press.

Glynn, I. M. (1993). All hands to the sodium pump. *J. Physiol.* **462**, 1–30.

Go, V. L. W. & Yaksh, T. L. (1987). Release of substance P from the cat spinal cord. *J. Physiol.* **391**, 141–67.

Godchaux, W. & Zimmerman, W. F. (1979). Membrane-dependent guanine nucleotide binding and GTPase activities of soluble proteins from bovine rod outer segments. *J. Biol. Chem.* **254**, 7874–84.

Gold, G. H. & Korenbrot, J. I. (1980). Light-induced calcium release by intact retinal rods. *Proc. Natl. Acad. Sci. USA.* **77**, 5557–61.

Goldman, D. E. (1943). Potential, impedance, and rectification in membranes. *J. Gen. Physiol.* **27**, 37–60.

Goldman, Y. E. (1987). Kinetics of the actomyosin ATPase in muscle fibres. *Ann. Rev. Physiol.* **49**, 637–54.

Goldman, Y. E. & Huxley, A F. (1994). Actin compliance: are you pulling my chain? *Biophys. J.* **67**, 2131–3.

Goldman, Y. E., Hibberd, M. G. & Trentham, D. R. (1984*a*). Relaxation of rabbit psoas muscle fibres from rigor by photochemical generation of adenosine–5′-triphosphate. *J. Physiol.* **354**, 577–604.

Goldman, Y. E., Hibberd, M. G. & Trentham, D. R. (1984*b*). Initiation of active contraction by photogeneration of adenosine–5′-triphosphate in rabbit psoas muscle fibres. *J. Physiol.* **354**, 605–24.

Goldstein, L. S. B. (1993). With apologies to Scheherazade: tales of 1001 kinesin motors. *Ann. Rev. Genetics* **27**, 319–51.

Goldstein, M. A., Michael, L. H., Schroeter, J. P. & Sass, R. L. (1986). The Z-band lattice in skeletal muscle before, during and after tetanic contraction. *J. Mus. Res. Cell Motil.* **7**, 527–36.

Gonzalez-Serratos, H. (1971). Inward spread of activation in vertebrate muscle fibres. *J. Physiol.* **212**, 777–99.

Goodson, H. V. (1994). Molecular evolution of the myosin superfamily: application of phylogenetic techniques to cell biological questions. In *Molecular Evolution of Physiological Processes*, ed. D. M. Fambrough. *Society of General Physiologists Series* **49**, 141–57. New York: Rockefeller University Press.

Göpfert, H. & Schaefer, H. (1938). Über den direkt und indirekt erregeten Aktionsstrom und die Funktion der motorischen Endplatte. *Pflügers Archiv* **239**, 597–619.

Gordon, A. M., Huxley, A. F. & Julian, F. J. (1966*a*). Tension development in highly stretched vertebrate muscle fibres. *J. Physiol.* **184**, 143–69.

Gordon, A. M., Huxley, A. F. & Julian, F. J. (1966*b*). The variation in isometric tension with sarcomere length in vertebrate muscle fibres. *J. Physiol.* **184**, 170–92.

Górecka, A., Aksoy, M. O. & Hartshorne, D. J. (1976). The effect of phosphorylation of gizzard myosin on actin activation. *Biochem. Biophys. Res. Comm.* **71**, 325–31.

Gorman, A. L. F. & Thomas, M. V. (1978). Changes in the intracellular concentration of the free calcium ions in a pace-maker neurone, measured with the metallochromic indicator dye arsenazo III. *J. Physiol.* **275**, 357–76.

Gorman, A. L. F. & Thomas, M. V. (1980). Intracellular calcium accumulation during depolarization in a molluscan neurone. *J. Physiol.* **308**, 259–85.

Gorodis, C., Virmaux, N., Cailla, H. L. & Delaage, A. (1974). Rapid, light-induced changes of retinal cyclic GMP levels. *FEBS Lett.* **49**, 167–9.

Gorter, E. & Grendel, F. (1925). On bimolecular layers of lipoids on the chromocytes of blood. *J. Exp. Med.* **41**, 439–43.

Goulding, E. H., Tibbs, G. R. & Siegelbaum, S. A. (1994). Molecular mechanism of cyclic-nucleotide-gated channel activation. *Nature* **372**, 369–74.

Grandy, D. K. & Civelli, O. (1992). G-protein-coupled receptors: the new dopamine receptor subtypes. *Curr. Opin. Neurobiol.* **2**, 275–81.

Granger, B. L. & Lazarides, E. (1978). The existence of an insoluble Z disc scaffold in chicken skeletal muscle. *Cell* **15**, 1253–68.

Granit, R. (1933). The components of the retinal action potential and their relation to the discharge in the optic nerve. *J. Physiol.* **77**, 207–40.

Granit, R. (1947). *Sensory Mechanisms of the Retina*. London: Oxford University Press.

Granit, R. (1962). The visual pathway. In *The Eye*, ed. H. Davson, vol. 2, pp. 535–763. London: Academic Press.

Gray, E. G. (1957). The spindle and extrafusal innervation of a frog muscle. *Proc. R. Soc. Lond.* B **146**, 416–30.

Gray, E. G. (1962). A morphological basis-for presynaptic inhibition? *Nature* **193**, 82–3.

Gray, E. G. (1978). Synaptic vesicles and microtubules in frog motor endplates. *Proc. R. Soc. Lond.* B **203**, 219–27.

Gray, J. A. B. (1962). Coding in systems of primary receptor neurons. *Symp. Soc. Exp. Biol.* **16**, 345–54.

Gray, J. A. B. & Matthews, P. B. C. (1951). A comparison of the adaptation of the Pacinian corpuscle with the accommodation of its own axon. *J. Physiol.* **114**, 454–64.

Graziadei, P. P. C. (1973). Cell dynamics in the olfactory mucosa. *Tiss. Cell* **5**, 113–31.

Graziadei, P. P. C. & Graziadei, G. A. M. (1979). Neurogenesis and neuron regeneration in the olfactory system of mammals. I. Morphological aspects of differentiation and structural organization of the olfactory sensory neurons. *J. Neurocytol.* **8**, 1–18.

Greaser, M. L. & Gergely, J. (1973). Purification and properties of the components from troponin. *J. Biol. Chem.* **248**, 2125–33.

Greaser, M. L., Yamaguchi, M., Brekke, C., Potter, J. & Gergely, J. (1973). Troponin subunits and their interactions. *Cold Spr. Harb. Symp. Quant. Biol.* **37**, 235–44.

Green, H. J. (1986). Muscle power: recruitment, metabolism and fatigue. In *Human Muscle Power*, ed. N. L. Jones, N. McCarmey & A. J. McComas, pp. 65–79. Champaign, IL: Human Kinetics Publishers.

Greenblatt, R. E., Blatt, Y. & Montal, M. (1985). The structure of the voltage-sensitive sodium channel. *FEBS Lett.* **193**, 125–34.

Greenspan, J. D. & Bolanowski, S. J. (1996). The psychophysics of tactile perception and its peripheral physiological basis. In *Pain and Touch*, ed. L. Kruger, pp. 25–103. San Diego, CA: Academic Press.

Greenspan, R. J. (1995). Flies, genes, learning, and memory. *Neuron* **15**, 747–50.

Greer Walker, M. & Pull, G. (1973). Skeletal muscle function and sustained swimming speeds in the coalfish *Gadus virens* L. *Comp. Biochem. Physiol.* A **44**, 495–501.

Gregorio, C. C., Weber, A., Bondad, M., Pennise, C. R. & Fowler, V. M. (1995). Requirement of pointed-end capping by tropomodulin to maintain actin filament length in embryonic chick cardiac myocytes. *Nature* **377**, 83–6.

Gregory, R. (1970). *The Intelligent Eye*. London: Weidenfeld.

Gregory, R. (1986). *Odd Perceptions*. London: Methuen.

Griffin, D. R. (1958). *Listening in the Dark*. New Haven, CN: Yale University Press.

Griffin, D. R. & Galambos, R. (1941). The sensory basis of obstacle avoidance by flying bats. *J. Exp. Zool.* **86**, 481–506.

Griffith, O. H., Dehlinger, P. J. & Van, S. P. (1974). Shape of the hydrophilic barrier of phospholipid bilayers. *J. Membrane Biol.* **15**, 159–92.

Griffith, O. W. & Stuehr, D. J. (1995). Nitric oxide synthases: properties and catalytic mechanism. *Ann. Rev. Physiol.* **57**, 707–36.

Guastella, J., Nelson, N., Nelson, H., Czyzyk, L., Keynan, S., Meidel, M. C., Davidson, N., Lester, H. A. & Kanner, B. I. (1990). Cloning and expression of a rat brain GABA transporter. *Science* **249**, 1303–6.

Gudermann, T., Kalkbrenner, F. & Schultz, G. (1996). Diversity and selectivity of receptor-G protein interaction. *Ann. Rev. Pharmacol. Toxicol.* **36**, 429–59.

Gurdon, J. B., Lane, C. D., Woodland, H. R. & Marbaix, G. (1971). Use of frog eggs and oocytes for the study of messenger RNA and its translation in living cells. *Nature* **233**, 177–82.

Guy, H. R. & Seetharamulu, P. (1986). Molecular model of the action potential sodium channel. *Proc. Natl. Acad. Sci. USA* **83**, 508–12.

Haga, K., Haga, T., Ichiyama, A., Katada, T., Kurose, H. & Ui, M. (1985). Functional reconstitution of purified muscarinic receptors and inhibitory guanine nucleotide regulatory protein. *Nature* **316**, 731–3.

Haga, K., Haga, T. & Ichiyama, A. (1986). Reconstitution of the muscarinic acetylcholine receptor. *J. Biol. Chem.* **261**, 10133–40.

Hagins, W. A. (1972). The visual process: excitatory mechanisms in the primary receptor cells. *Ann. Rev. Biophys.* **1**, 131–58.

Hagins, W. A. & Yoshikami. S. (1974). A role for Ca^{2+} in excitation of retinal rods and cones. *Exp. Eye Res.* **18**, 299–305.

Hagins, W. A., Penn, R. D. & Yoshikami, S. (1970). Dark current and photocurrent in retinal rods. *Biophys. J.* **10**, 380–412.

Hagiwara, S. & Byerly, L. (1981). Calcium channel. *Ann. Rev. Neurosci.* **4**. 69–125.

Hagiwara, S. & Morita, H. (1962). Electrotonic transmission between two nerve cells in leech ganglion. *J. Neurophysiol.* 25, 725–31.

Hagiwara, S. & Naka, K. (1964). The initiation of spike potential in barnacle muscle fibres under low intracellular Ca^{++}. *J. Gen. Physiol.* 48, 141–62.

Hagiwara, S., Watanabe, A. & Saito, N. (1959). Potential changes in syncytial neurons of lobster cardiac ganglion. *J. Neurophysiol.* 22, 554–72.

Hagiwara, S., Szabo, T. & Enger. P. S. (1965). Electroreceptor mechanisms in a high-frequency weakly electric fish, *Sternarchus albifrons*. *J. Neurophysiol.* 28, 784–99.

Hagopian, M. (1966). Myofilament arrangement in femoral muscle of the cockroach, *Leucophaea maderae* Fabricius. *J. Cell Biol.* 28, 545–62.

Hagopian, M. & Spiro, D. (1968). The filament lattice of cockroach thoracic muscle. *J. Cell Biol.* 36, 433–42.

Hai, C.-M. & Murphy, R. A. (1989). Ca^{2+}, cross-bridge phosphorylation, and contraction. *Ann. Rev. Physiol.* 51, 285–98.

Hall, Z. W. & Kelly, R. B. (1971). Enzymatic detachment of endplate acetylcholinesterase from muscle. *Nature New Biol.* 232, 62–3.

Halpern, M. (1987). The organization and function of the vomeronasal system. *Ann. Rev. Neurosci.* 10, 325–62.

Hamill, O. P. & McBride, D. W. (1996). A supramolecular complex underlying touch sensitivity. *Trends Neurosci.* 19, 258–61.

Hamill, O. P. & Sakmann, B. (1981). Multiple conductance states of single acetylcholine receptor channels in embryonic muscle cells. *Nature* 294, 462–4.

Hamill, O. P., Marty. A., Neher, E., Sakmann, B. & Sigworth. F. J. (1981). Improved patch-clamp techniques for high-resolution current recording from cells and cell-free membrane patches. *Pflügers Archiv* 391, 85–100.

Hammer, R. & Giachetti. A. (1982). Muscarinic receptor subtypes: M1 and M2. Biochemical and functional characterization. *Life Sci.* 31, 2991–8.

Hansen Bay, C. & Strichartz, G. R. (1980). Saxitoxin binding to sodium channels of rat skeletal muscle. *J. Physiol.* 300, 89–103.

Hanson, J. & Huxley, H. E. (1953). The structural basis of the cross-striations in muscle. *Nature* 172, 530–2.

Hanson, J. & Huxley, H. E. (1955). The structural basis of contraction in striated muscle. *Symp. Soc. Exp. Biol.* 9, 228–64.

Hanson, J. & Lowy, J. (1960). Structure and function of the contractile apparatus in the muscles of invertebrate animals. In *Structure and Function of Muscle*, ed. G. Bourne, vol. 1, pp. 263–365. New York: Academic Press.

Hanson, J. & Lowy, J. (1963). The structure of F-actin and of actin filaments isolated from muscle. *J. Molec. Biol.* 6, 46–60.

Harada, Y., Sakurada, K., Aoki, T., Thomas, D. D. & Yanagida, T. (1990). Mechanochemical coupling in actomyosin energy transduction studied by an *in vitro* movement assay. *J. Molec. Biol.* 216, 49–68.

Hardman, J. G., Limbird, L. E., Molinoff, P. B., Ruddon, R. W. & Gilman, A. G. (eds.) (1996). Goodman & Gilman's *The Pharmacological Basis of Therapeutics*, ninth edition. New York: McGraw-Hill.

Hardwicke, P. M. D. & Szent-Györgyi, A. G. (1985). Proximity of regulatory light chains in scallop myosin. *J. Molec. Biol.* 183, 203–11.

Hardwicke, P. M. D., Wallimann, T. & Szent-Györgyi, A. G. (1983). Light-chain movement and regulation in scallop myosin. *Nature* 301, 478–82.

Hargrave, P. A., McDowell, J. H., Curtis, D. R., Wang, J. K., Juszczak, E., Fong, S. L., Mohanna Rao, J. K. & Argos, P. (1983). The structure of bovine rhodopsin. *Biophys. Struct. Mech.* 9, 235–44.

Harrington, W. F. (1979). On the origin of the contractile force in skeletal muscle. *Proc. Natl. Acad. Sci. USA* 76, 5066–70.

Hartinger, J. & Jahn, R. (1993). An anion binding site that regulates the glutamate transporter of synaptic vesicles. *J. Biol. Chem.* 268, 23122–7.

Hartline, H. K. & Ratliff, F. (1957). Inhibitory interaction of receptor units in the eye of *Limulus*. *J. Gen. Physiol.* 40, 357–76.

Hartline, H. K., Wagner, H. G. & Ratliff, F. (1956). Inhibition in the eye of *Limulus*. *J. Gen. Physiol.* 39, 651–73.

Hartline, H. K., Ratliff, F. & Miller, W. H. (1961). Inhibitory interaction in the retina and its significance in vision. In *Nervous Inhibition*, ed. E. Florey, pp. 241–84. Oxford: Pergamon Press.

Hartmann, H. A., Kirsch, G. E., Drewe, J. A., Taglialatela, M., Joho, R. H. & Brown, A. M. (1991). Exchange of conduction pathways between two related K$^+$ channels. *Science* 251, 942–4.

Hartshorne, D. J. & Kawamura, T. (1992). Regulation of contraction–relaxation in smooth muscle. *News Physiol. Sci.* 7, 59–64.

Hartshorne, R. P. & Catterall, W. A. (1984). The sodium channel from rat brain. Purification and subunit composition. *J. Biol. Chem.* 259, 1667–75.

Hartshorne, R. P., Keller, B. U., Talvenheimo, J. A., Catterall, W. A. & Montal, M. (1985). Functional reconstitution of the purified brain sodium channel in planar lipid bilayers. *Proc. Natl. Acad. Sci. USA* 82, 240–4.

Hartzell, H. C., Méry, P.-F., Fischmeister, R. & Szabo, G. (1991). Sympathetic regulation of cardiac calcium current is due exclusively to cAMP-dependent phosphorylation. *Nature* 351, 573–6.

Haselgrove, J. C. (1973). X-ray evidence for a conformational change in the actin-containing filaments of vertebrate striated muscle. *Cold Spr. Harb. Symp. Quant. Biol.* 37, 341–52.

Haselgrove, J. C. (1983). Structure of vertebrate striated muscle as determined by X-ray-diffraction studies. In *Handbook of*

Physiology, section **10** *Skeletal Muscle*, ed. L. D. Peachey & R. H. Adrian, pp. 143–71. Bethesda, MD: American Physiological society.

Hasselbach, W. (1964). Relaxing factor and the relaxation of muscle. *Prog. Biophys.* **14**, 167–222.

Hasselbach, W. & Makinose, M. (1963). Über den mechanismus des Calciumtransportes durch die Membranen des Sarkoplasmatichen Reticulums. *Biochem. Z.* **339**, 94–111.

Hawkins, R. D. (1996). NO honey, I don't remember. *Neuron* **16**, 465–7.

Haydon, P. G., Henderson, E. & Stanley, E. F. (1994). Localization of individual calcium channels at the release face of a presynaptic nerve terminal. *Neuron* **13**, 1275–80.

Haynes, L. W. & Yau, K.-W. (1985). Cyclic GMP-sensitive conductance in outer segment of catfish cones. *Nature* **317**, 252–5.

Haynes, L. W., Kay, A. R. & Yau, K.-W. (1986). Single cyclic GMP-activated channel activity in excised patches of rod outer segment membrane. *Nature* **321**, 66–70.

Heath, J. K. (1993). *Growth Factors*. Oxford: IRL Press.

Hebb, D. O. (1949). *The Organization of Behaviour*. New York: John Wiley & Sons.

Hecht, S. (1937). Rods, cones, and the chemical basis of vision. *Physiol. Rev.* **17**, 239–90.

Hecht, S. & Schlaer, S. (1938). An adaptometer for measuring human dark adaptation. *J. Opt. Soc. Amer.* **28**, 269–75.

Hecht, S., Schlaer, S. & Pirenne, M. (1942). Energy, quanta and vision. *J. Gen. Physiol.* **25**, 819–40.

Heck, G. L., Mierson, S. & DeSimone, J. A. (1984). Salt taste transduction occurs through an amiloride-sensitive sodium transport pathway. *Science* **223**, 403–5.

Hecker, E. & Butenandt, A. (1984). Bombykol revisited. In *Techniques in Pheromone Research*, ed. H. E. Hummel & T. A. Miller, pp. 1–44. New York: Springer-Verlag.

Hegde, A. N., Goldberg, A. L. & Schwartz, J. H. (1993). Regulatory subunits of cAMP-dependent protein kinases are degraded after conjugation to ubiquitin: a molecular mechanism underlying long-term synaptic plasticity. *Proc. Natl. Acad. Sci. USA* **90**, 7436–40.

Heginbotham, L., Abramson, T. & MacKinnon, R. (1992). A functional connection between the pores of distantly related ion channels as revealed by mutant K^+ channels. *Science* **258**, 1152–5.

Heginbotham, L., Lu, Z., Abramson, T. & MacKinnon, R. (1994). Mutations in the K^+ channel signature sequence. *Biophys. J.* **66**, 1061–7.

Heidelberger, R., Heinemann, C., Neher, E. & Matthews, G. (1994). Calcium dependence of the rate of exocytosis in a synaptic terminal. *Nature* **371**, 513–5.

Heilbrunn, L. V. & Wiercinski, F. J. (1947). The action of various cations on muscle protoplasm. *J. Cell. Comp. Physiol.* **29**, 15–32.

Heiligenberg, W. (1977). *Principles of Electroreception and Jamming Avoidance in Electric Fish*. Berlin: Springer-Verlag.

Heinemann, C., Chow, R. H., Neher, E. & Zucker, R. S. (1994). Kinetics of the secretory response in bovine chromaffin cells following flash photolysis of caged Ca^{2+}. *Biophys. J.* **67**, 2546–57.

Heinemann, S. H., Terlau, H., Stühmer, W., Imoto, K. & Numa, S. (1992). Calcium channel characteristics conferred on the sodium channel by single mutations. *Nature* **356**, 441–3.

Heitler, W. J. (1992). *NeuroSim: a Neurophysiology Simulation Program*. Cambridge: Biosoft.

Hellam, D. C. & Podolsky, R. J. (1969). Force measurements in skinned muscle fibres. *J. Physiol.* **200**, 807–19.

Helmholtz, H. von (1863). *On the Sensations of Tone*, fourth edition, trans. A. J. Ellis (1885), reprinted 1954. New York: Dover.

Helmholtz, H. von (1866). *Handbook of Physiological Optics*, third edition, trans. J. P. C. Southall (1924), reprinted 1962. New York: Dover.

Hen, R. (1992). Of mice and flies: commonalities among 5-HT receptors. *Trends Pharmacol. Sci.* **13**, 160–5.

Henry, J. P., Botton, D., Sagne, C., Isambert, M. F., Desnos, C., Blanchard, V., Raisman-Vosari, R., Krejci, E., Massoulie, J. & Gasnier, B. (1994). Biochemistry and molecular biology of the vesicular monoamine transporter from chromaffin granules. *J. Exp. Biol.* **196**, 251–62.

Hensel, H. & Kenshalo, D. R. (1969). Warm receptors in the nasal region of cats. *J. Physiol.* **204**, 99–112.

Hensel, H. & Zotterman, Y. (1951). Quantitative Beziehungen zwischen der Entiadung einzelner Kältefasern und der Temperature. *Acta Physiol. Scand.* **23**, 291–319.

Hensel, H., Andres, K. H. & Düring, M. V. (1974). Structure and function of cold receptors. *Pflügers Archiv* **352**, 1–10.

Hepler, J. R. & Gilman, A. G. (1992). G proteins. *Trends Biochem. Sci.* **17**, 383–7.

Herman, R. K. (1996). Touch sensation in *Caenorhabditis elegans*. *BioEssays* **18**, 199–205.

Hermann, L. (1899). Zur Theorie der Erregungsleitung und der elektrischen Erregung. *Pflügers Archiv* **75**, 574.

Herzberg, O. & James, M. N. G. (1985). Structure of the calcium regulatory muscle protein troponin-C at 2.8 Å resolution. *Nature* **313**, 653–9.

Hestrin, S., Nicoll, R. A., Perkel, D. J. & Sah, P. (1990*a*). Analysis of excitatory synaptic action in pyramidal cells using whole-cell recording from rat hippocampal slices. *J. Physiol.* **422**, 203–25.

Hestrin, S., Sah, P. & Nicoll, R. A. (1990*b*). Mechanisms generating the time course of dual component excitatory synaptic currents recorded in hippocampal slices. *Neuron* **5**, 247–53.

Heuser, J. E. (1989). Review of electron microscopic evidence favouring vesicle exocytosis as the structural basis for quantal release during synaptic transmission. *Quart. J. Exp. Physiol.* **74**, 1051–69.

Heuser, J. E. & Reese, T. S. (1973). Evidence for recycling of synaptic vesicle membrane during transmitter release at the frog neuromuscular junction. *J. Cell Biol.* **57**, 315–44.

Heuser, J. E. & Reese, T. S. (1977). Structure of the synapse. In *Handbook of Physiology*, section 1, vol. 1, *Cellular Biology of Neurons* Part 1, ed. E. R. Kandel, pp. 261–94. Bethesda, MD: American Physiological Society.

Heuser, J. E. & Salpeter, S. R. (1979). Organization of acetylcholine receptors in quick-frozen deep-etched and rotary-replicated *Torpedo* postsynaptic membrane. *J. Cell Biol.* **82**, 150–73.

Heuser, J. E., Reese, T. S., Dennis, M. J., Jan, Y., Jan, L. & Evans, L. (1979). Synaptic vesicle exocytosis captured by quick-freezing and correlated with quantal transmitter release. *J. Cell Biol.* **81**, 275–300.

Hibberd, M. G. & Jewell, B. R. (1982). Calcium- and length-dependent force production in rat ventricular muscle. *J. Physiol.* **329**, 527–40.

Hibberd, M. G. & Trentham, D. R. (1986). Relationships between chemical and mechanical events during muscular contraction. *Ann. Rev. Biophys.* **15**, 119–61.

Hildebrand, J. G. & Shepherd, G. M. (1997). Mechanisms of olfactory discrimination: converging evidence for common principles across phyla. *Ann. Rev. Neurosci.* **20**, 595–631.

Hilgemann, D. W. (1986). Extracellular calcium transients at single excitations in rabbit atrium measured with tetramethylmurexide. *J. Gen. Physiol.* **87**, 707–35.

Hilgemann, D. W. & Noble, D. (1987). Excitation–contraction coupling and extracellular calcium transients in rabbit atrium: reconstruction of basic cellular mechanisms. *Proc. R. Soc. Lond.* B **230**, 163–205.

Hill, A. V. (1936). The strength–duration relation for electric excitation of medullated nerve. *Proc. R. Soc. Lond.* B **119**, 440–53.

Hill, A. V. (1938). The heat of shortening and the dynamic constants of muscle. *Proc. R. Soc. Lond.* B **126**, 136–95.

Hill, A. V. (1949). The abrupt transition from rest to activity in muscle. *Proc. R. Soc. Lond.* B **136**, 399–420.

Hill, A. V. (1950a). A challenge to biochemists. *Biochim. Biophys. Acta* **4**, 4–11.

Hill, A. V. (1950b). The dimensions of animals and their muscular dynamics. *Sci. Prog. Lond.* **38**, 209–30.

Hill, A. V. (1964). The effect of load on the heat of shortening of muscle. *Proc. R. Soc. Lond.* B **159**, 297–318.

Hill, A. V. (1965). *Trails and Trials in Physiology*. London: Edward Arnold.

Hill, A. V. & Hartree, W. (1920). The four phases of heat production of muscle. *J. Physiol.* **54**, 84–128.

Hill, A. V. & Howarth, J. V. (1957). The effect of potassium on the resting metabolism of the frog's sartorius. *Proc. R. Soc. Lond.* B **147**, 21–43.

Hill, R. B. & Usherwood, P. N. R. (1961). The action of 5-hydroxytryptamine and related compounds on neuromuscular transmission in the locust *Schistocerca gregaria*. *J. Physiol.* **157**, 393–401.

Hille, B. (1966). Common mode of action of three agents that decrease the transient change in sodium permeability in nerves. *Nature* **210**, 1220–2.

Hille, B. (1967). The selective inhibition of delayed potassium currents in nerve by tetraethylammonium ion. *J. Gen. Physiol.* **50**, 1287–1302.

Hille, B. (1968). Pharmacological modifications of the sodium channels of frog nerve. *J. Gen. Physiol.* **51**, 199–219.

Hille, B. (1971a). Voltage clamp studies on myelinated nerve fibres. In *Biophysics and Physiology of Excitable Membranes*, ed. W. J. Adelman, pp. 230–46. New York: Van Nostrand Reinhold.

Hille, B. (1971b). The permeability of the sodium channel to organic cations in myelinated nerve. *J. Gen. Physiol.* **58**, 599–619.

Hille, B. (1972). The permeability of the sodium channel to metal cations in myelinated nerve. *J. Gen. Physiol.* **59**, 637–58.

Hille, B. (1973). Potassium channels in myelinated nerve; selective permeability to small cations. *J. Gen. Physiol.* **61**, 669–86.

Hille, B. (1975). Ionic selectivity, saturation, and block in sodium channels: a four-barrier model. *J. Gen. Physiol.* **66**, 535–60.

Hille, B. (1984). *Ionic Channels of Excitable Membranes*. Sunderland, MA: Sinauer Associates.

Hille, B. (1992). *Ionic Channels of Excitable Membranes*, second edition. Sunderland, MA: Sinauer Associates.

Hille, B., Woodhull, A. M. & Shapiro, B. I. (1975). Negative surface charge near sodium channels of nerve: divalent ions, monovalent ions, and pH. *Phil. Trans. R. Soc. Lond.* B **270**, 301–18.

Hirano, A. (1981). Structure of normal central myelinated fibres. *Adv. Neurol.* **31**, 51–68.

Hirokawa, N. & Heuser, J. E. (1982). Internal and external differentiations of the postsynaptic membrane at the neuromuscular junction. *J. Neurocytol.* **11**, 487–510.

Hirose, K., Lenart, T. D., Murray, J. M., Franzini-Armstrong, C. & Goldman, Y. E. (1993). Flash and smash: rapid freezing of muscle fibers activated by photolysis of caged ATP. *Biophys. J.* **65**, 397–408.

Hirose, K., Franzini-Armstrong, C., Goldman, Y. E. & Murray, J. M. (1994). Structural changes in muscle cross-bridges accompanying force generation. *J. Cell Biol.* **127**, 763–78.

Ho, K., Nichols, C. G., Lederer, W. J., Lytton, J., Vassilev, P. M., Kanazirska, M. V. & Hebert, S. C. (1993). Cloning and expression of an inwardly rectifying ATP-regulated potassium channel. *Nature* **362**, 31–8.

Hochberg, J. (1984). Perception. In *Handbook of Physiology*, section, 1, vol. 3, *Sensory Processes*, ed. I. Darian-Smith, pp. 75–102. Bethesda, MD: American Physiological Society.

Hochner, B., Klein, M., Schacher, S. & Kandel, E. R. (1986). Action-potential duration and the modulation of transmitter release from the sensory neurons of *Aplysia* in presynaptic facilitation and behavioral sensitization. *Proc. Natl. Acad. Sci. USA* **83**, 8410–14.

Hodgkin, A. L. (1937). Evidence for electrical transmission in nerve. *J. Physiol.* **90**, 183–232.

Hodgkin, A. L. (1938). The subthreshold potentials in a crustacean nerve fibre. *Proc. R. Soc. Lond.* B **126**, 87–121.

Hodgkin, A. L. (1939). The relation between conduction velocity and the electrical resistance outside a nerve. *J. Physiol.* **94**, 560–70.

Hodgkin, A. L. (1951). The ionic basis of electrical activity in nerve and muscle. *Biol. Rev.* **26**, 339–409.

Hodgkin, A. L. (1954). A note on conduction velocity. *J. Physiol.* **125**, 221–4.

Hodgkin, A. L. (1958). Ionic movements and electrical activity in giant nerve fibres. *Proc. R. Soc. Lond.* B **148**, 1–37.

Hodgkin, A. L. (1964). *The Conduction of the Nervous Impulse*. Liverpool: Liverpool University Press.

Hodgkin, A. L. (1975). The optimum density of sodium channels in an unmyelinated nerve. *Phil. Trans. R. Soc. Lond.* B **270**, 297–300.

Hodgkin, A. L. (1992). *Chance and Design*. Cambridge: Cambridge University Press.

Hodgkin, A. L. (1996). In *The History of Science in Autobiography*, vol. 1, ed. L. R. Squire, pp. 254–91. Washington, DC: Society for Neuroscience.

Hodgkin, A. L. & Horowicz, P. (1957). The differential action of hypertonic solutions on the twitch and action potential of a muscle fibre. *J. Physiol.* **136**, 17P.

Hodgkin, A. L. & Horowicz, P. (1959). The influence of potassium and chloride ions on the membrane potential of single muscle fibres. *J. Physiol.* **148**, 127–60.

Hodgkin, A. L. & Horowicz, P. (1960). Potassium contractures in single muscle fibres. *J. Physiol.* **153**, 386–403.

Hodgkin, A. L. & Huxley, A. F. (1939). Action potentials recorded from inside a nerve fibre. *Nature* **140**, 710–11.

Hodgkin, A. L. & Huxley, A. F. (1945). Resting and action potentials in single nerve fibres. *J. Physiol.* **104**, 176–95.

Hodgkin, A. L. & Huxley, A. F. (1952a). Currents carried by sodium and potassium ions through the membrane of the giant axon of *Loligo*. *J. Physiol.* **116**, 449–72.

Hodgkin, A. L. & Huxley, A. F. (1952b). The components of membrane conductance in the giant axon of *Loligo*. *J. Physiol.* **116**, 473–96.

Hodgkin, A. L. & Huxley, A. F. (1952c). The dual effect of membrane potential on sodium conductance in the giant axon of *Loligo*. *J. Physiol.* **116**, 497–506.

Hodgkin, A. L. & Huxley, A. F. (1952d). A quantitative description of membrane current and its application to conduction and excitation in nerve. *J. Physiol.* **117**, 500–44.

Hodgkin, A. L. & Huxley, A. F. (1953). Movements of radioactive potassium and membrane current in a giant axon. *J. Physiol.* **121**, 403–14.

Hodgkin, A. L. & Katz, B. (1949). The effect of sodium ions on the electrical activity of the giant axon of the squid. *J. Physiol.* **108**, 37–77.

Hodgkin, A. L. & Keynes, R. D. (1953). The mobility and diffusion coefficient of potassium in giant axons from *Sepia*. *J. Physiol.* **119**, 513–28.

Hodgkin, A. L. & Keynes, R. D. (1955a). Active transport of cations in giant axons from *Sepia* and *Loligo*. *J. Physiol.* **128**, 28–60.

Hodgkin, A. L. & Keynes, R. D. (1955b). The potassium permeability of a giant nerve fibre. *J. Physiol.* **128**, 61–88.

Hodgkin, A. L. & Keynes, R. D. (1957). Movements of labelled calcium in squid giant axons. *J. Physiol.* **138**, 253–81.

Hodgkin, A. L. & Rushton, W. A. H. (1946). The electrical constants of a crustacean nerve fibre. *Proc. R. Soc. Lond.* B **133**, 444–79.

Hodgkin, A. L., Huxley, A. F. & Katz, B. (1952). Measurement of current–voltage relations in the membrane of the giant axon of *Loligo*. *J. Physiol.* **116**, 424–48.

Hodgkin, A. L., McNaughton, P. A., Nunn, B. J. & Yau, K.-W. (1984). Effect of ions on retinal rods from *Bufo marinus*. *J. Physiol.* **350**, 649–80.

Hodgkin, A. L., McNaughton, P. A. & Nunn, B. J. (1985). The ionic selectivity and calcium dependence of the light-sensitive pathway in toad rods. *J. Physiol.* **358**, 447–68.

Hodgkin, J., Plasterk, R. H. A. & Waterston, R. H. (1995). The nematode *Caenorhabditis elegans* and its genome. *Science* **270**, 410–14.

Hodgson, E. S. & Roeder, K. D. (1956). Electrophysiological studies of arthropod chemoreception. I. General properties of the labellar chemoreceptors of Diptera. *J. Cell. Comp. Physiol.* **48**, 51–76.

Hodgson, E. S., Lettvin, J. Y. & Roeder, K. D. (1955). Physiology of a primary chemoreceptor unit. *Science* **122**, 417–18.

Hoffman, E. P., Brown, R. H. & Kunkel, L. M. (1987). Dystrophin: the protein product of the Duchenne muscular dystrophy locus. *Cell* **51**, 919–28.

Hoffman, E. P., Lehmann-Horn, F. & Rüdel, R. (1995). Overexcited or inactive: ion channels in muscle disease. *Cell* **80**, 681–6.

Hofmann, F., Biel, M. & Flockerzi, V. (1994). Molecular basis for Ca^{2+} channel diversity. *Ann. Rev. Neurosci.* **17**, 399–418.

Hofmann, K. P., Jäger, S. & Ernst, O. P. (1995). Structure and function of activated rhodopsin. *Israel J. Chem.* **35**, 339–55.

Hökfelt, T. (1991). Neuropeptides in perspective: the last ten years. *Neuron* **7**, 867–79.

Hökfelt, T., Elfvin, L. G., Elde, R., Schultzberg, M., Goldstein, M. & Luft, R. (1977). Occurrence of somatostatin-like immunoreactivity in some peripheral sympathetic noradrenergic neurons. *Proc. Natl. Acad. Sci. USA* **74**, 3587–91.

Hökfelt, T., Johansson, O., Ljungdahl, A., Lundberg, J. M. & Schultzberg, M. (1980). Peptidergic neurones. *Nature* **284**, 515–21.

Hökfelt, T., Lundberg, J. M., Skirboll, L., Johansson, O.,

Schultzberg, M. & Vincent. S. R. (1982). Coexistence of classical transmitters and peptides in neurones. In *Co-transmission*, ed. A. C. Cuello, pp. 77–125. London: Macmillan.

Hökfelt, T., Everitt, B., Holets, V. R., Meister, B., Melander, T., Schalling, M., Staines, W. & Lundberg, J. M. (1986). Coexistence of peptides and other active molecules in neurons: diversity of chemical signalling potential. In *Fast and Slow Chemical Signalling in the Nervous System*, ed. L. L. Iversen & E. Gootiman, pp. 205–31. Oxford: Oxford University Press.

Holley, M. C. & Ashmore, J. F. (1988). A cytoskeletal spring in cochlear outer hair cells. *Nature* 335, 635–7.

Holley, M. C., Kalinec, F. & Kachar, B. (1992). Structure of the cortical cytoskeleton in mammalian outer hair cells. *J. Cell Sci.* 102, 569–80.

Hollmann, M. & Heinemann, S. (1994). Cloned glutamate receptors. *Ann. Rev. Neurosci.* 17, 31–108.

Hollmann, M., Maron, C. & Heinemann, S. (1994). N-glycosylation site tagging suggests a three transmembrane domain topology for the glutamate receptor GluR1. *Neuron* 13, 1331–43.

Holman, M. E. & Neild, T. O. (1979). Membrane properties. *Brit. Med. Bull.* 35, 235–41.

Holmes, K. C. (1997). The swinging lever-arm hypothesis of muscle contraction. *Current Biol.* 7, R112–R118.

Holmes, K. C., Popp, D., Gebhard, W. & Kabsch, W. (1990). Atomic model of the actin filament. *Nature* 347, 44–9.

Holmes, W. (1942). The giant myelinated nerve fibres of the prawn. *Phil. Trans. R. Soc. Lond.* B 231, 293–311.

Holmgren, F. (1865). Method att objectivera effecten av ljusintryck på retina. *Upsala Läkareförh.* 1, 177–91.

Holtzman, E., Freeman, A. R. & Kashnev, L. A. (1971). Stimulation-dependent alterations in peroxidase uptake at lobster neuromuscular junctions. *Science* 173, 733–6.

Holzbauer, E. L. F. & Vallee, R. B. (1994). Dyneins: molecular structure and cellular function. *Ann. Rev. Cell. Biol.* 10, 339–72.

Holzer, P., Wachter, C., Heinemann, A., Jocic, M., Lippe, I. T. & Herbert, M. K. (1995). Sensory nerve, nitric oxide and NANC vasodilation. *Arch. Int. Pharmacodyn. Ther.* 329, 67–79.

Homsher, E. (1987). Muscle enthalpy production and its relationship to actomyosin ATPase. *Ann. Rev. Physiol.* 49, 673–90.

Homsher, E. & Kean, C. J. C. (1982). Unexplained enthalpy production in contracting skeletal muscles. *Fed Proc.* 41, 149–54.

Homsher, E. & Rall, J. A. (1973). Energetics of shortening muscles in twitches and tetanic contractions. I. A reinvestigation of Hill's concept of shortening heat. *J. Gen. Physiol.* 62, 663–76.

Homsher, E., Irving, M. & Wallner, A. (1981). High-energy phosphate metabolism and energy liberation associated with

rapid shortening in frog skeletal muscle. *J. Physiol.* 321, 423–36.

Homsher, E., Lacktis, J., Yamada, T. & Zohman, G. (1987). Repriming and reversal of the isometric unexplained enthalpy in frog skeletal muscle. *J. Physiol.* 393, 157–70.

Hong, K. & Driscoll, M. (1994). A transmembrane domain of the putative channel subunit MEC–4 influences mechanotransduction and neurodegeneration in *C. elegans*. *Nature* 367, 470–3.

Hornykiewicz, O. (1973). Parkinsons' disease: from brain homogenate to treatment. *Fedn Proc.* 32, 183–90.

Horowitz, A., Menice, C. B., Laporte, R. & Morgan, K. G. (1996). Mechanisms of smooth muscle contraction. *Physiol. Rev.* 76, 967–1003.

Hoshi, T., Zagotta, W. N. & Aldrich, R. W. (1990). Biophysical and molecular mechanisms of *Shaker* potassium channel inactivation. *Science* 250, 533–8.

Hoshi, T., Zagotta, W. N. & Aldrich, R. W. (1991). Two types of inactivation in *Shaker* K^+ channels: effects of alterations in the carboxy-terminal region. *Neuron* 7, 547–56.

Hoshi, T., Zagotta, W. N. & Aldrich, R. W. (1994). *Shaker* potassium channel gating I: transitions near the open state. *J. Gen. Physiol.* 103, 249–78.

Houamed, K. M., Kuijper, J. L., Gilbert, T. L., Haldeman, B. A., O'Hara, P. J., Mulvihill, E. R., Almers, W. & Hagen, F. S. (1991). Cloning, expression, and gene structure of a G protein-coupled glutamate receptor from rat brain. *Science* 252, 1318–21.

Houmeida, A., Holt, J., Tskhovrebova, L. & Trinick, J. (1995). Studies of the interaction between titin and myosin. *J. Cell Biol.* 131, 1471–81.

Hoyer, D., Clarke, D. E., Fozard, J. R., Hartig, P. R., Martin, G. R., Mylecharane, E. J., Saxena, P. R. & Humphrey, P. P. A. (1994). International Union of Pharmacology classification of receptors for 5-hydroxytryptamine (serotonin). *Pharmacol. Rev.* 46, 157–203.

Hoyle, G. (1955a). The anatomy and innervation of locust skeletal muscle. *Proc. R. Soc. Lond.* B 143, 281–92.

Hoyle, G. (1955b). Neuromuscular mechanisms of a locust skeletal muscle. *Proc. R. Soc. Lond.* B 143, 343–67.

Hoyle, G. (1959). The neuromuscular mechanism of an insect spiracular muscle. *J. Insect Physiol.* 3, 378–94.

Hoyle, G. (1978). Distributions of nerve and muscle fibre types in locust jumping muscle. *J. Exp. Biol.* 73, 205–33.

Hoyle, G. & Wiersma, C. A. G. (1958). Excitation at neuromuscular junctions in Crustacea. *J. Physiol.* 143, 403–25.

Hoyle, G., Dagan, D., Moberly, B. & Colquhoun, W. (1974). Dorsal unpaired median insect neurons make neurosecretory endings on skeletal muscle. *J. Exp. Zool.* 187, 159–65.

Hsu, Y.-T. & Molday, R. S. (1993). Modulation of the cGMP-gated channel of rod photoreceptor cells by calmodulin. *Nature* 361, 76–9.

Hsu, Y.-T. & Molday, R. S. (1994). Interaction of calmodulin with the cyclic GMP-gated channel of rod photoreceptor cells: modulation of activity, affinity purification, and localization. *J. Biol. Chem.* **269**, 29765–70.

Hu, Y., Barzilai, A., Chen., M., Bailey, C. H. & Kandel, E. R. (1993). 5-HT and cAMP induce the formation of coated pits and vesicles and increase the expression of clathrin light chain in sensory neurons of *Aplysia. Neuron* **10**, 921–9.

Huang, C. L.-H. (1988). Intramembrane charge movements in skeletal muscle. *Physiol. Rev.* **68**, 1197–247.

Huang, C. L.-H. (1993). *Intramembrane Charge Movements in Skeletal Muscle.* Oxford: Oxford University Press.

Huang, M. & Chalfie, M. (1994). Gene interactions affecting mechanosensory transduction in *Caenorhabditis elegans. Nature* **367**, 467–70.

Huang, Y.-Y., Li, X.-C. & Kandel, E. R. (1994). cAMP contributes to mossy fibre LTP by initiating both a covalently mediated early phase and macromolecular synthesis-dependent late phase. *Cell* **79**, 69–79.

Hubbard, J. I. (1963). Repetitive stimulation at the mammalian neuromuscular junction, and the mobilisation of transmitter. *J. Physiol.* **169**, 641–62.

Hubbard, R. (1959). The thermal stability of rhodopsin and opsin. *J. Gen. Physiol.* **42**, 259–80.

Hubbard, R. & Kropf, A. (1958). The action of light on rhodopsin. *Proc. Natl. Acad. Sci. USA* **44**, 130–9.

Hubbard, R. & Wald, G. (1952). *Cis–trans* isomers of vitamin A and retinene in the rhodopsin system. *J. Gen. Physiol.* **36**, 269–315.

Hubbard, S. J. (1958). A study of rapid mechanical events in a mechanoreceptor. *J. Physiol.* **141**, 198–218.

Hucho, F., Görne-Tschelnokow, U. & Strecker, A. (1994). β-structure in the membrane-spanning part of the nicotinic acetylcholine receptor (or how helical are transmembrane helices?). *Trends Biochem. Sci.* **19**, 383–7.

Hudspeth, A. J. (1982). Extracellular current flow and the site of transduction by vertebrate hair cells. *J. Neurosci.* **2**, 1–10.

Hudspeth, A. J. (1985). The cellular basis of hearing: the biophysics of hair cells. *Science* **230**, 745–52.

Hudspeth, A. J. (1989). How the ear's works work. *Nature* **341**, 397–404.

Hudspeth, A. J. & Corey, D. P. (1977). Sensitivity, polarity, and conductance change in the response of vertebrate hair cells to controlled mechanical stimuli. *Proc. Natl. Acad. Sci. USA* **74**, 2407–11.

Hudspeth, A. J. & Gillespie, P. G. (1994). Pulling strings to tune transduction: adaptation by hair cells. *Neuron* **12**, 1–9.

Hudspeth, A. J. & Jacobs, R. (1979). Stereocilia mediate transduction in vertebrate hair cells. *Proc. Natl. Acad. Sci. USA* **76**, 1506–9.

Hughes, J., Smith, T. W., Kosterlitz, H. W., Fothergill, L. A., Morgan, B. A. & Morris, H. R. (1975). Identification of two related pentapeptides from the brain with potent opiate agonist activity. *Nature* **258**, 577–9.

Hulme, E. C., Birdsall, N. J. M. & Buckley, N. J. (1990). Muscarinic receptor subtypes. *Ann. Rev. Pharmacol. Toxicol.* **30**, 633–73.

Hume, R. I., Dingledine, R. & Heinemann, S. F. (1991). Identification of a site in glutamate receptor subunits that controls calcium permeability. *Science* **253**, 1028–31.

Hunt, C. C. (1954). Relation of function to diameter in afferent fibers of muscle nerves. *J. Gen. Physiol.* **38**, 117–31.

Hunt, C. C. (1990). Mammalian muscle spindle: peripheral mechanisms. *Physiol. Rev.* **70**, 643–63.

Hunt, C. C. & Ottoson, D. (1975). Impulse activity and receptor potential of primary and secondary endings of isolated mammalian muscle spindles. *J. Physiol.* **252**, 259–81.

Hunt, D. M., Dulai, K. S., Bowmaker, J. K. & Mollon, J. D. (1995). The chemistry of John Dalton's color blindness. *Science* **267**, 984–8.

Hurlbut, W. P., Iezzi, N., Fesce, R. & Ceccarelli, B. (1990). Correlation between quantal secretion and vesicle loss at the frog neuromuscular junction. *J. Physiol.* **425**, 501–26.

Hurley, J. B. & Stryer, L. (1982). Purification and characterization of the γ subunit of the cyclic GMP phosphodiesterase from retinal rod outer segments. *J. Biol. Chem.* **257**, 11094–9.

Hursh, J. B. (1939). Conduction velocity and diameter of nerve fibres. *Amer. J. Physiol.* **127**, 131–9.

Hutter, O. F. & Noble, D. (1960). Rectifying properties of heart muscle. *Nature* **188**, 495.

Hutter, O. F. & Trautwein, W. (1956). Vagal and sympathetic effects on the pacemaker fibers in the sinus venosus of the heart. *J. Gen. Physiol.* **39**, 715–33.

Huxley, A. F. (1957). Muscle structure and theories of contraction. *Prog. Biophys.* **7**, 255–318.

Huxley, A. F. (1974). Muscular contraction. *J. Physiol.* **243**, 1–43.

Huxley, A. F. (1980). *Reflections on Muscle.* Liverpool: Liverpool University Press.

Huxley, A. F. & Niedergerke, R. (1954). Structural changes in muscle during contraction. Interference microscopy of living muscle fibres. *Nature* **173**, 971–3.

Huxley, A. F. & Peachey, L. D. (1961). The maximum length for contraction in vertebrate striated muscle. *J. Physiol.* **156**, 150–65.

Huxley, A. F. & Simmons, R. M. (1971). Proposed mechanism of force generation in striated muscle. *Nature* **233**, 533–8.

Huxley, A. F. & Simmons, R. M. (1973). Mechanical transients and the origin of muscular force. *Cold Spr. Harb. Symp. Quant. Biol.* **37**, 669–80.

Huxley, A. F. & Stämpfli, R. (1949). Evidence for saltatory conduction in peripheral myelinated nerve fibres. *J. Physiol.* **108**, 315–39.

Huxley, A. F. & Stämpfli, R. (1951). Effect of potassium and sodium on resting and action potentials of single myelinated nerve fibres. *J. Physiol.* **122**, 496–508.

Huxley, A. F. & Straub, R. W. (1958). Local activation and

interfibrillar structures in striated muscle. *J. Physiol.* **143**, 40P–41P.

Huxley, A. F. & Taylor, R. E. (1955). Function of Krause's membrane. *Nature* **176**, 1068.

Huxley, A. F. & Taylor, R. E. (1958). Local activation of striated muscle fibres. *J. Physiol.* **144**, 426–41.

Huxley, A. F. & Tideswell, S. (1996). Filament compliance and tension transients in muscle. *J. Mus. Res. Cell Motil.* **17**, 507–11.

Huxley, H. E. (1953a). Electron microscope studies of the organisation of the filaments in striated muscle. *Biochim. Biophys. Acta* **12**, 387–94.

Huxley, H. E. (1953b). X-ray analysis and the problem of muscle. *Proc. R. Soc. Lond.* B **141**, 59–66.

Huxley, H. E. (1957). The double array of filaments in cross-striated muscle. *J. Biophys. Biochem. Cytol.* **3**, 631–48.

Huxley, H. E. (1963). Electron microscope studies on the structure of natural and synthetic protein filaments from striated muscle. *J. Molec. Biol.* **7**, 281–308.

Huxley, H. E. (1964). Evidence for continuity between the central elements of the triads and extracellular space in frog sartorius muscle. *Nature* **202**, 1067–71.

Huxley, H. E. (1969). The mechanism of muscular contraction. *Science* **164**, 1356–66.

Huxley, H. E. (1971). The structural basis of muscular contraction. *Proc. R. Soc. Lond.* B **178**, 131–49.

Huxley, H. E. (1973). Structural changes in the actin- and myosin-containing filaments during contraction. *Cold Spr. Harb. Symp. Quant. Biol.* **37**, 361–76.

Huxley, H. E. (1976). The structural basis of contraction and regulation in skeletal muscle. In *Molecular Basis of Motility*, ed. L. M. G. Heilmeyer, J. C. Rüegg & T. Wieland. Berlin: Springer-Verlag.

Huxley, H. E. (1990). Sliding filaments and molecular motile systems. *J. Biol. Chem.* **265**, 8347–50.

Huxley, H. E. (1996). A personal view of muscle and motility mechanisms. *Ann. Rev. Physiol.* **58**, 1–19.

Huxley, H. E. & Brown, W. (1967). The low-angle X-ray diagram of vertebrate striated muscle and its behaviour during contraction and rigor. *J. Molec. Biol.* **30**, 383–434.

Huxley, H. E. & Faruqi, A. R. (1983). Time-resolved X-ray diffraction studies on vertebrate striated muscle. *Ann. Rev. Biophys.* **12**, 381–417.

Huxley, H. E. & Hanson, J. (1954). Changes in the cross-striations of muscle during contraction and stretch and their structural interpretation. *Nature* **173**, 973–6.

Huxley, H. E., Brown, W. & Holmes, K. C. (1965). Constancy of axial spacings in frog sartorius muscle during contraction. *Nature* **206**, 1358.

Huxley, H. E., Faruqi, A. R., Kress, M., Borda, J. & Koch, M. H. J. (1982). Time-resolved X-ray diffraction studies of the myosin layer-line reflections during muscular contraction. *J. Molec. Biol.* **158**, 637–84.

Huxley, H. E., Stewart, A., Sosa, H. & Irving, T. (1994). X-ray diffraction measurements on the extensibility of actin and myosin filaments in contracting muscle. *Biophys. J.* **67**, 2411–21.

Iggo, A. (1964). Temperature discrimination in the skin. *Nature* **204**, 481–3.

Iggo, A. (1969). Cutaneous thermoreceptors in primates and sub-primates. *J. Physiol.* **200**, 403–30.

Iggo, A. (1990). Whither sensory specificity? In *Thermoreception and Temperature Regulation*, ed. J. Bligh & K. Voigt, pp. 9–18. Berlin: Springer-Verlag.

Iggo, A. & Andres, K. H. (1982). Morphology of cutaneous receptors. *Ann. Rev. Neurosci.* **5**, 1–31.

Iggo, A. & Muir, A. R. (1969). The structure and function of a slowly adapting touch corpuscle in hairy skin. *J. Physiol.* **200**, 763–96.

Ignarro, L. J., Buga, G. M., Wood, K. S., Byrns, R. E. & Chaudhuri, G. (1987). Endothelium-derived relaxing factor produced and released from artery and vein is nitric oxide. *Proc. Natl. Acad. Sci. USA* **84**, 9265–9.

Ikeda, I., Yamashita, Y., Ono, T. & Ogawa, H. (1994). Selective phototoxic destruction of rat Merkel cells abolishes responses of slowly adapting type I mechanoreceptor units. *J. Physiol.* **479**, 247–56.

Imoto, K. (1993). Molecular aspects of ion permeation through channels. *Annals N. Y. Acad. Sci.* **707**, 38–50.

Imoto, K., Methfessel, C., Sakmann, B., Mishina, M., Mori, Y., Konno, T., Fukuda, K., Kurasaki, M., Bujo, H., Fujita, Y. & Numa, S. (1986). Location of a subunit region determining ion transport through the acetylcholine receptor channel. *Nature* **324**, 670–4.

Imoto, K., Busch, C., Sakmann, B., Mishina, M., Konno, T., Nakai, J., Bujo, H., Mori, Y., Fukuda, K. & Numa, S. (1988). Rings of negatively charged amino acids determine the acetylcholine receptor channel conductance. *Nature* **335**, 645–8.

Imoto, K., Konno, T., Nakai, J., Wang, F., Mishina, M. & Numa, S. (1991). A ring of uncharged polar amino acids as a component of channel constriction in the nicotinic acetylcholine receptor. *FEBS Letters* **289**, 193–200.

Impey, S., Mark, M., Villacres, E. C., Poser, S., Chavkin, C. & Storm, D. R. (1996). Induction of CRE-mediated gene expression by stimuli that generate long-lasting LTP in area CA1 of the hippocampus. *Neuron* **16**, 973–82.

Infante, A. A. & Davies, R. E. (1962). Adenosine triphosphate breakdown during a single isotonic twitch of frog sartorius muscle. *Biochem. Biophys. Res. Comm.* **9**, 410–15.

Ingi, T., Cheng, J. & Ronnett, G. V. (1996). Carbon monoxide: an endogenous modulator of the nitric oxide-cyclic GMP signaling system. *Neuron* **16**, 835–42.

Inui, M., Saito, A. & Fleischer, S. (1987). Purification of the ryanodine receptor and identity with feet structures of the junctional terminal cisternae of sarcoplasmic reticulum from fast skeletal muscle. *J. Biol. Chem.* **262**, 1740–7.

Irisawa, H., Brown, H. F. & Giles, W. (1993). Cardiac

pacemaking in the sinuatrial node. *Physiol. Rev.* **73**, 197–227.

Irving, M., Allen, T. St C., Sabido-David, C., Craik, J. S., Brandmeier, B., Kendrick-Jones, J., Corrie, J. E. T., Trentham, D. R. & Goldman, Y. E. (1995). Tilting of the light-chain region of myosin during step length changes and active force generation in skeletal muscle. *Nature* **375**, 688–91.

Isaacson, J. S. & Walmsley, B. (1995). Counting quanta: direct measurements of transmitter release at a central synapse. *Neuron* **15**, 875–84.

Isacoff, E. Y., Jan, Y. N. & Jan, L. Y. (1991). Putative receptor for the cytoplasmic inactivation gate in the *Shaker* K$^+$ channel. *Nature* **353**, 86–90.

Ishihara, T., Shigemoto, R., Mori, K., Takahashi, K. & Nagata, S. (1992). Functional expression and tissue distribution of a novel receptor for vasoactive intestinal polypeptide. *Neuron* **8**, 811–19.

Ishijima, A., Kojima, H., Higuchi, H., Harada, Y., Funatsu, T. & Yanagida, T. (1996). Multiple- and single-molecule analysis of the actomyosin motor by nanometer-piconewton manipulation with a microneedle: unitary steps and forces. *Biophys. J.* **70**, 383–400.

Isom, L. L., de Jongh, K. S., Patton, D. E., Reber, B. F. X., Offord, J., Charbonneau, H., Walsh, K., Goldin, A. L. & Catterall, W. A. (1992). Primary structure and functional expression of the β$_1$ subunit of the rat brain sodium channel. *Science* **256**, 839–42.

Iversen, L. L. (1963). Uptake of noradrenaline by isolated perfused rat heart. *Brit. J. Pharmacol.* **21**, 523–37.

Iversen, L. L. (1971). Role of transmitter uptake mechanisms in synaptic neurotransmission. *Brit. J. Pharmacol.* **41**, 571–91.

Iversen, L. L. (1995). Neuropeptides: promise unfulfilled? *Trends Neurosci.* **18**, 49–50.

Jack, J. J. B., Noble, D. & Tsien, R. W. (1975). *Electric Current Flow in Excitable Cells.* Oxford: Oxford University Press.

Jack, J. J. B., Redman, S. J. & Wong, K. (1981). The components of synaptic potentials evoked in cat spinal motoneurones by impulses in single group Ia afferents. *J. Physiol.* **321**, 65–96.

Jack, J. J. B., Larkman, A. U., Major, G. & Stratford, K. (1994). Quantal analysis of the synaptic excitation of CA1 hippocampal pyramidal cells. In *Molecular and Cellular Mechanisms of Neurotransmitter Release* (*Adv. Second Mess. Phosphoprotein Res.* **29**), ed. L. Stjärne, P. Greengard, S. E. Grillner, T. G. M. Hökfelt & D. R. Ottoson, pp. 275–99. New York: Raven Press.

Jackson, M. B. (1988). Dependence of acetylcholine receptor channel kinetics on agonist concentration in cultured mouse muscle fibres. *J. Physiol.* **397**, 555–83.

Jackson, M. B. (1989). Perfection of a synaptic receptor: kinetics and energetics of the acetylcholine receptor. *Proc. Natl. Acad. Sci. USA* **86**, 2199–203.

Jackson, M. B. (1994). Single channel currents in the nicotinic acetylcholine receptor: a direct demonstration of allosteric transitions. *Trends Biochem. Sci.* **19**, 396–9.

Jacobs, G. H., Neitz, J. & Crognale, M. (1985). Spectral sensitivity of ground-squirrel cones measured with ERG flicker photometry. *J. Comp. Physiol.* A **156**, 503–9.

Jaffrey, S. R. & Snyder, S. H. (1995). Nitric-oxide – a neural messenger. *Ann. Rev. Cell Dev. Biol.* **11**, 417–40.

Jahn, R. & Südhof, T. C. (1994). Synaptic vesicles and exocytosis. *Ann. Rev. Neurosci.* **17**, 219–46.

Jami, L., Murthy, K. S. K. & Petit, J. (1982). A quantitative study of skeletofusimotor innervation in the rat peroneus tertius muscle. *J. Physiol.* **325**, 125–44.

Jan, L. Y. & Jan, Y. N. (1982). Peptidergic transmission in sympathetic ganglia of the frog. *J. Physiol.* **327**, 219–46.

Jan, L. Y. & Jan, Y. N. (1990). A superfamily of ion channels. *Nature* **345**, 672.

Jan, L. Y. & Jan, Y. N. (1992). Tracing the roots of ion channels. *Cell* **69**, 715–18.

Jaslove, S. W. & Brink, P. R. (1986). The mechanism of rectification at the electronic motor giant synapse of the crayfish. *Nature* **323**, 63–5.

Jentsch, T. J., Steinmeyer, K. & Schwarz, G. (1990). Primary structure of *Torpedo marmorata* chloride channel isolated by expression cloning in *Xenopus* oocytes. *Nature* **348**, 510–14.

Jewell, B. R. (1959). The nature of the phasic and tonic responses of the anterior byssal retractor of *Mytilus*. *J. Physiol.* **149**, 154–77.

Jewell, B. R. & Rüegg, J. C. (1966). Oscillatory contraction of insect fibrillar muscle after glycerol extraction. *Proc. R. Soc. Lond.* B **164**, 428–59.

Jewell, B. R. & Wilkie, D. R. (1958). An analysis of the mechanical components in frog striated muscle. *J. Physiol.* **143**, 515–40.

Jöbsis, F. F. & O'Connor, M. J. (1966). Calcium release and reabsorption in the sartorius muscle of the toad. *Biochem. Biophys. Res. Comm.* **25**, 246–52.

Johansson, R. S. (1978). Tactile sensibility in the human hand: receptive field characteristics of mechanoreceptive units in the glabrous skin area. *J. Physiol.* **281**, 101–23.

Johnson, J. W. & Ascher, P. (1987). Glycine potentiates the NMDA response in cultured mouse brain neurons. *Nature* **325**, 529–31.

Johnson, R. (1988). Accumulation of biological amines in chromaffin granules: a model for hormone and neurotransmitter transport. *Physiol. Rev.* **68**, 232–307.

Johnston, I. A. (1980). Specialization of fish muscle. In *Development and Specialization of Skeletal Muscle*, ed. D. F. Goldspink, pp. 123–48. Cambridge: Cambridge University Press.

Johnston, I. A., Davison, W. & Goldspink, G. (1977). Energy metabolism of carp swimming muscles. *J. Comp. Physiol.* **114**, 203–16.

Johnston, M. F. & Ramon, F. (1982). Voltage independence of an electrotonic synapse. *Biophys. J.* **39**, 115–17.

Johnstone, B. M. & Boyle, A. J. F. (1967). Basilar membrane vibration examined with the Mössbauer technique. *Science* **158**, 389–90.

Johnstone, B. M., Taylor, J. J. & Boyle, A. J. (1970). Mechanics of guinea-pig cochlea. *J. Acoust. Soc. Amer.* **47**, 504–9.

Jones, D. T. & Reed, R. R. (1989). G_{olf}: an olfactory neuron specific-G protein involved in odorant signal transduction. *Science* **244**, 790–5.

Jones, S. W. (1985). Muscarinic and peptidergic excitation of bull-frog sympathetic neurones. *J. Physiol.* **366**, 63–87.

Jorgensen, A. O., Shen, A. C.-Y., Campbell, K. P. & MacLennan, D. H. (1983). Ultrastructural localization of calsequestrin in rat skeletal muscle by immunoferritin labeling of ultrathin frozen sections. *J. Cell Biol.* **92**, 1573–81.

Jørgensen, P. L. (1975). Purification and characterization of (Na^+, K^+)-ATPase. V. Conformational changes in the enzyme. Transitions between the Na-form and the K-form studied with tryptic digestion as a tool. *Biochim. Biophys. Acta* **401**, 399–415.

Jørgensen, P. L. (1992). Na,K-ATPase, structure and transport mechanism. In *Molecular Aspects of Transport Proteins*, ed. J. J. H. H. M. de Pont (*New Comprehensive Biochemistry*, ed. A. Neuberger & L. L. M. van Deenan, vol. 21), pp. 1–26. Amsterdam: Elsevier.

Josephson, R. K. (1993). Contraction dynamics and power output of skeletal muscle. *Ann. Rev. Physiol.* **55**, 527–46.

Kabsch, W., Mannherz, H. G., Suck, D., Pai, E. F. & Holmes, K. C. (1990). Atomic structure of the actin:DNase I complex. *Nature* **347**, 37–44.

Kaissling, K.-E. (1971). Insect olfaction. In *Handbook of Sensory Physiology*, vol. 4, part 1, ed. L. M. Beidler, pp. 351–431. Berlin: Springer-Verlag.

Kaissling, K.-E. (1986). Chemo-electrical transduction in insect olfactory receptors. *Ann. Rev. Neurosci.* **9**, 121–45.

Kaissling, K.-E. (1996). Peripheral mechanisms of pheromone reception in moths. *Chem. Senses* **21**, 257–68.

Kaissling, K.-E. & Priesner, E. (1970). Die Riechschwelle des Seidenspinners. *Naturwissenschaften* **57**, 23–8.

Kaissling, K.-E., Kasang, G., Bestmann, H. J., Stransky, & Vostrowsky, O. (1978). A new pheromone of the silkworm moth *Bombyx mori*. Sensory pathway and behavioural effects. *Naturwissenschaften* **65**, 382–4.

Kalinec, F., Holley, M. C., Iwasa, K. H., Lim, D. J. & Kachar, B. (1992). A membrane-based force generation mechanism in auditory sensory cells. *Proc. Natl. Acad. Sci. USA* **89**, 8671–5.

Kalmijn, A. J. (1971). The electric sense of sharks and rays. *J. Exp. Biol.* **55**, 371–83.

Kalmijn, A. J. (1982). Electric and magnetic field detection in elasmobranch fishes. *Science* **218**, 916–18.

Kalmijn, A. J. (1984). Theory of electromagnetic orientation: a further analysis. In *Comparative Physiology of Sensory Systems*, ed. L. Bolis, R. D. Keynes & S. H. P. Maddrell, pp. 525–60. Cambridge: Cambridge University Press.

Kamm, K. E. & Stull, J. T. (1985). The function of myosin and myosin light chain kinase phosphorylation in smooth muscle. *Ann. Rev. Pharmacol. Toxicol.* **25**, 593–620.

Kanaujia, S. & Kaissling, K.-E. (1985). Interactions of pheromone with moth antennae: adsorption, desorption and transport. *J. Insect Physiol.* **31**, 71–81.

Kandel, E. R. (1976). *Cellular Basis of Behaviour*. San Francisco: W. H. Freeman and Company.

Kandel, E. R. (1991). Cellular mechanisms of learning and the biological basis of individuality. In *Principles of Neural Science*, third edition, ed. E. R. Kandel, J. H. Schwartz & T. M. Jessell, pp. 1009–31. New York: Elsevier.

Kandel, E. R. & Schwartz, J. H. (1982). Molecular biology of learning: modulation of transmitter release. *Science* **218**, 433–43.

Kandel, E. R., Klein, M., Hochner, B., Shuster, M., Siegelbaum, S. A., Hawkins, R. D., Glanzman, D. L., Castellucci, V. F. & Abrams, T. W. (1987). Synaptic modulation and learning: new insights into synaptic transmission from the study of behaviour. In *Synaptic Function*, ed. G. M. Edelman, W. E. Gall & W. M. Cowan, pp. 471–518. New York: Wiley.

Kang, H. J. & Schuman, E. M. (1995). Long-lasting neurotrophin-induced enhancement of synaptic transmission in the adult hippocampus. *Science* **267**, 1658–62.

Kanner, B. I. (1993a). Structure and function of sodium-coupled neurotransmitter transporters. In *Molecular Biology and Function of Carrier Proteins*, ed. L. Reuss, J. M. Russell & M. L. Jennings (*Society of General Physiologists Series* **48**), pp. 243–50. New York: Rockefeller University Press.

Kanner, B. I. (1993b). Glutamate transporters from brain. *FEBS Lett.* **325**, 95–9.

Kanno, Y. & Loewenstein, W. R. (1964). Intercellular diffusion. *Science* **143**, 959–60.

Karn, J., Brenner, S. & Barnett, L. (1983). Protein structural domains in the *Caenorhabditis elegans unc*–54 myosin heavy chain gene are not separated by introns. *Proc. Natl. Acad. Sci. USA* **80**, 4253–7.

Karpinski, B. A., Morle, G. D., Huggenvik, J., Uhler, M. D. & Leiden, J. M. (1992). Molecular-cloning of human CREB-2 – an ATF/CREB transcription factor that can negatively regulate transcription from the cAMP response element. *Proc. Natl. Acad. Sci. USA* **89**, 4820–4.

Kato, G. (1936). On the excitation, conduction, and narcotization of single nerve fibers. *Cold Spr. Harb. Symp. Quant. Biol.* **4**, 202–13.

Katz, A. M. (1977). *Physiology of the Heart*. New York: Raven Press.

Katz, B. (1937). Experimental evidence for a non-conducted response of nerve to subthreshold stimulation. *Proc. R. Soc. Lond.* B **124**, 244–76.

Katz, B. (1949). Les constants électriques de la membrane du muscle. *Arch. Sci. Physiol.* **3**, 285.

Katz, B. (1950). Depolarization of sensory terminals and the initiation of impulses in the muscle spindle. *J. Physiol.* **111**, 261–82.

Katz, B. (1962). The transmission of impulses from nerve to muscle, and the subcellular unit of synaptic action. *Proc. R. Soc. Lond.* B **155**, 455–79.

Katz, B. (1969). *The Release of Neural Transmitter Substances.* Liverpool: Liverpool University Press.

Katz, B. & Miledi, R. (1965*a*). Propagation of electrical activity in motor nerve terminals. *Proc. R. Soc. Lond.* B **161**, 453–82.

Katz, B. & Miledi, R. (1965*b*). The measurement of synaptic delay, and the time course of acetylcholine release at the neuromuscular junction. *Proc. R. Soc. Lond.* B **161**, 483–95.

Katz, B. & Miledi, R. (1966). Input/output relation of a single synapse. *Nature* **212**, 1242–5.

Katz, B. & Miledi, R. (1967*a*). Tetrodotoxin and neuromuscular transmission. *Proc. R. Soc. Lond.* B **167**, 8–22.

Katz, B. & Miledi, R. (1967*b*). The timing of calcium action during neuromuscular transmission. *J. Physiol.* **189**, 535–44.

Katz, B. & Miledi, R. (1968). The role of calcium in neuromuscular facilitation. *J. Physiol.* **195**, 481–92.

Katz, B. & Miledi, R. (1970). Membrane noise produced by acetylcholine. *Nature* **226**, 962–3.

Katz, B. & Miledi, R. (1972). The statistical nature of the acetylcholine potential and its molecular components. *J. Physiol.* **224**, 665–700.

Katz, B. & Miledi, R. (1977). Transmitter leakage from motor nerve endings. *Proc. R. Soc. Lond.* B **196**, 59–72.

Katz, B. & Thesleff, S. (1957*a*). A study of the 'desensitization' produced by acetylcholine at the motor endplate. *J. Physiol.* **138**, 63–80.

Katz, B. & Thesleff, S. (1957*b*). On the factors which determine the amplitude of the 'miniature end-plate potential'. *J. Physiol.* **137**, 267–78.

Katzung, B. G. (1995). *Basic & Clinical Pharmacology*, sixth edition. Norwalk, CN: Appleton & Lange.

Kaupp, U. B., Niidome, T., Tanabe, T., Terada, S., Bönigk, W., Stühmer, W., Cook, N. J., Kangawa, K., Matsuo, H., Hirose, T., Miyata, T. & Numa, S. (1989). Primary structure and functional expression from complementary DNA of the rod photoreceptor cyclic GMP-gated channel. *Nature* **342**, 762–6.

Kawakami, K., Noguchi, S., Noda, M., Takahashi, H., Ohta, T., Kawamura, M., Nojima, H., Nagano, K., Hirose, T., Inayama, S., Hayashida, H., Miyata, T. & Numa, S. (1985). Primary structure of the α subunit of *Torpedo californica* ($Na^+ + K^+$) ATPase deduced from cDNA sequence. *Nature* **316**, 733–6.

Kebabian, J. W. & Calne, D. B. (1979). Multiple receptors for dopamine. *Nature* **277**, 93–6.

Keeton, W. T. (1981). The orientation and navigation of birds. In *Animal Migration*, ed. D. J. Aidley, pp. 81–104. Cambridge: Cambridge University Press.

Kelly, A. M. & Rubinstein, N. A. (1994). The diversity of muscle fiber types and its origin during development. In *Myology: Basic and Clinical*, second edition, ed. A. G. Engel & C. Franzini-Armstrong, vol. 1, pp. 119–33. New York: McGraw-Hill.

Kelso, S. R., Ganong, A. H. & Brown, T. H. (1986). Hebbian synapses in hippocampus. *Proc. Natl. Acad. Sci. USA* **83**, 5326–30.

Kemp, D. T. (1978). Stimulated acoustic emissions from within the human auditory system. *J. Acoust. Soc. Amer.* **64**, 1386–91.

Kendrick-Jones, J., Lehman, W. & Szent-Györgyi, A. G. (1970). Regulation in molluscan muscles. *J. Molec. Biol.* **54**, 313–26.

Kennedy, T. E., Hawkins, R. D. & Kandel, E. R. (1992). Molecular interrelationships between short- and long-term memory. In *Neuropsychology of Memory*, second edition, ed. L. R. Squire & N. Butters, pp. 557–74. New York: Guilford Press.

Kerjaschki, D. & Hörander, H. (1976). The development of mouse olfactory vesicles and their cell contacts: a freeze-etching study. *J. Ultrastruct. Res.* **54**, 420–44.

Keynan, S. & Kanner, B. I. (1988). γ-Aminobutyric acid transport in reconstituted preparations from rat brain: coupled sodium and chloride fluxes. *Biochemistry* **27**, 12–17.

Keynes, R. D. (1951). The ionic movements during nervous activity. *J. Physiol.* **114**, 119–50.

Keynes, R. D. (1963). Chloride in the squid giant axon. *J. Physiol.* **169**, 696–705.

Keynes, R. D. (1983). Voltage-gated ion channels in the nerve membrane. *Proc. R. Soc. Lond.* B **220**, 1–30.

Keynes, R. D. (1990). A series-parallel model of the voltage-gated sodium channel. *Proc. R. Soc. Lond.* B **240**, 425–32.

Keynes, R. D. (1994). The kinetics of voltage-gated ion channels. *Quart. Rev. Biophys.* **27**, 339–434.

Keynes, R. D. & Aidley, D. J. (1991). *Nerve and Muscle*, second edition. Cambridge: Cambridge University Press.

Keynes, R. D. & Lewis, P. R. (1951). The sodium and potassium content of cephalopod nerve fibres. *J. Physiol.* **114**, 151–82.

Keynes, R. D. & Martins-Ferreira, H. (1953). Membrane potentials in the electroplates of the electric eel. *J. Physiol.* **119**, 315–51.

Keynes, R. D. & Ritchie, J. M. (1984). On the binding of labelled saxitoxin to the squid giant axon. *Proc. R. Soc. Lond.* B **222**, 147–53.

Keynes, R. D. & Rojas, E. (1974). Kinetics and steady-state properties of the charged system controlling sodium conductance in the squid giant axon. *J. Physiol.* **239**, 393–434.

Kiang, N. Y.-S. (1965). *Discharge Patterns of Single Fibers in the Cat's Auditory Nerve.* Cambridge, MA: MIT Press.

Kiang, N. Y.-S., Moxon, E. C. & Levine, R. A. (1970). Auditory nerve activity in cats with normal and abnormal cochleas. In *Sensorineural Hearing Loss*, ed. G. E. Wolstenholme & J. Knight, pp. 241–68. London: Churchill.

Kinnamon, S. C. (1992). Role of K^+ channels in taste transduction. In *Sensory Transduction (Society of General*

Physiologists Series, vol. 47), ed. D. P. Corey & S. D. Roper, pp. 261–70. New York: Rockefeller University Press.

Kinnamon, S. C. & Margolskee, R. F. (1996). Mechanisms of taste transduction. *Curr. Opin. Neurobiol.* **6**, 506–13.

Kinnamon, S. C., Dionne, V. E. & Beam, K. G. (1988). Apical localization of K^+ channels in taste cells provides the basis for sour taste transduction. *Proc. Natl. Acad. Sci. USA* **85**, 7023–7.

Klein, M. & Kandel, E. R. (1980). Mechanism of calcium current modulation underlying presynaptic facilitation and behavioral sensitization in *Aplysia*. *Proc. Natl. Acad. Sci. USA* **77**, 6912–16.

Klein, M. G., Simon, B. J., Szucs, G. & Schneider, M. F. (1988). Simultaneous recording of calcium transients in skeletal muscle using high- and low-affinity calcium indicators. *Biophys. J.* **53**, 971–88.

Klein, M. G., Cheng, H., Santana, L. F., Jiang, Y. H., Lederer, W. J. & Schneider, M. F. (1996). Two mechanisms of quantized calcium release in skeletal muscle. *Nature* **379**, 455–8.

Klein, R. L. & Lagercrantz, H. (1981). Noradrenergic vesicles: composition and function. In *Chemical Neurotransmission: 75 Years*, ed. L. Stjärne, P. Hedqvist, H. Lagercrantz & Å. Wennmalm, pp. 69–83.

Klenchin, V. A., Calvert, P. D. & Bownds, M. D. (1995). Inhibition of rhodopsin kinase by recoverin: further evidence for a negative feedback-system in phototransduction. *J. Biol. Chem.* **270**, 16147–52.

Knappeis, G. G. & Carlsen, F. (1962). The ultrastructure of the Z disc in skeletal muscle. *J. Cell Biol.* **13**, 323–35.

Knappeis, G. G. & Carlsen, F. (1968). The ultrastructure of the M line in skeletal muscle. *J. Cell Biol.* **38**, 202–11.

Knibestöl, M. & Vallbo, A. B. (1970). Single unit analysis of mechanoreceptor activity from the human glabrous skin. *Acta Physiol. Scand.* **80**, 178–95.

Knight, P. J., Erickson, M. A., Rodgers, M. E., Beer, M. & Wiggins, J. W. (1986). Distribution of mass within native thick filaments of vertebrate skeletal muscle. *J. Molec. Biol.* **189**, 167–77.

Knowles, A. & Dartnall, H. J. A. (1977). *The Photobiology of Vision*. Vol. 2B of *The Eye*, second edition, ed. H. Davson. New York: Academic Press.

Kobayashi, H. & Libet, B. (1970). Actions of noradrenaline and acetylcholine on sympathetic ganglion cells. *J. Physiol.* **208**, 353–72.

Koch, K.-W. & Stryer, L. (1988). Highly cooperative feedback control of retinal rod guanylate cyclase by calcium ion. *Nature* **334**, 64–6.

Koenig, M., Monaco, A. P. & Kunkel, L. M. (1988). The complete sequence of dystrophin predicts a rod-shaped cytoskeletal protein. *Cell* **53**, 219–28.

Kolesnikov, S. S. & Margolskee, R. F. (1995). A cyclic nucleotide-suppressible conductance activated by transducin in taste cells. *Nature* **376**, 85–8.

Kolston, P. J. & Ashmore, J. F. (1996). Finite element micromechanical modeling of the cochlea in 3 dimensions. *J. Acoust. Soc. Amer.* **99**, 455–67.

Konno, T., Busch, C., von Kitzing, E., Imoto, K., Wang, F., Nakai, J., Mishina, M., Numa, S. & Sakmann, B. (1991). Rings of anionic acids as structural determinants of ion selectivity in the acetylcholine receptor channel. *Proc. R. Soc. Lond.* B **244**, 69–79.

Korenbrot, J. I. & Cone, R. A. (1972). Dark ionic flux and effects of light in isolated rod outer segments. *J. Gen. Physiol.* **60**, 20–45.

Korn, H. & Faber, D. S. (1975). An electrically mediated inhibition in goldfish medulla. *J. Neurophysiol.* **38**, 452–71.

Korn, H., Sur, C., Charpier, S., Legendre, P. & Faber, D. S. (1994). The one-vesicle hypothesis and multivesicular release. In *Molecular and Cellular Mechanisms of Neurotransmitter Release* (*Adv. Second Mess. Phosphoprotein Res.* **29**), ed. L. Stjärne, P. Greengard, S. E. Grillner, T. G. M. Hökfelt & D. R. Ottoson, pp. 301–22. New York: Raven Press.

Körschen, H. G., Illing, M., Seifert, R., Sesti, F., Williams, A., Gotzes, S., Colville, C., Müller, F., Dosé, A., Godde, M., Molday, L., Kaupp, B. & Molday, R. S. (1995). A 240 kDa protein represents the complete β subunit of the cyclic-nucleotide-gated channel from rod photoreceptor. *Neuron* **15**, 627–36.

Korte, M., Carroll, P., Wolf, E., Brem, G., Thoenen, H. & Bonhoeffer, T. (1995). Hippocampal long-term potentiation is impaired in mice lacking brain-derived neurotrophic factor. *Proc. Natl. Acad. Sci. USA* **92**, 8856–60.

Koutalos, Y. & Yau, K.-W. (1996). Regulation of sensitivity in vertebrate rod photoreceptors by calcium. *Trends Neurosci.* **19**, 73–81.

Kress, M., Huxley, H. E., Faruqi, A. R. & Hendrix, J. (1986). Structural changes during activation of frog muscle studied by time-resolved X-ray diffraction. *J. Molec. Biol.* **188**, 325–42.

Kretschmar, K. M. & Wilkie, D. R. (1969). A new approach to freezing tissues rapidly. *J. Physiol.* **202**, 66P–67P.

Kreuger, B. K., Worley, J. F. & French, R. J. (1983). Single sodium channels from rat brain incorporated into planar lipid bilayers. *Nature* **303**, 172–5.

Kriebel, M. E. & Gross, C. E. (1974). Multimodal distribution of frog miniature endplate potentials in adult, denervated, and tadpole leg muscle. *J. Gen. Physiol.* **64**, 85–103.

Krnjević, K. (1986). Amino acid transmitters: 30 years progress in research. In *Fast and Slow Chemical Signalling in the Nervous System*, ed. L. L. Iversen & E. C. Goodman, pp. 3–15. Oxford: Oxford University Press.

Krnjević, K. & Mitchell, J. F. (1961). The release of acetylcholine in the isolated rat diaphragm. *J. Physiol.* **155**, 246–62.

Krnjević, K. & Schwarz, S. (1967). The action of gamma-aminobutyric acid on cortical neurones. *Exp. Brain Res.* **3**, 320–36.

Kron, S. J. & Spudich, J. A. (1986). Fluorescent actin filaments move on myosin fixed to a glass surface. *Proc. Natl. Acad. Sci. USA* **83**, 6272–6.

Kroner, C., Breer, H., Singer, A. G. & O'Connell, R. J. (1996). Pheromone-induced second messenger signaling in hamster vomeronasal organ. *Neuroreport* **7**, 2989–92.

Kruger, M., Wright, J. & Wang, K. (1991). Nebulin as a length regulator of thin filaments of vertebrate skeletal muscles: correlation of thin filament length, nebulin size, and epitope profile. *J. Cell Biol.* **115**, 97–107.

Kubo, T., Fukuda, K., Mikami, A., Maeda, A., Takahashi, H., Mishina, M., Haga, T., Haga, K., Ichiyama, A., Kangawa, K., Kojima, M., Matsuo, H., Hirose, T. & Numa, S. (1986). Cloning, sequencing and expression of complementary DNA encoding the muscarinic acetylcholine receptor. *Nature* **323**, 411–16.

Kubo, Y., Baldwin, T. J., Jan, Y. N. & Jan, L. Y. (1993). Primary structure and functional expression of a mouse inward rectifier potassium channel. *Nature* **362**, 127–33.

Kuffler, S. W. (1942). Electrical potential changes in an isolated nerve–muscle junction. *J. Neurophysiol.* **5**, 18–26.

Kuffler, S. W. (1946). The relation of electrical potential changes to contracture in skeletal muscle. *J. Neurophysiol.* **9**, 367–77.

Kuffler, S. W. (1980). Slow synaptic responses in the autonomic ganglia and the pursuit of a peptidergic transmitter. *J. Exp. Biol.* **89**, 257–86.

Kuffler, S. W. & Vaughan Williams, E. M. (1953a). Small-nerve junctional potentials. The distribution of small motor nerves to frog skeletal muscle, and the membrane characteristics of the fibres they innervate. *J. Physiol.* **121**, 289–317.

Kuffler, S. W. & Vaughan Williams, E. M. (1953b). Properties of the 'slow' skeletal muscle fibres of the frog. *J. Physiol.* **121**, 318–40.

Kuffler, S. W. & Yoshikami, D. (1975a). The distribution of acetylcholine sensitivity at the post-synaptic membrane of vertebrate skeletal twitch muscles: iontophoretic mapping in the micron range. *J. Physiol.* **244**, 703–30.

Kuffler, S. W. & Yoshikami, D. (1975b). The number of transmitter molecules in a quantum: an estimate from iontophoretic application of acetylcholine at the neuromuscular synapse. *J. Physiol.* **251**, 465–82.

Kuffler, S. W., Hunt, C. C. & Quilliam, J. P. (1951). Function of medullated small-nerve fibres in mammalian ventral roots: efferent muscle spindle innervation. *J. Neurophysiol.* **14**, 29–54.

Kühne, W. (1864). *Untersuchungen über das Protoplasma und die Contractilität*. Leipzig: W. Engelmann.

Kühne, W. (1878). *On the Photochemistry of the Retina and on Visual Purple*, edited with notes by M. Foster. London: Macmillan.

Kullmann, D. M. & Siegelbaum, S. A. (1995). The site of expression of NMDA receptor-dependent LTP: new fuel for an old fire. *Neuron* **15**, 997–1002.

Kumar, N. M. & Gilula, N. B. (1996). The gap junction communication channel. *Cell* **84**, 381–8.

Kuno, M. (1964). Quantal components of excitatory synaptic potentials in spinal motoneurones. *J. Physiol.* **175**, 81–99.

Kuno, M. (1995). *The Synapse: Function, Plasticity, and Neurotrophism*. Oxford: Oxford University Press.

Kuo, S. C. & Sheetz, M. P. (1992). Optical tweezers in cell biology. *Trends Cell Biol.* **2**, 116–18.

Kurahashi, T. & Kaneko, A. (1993). Gating properties of the cAMP-gated channel in toad olfactory receptor cells. *J. Physiol.* **466**, 287–302.

Kürz, L. L., Zuhlke, R. D., Zhang, H. J. & Joho, R. H. (1995). Side-chain accessibilities in the pore of a K channel probed by sulfhydryl-specific reagents after cysteine-scanning mutagenesis. *Biophys. J.* **68**, 900–5.

Kushmerick, M. J. (1983). Energetics of muscle contraction. In *Handbook of Physiology*, section 10, *Skeletal Muscle*, ed. L. D. Peachey. Bethesda, MD: American Physiological Society.

Kushmerick, M. J. & Davies, R. E. (1969). The chemical energetics of muscle contraction. II. The chemistry, efficiency and power of maximally working sartorius muscles. *Proc. R. Soc. Lond.* B **174**, 315–53.

Kyte, J. & Doolittle, R. F. (1982). A simple method for displaying the hydropathic character of a protein. *J. Molec. Biol.* **157**, 105–32.

Labeit, S. & Kolmerer, B. (1995a). Titins: giant proteins in charge of muscle ultrastructure and elasticity. *Science* **270**, 293–6.

Labeit, S. & Kolmerer, B. (1995b). The complete primary structure of human nebulin and its correlation to muscle structure. *J. Molec. Biol.* **248**, 308–15.

Lagnado, L. & Baylor, D. A. (1994). Calcium controls light-triggered formation of catalytically active rhodopsin. *Nature* **367**, 273–7.

Lai, F. A., Erickson, H. P., Rousseau, E., Liu, Q.-Y. & Meissner, G. (1988). Purification and reconstitution of the calcium release channel from skeletal muscle. *Nature* **331**, 315–19.

Lamb, T. D. (1986). Transduction in vertebrate photoreceptors: the roles of cyclic GMP and calcium. *Trends Neurosci.* **9**, 224–8.

Lamb, T. D. (1994). Stochastic simulation of activation in the G-protein cascade of phototransduction. *Biophys. J.* **67**, 1439–54.

Lamb, T. D. (1996). Gain and kinetics in the G-protein cascade of phototransduction. *Proc. Natl. Acad. Sci. USA* **93**, 566–70.

Lamb, T. D. & Pugh, E. N. (1992). A quantitative account of the activation steps involved in phototransduction in amphibian photoreceptors. *J. Physiol.* **449**, 719–58.

LaMotte, R. H., Lundberg, L. E. R. & Torebjörk, H. E. (1992). Pain, hyperalgesia and activity in nociceptive C units in humans after intradermal injection of capsaicin. *J. Physiol.* **448**, 749–64.

Lamvic, M. K. (1978). Muscle thick filament mass measured by electron scattering. *J. Molec. Biol.* **122**, 55–68.

Lancet, D. (1986). Vertebrate olfactory reception. *Ann. Rev. Neurosci.* **9**, 329–55.

Langer, S. Z. (1974). Presynaptic regulation of catecholamine release. *Biochem. Pharmacol.* **23**, 1793–800.

Langer, S. Z. (1981). Presynaptic regulation of the release of catecholamines. *Pharmacol. Rev.* **32**, 337–62.

Langosch, D., Thomas, L. & Betz, H. (1988). Conserved quaternary structure of ligand-gated ion channels: the postsynaptic glycine receptor is a pentamer. *Proc. Natl. Acad. Sci. USA* **85**, 7394–8.

Lännergren, J. (1979). An intermediate type of muscle fibre in *Xenopus laevis*. *Nature* **279**, 254–6.

Lanzavecchia, G. (1968). Studi sulla muscolatura elicoidale e paramiosinica. II. Meccanismo di contrazione dei muscoli elicoidale. *Atti Accad. Naz. Lincei* **44**, 575–83.

Lapique, L. (1907). Recherches quantitatifs sur l'excitation électrique des nerfs traitée comme une polarisation. *J. Physiol. Paris* **9**, 622–35.

Larkman, A. U. & Jack, J. J. B. (1995). Synaptic plasticity: hippocampal LTP. *Curr. Opin. Neurobiol.* **5**, 324–34.

Larsson, H. P., Baker, O. S., Dhillon, D. S. & Isacoff, E. Y. (1996). Transmembrane movement of the *Shaker* K$^+$ channel S4. *Neuron* **16**, 387–97.

Latorre, R. (1994). Molecular workings of large conductance (maxi) Ca^{2+}-activated K$^+$ channels. In *Handbook of Membrane Channels*, ed. C. Peracchia, pp. 79–102. San Diego, CA: Academic Press.

Latorre, R., Oberhauser, A., Labarca, P. & Alvarez, O. (1989). Varieties of calcium-activated potassium channels. *Ann. Rev. Physiol.* **51**, 385–99.

Läuger, P. (1991). *Electrogenic Ion Pumps*. Sunderland, MA: Sinauer.

Lee, C. Y. (1970). Elapid neurotoxins and their mode of action. *Clinical Toxicol.* **3**, 457–72.

Lehman, W. (1976). Phylogenetic diversity of the proteins regulating muscular contraction. *Int. Rev. Cytol.* **44**, 55–92.

Lehman, W., Craig, R. & Vibert, P. (1994). Ca^{2+}-induced tropomyosin movement in *Limulus* thin filaments revealed by three-dimensional reconstruction. *Nature* **368**, 65–7.

Lehman, W., Vibert, P., Uman, P. & Craig, R. (1995). Steric-blocking by tropomyosin visualized in relaxed vertebrate muscle thin filaments. *J. Molec. Biol.* **251**, 191–6.

Lehman, W., Vibert, P., Craig, R & Bárány, M. (1996). Actin and the structure of smooth muscle thin filaments. In *Biochemistry of Smooth Muscle Contraction*, ed. M. Bárány, pp. 47–60. San Diego, CA: Academic Press.

Leksell, L. (1945). The action potential and excitatory effects of the small ventral root fibres to skeletal muscle. *Acta Physiol. Scand. Suppl.* **31**, 1–84.

Lester, H. A. (1992). The permeation pathway of neurotransmitter-gated ion channels. *Ann. Rev. Biophys.* **21**, 267–92.

Lester, R. A. J., Clements, J. D., Westbrook, G. L. & Jahr, C. E. (1990). Channel kinetics determine the time course of NMDA receptor-mediated synaptic currents. *Nature* **346**, 565–7.

Levi-Montalcini, R. (1987). The nerve growth-factor 35 years later. Nobel Lecture. *Science* **237**, 1154–62.

Levi-Montalcini, R. & Hamburger, V. (1951). Selective growth stimulating effects of mouse sarcoma on the sensory and sympathetic nervous system of the chick embryo. *J. Exp. Zool.* **116**, 321–61.

Levick, J. R. (1995). *An Introduction to Cardiovascular Physiology*, second edition. Oxford: Butterworth-Heinemann.

Levin, A. & Wyman, J. (1927). The viscous elastic properties of muscle. *Proc. R. Soc. Lond.* B **101**, 218–43.

Levin, L. R., Han, P.-L., Hwang, P. M., Feinstein, P. G., Davis, R. L. & Reed, R. R. (1992). The *Drosophila* learning and memory gene *rutabaga* encodes a Ca$^+$/calmodulin-responsive adenylyl cyclase. *Cell* **68**, 479–89.

Levine, J. & Taiwo, Y. (1994). Inflammatory pain. In *Textbook of Pain*, third edition, ed. P. D. Wall & R. Melzack, pp. 45–56. Edinburgh: Churchill Livingstone.

Levitzki, A. (1986). β-Adrenegic receptors and their mode of coupling to adenylate cyclase. *Physiol. Rev.* **66**, 819–54.

Lewin, G. R. & Barde, Y.-A. (1996). Physiology of the neurotrophins. *Ann. Rev. Neurosci.* **19**, 289–317.

Lewis, C., Neidhart, S., Holy, C., North, R. A., Buell, G. & Surprenant, A. (1995). Coexpression of P$_{2X2}$ and P$_{2X3}$ receptor subunits can account for ATP-gated currents in sensory neurons. *Nature* **377**, 432–5.

Lewis, C. A. (1979). Ion-concentration dependence of the reversal potential and the single-channel conductance of ion channels at the frog neuromuscular junction. *J. Physiol.* **286**, 417–45.

Li, M., West, J. W., Lai, Y., Scheuer, T. & Catterall, W. A. (1992). Functional modulation of brain sodium channels by cAMP-dependent phosphorylation. *Neuron* **8**, 1151–9.

Liberman, M. C. (1982). Single-neuron labelling in the cat auditory nerve. *Science* **216**, 1239–41.

Liebman, P. A. (1962). In situ microspectrophotometric study of the pigments of single retinal rods. *Biophys. J.* **2**, 161–78.

Liebman, P. A. (1972). Microspectrophotometry of photoreceptors. In *Handbook of Sensory Physiology*, vol. VII/1 *Photochemistry of Vision*, ed. H. J. A. Dartnall, pp. 481–528. Berlin: Springer-Verlag.

Liley, A. W. (1956). The effects of presynaptic polarization on the spontaneous activity at the mammalian neuromuscular junction. *J. Physiol.* **134**, 427–43.

Liman, E. R. (1996). Pheromone transduction in the vomeronasal organ. *Curr. Opin. Neurobiol.* **6**, 487–93.

Liman, E. R. & Corey, D. P. (1996). Electrophysiological characterization of chemosensory neurons from the mouse vomeronasal organ. *J. Neurosci.* **16**, 4625–37.

Liman, E. R., Tytgat, J. & Hess, P. (1992). Subunit stoichiometry of a mammalian K$^+$ channel determined by construction of multimeric cDNAs. *Neuron* **9**, 861–71.

Lindauer, M. & Martin, H. (1972). Magnetic effect in dancing bees. In *Animal Orientation and Navigation*, ed. S. R. Galler,

K. Schmidt-Koenig, G. J. Jacobs & R. Belleville, pp. 559–567. Washington, DC: National Aeronautics and Space Administration.

Lindblom, U. (1965). Properties of touch receptors in distal glabrous skin of the monkey. *J. Neurophysiol.* **28**, 966–85.

Lindemann, B. (1996). Taste reception. *Physiol. Rev.* **76**, 719–66.

Linder, T. M. & Quastel, D. M. J. (1978). A voltage-clamp study of the permeability change induced by quanta of transmitter at the mouse end-plate. *J. Physiol.* **281**, 535–56.

Ling, G. & Gerard, R. W. (1949). The normal membrane potential of frog sartorius fibers. *J. Cell. Comp. Physiol.* **34**, 383–96.

Linke, W. A., Ivemeyer, M., Olivieri, N., Kolmerer, B., Rüegg, J. C. & Labeit, S. (1996). Towards a molecular understanding of the elasticity of titin. *J. Molec. Biol.* **261**, 62–71.

Lipmann, F. (1941). Metabolic generation and utilization of phosphate bond energy. *Adv. Enzymol.* **1**, 99–162.

Lisman, J., Erickson, M. A., Richard, E. A., Cote, R. H., Bacigalupo, J., Johnson, E. & Kirkwood, A. (1992). Mechanisms of amplification, deactivation, and noise reduction in invertebrate photoreceptors. In *Sensory Transduction* (*Society of General Physiologists Series*, vol. 47), ed. D. P. Corey & S. D. Roper, pp. 175–99. New York: Rockefeller University Press.

Lissmann, H. W. (1951). Continuous electrical signals from the tail of a fish, *Gymnarchus niloticus* Cuv. *Nature* **167**, 201.

Lissmann, H. W. (1958). On the function and evolution of electric organs in fish. *J. Exp. Biol.* **35**, 156–91.

Lissmann, H. W. & Machin, K. E. (1958). The mechanism of object location in *Gymnarchus niloticus* and similar fish. *J. Exp. Biol.* **35**, 451–86.

Llinas, R., Steinberg, I. Z. & Walton, K. (1981). Presynaptic calcium currents in squid giant synapse. *Biophys. J.* **33**, 289–322.

Llinas, R., Sugimori, M & Silver, R. B. (1995). The concept of calcium-concentration microdomains in synaptic transmission. *Neuropharmacology* **34**, 1443–51.

Lloyd, D. P. C. (1943). Neuron patterns controlling transmission of ipsilateral hind limb reflexes in cat. *J. Neurophysiol.* **6**, 293–315.

Lloyd, D. P. C. (1949). Post-tetanic potentiation of response in monosynaptic reflex pathways of the spinal cord. *J. Gen. Physiol.* **33**, 147–70.

Lo, D. C. (1995). Neurotrophic factors and synaptic plasticity. *Neuron* **15**, 979–81.

Lodish, H., Baltimore, D., Berk, A., Zipursky, S. L., Matsudaira, P. & Darnell, J. (1995). *Molecular Cell Biology*, third edition. New York: Scientific American

Loewi, O. (1921). Über humorale Übertragbarkeit der Herznervenwirkung. *Pflügers Archiv* **189**, 239–42.

Logothetis, D. E., Kurachi, Y., Galper, J., Neer, E. J. & Clapham, D. E. (1987). The βγ subunits of GTP-binding proteins activate the muscarinic K$^+$ channel in heart. *Nature* **325**, 321–6.

Lohmann, K. (1929). Über die Pyrophosphatfraktion im Muskel. *Naturwissenschaft* **17**, 624–5.

Lohmann, K. (1934). Über die enzymatische Aufspaltung der Kreatinphosphorsäure: zugleich ein Beitrag zum Chemismus der Muskelkontraktion. *Biochem. Z.* **271**, 264–77.

Lombardi, V., Piazzesi, G. & Linari, M. (1992). Rapid regeneration of the actin-myosin power stroke in contracting muscle. *Nature* **355**, 638–41.

López-García, J. C., Arancio, O., Kandel, E. R. & Baranes, D. (1996). A presynaptic locus for long-term potentiation of elementary synaptic transmission at mossy fiber synapses in culture. *Proc. Natl. Acad. Sci. USA* **93**, 4712–17.

Lorenz, M., Popp, D. & Holmes, K. C. (1993). Refinement of the F-actin model against X-ray fiber diffraction data by use of a directed mutation algorithm. *J. Molec. Biol.* **234**, 826–36.

Lowenstein, O. & Roberts, T. D. M. (1950). The equilibrium function of the otolith organs of the thornback ray (*Raja clavata*). *J. Physiol.* **110**, 392–415.

Lowenstein, O. & Roberts, T. D. M. (1951). The localization and analysis of the responses to vibration from the isolated elasmobranch labyrinth. A contribution to the problem of the evolution of hearing in vertebrates. *J. Physiol.* **114**, 471–89.

Lowenstein, O. & Sand, A. (1936). The activity of the horizontal semicircular canal of the dogfish, *Scyllium canicula*. *J. Exp. Biol.* **13**, 416–28.

Lowenstein, O. & Sand, A. (1940a). The mechanism of the semicircular canal. A study of the responses of single-fibre preparations to angular accelerations and to rotation at constant speed. *Proc. R. Soc. Lond.* B **129**, 256–75.

Lowenstein, O. & Sand, A. (1940b). The individual and integrated activity of the semicircular canals of the elasmobranch labyrinth. *J. Physiol.* **99**, 89–101.

Lowenstein, O. & Wersäll, J. (1959). A functional interpretation of the electronmicroscopic structure of the sensory hairs in the cristae of the elasmobranch *Raja clavata* in terms of directional sensitivity. *Nature* **184**, 1807.

Lowenstein, O., Osborne, P. & Wersäll, J. (1964). Structure and innervation of the sensory epithelia in the labyrinth of the thornback ray (*Raja clavata*). *Proc. R. Soc. Lond.* B **160**, 1–12.

Lowey, S. (1994). The structure of vertebrate muscle myosin. In *Myology: Basic and Clinical*, second edition, ed. A.G. Engel & C. Franzini-Armstrong, vol. 1, pp. 485–505. New York: McGraw-Hill.

Lowey, S., Slayter, H. S., Weeds, A. G. & Baker, H. (1969). Substructure of the myosin molecule. I. Subfragments of myosin by enzymic degradation. *J. Molec. Biol.* **42**, 1–29.

Lowy, J. & Millman, B. M. (1963). The contractile mechanism of the anterior byssus retractor muscle of *Mytilus edulis*. *Phil. Trans. R. Soc. Lond.* B **246**, 105–48.

Lowy, J., Poulson, F. R. & Vibert, P. J. (1970). Myosin filaments in vertebrate smooth muscle. *Nature* **225**, 1053–4.

Lü, Q. & Miller, C. (1995). Silver as a probe of pore-forming residues in a potassium channel. *Science* **268**, 304–7.

Lucas, K. (1917). *The Conduction of the Nervous Impulse*, ed. E. D. Adrian. London: Longmans.

Luvigh, E. & McCarthy, E. F. (1938). Absorption of visible light by the refractive media of the human eye. AMA *Arch. Ophthalmol. N Y* **20**, 37–51.

Lumpkin, E. A. & Hudspeth, A. J. (1995). Detection of Ca^{2+} entry through mechanosensitive channels localizes the site of mechanoelectrical transduction in hair cells. *Proc. Natl. Acad. Sci. USA* **92**, 10297–301.

Lundberg, A. & Quilisch, H. (1953). On the effect of calcium on presynaptic potentiation and depression at the neuromuscular junction. *Acta Physiol. Scand. Suppl.* **111**, 121–9.

Lundsgaard, E. (1930a). Untersuchungen über Muskelkontraktionen ohne Milchsäurebildung. *Biochem. Z.* **217**, 162–77.

Lundsgaard, E. (1930b). Weitere Untersuchungen über Muskelkontraktionen ohne Milchsäurebildung. *Biochem. Z.* **227**, 51–83.

Luther, P. K. (1991). Three-dimensional reconstruction of a simple Z-band in fish muscle. *J. Cell Biol.* **113**, 1043–55.

Luther, P. K. & Squire, J. M. (1978). Three dimensional structure of the vertebrate muscle M-region. *J. Molec. Biol.* **125**, 313–24.

Luvigh, E. & McCarthy, E. F. (1938). Absorption of visible light by the refractive media of the human eye. *AMA Arch. Ophthalmol.* NY **20**, 37–51.

Lymn, R. W. & Taylor, E. W. (1970). Transient state phosphate production in the hydrolysis of nucleoside triphosphates by myosin. *Biochemistry* **9**, 2975–83.

Lymn, R. W. & Taylor, E. W. (1971). Mechanism of adenosine triphosphate hydrolysis by actomyosin. *Biochemistry* **10**, 4617–24.

Lynch, D. R. & Snyder, S. H. (1986). Neuropeptides: multiple molecular forms, metabolic pathways, and receptors. *Ann. Rev. Biochem.* **55**, 773–99.

Lynch, G., Larson, J., Kelso, S., Barrionuevo, G. & Schottler, F. (1983). Intracellular injections of EGTA block induction of hippocampal long-term potentiation. *Nature* **305**, 719–21.

Ma, W.-C. (1972). Dynamics of feeding responses in *Pieris brassicae* (Linn.) as a function of chemosensory input: a behavioural, ultrastructural and electrophysiological study. *Meded. Land. Wageningen* **72**, 1–162.

MacDermott, A. B., Mayer, M. L., Westbrook, M., Smith, S. J. & Barker, J. L. (1986). NMDA-receptor activation increases cytoplasmic calcium concentration in cultured spinal cord neurones. *Nature* **321**, 519–22.

MacDonald, J. F., Porietis, A. V. & Wojtowicz, J. (1982). L-Aspartic acid induces a region of negative slope conductance in the current–voltage relationship of cultured spinal neurons. *Brain Res.* **237**, 248–53.

Macdonald, R. L. & Olsen, R. W. (1994). $GABA_A$ receptor channels. *Ann. Rev. Neurosci.* **17**, 569–602.

Macdonald, R. L., Rogers, C. J. & Twyman, R. E. (1989). Barbiturate regulation of kinetic properties of the $GABA_A$ receptor channel of mouse spinal neurones in culture. *J. Physiol.* **417**, 483–500.

Machin, K. E. (1962). Electric receptors. *Symp. Soc. Exp. Biol.* **16**, 227–44.

Machin, K. E. & Pringle, J. W. S. (1959). The physiology of insect fibrillar muscle. II. Mechanical properties of a beetle flight muscle. *Proc. R. Soc. Lond.* B **151**, 204–25.

Machin, K. E. & Pringle, J. W. S. (1960). The physiology of insect fibrillar muscle. III. The effect of sinusoidal changes of length on a beetle flight muscle. *Proc. R. Soc. Lond.* B **152**, 311–30.

MacKinnon, R. (1991). Determination of the subunit stoichiometry of a voltage-activated potassium channel. *Nature* **350**, 232–35.

MacKinnon, R. & Yellen, G. (1990). Mutations affecting TEA blockade and ion permeation in voltage-activated K^+ channels. *Science* **250**, 276–79.

MacKinnon, R., Heginbotham, L. & Abramson, T. (1990). Mapping the receptor site for charybdotoxin, a pore-blocking potassium channel inhibitor. *Neuron* **5**, 767–71.

MacLennan, D. H. & Wong, P. T. S. (1971). Isolation of a calcium-sequestering protein from sarcoplasmic reticulum. *Proc. Natl. Acad. Sci. USA* **68**, 1231–5.

MacLennan, D. H., Brandl, C. J., Korczak, B. & Green, N. M. (1985). Amino-acid sequence of a $Ca^{2+} + Mg^{2+}$-dependent ATPase from rabbit muscle sarcoplasmic reticulum, deduced from its complementary DNA sequence. *Nature* **316**, 696–700.

MacVicar, B. & Dudek, F. E. (1981). Electrotonic coupling between pyramidal cells: a direct demonstration in rat hippocampal slices. *Science* **213**, 782–4.

Maggi, C. A. & Giuliani, S. (1995). Role of tachykinins as excitatory mediators of NANC contraction in the circular muscle of rat small intestine. *J. Autonom. Pharmacol.* **15**, 335–50.

Magleby, K. L. & Zengel, J. E. (1982). A quantitative description of stimulation-induced changes in transmitter release at the frog neuromuscular junction. *J. Gen. Physiol.* **80**, 613–38.

Makowski, L., Caspar, D. L. D., Phillips, W. C., Baker, T. S. & Goodenough, D. A. (1984). Gap junction structures. VI. Variation and conservation in connexon conformation and packing. *Biophys. J.* **45**, 208–18.

Malenka, R. C., Kauer, J. A., Zucker, R. S. & Nicoll, R. A. (1988). Postsynaptic calcium is sufficient for potentiation of hippocampal synaptic transmission. *Science* **242**, 81–4.

Malinow, R., Schulman, H. & Tsien, R. W. (1989). Inhibition of postsynaptic PKC or CaMKII blocks induction but not expression of LTP. *Science* **245**, 862–6.

Manabe, T. & Nicoll, R. A. (1994). Long-term potentiation: evidence against an increase in transmitter release probability in the CA1 region of the hippocampus. *Science* **265**, 1888–92.

Mansour, A., Fox, C. A., Akil, H. & Watson, S. J. (1995). Opioid-receptor mRNA expression in the rat CNS: anatomical and functional implications. *Trends Neurosci.* **18**, 22–9.

Margolskee, R. F. (1993). The molecular biology of taste transduction. *BioEssays* **15**, 645–50.

Maricq, A. V., Peterson, A. S., Brake, A. J., Myers, R. M. & Julius, D. (1991). Primary structure and functional expression of the $5HT_3$ receptor, a serotonin-gated ion channel. *Science* **254**, 432–6.

Markin, V. S. & Hudspeth, A. J. (1995). Modeling the active process of the cochlea: phase relations, amplification, and spontaneous oscillation. *Biophys. J.* **69**, 138–47.

Marks, W. B. (1965). Visual pigments of single goldfish cones. *J. Physiol.* **178**, 14–32.

Marks, W. B., Dobelle, W. H. & MacNichol, E. (1964). Visual pigments of single primate cones. *Science* **143**, 1181–3.

Marmont, G. (1949). Studies on the axon membrane: I. A new method. *J. Cell Comp. Physiol.* **34**, 351–82.

Marnay, A. & Nachmansohn, D. (1938). Choline esterase in voluntary muscle. *J. Physiol.* **92**, 37–47.

Marsh, B. B. (1952). The effects of ATP on the fibre volume of a muscle homogenate. *Biochim. Biophys. Acta* **9**, 247–60.

Marsh, D. (1975). Spectroscopic studies of membrane structure. *Essays Biochem.* **11**, 139–80.

Marston, S. B. (1982). The regulation of smooth muscle contractile proteins. *Prog. Biophys. Molec. Biol.* **41**, 1–41.

Marston, S. B. & Huber, P. A. J. (1996). Caldesmon. In *Biochemistry of Smooth Muscle Contraction*, ed. M. Bárány, pp. 77–90. San Diego, CA: Academic Press.

Marston, S. B. & Redwood, C. S. (1991). The molecular anatomy of caldesmon. *Biochem. J.* **279**, 1–16.

Marston, S. B. & Redwood, C. S. (1992). Inhibition of actin-tropomyosin activation of myosin MgATPase activity by the smooth muscle regulatory protein caldesmon. *J. Biol. Chem.* **267**, 16796–800.

Martin, A. R. (1955). A further study of the statistical composition of the end-plate potential. *J. Physiol.* **130**, 114–22.

Martinez, J. L. & Derrick, B. E. (1996). Long-term potentiation and learning. *Ann. Rev. Psychol.* **47**, 173–203.

Martonosi, A. N. & Beeler, T. J. (1983). Mechanism of Ca^{2+} transport by sarcoplasmic reticulum. In *Handbook of Physiology*, section 10 *Skeletal Muscle*, ed. L. Peachey, pp. 417–85. Bethesda, MD: American Physiological Society.

Martynov, V. I., Kostina, M. B., Feigina, M. Y., Miroshnikov, A. I. (1983). Limited proteolysis studies on the molecular organization of bovine rhodopsin in the photoreceptor membrane. *Bioorganicheskaya Khimiya* **9**, 734–45.

Maruyama, K. (1986). Connectin, an elastic filamentous protein of striated muscle. *Int. Rev. Cytol.* **104**, 81–114.

Maruyama, K. (1995). Birth of the sliding filament concept in muscle contraction. *J. Biochem.* **117**, 1–6.

Maruyama, K., Kimura, S., Ishii, T., Kuroda, M., Ohashi, K. & Muramatsu, S. (1977a). β-Actinin, a regulatory protein of muscle. *J. Biochem.* **81**, 215–32.

Maruyama, K., Matsubara, S., Natori, S. R., Nonomura, Y., Kimura, S., Ohashi, K., Murakami, F., Handa, S. & Eguchi, G. (1977b). Connectin, an elastic protein in muscle. *J. Biochem.* **82**, 317–37.

Maruyama, K., Kurokawa, H., Oosawa, M., Shimaoka, S., Yamamoto, H., Ito, M. & Maruyama, K. (1990). β-Actinin is equivalent to Cap Z protein. *J. Biol. Chem.* **265**, 8712–15.

Masaki, T., Endo, M. & Ebashi, S. (1967). Localization of 6S component of α-actinin at Z-band. *J. Biochem.* **62**, 630–1.

Massoulié, J., Pezzementi, L., Bon, S., Krejci, E. & Vallette, F.-M. (1993). Molecular and cellular biology of cholinesterases. *Prog. Neurobiol.* **41**, 31–91.

Matsuda, H., Saigusa, A. & Irisawa, H. (1987). Ohmic conductance through the inwardly rectifying K channel and blocking by internal Mg^{2+}. *Nature* **325**, 156–9.

Matsuda, L. A., Lolait, S. J., Brownstein, M. J., Young, A. C. & Bonner, T. I. (1990). Structure of a cannabinoid receptor and functional expression of the cloned cDNA. *Nature* **346**, 561–4.

Matthews, B. H. C. (1933). Nerve endings in mammalian muscle. *J. Physiol.* **78**, 1–53.

Matthews, H. R., Torre, V. & Lamb, T. D. (1985). Effects on the photoresponse of calcium buffers and cyclic GMP incorporated into the cytoplasm of retinal rods. *Nature* **313**, 582–5.

Matthews, H. R., Fain, G. L. & Cornwall, M. C. (1996). Role of cytoplasmic calcium concentration in the bleaching adaptation of salamander cone photoreceptors. *J. Physiol.* **490**, 293–303.

Matthews, P. B. C. (1962). The differentiation of two types of fusimotor fibre by their effects on the dynamic response of muscle spindle primary endings. *Quart. J. Exp. Physiol.* **47**, 324–33.

Matthews, P. B. C. (1964). Muscle spindles and their motor control. *Physiol. Rev.* **44**, 219–88.

Matthews, P. B. C. (1981). Evolving views on the internal operation and functional role of the muscle spindle. *J. Physiol.* **320**, 1–20.

Maxam, A. M. & Gilbert, W. (1980). Sequencing end-labelled DNA with base-specific chemical cleavages. *Methods Enzymol.* **65**, 499–560.

Maxwell, J. C. (1860). On the theory of compound colours and the relations of the colours of the spectrum. *Phil. Trans. R. Soc. Lond.* **150**, 57–84.

May, D. C., Ross, E. M., Gilman, A. G. & Smigel, M. D. (1985). Reconstitution of catecholamine-stimulated adenylate cyclase activity using three purified proteins. *J. Biol. Chem.* **260**, 15829–33.

Mayford, M., Barzilai, A., Keller, F., Schacher, S. & Kandel, E. R. (1992). Modulation of an NCAM-related adhesion molecule with long-term synaptic plasticity in *Aplysia*. *Science* **256**, 638–44.

McAllister, R. E., Noble, D. & Tsien, R. W. (1975). Reconstruction of the electrical activity of cardiac Purkinje fibres. *J. Physiol.* **251**, 1–59.

McBain, C. J. & Mayer, M. L. (1994). *N*-methyl-D-aspartic acid receptor structure and function. *Physiol. Rev.* **74**, 723–60.

McCleskey, E. W. (1994). Calcium channels: cellular roles and molecular mechanisms. *Curr. Opin. Neurobiol.* **4**, 304–312.

McDonald, K. S. & Moss, R. L. (1995). Osmotic compression of single cardiac myocytes eliminates the reduction in Ca^{2+} sensitivity of tension at short sarcomere length. *Circulation Res.* **77**, 199–205.

McLachlan, A. D. (1984). Structural implications of the myosin amino acid sequence. *Ann. Rev. Biophys.* **13**, 167–89.

McLachlan, A. D. & Karn, J. (1982). Periodic charge distributions in the myosin rod amino acid sequence match cross-bridge spacings in muscle. *Nature* **299**, 226–31.

McLachlan, A. D. & Stewart, M. (1976). The 14-fold periodity in α-tropomyosin and the interaction with actin. *J. Molec. Biol.* **103**, 271–98.

McLaughlin, S. & Margolskee, R. F. (1994). The sense of taste. *Amer. Sci.* **82**, 538–45.

McLaughlin, S., McKinnon, P. J. & Margolskee, R. F. (1992). Gustducin is a taste-cell-specific G protein closely related to the transducins. *Nature* **357**, 563–9.

McLennan, H. (1963). *Synaptic Transmission.* Philadelphia: W. B. Saunders.

McNaughton, P. A. (1990). Light response of vertebrate photoreceptors. *Physiol. Rev.* **70**, 847–83.

McNaughton, P. A., Cervetto, L. & Nunn, B. J. (1986). Measurement of the intracellular free calcium concentration in salamander rods. *Nature* **322**, 261–3.

McNicol, D. (1972). *A Primer of Signal Detection Theory.* London: Allen & Unwin.

Meech, R. W. (1972). Intracellular calcium injection causes increased potassium conductance in *Aplysia* nerve cells. *Comp. Biochem. Physiol.* **42A**, 493–9.

Meijer, G. M., Ritter, F. J., Persoons, C. J., Minks, A. & Voerman, S. (1972). Sex pheromones of summer fruit tortrix moth *Adoxophyes orana*: two synergistic isomers. *Science* **175**, 1469–70.

Meissner, G. (1994). Ryanodine receptor/Ca^{2+} release channels and their regulation by endogenous effectors. *Ann. Rev. Physiol.* **56**, 485–508.

Melzack, R. & Wall, P. D. (1996). *The Challenge of Pain*, updated second edition. London: Penguin Books.

Melzer, W., Schneider, M. F., Simon, B. J. & Szucs, G. (1986). Intramembrane charge movement and calcium release in frog skeletal muscle. *J. Physiol.* **373**, 481–511.

Melzer, W., Herrmann-Frank, A. & Lüttgau, H. C. (1995). The role of Ca^{2+} ions in excitation–contraction coupling of skeletal muscle fibres. *Biochim. Biophys. Acta* **1241**, 59–116.

Menco, B. P. M., Bruch, R. C., Dau, B. & Danho, W. (1992). Ultrastructural localization of olfactory transduction components: the G protein subunit $G_{olf\alpha}$ and type III adenylyl cyclase. *Neuron* **8**, 441–53.

Menini, A., Picco, C. & Firestein, S. (1995). Quantal-like current fluctuations induced by odorants in olfactory receptor cells. *Nature* **373**, 435–7.

Merbs, S. L. & Nathans, J. (1992*a*). Absorption spectra of human cone pigments. *Nature* **356**, 433–5.

Merbs, S. L. & Nathans, J. (1992*b*). Absorption spectra of the hybrid pigments responsible for anomalous color vision. *Science* **258**, 464–6.

Metcalf, A., Chelliah, Y. & Hudspeth, A. J. (1994). Molecular cloning of a myosin Iβ isozyme that may mediate adaptation by hair cells of the bullfrog's internal ear. *Proc. Natl. Acad. Sci. USA* **91**, 11821–5.

Meyerhof, O. & Lohmann, K. (1932). Über energetische Wechselbeziehungen zwischem dem Umstatz der Phosphorsäure ester im Muskelextrakt. *Biochem. Z.* **253**, 431–61.

Mihalyi, E. & Szent-Györgyi, A. G. (1953). Trypsin digestion of muscle proteins. III. Adenosine triphosphatase activity and actin-binding capacity of the digested myosin. *J. Biol. Chem.* **201**, 211–19.

Mikami, A., Imoto, K., Tanabe, T., Niidome, T., Mori, Y., Takeshima, H., Narumiya, S. & Numa, S. (1989). Primary structure and functional expression of the cardiac dihydropyridine-sensitive calcium channel. *Nature* **340**, 230–3.

Miledi, R. (1973). Transmitter release induced by injection of calcium ions into nerve terminals. *Proc. R. Soc. Lond.* B **183**, 421–5.

Miledi, R., Parker, I. & Zhu, P. H. (1982). Calcium transients evoked by action potentials in frog twitch muscle fibres. *J. Physiol.* **333**, 655–79.

Miledi, R., Molenaar, P. C. & Polak, R. L. (1983). Electrophysiological and chemical determination of acetylcholine release at the frog neuromuscular junction. *J. Physiol.* **334**, 245–54.

Mill, P. J. & Knapp, M. F. (1970). The fine structure of obliquely striated body wall muscles in the earthworm *Lumbricus terrestris* Linn. *J. Cell Sci.* **7**, 233–61.

Miller, C. (1984). Integral membrane channels: studies on model membranes. *Physiol. Rev.* **63**, 1209–42.

Miller, C. (1986). Ion channel reconstitution: why bother? In *Ionic Channels in Cell and Model Systems*, ed. R. Latorre, pp. 257–71. New York: Plenum Press.

Miller, C. (1988). *Shaker* shakes out potassium channels. *Trends Neurosci.* **11**, 185–6.

Miller, C. (1995). The charybdotoxin family of K^+ channel-blocking peptides. *Neuron* **15**, 5–10.

Miller, J. A., Agnew, W. S. & Levinson, S. R. (1983). Principal glycopeptide of the tetrodotoxin/saxitoxin binding protein from *Electrophorus electricus*: isolation and partial chemical and physical characterization. *Biochemistry* **22**, 462–70.

Miller, R. F. & Dowling, J. E. (1970). Intracellular responses of the Müller (glial) cells of the mudpuppy retina: their relation to the b-wave of the electroretinogram. *J. Neurophysiol.* **33**, 323–41.

Miller, T. M. & Heuser, J. E. (1984). Endocytosis of synaptic vesicle membrane at the frog neuromuscular junction. *J. Cell Biol.* **98**, 685–98.

Miller, W. H. (1990). Dark mimic. *Invest. Ophthalmol.* **31**, 1664–73.

Miller, W. H. & Nicol, G. D. (1979). Evidence that cyclic GMP regulates membrane potential in rod photoreceptors. *Nature* **280**, 64–6.

Milligan, R. A. & Flicker, P. F. (1987). Structural relationships of actin, myosin and tropomyosin revealed by cryo-electron microscopy. *J. Cell Biol.* **105**, 29–39.

Milligan, R. A., Whittaker, M. & Safer, D. (1990). Molecular structure of F-actin and location of surface binding sites. *Nature* **348**, 217–21.

Mishina, M., Kurosaki, T., Tobimatsu, T., Morimoto, Y., Noda, M., Yamamoto, T., Tereo, M., Lindstrom, J., Takahashi, T., Kuno, M. & Numa, S. (1984). Expression of functional acetylcholine receptor from cloned cDNAs. *Nature* **307**, 604–8.

Mishina, M., Takai, T., Imoto, K., Noda, M., Takahashi, T., Numa, S., Methfessel, C. & Sakmann, B. (1986). Molecular distinction between fetal and adult forms of muscle acetylcholine receptor. *Nature* **321**, 406–11.

Mitchell, J. F. & Silver, A. (1963). The spontaneous release of acetylcholine from the denervated hemidiaphragm of the rat. *J. Physiol.* **165**, 117–29.

Miyata, H., Yoshikawa, H., Hakozaki, H., Suzuki, N., Furuno, T., Ikegami, A., Kinosita, K., Nishizaka, T & Ishiwata, S. (1995). Mechanical measurements of single actomyosin motor force. *Biophys. J.* **68**, 286s–290s.

Molday, R. S. (1996). Calmodulin regulation of cyclic-nucleotide-gated channels. *Curr. Opin. Neurobiol.* **6**, 445–52.

Mollon, J. D. & Sharpe, L. T. (1983). *Colour Vision: Physiology and Psychophysics*. London: Academic Press.

Molloy, J. E., Kyrtatas, V., Sparrow, J. C. & White, D. C. S. (1987). Kinetics of flight muscles from insects with different wingbeat frequencies. *Nature* **328**, 449–51.

Molloy, J. E., Burns, J. E., Kendrick-Jones, J., Tregear, R. T. & White, D. C. S. (1995). Movement and force produced by a single myosin head. *Nature* **378**, 209–12.

Mombaerts, P. (1996). Targeting olfaction. *Curr. Opin. Neurobiol.* **6**, 481–6.

Mombaerts, P., Wang, F., Dulac, C., Chao, S. K., Nemes, A., Medelsohn, M., Edmondson, J. & Axel, R. (1996). Visualizing an olfactory sensory map. *Cell* **87**, 675–86.

Monaghan, D. T., Bridges, R. J. & Cotman, C. W. (1989). The excitatory amino acid receptors. *Ann. Rev. Pharmacol. Toxicol.* **29**, 365–402.

Monck, J. R. & Fernandez, J. M. (1994). The exocytotic fusion pore and neurotransmitter release. *Neuron* **12**, 707–16.

Montarolo, P. G., Goelet, P., Castellucci, V. F., Morgan, J.,

Kandel, E. R. & Schacher, S. (1986). A critical period for macromolecular synthesis in long-term heterosynaptic facilitation in *Aplysia. Science* **234**, 1249–54.

Monyer, H., Sprengel, R., Schoepfer, R., Herb, A., Higuchi, M., Lomeli, H., Burnashev, N., Sakmann, B. & Seeburg, P. H. (1992). Heteromeric NMDA receptors: molecular and functional distinction of subtypes. *Science* **256**, 1217–21.

Moore, B. C. J. (1997). *Introduction to the Psychology of Hearing*, fourth edition. London: Academic Press.

Moore, P. B., Huxley, H. E. & DeRosier, D. J. (1970). Three-dimensional reconstruction of F-actin, thin filaments and decorated thin filaments. *J. Molec. Biol.* **50**, 279–95.

Mooseker, M. S. & Cheney, R. E. (1995). Unconventional myosins. *Ann. Rev. Cell Biol.* **11**, 633–75.

Morgan, J. P. & Blinks, J. R. (1982). Intracellular Ca^{2+} transients in the cat papillary muscle. *Can. J. Physiol. Pharmacol.* **60**, 524–9.

Morita, H. (1992). Transduction process and impulse initiation in insect contact chemoreceptor. *Zool. Sci.* **9**, 1–16.

Morita, H. & Shiraishi, A. (1985). Chemoreception physiology. In *Comparative Insect Physiology, Biochemistry and Pharmacology*, ed. G. A. Kerkut & I. Gilbert, vol. 6, pp. 133–70. Oxford: Pergamon Press.

Morita, H. & Yamashita, S. (1959). Generator potential of an insect chemoreceptor. *Science* **130**, 922.

Moriyoshi, K., Masu, M., Ishii, T., Shigemoto, R., Mizuno, N. & Nakanishi, S. (1991). Molecular cloning and characterization of the rat NMDA receptor. *Nature* **354**, 31–7.

Morton, R. A. (1944). Chemical aspects of the visual process. *Nature* **153**, 69–71.

Morton, R. A. & Goodwin, T. W. (1944). Preparation of retinene in vitro. *Nature* **153**, 405–6.

Morton, R. A. & Pitt, G. A. J. (1955). Studies on rhodopsin 9. pH and the analysis of indicator yellow. *Biochem. J.* **59**, 128–34.

Mueller, H. & Perry, S. V. (1962). The degradation of heavy meromyosin by trypsin. *Biochem. J.* **85**, 431–9.

Mullen, E. & Akhtar, M. (1983). Structural studies on membrane-bound bovine rhodopsin. *Biochem. J.* **211**, 45–54.

Müller, J. (1826). *Zur vergleichenden Physiologie des Gesichtssinnes*. Leipzig.

Murphy, R. A. (1994). What is special about smooth muscle? The significance of covalent cross-bridge regulation. *FASEB J.* **8**, 311–17.

Murray, R. W. (1962). The response of the ampullae of Lorenzini of elasmobranchs to electrical stimulation. *J. Exp. Biol.* **39**, 119–28.

Myers, R. A., Cruz, L. J., Rivier, J. E. & Olivera, B. M. (1993). *Conus* peptides as chemical probes for receptors and ion channels. *Chemical Rev.* **93**, 1923–36.

Nachtigall, P. E. & Moore, P. W. B. (eds.) (1988). *Animal Sonar*. NATO Advanced Study Institute series. New York: Plenum Press.

Nafe, J. P. (1929). A quantitative theory of feeling. *J. Gen. Psychol.* **2**, 199–211.

Nagai, T., Makinose, M. & Hasselbach, W. (1960). Der physiologische Erschlaffungsfaktor und die Muskelgrana. *Biochim. Biophys. Acta* **54**, 338–44.

Nakamura, T. & Gold, G. H. (1987). A cyclic nucleotide-gated conductance in olfactory receptor cilia. *Nature* **325**, 442–4.

Nakanishi, S. & Masu, M. (1994). Molecular diversity and functions of glutamate receptors. *Ann. Rev. Biophys.* **23**, 319–48.

Narahashi, T. (1963). The properties of insect axons. *Adv. Insect Physiol.* **1**, 176–256.

Narahashi, T., Moore, J. W. & Scott, W. R. (1964). Tetrodotoxin blockage of sodium conductance increase in lobster giant axons. *J. Gen. Physiol.* **47**, 965–74.

Nastuk, W. L. (1953). The electrical activity of the muscle cell membrane at the neuromuscular junction. *J. Cell. Comp. Physiol.* **42**, 249–72.

Nastuk, W. L. & Hodgkin, A. L. (1950). The electrical activity of single muscle fibres. *J. Cell. Comp. Physiol.* **35**, 39–73.

Nathans, J. (1992). Rhodopsin: structure, function, and genetics. *Biochemistry* **31**, 4923–31.

Nathans, J. (1994). In the eye of the beholder: visual pigments and inherited variation in human vision. *Cell* **78**, 357–60.

Nathans, J. & Hogness, D. S. (1983). Isolation, sequence analysis, and intron–exon arrangement of the gene encoding bovine rhodopsin. *Cell* **34**, 807–14.

Nathans, J. & Hogness, D. S. (1984). Isolation and nucleotide sequence of the gene encoding human rhodopsin. *Proc. Natl. Acad. Sci. USA* **81**, 4851–5.

Nathans, J., Thomas, D. & Hogness, D. S. (1986a). Molecular genetics of human colour vision: the genes encoding blue, green and red pigments. *Science* **232**, 193–202.

Nathans, J., Piantanida, T. P., Eddy, R., Shows, T. B. & Hogness, D. S. (1986b). Molecular genetics of inherited variation in human colour vision. *Science* **232**, 203–10.

Nathans, J., Merbs, S. L., Sung, C.-H., Weitz, C. J. & Wang, Y. (1992). Molecular genetics of human visual pigments. *Ann. Rev. Genetics* **26**, 403–24.

Nave, R., Fürst, D. O. & Weber, K. (1989). Visualization of the polarity of isolated titin molecules – a single globular head on a long thin rod as the M-band anchoring domain. *J. Cell Biol.* **109**, 2177–87.

Nawa, H., Kotani, H. & Nakanishi, S. (1984). Tissue-specific generation of two preprotachykinin mRNAs from one gene by alternative RNA splicing. *Nature* **312**, 729–34.

Needham, D. M. (1971). *Machina Carnis.* Cambridge: Cambridge University Press.

Neher, E. (1992). Ion channels for communication between and within cells. *Science* **256**, 498–502. Also published in *Les Prix Nobel 1991*, pp. 120–35. Stockholm: Almquist & Wiksell.

Neher, E. & Sakmann, B. (1976). Single-channel currents recorded from membrane of denervated frog muscle cells. *Nature* **260**, 799–802.

Neher, E. & Stevens, C. F. (1977). Conductance fluctuations and ionic pores in membranes. *Ann. Rev. Biophys.* **6**, 345–81.

Neitz, M., Neitz, J. & Jacobs, G. H. (1991). Spectral tuning of pigments underlying red-green color vision. *Science* **252**, 971–4.

Nemeth, P. M., Pette, D. & Vrbova, G. (1981). Comparison of enzyme activities among single muscle fibres within defined motor units. *J. Physiol.* **311**, 489–95.

Neumcke, B. & Stämpfli, R. (1982). Sodium currents and sodium-current fluctuations in rat myelinated nerve fibres. *J. Physiol.* **329**, 163–84.

Neville, A. C. (1963). Motor unit distribution of the dorsal longitudinal flight muscles in locusts. *J. Exp. Biol.* **40**, 123–36.

Ngai, J., Chess, A., Dowling, M. M., Necles, N., Macagno, E. R. & Axel, R. (1993). Coding of olfactory information: topography of odorant receptor expression in the catfish epithelium. *Cell* **72**, 667–80.

Nguyen, P. V., Abel, T. & Kandel, E. R. (1994). Requirement of a critical period of transcription for induction of a late phase of LTP. *Science* **265**, 1104–7.

Nicoll, D. A., Longoni, S. & Philipson, K. D. (1990). Molecular cloning and functional expression of the cardiac sarcolemmal Na^+–Ca^{2+} exchanger. *Science* **250**, 562–5.

Nicoll, R. A. & Alger, B. E. (1979). Presynaptic inhibition: transmitter and ionic mechanisms. *Int. Rev. Neurobiol.* **21**, 217–58.

Nicoll, R. A. & Malenka, R. C. (1995). Contrasting properties of two forms of long-term potentiation in the hippocampus. *Nature* **377**, 115–18.

Nicoll, R. A., Kauer, J. A. & Malenka, R. C. (1988). The current excitement in long-term potentiation. *Neuron* **1**, 97–103.

Nicoll, R. A., Malenka, R. C. & Kauer, J. A. (1990). Functional comparison of neurotransmitter receptor subtypes in mammalian central nervous system. *Physiol. Rev.* **70**, 513–65.

Niedergerke, R. (1956). The potassium chloride contracture of the heart and its modification by calcium. *J. Physiol.* **134**, 584–99.

Niemann, H., Blasi, J. & Jahn, R. (1994). Clostridial neurotoxins: new tools for dissecting exocytosis. *Trends Cell Biol.* **4**, 179–85.

Nishi, S. & Koketsu, K. (1960). Electrical properties and activities of single sympathetic neurons in frogs. *J. Cell. Comp. Physiol.* **55**, 15–30.

Nishi, S. & Koketsu, K. (1968). Early and late afterdischarges of amphibian sympathetic ganglion cells. *J. Neurophysiol.* **31**, 109–21.

Nishizuka, Y. (1984). The role of protein kinase C in cell surface signal transduction and tumour promotion. *Nature* **308**, 693–8.

Nishizuka, Y. (1992). Intracellular signalling by hydrolysis of phospholipids and activation of protein kinase C. *Science* **258**, 607–14.

Noble, D. (1962). A modification of the Hodgkin–Huxley equations applicable to Purkinje fibre action and pacemaker potentials. *J. Physiol.* **160**, 317–52.

Noble, D. (1966). Applications of Hodgkin–Huxley equations to excitable tissues. *Physiol. Rev.* **46**, 1–50.

Noble, D. (1984). The surprising heart: a review of recent progress in cardiac electrophysiology. *J. Physiol.* **353**, 1–50.

Noble, D. (1994). Ionic mechanisms in cardiac electrical activity. In *Cardiac Electrophysiology: From Cell to Bedside*, 2nd edition, ed. D. P. Zipes & J. Jalife, pp. 305–13. Philadelphia: W. B. Saunders.

Noble, D. (1995). The development of mathematical models of the heart. *Chaos, Solitons & Fractals* **5**, 321–33.

Noble, D. & Powell, T. (eds.) (1987). *Electrophysiology of Single Cardiac Cells*. London: Academic Press.

Noda, M., Takahashi, H., Tanabe, T., Toyosato, M., Furutani, Y., Hirose, T., Asai, M., Inayama, S., Miyata, T. & Numa, S. (1982a). Primary structure of α-subunit precursor of *Torpedo californica* acetylcholine receptor deduced from cDNA sequence. *Nature* **299**, 793–7.

Noda, M., Furutani, Y., Takahashi, H., Toyosato, M., Hirose, T., Inayama, S., Nakanishi, S. & Numa, S. (1982b). Cloning and sequence analysis of cDNA for bovine adrenal preproenkephalin. *Nature* **295**, 202–6.

Noda, M., Takahashi, H., Tanabe, T., Toyosato, M., Kikyotani, S., Hirose, T., Asai, M., Takashima, H., Inayama, S., Miyata, T. & Numa, S. (1983a). Primary structures of β- and δ-subunit precursors of *Torpedo californica* acetylcholine receptor deduced from cDNA sequences. *Nature* **301**, 251–5.

Noda, M., Takahashi, H., Tanabe, T., Toyosato, M., Kikyotani, S., Furutani, Y., Hirose, T., Takashima, H., Inayama, S., Miyata, T. & Numa, S. (1983b). Structural homology of *Torpedo californica* acetylcholine receptor units. *Nature* **302**, 528–32.

Noda, M., Shimizu, S., Tanabe, T., Takai, T., Kayano, T., Ikeda, T., Takahashi, H., Nakayama, H., Kanaoka, Y., Minamino, N., Kangawa, K., Matsuo, H., Raftery, M. A., Hirose, T., Inayama, S., Hayashida, H., Miyati, T. & Numa, S. (1984). Primary structure of *Electrophorus electricus* sodium channel deduced from cDNA sequence. *Nature* **312**, 121–7.

Noda, M., Ikeda, T., Kayano, T., Suzuki, H., Takeshima, H., Kurasaki, M., Takahashi, H. & Numa, S. (1986a). Existence of distinct sodium channel messenger RNAs in rat brain. *Nature* **320**, 188–92.

Noda, M., Ikeda, T., Suzuki, H., Takeshima, H., Takahashi, T., Kuno, M. & Numa, S. (1986b). Expression of functional sodium channels from cloned cDNA. *Nature* **322**, 826–8.

Noguchi, J., Yanagisawa, M., Imamura, M., Kasuya, Y., Sakurai, T., Tanaka, T. & Masaki, T. (1992). Complete primary structure and tissue expression of chicken pectoralis M-protein. *J. Biol. Chem.* **267**, 20302–10.

Nonner, W., Rojas, E. & Stämpfli, R. (1975a). Gating currents in the node of Ranvier: voltage and time dependence. *Phil. Trans. R. Soc. Lond.* B **270**, 483–92.

Nonner, W., Rojas, E. & Stämpfli, R. (1975b). Displacement currents in the node of Ranvier. *Pflügers Archiv* **354**, 1–18.

North, R. A. (1986). Opioid receptor types and membrane ion channels. *Trends Neurosci.* **9**, 114–17.

Norton, S. J. & Neely, S. T. (1987). Tone-burst-evoked otoacoustic emissions from normal-hearing subjects. *J. Acoust. Soc. Amer.* **81**, 1860–72.

Nowak, L., Bregestovski, P., Ascher, P., Herbert, A. & Prochiantz, A. (1984). Magnesium gates glutamate-activated channels in mouse central neurones. *Nature* **307**, 462–5.

Nowycky, M. C., Fox, A. P. & Tsien, R. W. (1985). Three types of neuronal calcium channel with different calcium agonist sensitivity. *Nature* **316**, 440–3.

Numa, S. (1986). Evolution of ionic channels. *Chemica Scripta* **26B**, 173–8.

Numa, S., Noda, M., Takahashi, H., Tanabe, T., Toyosato, M., Furutani, Y. & Kikyotani, S. (1983). Molecular structure of the nicotinic acetylcholine receptor. *Cold Spr. Harb. Symp. Quant. Biol.* **48**, 57–69.

Nuttall, A. L., Dolan, D. F. & Avinish, G. (1991). Laser Doppler velocimetry of basilar membrane vibration. *Hearing Res.* **51**, 203–13.

Obara, S. & Sugawara, Y. (1984). Electroreceptor mechanisms in teleost and non-teleost fishes. In *Comparative Physiology of Sensory Systems*, ed. L. Bolis, R. D. Keynes & S. H. P. Maddrell, pp. 509–23. Cambridge: Cambridge University Press.

Obermann, W. M. J., Gautel, M., Steiner, F., Vanderven, P. F. M., Weber, K. & Furst, D. O. (1996). The structure of the sarcomeric M-band – localization of defined domains of myomesin, M-protein, and the 250–kD carboxy-terminal region of titin by immunoelectron microscopy. *J. Cell. Biol.* **134**, 1441–53.

Obermann, W. M. J., Gautel, M., Weber, K. & Fürst, D. O. (1997). Molecular structure of the sarcomeric M band: mapping of titin and myosin binding domains in myomesin and the identification of a potential regulatory phosphorylation site in myomesin. *EMBO J.* **16**, 211–20.

Ochoa, J. & Torebjörk, E. (1983). Sensations evoked by intraneural microstimulation of single mechanoreceptor units innervating the human hand. *J. Physiol.* **342**, 633–45.

Ochoa, J. & Torebjörk, E. (1989). Sensations evoked by intraneural microstimulation of C nociceptor fibres in human skin nerves. *J. Physiol.* **415**, 583–99.

O'Dell, T. J., Huang, P. L., Dawson, T. M., Dinerman, J. L., Snyder, S. H., Kandel, E. R. & Fishman, M. C. (1994). Endothelial NOS and the blockade of LTP by NOS inhibitors in mice lacking neuronal NOS. *Science* **265**, 542–6.

O'Dowd, B. F., Hnatowich, M., Caron, M. G., Lefkowitz, R. J. &

Bouvier, M. (1989). Palmitoylation of the human β_2-adrenergic receptor. *J. Biol. Chem.* **264**, 7564–9.

Offer, G. (1974). The molecular basis of muscular contraction. In *Companion to Biochemistry*, ed. A. T. Bull, J. R. Lagnado, J. O. Thomas & K. F. Tipton, pp. 623–71. London: Longman.

Offer, G., Moos, C. & Starr, R. (1973). A new protein of the thick filaments of vertebrate skeletal myofibrils. *J. Molec. Biol.* **74**, 653–76.

Ogawa, H. (1996). The Merkel cell as a possible mechanoreceptor cell. *Prog. Neurobiol.* **49**, 317–34.

Ogden, D. (1994). *Microelectrode Techniques. The Plymouth Workshop Handbook*, second edition. Cambridge: Company of Biologists.

Ogden, D. & Stanfield, P. R. (1994). Patch clamp techniques for single channel and whole-cell recording. In *Microelectrode Techniques. The Plymouth Workshop Handbook*, second edition, ed. D. Ogden, pp. 53–78. Cambridge: Company of Biologists.

O'Hara, P. J., Sheppard, P. O., Thøgersen, H., Venezia, D., Haldeman, B. A., McGrane, V., Houamed, K. M., Thomsen, C., Gilbert, T. L. & Mulvihill, E. R. (1993). The ligand-binding domain in metabotropic glutamate receptors is related to bacterial periplasmic binding proteins. *Neuron* **11**, 41–52.

Ohmori, H., Yoshida, S. & Hagiwara, S. (1981). Single K channel currents of anomalous rectification in cultured rat myotubes. *Proc. Natl. Acad. Sci. USA* **78**, 4960–4.

Ohtsuki, I., Masaki, T., Nonomura, Y. & Ebashi, S. (1967). Periodic distribution of troponin along the thin filament. *J. Biochem.* **61**, 817–19.

Okada, Y., Miyamoto, T. & Sato, T. (1994). Activation of a cation conductance by acetic acid in taste cells isolated from bullfrog. *J. Exp. Biol.* **187**, 19–32.

Olivera, B. M., Rivier, J., Clark, C., Ramilo, C., Corpuz, G. P., Abogadie, F. C., Mena, E. E., Woodward, S. R., Hillyard, D. R. & Cruz, L. J. (1990). Diversity of *Conus* neuropeptides. *Science* **249**, 257–63.

Olney, J. W. (1969). Brain lesions, obesity and other disturbances in mice treated with monosodium glutamate. *Science* **164**, 719–21.

Olney, J. W. (1995). Glutamate receptor-mediated neurotoxicity. In *Neurotoxicology: Approaches and Methods*, ed. L. W. Chang & W. Slikker, pp. 455–63. San Diego, CA: Academic Press.

Østerberg, E. (1935). Topography of the layer of rods and cones in the human retina. *Acta Ophthal. Kbh.* Suppl. **6**, 1–103.

Ostrowski, J., Kjelsberg, M. A., Caron, M. G. & Lefkowitz, R. J. (1992). Mutagenesis of the β_2-adrenergic receptor: how structure elucidates function. *Ann. Rev. Pharmacol. Toxicol.* **32**, 167–83.

Ottoson, D. (1983). *Physiology of the Nervous System*. London: Macmillan.

Ovalle, W. K. & Smith, R. S. (1972). Histochemical identification of three types of intrafusal muscle fibres in the cat and monkey based on myosin ATPase reaction. *Can. J. Physiol. Pharmacol.* **50**, 195–202.

Ovchinnikov, Y. A., Abdulaev, N. G., Feigma, M. Y. Artamonov, I. D., Bogachuk, A. S., Zolotarev, A., Eganyan, E. R. & Kostetskii, P. V. (1983). Visual rhodopsin. III. Complete amino acid sequence and topography in the membrane. *Bioorganicheskaya Khimiya* **9**, 1331–40.

Ovchinnikov, Y. A., Luneva, N. M., Arystarkhova, E. A., Gvondyan, N. M., Arzamazova, N. M., Kozhich, A. T., Nesmayanov, V. A. & Modyanov, N. N. (1988). Topology of Na^+,K^+-ATPase: identification of the extra- and intracellular hydrophilic loops of the catalytic subunit by specific antibodies. *FEBS Lett.* **227**, 230–4.

Owen, F., Crawley, J., Cross, A. J., Crow, T. J., Oldland, R., Poulter, M., Veall, N. & Zanelli, G. D. (1985). Dopamine D_2 receptors and schizophrenia. In *Psychopharmacology: Recent Advances and Future Prospects*, ed. S. D. Iversen, pp. 21–7. Oxford: Oxford University Press.

Pace, U., Hanski, E., Salomon, Y. & Lancet, D. (1985). Odorant-sensitive adenylate cyclase may mediate olfactory reception. *Nature* **316**, 255–8.

Pacholczyk, T., Blakely, R. D. & Amara, S. G. (1991). Expression cloning of a cocaine- and antidepressant-sensitive human noradrenaline transporter. *Nature* **350**, 350–4.

Page, S. G. (1964). Filament lengths in resting and excited muscles. *Proc. R. Soc. Lond.* B **160**, 460–6.

Page, S. G. (1965). A comparison of the fine structures of frog slow and twitch muscle fibres. *J. Cell Biol.* **26**, 477–97.

Page, S. G. & Huxley, H. E. (1963). Filament lengths in striated muscle. *J. Cell Biol.* **19**, 369–90.

Pak, W. L. (1968). Rapid photoresponses in the retina and their relevance to vision research. *Photochem. Photobiol.* **8**, 495–503.

Palade, G. E. & Palay, S. L. (1954). Electron microscope observations of interneuronal and neuromuscular synapses. *Anat. Rec.* **118**, 335.

Palmer, R. M. J., Ferrige, A. G. & Moncada, S. (1987). Nitric oxide release accounts for the biological activity of endothelium-derived relaxing factor. *Nature* **327**, 524–6.

Paniagua, R., Royuela, M., García-Anchuelo, R. M. & Fraile, B. (1996). Ultrastructure of invertebrate muscle-cell types. *Histol. Histopathol.* **11**, 181–201.

Pannbacker, R. G. (1973). Control of guanylate cyclase activity in the rod outer segment. *Science* **182**, 1138–40.

Pannbacker, R. G., Fleischman, D. E. & Reed, D. (1972). Cyclic nucleotide phosphodiesterase: high activity in a mammalian photoreceptor. *Science* **175**, 757–8.

Papazian, D. M., Schwarz, T. L., Tempel, B. L., Jan, Y. N. & Jan, L. Y. (1987). Cloning of genomic and complementary DNA from *Shaker*, a putative potassium channel gene from *Drosophila*. *Science* **237**, 749–53.

Papazian, D. M., Timpe, L. C., Jan, Y. N. & Jan, L. Y. (1991). Alteration of voltage-dependence of *Shaker* potassium channel by mutations in the S4 sequence. *Nature* **349**, 305–10.

Parry, D. A. D. (1975). Analysis of the primary sequence of α-tropomyosin from rabbit skeletal muscle. *J. Molec. Biol.* **98**, 519–35.

Parry, D. A. D. & Squire, J. M. (1973). Structural role of tropomysin in muscle regulation: analysis of the X-ray diffraction patterns from relaxed and contracting muscles. *J. Molec. Biol.* **75**, 33–55.

Parsons, S. M., Prior, C. & Marshall, I. G. (1993). Acetylcholine transport, storage, and release. *Int. Rev. Neurobiol.* **35**, 279–390.

Pascual, J. M., Shieh, C.-C., Kirsch, G. E. & Brown, A. M. (1995). K^+ pore structure revealed by reporter cysteines at inner and outer surfaces. *Neuron* **14**, 1055–63.

Patterson, S. & Starling, E. H. (1914). On the mechanical factors which determine the output of the ventricles. *J. Physiol.* **48**, 357–79.

Patterson, S. L., Grover, L. M., Schwartzkroin, P. A. & Bothwell, M. (1992). Neurotrophin expression in rat hippocampal slices: a stimulus paradigm producing LTP in CA1 evokes increases in BDNF and NT–3 mRNAs. *Neuron* **9**, 1081–8.

Patterson, S. L., Abel, T., Deuel, T. A. S., Martin, K. C., Rose, J. C. & Kandel, E. R. (1996). Recombinant BDNF rescues deficits in basal synaptic transmission and hippocampal LTP in BDNF knockout mice. *Neuron* **16**, 1137–45.

Patuzzi, R. (1996). Cochlear micromechanics and macromechanics. In *The Cochlea*, ed. P. Dallos, A. N. Popper & R. R. Fay, pp. 186–257. New York: Springer-Verlag.

Paul, D. L. (1986). Molecular cloning of cDNA for rat liver gap junction protein. *J. Cell. Biol.* **103**, 123–34.

Paul, R. J. (1983). Physical and biochemical energy balance during an isometric tetanus and steady state recovery in frog sartorius at 0°C. *J. Gen. Physiol.* **81**, 337–54.

Paul, R. J. (1989). Smooth muscle energetics. *Ann. Rev. Physiol.* **51**, 331–49.

Pavlov, I. P. (1906). The scientific investigation of the psychical faculties or processes in the higher animals. *Science* **24**, 613–19.

Peachey, L. D. (1965). The sarcoplasmic reticulum and transverse tubules of the frog's sartorius. *J. Cell Biol.* **25** (part 2), 209–32.

Peachey, L. D. & Huxley, A. F. (1962). Structural identification of twitch and slow striated muscle fibres of the frog. *J. Cell Biol.* **13**, 117–80.

Pelc, R., Smith, P. J. S. & Ashley, C. C. (1996). *In vivo* recording of calcium fluxes accompanying 'catch' contraction of molluscan smooth muscle. *J. Physiol.* **497**, 41P.

Pellegrino, R. G. & Ritchie, J. M. (1984). Sodium channels in the axolemma of normal and degenerating rabbit optic nerve. *Proc. R. Soc. Lond.* B **222**, 155–60.

Penn, R. D. & Hagins, W. A. (1969). Signal transmission along retinal rods and the origin of the electroretinographic a-wave. *Nature* **223**, 201–5.

Penn, R. D. & Hagins, W. A. (1972). Kinetics of the photocurrent of retinal rods. *Biophys. J.* **12**, 1073–94.

Peracchia, C., Lazrak, A. & Peracchia, L. L. (1994). Molecular models of channel interaction and gating in gap junctions. In *Handbook of Membrane Channels*, ed. C. Peracchia, pp. 361–77. San Diego, CA: Academic Press.

Perkel, D. J., Hestrin, S., Sah, P & Nicoll, R. A. (1990). Excitatory synaptic currents in cerebellar Purkinje cells. *Proc. R. Soc. Lond.* B **241**, 116–21.

Perl, E. R. (1996). Pain and the discovery of nociceptors. In *Neurobiology of Nociceptors*, ed. C. Belmonte & F. Cervero, pp. 5–36. Oxford: Oxford University Press.

Peroutka, S. J., Lebowitz, R. M. & Snyder, S. H. (1981). Two distinct central serotonin receptors with different physiological functions. *Science* **212**, 827–8.

Perry, S. V. (1994). Activation of the contractile mechanism by calcium. In *Myology: Basic and Clinical*, second edition, ed. A. G. Engel & C. Franzini-Armstrong, vol. 1, pp. 529–52. New York: McGraw-Hill.

Perry, S. V. (1996). *Molecular Mechanisms in Striated Muscle*. Cambridge: Cambridge University Press.

Pert, C. B. & Snyder, S. H. (1973). Opiate receptor: demonstration in nervous tissue. *Science* **179**, 1011–14.

Peter, D., Liu, Y., Sternini, C., de Giorgio, R., Brecha, N. & Edwards, R. H. (1995). Differential expression of two monoamine transporters. *J. Neurosci.* **15**, 6179–88.

Peter, J. B., Barnard, R. J., Edgerton, V. R., Gillespie, A. & Stempel, K. E. (1972). Metabolic profiles of three fibre types of skeletal muscle in guinea pigs and rabbits. *Biochemistry* **11**, 2627–33.

Petrozzino, J. J., Miller, L. D. P. & Connor, J. A. (1995). Micromolar Ca^{2+} transients in dendritic spines of hippocampal pyramidal neurons in brain slice. *Neuron* **14**, 1223–31.

Pette, D. & Staron, R. S. (1990). Cellular and molecular diversities of mammalian skeletal muscle fibres. *Rev. Physiol. Biochem. Pharmacol.* **116**, 1–76.

Pevsner, J. & Snyder, S. (1990). Odorant binding protein: localization to nasal glands. *Chem. Senses* **15**, 217–22.

Pfaffinger, P. J., Martin, J. M. Hunter, D. D., Nathanson, N. M. & Hille, B. (1985). GTP-binding proteins couple cardiac muscarinic receptors to a K channel. *Nature* **317**, 536–8.

Pfaffmann, C. (1941). Gustatory afferent impulses. *J. Cell. Comp. Physiol.* **17**, 243–58.

Pfaffmann, C. (1984). Taste electrophysiology, sensory coding, and behaviour. In *Foundations of Sensory Science*, ed. W. W. Dawson & J. M. Enoch, pp. 325–49. Berlin: Springer-Verlag.

Pfuhl, M., Winder, S. J. & Pastore, A. (1994). Nebulin, a helical actin binding protein. *EMBO J.* **13**, 1782–9.

Philipson, K. D., Nicoll, D. A. & Li, Z. (1993). The cardiac sodium–calcium exchanger. In *Molecular Biology and Function of Carrier Proteins*, ed. L. Reuss, J. M. Russell &

M. L. Jennings (*Society of General Physiologists Series* **48**), pp. 187–91. New York: Rockefeller University Press.

Pickles, J. O. & Corey, D. P. (1992). Mechanoelectrical transduction by hair cells. *Trends Neurosci.* **15**, 254–9.

Pickles, J. O., Comis, S. D. & Osborne, M. P. (1984). Cross-links between stereocilia in the guinea pig organ of Corti, and their possible relation to sensory transduction. *Hearing Res.* **15**, 103–12.

Pierau, F.-K. & Wurster, R. D. (1981). Primary afferent input from cutaneous thermoreceptors. *Fedn Proc.* **40**, 2819–24.

Pierau, F.-K., Torrey, P. & Carpenter, D. (1975). Effect of ouabain and potassium-free solution on mammalian thermosensitive afferents in vitro. *Pflügers Archiv* **359**, 349–56.

Pierce, U. W. & Griffin, D. R. (1938). Experimental determination of supersonic notes emitted by bats. *J. Mammal.* **19**, 454–5.

Pin, J.-P. & Bockaert, J. (1995). Get receptive to metabotropic glutamate receptors. *Curr. Opin. Neurobiol.* **5**, 342–9.

Pines, G., Danbolt, N. C., Bjoras, M., Zhang, Y., Bendahan, A., Eide, L., Koepsell, H., Storm-Mathisen, J., Seeberg, E. & Kanner, B. I. (1992). Cloning and expression of a rat brain L-glutamate transporter. *Nature* **360**, 464–7.

Pinsker, H. M., Hening, W. A., Carew, T. J. & Kandel, E. R. (1973). Long-term sensitization of a defensive withdrawal reflex in *Aplysia*. *Science* **182**, 1039–42.

Pirenne, M. H. (1943). Binocular and monocular threshold of vision. *Nature* **152**, 698–9.

Pirenne, M. H. (1944). Rods and cones, and Thomas Young's theory of colour vision. *Nature* **154**, 741–2.

Pirenne, M. H. (1962). Visual functions in man. In *The Eye*, ed. H. Davson, vol. 2, chapters 1–11, pp. 1–204. London: Academic Press.

Pitchford, S. & Levine, J. D. (1991). Prostaglandins sensitize nociceptors in cell-culture. *Neurosci. Lett.* **132**, 105–8.

Podolsky, R. J. (1960). Kinetics of muscular contraction: the approach to the steady state. *Nature* **188**, 666–8.

Pollack (1990). *Muscles and Molecules: Uncovering the Principles of Biological Motion.* Seattle, WA: Ebner and Sons.

Polyak, S. (1941). *The Retina.* Chicago, IL: Chicago University Press.

Pongs, O. (1992). Molecular biology of voltage-dependent potassium channels. *Physiol. Rev.* **72**, S69–S88.

Popot, J.-L. (1993). Integral membrane protein structure: transmembrane α-helices as autonomous folding domains. *Curr. Biol.* **3**, 532–40.

Popot, J.-L. & Changeux, J.-P. (1984). Nicotinic receptor of acetylcholine: structure of an oligomeric integral membrane protein. *Physiol. Rev.* **64**, 1162–239.

Poppele, R. E. & Quick, D. C. (1981). Stretch-induced contraction of intrafusal muscle in cat muscle spindle. *J. Neurosci.* **1**, 1069–74.

Popper, K. R. (1963). *Conjectures aad Refutations: The Growth of Scientific Knowlege.* London: Routledge and Kegan Paul.

Porter, K. R. & Palade, G. E. (1957). Studies on the endoplasmic reticulum. III. Its form and distribution in striated muscle cells. *J. Biophys. Biochem. Cytol.* **3**, 269–300.

Portzehl, H. (1957). Die Bindung des Erschlaffungsfaktors von Marsh an die Muskelgrana. *Biochim. Biophys. Acta* **26**, 373–7.

Portzehl, H., Caldwell, P. C. & Rüegg, J. C. (1964). The dependence of contraction and relaxation of muscle fibres from the crab *Maia squinado* on the internal concentration of free calcium ions. *Biochim. Biophys. Acta* **79**, 581–99.

Post, R. L. & Jolly, P. C. (1957). The linkage of sodium, potassium and ammonium active transport across the human erythrocyte membrane. *Biochim. Biophys. Acta* **25**, 118–28.

Post, R. L., Sen, A. K. & Rosenthal, A. S. (1965). A phosphorylated intermediate in adenosine triphosphate-dependent sodium and potassium transport across kidney membranes. *J. Biol. Chem.* **240**, 1437–45.

Poyner, D. R. (1992). Calcitonin gene-related peptide – multiple actions, multiple receptors. *Pharmacol. Ther.* **56**, 23–51.

Priesner, E. (1979). Specificity studies on pheromone receptors of noctuid and tortricid lepidoptera. In *Chemical Ecology: Odour Communication in Animals*, ed. F. J. Ritter, pp. 57–71. Amsterdam: Elsevier/North-Holland.

Pringle, J. W. S. (1949). The excitation and contraction of the flight muscles of insects. *J. Physiol.* **108**, 226–32.

Pringle, J. W. S. (1954). The mechanism of the myogenic rhythm of certain insect striated muscles. *J. Physiol.* **124**, 269–91.

Pringle, J. W. S. (1979). Stretch activation of muscle: function and mechanism. *Proc. R. Soc. Lond.* B **201**, 107–30.

Pringle, J. W. S. (1981). The evolution of fibrillar muscle in insects. *J. Exp. Biol.* **94**, 1–14.

Proske, U. (1997). The mammalian muscle spindle. *News Physiol. Sci.* **12**, 37–42.

Pugh, E. N. & Cobbs, W. H. (1986). Visual transduction in vertebrate rods and cones: a tale of two transmitters, calcium and cyclic GMP. *Vision Res.* **26**, 1613–43.

Pugh, E. N. & Lamb, T. D. (1993). Amplification and kinetics of the activation steps in phototransduction. *Biochim. Biophys. Acta* **1141**, 111–49.

Pumphrey, R. J. & Young, J. Z. (1938). The rates of conduction of nerve fibres of various diameters of cephalopods. *J. Exp. Biol.* **15**, 453–66.

Pusch, M. & Jentsch, T. J. (1994). Molecular physiology of voltage-gated chloride channels. *Physiol. Rev.* **74**, 813–27.

Qian, Z., Gilbert, M. E., Colicos, M. A., Kandel, E. R. & Kuhl, D. (1993). Tissue-plasminogen activator is induced as an immediate-early gene during seizure, kindling and long-term potentiation. *Nature* **361**, 453–7.

Qiu, Y., Chen, C.-N., Malone, T., Richter, L., Beckendorf, S. K. & Davis, R. L. (1991). Characterization of the memory gene *dunce* of *Drosophila melanogaster*. *J. Molec. Biol.* **222**, 553–65.

Quilliam, T. A. & Sato, M. (1955). The distribution of myelin on nerve fibres from Pacinian corpuscles. *J. Physiol.* **129**, 167–76.

Quinn, D. M. (1987). Acetylcholinesterase: enzyme structure, reaction dynamics, and virtual transition states. *Chem. Rev.* **87**, 955–79.

Quinn, W. G., Harris, W. A. & Benzer, S. (1974). Conditioned behaviour in *Drosophila melanogaster*. *Proc. Natl. Acad. Sci. USA* **71**, 708–12.

Quinta-Ferreira, M. E., Rojas, E. & Arispe, N. (1982). Potassium currents in the giant axon of the crab *Carcinus maenas*. *J. Membrane Biol.* **66**, 171–81.

Radermacher, M., Rao, V., Grassucci, R., Frank, J., Timerman, A. P., Fleischer, S. & Wagenknecht, T. (1994). Cryo-electronmicroscopy and three-dimensional reconstruction of the calcium release channel/ryanodine receptor from skeletal muscle. *J. Cell Biol.* **127**, 411–23.

Raftery, M. A., Vandlen, R. L., Reed, K. L. & Lee, T. (1976). Characterization of *Torpedo californica* acetylcholine receptor: its subunit composition and ligand-binding properties. *Cold Spr. Harb. Symp. Quant. Biol.* **40**, 193–202.

Raftery, M. A., Hunkapiller, M. W., Strader, C. D. & Hood, L. E. (1980). Acetylcholine receptor: complex of homologous subunits. *Science* **208**, 1454–7.

Raming, K., Krieger, J., Strotmann, J., Boekhoff, I., Kubick, S., Baumstark, C. & Breer, H. (1993). Cloning and expression of odorant receptors. *Nature* **361**, 353–6.

Ramón y Cajal, S. (1911). *Histologie du Système Nerveux de l'Homme et des Vertébrés.* Paris: Maloine.

Randall, J. (1975). Emmeline Jean Hanson. *Biog. Mem. Fellows R. Soc. Lond.* **21**, 313–44.

Rash, J. E., Dillman, R. K., Bilhartz, B. L., Duffy, H. S., Whalen, L. R. & Yasumura, T. (1996). Mixed synapses discovered and mapped throughout mammalian spinal cord. *Proc. Natl. Acad. Sci. USA* **93**, 4235–9.

Rasmussen, G. L. (1953). Further observations on the efferent cochlear bundle. *J. Comp. Neurol* **99**, 61–74.

Ratto, G. M., Payne, R., Owen, W. G. & Tsien, R. Y. (1988). The concentration of cytosolic free calcium in vertebrate rod outer segments measured with Fura-2. *J. Neurosci.* **8**, 3240–6.

Rayment, I. & Holden, H. M. (1994). The three-dimensional structure of a molecular motor. *Trends Biochem. Sci.* **19**, 129–34.

Rayment, I., Rypniewski, W. R., Schmidt-Bäse, K., Smith, R., Tomchick, D. R., Benning, M. M., Winkelmann, D. A., Wesenberg, G. & Holden, H. M. (1993a). Three-dimensional structure of myosin subfragment-1: a molecular motor. *Science* **261**, 50–8.

Rayment, I., Holden, H. M., Whittaker, M., Yohn, C. B., Lorenz, M., Holmes, K. C. & Milligan, R. A. (1993b). Structure of the actin–myosin complex and its implications for muscle contraction. *Science* **261**, 58–65.

Rayment, I., Smith, C. & Yount, R. G. (1996). The active site of myosin. *Ann. Rev. Physiol.* **58**, 671–702.

Rayner, M. D., Starkus, J. G., Ruben, P. C. & Alicata, D. A. (1992). Voltage-sensitive and solvent-sensitive processes in ion channel gating. *Biophys. J.* **61**, 96–108.

Redman, S. (1990). Quantal analysis of synaptic potentials in neurones of the central nervous system. *Physiol. Rev.* **70**, 165–98.

Redman, S. & Walmsley, B. (1983a). The time course of synaptic potentials evoked in cat spinal motoneurones at identified group Ia synapses. *J. Physiol.* **343**, 117–33.

Redman, S. & Walmsley, B. (1983b). Amplitude fluctuations in synaptic potentials evoked in cat spinal motoneurones at identified group Ia synapses. *J. Physiol.* **343**, 135–45.

Reedy, M. K., Holmes, K. C. & Tregear, R. T. (1965). Induced changes in orientation of the cross-bridges of glycerinated insect flight muscle. *Nature* **207**, 1276–80.

Reedy, M., Leonard, K. R., Freeman, R. & Arad, (1981). Thick filament mass determination by electron scattering measurements with the scanning transmission electron microscope. *J. Mus. Res. Cell Motil.* **2**, 45–64.

Rees, C. J. C. (1969). Chemoreceptor specificity associated with the choice of feeding site by the beetle *Chrysolina brunsvicensis* on its foodplant *Hypericum hirsutum*. *Entomol. Exp. Appl.* **12**, 565–83.

Regehr, W. G., Connor, J. A. & Tank, D. W. (1989). Optical imaging of calcium accumulation in hippocampal pyramidal cells during synaptic activation. *Nature* **341**, 533–6.

Reichardt, W. (1962). Theoretical aspects of neural inhibition in the lateral eye of *Limulus*. *Proc. Int. Union Physiol. Sci.* **3**, 65–84.

Reilander, H., Achilles, A., Friedel, U., Maul, G, Lottspeich, F. & Cook, N. J. (1992). Primary structure and functional expression of the Na/Ca,K-exchanger from bovine rod photoreceptors. *EMBO J.* **11**, 1689–95.

Reisine, T. & Bell, G. I. (1993). Molecular biology of opioid receptors. *Trends Neurosci.* **16**, 506–10.

Rembold, C. M. (1996). Electromechanical and pharmacomechanical coupling. In *Biochemistry of Smooth Muscle Contraction*, ed. M. Bárány, pp. 227–39. San Diego, CA: Academic Press.

Requena, J., Mullins, L. J., Whittembury, J. & Brinley, F. J. (1986). Dependence of ionized and total Ca in squid axons on Na_o-free or high K_o conditions. *J. Gen. Physiol.* **87**, 143–59.

Ressler, K. J., Sullivan, S. L. & Buck, L. (1993). A zonal organization of odorant receptor gene expression in the olfactory epithelium. *Cell* **73**, 597–609.

Rettig, J., Heinemann, S. H., Wunder, F., Lorra, C., Parcej, D. N., Dolly, J. O. & Pongs, O. (1994). Inactivation properties of voltage-gated K^+ channels altered by presence of β-subunit. *Nature* **369**, 289–94.

Reuter, H. (1984). Ion channels in cardiac cell membranes. *Ann. Rev. Physiol.* **46**, 473–84.

Reuter, H. & Seitz, H. (1968). The dependence of calcium efflux from cardiac muscle on temperature and external ion composition. *J. Physiol.* **195**, 451–70.

Reuveny, E., Slesinger, P. A., Inglese, J., Morales, J. M., Iñiguez-Lluhi, J. A., Lefkowitz, R. J., Bourne, H. R., Jan, Y. N. & Jan, L. Y. (1994). Activation of the cloned muscarinic potassium channel by G protein βγ subunits. *Nature* **370**, 143–6.

Revah, F., Galzi, J.-L., Giraudat, J., Haumont, P.-Y., Lederer, F. & Changeux, J.-P. (1990). The noncompetitive blocker [³H]chlorpromazine labels three amino acids of the acetylcholine receptor gamma subunit: implications for the α-helical organization of regions MII and for the structure of the ion channel. *Proc. Natl. Acad. Sci. USA* **87**, 4675–9.

Revel, J. P. & Karnovsky, M. J. (1967). Hexagonal array of subunits in intercellular junctions of the mouse heart and liver. *J. Cell Biol.* **33**, C7–C12.

Rhode, W. S. & Robles, L. (1974). Evidence from Mössbauer experiments for nonlinear vibration in the cochlea. *J. Acoust. Soc. Amer.* **55**, 588–96.

Richardson, C. L., Tate, W. P., Mason, S. E., Lawlor, P. A., Dragunow, M. & Abraham, W. C. (1992). Correlation between the induction of an immediate early gene, *zif/268*, and long-term potentiation in the dentate gyrus. *Brain Res.* **580**, 147–54.

Riedel, G. & Reymann, K. G. (1996). Metabotropic glutamate receptors in hippocampal long-term potentiation and learning and memory. *Acta Physiol. Scand.* **157**, 1–19.

Ringer, S. (1883). A further contribution regarding the influence of different constituents of the blood on the contraction of the heart. *J. Physiol.* **4**, 29–42.

Ríos, E., Ma, J. & González, A. (1991). The mechanical hypothesis of excitation-contraction (EC) coupling in skeletal muscle. *J. Mus. Res. Cell Motil.* **12**, 127–35.

Ritchie, J. M. (1971). Electrogenic ion pumping in nervous tissue. In *Current Topics in Bioenergetics* vol. 4, pp. 327–56. New York: Academic Press.

Ritchie, J. M. (1982). On the relation between fibre diameter and conduction velocity in myelinated nerve fibres. *Proc. R. Soc. Lond.* B **217**, 29–35.

Ritchie, J. M. & Straub, R. W. (1975). The movement of potassium ions during electrical activity, and the kinetics of the recovery process, in the non-myelinated fibres of the garfish olfactory nerve. *J. Physiol.* **249**, 327–48.

Ritchie, J. M., Rogart, R. B. & Strichartz, G. R. (1976). A new method for labelling saxitoxin and its binding to nonmyelinated fibres of the rabbit vagus, lobster walking leg, and garfish olfactory nerves. *J. Physiol.* **261**, 477–94.

Robertson, J. D. (1960). The molecular structure and contact relationships of cell membranes. *Prog. Biophys.* **10**, 343–418.

Robertson, J. D. (1989). Membranes, molecules, nerves, and people. In *Membrane Transport: People and Ideas*, ed. D. C. Tosteson, pp. 51–124. Bethesda, MD: American Physiological Society.

Rodbell, M. (1995). Signal transduction: evolution of an idea (Nobel lecture). *Angew. Chem. Int. Ed. Eng.* **34**, 1420–8.

Rodbell, M., Birnbaumer, L., Pohl, S. L. & Krans, M. J. (1971). The glucagon-sensitive adenyl cyclase system in plasma membranes of rat liver. *J. Biol. Chem.* **246**, 1877–82.

Roeder, K. D. (1951). Movements of the thorax and potential changes in the thoracic muscles of insects during flight. *Biol. Bull.* **100**, 95–106.

Roelofs, W. L. (1995). Chemistry of sex attraction. *Proc. Natl. Acad. Sci. USA* **92**, 44–9.

Rogawski, M. A. (1993). Therapeutic potential of excitatory amino acid antagonists: channel blockers and 2, 3-benzodiazepines. *Trends Pharmacol. Sci.* **14**, 325–31.

Rogers, C. J., Twyman, R. E. & Macdonald, R. L. (1994). Benzodiazepine and β-carboline regulation of single GABA_A receptor channels of mouse spinal neurones in culture. *J. Physiol.* **475**, 69–82.

Romanes, G. J. (1951). The motor cell columns of the lumbosacral spinal cord of the cat. *J. Comp. Neurol.* **94**, 313–63.

Roper, S. (1983). Regenerative impulses in taste cells. *Science* **220**, 1311–12.

Roper, S. D. (1992). The microphysiology of peripheral taste organs. *J. Neurosci.* **12**, 1127–34.

Rosenberg, R. L., Tomiko, S. A. & Agnew, W. (1984). Single-channel properties of the reconstituted voltage-regulated Na channel isolated from the electroplax of *Electrophorus electricus*. *Proc. Natl. Acad. Sci. USA* **81**, 5594–8.

Rosenberry, J. (1975). Acetylcholine esterase. *Adv. Enzymol.* **43**, 103–218.

Rosenbluth, J. (1968). Obliquely striated muscle. *J. Cell Biol.* **36**, 245–59.

Rosenbluth, J. (1974). Substructure of the amphibian motor end plate. Evidence for a granular component projecting from the outer surface of the receptive membrane. *J. Cell Biol.* **62**, 755–66.

Rosenfeld, M. G., Mermod, J.-J., Amara, S. G., Swanson, L. W., Sawchenko, P. E., Rivier, J., Vale, W. W. & Evans, R. M. (1983). Production of a novel neuropeptide encoded by the calcitonin gene via tissue-specific RNA processing. *Nature* **304**, 129–35.

Rothman, J. E. & Lenard, J. (1977). Membrane asymmetry. *Science* **195**, 743–53.

Rothman, S. J. & Olney, J.W. (1987). Excitotoxicity and the NMDA receptor. *Trends Neurosci.* **10**, 299–302.

Rothman, S. M. & Olney, J. W. (1995). Excitotoxicity and the NMDA receptor – still lethal after eight years. *Trends Neurosci.* **18**, 57–8.

Rowlerson, A., Gorza, L. & Schiaffino, S. (1985). Immunohistochemical identification of spindle fibre types in mammalian muscle using type-specific antibodies to isoforms of myosin. In *The Muscle Spindle*, ed. I. A. Boyd & M. H. Gladden, pp. 29–34. New York: Macmillan.

Rüegg, J. C. (1992). *Calcium in Muscle Contraction*, second edition. Berlin: Springer-Verlag.

Rüegg, J. C. & Tregear, R. T. (1966). Mechanical factors affecting the ATPase activity of glycerol-extracted insect fibrillar flight muscle. *Proc. R. Soc. Lond.* B **165**, 497–512.

Ruffini, A. (1898). On the minute anatomy of the neuromuscular spindles of the cat, and on their physiological significance. *J. Physiol.* **23**, 190–208.

Ruggero, M. A. (1992). Responses to sound of the basilar membrane of the mammalian cochlea. *Curr. Opin. Neurobiol.* **2**, 449–56.

Ruggero, M. A. & Rich, N. C. (1991). Application of a commercially-manufactured Doppler-shift laser velocimeter to the measurement of basilar membrane vibration. *Hearing Res.* **51**, 215–30.

Ruiz-Avila, L., McLaughlin, S. K., Wildman, D., McKinnon, P. J., Robichon, A., Spickofsky, N. & Margolskee, R. F. (1995). Coupling of bitter receptor to phosphodiesterase through transducin in taste receptor cells. *Nature* **376**, 80–5.

Ruppersberg, J. P., Schröter, K. H., Sakmann, B., Stocker, M., Sewing, S. & Pongs, O. (1990). Heteromultimeric channels formed by rat brain potassium-channel proteins. *Nature* **345**, 535–7.

Rushton, W. A. H. (1945). Action potentials from the isolated nerve cord of the earthworm. *Proc. Roy. Soc. Lond.* B **132**, 423–37.

Rushton, W. A. H. (1951). A theory of the effects of fibre size in medullated nerve. *J. Physiol.* **115**, 101–22.

Rushton, W. A. H. (1963). A cone pigment in the protanope. *J. Physiol.* **168**, 345–59.

Rushton, W. A. H. (1965). Visual adaptation. *Proc. R. Soc. Lond.* B **162**, 20–46.

Rushton, W. A. H. (1972). Visual pigments in man. In *Handbook of Sensory Physiology*, vol. VIII/1 *Photochemistry of Vision*, ed. H. J. A. Darnall, pp. 364–94. Berlin: Springer-Verlag. .

Russell, I. J. (1971). The role of the lateral-line efferent system in *Xenopus laevis*. *J. Exp. Biol.* **54**, 621–41.

Russell, I. J. & Sellick, P. M. (1978). Intracellular studies of hair cells in the mammalian cochlea. *J. Physiol.* **284**, 261–90.

Russell, J. M. (1983). Cation-coupled chloride influx in squid axon: role of potassium and stoichiometry of the transport processss. *J. Gen. Physiol.* **81**, 909–25.

Russell, J. M. & Boron, W. F. (1976). Role of chloride transport in regulation of intracellular pH. *Nature* **264**, 73–4.

Russell, J. M. & Boron, W. F. (1990). Chloride transport in the squid giant axon. In *Chloride Channels and Carriers in Nerve, Muscle, and Glial Cells*, ed. F. J. Alvarez-Leefmans & J. M. Russell, pp. 85–107. New York: Plenum Press.

Sakitt, B. (1972). Counting every quantum. *J. Physiol.* **223**, 131–50.

Sakmann, B. (1992). Elementary steps in synaptic transmission revealed by currents through single ion channels. In *Les Prix Nobel 1991*, pp. 141–69. Stockholm: Almquist & Wiksell. Also published in *Science* **256**, 503–12.

Sakmann, B. & Neher, E. (eds.) (1995). *Single-Channel Recording*, second edition. New York: Plenum Press.

Sakmann, B., Noma, A. & Trautwein, W. (1983). Acetylcholine activation of single muscarinic K$^+$ channels in isolated pacemaker cells of the mammalian heart. *Nature* **303**, 250–3.

Sakmann, B., Methfessel, C., Mishina, M., Takahashi, T., Takai, T., Kurasaki, M., Fukuda, K. & Numa, S. (1985). Role of acetylcholine receptor subunits in gating of the channel. *Nature* **318**, 538–43.

Sakmann, B., Edwards, F., Konnerth, A. & Takahashi, T. (1989). Patch clamp techniques used for studying synaptic transmission in slices of mammalian brain. *Quart. J. Exp. Physiol.* **74**, 1107–18.

Salkoff, L. & Wyman, R. (1981). Genetic modification of potassium channels in *Drosophila* Shaker mutants. *Nature* **293**, 228–30.

Salkoff, L., Butler, A., Wei, A., Scavarda, N., Giffen, K., Ifune, C., Goodman, R. & Mandel, G. (1987). Genomic organization and deduced amino acid sequence of a putative sodium channel gene in *Drosophila*. *Science* **237**, 744–9.

Salkoff, L., Baker, K., Butler, A., Covarrubias, M., Pak, M. D. & Wei, A. (1992). An essential 'set' of K$^+$ channels conserved in flies, mice and humans. *Trends Neurosci.* **15**, 161–6.

Salpeter, M. M. (1987). Vertebrate neuromuscular junctions: general morphology, molecular organization, and functional consequences. In *The Vertebrate Neuromuscular Junction*, ed. M. M. Salpeter, pp. 1–54. New York: Alan R. Liss.

Salpeter, M. M., Rogers, A. W., Kasprzak, H. & McHenry, F. A. (1978). Acetylcholinesterase in the fast extraocular muscle of the mouse by light and electron microscope autoradiography. *J. Cell Biol.* **78**, 274–85.

Sand, A. (1937). The mechanism of the lateral sense organs of fishes. *Proc. R. Soc. Lond.* B **123**, 472–95.

Sandow, A. (1965). Excitation–contraction coupling in skeletal muscle. *Pharmacol. Rev.* **17**, 265–320.

Santos-Sacchi, J. (1991). Reversible inhibition of voltage-dependent outer hair cell motility and capacitance. *J. Neurosci.* **11**, 3096–110.

Sargent, P. B. (1993). The diversity of neuronal nicotinic acetylcholine receptors. *Ann. Rev. Neurosci.* **16**, 403–43.

Sastry, B. R., Goh, J. W. & Auyeung, A. (1986). Associative induction of post-tetanic and long-term potentiation in CA1 neurons of rat hippocampus. *Science* **232**, 988–90.

Schacher, S., Castellucci, V. F. & Kandel, E. R. (1988). cAMP evokes long-term facilitation in *Aplysia* sensory neurons that requires new protein synthesis. *Science* **240**, 1667–9.

Scharrer, E. & Scharrer, B. (1945). Neurosecretion. *Physiol. Rev.* **25**, 171–81.

Scheich, H., Langner, G., Tidemann, C., Coles, R. B. & Gupp, A. (1986). Electroreception and electrolocation in platypus. *Nature* **319**, 401–2.

Scheonlein, R. W., Peteanu, L. A., Mathies, R. A. & Shank, C. V. (1991). The first step in vision: femtosecond isomerization of rhodopsin. *Science* **254**, 412–15.

Scher, A. M. (1965). Mechanical events in the cardiac cycle. In *Physiology and Biophysics*, nineteenth edition, ed. T. C. Ruch & H. D. Patton, pp. 550–9. Philadelphia: W. B. Saunders.

Schiaffino, S. & Reggiani, C. (1996). Molecular diversity of myofibrillar proteins: gene regulation and functional significance. *Physiol. Rev.* **76**, 371–423.

Schiffman, S. S., Lockhead, E. & Maes, F. W. (1983). Amiloride reduces the taste intensity of Na$^+$ and Li$^+$ salts and sweeteners. *Proc. Natl. Acad. Sci. USA* **80**, 6136–40.

Schmidt, W. J. (1938). Polarizations-optische Analyse eines Eiweiss-Lipoid-Systems erläutert am Aussenglied der Schzellen. *Kolloidzeitschrift* **85**, 137–48.

Schmidt-Nielsen, K. (1983). *Animal Physiology*, third edition. Cambridge: Cambridge University Press.

Schnapf, J. L., Kraft, T. W. & Baylor, D. A. (1987). Spectral sensitivity of human cone photoreceptors. *Nature* **325**, 439–41.

Schneider, D. (1965). Chemical sense communication in insects. *Symp. Soc. Exp. Biol.* **20**, 273–97.

Schneider, D. (1984). Insect olfaction – our research endeavour. In *Foundations of Sensory Science*, ed. W. W. Dawson & J. M. Enoch, pp. 381–418. Berlin: Springer-Verlag.

Schneider, D., Lacher, V. & Kaissling, K.-E. (1964). Die Reacktionweise und das Reacktionsspektrum von Reichzellen bei *Antherea pernyi* (Lepidoptera, Saturniidae). *Z. Vergl. Physiol.* **48**, 632–62.

Schneider, M. F. (1994). Control of calcium release in functioning skeletal muscle fibers. *Ann. Rev. Physiol.* **56**, 463–84.

Schneider, M. F. & Chandler, W. K. (1973). Voltage dependent charge movement in skeletal muscle: a possible step in excitation-contraction coupling. *Nature* **242**, 244–6.

Schoenlein, R. W., Peteanu, L. A., Mathies, R. A. & Shank, C. V. (1991). The 1st step in vision – femtosecond isomerization of rhodopsin. *Science* **254**, 412–15.

Schoffeniels, E. (1959). Ion movements studied with single isolated electroplax. *Ann. N. Y. Acad. Sci.* **81**, 285–306.

Schofield, P. R., Darlison, M. G., Fujita, N., Burt, D., Stephenson, F. A., Rodriguez, H., Rhee, L. M., Ramachandran, J., Reale, V., Glencorse, T. A., Seeburg, P. H. & Barnard, E. A. (1987). Sequence and functional expression of the GABA$_A$ receptor shows a ligand-gated receptor super-family. *Nature* **328**, 221–7.

Schuldiner, S., Shirvan, A. & Linial, M. (1995). Vesicular neurotransmitter transporters: from bacteria to humans. *Physiol. Rev.* **75**, 369–92.

Schultze, M. (1866). Zur Anatomie und Physiologie der Retina. *Arch. Mikrosk. Anat. Entw. Mech.* **2**, 175–286.

Schulz, J. B., Henshaw, D. R., Siwek, D., Jenkins, B. G., Ferrante, R. J., Cipolloni, P. B., Kowall, N. W., Rosen, B. R. & Beal, M. F. (1995). Involvement of free radicals in excitotoxicity in vivo. *J. Neurochem.* **64**, 2239–47.

Schumacher, M., Camp, S., Maulet, Y., Newton, M., MacPhee-Quigley, K., Taylor, S. S., Friedmann, T. & Taylor, P. (1986).

Primary structure of *Torpedo californica* acetylcholinesterase deduced from its cDNA sequence. *Nature* **319**, 407–9.

Schuman, E. M. & Madison, D. V. (1994). Nitric oxide and synaptic function. *Ann. Rev. Neurosci.* **17**, 153–83.

Schwartz, J.-C., Arrang, J.-M., Garbarg, M., Pollard, H. & Ruat, M. (1991). Histaminergic transmission in the mammalian brain. *Physiol. Rev.* **71**, 1–51.

Schwarz, W., Palade, P. T. & Hille, B. (1977). Local anaesthetics: effect of pH on use-dependent block of sodium channels in frog muscle. *Biophys. J.* **20**, 343–68.

Schwarzmann, G., Wiegand, H., Rose, B., Zimmerman, D., Ben-Haim, D. & Loewenstein, W. R. (1981). Diameter of cell-to-cell junctional membrane channels as probed with neutral molecules. *Science* **213**, 551–3.

Seeburg, P. H. (1993). The molecular biology of mammalian glutamate receptor channels. *Trends Neurosci.* **16**, 359–65.

Seizinger, B. R., Liebisch, D. C., Gramsch, C., Herz, A., Weber, E., Evans, C. J., Esch, F. S. & Böhlen, P. (1985). Isolation and structure of a novel C-terminally amidated opioid peptide, amidorphin, from bovine adrenal medulla. *Nature* **313**, 57–9.

Selby, C. C. & Bear, R. S. (1956). The structure of the actin-rich filaments of muscles according to X-ray diffraction. *J. Biophys. Biochem. Cytol.* **2**, 71–85.

Sellick, P. M., Patuzzi, R. & Johnstone, B. M. (1982). Measurement of basilar membrane motion in the guinea pig using the Mössbauer technique. *J. Acoust. Soc. Amer.* **72**, 131–41.

Shenker, A. (1995). G protein-coupled receptor structure and function: the impact of disease-causing mutations. *Ballière's Clin. Endocrinol. Metab.* **9**, 427–51.

Shepherd, G. M. (1991). *Foundations of the Neuron Doctrine*. New York: Oxford University Press.

Shepherd, G. M. (1994). Discrimination of molecular signals by the olfactory receptor neuron. *Neuron* **13**, 771–90.

Sherrington, C. S. (1897). The central nervous system. Vol. 3 of *A Textbook of Physiology*, seventh edition, ed. M. Foster, London: Macmillan.

Sheterline, P. & Sparrow, J. C. (1994). Actin. *Protein Profile* **1**, 1–121.

Shimizu, H., Miyahara, M., Ichikawa, K., Okubu, S., Konishi, T., Naka, M., Tanaka, T., Hirano, K., Harshorne, D. J. & Nakano, T. (1994). Characterization of the myosin-binding subunit of smooth muscle myosin phosphatase. *J. Biol. Chem.* **269**, 30407–11.

Shirazi, A., Iizuka, K., Fadden, P., Mosse, C., Somlyo, A. P., Somlyo, A. V. & Haystead, T. A. (1994). Purification and characterization of the mammalian myosin light chain phosphatase holoenzyme. *J. Biol. Chem.* **269**, 31598–606.

Shoenberg, C. F. & Needham, D. M. (1976). A study of the mechanism of contraction in vertebrate smooth muscle. *Biol. Rev.* **51**, 53–104.

Shull, G. A., Schwarz, A. & Lingrel (1985). Aminoacid sequence of the catalytic subunit of the (Na$^+$ + K$^+$)ATPase deduced from a complementary DNA. *Nature* **316**, 691–5.

Sicard, G. & Holley, A. (1984). Receptor cell responses to odorants: similarities and differences among odorants. *Brain Res.* **292**, 283–96.

Siegelbaum, S. A., Camardo, J. S. & Kandel, E. R. (1982). Serotonin and cyclic AMP close single K$^+$ channels in *Aplysia* sensory neurones. *Nature* **299**, 413–17.

Sigworth, F. J. (1980a). The variance of sodium current fluctuations at the node of Ranvier. *J. Physiol.* **307**, 97–129.

Sigworth, F. J. (1980b). The conductance of sodium channels under conditions of reduced current at the node of Ranvier. *J. Physiol.* **307**, 131–42.

Sigworth, F. J. (1994). Voltage gating of ion channels. *Quart. Rev. Biophys.* **27**, 1–40.

Sigworth, F. J. & Neher, E. (1980). Single Na$^+$ channel currents observed in cultured rat muscle cells. *Nature* **287**, 447–9.

Silinsky, E. M. (1975). On the association between transmitter secretion and the release of adenine nucleotides from mammalian motor nerve terminals. *J. Physiol.* **247**, 145–62.

Silinsky, E. M. (1985). The biophysical pharmacology of calcium-dependent acetylcholine secretion. *Pharmacol. Rev.* **37**, 81–132.

Silinsky, E. M. & Redman, R. S. (1996). Synchronous release of ATP and neurotransmitter within milliseconds of a motor nerve impulse in the frog. *J. Physiol.* **492**, 815–22.

Sillman, A. J., Ito, H. & Tomita, T. (1969). Studies on the mass receptor potential of the isolated frog retina. II. On the basis of the ionic mechanism. *Vision Res.* **9**, 1443–51.

Silva, A. J., Stevens, C. F., Tonegawa, S. & Wang, Y. Y. (1992a). Deficient hippocampal long-term potentiation in α-calcium-calmodulin kinase-II mutant mice. *Science* **257**, 201–6.

Silva, A. J., Paylor, R., Wehner, J. M. & Tonegawa, S. (1992b). Impaired spatial-learning in α-calcium-calmodulin kinase-II mutant mice. *Science* **257**, 206–11.

Simantov, R. & Snyder, S. H. (1976). Morphine-like peptides in mammalian brain: isolation, structure elucidation, and interactions with the opiate receptor. *Proc. Natl. Acad. Sci. USA* **73**, 2515–19.

Simmons, R. M. (1992). A. F. Huxley's research on muscle. In *Muscular Contraction*, ed. R. M. Simmons, pp. 19–42. Cambridge: Cambridge University Press.

Simmons, R. M. & Szent-Györgyi, A. G. (1985). A mechanical study of regulation in the striated adductor muscle of the scallop. *J. Physiol.* **358**, 47–64.

Simmons, R. M., Finer, J. T., Chu, S. & Spudich, J. A. (1996). Quantitative measurements of force and displacement using an optical trap. *Biophys. J.* **70**, 1813–22.

Simon, M. I., Strathmann, M. P. & Gautam, N. (1991). Diversity of G proteins in signal transduction. *Science* **252**, 802–8.

Singer, S. J. (1990). The structure and insertion of integral proteins in membranes. *Ann. Rev. Cell Biol.* **6**, 247–96.

Singer, S. J. (1992). The structure and function of membranes – a personal memoir. *J. Membrane Biol.* **129**, 3–12.

Singer, S. J. & Nicolson, G. L. (1972). The fluid mosaic model of the structure of cell membranes. *Science* **175**, 720–31.

Skou, J. C. (1957). The influence of some cations on an adenosine triphosphatase from peripheral nerves. *Biochim. Biophys. Acta* **23**, 394–401.

Skou, J. C. (1989). Sodium-potassium pump. In *Membrane Transport: People and Ideas*, ed. D. C. Tosteson, pp. 155–85. Bethesda, MD: American Physiological Society.

Slayter, H. S. & Lowey, S. (1967). Substructure of the myosin molecule as visualized by electron microscopy. *Proc. Natl. Acad. Sci. USA* **58**, 1611–18.

Slepecky, N. B. (1996). Structure of the mammalian cochlea. In *The Cochlea*, ed. P. Dallos, A. N. Popper & R. R. Fay, pp. 44–129. New York: Springer-Verlag.

Slesinger, P. A., Jan, Y. N. & Jan, L. Y. (1993). The S4–S5 loop contributes to the ion-selective pore of potassium channels. *Neuron* **11**, 739–49.

Small, J. V. & Sobieszek, A. (1977). Ca-regulation of mammalian smooth muscle actomyosin via a kinase-phosphatase-dependent phosphorylation and dephosphorylation of the 20000-M_r light chain of myosin. *Eur. J. Biochem.* **77**, 521–30.

Small, J. V. & Squire, J. M. (1972). Structural basis of contraction in vertebrate smooth muscle. *J. Molec. Biol.* **67**, 117–49.

Small, J. V., Fürst, D. O. & Thornell, L.-A. (1992). The cytoskeletal lattice of muscle cells. *Eur. J. Biochem.* **208**, 559–72.

Smith, D. A. & Geeves, M. A. (1995). Strain-dependent cross-bridge cycle for muscle. *Biophys. J.* **69**, 524–37.

Smith, D. P. (1996). Olfactory mechanisms in *Drosophila melanogaster*. *Curr. Opin. Neurobiol.* **6**, 500–5.

Smith, D. S., Gupta, B. L. & Smith, U. (1966). The organization and myofilament array of insect visceral muscles. *J. Cell Sci.* **1**, 49–57.

Smith, M. (1994). Synthetic DNA and biology. Nobel lecture. *Angew. Chemie Int. Ed.* **33**, 1214–21.

Smith, S. J. & Augustine, G. J. (1988). Calcium ions, active zones and synaptic transmitter release. *Trends Neurosci.* **11**, 458–64.

Smith, S. J., Buchanan, J., Osses, L. R., Charlton, M. P. & Augustine, G. J. (1993). The spatial distribution of calcium signals in squid presynaptic terminals. *J. Physiol.* **472**, 573–93.

Sneddon, P. & Westfall, D. P. (1984). Pharmacological evidence that adenosine triphosphate and noradrenaline are cotransmitters in the guinea-pig vas deferens. *J. Physiol.* **347**, 561–80.

Sobieszek, A. (1973). The fine structure of the contractile apparatus of the anterior byssus retractor muscle of *Mytilus edulis*. *J. Ultrastruct. Res.* **43**, 313–43.

Sobieszek, A. (1977). Ca-linked phosphorylation of a light chain of vertebrate smooth-muscle myosin. *Eur. J. Biochem.* **73**, 477–83.

Sobue, K., Muramoto, Y., Fujita, M. & Kakiuchi, (1981).

Purification of a calmodulin binding protein from chicken gizzard that interacts with F-actin. *Proc. Natl. Acad. Sci. USA* **78**, 5652–5.

Somlyo, A. P. & Somlyo, A. V. (1994). Signal transduction and regulation in smooth muscle. *Nature* **372**, 231–6.

Somlyo, A. V. & Somlyo, A. P. (1968). Electromechanical and pharmacomechanical coupling in vascular smooth muscle. *J. Pharmacol. Exp. Ther.* **159**, 129–45.

Sommer, B., Keinänen, K., Verdoorn, T. A., Wisden, W., Burnashev, N., Herb, A., Köhler, M., Takagi, T., Sakmann, B. & Seeburg, P. H. (1990). Flip and flop: a cell-specific functional switch in glutamate-operated channels of the CNS. *Science* **249**, 1580–5.

Sommer, B., Köhler, M., Sprengel, R. & Seeburg, P. H. (1991). RNA editing in brain controls a determinant of ion flow in glutamate-gated channels. *Cell* **67**, 11–19.

Song, Y. & Fambrough, D. (1994). Molecular evolution of the calcium-transporting ATPases analyzed by the maximum parsimony method. In *Molecular Evolution of Physiological Processes*, ed. D. M. Fambrough (*Society of General Physiologists Series* **49**), pp. 271–83. New York: Rockefeller University Press.

Sosa, H., Popp, D., Ouyang, G. & Huxley, H. E. (1994). Ultrastructure of skeletal muscle fibers studied by a plunge quick-freezing method – myofilament lengths. *Biophys. J.* **67**, 283–92.

Sossin, W. S., Fisher, J. M. & Scheller, R. H. (1989). Cellular and molecular biology of neuropeptide processing and packaging. *Neuron* **2**, 1407–17.

Sotavalta, O. (1953). Recordings of high wing-stroke and thoracic vibration frequency in some midges. *Biol. Bull.* **104**, 439–44.

Southam, E. & Garthwaite, J. (1993). The nitric oxide-cyclic GMP signalling pathway in rat brain. *Neuropharmacology* **32**, 1267–77.

Spielman, A. I., Huque, T., Nagai, H., Whitney, G. & Brand, J. G. (1994). Generation of inositol phosphates in bitter taste transduction. *Physiol. Behav.* **56**, 1149–55.

Spoendlin, H. (1975). Neuranatomical basis of cochlear coding mechanisms. *Audiology* **14**, 383–407.

Spoendlin, H. (1984). Efferent innervation of the cochlea. In *Comparative Physiology of Sensory Systems*, ed. L. Bolis, R. D. Keynes & S. H. P. Maddrell, pp. 163–88. Cambridge: Cambridge University Press.

Spray, D. C. (1974). Metabolic dependence of frog cold receptor sensitivity. *Brain Res.* **72**, 354–9.

Spray, D. C. (1986). Cutaneous temperature receptors. *Ann. Rev. Physiol.* **48**, 625–38.

Spray, D. C., Harris, A. L. & Bennett, M. V. L. (1979). Voltage dependence of junctional conductance in early amphibian embryos. *Science* **204**, 432–4.

Spudich, J. A. (1994). How molecular motors work. *Nature* **372**, 515–18.

Spudich, J. A., Finer, J., Simmons, B., Ruppel, K., Patterson, B.

& Uyeda, T. (1995). Myosin structure and function. *Cold Spr. Harb. Symp. Quant. Biol.* **60**, 783–91.

Squire, J. M. (1974). Symmetry and three-dimensional arrangement of filaments in vertebrate striated muscle. *J. Molec. Biol.* **90**, 153–60.

Squire, J. M. (1981). *The Structural Basis of Muscular Contraction*. New York: Plenum Press.

Squire, J. M. (1986). *Muscle: Design, Diversity and Disease*. Menlo Park, CA: Benjamin/Cummings.

Squire, J. M. (1992). Muscle filament lattices and stretch-activation: the match-mismatch model reassessed. *J. Mus. Res. Cell Motil.* **13**, 183–9.

Squire, J. M. (1997). Architecture and function in the muscle sarcomere. *Curr. Opin. Struct. Biol.* **7**, 247–57.

Squire, J. M. & Al-Khayat, H. A. (1993). Muscle thin filament structure and regulation: actin sub-domain movements and tropomyosin shift modelled from low-angle X-ray diffraction. *J. Chem. Soc. Faraday Trans.* **89**, 2717–26.

Squire, L. R. (1992). Memory and the hippocampus: a synthesis from findings with rats, monkeys, and humans. *Psychol. Rev.* **99**, 195–231.

Squire, L. R. & Zola-Morgan, S. (1988). Memory: brain systems and behaviour. *Trends Neurosci.* **11**, 170–5.

Städler, E. (1984). Contact chemoreception. In *Chemical Ecology of Insects*, ed. W. J. Bell & R. T. Cardé, pp. 3–35. London: Chapman and Hall.

Standen, N. B. (1992). Potassium channels, metabolism and muscle. *Exp. Physiol.* **77**, 1–25.

Standen, N. B., Stanfield, P. R. & Ward, T. A. (1985). Properties of single potassium channels in vesicles formed from the sarcolemma of frog skeletal muscle. *J. Physiol.* **364**, 339–58.

Stanfield, P. R. (1986). Voltage-dependent calcium channels of excitable membranes. *Brit. Med. Bull.* **42**, 359–67.

Stanfield, P. R., Davies, N. W., Shelton, P. A., Khan, I. A., Brammar, W. J., Standen, N. B. & Conley, E. C. (1994). The intrinsic gating of inward rectifier K^+ channels expressed from the murine IRK1 gene depends on voltage, K^+ and Mg^{2+}. *J. Physiol.* **475**, 1–7.

Stanley, E. F. (1991). Single calcium channels on a cholinergic presynaptic nerve terminal. *Neuron* **7**, 585–91.

Stanley, E. F. (1993). Single calcium channels and acetylcholine release at a presynaptic nerve terminal. *Neuron* **11**, 1007–11.

Starr, R. & Offer, G. (1983). H-protein and X-protein: two new components of the thick filaments of vertebrate skeletal muscle. *J. Molec. Biol.* **170**, 673–98.

Steele, P. A. & Costa, M. (1990). Opioid-like immunoreactive neurons in secretomotor pathways of the guinea-pig ileum. *Neuroscience* **38**, 771–86.

Stein, L. A., Schwarz, R. P., Chock, P. B. & Eisenberg, E. (1979). Mechanism of actomyosin adenosine triphospatase. Evidence that adenosine 5′-triphosphate hydrolysis can occur without dissociation of the actomyosin complex. *Biochemistry* **18**, 3895–909.

Stein, W. D. (1990). *Channels, Carriers, and Pumps: An Introduction to Membrane Transport.* San Diego, CA: Academic Press.

Steinbrecht, R. A. & Schneider, D. (1980). Pheromone communication in moths. In *Insect Biology in the Future*, ed. M. Locke & D. S. Smith, pp. 685–703. New York: Academic Press.

Steinhausen, W. (1931). Über den Nachwels der Bewegung der Cupula in der intakten Bogengangsampullen des Labyrinths bei der natürlichen rotatorischen und calorischen Reizung. *Pflügers Archiv* **228**, 322–8.

Steinhausen, W. (1933). Über die Beobachtung der Cupula in den Begengangsampullen des Labyrinths des lebenden Hechets. *Pflügers Archiv* **232**, 500–12.

Stephenson, D. G. & Williams, D. A. (1981). Calcium-activated force responses in fast- and slow-twitch skinned muscle fibres of the rat at different temperatures. *J. Physiol.* **317**, 281–302.

Stephenson, D. G. & Williams, D. A. (1982). Effects of sarcomere length on the force-pCa relation in fast- and slow-twitch skinned muscle fibres from the rat. *J. Physiol.* **333**, 637–53.

Stern, M. D. (1992). Theory of excitation–contraction coupling in cardiac muscle. *Biophys. J.* **63**, 497–517.

Stern, P., Edwards, F. A. & Sakmann, B. (1992). Fast and slow components of unitary EPSCs on stellate cells elicited by focal stimulation in slices of rat visual cortex. *J. Physiol.* **449**, 247–78.

Stevens, C. F. (1972). Inferences about membrane properties from electrical noise measurements. *Biophys. J.* **12**, 1028–47.

Stevens, C. F. (1993). Quantal release of neurotransmitter and long-term potentiation. *Cell* **72** (*Neuron* **10**), supplement, 55–63.

Stevens, C. F. & Wang, Y. Y. (1994). Changes in reliability of synaptic function as a mechanism for plasticity. *Nature* **371**, 704–7.

Stevens, J. C. & Green, B. G. (1996). History of research on touch. In *Pain and Touch*, ed. L. Kruger, pp. 1–23. San Diego, CA: Academic Press.

Stevens, S. S. (1961*a*). To honor Fechner and repeal his law. *Science* **133**, 80–6.

Stevens, S. S. (1961*b*). The psychophysics of sensory function. In *Sensory Communication*, ed. W. A. Rosenblith, pp. 1–33. Cambridge, MA: MIT Press.

Stevens, S. S. (1970). Neural events and the psychophysical law. *Science* **170**, 1043–50.

Stiles, W. S. (1939). The directional sensitivity of the retina and the spectral sensitivities of the rods and cones. *Proc. R. Soc. Lond.* B **127**, 64–105.

Stiles, W. S. (1978). *Mechanisms of Colour Vision.* London: Academic Press.

Stocker, M., Stühmer, W., Wittka, R., Wang, S., Müller, R., Ferrus, A. & Pongs, O. (1990). Alternative *Shaker* transcripts express either rapidly inactivating or noninactivating K$^+$ channels. *Proc. Natl. Acad. Sci. USA* **87**, 8903–7.

Stocker, R. F. (1994). The organization of the chemosensory system in *Drosophila melanogaster*: a review. *Cell Tiss. Res.* **275**, 3–26.

Stone, D., Sodek, J., Johnson, P. & Smillie, L. (1974). Tropomyosin: correlation of amino acid sequence and structure. *Proc. IX FEBS Meeting (Budapest)* **31**, 125–36.

Strader, C. D., Sigal, I. S., Candelore, M. R., Rands, E., Hill, W. S. & Dixon, R. A. F. (1988). Conserved aspartic acid residues 79 and 113 of the β-adrenergic receptor have different roles in receptor function. *J. Biol. Chem.* **263**, 10267–71.

Strader, C. D., Candelore, M. R., Hill, W. S., Sigal, I. S. & Dixon, R. A. F. (1989). Identification of two serine residues involved in agonist action of the β-adrenergic receptor. *J. Biol. Chem.* **264**, 13572–8.

Straub, F. B. (1943). Actin. *Stud. Inst. Med. Chem. Univ. Szeged* **2**, 3–15.

Streb, H., Irvine, R. F., Berridge, M. J. & Schultz, I. (1983). Release of Ca^{2+} from a nonmitochondrial intracellular store in pancreatic acinar cells by inositol 1,4,5-trisphosphate. *Nature* **306**, 67–8.

Strehler, E. E., Carlsson, E., Eppenberger, H. M. & Thornell, L.-E. (1983). Ultrastructural localization of band proteins in chicken breast muscle as revealed by combined immunocytochemistry and ultramicroscopy. *J. Molec. Biol.* **166**, 141–58.

Strehler, E. E., Strehler-Page, M.-A., Perriard, J.-C., Periasamy, M. & Nadal-Ginard, B. (1986). Complete nucleotide and encoded amino acid sequence of a mammalian myosin heavy chain gene. *J. Molec. Biol.* **190**, 291–317.

Strichartz, G. R. (1973). The inhibition of sodium currents in myelinated nerve by quaternary derivatives of lidocaine. *J. Gen. Physiol.* **62**, 37–57.

Strichartz, G. R., Rando, T. & Wang, G. K. (1987). An integrated view of the molecular toxinology of sodium channel gating in excitable cells. *Ann. Rev. Neurosci.* **10**, 237–67.

Striem, B. J., Pace, U., Zehavi, U., Naim, M. & Lancet,D. (1989). Sweet tastants stimulate adenylate cyclase coupled to GTP-binding protein in rat tongue membranes. *Biochem. J.* **260**, 121–6.

Strong, M., Chandy, K. G. & Gutman, G. A. (1993). Molecular evolution of voltage-sensitive ion channel genes: on the origins of electrical excitability. *Molec. Biol. Evol.* **10**, 221–42.

Strosberg, A. D. (1991). Structure/function relationship of proteins belonging to the family of receptors coupled to GTP-binding proteins. *Eur. J. Biochem.* **196**, 1–10.

Stryer, L. (1986). Cyclic GMP cascade of vision. *Ann. Rev. Neurosci.* **9**, 87–119.

Stryer, L. (1991). Visual excitation and recovery. *J. Biol. Chem.* **266**, 10711–14.

Stryer, L. (1993). Molecular mechanism of visual excitation. *Harvey Lectures* **87**, 129–43.

Stühmer, W., Conti, F., Suzuki, H., Wang, X., Noda, M., Yahagi,

N., Kubo, H. & Numa, S. (1989). Structural parts involved in activation and inactivation of the sodium channel. *Nature* **339**, 597–603.

Stull, J. T., Krueger, J. K., Kamm, K. E., Gao, Z.-H., Zhi, G. & Padre, R. (1996). Myosin light chain kinase. In *Biochemistry of Smooth Muscle Contraction*, ed. M. Bárány, pp. 119–30. San Diego, CA: Academic Press.

Südhof, T. C. (1995). The synaptic vesicle cycle: a cascade of protein–protein interactions. *Nature* **375**, 645–53.

Sugawara, Y. & Obara, S. (1989). Receptor Ca current and Ca-gated K current in tonic electroreceptors of the marine catfish *Plotosus*. *J. Gen. Physiol.* **93**, 343–64.

Sugiyama, H., Ito, I. & Hirono, C. (1987). A new type of glutamate receptor linked to inositol phospholipid metabolism. *Nature* **325**, 531–3.

Sumikawa, K., Houghton, M., Smith, J. C., Bell, L., Richards, B. M. & Barnard, E. A. (1982). The molecular cloning and characterisation of cDNA coding for the α subunit of the acetylcholine receptor. *Nucleic Acids Res.* **10**, 5802–22.

Summers, R. D., Morgan, R. & Relmann, S. P. (1943). The semicircular canals as a device for vectorial resolution. *Arch. Otolaryng.* **37**, 219–37.

Sun, X. H., Protasi, F., Takahashi, M., Takeshima, H., Ferguson, D. G. & Franzini-Armstrong, C. (1995). Molecular architecture of membranes involved in excitation–contraction coupling of cardiac muscle. *J. Cell. Biol.* **129**, 659–71.

Sunahara, R. K., Dessauer, C. W. & Gilman, A. G. (1996). Complexity and diversity of mammalian adenylyl cyclases. *Ann. Rev. Pharmacol. Toxicol.* **36**, 461–80.

Surprenant, A., Buell, G. & North, R. A. (1995). P_{2X} receptors bring new structure to ligand-gated ion channels. *Trends Neurosci.* **18**, 224–9.

Sussman, J. L., Harel, M., Frolow, F., Oefner, C., Goldman, A., Toker, L. & Silman, I. (1991). Atomic structure of acetylcholinesterase from *Torpedo californica*: a prototypic acetylcholine-binding protein. *Science* **253**, 872–9.

Sutcliffe, M. J., Wo, Z. G. & Oswald, R. E. (1996). Three-dimensional models of non-NMDA glutamate receptors. *Biophys. J.* **70**, 1575–89.

Sutherland, E. W. (1971). Studies on the mechanism of hormone action. In *Les Pris Nobel 1971*, pp. 240–57. Stockholm: Almquist & Wiksell.

Sutherland, E. W. & Rall, T. W. (1960). Relation of adenosine 3′,5′ phosphate and phosphorylase to the action of catecholamines and other hormones. *Pharmacol. Rev.* **12**, 265–99.

Sweatt, J. D. & Kandel, E. R. (1989). Persistent and transcriptionally-dependent increase in protein phosphorylation in long-term facilitation of *Aplysia* sensory neurons. *Nature* **339**, 51–4.

Swenson, R. P. (1982). Inactivation of potassium current in a squid axon by a variety of quaternary ammonium ions. *J. Gen. Physiol.* **77**, 255–71.

Szabo, T. (1965). Sense organs of the lateral line system in some electric fish of the Gymnotidae, Mormyridae and Gymnarchidae. *J. Morphol.* **117**, 229–50.

Szabo, T. (1974). Anatomy of the specialized lateral line organs of electroreception. In *Handbook of Sensory Physiology*, vol. III/3, ed. A. Fessard, pp. 13–58. Berlin: Springer-Verlag.

Szent-Györgyi, A. (1949). Free energy relations and contraction of actomyosin. *Biol. Bull.* **96**, 140–61.

Szent-Györgyi, A. G. (1951). The reversible depolymerisation of actin by potassium iodide. *Arch. Biochem. Biophys.* **31**, 97.

Szent-Györgyi, A. G., Cohen, C. & Kendrick-Jones, J. (1971). Paramyosin and the filaments of molluscan 'catch' muscles. II. Native filaments: isolation and characterization. *J. Molec. Biol.* **56**, 239–58.

Szent-Györgyi, A. G., Szentkiralyi, E. M. & Kendrick-Jones, J. (1973). The light chains of scallop myosin as regulatory subunits. *J. Molec. Biol.* **74**, 179–203.

Szerekes, L. & Papp, J. G. (eds.) (1994). *Pharmacology of Smooth Muscle (Handb. Exp. Pharmacol.* **111**). Berlin: Springer-Verlag.

Takai, T., Noda, M., Mishina, M., Shimizu, S., Furutani, Y., Kayano, T., Ikeda, T., Kubo, T., Takahashi, H., Takahashi, T., Kuno, M. & Numa, S. (1985). Cloning, sequencing and expression of cDNA for a novel subunit of acetylcholine receptor from calf muscle. *Nature* **315**, 761–4.

Takeshima, H., Nishimura, S., Matsumoto, T., Ishida, H., Kangawa, K., Minamino, N., Matsuo, H., Ueda, M., Hanaoka, M., Hirose, T., Numa, S. (1989). Primary structure and expression from complementary DNA of skeletal muscle ryanodine receptor. *Nature* **339**, 439–45.

Takeuchi, A. & Takeuchi, N. (1959). Active phase of frog's end-plate potential. *J. Neurophysiol.* **22**, 395–411.

Takeuchi, A. & Takeuchi, N. (1960). On the permeability of the end-plate membrane during the action of the transmitter. *J. Physiol.* **154**, 52–67.

Takeuchi, A. & Takeuchi, N. (1966). On the permeability of the presynaptic terminal of the crayfish neuromuscular junction during presynaptic inhibition and the action of γ-aminobutyric acid. *J. Physiol.* **183**, 433–49.

Takeuchi, N. (1963a). Some properties of conductance changes at the end-plate membrane during the action of acetylcholine. *J. Physiol.* **167**, 128–40.

Takeuchi, N. (1963b). Effects of calcium on the conductance change of the end-plate membrane during the action of the transmitter. *J. Physiol.* **167**, 141–55.

Talbot, W. H., Darian-Smith, I., Kornhuber, H. H. & Mountcastle, V. B. (1968). The sense of flutter-vibration: comparison of the human capacity with response patterns of mechanoreceptive afferents from the monkey hand. *J. Neurophysiol.* **31**, 301–34.

Tanabe, T., Takeshima, H., Mikami, A., Flockerzi, V., Takahashi, H., Kangawa, K., Kojima, M., Matsuo, H., Hirose, T. & Numa, S. (1987). Primary structure of the receptor for calcium channel blockers from skeletal muscle. *Nature* **328**, 313–18.

Tanabe, T., Beam, K. G., Powell, J. A. & Numa, S. (1988). Restoration of excitation–contraction coupling and slow calcium current in dysgenic muscle by dihydropyridine receptor complementary DNA. *Nature* **336**, 134–9.

Tanaka, C. & Nishizuka, Y. (1994). The protein kinase C family for neuronal signalling. *Ann. Rev. Neurosci.* **17**, 551–67.

Tanaka, J. C., Eccleston, J. F. & Barchi, R. L. (1983). Cation selectivity characteristics of the reconstituted voltage-dependent sodium channel purified from rat skeletal muscle sarcolemma. *J. Biol. Chem.* **258**, 7519–26.

Tanner, W. P. & Swets, J. A. (1954). A decision-making theory of visual detection. *Psychol. Rev.* **61**, 401–9.

Tasaki, I. (1953). *Nervous Transmission*. Springfield, IL: Charles C. Thomas.

Tasaki, I. (1954). Nerve impulses in individual auditory nerve fibres of guinea pig. *J. Neurophysiol.* **17**, 97–122.

Tasaki, I. & Shimamura, M. (1962). Further observations on resting and action potential of intracellularly perfused squid giant axon. *Proc. Natl. Acad. Sci. USA* **48**, 1571–7.

Tasaki, I. & Takeuchi, T. (1942). Weitere Studien über den Aktionsstrom der markhaltigen Nerfenvaser und über die elektrosaltorische Übertragung des Nervenimpulses. *Pflügers Archiv* **245**, 764–82.

Tasaki, I., Ishii, K. & Ito, H. (1943). On the relation between the conduction-rate, the fiber diameter and the internodal distance of the myelinated nerve fiber. *Jap. J. Med. Sci. Biophys.* **9**, 189–99.

Taussig, R. & Gilman, A. G. (1995). Mammalian membrane-bound adenylyl cyclases. *J. Biol. Chem.* **270**, 1–4.

Tavernarakis, N. & Driscoll, M. (1997). Molecular modeling of mechanotransduction in the nematode *Caenorhabditis elegans*. *Ann. Rev. Physiol.* **59**, 659–89.

Taylor, P. (1991). The cholinesterases. *J. Biol. Chem.* **266**, 4025–8.

Taylor, P. & Radic, Z. (1994). The cholinesterases: from genes to proteins. *Ann. Rev. Pharmacol. Toxicol.* **34**, 281–320.

Taylor, R. E. (1963). Cable theory. In *Physical Techniques in Biological Research*, ed. W. L. Nastuk, vol. 6, pp. 219–62. New York: Academic Press.

Taylor, S. S. (1989). cAMP-dependent protein kinase. *J. Biol. Chem.* **264**, 8443–6.

Tempel, B. L., Papazian, D. M., Schwarz, T. L., Jan, Y. N. & Jan, L. Y. (1987). Sequence of a probable potassium channel component encoded at *Shaker* locus of *Drosphila*. *Science* **237**, 770–5.

Thesleff, S. (1955). The mode of neuromuscular block caused by acetylcholine, nicotine, decamethonium and succinylcholine. *Acta Physiol. Scand.* **34**, 218–31.

Thoenen, H. (1995). Neurotrophins and neuronal plasticity. *Science* **270**, 593–8.

Thomas, D. D. (1987). Spectroscopic probes of muscle cross-bridge rotation. *Ann. Rev. Physiol.* **49**, 691–709.

Thomas, D. D. & Cooke, R. (1980). Orientation of spin-labelled myosin heads in glycerinated muscle fibres. *Biophys. J.* **32**, 891–906.

Thomas, R. C. (1969). Membrane current and intracellular sodium changes in a snail neurone during extrusion of injected sodium. *J. Physiol.* **201**, 495–514.

Thomas, R. C. (1972). Intracellular sodium activity and the sodium pump in snail neurones. *J. Physiol.* **220**, 55–71.

Thomas, R. C. (1977). The role of bicarbonate, chloride and sodium ions in the regulation of intracellular pH in snail neurones. *J. Physiol.* **273**, 317–38.

Thompson, S. M. & Gähwiler, B. H. (1989). Activity-dependent disinhibition. II. Effects of extracellular potassium, furosemide, and membrane potential on E_{Cl^-} in hippocampal CA3 neurons. *J. Neurophysiol.* **61**, 512–23.

Thompson, S. M., Deisz, R. A. & Prince, D. A. (1988). Outward chloride/cation cotransport in mammalian cortical neurons. *Neurosci. Letters* **89**, 49–54.

Thorson, J. & White, D. C. S. (1969). Distributed representations for actin–myosin interaction in the oscillatory contraction of muscle. *Biophys. J.* **9**, 360–90.

Thorson, J. & White, D. C. S. (1983). Role of cross-bridge distortion in the small-signal mechanical dynamics of insect and rabbit skeletal muscle. *J. Physiol.* **343**, 59–84.

Thurm, U. (1965). An insect mechanoreceptor. I. Fine structure and adequate stimulus. *Cold Spr. Harb. Symp. Quant. Biol.* **30**, 75–82.

Tilney, L. G. & Saunders, J. C. (1983). Actin-filaments, stereocilia, and hair-cells of the bird cochlea. 1. Length, number, width, and distribution of stereocilia of each hair cell are related to the position of the hair cell on the cochlea. *J. Cell Biol.* **96**, 807–21.

Tilney, L. G., DeRosier, D. J. & Mulroy, M. J. (1980). The organization of actin filaments in the stereocilia of cochlear hair cells. *J. Cell Biol.* **86**, 244–59.

Tinsley, J. M., Blake, D. J., Zuellig, R. A. & Davies, K. E. (1994). Increasing complexity of the dystrophin-associated protein complex. *Proc. Natl. Acad. Sci. USA* **91**, 8307–13.

Tomita, T. (1965). Electrophysiological study of the mechanisms subserving color coding in the fish retina. *Cold Spr. Harb. Symp. Quant. Biol.* **30**, 559–66.

Tomita, T. (1970). Electrical activity of vertebrate photoreceptors. *Quart. Rev. Biophys.* **3**, 179–222.

Tomita, T. (1981). Electrical activity (spikes and slow waves) in gastrointestinal smooth muscles. In *Smooth Muscle: An Assessment of Current Knowledge*, ed. E. Bülbring, A. F. Brading, A. W. Jones & T. Tomita, pp. 127–56. London: Edward Arnold.

Tomita, T. (1984). Neurophysiology of the retina. In *Foundations of Sensory Science*, ed. W. W. Dawson & J. M. Enoch, pp. 151–90. Berlin: Springer–Verlag.

Tomita, T., Kaneko, A., Murakami, M. & Pautter, E. L. (1967). Spectral response curves of single cones in the carp. *Vision Res.* **7**, 519–31.

Torebjörk, H. E., Lundberg, L. E. R. & LaMotte, R. H. (1992). Central changes in processing of mechanoreceptive input in capsaicin-induced secondary hyperalgesia in humans. *J. Physiol.* **448**, 765–80.

Toro, L., Stefani, E. & Latorre, R. (1992). Internal blockade of a Ca⁺-activated K⁺ channel by *Shaker* B inactivating "ball" peptide. *Neuron* **9**, 237–45.

Torre, V., Matthews, H. R. & Lamb, T. D. (1986). Role of calcium in regulating the cyclic GMP cascade of phototransduction in retinal rods. *Proc. Natl. Acad. Sci. USA* **83**, 7109–13.

Torre, V., Ashmore, J. F., Lamb, T. D. & Menini, A. (1995). Transduction and adaptation in sensory receptor cells. *J. Neurosci.* **15**, 7757–68.

Tota, M. R. & Strader, C. D. (1990). Characterization of the binding domain of the β-adrenergic receptor with the fluorescent antagonist carazolol: evidence for a buried ligand-binding site. *J. Biol. Chem.* **265**, 16891–7.

Tovée, M. J. (1994). The molecular genetics and evolution of primate colour vision. *Trends Neurosci.* **17**, 30–7.

Toyoda, J., Nosaki, H. & Tomita, T. (1969). Light-induced resistance changes in single photoreceptors of *Necturus* and *Gekko*. *Vision Res.* **9**, 453–63.

Toyoshima, C. & Unwin, P. N. T. (1988). Ion channel of acetylcholine receptor reconstructed from images of postsynaptic membranes. *Nature* **336**, 247–50.

Toyoshima, Y. Y., Kron, S. J., McNally, E. M., Niebling, K. R., Toyoshima, C. & Spudich, J. A. (1987). Myosin subfragment–1 is sufficient to move actin filaments in vitro. *Nature* **328**, 536–9.

Toyoshima, Y. Y., Kron, S. J. & Spudich, J. A. (1990). The myosin step size: measurement of the unit displacement per ATP hydrolyzed in an *in vitro* assay. *Proc. Natl. Acad. Sci. USA* **87**, 7130–4.

Trautwein, W. & Hescheler, J. (1990). Regulation of cardiac L-type calcium current by phosphorylation and G proteins. *Ann. Rev. Physiol.* **52**, 257–74.

Tregear, R. T. (1983). Insect flight muscle. In *Handbook of Physiology*, section 10 *Skeletal Muscle*, ed. L. Peachey & R. H. Adrian, pp. 487–506. Bethesda, MD: American Physiological Society.

Tregear, R. T. & Marston, S. B. (1979). The cross-bridge theory. *Ann. Rev. Physiol.* **41**, 723–36.

Trinick, J. (1994). Titin and nebulin: protein rulers in muscle? *Trends Biochem. Sci.* **19**, 405–9.

Tsien, R. W. & Tsien, R. Y. (1990). Calcium channels, stores, and oscillations. *Ann. Rev. Cell Biol.* **6**, 715–60.

Tsien, R. W., Hess, P., McCleskey, E. W. & Rosenberg, R. W. (1987). Calcium channels: mechanisms of selectivity, permeation and block. *Ann. Rev. Biophys.* **16**, 265–90.

Tsugorka, A., Ríos, E. & Blatter, L. A. (1995). Imaging elementary events of calcium release in skeletal muscle cells. *Science* **269**, 1723–6.

Tully, T. & Quinn, W. G. (1985). Classical conditioning and retention in normal and mutant *Drosophila melanogaster*. *J. Comp. Physiol.* **157**, 263–77.

Tully, T., Preat, T., Boynton, S. C. & Del Vecchio, M. (1994). Genetic dissection of consolidated memory in *Drosophila*. *Cell* **79**, 35–47.

Turner, D. C., Wallimann, T. & Eppenberger, H. M. (1973). A protein that binds specifically to the M-line of skeletal muscle is identified as the muscle form of creatine kinase. *Proc. Natl. Acad. Sci. USA* **70**, 702–5.

Twarog, B. M. (1954). Responses of a molluscan smooth muscle to acetylcholine and 5-hydroxytryptamine. *J. Cell. Comp. Physiol.* **44**, 141–63.

Twarog, B. M. & Page, I. H. (1953). Serotonin content of some mammalian tissues and urine and a method for its determination. *Amer. J. Physiol.* **175**, 157–61.

Twyman, R. E. & Macdonald, R. L. (1992). Neurosteroid regulation of GABA_A receptor single-channel kinetic properties of mouse spinal cord neurons in culture. *J. Physiol.* **456**, 215–45.

Unger, V. M. & Schertler, G. F. X. (1995). Low resolution structure of bovine rhodopsin determined by electron cryo-microscopy. *Biophys. J.* **68**, 1776–86.

Unwin, N. (1993). Nicotinic acetylcholine receptor at 9 Å resolution. *J. Molec. Biol.* **229**, 1101–24.

Unwin, N. (1995). Acetylcholine receptor channel imaged in the open state. *Nature* **373**, 37–43.

Unwin, P. N. T. & Ennis, P. D. (1984). Two configurations of a channel-forming membrane protein. *Nature* **307**, 609–13.

Unwin, P. N. T. & Zampighi, G. (1980). Structure of the junction between communicating cells. *Nature* **283**, 545–9.

Usdin, T. B., Eiden, L. E., Bonner, T. I. & Erickson, J. D. (1995). Molecular biology of the vesicular ACh transporter. *Trends Neurosci.* **18**, 218–24.

Usherwood, P. N. R. (1967). Insect neuromuscular mechanisms. *Amer. Zoologist* **7**, 553–82.

Usherwood, P. N. R. & Blagbrough, I. S. (1991). Spider toxins affecting glutamate receptors: polyamines in therapeutic neurochemistry. *Pharmacol. Therapeut.* **52**, 245–68.

Usherwood, P. N. R. & Grundfest, H. (1965). Peripheral inhibition in skeletal muscle of insects. *J. Neurophysiol.* **28**, 497–518.

Ussing, H. H. (1949). The distinction by means of tracers between active transport and diffusion. *Acta Physiol. Scand.* **19**, 43–56.

Uyeda, T. Q. P. (1994). Three recent breakthroughs in molecular motor research: recombinant myosin, monomolecular in vitro assay and atomic structure of S1. *Materials Sci. Eng.* **C2**, 1–11.

Uyeda, T. Q. P., Abramson, P. D. & Spudich, J. A. (1996). The neck region of the myosin motor domain acts as a lever arm to generate movement. *Proc. Natl. Acad. Sci. USA* **93**, 4459–64.

Valdeolmillos, M., O'Neill, S. C., Smith, G. L. & Eisner, D. A.

(1989). Calcium-induced Ca^{2+} release activates contraction in intact cardiac cells. *Pflügers Archiv* **413**, 676–8.

Valera, S., Hussy, N., Evans, R. J., Adami, N., North, R. A., Surprenant, A. & Buell, G. (1994). A new class of ligand-gated ion channel defined by P_{2X} receptor for extracellular ATP. *Nature* **371**, 516–19.

van der Kloot, W. & Molgó, J. (1994). Quantal acetylcholine release at the vertebrate neuromuscular junction. *Physiol. Rev.* **74**, 899–991.

van Egmond, A. A. J., Groen, J. J. & Jongkees, L. (1949). The mechanics of the semicircular canal. *J. Physiol.* **110**, 1–17.

VanBuren, P., Guilford, W. H., Kennedy, G., Wu, J. & Warshaw, D. M. (1995). Smooth muscle myosin: a high force generating molecular motor. *Biophys. J.* **68**, suppl., 256s–9s.

Vandekerckhove, J. & Weber, K. (1978a). Actin amino-acid sequences. *Eur. J. Biochem.* **90**, 451–62.

Vandekerckhove, J. & Weber, K. (1978b). At least six different actins are expressed in a higher mammal: an analysis based on the amino acid sequence of the amino terminal tryptic peptide. *J. Molec. Biol.* **126**, 783–802.

Vassar, R., Ngai, J. & Axel, R. (1993). Spatial segregation of odorant receptor expression in the mammalian olfactory epithelium. *Cell* **74**, 309–18.

Vaughan, K. T., Weber, F. E., Einheber, S. & Fischman, D. A. (1993). Molecular cloning of chicken myosin-binding protein (MyBP) H (86-kDa protein) reveals extensive homology with MyBP-C (C-protein) with conserved immunoglobulin C2 and fibronectin type III motifs. *J. Biol. Chem.* **268**, 3670–6.

Vega-Saenz de Meira, E., Weiser, M., Kentros, C., Lau, D., Moreno, H., Serodio, P. & Rudy, B. (1994). *Shaw*-related K$^+$ channels in mammals. In *Handbook of Membrane Channels*, ed. C. Peracchia, pp. 41–78. San Diego, CA: Academic Press.

Verkleij, A. J., Zwani, R. F. A., Roelofsen, B., Cumfurius, P., Kastelijn, D. & von Deenan, L. L. M. (1973). The asymmetric distribution of phospholipids in the human red cell membrane. *Biochim. Biophys. Acta* **323**, 178–93.

Verma, A., Hirsch, D. J., Glatt, C. E., Ronnett, G. V. & Snyder, S. H. (1993). Carbon monoxide: a putative neural messenger. *Science* **259**, 381–4.

Vibert, P. J., Haselgrove, J. C., Lowy, J. & Poulsen, F. (1972). Structural changes in actin-containing filaments in muscle. *Nature New Biol.* **236**, 182–3.

Vigoreaux, J. O. (1994). The muscle Z band: lessons in stress management. *J. Mus. Res. Cell Motil.* **15**, 237–55.

Villaroel, A. & Sakmann, B. (1992). Threonine in the selectivity filter of the acetylcholine receptor channel. *Biophys. J.* **62**, 196–208.

Vincent, S. R. (1995). *Nitric Oxide in the Nervous System*. London: Academic Press.

Vizi, E. S. & Vyskocil, F. (1979). Changes in total and quantal release of acetylcholine in the mouse diaphragm during activation and inhibition of membrane ATPase. *J. Physiol.* **286**, 1–14.

Vogt, R. G. & Riddiford, L. M. (1981). Pheromone binding and inactivation by moth antennae. *Nature* **293**, 161–3.

Vogt, R. G., Riddiford, L. M. & Prestwich, G. D. (1985). Kinetic properties of a sex pheromone-degrading enzyme: the sensillar esterase of *Antheraea polyphemus*. *Proc. Natl. Acad. Sci. USA* **82**, 8827–31.

von Békésy, G. (1960). *Experiments in Hearing*. New York: McGraw-Hill.

von Békésy, G. (1962). The gap between the hearing of internal and external sounds. *Symp. Soc. Exp. Biol.* **16**, 267–88.

von Békésy, G. (1964). Concerning the pleasures of observing, and the mechanics of the inner ear. Nobel Lecture 1961. Reprinted in *Nobel Lectures, Physiology or Medicine 1942–1962*, pp. 722–46. Amsterdam: Elsevier for the Nobel Foundation.

von Békésy, G. & Rosenblith, W. A. (1951). The mechanical properties of the ear. In *Handbook of Experimental Psychology*, ed. S. S. Stevens, pp. 1075–115. New York: John Wiley.

von Euler, U. S. (1955). *Noradrenaline*. Springfield, IL: Charles C. Thomas.

von Euler, U. S. & Gaddum, J. H. (1931). An unidentified depressor substance in certain tissue extracts. *J. Physiol.* **72**, 74–87.

von Frisch, K. (1936). Über den Gehörsinn der Fische. *Biol. Rev.* **11**, 210–46.

Wada, K., Ballivet, M., Boulter, J., Connolly, J., Wada, E., Deneris, E. S., Swanson, L. W., Heinemann, S. & Patrick, J. (1988). Functional expression of a new pharmacological subtype of brain nicotinic acetylcholine receptor. *Science* **240**, 330–4.

Wakabayashi, K., Sugimoto, Y., Tanaka, H., Ueno, Y., Takezawa, Y. & Amemiya, Y. (1994). X-ray diffraction evidence for the extensibility of actin and myosin filaments during muscle contraction. *Biophys. J.* **67**, 2422–35.

Wald, G. (1933). Vitamin A in the retina. *Nature* **132**, 316–17.

Wald, G. (1934). Carotenoids and the vitamin A cycle in vision. *Nature* **134**, 65.

Wald, G. (1965). Visual excitation and blood clotting. *Science* **150**, 1028–30.

Wald, G. (1968). Molecular basis of visual excitation. *Science* **162**, 230–9.

Wald, G., Brown, P. K. & Gibbons, I. R. (1962). Visual excitation: a chemo-anatomical study. *Symp. Soc. Exp. Biol.* **16**, 32–57.

Wall, P. D. & Melzack, R. (1994). *Textbook of Pain*, third edition. Edinburgh: Churchill Livingstone.

Walls, G. L. (1942). *The Vertebrate Eye and its Adaptive Radiation*. Bloomfield Hills, MI: Cranbrook Institute of Science.

Walmsley, B. (1995). Interpretation of 'quantal' peaks in distributions of evoked synaptic transmission at central synapses. *Proc. R. Soc. Lond.* B **261**, 245–50.

Wang, H., Kunkel, D. D., Martin, T. M., Schwartzkroin, P. A. &

Tempel, B. L. (1993). Heteromultimeric K$^+$ channels in terminal and juxtaparanodal regions of neurons. *Nature* **365**, 75–9.

Wang, H.-Y., Lipfert, L., Malbon, C. C. & Bahouth, S. (1989). Site-directed anti-peptide antibodies define the topography of the β-adrenergic receptor. *J. Biol. Chem.* **264**, 14424–31.

Wang, K. (1984). Cytoskeletal matrix in striated muscle: the role of titin, nebulin and intermediate filaments. In *Contractile Mechanisms in Muscle*, ed. G. H. Pollack & H. Sugi, pp. 285–302. New York: Plenum Press.

Wang, K. & Ramirez-Mitchell, R. (1983). A network of transverse and longitudinal intermediate filaments is associated with sarcomeres of adult vertebrate skeletal muscle. *J. Cell Biol.* **96**, 562–70.

Wang, K., McClure, J. & Tu, A. (1979). Titin: major myofibrillar component of striated muscle. *Proc. Natl. Acad. Sci. USA* **76**, 3698–702.

Wang, K., McCarter, R., Wright, J., Beverly, J. & Ramirez-Mitchell, R. (1993). Viscoelasticity of the sarcomere matrix of skeletal muscles. *Biophys. J.* **64**, 1161–77.

Warrick, H. M. & Spudich, J. A. (1987). Myosin structure and function in cell motility. *Ann. Rev. Cell Biol.* **3**, 379–421.

Warshaw, D. M. (1996). The in vitro motility assay: a window into the myosin molecular motor. *News Physiol. Sci.* **11**, 1–7.

Watanabe, A. & Grundfest, H. (1961). Impulse propagation at the septal and commissural junctions of crayfish lateral giant axons. *J. Gen. Physiol.* **45**, 267–308.

Watson, J. D., Gilman, M., Witkowski, J. & Zoller, M. (1992). *Recombinant DNA*, second edition. New York: Scientific American Books.

Watson, S. & Arkinstall, S. (1994). *The G-protein Linked Receptor FactsBook*. London: Academic Press.

Waxman, S. G. & Bennett, M. V. L. (1972). Relative conduction velocities of small myelinated and nonmyelinated fibres in the central nervous system. *Nature New Biol.* **238**, 217–19.

Waxman, S. G., Pappas, G. D. & Bennett, M. V. L. (1972). Morphological correlates of functional differentiation of nodes of Ranvier along single fibers in the neurogenic electric organ of the knife fish *Sternarchus*. *J. Cell Biol.* **53**, 210–24.

Weber, A. & Herz, R. (1963). The binding of calcium to actomyosin systems in relation to their biological activity. *J. Biol. Chem.* **238**, 599–605.

Weber, A., Pennise, C. R., Bondad, M. & Fowler, V. M. (1994). Tropomodulin caps the pointed ends of actin filaments. *J. Cell Biol.* **127**, 1627–35.

Weber, E., Esch, F. S., Böhlen, P., Paterson, S., Corbett, A. D., McNight, A. T., Kosterlitz, H. W., Barchas, J. & Evans, C. J. (1983). Metorphamide: isolation, structure and biological activity of an amidated opioid octapeptide from mammalian brain. *Proc. Natl. Acad. Sci. USA* **80**, 7362–6.

Weber, E. H. (1846). Der Tastsinn und das Gemeingefühl. *Handwörterbuch d. Physiologie* **3**, no. 2, 481–588.

Weddell, G. (1955). Somesthesis and chemical senses. *Ann. Rev. Psychol.* **6**, 119–36.

Weeds, A. G. & Lowey, S. (1971). Substructure of the myosin molecule. II. The light chains of myosin. *J. Molec. Biol.* **61**, 701–25.

Wei, A., Covarrubias, M., Butler, A., Baker, K., Pak, M. & Salkoff, L. (1990). K$^+$ current diversity is produced by an extended gene family conserved in *Drosophila* and the mouse. *Science* **248**, 599–603.

Wei, A., Solaro, C., Lingle, C. & Salkoff, L. (1994). Calcium sensitivity of BK-type K_{Ca} channels determined by a separable domain. *Neuron* **13**, 671–81.

Weidmann, S. (1951). Effect of current flow on the membrane potential of cardiac muscle. *J. Physiol.* **115**, 227–36.

Weight, F. F. & Votova, J. (1970). Slow synaptic excitation in sympathetic ganglion cells: evidence for synaptic inactivation of potassium conductance. *Science* **170**, 755–8.

Weiser, M., Vega-Saenz de Meira, E., Kentros, C., Moreno, H., Franzen, L., Hillman, D., Baker, H. & Rudy, B. (1994). Differential expression of *Shaw*-related K$^+$ channels in the rat central nervous system. *J. Neurosci.* **14**, 949–72.

Weisskopf, M. G., Castillo, P. E., Zalutsky, R. A. & Nicoll, R. A. (1994). Mediation of hippocampal mossy fiber long-term potentiation by cyclic AMP. *Science* **265**, 1878–82.

Wells, C. & Bagshaw, C. R. (1985). Calcium regulation of molluscan myosin ATPase in the absence of actin. *Nature* **313**, 696–7.

Wernig, A. (1975). Estimates of statistical release parameters from crayfish and frog neuromuscular junctions. *J. Physiol.* **244**, 207–21.

Wernig, A. & Stirner, H. (1977). Quantum amplitude distributions point to the functional unity of the synaptic 'active zone'. *Nature* **269**, 820–2.

Wersäll, J., Flock, Å. & Lundquist, P.-G. (1965). Structural basis for directional sensitivity in cochlear and vestibular sensory receptors. *Cold Spr. Harb. Symp. Quant. Biol.* **30**, 115–32.

Wess, J. (1993). Molecular basis of muscarinic acetylcholine receptor function. *Trends Pharmacol. Sci.* **14**, 308–13.

Wess, J., Gdula, D. & Brann, M. R. (1991). Site-directed mutagenesis of the m3 muscarinic receptor: identification of a series of threonine and tyrosine residues involved in agonist but not anatagonist binding. *EMBO J.* **10**, 3729–34.

West, J. W., Numann, R., Murphy, B. J., Scheuer, T. & Catterall, W. A. (1991). A phosphorylation site in the Na$^+$ channel required for modulation by protein kinase C. *Science* **254**, 866–8.

West, J. W., Patton, D. E., Scheuer, T., Wang, Y., Goldin, A. L. & Catterall, W. A. (1992). A cluster of hydrophobic amino acid residues required for fast Na$^+$ channel inactivation. *Proc. Natl. Acad. Sci. USA* **89**, 10910–14.

Wheeler, G. L., Matuo, Y. & Bitensky, M. W. (1977). Light-activated GTPase in vertebrate photoreceptors. *Nature* **269**, 822–4.

Whitby, L. G., Axelrod, J. & Weil-Malherbe, H. (1961). The fate of ^{3}H-norepinephrine in animals. *J. Pharmacol. Exp. Ther.* **132**, 193–201.

White, D. C. S. (1983). The elasticity of relaxed insect fibrillar flight muscle. *J. Physiol.* **343**, 31–57.

White, J. G., Southgate, E., Thomson, J. N. & Brenner, S. (1986). The structure of the nervous system of *Caenorhabditis elegans. Phil. Trans. R. Soc. Lond.* B **314**, 1–340.

White, M. M. & Bezanilla, F. (1985). Activation of squid axon K channels. *J. Gen. Physiol.* **85**, 539–54.

White, M. M. & Miller, C. (1979). A voltage-gated anion channel from electric organ of *Torpedo californica. J. Biol. Chem.* **254**, 10161–6.

White, T. W., Bruzzone, R., Wolfram, S., Paul, D. L. & Goodenough, D. A. (1994). Selective interactions among the multiple connexin proteins expressed in the vertebrate lens: the extracellular domain is a determinant of compatibility between connexins. *J. Cell Biol.* **125**, 879–92.

Whittaker, V. P. (1984). The structure and function of cholinergic synaptic vesicles. *Biochem. Soc. Trans.* **12**, 561–76.

Whittaker, V. P. (1993). Thirty years of synaptosome research. *J. Neurocytol.* **22**, 735–42.

Whittaker, V. P., Michaelson, J. A. & Kirkland, R. (1964). The separation of synaptic vesicles from nerve-ending particles ('synaptosomes'). *Biochem. J.* **90**, 293.

Wickman, K. D., Iñiguez-Lluhi, J. A., Davenport, P. A., Taussig, R., Krapivinsky, G. B., Linder, M. E., Gilman, A. G. & Clapham, D. E. (1994). Recombinant G-protein βγ-subunits activate the muscarinic-gated atrial potassium channel. *Nature* **368**, 255–7.

Wier, W. G. (1990). Cytoplasmic [Ca^{2+}] in mammalian ventricle: dynamic control by cellular processes. *Ann. Rev. Physiol.* **52**, 467–85.

Wilden, U., Hall, S. W. & Kuhn, H. (1986). Phosphodiesterase activation by photoexcited rhodopsin is quenched when rhodopsin is phosphorylated and binds the intrinsic 48-kDa protein of rod outer segments. *Proc. Natl. Acad. Sci. USA* **83**, 1174–8.

Wilkie, D. R. (1968). Heat work and phosphorylcreatine breakdown in muscle. *J. Physiol.* **195**, 157–83.

Wilkins, M. H. F., Blaurock, A. E. & Engelman, D. (1971). Bilayer structure in membranes. *Nature New Biol.* **230**, 72–6.

Wiltschko, W. & Wiltschko, R. (1996). Magnetic orientation in birds. *J. Exp. Biol.* **199**, 29–38.

Winderickx, J., Lindsey, D. T., Sanocki, E., Teller, D. Y., Motulsky, A. G. & Deeb, S. S. (1992). Polymorphism in red photopigment underlies variation in colour matching. *Nature* **356**, 431–3.

Winton, F. R. (1937). The changes in viscosity of an unstriated muscle (*Mytilus edulis*) during and after stimulation with alternating, interrupted and uninterrupted direct currents. *J. Physiol.* **88**, 492–511.

Withering, W. (1785). *An Account of the Foxglove and some of its Medical Uses, with Practical Remarks on Dropsy, and other Disease.* Birmingham. Reprinted in 1948, London: Broomsleigh Press.

Wnuk, W., Cox, J. A. & Stein, E. A. (1982). Parvalbumins and other soluble high-affinity calcium-binding proteins from muscle. In *Calcium and Cell Function*, ed. W. Y. Cheung, vol. 2, pp. 243–78. New York: Academic Press.

Wo, Z. G. & Oswald, R. E. (1995). Unraveling the modular design of glutamate-gated ion channels. *Trends Neurosci.* **18**, 161–8.

Wolbarsht, M. L. (1965). Receptor sites in insect chemoreceptors. *Cold Spr. Harb. Symp. Quant. Biol.* **30**, 281–8.

Wolbarsht, M. L. & Dethier, V. G. (1958). Electrical activity in the chemoreceptors of the blowfly. I. Responses to chemical and mechanical stimuli. *J. Gen. Physiol.* **42**, 393–412.

Woledge, R. C. (1961). The thermoelastic effect of change of tension in active muscle. *J. Physiol.* **155**, 187–208.

Woledge, R. C. (1973). In vitro calorimetric studies relating to the interpretation of muscle heat experiments. *Cold Spr. Harb. Symp. Quant. Biol.* **37**, 629–34.

Woledge, R. C., Curtin, N. A. & Homsher, E. (1985). *Energetic Aspects of Muscular Contraction.* London: Academic Press.

Wong, G. T., Gannon, K. S. & Margolskee, R. F. (1996). Transduction of bitter and sweet by gustducin. *Nature* **381**, 796–800.

Wood, J. & Garthwaite, J. (1994). Models of the diffusional spread of nitric oxide: implications for neural nitric oxide signalling and its pharmacological properties. *Neuropharmacology* **33**, 1235–44.

Woodhull, A. M. (1973). Ionic blockage of sodium channels in nerve. *J. Gen. Physiol.* **61**, 687–708.

Woodruff, M. L. & Bownds, M. D. (1979). Amplitude, kinetics and reversibility of a light-induced decrease in guanosine 3′, 5′-cyclic monophosphate in frog photoreceptor membranes. *J. Gen. Physiol.* **73**, 629–53.

Worrall, D. M. & Williams, D. C. (1994). Sodium ion-dependent transporters for neurotransmitters: a review of recent developments. *Biochem. J.* **297**, 425–36.

Worthington, C. R. (1959). Large axial spacings and striated muscle. *J. Molec. Biol.* **1**, 398–401.

Worton, R. G. & Brooke, M. H. (1995). The X-linked muscular dystrophies. In *The Metabolic and Molecular Bases of Inherited Disease*, ed. C. R. Scriver, A. L. Beaudet, W. S. Sly & D. Valle, vol. 3, pp. 4195–226. New York: McGraw-Hill.

Wray, J. S. (1979). Filament geometry and the activation of insect flight muscles. *Nature* **280**, 325–6.

Wright, W. D. (1967). *The Rays are not Coloured.* London: Adam Hilger.

Wu, C. H. (1984). Electric fish and the discovery of animal electricity. *Amer. Sci.* **72**, 598–607.

Wu, L.-G. & Saggau, P. (1995). GABA$_B$ receptor-mediated presynaptic inhibition in guinea-pig hippocampus is caused by reduction of presynaptic Ca^{2+} influx. *J. Physiol.* **485**, 649–57.

Wu, L.-G. & Saggau, P. (1997). Presynaptic inhibition of elicited neurotransmitter release. *Trends Neurosci.* **20**, 204–12.

Yamaguchi, N., de Champlain, J. & Nadeau, R. A. (1977). Regulation of norepinephrine release from cardiac sympathetic fibres in the dog by presynaptic α- and β-receptors. *Circ. Res.* **41**, 108–17.

Yamazaki, A., Stein, P. J., Chernoff, N. & Bitensky, M. W. (1983). Activation mechanism of rod outer segment cyclic GMP phosphodiesterase. Release of inhibitor by the GTP/GDP-binding protein. *J. Biol. Chem.* **258**, 8188–94.

Yanagida, T. (1990). Loose coupling between chemical and mechanical reactions in actomyosin energy transduction. *Adv. Biophys.* **26**, 75–95.

Yanagida, T. & Ishijima, A. (1995). Forces and steps generated by single myosin molecules. *Biophys. J.* **68**, 312s–320s.

Yanagida, T., Nakase, M., Nishiyama, K. & Oosawa, F. (1984). Direct observation of motion of single F-actin filaments in the presence of myosin. *Nature* **307**, 58–60.

Yanagida, T., Harada, Y. & Ishijima, A. (1993). Nano-manipulation of actomyosin molecular motors *in vitro*: a new working principle. *Trends Biochem. Sci.* **18**, 319–24.

Yanagisawa, M., Hamada, Y., Katsuragawa, Y., Imamura, M., Mikawa, T. & Misaki, T. (1987). Complete primary structure of vertebrate smooth-muscle myosin heavy-chain deduced from its complementary-DNA sequence. Implications on topography and function of myosin. *J. Molec. Biol.* **198**, 143–57.

Yang, J., Ellinor, P. T., Sather, W. A., Zhang, J.-F. & Tsien, R. W. (1993). Molecular determinants of Ca^{2+} selectivity and ion permeation in L-type Ca^{2+} channels. *Nature* **366**, 158–61.

Yang, N. & Horn, R. (1995). Evidence for voltage-dependent S4 movement in sodium channels. *Neuron* **15**, 213–18.

Yang, N., George, A. L. & Horn, R. (1996). Molecular basis of charge movement in voltage-gated sodium channels. *Neuron* **16**, 113–22.

Yarfitz, S. & Hurley, J. B. (1994). Transduction mechanisms of vertebrate and invertebrate photoreceptors. *J. Biol. Chem.* **269**, 14329–32.

Yatani, A., Codina, J., Brown, A. M. & Birnbaumer, L. (1987). Direct activation of mammalian atrial muscarinic potassium channels by GTP regulatory protein G_k. *Science* **235**, 207–11.

Yatani, A., Okabe, K., Codina, J., Birnbaumer, L. & Brown, A. M. (1990). Heart rate regulation by G proteins acting on the cardiac pacemaker channel. *Science* **249**, 1163–6.

Yau, K.-W. (1994*a*). Cyclic nucleotide-gated channels: an expanding new family of ion channels. *Proc. Natl. Acad. Sci. USA* **91**, 3481–3.

Yau, K.-W. (1994*b*). Phototransduction mechanism in retinal rods and cones. *Invest. Ophthalmol.* **35**, 9–32.

Yau, K.-W. & Baylor, D. A. (1989). Cyclic GMP-activated conductance of retinal photoreceptor cells. *Ann. Rev. Neurosci.* **12**, 289–327.

Yau, K.-W. & Nakatani, K. (1984*a*). Cation selectivity of light-sensitive conductance in retinal rods. *Nature* **309**, 352–4.

Yau, K.-W. & Nakatani, K. (1984*b*). Electrogenic Na–Ca exchange in retinal rod outer segment. *Nature* **311**, 661–3.

Yau, K.-W. & Nakatani, K. (1985). Light-suppressible, cyclic GMP-sensitive conductance in the plasma membrane of a truncated rod outer segment. *Nature* **317**, 252–5.

Ye, Q., Heck, G. L. & DeSimone, J. A. (1991). The anion paradox in sodium taste reception: resolution by voltage-clamp studies. *Science* **254**, 724–6.

Yeager, M. & Gilula, N. B. (1992). Membrane topology and quaternary structure of cardiac gap junction ion channels. *J. Molec. Biol.* **223**, 929–48.

Yeagle, P. L. (1993). *The Membranes of Cells*, second edition. San Diego, CA: Academic Press.

Yee, R. & Liebman, P. A. (1978). Light-activated phosphodiesterase of the rod outer segment. *J. Biol. Chem.* **253**, 8902–9.

Yellen, G., Jurman, M. E., Abramson, T. & MacKinnon, R. (1991). Mutations affecting internal TEA blockade identify the probable pore-forming region of a K^+ channel. *Science* **251**, 939–42.

Yin, J. C. P., Wallach, J. S., Del Vecchio, M., Wilder, E. L., Zhou, H., Quinn, W. G. & Tully, T. (1994). Induction of a dominant-negative CREB transgene specifically blocks long-term-memory in *Drosophila*. *Cell* **79**, 49–58.

Yin, J. C. P., Del Vecchio, M., Zhou, H. & Tully, T. (1995). CREB as a memory modulator: induced expression of a dCREB2 activator isoform enhances long-term-memory in *Drosophila*. *Cell* **81**, 107–15.

Yokoi, M., Mori, K. & Nakanishi, S. (1995). Refinement of odor molecule tuning by dendrodendritic synaptic inhibition in the olfactory bulb. *Proc. Natl. Acad. Sci. USA* **92**, 3371–5.

Yoshikami, S. & Hagins, W. A. (1973). Control of the dark current in vertebrate rods and cones. In *Biochemistry aad Physiology of Visual Pigments*, ed. H. Langer, pp. 245–56. Berlin: Springer-Verlag.

Yoshizawa, T., Shichida, Y. & Matuoka, S. (1984). Primary intermediates of rhodopsin studied by low temperature spectrophotometry and laser photolysis. *Vision Res.* **24**, 1455–63.

Young, J. Z. (1936). The giant nerve fibres and epistellar body of cephalopods. *Quart. J. Microsc. Sci.* **78**, 367–86.

Young, T. (1802). On the theory of light and colours. *Phil. Trans. R. Soc. Lond.* **92**, 12–48.

Zagorodnyuk, V., Santicioli, P., Maggi, C. A. & Giachetti, A. (1996). The possible role of ATP and PACAP as mediators of apamin-sensitive NANC inhibitory junction potentials in circular muscle of guinea-pig colon. *Brit. J. Pharmacol.* **119**, 779–86.

Zagotta, W. N., Hoshi, T. & Aldrich, R. W. (1990). Restoration of inactivation in mutants of *Shaker* potassium channels by a peptide derived from ShB. *Science* **250**, 568–71.

Zagotta, W. N., Hoshi, T. Aldrich, R. W. (1994). *Shaker*

potassium channel gating. III. Evaluation of kinetic models for activation. *J. Gen. Physiol.* **103**, 321–62.

Zaki, P. A., Bilsky, E. J., Vandereh, T. W., Lai, J., Evans, C. J. & Porreca, F. (1996). Opioid receptor subtypes: the δ receptor as a model. *Ann. Rev. Pharmacol. Toxicol.* **36**, 379–401.

Zaza, A., Robinson, R. B. & DiFrancesco, D. (1996). Basal responses of the L-type Ca^{2+} and hyperpolarization-activated currents to autonomic agonists in the rabbit sinoatrial node. *J. Physiol.* **491**, 347–55.

Zerangue, N. & Kavanaugh, M. P. (1996). Flux coupling in a neuronal glutamate transporter. *Nature* **383**, 634–7.

Zhang, J. F., Ellinor, P. T., Aldrich, R. W. & Tsien, R. W. (1994). Molecular determinants of voltage-dependent inactivation in calcium channels. *Nature* **372**, 97–100.

Zhong, Y. & Wu, C.-F. (1991). Altered synaptic plasticity in *Drosophila* memory mutants with a defective cyclic AMP cascade. *Science* **251**, 198–201.

Ziegelberger, G. (1995). Redox shift of the pheromone-binding protein in the silkmoth *Antheraea polyphemus. Eur. J. Biochem.* **232**, 706–11.

Ziegelberger, G., van den Berg, M. J., Kaissling, K.-E., Klumpp, S. & Schultz, J. E. (1990). Cyclic GMP levels and guanylate cyclase activity in pheromone sensitive antennae of the silkmoths *Antheraea polyphemus* and *Bombyx mori. J. Neurosci.* **10**, 1217–25.

Zimmerman, A. L. & Baylor, D. A. (1986). Cyclic GMP-sensitive conductance of retinal rods consists of aqueous pores. *Nature* **321**, 70–2.

Zimmerman, A. L. & Baylor, D. A. (1992). Cation interactions within the cyclic GMP-gated channel of retinal rods from the tiger salamander. *J. Physiol.* **449**, 759–83.

Zorzato, F., Fujii, J., Otsu, K., Phillips, M., Green, N. M., Lai, F. A., Meissner, G., & MacLennan, D. H. (1990). Molecular cloning of cDNA encoding human and rabbit forms of the Ca^{2+} release channel (ryanodine receptor) of skeletal muscle sarcoplasmic reticulum. *J. Biol. Chem.* **265**, 2244–56.

Zuckerman, R. (1973). Ionic analysis of photoreceptor membrane currents. *J. Physiol.* **235**, 333–54.

Zufall, F., Firestein, S. & Shepherd, G. M. (1994). Cyclic nucleotide-gated ion channels and sensory transduction in olfactory receptor neurons. *Ann. Rev. Biophys.* **23**, 577–607.

Zuker, C. S. (1996). The biology of vision in *Drosophila. Proc. Natl. Acad. Sci. USA* **93**, 571–6.

Zukin, R. S. & Bennett, M. V. L. (1995). Alternatively spliced isoforms of the NMDAR1 receptor subunit. *Trends Neurosci.* **18**, 306–13.

Index

8256699R0

Made in the USA
Lexington, KY
18 January 2011